de Gruyter Studies in Mathematics 4

Editors: Heinz Bauer · Peter Gabriel

Corneliu Constantinescu

Spaces of Measures

Walter de Gruyter · Berlin · New York 1984

Author

Dr. Corneliu Constantinescu
Professor at the Eidgenössische
Technische Hochschule Zürich
Department of Mathematics

CIP-Kurztitelaufnahme der Deutschen Bibliothek

Constantinescu, Corneliu:
Spaces of measures / Corneliu Constantinescu. –
Berlin ; New York : de Gruyter, 1984.
(De Gruyter studies in mathematics ; 4)
ISBN 3-11-008784-7

Library of Congress Cataloging in Publication Data

Constantinescu, Corneliu.
Spaces of measures.
(De Gruyter studies in mathematics ; 4)
Bibliography: p.
Includes index.
1. Spaces of measures. I. Title. II. Series.
QA312.C578 1984 515.4'2 84–5815
ISBN 3-11-008784-7

Cover design: Rudolf Hübler, Berlin. Typesetting and Printing: Tutte Druckerei GmbH, Salzweg-Passau.
Binding: Lüderitz & Bauer, Berlin.

Preface

The theory of spaces of measures may be considered as a chapter of measure and integration theory, if we accept a rather comprehensive definition for the latter theory. In contrast to the other chapters, in which attention is concentrated generally on isolated measures, the properties of sets of measures are of interest for the theory of spaces of measures. Nikodym's convergence theorem may illustrate this situation quite well. This theorem states that if $(\mu_n)_{n\in\mathbb{N}}$ is a sequence of real (σ-additive) measures defined on a σ-ring $\mathfrak{R}$ such that $(\mu_n(A))_{n\in\mathbb{N}}$ converges for every $A\in\mathfrak{R}$, then the limit function $A\mapsto\lim_{n\to\infty}\mu_n(A)$ is a measure as well; moreover, the set $\{\mu_n \mid n\in\mathbb{N}\}$ is equi-σ-additive. Here we have a typical example of a topological property of the set of all real measures of $\mathfrak{R}$. The greatest part of the properties of the spaces of measures is of topological nature and so the topological methods are the main tools in this theory.

Besides the σ-additivity there is another weaker property, called exhaustivity, which compels attention in this field. A function μ defined on a σ-ring $\mathfrak{R}$ with values in a topological commutative group is called exhaustive if $(\mu(A_n))_{n\in\mathbb{N}}$ converges to 0 for any disjoint sequence $(A_n)_{n\in\mathbb{N}}$ in $\mathfrak{R}$. The greatest part of the theorems of the theory of spaces of measures consists of two parallel formulations, one for the measures, the other for the exhaustive finitely additive functions, the latter one being in general weaker. An important case occurs when the measures in question are defined on a ring of subsets of a Hausdorff space and when they satisfy some compatibility conditions with the topology; in this case the theorems may be strongly reinforced.

As an application of this theory we mention integration theory itself (e.g. the theorem stating that L^1 is weakly sequentially complete is a consequence of Nikodym's convergence theorem quoted above) and the theory of vector lattices, especially of M-spaces. An M-space is a vector lattice E endowed with a topology generated by a set of semi-norms p satisfying the conditions

$$x, y\in E,\ |x|\le|y| \Rightarrow p(x)\le p(y),$$
$$x, y\in E,\ x\ge 0,\ y\ge 0 \Rightarrow p(x\vee y)=\sup(p(x),p(y)).$$

The origins of the theory of spaces of measures can be traced back to the period between the two world wars. The theorems, classical ones nowadays, are associated with the names Orlicz-Pettis, Vitali-Hahn-Saks, Nikodym (the convergence theorem and the boundedness theorem), and Phillips. After the war, especially during the seventies, an almost explosive development took place in this field, the above mentioned theorems being generalized in all kinds of directions. Up to the present time no unified bibliography on this subject has existed. The literature is spread out and, in part, difficult to trace. One aim of this book is to collect a part of this literature and to present it in a unified manner. The book also contains the applications of this theory to the study of vector lattices and of the M-spaces.

The author would like to thank the editors of the "de Gruyter Studies in Mathematics" for accepting this book as part of their series. My special thanks go to Wolfgang Filter and to Peter von Siebenthal for reading the manuscript and for contributing improvements.

Corneliu Constantinescu
Zurich, March 29, 1983

Contents

Chapter 4: Spaces of measures

Chapter 5: Locally convex lattices

Introduction

The theory of spaces of measures developed from the following classical theorems.

1. Nikodym's convergence theorem (O. Nikodym (1931) [1] II, (1933) [3] page 427).
Let $\mathfrak{R}$ be a σ-ring, and let $(\mu_n)_{n \in \mathbb{N}}$ be a sequence of real-valued (σ-additive) measures on $\mathfrak{R}$ such that $(\mu_n(A))_{n \in \mathbb{N}}$ converges for every $A \in \mathfrak{R}$. Then the measures μ_n $(n \in \mathbb{N})$ are equi-σ-additive and the map

$$\mathfrak{R} \to \mathbb{R}, \quad A \mapsto \lim_{n \to \infty} \mu_n(A)$$

is also a measure.

2. Orlicz-Pettis theorem (W. Orlicz (1929) [1] Satz 2; B.J. Pettis (1938) [1] Theorem 2.32).
Let $\mathfrak{R}$ be a σ-ring, E a Banach space, and μ a map of $\mathfrak{R}$ into E. If μ is a measure with respect to the weak topology of E then μ is also a measure with respect to the norm topology of E.

3. Boundedness theorem of Nikodym (O. Nikodym (1931) [1] I, (1933) [2] page 418).
Let $\mathfrak{R}$ be a σ-ring, and let $\mathcal{M}$ be a set of real-valued measures on $\mathfrak{R}$. If

$$\sup_{\mu \in \mathcal{M}} |\mu(A)| < \infty$$

for every $A \in \mathfrak{R}$ then

$$\sup_{\substack{\mu \in \mathcal{M} \\ A \in \mathfrak{R}}} |\mu(A)| < \infty .$$

4. Vitali-Hahn-Saks theorem (G. Vitali (1907) [1] Teorema, page 147; H. Hahn (1922) [1] XXI; S. Saks (1933) [1] Theorem 5).
Let $\mathfrak{R}$ be a σ-ring, and let μ be a positive real-valued measure on $\mathfrak{R}$. Let $(\mu_n)_{n \in \mathbb{N}}$ be a sequence of μ-absolutely continuous, real-valued measures on $\mathfrak{R}$ such that $(\mu_n(A))_{n \in \mathbb{N}}$ converges for every $A \in \mathfrak{R}$. Then $(\mu_n)_{n \in \mathbb{N}}$ is equi-μ-absolutely continuous, and the limit function is also μ-absolutely continuous.

5. Phillips' lemma (R.S. Phillips (1940) [1] Lemma 3.3).
Let $(\mu_n)_{n \in \mathbb{N}}$ be a sequence of finitely additive bounded real-valued functions on the power set of $\mathbb{N}$ such that $(\mu_n(M))_{n \in \mathbb{N}}$ converges to 0 for every $M \subset \mathbb{N}$. Then

$$\lim_{n \to \infty} \sum_{m \in \mathbb{N}} |\mu_n(\{m\})| = 0 .$$

The first four theorems give information about spaces of measures, the last one about a space of exhaustive additive maps. For instance let us reformulate Nikodym's convergence theorem in the language of spaces of measures. Let $\mathbb{R}^{\mathfrak{R}}$ be the space of all real-valued functions on a σ-ring $\mathfrak{R}$ endowed with the product topology, i.e. with the topology of pointwise convergence. Any real-valued measure on $\mathfrak{R}$ may be considered as a point in the space $\mathbb{R}^{\mathfrak{R}}$. In general the set of all real-valued measures on $\mathfrak{R}$ is not a closed set in $\mathbb{R}^{\mathfrak{R}}$, but Nikodym's convergence theorem says that it is sequentially closed (in fact a stronger result holds). The other theorems can also be formulated in a similar way.

The results concerning spaces of measures rely primarily on two theorems. In order to present these two theorems we denote by G a Haudorff topological additive (i.e. commutative) group, and we call a sequence in G supersummable if every one of its subsequences is summable. The announced theorems read as follows.

a) If $(x_n)_{n \in \mathbb{N}}$ is a supersummable sequence from G, then the map

$$\mathfrak{P}(\mathbb{N}) \to G, \quad A \mapsto \sum_{n \in A} x_n$$

is continuous. Here $\mathfrak{P}(\mathbb{N})$ denotes the power set of $\mathbb{N}$ endowed with the compact topology obtained by identifying $\mathfrak{P}(\mathbb{N})$ with $\{0,1\}^{\mathbb{N}}$ via the map

$$\{0,1\}^{\mathbb{N}} \to \mathfrak{P}(\mathbb{N}), \quad f \mapsto \{n \in \mathbb{N} \mid f(n) = 1\}.$$

b) If $((x_{mn})_{n \in \mathbb{N}})_{m \in \mathbb{N}}$ is a sequence of supersummable sequences from G such that $(\sum_{n \in A} x_{mn})_{m \in \mathbb{N}}$ converges for every $A \subset \mathbb{N}$, then the convergence is uniform with respect to A.

These two theorems hold even with $\mathbb{N}$ replaced by an arbitrary set. These results open the way to a study of spaces of supersummable families to which chapter 3 of this book is devoted.

The classical results about spaces of measures are presented in chapter 4. In this sense chapter 4 is somehow the core of the book. Besides spaces of measures this chapter also discusses spaces of additive exhaustive maps. A part of this chapter is dedicated to spaces of measures on topological spaces; such measures possess additional properties which reflect the compatibility of the measure with the topology.

The most convenient way to state the results presented in this book is in the context of spaces of functions. Certain notions involving nets, filters, and associated maps of uniform spaces play a crucial role in such a presentation. These objects are defined and their properties are studied in the first chapter. It is somehow astonishing to find that this theory is so rich in properties that it forms a whole world, deserving to be studied for itself. Thus chapter 1 furnishes the language in which the main results of the book are formulated.

The spaces of supersummable families and various spaces of measures and of exhaustive additive maps possess many remarkable properties. Similar properties, it turns out, are possessed by many other important spaces appearing in mathematics e.g. the spaces of order σ-continuous group homomorphisms from a σ-complete lattice ordered group in a topological group, (these are treated in the first part of chapter 5), the duals of M-spaces, (an M-space is a vector lattice E endowed with a topology generated by a set $\mathscr{P}$ of seminorms such that

$$p \in \mathscr{P},\ x, y \in E,\ |x| \leq |y| \Rightarrow p(x) \leq p(y),$$
$$p \in \mathscr{P},\ x, y \in E,\ x \geq 0,\ y \geq 0 \Rightarrow p(x \vee y) = \sup(p(x), p(y))),$$

(these spaces are discussed in the second part of chapter 5), etc. In all of these spaces the methods for establishing the properties in question are similar. Certain constructions on these spaces of functions appear again and again. Repeating these constructions would not only be irritating but it would also be unproductive since the number of repetitions is considerable. Therefore it seemed reasonable to extract the common features of all these constructions and formulate them as theorems that can be quoted later. The resulting abstract construction is carried out in chapter 2.

The author is aware of the difficulties connected with any step in the direction of abstractness, especially when the reason for the abstraction or the way in which this abstract theory applies is not clear. We advise the readers who find it hard to study chapter 2 because of such motives (and the same advice holds for chapter 1) to start by searching the historical remarks for the more or less classical results (the list of references can be used for a purposeful tracking of these historical remarks, since the reader will find for each paper the places where it is cited). The historical remarks will lead back to the corresponding theorems, the formulations and the proofs of which will lead rapidly to the first two chapters of the book. Via this backtracking the reader will find out where and in what connection use is made of the various notions and deliberations that are presented abstractly in chapters 1 and 2, and this may facilitate the study of these chapters. In the present context no other choice seemed practicable.

Some results of chapter 5 deserve special mention since they offer a nice example of applications of the theory of spaces of measures. The main result of Grothendieck's thesis (1953, [2], Théorème 1 and Théorème 6) consists of the proof that the Banach space of continuous real functions on a compact space possesses the DP-property and the D-property. In fact these properties (and even stronger ones) are possessed by any M-space. The proof given in this book uses the fact that large parts of the dual of an M-space may be identified (via Choquet's theorem on simplexes) with spaces of measures. The above mentioned properties of the M-spaces are nothing other than the reflection of the corresponding properties of the spaces of measures.

It is assumed that, besides possessing the usual knowledge of a graduate student, the reader is also familiar with some special topological objects, namely filters, nets, special kinds of topological spaces (Hausdorff, regular, completely regular and so on: reference book N. Bourbaki [1]), knows elementary facts abort topological groups (reference book N. Bourbaki [1]) and has some knowledge about locally convex spaces (reference book H.H. Schaefer [1]). Section 4.9 and all the other sections which use it require (real) integration theory. Starting with section 5.5 we use the theory of vector lattices and of locally convex vector lattices (reference book Ch.D. Aliprantis, O. Burkinshaw [1]), and from section 5.6 on the theory of duality in measure theory (reference book C. Constantinescu [5]).

There are certain sections in which some notation occurs very frequently. In order to avoid too many repetitions we have preferred to adopt the following convention: throughout the book it is assumed that, besides the hypotheses explicitely formulated, the hypotheses stated at the beginning of the section also hold.

Logical connections between sections

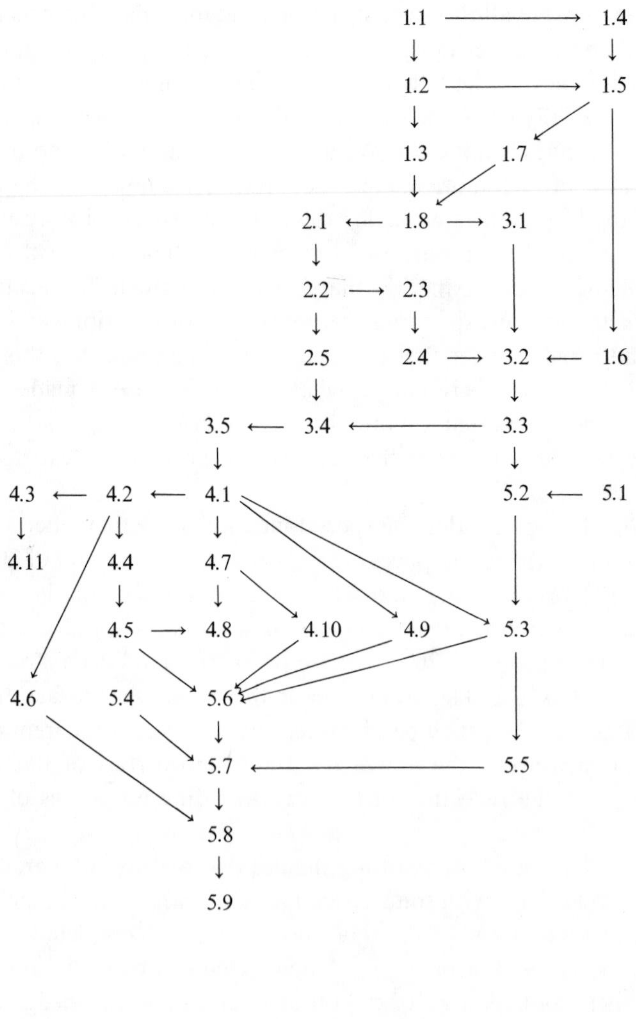

Notations and Terminology

In general we shall use the notation and terminology of N. Bourbaki.

1. Set theory

If X is a set, then $\mathfrak{P}(X)$ will denote the *power set of* X, i.e. the set of subsets of X. We will denote by Y^X the set of maps of X into Y for any sets X, Y. Let X, Y be sets, S be a subset of X, $\mathfrak{F}$ be a filter on X, and $\varphi \in Y^X$. We shall denote by $\varphi|S$ the *restriction of* φ *to* S, i.e. the map

$$S \to Y, \quad x \mapsto \varphi(x),$$

and by $\varphi(\mathfrak{F})$ the *image of* $\mathfrak{F}$ *with respect to* φ, i.e. the filter on Y generated by the filter base $\{\varphi(A) | A \in \mathfrak{F}\}$. Let $(X_\iota)_{\iota \in I}$ be a family of sets and, for any $\iota \in I$, let $\mathfrak{F}_\iota$ be a filter on $X_\iota (I \neq \emptyset)$. We shall denote by $\prod_{\iota \in I} \mathfrak{F}_\iota$ *the product filter of the family* $(\mathfrak{F}_\iota)_{\iota \in I}$, i.e. the filter on $\prod_{\iota \in I} X_\iota$ generated by the filter base

$$\{\prod_{\iota \in I} A_\iota | (A_\iota)_{\iota \in I} \in \prod_{\iota \in I} \mathfrak{F}_\iota, \ \{\iota \in I | A_\iota \neq X_\iota\} \text{ finite}\}.$$

We set $\mathfrak{F}_1 \times \mathfrak{F}_2 := \prod_{\iota \in I} \mathfrak{F}_\iota$ if $I = \{1, 2\}$. Let $(x_n)_{n \in \mathbb{N}}$ be a sequence. A *subsequence of* $(x_n)_{n \in \mathbb{N}}$ is a sequence of the form $(x_{k_n})_{n \in \mathbb{N}}$, where $(k_n)_{n \in \mathbb{N}}$ denotes a strictly increasing sequence in $\mathbb{N}$. An ultrafilter is called *free* or *non-trivial* if it contains no one-point set. We denote for any set X by $\mathrm{Card}(X)$ the cardinal number of X.

2. Order relations

A *preorder relation* on a set I is a binary relation $\leq$ on I such that:

a) $\iota \in I \Rightarrow \iota \leq \iota$;

b) $\iota, \iota', \iota'' \in I, \ \iota \leq \iota', \iota' \leq \iota'' \Rightarrow \iota \leq \iota''$.

We set

$$\iota' \geq \iota'' : \Leftrightarrow \iota'' \leq \iota'$$

for any $\iota', \iota'' \in I$. We set $\iota' < \iota''$ for any different $\iota', \iota'' \in I$ with $\iota' \leq \iota''$. If

c) $\iota', \iota'' \in I, \ \iota' \leq \iota'', \ \iota'' \leq \iota' \Rightarrow \iota' = \iota''$

then the preorder relation is called an *order relation*. A *preordered* (*ordered*) *set* is a set endowed with a preorder (order) relation. A preordered set I is called *upper* (*lower*) *directed* if

$$\iota', \iota'' \in I \Rightarrow \exists \iota \in I,\ \iota' \leq \iota, \iota'' \leq \iota$$

$$(\iota', \iota'' \in I \Rightarrow \exists \iota \in I,\ \iota \leq \iota', \iota \leq \iota'').$$

A preordered set is called *directed* if it is both upper and lower directed. A *directed* (*upper directed, lower directed*) *set* of a preordered set I is a subset of I which possesses the corresponding property with respect to the induced order relation. Let I be a nonempty upper directed preordered set. Then

$$\{\{\lambda \in I \mid \lambda \geq \iota\} \mid \iota \in I\}$$

is a filter base on I; the filter on I generated by this filter base is called the *section filter of I*. A *net in a set* X is a pair (I, f) such that I is an upper directed preordered set ($I \neq \emptyset$) and f is a map of I into X. If X, Y are sets and if (I, f) is a net in a subset of Y^X, then we will use the indexed notation rather than the functional one, i.e. we set $f_\iota := f(\iota)$ for any $\iota \in I$.

3. Topological spaces

Let X be a topological space and A be a subset of X. A is called *sequentially closed* if any point x of X belongs to A if there exists a sequence in A converging to x. A is called *σ-compact* (*σ-quasicompact*) if there exists a sequence of compact (of quasi-compact) sets of X the union of which is equal to A. A map φ of X into a topological space is called *sequentially continuous* if $(\varphi(x_n))_{n \in \mathbb{N}}$ converges to $\varphi(x)$ for any $x \in X$ and for any sequence $(x_n)_{n \in \mathbb{N}}$ in X, which converges to x. A topological space is called *countably compact* if any sequence in it possesses an adherent point.

4. Uniform spaces

A uniform space X is called *sequentially complete* if any Cauchy sequence in X converges.

Let X be a set. A *pseudometric on* X is a map

$$d : X \times X \to \overline{\mathbb{R}}_+ (:= [0, \infty])$$

such that we have for any $x, y, z \in X$

a) $d(x, x) = 0$;

b) $d(x, y) = d(y, x)$;

c) $d(x, y) \leq d(x, z) + d(z, y)$.

Let X be a set and let $\mathscr{D}$ be a set of pseudometrics on X. Then the set of finite intersections of the sets of the form

$$\{(x, y) \in X \times X \mid d(x, y) < \alpha\},$$

where (d, α) runs through $\mathscr{D} \times]0, \infty[$, is a fundamental system of a uniformity on

X which will be called the *uniformity (on X) generated by $\mathcal{D}$*. Any uniformity is generated by a set of pseudometrics. If d is a pseudometric on a set X, then we will call the *uniformity (on X) generated by d* the uniformity on X generated by $\{d\}$ and the *topology (on X) generated by d* the corresponding topology. A uniform (topological) space is called *pseudometrizable* if its uniformity (topology) is generated by a pseudometric.

Let X be a set and d be a pseudometric on X. Then

$$x \sim y :\Leftrightarrow (x, y \in X, d(x, y) = 0)$$

is an equivalence relation on X. Let $X/\sim$ be the quotient set of X with respect to $\sim$ (i.e. the set of equivalence classes with respect to $\sim$) and let φ be the canonical map $X \to X/\sim$ (i.e. the map which associates with each point of X its equivalence class with respect to $\sim$). Then we have

$$d(x, y) = d(x', y')$$

for any $A, B \in X/\sim$ and for any $x, x' \in A$, $y, y' \in B$. We set

$$d_0(A, B) := d(x, y)$$

for any $A, B \in X/\sim$, where $(x, y) \in A \times B$. d_0 is a metric on $X/\sim$. $X/\sim$ endowed with this metric will be called the *canonical metric space associated with (X, d)* and φ will be called the *canonical map associated with (X, d)*.

Let X be a pseudometrizable uniform (topological) space and d be a pseudometric on X generating its uniformity (topology). Further, let Y be the canonical metric space associated with (X, d) and φ be the canonical map associated with (X, d). Y endowed with its uniformity (topology) will be called the *canonical metrizable uniform (topological) space associated with X*. The uniformity (topology) of X is the initial uniformity (topology) on X with respect to φ.

Chapter 1: Topological preliminaries

§ 1.1 Sets of filters

Definition 1.1.1. *Let X be a set and Φ be a set of filters on X. The expression* $\mathfrak{F}$ is a Φ-filter *will mean $\mathfrak{F} \in \Phi$. A subset A of X is called a* Φ-set *if any filter on X containing A belongs to Φ. We denote by $\underline{\hat{\Phi}}$ the set of filters $\mathfrak{F}$ on X possessing the following property: for any map $M : \Phi \to \mathfrak{P}(X)$ with $M(\mathfrak{G}) \in \mathfrak{G}$ for any $\mathfrak{G} \in \Phi$ there exists a finite subset Ψ of Φ such that $\bigcup_{\mathfrak{G} \in \Psi} M(\mathfrak{G}) \in \mathfrak{F}$.*

We have $\Phi \subset \hat{\Phi}$ and any filter on X finer than a filter of $\hat{\Phi}$ belongs to $\hat{\Phi}$.

Proposition 1.1.2. *Let X be a set and let Φ, Φ' be sets of filters on X such that for any $\mathfrak{G} \in \Phi$ there exists $\mathfrak{G}' \in \Phi'$ with $\mathfrak{G}' \subset \mathfrak{G}$. Then $\hat{\Phi} \subset \hat{\Phi}'$.*

Let $\mathfrak{F} \in \hat{\Phi}$ and M be a map of Φ' into $\mathfrak{P}(X)$ with $M(\mathfrak{G}') \in \mathfrak{G}'$ for any $\mathfrak{G}' \in \Phi'$. Let Λ be a map of Φ into Φ' such that $\Lambda(\mathfrak{G}) \subset \mathfrak{G}$ for any $\mathfrak{G} \in \Phi$. Then $M \circ \Lambda(\mathfrak{G}) \in \mathfrak{G}$ for any $\mathfrak{G} \in \Phi$ and therefore there exists a finite subset Ψ of Φ such that

$$\bigcup_{\mathfrak{G} \in \Psi} M \circ \Lambda(\mathfrak{G}) \in \mathfrak{F}.$$

Hence, $\mathfrak{F} \in \hat{\Phi}'$. □

Proposition 1.1.3. *Let X be a set, Φ be a set of filters on X, $\mathfrak{F}$ be a filter on X, and Φ_0 be the set of ultrafilters $\mathfrak{G}$ on X finer than $\mathfrak{F}$ such that there exists a filter of Φ coarser than $\mathfrak{G}$. The following assertions are equivalent:*

a) $\mathfrak{F} \in \hat{\Phi}$;

b) any ultrafilter on X finer than $\mathfrak{F}$ belongs to Φ_0;

c) $\mathfrak{F} \in \hat{\Phi}_0$.

In particular $\hat{\hat{\Phi}} = \hat{\Phi}$.

a ⇒ b. Let $\mathfrak{F}_0$ be an ultrafilter on X finer than $\mathfrak{F}$. Assume $\mathfrak{F}_0 \notin \Phi_0$. Then there exists a map $M : \Phi \to \mathfrak{P}(X)$ with $M(\mathfrak{G}) \in \mathfrak{G} \backslash \mathfrak{F}_0$ for any $\mathfrak{G} \in \Phi$. By a) there exists a finite subset Ψ of Φ such that $\bigcup_{\mathfrak{G} \in \Psi} M(\mathfrak{G}) \in \mathfrak{F}$. Since $\mathfrak{F}_0$ is an ultrafilter finer than $\mathfrak{F}$ there exists $\mathfrak{G} \in \Psi$ with $M(\mathfrak{G}) \in \mathfrak{F}_0$, which is a contradiction.

b ⇒ c. Let M be a map of Φ_0 into $\mathfrak{P}(X)$ with $M(\mathfrak{G}) \in \mathfrak{G}$ for any $\mathfrak{G} \in \Phi_0$. Assume that for any finite subset Ψ of Φ_0 we have $\bigcup_{\mathfrak{G} \in \Psi} M(\mathfrak{G}) \notin \mathfrak{F}$. Then

$$\mathfrak{F}' := \{A \backslash \bigcup_{\mathfrak{G} \in \Psi} M(\mathfrak{G}) \mid A \in \mathfrak{F},\ \Psi \text{ finite} \subset \Phi_0\}$$

is a filter base. Let $\mathfrak{F}_0$ be an ultrafilter on X finer than the filter on X generated by $\mathfrak{F}'$. Since $\mathfrak{F} \subset \mathfrak{F}_0$ we get by b) $\mathfrak{F}_0 \in \Phi_0$. Hence, $X \backslash M(\mathfrak{F}_0) \in \mathfrak{F}' \subset \mathfrak{F}_0$, which contradicts the relation $M(\mathfrak{F}_0) \in \mathfrak{F}_0$.

c $\Rightarrow$ a follows from Proposition 1.1.2.

The relation $\hat{\Phi} = \hat{\hat{\Phi}}$ follows from a $\Leftrightarrow$ b. □

Corollary 1.1.4. *Let X be a set, Φ be a set of filters on X and Φ' (resp. Φ'') be the set of filters (resp. ultrafilters) $\mathfrak{G}$ on X for which there exists a filter of Φ coarser than $\mathfrak{G}$. Then $\hat{\Phi} = \hat{\Phi}' = \hat{\Phi}''$.*

By Proposition 1.1.2 $\hat{\Phi}'' \subset \hat{\Phi}' \subset \hat{\Phi}$. The inclusion $\hat{\Phi} \subset \hat{\Phi}''$ follows from Proposition 1.1.2 and Proposition 1.1.3 a $\Rightarrow$ c. □

Corollary 1.1.5. *Let X be a set, Φ be a set of filters on X such that any ultrafilter on X belongs to Φ if there exists a filter of Φ coarser than it, and let $\mathfrak{F}$ be a filter on X. The following assertions are equivalent:*

a) $\mathfrak{F} \in \hat{\Phi}$;

b) any ultrafilter on X finer than $\mathfrak{F}$ belongs to Φ;

c) $\mathfrak{F} \in \hat{\Phi}_0$, where Φ_0 denotes the set of ultrafilters of Φ finer than $\mathfrak{F}$.

The assertion follows immediately from Proposition 1.1.3 if we remark that Φ_0 defined there coincides with Φ_0 defined in c). □

Corollary 1.1.6. *Let X be a set and Φ be a set of filters on X. If $\mathfrak{F}$, $\mathfrak{G} \in \hat{\Phi}$, then $\mathfrak{F} \cap \mathfrak{G} \in \hat{\Phi}$.*

Let $\mathfrak{H}$ be an ultrafilter on X finer than $\mathfrak{F} \cap \mathfrak{G}$. Then either $\mathfrak{F} \subset \mathfrak{H}$ or $\mathfrak{G} \subset \mathfrak{H}$. By Proposition 1.1.3 a $\Leftrightarrow$ b $\mathfrak{F} \cap \mathfrak{G} \in \hat{\Phi}$. □

Proposition 1.1.7. *Let X be a set, A be a subset of X and Φ be a set of filters on X such that any filter on X finer than a Φ-filter belongs to Φ. Then A is a Φ-set iff $\{B \mid A \subset B \subset X\} \in \Phi$.*

If A is a Φ-set then $\{B \mid A \subset B \subset X\} \in \Phi$. Conversely assume $\{B \mid A \subset B \subset X\} \in \Phi$ and let $\mathfrak{F}$ be a filter on X with $A \in \mathfrak{F}$. Let B be a subset of X containing A. Then $B \in \mathfrak{F}$. Hence, $\mathfrak{F}$ is finer than $\{B \mid A \subset B \subset X\}$ and by the hypothesis $\mathfrak{F} \in \Phi$. □

Corollary 1.1.8. *Let X be a set, A be a subset of X and Φ be a set of filters on X. Then A is a $\hat{\Phi}$-set iff $\{B \mid A \subset B \subset X\} \in \hat{\Phi}$.* □

Proposition 1.1.9. *Let X, Y be sets, Φ, Ψ be sets of filters on X and Y respectively and let φ be a map of X in Y such that $\varphi(\mathfrak{F}) \in \hat{\Psi}$ for any $\mathfrak{F} \in \Phi$. Then $\varphi(\mathfrak{F}) \in \hat{\Psi}$ for any $\mathfrak{F} \in \hat{\Phi}$.*

Let $\mathfrak{F} \in \hat{\Phi}$ and $\mathfrak{G}$ be an ultrafilter on Y finer than $\varphi(\mathfrak{F})$. Then there exists an ultrafilter $\mathfrak{H}$ on X finer than $\mathfrak{F}$ and such that $\varphi(\mathfrak{H}) = \mathfrak{G}$. By Proposition 1.1.3 a $\Rightarrow$ b there exists $\mathfrak{F}_0 \in \Phi$ with $\mathfrak{F}_0 \subset \mathfrak{H}$. We get $\varphi(\mathfrak{F}_0) \subset \mathfrak{G}$ and by the hypothesis $\varphi(\mathfrak{F}_0) \in \hat{\Psi}$. Hence, $\mathfrak{G} \in \hat{\Psi}$ and by Proposition 1.1.3 a $\Rightarrow$ b there exists a filter of Ψ coarser than $\mathfrak{G}$. By Proposition 1.1.3 b $\Rightarrow$ a we get $\varphi(\mathfrak{F}) \in \hat{\Psi}$. □

Proposition 1.1.10. *Let $(X_\iota)_{\iota \in I}$ be a family of sets. For each $\iota \in I$ let Φ_ι be a set of filters on X_ι, and let Φ be the set of filters on $\prod_{\iota \in I} X_\iota$ of the form $\prod_{\iota \in I} \mathfrak{F}_\iota$, where $(\mathfrak{F}_\iota)_{\iota \in I} \in \prod_{\iota \in I} \Phi_\iota$. Then a filter on $\prod_{\iota \in I} X_\iota$ belongs to $\hat{\Phi}$ iff for any $\iota \in I$ its projection on X_ι belongs to $\hat{\Phi}_\iota$.*

Let $\mathfrak{F} \in \hat{\Phi}$ and $\iota \in I$. The projection of any filter of Φ on X_ι belongs to Φ_ι. By Proposition 1.1.9 the projection of $\mathfrak{F}$ on X_ι belongs to $\hat{\Phi}_\iota$.

Now let $\mathfrak{F}$ be a filter on $\prod_{\iota \in I} X_\iota$ such that for any $\iota \in I$ its projection $\mathfrak{F}_\iota$ on X_ι belongs to $\hat{\Phi}_\iota$. Let $\mathfrak{G}$ be an ultrafilter on $\prod_{\iota \in I} X_\iota$ finer than $\mathfrak{F}$ and for any $\iota \in I$ let $\mathfrak{G}_\iota$ be its projection on X_ι. Then for any $\iota \in I$, $\mathfrak{G}_\iota$ is finer than $\mathfrak{F}_\iota$ and by Proposition 1.1.3 a $\Rightarrow$ b there exists $\mathfrak{H}_\iota \in \Phi_\iota$ with $\mathfrak{H}_\iota \subset \mathfrak{G}_\iota$. We get $\prod_{\iota \in I} \mathfrak{H}_\iota \subset \mathfrak{G}$ and therefore by Proposition 1.1.3 b $\Rightarrow$ a, $\mathfrak{F} \in \hat{\Phi}$. □

Proposition 1.1.11. *Let X be a set, Φ be a set of filters on X and $(X_\iota)_{\iota \in I}$ be a family of sets. For each $\iota \in I$ let Φ_ι be a set of filters on X_ι, and let φ be a map of $\prod_{\iota \in I} X_\iota$ in X such that $\varphi(\prod_{\iota \in I} \mathfrak{F}_\iota) \in \hat{\Phi}$ for any $(\mathfrak{F}_\iota)_{\iota \in I} \in \prod_{\iota \in I} \Phi_\iota$. Then $\varphi(\prod_{\iota \in I} \mathfrak{F}_\iota) \in \hat{\Phi}$ for any $(\mathfrak{F}_\iota)_{\iota \in I} \in \prod_{\iota \in I} \hat{\Phi}_\iota$.*

Let Ψ be the set of filters on $\prod_{\iota \in I} X_\iota$ of the form $\prod_{\iota \in I} \mathfrak{F}_\iota$ with $(\mathfrak{F}_\iota)_{\iota \in I} \in \prod_{\iota \in I} \Phi_i$ and let $(\mathfrak{F}_\iota)_{\iota \in I} \in \prod_{\iota \in I} \hat{\Phi}_\iota$. By Proposition 1.1.10 $\prod_{\iota \in I} \mathfrak{F}_\iota \in \hat{\Psi}$ and by Proposition 1.1.9 $\varphi(\prod_{\iota \in I} \mathfrak{F}_\iota) \in \hat{\Phi}$. □

§ 1.2 Sets of filters on topological and uniform spaces

Definition 1.2.1. *Let X be a uniform (topological) space. We denote by $\underline{\Phi_C(X)}$ $(\underline{\Phi_c(X)})$ or simply by $\underline{\Phi_C}$ $(\underline{\Phi_c})$ the set of Cauchy (convergent) filters on X.*

Proposition 1.2.2. *A filter on a uniform (topological) space belongs to $\hat{\Phi}_C$ $(\hat{\Phi}_c)$ if and only if any ultrafilter finer than it is a Cauchy (convergent) filter.*

The assertion follows immediately from Corollary 1.1.5 a $\Leftrightarrow$ b. □

Corollary 1.2.3. *On uniform spaces the precompact sets are exactly the $\hat{\Phi}_C$-sets. A subset A of a topological space X is a $\hat{\Phi}_c$-set iff any ultrafilter on X containing A is convergent. In particular, on regular topological spaces the $\hat{\Phi}_c$-sets and the relatively compact sets coincide.* □

Proposition 1.2.4. *Let X be a uniform space and $\mathfrak{F}$ be a filter on X. The following assertions are equivalent:*

a) $\mathfrak{F} \in \hat{\Phi}_C$;

b) for any entourage U of X there exists a finite set $\mathfrak{A}$ of subsets of X such that

$$\bigcup_{A \in \mathfrak{A}} (A \times A) \subset U, \quad \bigcup_{A \in \mathfrak{A}} A \in \mathfrak{F}.$$

a $\Rightarrow$ b. Let U be an entourage of X. For any $\mathfrak{G} \in \Phi_C$ let $M(\mathfrak{G})$ be a set of $\mathfrak{G}$ with $M(\mathfrak{G}) \times M(\mathfrak{G}) \subset U$. By a) there exists a finite subset Ψ of Φ_C such that $\bigcup_{\mathfrak{G} \in \Psi} M(\mathfrak{G}) \in \mathfrak{F}$.

b $\Rightarrow$ a. Let $\mathfrak{G}$ be an ultrafilter on X finer than $\mathfrak{F}$ and let U be an entourage of X. By b) there exists a finite set $\mathfrak{A}$ of subsets of X such that

$$\bigcup_{A \in \mathfrak{A}} (A \times A) \subset U, \quad \bigcup_{A \in \mathfrak{A}} A \in \mathfrak{F} \subset \mathfrak{G}.$$

Since $\mathfrak{G}$ is an ultrafilter there exists $A \in \mathfrak{A}$ with $A \in \mathfrak{G}$. Hence $\mathfrak{G}$ is a Cauchy filter. By Proposition 1.2.2 $\mathfrak{F} \in \hat{\Phi}_C$. □

The following result will not be used later.

Corollary 1.2.5. *The adherence of a filter of $\hat{\Phi}_C$ is precompact.*

Let X be a uniform space, $\mathfrak{F} \in \hat{\Phi}_C(X)$ and F be its adherence. Let further U be a closed entourage of X. By Proposition 1.2.4 there exists a finite set $\mathfrak{A}$ of subsets of X such that

$$\bigcup_{A \in \mathfrak{A}} (A \times A) \subset U, \quad \bigcup_{A \in \mathfrak{A}} A \in \mathfrak{F}.$$

Then

$$F \subset \overline{\bigcup_{A \in \mathfrak{A}} A} = \bigcup_{A \in \mathfrak{A}} \bar{A}, \qquad \bigcup_{A \in \mathfrak{A} \times \mathfrak{A}} (\bar{A} \times \bar{A}) = \bigcup_{A \in \mathfrak{A}} \overline{A \times A} \subset U.$$

Hence, F is precompact. □

Proposition 1.2.6. *Let X be a topological space and $\mathfrak{F}$ be a filter on X. The following assertions are equivalent:*

a) $\mathfrak{F} \in \hat{\Phi}_c$;

b) for any open covering $\mathfrak{U}$ of X there exists a finite subset $\mathfrak{V}$ of $\mathfrak{U}$ with $\bigcup_{V \in \mathfrak{V}} V \in \mathfrak{F}$.

a $\Rightarrow$ b. Let $\mathfrak{U}$ be an open covering of X and for any $\mathfrak{G} \in \Phi_c$ let $U(\mathfrak{G})$ be a set of $\mathfrak{G} \cap \mathfrak{U}$. By a) there exists a finite subset Ψ of Φ_c with $\bigcup_{\mathfrak{G} \in \Psi} U(\mathfrak{G}) \in \mathfrak{F}$.

b $\Rightarrow$ a. Let $\mathfrak{G}$ be an ultrafilter on X finer than $\mathfrak{F}$. Assume $\mathfrak{G}$ is not convergent. Then for any $x \in X$ there exists an open set U_x of X with $x \in U_x \notin \mathfrak{G}$. By b) there exists a finite subset A of X such that $\bigcup_{x \in A} U_x \in \mathfrak{F} \subset \mathfrak{G}$. Since $\mathfrak{G}$ is an ultrafilter there exists $x \in A$ with $U_x \in \mathfrak{G}$, which is a contradiction. Hence, $\mathfrak{G}$ is convergent and by Proposition 1.2.2 $\mathfrak{F} \in \hat{\Phi}_c$. □

Proposition 1.2.7. *Let X be a topological space such that any point of X possesses a fundamental system of closed neighbourhoods and let $\mathfrak{F}$ be a filter on X. The following assertions are equivalent:*

a) $\mathfrak{F} \in \hat{\Phi}_c$;

b) the adherence of $\mathfrak{F}$ is quasicompact and nonempty and $\mathfrak{F}$ is finer than the neighbourhood filter of its adherence;

c) there exists a quasicompact nonempty set K of X such that $\mathfrak{F}$ is finer than the neighbourhood filter of K.

a ⇒ b. Let F be the adherence of $\mathfrak{F}$ and $\mathfrak{U}$ be an open covering of X. For each $x \in X$ let V_x be an open neighbourhood of x whose closure is contained in a set $U_x \in \mathfrak{U}$. By Proposition 1.2.6 there exists a finite subset A of X such that $\bigcup_{x \in A} V_x \in \mathfrak{F}$. We get

$$F \subset \overline{\bigcup_{x \in A} V_x} = \bigcup_{x \in A} \overline{V_x} \subset \bigcup_{x \in A} U_x .$$

Hence, F is quasicompact. Let U be an open set of X containing F. If $U \notin \mathfrak{F}$ then there exists an ultrafilter $\mathfrak{G}$ on X finer than $\mathfrak{F}$ and such that $U \notin \mathfrak{G}$. By Proposition 1.2.2 $\mathfrak{G}$ is convergent. Let x be a limit point of $\mathfrak{G}$. Then $x \in F$ and therefore $U \in \mathfrak{G}$. This contradiction shows that $U \in \mathfrak{F}$. We deduce that F is nonempty and $\mathfrak{F}$ is finer than the neighbourhood filter of F.

b ⇒ c is trivial.

c ⇒ a. Let $\mathfrak{U}$ be an open covering of X. There exists a finite subset $\mathfrak{V}$ of $\mathfrak{U}$ such that $K \subset \bigcup_{V \in \mathfrak{V}} V$. Hence, $\bigcup_{V \in \mathfrak{V}} V \in \mathfrak{F}$. By Proposition 1.2.6 b ⇒ a $\mathfrak{F} \in \hat{\Phi}_c$. □

Proposition 1.2.8. *Let $(X_\iota)_{\iota \in I}$ be a family of uniform (topological) spaces. A filter $\mathfrak{F}$ on $\prod_{\iota \in I} X_\iota$ belongs to $\hat{\Phi}_C(\prod_{\iota \in I} X_\iota)$ ($\hat{\Phi}_c(\prod_{\iota \in I} X_\iota)$) iff for any $\iota \in I$ the projection of $\mathfrak{F}$ on X_ι belongs to $\hat{\Phi}_C(X_\iota)$ ($\hat{\Phi}_c(X_\iota)$). In particular the product of a family of $\hat{\Phi}_C$-filters ($\hat{\Phi}_c$-filters) is a $\hat{\Phi}_C$-filter ($\hat{\Phi}_c$-filter).*

The product of a family of Cauchy (convergent) filters being a Cauchy (convergent) filter, the assertion follows from Proposition 1.1.10. □

Definition 1.2.9. *Let X be a topological space and let Φ be a set of filters on X. We say that X is Φ-compact if the adherence of any filter of Φ is nonempty.*

If the ultrafilters on X finer than the filters of Φ belong to Φ then X is Φ-compact iff any ultrafilter of Φ converges.

Proposition 1.2.10. *Let X be a regular topological space, Φ be a set of filters on X and A be a Φ-set. If X is Φ-compact then $\bar{A}$ is compact.*

Let $\mathfrak{F}$ be an ultrafilter on $\bar{A}$. We denote by $\mathfrak{G}$ the filter on X generated by the filter base

$$\{A \cap U \mid U \quad \text{open set of } X, \quad U \cap \bar{A} \in \mathfrak{F}\} .$$

Then $A \in \mathfrak{G}$ and therefore $\mathfrak{G} \in \Phi$. Since X is Φ-compact, $\mathfrak{G}$ possesses an adherent point $x \in \bar{A}$. Let F be a closed neighbourhood of x in X. Then $X \backslash F \notin \mathfrak{G}$ and therefore $\bar{A} \backslash F \notin \mathfrak{F}$. We get $F \cap \bar{A} \in \mathfrak{F}$ and therefore $\mathfrak{F}$ converges to x. Hence, $\bar{A}$ is compact. □

§ 1.3 Φ-continuous and uniformly Φ-continuous maps

Definition 1.3.1. *Let X, Y be uniform spaces and Φ be a set of filters on X. We say that a map $\varphi : X \to Y$ is* uniformly Φ-continuous *if the image with respect to φ of any Cauchy filter of $\hat{\Phi}$ is a Cauchy filter.*

Any uniformly continuous map $X \to Y$ is uniformly Φ-continuous. From $\hat{\Phi} = \hat{\hat{\Phi}}$ (Proposition 1.1.3) it follows that the uniformly Φ-continuous maps and the uniformly $\hat{\Phi}$-continuous maps coincide.

Definition 1.3.2. *Let X, Y be topological spaces and Φ be a set of filters on X. We say that a map $\varphi : X \to Y$ is* Φ-continuous *if for any $x \in X$ and for any filter $\mathfrak{F}$ of Φ converging to x, $\varphi(\mathfrak{F})$ converges to $\varphi(x)$.*

Any continuous map $X \to Y$ is Φ-continuous. If any convergent filter belongs to Φ, then a map $X \to Y$ is continuous iff it is Φ-continuous.

Proposition 1.3.3. *Let X, Y be uniform spaces, Φ be a set of filters on X and φ be a uniformly Φ-continuous map of X into Y. Then φ is continuous at any point of X for which the neighbourhood filter belongs to $\hat{\Phi}$.*

Let $x \in X$ and let $\mathfrak{F}$ be the neighbourhood filter of x. If $\mathfrak{F} \in \hat{\Phi}$ then $\varphi(\mathfrak{F})$ is a Cauchy filter. Since $\varphi(x)$ is an adherent point of $\varphi(\mathfrak{F})$, it converges to $\varphi(x)$. Hence, φ is continuous at x. □

Proposition 1.3.4. *Let X be a topological space, Φ be a set of filters on X such that $\{A \mid x \in A \subset X\} \in \Phi$ for any $x \in X$, Y be a uniform space and φ be a map of X into Y such that $\varphi(\mathfrak{F})$ is a Cauchy filter for any convergent filter $\mathfrak{F}$ of $\hat{\Phi}$. Then φ is Φ-continuous.*

Let $x \in X$ and $\mathfrak{F}$ be a filter of Φ converging to x. By Corollary 1.1.6

$$\mathfrak{G} := \mathfrak{F} \cap \{A \mid x \in A \subset X\} \in \hat{\Phi}.$$

By the hypothesis $\varphi(\mathfrak{G})$ is a Cauchy filter. Since $\varphi(x)$ is an adherent point of $\varphi(\mathfrak{G})$, it is a limit point too. Hence $\varphi(\mathfrak{F})$ converges to $\varphi(x)$ and φ is Φ-continuous. □

Corollary 1.3.5. *Let X be a uniform space and Φ be a set of filters on X such that $\{A \mid x \in A \subset X\} \in \Phi$ for any $x \in X$. Then any uniformly Φ-continuous map of X into a uniform space is Φ-continuous.* □

Proposition 1.3.6. *Let X, Y be topological spaces and Φ be a set of filters on X such that any ultrafilter on X belongs to Φ if there exists a filter of Φ coarser than it. Then a map of X into Y is Φ-continuous iff it is $\hat{\Phi}$-continuous. In particular, a Φ-continuous map of X into Y is continuous at any point of X whose neighbourhood filter belongs to $\hat{\Phi}$.*

Let $\varphi : X \to Y$ be a Φ-continuous map, $\mathfrak{F} \in \hat{\Phi}$ and x be a limit point of $\mathfrak{F}$. Further, let $\mathfrak{G}$ be an ultrafilter on X finer than $\mathfrak{F}$. By Corollary 1.1.5 a $\Rightarrow$ b $\mathfrak{G} \in \Phi$. Hence, $\varphi(\mathfrak{G})$ converges to $\varphi(x)$. It follows that $\varphi(\mathfrak{F})$ converges to $\varphi(x)$. □

Proposition 1.3.7. *Let X, Y be topological spaces, Φ be a set of filters on X such that any ultrafilter on X belongs to Φ if it is finer than a filter of Φ, let A be a $\hat{\Phi}$-set and φ* be *a Φ-continuous map of X into Y. Then the restriction of φ to A is continuous. If any point of Y possesses a fundamental system of closed neighbourhoods, then the restriction of φ to $\bar{A}$ is continuous.*

Let $x \in \bar{A}$, let $\mathfrak{V}$ be the neighbourhood filter of x, and let $\mathfrak{F}$ be the filter on X generated by the filter base $\{A \cap V \mid V \in \mathfrak{V}\}$. Then $\mathfrak{F}$ is a $\hat{\Phi}$-filter on X converging to x. By Proposition 1.3.6 $\varphi(\mathfrak{F})$ converges to $\varphi(x)$. Both assertions follow immediately from this result. □

The following corollary will not be used in the sequel.

Corollary 1.3.8. *Let X', X'' be topological spaces defined on the same set X and let Φ be a set of filters on X such that any ultrafilter on X belongs to Φ if it is finer than a filter of Φ and such that the identity maps $X' \to X''$, $X'' \to X'$ are Φ-continuous. If A is a $\hat{\Phi}$-set, then X' and X'' induce the same topology on A. If in addition any point of X possesses a fundamental system of closed neighbourhoods in X' and X'', then the closures of A in X' and X'' coincide and X' and X'' induce the same topology on this closure of A.* □

Proposition 1.3.9. *Let X, Y be uniform spaces and Φ be a set of filters on X such that any ultrafilter on X finer than a filter of Φ belongs to Φ and such that $\mathfrak{F} \cap \mathfrak{G} \in \Phi$ for any ultrafilters $\mathfrak{F}$, $\mathfrak{G}$ of Φ. If φ is a map $X \to Y$ such that $\varphi(\mathfrak{F})$ is a Cauchy filter for any Cauchy filter $\mathfrak{F}$ of Φ, then φ is uniformly Φ-continuous.*

Let $\mathfrak{F}$ be a Cauchy filter of $\hat{\Phi}$ such that $\varphi(\mathfrak{F})$ is not a Cauchy filter. Then there exists an entourage V of Y such that $\varphi(A) \times \varphi(A) \not\subset V$ for any $A \in \mathfrak{F}$. Hence, there exist two maps $x, y : \mathfrak{F} \to X$ such that $x(A), y(A) \in A$ and $(\varphi(x(A)), \varphi(y(A))) \notin V$ for any $A \in \mathfrak{F}$. Let us endow $\mathfrak{F}$ with the order relation defined by the converse inclusion relation and let $\mathfrak{G}$ be an ultrafilter on $\mathfrak{F}$ finer than the section filter of $\mathfrak{F}$. Then $x(\mathfrak{G})$, $y(\mathfrak{G})$ are ultrafilters on X finer than $\mathfrak{F}$ and therefore belong to Φ (Corollary 1.1.5 a $\Rightarrow$ b). By the hypothesis $x(\mathfrak{G}) \cap y(\mathfrak{G}) \in \Phi$. We get $\varphi(x(\mathfrak{G}) \cap y(\mathfrak{G}))$ is a Cauchy filter and this is a contradiction. □

Proposition 1.3.10. *Let X, Y be uniform spaces, Φ be a set of filters on X, φ be a uniformly Φ-continuous map of X into Y and let $\mathfrak{F} \in \hat{\Phi}$. Then for any entourage V of Y*

there exist $A \in \mathfrak{F}$ *and an entourage* U *of* X *such that* $(\varphi(x), \varphi(y)) \in V$ *for any* $(x, y) \in (A \times A) \cap U$.

Let $\mathfrak{U}$ be the uniformity of X and I be the set $\mathfrak{U} \times \mathfrak{F}$ endowed with the upper directed order relation $\leq$ defined by

$$(U, A) \leq (U', A') :\Leftrightarrow (U' \subset U \quad \text{and} \quad A' \subset A).$$

Assume the assertion does not hold. Then there exists an entourage V of Y such that for any $\iota := (U, A) \in I$ there exists $(x(\iota), y(\iota)) \in (A \times A) \cap U$ with $(\varphi(x(\iota)), \varphi(y(\iota))) \notin V$. Let $\mathfrak{G}$ be the section filter of I. It is easy to see that $x(\mathfrak{G}) \cap y(\mathfrak{G})$ is a Cauchy filter finer than $\mathfrak{F}$, so it belongs to $\hat{\Phi}$. Hence, $\varphi(x(\mathfrak{G}) \cap y(\mathfrak{G}))$ is a Cauchy filter and this is a contradiction. □

Proposition 1.3.11. *Let* X, Y *be uniform spaces,* Φ *be a set of filters on* X, φ *be a uniformly* Φ*-continuous map of* X *into* Y *and let* A *be a* $\hat{\Phi}$*-set. Then the restriction of* φ *to* A *is uniformly continuous. If in addition* $\{B \mid x \in B \subset X\} \in \Phi$ *for any* $x \in X$, *then the restriction of* φ *to* $\bar{A}$ *is uniformly continuous.*

The first assertion follows immediately from Proposition 1.3.10. Assume now $\{B \mid x \in B \subset X\} \in \Phi$ for any $x \in X$. By Corollary 1.3.5 φ is $\hat{\Phi}$-continuous. By Proposition 1.3.7 the restriction of φ to $\bar{A}$ is continuous. Since the restriction of φ to A is uniformly continuous its restriction to $\bar{A}$ is uniformly continuous too. □

Corollary 1.3.12. *Let* X', X'' *be uniform spaces defined on the same set* X *and* Φ *be a set of filters on* X *such that the identity maps* $X' \to X''$, $X'' \to X'$ *are uniformly* Φ*-continuous. If* A *is a* $\hat{\Phi}$*-set, then* X' *and* X'' *induce the same uniformity on* A. *If in addition* $\{B \mid x \in B \subset X\} \in \Phi$ *for any* $x \in X$ *then the closures of* A *in* X' *and* X'' *coincide and* X' *and* X'' *induce the same uniformity on this closure of* A.

The assertion follows immediately from Proposition 1.2.2 and Proposition 1.3.11. □

Proposition 1.3.13. *Let* X, Y, Z *be uniform spaces,* Φ, Ψ *be sets of filters on* X *and* Y *respectively, let* Λ *be the set of filters on* $X \times Y$ *of the form* $\mathfrak{F} \times \mathfrak{G}$ *with* $(\mathfrak{F}, \mathfrak{G}) \in \Phi \times \Psi$ *and* φ *be a uniformly* Λ*-continuous map of* $X \times Y$ *into* Z. *Then for any* $\mathfrak{F} \in \hat{\Phi} \cap \Phi_c(X)$, *for any* $\mathfrak{G} \in \hat{\Psi}$, *and for any entourage* V *of* Z *there exist* $(A, B) \in \mathfrak{F} \times \mathfrak{G}$ *and an entourage* U *of* Y *such that* $(\varphi(x, y), \varphi(x', y')) \in V$ *for any* $x, x' \in A$ *and* $(y, y') \in (B \times B) \cap U$.

By Proposition 1.1.10 $\mathfrak{F} \times \mathfrak{G} \in \hat{\Lambda}$. By Proposition 1.3.10 there exist $(A', B) \in \mathfrak{F} \times \mathfrak{G}$, an entourage U' of X and an entourage U of Y such that $(\varphi(x, y), \varphi(x', y')) \in V$ for any $(x, x') \in (A' \times A') \cap U'$ and $(y, y') \in (B \times B) \cap U$. Since $\mathfrak{F}$ is a Cauchy filter there exists $A'' \in \mathfrak{F}$ with $A'' \times A'' \subset U'$. If we set $A := A' \cap A''$, then $(\varphi(x, y), \varphi(x', y')) \in V$ for any $x, x' \in A$ and $(y, y') \in (B \times B) \cap U$. □

The next proposition will not be used later.

Proposition 1.3.14. *Let X, Y be uniform spaces, φ be the map*

$$Y^X \times X \to Y, \quad (f, x) \mapsto f(x),$$

$\mathfrak{F}$ be a filter on Y^X and Φ be a set of filters on X such that $\varphi(\mathfrak{F} \times \mathfrak{G})$ is a Cauchy filter for any Cauchy filter $\mathfrak{G}$ on X belonging to $\hat{\Phi}$. Then for any entourage V of Y and any $\mathfrak{G} \in \hat{\Phi}$ there exist $(\mathscr{F}, A) \in \mathfrak{F} \times \mathfrak{G}$ and an entourage U of X such that $(f(x), g(y)) \in V$ for any $f, g \in \mathscr{F}$ and any $(x, y) \in (A \times A) \cap U$.

Let us endow Y^X with the coarsest uniformity and let Ψ be the set of filters on $Y^X \times X$ of the form $\mathfrak{F} \times \mathfrak{G}$, with $\mathfrak{G} \in \Phi$. Further, let $\mathfrak{H}$ be a Cauchy filter of $\hat{\Psi}$ and $\mathfrak{H}_1$, $\mathfrak{H}_2$ be its projections on Y^X and X respectively. By Proposition 1.1.10 $\mathfrak{F} \subset \mathfrak{H}_1$ and $\mathfrak{H}_2 \in \hat{\Phi}$ and therefore, by the hypothesis of the proposition, $\varphi(\mathfrak{F} \times \mathfrak{H}_2)$ is a Cauchy filter. We deduce that $\varphi(\mathfrak{H})$ is a Cauchy filter. Hence, φ is uniformly Ψ-continuous and the assertion follows from Proposition 1.3.13. □

§ 1.4 Filters defined by sets of sequences

Definition 1.4.1. *Let X be a set and Θ be a set of sequences in X. A* <u>Θ-net in X</u> *is a net (I, f) in X such that for any increasing sequence $(\iota_n)_{n \in \mathbb{N}}$ in I the sequence $(f(\iota_n))_{n \in \mathbb{N}}$ belongs to Θ. A* <u>Θ-sequence in X</u> *is a sequence $(x_n)_{n \in \mathbb{N}}$ in X such that $(\mathbb{N}, f)$ is a Θ-net, where*

$$f : \mathbb{N} \to X, \quad n \mapsto x_n .$$

We shall denote by <u>$\Phi(\Theta, X)$</u>*, or simply by* <u>$\Phi(\Theta)$</u>*, the set of filters on X of the form $f(\mathfrak{F})$, where (I, f) is a Θ-net in X and $\mathfrak{F}$ is the section filter of I. We set*

$$\underline{\hat{\Phi}(\Theta, X)} := \underline{\hat{\Phi}(\Theta)} := \widehat{\Phi(\Theta)} .$$

If (I, f) is a Θ-net in X, then $(f(\iota_n))_{n \in \mathbb{N}}$ is a Θ-sequence in X for any increasing sequence $(\iota_n)_{n \in \mathbb{N}}$ in I.

Proposition 1.4.2. *Let X be a set and Θ be a set of sequences in X. Any filter on X finer than a filter of $\Phi(\Theta)$ belongs to $\Phi(\Theta)$. A filter on X belongs to $\hat{\Phi}(\Theta)$ iff any ultrafilter on X finer than it belongs to $\Phi(\Theta)$. A subset A of X is a $\Phi(\Theta)$-set iff $\{B \mid A \subset B \subset X\} \in \Phi(\Theta)$.*

Let (I, f) be a Θ-net in X, $\mathfrak{F}$ be the section filter of I and let $\mathfrak{G}$ be a filter on X finer than $f(\mathfrak{F})$. We denote by J the set

$$\{(\iota, A) \in I \times \mathfrak{G} \mid f(\iota) \in A\} ,$$

by g the map

$$J \to X, \quad (\iota, A) \mapsto f(\iota) ,$$

and by $\leq$ the binary relation on J defined by

$$(\iota, A) \le (\iota', A'): \Leftrightarrow (\iota \le \iota' \quad \text{and} \quad A' \subset A).$$

It is easy to check that $\le$ is an upper directed preorder relation on J and that (J, g) is a Θ-net in X such that $\mathfrak{G} = g(\mathfrak{F}')$, where $\mathfrak{F}'$ denotes the section filter of J. Hence, $\mathfrak{G} \in \Phi(\Theta)$.

The last assertions follow from the first one and from Corollary 1.1.5 and Proposition 1.1.7. □

Proposition 1.4.3. *Let X be a set, A be a subset of X and Θ be a set of sequences in X. Then A is a $\Phi(\Theta)$-set iff any sequence in A belongs to Θ.*

Assume first that A is a $\Phi(\Theta)$-set and let $(x_n)_{n \in \mathbb{N}}$ be a sequence in A. Since $\{B \mid A \subset B \subset X\}$ is a Θ-filter there exists a Θ-net (I, f) in X such that

$$f(\mathfrak{F}) = \{B \mid A \subset B \subset X\},$$

where $\mathfrak{F}$ denotes the section filter of I. We construct inductively an increasing sequence $(\iota_n)_{n \in \mathbb{N}}$ in I such that $f(\iota_n) = x_n$ for any $n \in \mathbb{N}$. Let $n \in \mathbb{N}$ and suppose the sequence was constructed up to $n-1$. Then

$$A \subset \{f(\iota) \mid \iota \in I, \quad \iota \ge \iota_{n-1}\}.$$

Hence, there exists $\iota_n \in I$ with $\iota_n \ge \iota_{n-1}$ and $f(\iota_n) = x_n$. It follows that $(x_n)_{n \in \mathbb{N}}$ belongs to Θ.

Assume now that any sequence in A belongs to Θ. Let $\le$ be the coarsest preorder relation on A (i.e. $x \le y$ for any $x, y \in A$) and f be the inclusion map $A \to X$. Then (A, f) is a Θ-net in X and

$$f(\mathfrak{F}) = \{B \mid A \subset B \subset X\},$$

where $\mathfrak{F}$ denotes the section filter of A. By Proposition 1.4.2 A is a $\Phi(\Theta)$-set. □

Proposition 1.4.4. *Let X be a set, Θ be a set of sequences in X and $(x_n)_{n \in \mathbb{N}}$ be a sequence in X such that there exists a filter of $\hat{\Phi}(\Theta)$ finer than the elementary filter of $(x_n)_{n \in \mathbb{N}}$. Then there exists a subsequence of $(x_n)_{n \in \mathbb{N}}$ belonging to Θ.*

By Proposition 1.4.2 we may assume there exists a filter $\mathfrak{F}$ of $\Phi(\Theta)$ finer than the elementary filter of $(x_n)_{n \in \mathbb{N}}$. Let (I, f) be a Θ-net in X such that $f(\mathfrak{G}) = \mathfrak{F}$, where $\mathfrak{G}$ is the section filter of I. We construct inductively an increasing sequence $(\iota_n)_{n \in \mathbb{N}}$ in I and a strictly increasing sequence $(k_n)_{n \in \mathbb{N}}$ in $\mathbb{N}$ such that $f(\iota_n) = x_{k_n}$ for any $n \in \mathbb{N}$. Let $n \in \mathbb{N}$ and assume the sequences were constructed up to $n-1$. Then

$$\{f(\iota) \mid \iota \in I, \ \iota \ge \iota_{n-1}\} \cap \{x_m \mid m > k_{n-1}\} \in \mathfrak{F}.$$

Hence there exists $(\iota_n, k_n) \in I \times \mathbb{N}$ with

$$\iota_n \ge \iota_{n-1}, \ k_n > k_{n-1}, \ f(\iota_n) = x_{k_n}.$$

Then $(x_{k_n})_{n \in \mathbb{N}}$ is a subsequence of $(x_n)_{n \in \mathbb{N}}$ belonging to Θ. □

Proposition 1.4.5. *Let X be a set, A be a subset of X and Θ be a set of sequences in X such that any sequence in X belongs to Θ if it possesses a subsequence belonging to Θ. Then the following assertions are equivalent:*

a) A is a $\Phi(\Theta)$-set;

b) A is a $\hat{\Phi}(\Theta)$-set;

c) any sequence in A belongs to Θ.

a $\Rightarrow$ b is trivial;

b $\Rightarrow$ c. Let $(x_n)_{n\in\mathbb{N}}$ be a sequence in A. Then the elementary filter of $(x_n)_{n\in\mathbb{N}}$ belongs to $\hat{\Phi}(\Theta)$ and by Proposition 1.4.4 it possesses a subsequence belonging to Θ. By the hypothesis $(x_n)_{n\in\mathbb{N}}\in\Theta$.

c $\Rightarrow$ a follows from Proposition 1.4.3. $\square$

Proposition 1.4.6. *Let X be a set, Θ be a set of sequences in X, A be the set of $x\in X$ such that the constant sequence $(x)_{n\in\mathbb{N}}\in\Theta$, (I,f) be a Θ-net in X and $\mathfrak{F}\in\hat{\Phi}(\Theta)$. Then $f(I)\subset A$ and $A\in\mathfrak{F}$.*

Let $x\in f(I)$. Then there exists $\iota\in I$ so that $f(\iota)=x$. The constant sequence $(\iota)_{n\in\mathbb{N}}$ in I being increasing it follows that the constant sequence $(x)_{n\in\mathbb{N}}$ belongs to Θ. Hence, $x\in A$ and $f(I)\subset A$.

Assume $A\notin\mathfrak{F}$. Then there exists an ultrafilter $\mathfrak{G}$ on X finer than $\mathfrak{F}$ with $X\setminus A\in\mathfrak{G}$. By Proposition 1.4.2 $\mathfrak{G}\in\Phi(\Theta)$. Hence, there exists a Θ-net (J,g) in X with $g(\mathfrak{H})=\mathfrak{G}$, where $\mathfrak{H}$ denotes the section filter of J. By the above proof $g(J)\subset A$ and therefore $A\in\mathfrak{G}$, which is a contradiction. Hence, $A\in\mathfrak{F}$. $\square$

Proposition 1.4.7. *Let X be a set, Θ be a set of sequences in X and $(x_n)_{n\in\mathbb{N}}$ be a sequence in X. If $\{x_n \mid n\in\mathbb{N}\}$ is a $\Phi(\Theta)$-set, then $(x_n)_{n\in\mathbb{N}}$ is a Θ-sequence. If $(x_n)_{n\in\mathbb{N}}$ is a Θ-sequence, then $\{x_n \mid n\in\mathbb{N}\}$ is a $\hat{\Phi}(\Theta)$-set. If in addition any sequence in X, which possesses a subsequence belonging to Θ belongs itself to Θ, then $\{x_n \mid n\in\mathbb{N}\}$ is a $\Phi(\Theta)$-set.*

By Proposition 1.4.3 if $\{x_n \mid n\in\mathbb{N}\}$ is a $\Phi(\Theta)$-set, then $(x_n)_{n\in\mathbb{N}}$ is a Θ-sequence. Assume that $(x_n)_{n\in\mathbb{N}}$ is a Θ-sequence and let $\mathfrak{F}$ be an ultrafilter on X with $\{x_n \mid n\in\mathbb{N}\}\in\mathfrak{F}$. If there exists $n\in\mathbb{N}$ with $\{x_n\}\in\mathfrak{F}$, then $\mathfrak{F}\in\Phi(\Theta)$. If $\{x_n\}\notin\mathfrak{F}$ for any $n\in\mathbb{N}$, then $\mathfrak{F}$ is finer than the elementary filter of $(x_n)_{n\in\mathbb{N}}$ and therefore $\mathfrak{F}\in\Phi(\Theta)$ (Proposition 1.4.2). Hence, $\{x_n \mid n\in\mathbb{N}\}$ is a $\hat{\Phi}(\Theta)$-set (Proposition 1.4.2) and if the supplementary hypothesis is fulfilled it is even a $\Phi(\Theta)$-set (Proposition 1.4.5). $\square$

Proposition 1.4.8. *Let X, X' be sets, Θ, Θ' be sets of sequences in X and X' respectively and φ be a map of X into X'. Let us consider the following assertions:*

a) $(x_n)_{n\in\mathbb{N}}$ Θ-sequence in $X \Rightarrow (\varphi(x_n))_{n\in\mathbb{N}}$ is a Θ'-sequence in X';

b) (I,f) Θ-net in $X \Rightarrow (I,\varphi\circ f)$ is a Θ'-net in X';

c) $\mathfrak{F}\in\Phi(\Theta)\Rightarrow\varphi(\mathfrak{F})\in\Phi(\Theta')$;

d) $A\subset X$, A is a $\Phi(\Theta)$-set $\Rightarrow \varphi(A)$ is a $\Phi(\Theta')$-set:

e) $\mathfrak{F} \in \hat{\Phi}(\Theta) \Rightarrow \varphi(\mathfrak{F}) \in \hat{\Phi}(\Theta')$:

f) $A \subset X$, *A is a* $\hat{\Phi}(\Theta)$*-set* $\Rightarrow$ $\varphi(A)$ *is a* $\hat{\Phi}(\Theta')$*-set.*

Then:

1) $a \Leftrightarrow b \Rightarrow c \Rightarrow d$ *and* $c \Rightarrow e \Rightarrow f$;

2) if every sequence in X which possesses a subsequence belonging to Θ *belongs itself to* Θ *then* $d \Rightarrow a \& f$;

3) if every sequence in X′ which possesses a subsequence belonging to Θ' *belongs itself to* Θ' *then* $f \Rightarrow d$;

4) if the hypotheses of 2) and 3) are fulfilled, then the assertions a), b), c), d), e), f) coincide.

1) $a \Leftrightarrow b \Rightarrow c$ is trivial, $c \Rightarrow d$ follows from Proposition 1.4.2. $c \Rightarrow e$ follows from Proposition 1.1.9. $e \Rightarrow f$ follows from Corollary 1.1.8.

2) $d \Rightarrow a$. By Proposition 1.4.7 $\{x_n \mid n \in \mathbb{N}\}$ is a $\Phi(\Theta)$-set and by d) $\{\varphi(x_n) \mid n \in \mathbb{N}\}$ is a $\Phi(\Theta')$-set. By Proposition 1.4.3 $(\varphi(x_n))_{n \in \mathbb{N}}$ is a Θ'-sequence. $d \Rightarrow f$ follows from Proposition 1.4.5.

3) follows from Proposition 1.4.5.

4) follows from 1), 2), and 3). □

Proposition 1.4.9. *Let X be a set,* Θ *be a set of sequences in X and let* $\mathfrak{F}, \mathfrak{G} \in \Phi(\Theta)$. *We assume that any sequence* $(x_n)_{n \in \mathbb{N}}$ *in X belongs to* Θ *if there exist strictly increasing maps*

$$f: \mathbb{N} \to \mathbb{N}, \quad g: \mathbb{N} \to \mathbb{N}$$

with

$$f(\mathbb{N}) \cup g(\mathbb{N}) = \mathbb{N}, \quad f(\mathbb{N}) \cap g(\mathbb{N}) = \emptyset$$

for which the sequences $(x_{f(n)})_{n \in \mathbb{N}}, (x_{g(n)})_{n \in \mathbb{N}}$ *belong to* Θ. *Then* $\mathfrak{F} \cap \mathfrak{G} \in \Phi(\Theta)$.

Let $(I, f), (J, g)$ be Θ-nets in X such that $f(\mathfrak{F}') = \mathfrak{F}$, $g(\mathfrak{G}') = \mathfrak{G}$, where $\mathfrak{F}', \mathfrak{G}'$ denote the section filters of I and J respectively. Let $\leq$ be the binary relation on $I \times J \times \{0, 1\}$ defined by

$$(\iota, \kappa, \lambda) \leq (\iota', \kappa', \lambda') :\Leftrightarrow (\iota \leq \iota', \kappa \leq \kappa')$$

and let h be the map of $I \times J \times \{0, 1\}$ into X defined by

$$h(\iota, \kappa, 0) := f(\iota), \quad h(\iota, \kappa, 1) := g(\kappa)$$

for any $(\iota, \kappa) \in I \times J$. Then $\leq$ is an upper directed preorder relation on $I \times J \times \{0, 1\}$, $(I \times J \times \{0, 1\}, h)$ is a Θ-net in X, and $h(\mathfrak{H}) = \mathfrak{F} \cap \mathfrak{G}$, where $\mathfrak{H}$ denotes the section filter of $I \times J \times \{0, 1\}$. Hence, $\mathfrak{F} \cap \mathfrak{G} \in \Phi(\Theta)$. □

Proposition 1.4.10. *Let X, Y be sets, Z be a uniform space,* φ *be a map of* $X \times Y$ *into Z,* Φ *be the set of filters* $\mathfrak{F}$ *on* $X \times Y$ *such that for any entourage U of Z there exists*

$A \in \mathfrak{F}$ with $(\varphi(x, y), \varphi(x', y)) \in U$ for any $(x, y), (x', y) \in A$, and let Ψ be a set of ultrafilters of Φ such that $\mathfrak{F} \cap \mathfrak{G} \in \Phi$ for any $\mathfrak{F}, \mathfrak{G} \in \Psi$. Then $\hat{\Psi} \subset \Phi$.

Let $\mathfrak{F} \in \hat{\Psi}$ and assume $\mathfrak{F} \notin \Phi$. Then there exists an entourage U of Z such that for any $A \in \mathfrak{F}$ there exist $x(A), x'(A) \in X$, $y(A) \in Y$ with

$$z(A) := (x(A), y(A)) \in A, \quad z'(A) := (x'(A), y(A)) \in A, \quad (\varphi(z(A)), \varphi(z'(A))) \notin U.$$

Let us order $\mathfrak{F}$ by the converse inclusion relation and let $\mathfrak{G}$ be an ultrafilter on $\mathfrak{F}$ finer than the section filter of $\mathfrak{F}$. Then $z(\mathfrak{G})$, $z'(\mathfrak{G})$ are ultrafilters on $X \times Y$ finer than $\mathfrak{F}$. By Proposition 1.1.3 a $\Rightarrow$ b they belong to Ψ. Thus, $z(\mathfrak{G}) \cap z'(\mathfrak{G}) \in \Phi$. Hence, there exists $B \in z(\mathfrak{G}) \cap z'(\mathfrak{G})$ such that $(\varphi(x, y), \varphi(x', y)) \in U$ for any $(x, y), (x', y) \in B$. Let $C, C' \in \mathfrak{G}$ such that $z(C) \subset B$, $z'(C') \subset B$. Then $C \cap C' \in \mathfrak{G}$ and therefore there exists $A \in C \cap C'$. We get $z(A), z'(A) \in B$ and therefore $(\varphi(z(A)), \varphi(z'(A))) \in U$ and this is a contradiction. Hence, $\mathfrak{F} \in \Phi$ and $\hat{\Psi} \subset \Phi$. □

Proposition 1.4.11. *Let X, Y be sets, Z be a uniform space, φ be a map of $X \times Y$ into Z and let Θ be the set of sequences $(x_n, y_n)_{n \in \mathbb{N}}$ in $X \times Y$ such that for any entourage U of Z there exist $m, n \in \mathbb{N}$ with $m < n$ and*

$$(\varphi(x_m, y_m), \varphi(x_n, y_m)) \in U.$$

Let $\mathfrak{F} \in \hat{\Phi}(\Theta)$ and $\mathfrak{G}, \mathfrak{H}$ be the projections of $\mathfrak{F}$ on X and Y respectively. If

$$\{y \in Y \mid \varphi(\mathfrak{G}, y) \quad \text{is a Cauchy filter}\} \in \mathfrak{H},$$

then for any entourage U of Z there exists $A \in \mathfrak{F}$ such that

$$(\varphi(x, y), \varphi(x', y)) \in U$$

for any $(x, y), (x', y) \in A$.

Assume first $\mathfrak{F} \in \Phi(\Theta)$ and let (I, f) be a Θ-net in $X \times Y$ such that $f(\mathfrak{F}') = \mathfrak{F}$, where $\mathfrak{F}'$ denotes the section filter of I. We set

$$\begin{cases} p: X \times Y \to X, & (x, y) \mapsto x; \\ q: X \times Y \to Y, & (x, y) \mapsto y; \end{cases} \quad \begin{cases} g := p \circ f, \\ h := q \circ f. \end{cases}$$

Let U be an entourage of Z. Assume that for any $A \in \mathfrak{F}$ there exist $(x, y), (x', y) \in A$ with

$$(\varphi(x, y), \varphi(x', y)) \notin U.$$

Let V be an entourage of Z such that $\overset{-1}{V} \circ V \circ V \subset U$. We construct inductively an increasing sequence $(\iota_n)_{n \in \mathbb{N}}$ in I such that

$$(\varphi(f(\iota_{2n})), \varphi(g(\iota), h(\iota_{2n}))) \notin V$$

for any $n \in \mathbb{N}$ and for any $\iota \in I$ with $\iota \geq \iota_{2n+1}$. Let $n \in \mathbb{N}$ and assume the sequence was constructed up to $2n - 1$. We set

$$B := \{y \in Y \mid \varphi(\mathfrak{G}, y) \quad \text{is a Cauchy filter}\}.$$

Since $B \in \mathfrak{H}$ there exists $\kappa \in I$ such that $\kappa \geq \iota_{2n-1}$ and

$$\{h(\iota) \mid \iota \in I,\ \iota \geq \kappa\} \subset B$$

(ι_{-1} is an arbitrary element of I if $n = 0$). By the above hypothesis there exist $\iota', \iota'' \in I$ such that $\iota' \geq \kappa, \iota'' \geq \kappa, h(\iota') = h(\iota'')$ and $(\varphi(f(\iota')), \varphi(f(\iota''))) \notin U$. Since $\varphi(\mathfrak{G}, h(\iota'))$ is a Cauchy filter there exists $\iota_{2n+1} \in I$ such that $\iota_{2n+1} \geq \iota', \iota_{2n+1} \geq \iota''$, and

$$(\varphi(g(\iota), h(\iota')), \varphi(g(\lambda), h(\iota'))) \in V$$

for any $\iota, \lambda \in I$ with $\iota \geq \iota_{2n+1}, \lambda \geq \iota_{2n+1}$. Hence, either

$$(\varphi(f(\iota')), \varphi(g(\iota), h(\iota'))) \notin V$$

for any $\iota \in I$ with $\iota \geq \iota_{2n+1}$ or

$$(\varphi(f(\iota'')), \varphi(g(\iota), h(\iota'')) \notin V$$

for any $\iota \in I$ with $\iota \geq \iota_{2n+1}$. We set $\iota_{2n} := \iota'$ in the first case and $\iota_{2n} := \iota''$ in the second case. This proves the existence of the sequence $(\iota_n)_{n \in \mathbb{N}}$. Since $(f(\iota_{2n}))_{n \in \mathbb{N}} \in \Theta$ there exist $m, n \in \mathbb{N}$ such that $m < n$ and

$$(\varphi(f(\iota_{2m})), \varphi(g(\iota_{2n}), h(\iota_{2m})) \in V,$$

which is the expected contradiction.

Now let $\mathfrak{F}$ belong to $\hat{\Phi}(\Theta)$. Let Ψ be the set of ultrafilters on $X \times Y$ finer than $\mathfrak{F}$ and Φ_0 be the set of filters $\mathfrak{F}_0$ on $X \times Y$ such that for any entourage U of Z there exists $A_0 \in \mathfrak{F}_0$ with $(\varphi(x, y), \varphi(x', y)) \in U$ for any $(x, y), (x', y) \in A_0$. By Proposition 1.4.2 $\Psi \subset \Phi(\Theta)$. Let $\mathfrak{F}_1, \mathfrak{F}_2 \in \Psi$. By Proposition 1.4.9 $\mathfrak{F}_1 \cap \mathfrak{F}_2 \in \Phi(\Theta)$ and by the above proof $\mathfrak{F}_1 \cap \mathfrak{F}_2 \in \Phi_0$. By Proposition 1.4.10 $\hat{\Psi} \subset \Phi_0$. Since $\mathfrak{F} \in \hat{\Psi}$ (Corollary 1.1.5) we get $\mathfrak{F} \in \Phi_0$. □

Proposition 1.4.12. *Let $(X_\iota)_{\iota \in I}$ be a countable family of sets and for any $\iota \in I$ let Θ_ι be a set of sequences in X_ι containing all constant sequences, and Θ'_ι be the set of sequences in X_ι possessing subsequences belonging to Θ_ι. We denote by Θ the set of sequences $(x_n)_{n \in \mathbb{N}}$ in $\prod_{\iota \in I} X_\iota$ such that for any $\iota \in I$ the projection of $(x_{n+m})_{n \in \mathbb{N}}$ in X_ι belongs to Θ_ι for an $m \in \mathbb{N}$ and by Θ' the set of sequences in $\prod_{\iota \in I} X_\iota$ possessing subsequences belonging to Θ. Then:*

a) $(\mathfrak{F}_\iota)_{\iota \in I} \in \prod_{\iota \in I} \Phi(\Theta'_\iota) \Rightarrow \prod_{\iota \in I} \mathfrak{F}_\iota \in \Phi(\Theta')$;

b) $(\mathfrak{F}_\iota)_{\iota \in I} \in \prod_{\iota \in I} \hat{\Phi}(\Theta'_\iota) \Rightarrow \prod_{\iota \in I} \mathfrak{F}_\iota \in \hat{\Phi}(\Theta')$.

a) Without loss of generality we may assume $I = \mathbb{N}$. For any $n \in \mathbb{N}$ let (I_n, f_n) be a Θ'_n-net in X_n such that $f_n(\mathfrak{G}_n) = \mathfrak{F}_n$, where $\mathfrak{G}_n$ denotes the section filter of I_n, and such that $I_n \cap X_n = \emptyset$. We set

$$J := \bigcup_{n\in\mathbb{N}} ((\prod_{m\le n} I_m) \times (\prod_{m>n} X_m)).$$

For any $g \in J$ there exists exactly one $n(g) \in \mathbb{N}$ such that

$$g \in (\prod_{m\le n(g)} I_m) \times (\prod_{m>n(g)} X_m).$$

We define an order relation $\le$ on J as follows. Let $g, h \in J$; we set $g \le h$ if $g = h$ or if $n(g) < n(h)$ and $g(m) \le h(m)$ for any $m \le n(g)$. Let f be the map of J into $\prod_{n\in\mathbb{N}} X_n$ defined as follows. For any $g \in J$ we set

$$f(g)_n := f_n(g(n))$$

if $n \le n(g)$ and

$$f(g)_n := g(n)$$

if $n > n(g)$. Then (J, f) is a net in $\prod_{n\in\mathbb{N}} X_n$ such that $f(\mathfrak{G}) = \prod_{n\in\mathbb{N}} \mathfrak{F}_n$, where $\mathfrak{G}$ denotes the section filter of J. We want to show that (J, f) is a Θ'-net. Let $(g_n)_{n\in\mathbb{N}}$ be an increasing sequence in J. By the diagonal method we may construct a subsequence $(g_{k_m})_{m\in\mathbb{N}}$ of $(g_n)_{n\in\mathbb{N}}$ such that for any $n \in \mathbb{N}$ the projection of $(f(g_{k_{m+n}}))_{m\in\mathbb{N}}$ on X_n belongs to Θ_n. Hence, $(f(g_{k_m}))_{m\in\mathbb{N}} \in \Theta$ and therefore $(g_n)_{n\in\mathbb{N}} \in \Theta'$. We get $f(\mathfrak{G}) \in \Phi(\Theta')$.

b) follows from a) and Proposition 1.1.10. □

§ 1.5 Sets of sequences in topological and uniform spaces

Definition 1.5.1. *Let X be a topological space. We denote by* $\underline{\Theta_1(X)}$ *the set of sequences in X possessing convergent subsequences and by* $\underline{\Theta_2(X)}$ *the set of sequences in X possessing adherent points. We set* $\underline{\Phi_i(X)} := \Phi(\Theta_i(X))$ *and* $\underline{\hat{\Phi}_i(X)} := \widehat{\Phi_i(X)}$ *for any* $i \in \{1, 2\}$.

We have $\Theta_1(X) \subset \Theta_2(X)$, $\Phi_1(X) \subset \Phi_2(X)$.

Definition 1.5.2. *Let X be a uniform space. We denote by* $\underline{\Theta_3(X)}$ *the set of sequences in X possessing Cauchy subsequences, by* $\underline{\Theta_4(X)}$ *the set* $\Theta_2(X) \cup \Theta_3(X)$, *and by* $\underline{\Theta_5(X)}$ *the set of sequences* $(x_n)_{n\in\mathbb{N}}$ *in X such that for any entourage U of X there exist different* $m, n \in \mathbb{N}$ *with* $(x_m, x_n) \in U$. *We set* $\underline{\Phi_i(X)} := \Phi(\Theta_i(X))$ *and* $\underline{\hat{\Phi}_i(X)} := \widehat{\Phi_i(X)}$ *for any* $i \in \{3, 4, 5\}$.

We have

$$\begin{array}{ccccc} \Theta_1(X) & \subset & \Theta_2(X) & & \\ \cap & & \cap & & \\ \Theta_3(X) & \subset & \Theta_4(X) & \subset & \Theta_5(X). \end{array}$$

In general we shall write Θ_i, Φ_i, and $\hat{\Phi}_i$ instead of $\Theta_i(X)$, $\Phi_i(X)$, and $\hat{\Phi}_i(X)$ respectively for any $i \in \{1, 2, 3, 4, 5\}$. For sequentially complete uniform spaces X we have

$$\Theta_1(X) = \Theta_3(X), \quad \Theta_2(X) = \Theta_4(X),$$
$$\Phi_1(X) = \Phi_3(X), \quad \Phi_2(X) = \Phi_4(X).$$

Definition 1.5.3. *Let X be a topological (uniform) space and let $i \in \{1, 2\}$ ($i \in \{1, 2, 3, 4, 5\}$). The expression* $\mathfrak{F}$ is a Φ_i-filter on X *and* $\mathfrak{F}$ is a $\hat{\Phi}_i$-filter on X *means $\mathfrak{F} \in \Phi_i(X)$ and $\mathfrak{F} \in \hat{\Phi}_i(X)$ respectively. Similarly, a* Φ_i-set *(a* $\hat{\Phi}_i$-set) of X *means a $\Phi_i(X)$-set (a $\hat{\Phi}_i(X)$-set) and a* Θ_i-sequence *(a* Θ_i-net) in X *means a $\Theta_i(X)$-sequence (a $\Theta_i(X)$-net). X is called* Φ_i-compact *if it is $\Phi_i(X)$-compact.*

Proposition 1.5.4. *Let X be a topological (uniform) space and A be a subset of X. Then the following assertions are equivalent:*

a) A is a Φ_1-set (Φ_3-set);

b) A is a $\hat{\Phi}_1$-set ($\hat{\Phi}_3$-set);

c) any sequence in A possesses a convergent (Cauchy) subsequence.

The proposition follows immediately from Proposition 1.4.5. □

Remark. The sets that we call Φ_3-sets appear, in case of weak uniformity of a Banach space, in a paper of P. Meyer-Nieberg (1973) ([1] p. 304), where they are called „schwach folgenpräkompakt".

Proposition 1.5.5. *Let X be a topological (uniform) space and A be a subset of X. Then the following assertions are equivalent:*

a) A is a Φ_2-set (Φ_4-set);

b) A is a $\hat{\Phi}_2$-set ($\hat{\Phi}_4$-set);

c) any sequence in A possesses adherent points in X (or a Cauchy subsequence).

The proposition follows immediately from Proposition 1.4.5. □

Proposition 1.5.6. *Let X be a uniform space and A be a subset of X. Then the following assertions are equivalent:*

a(A is a Φ_5-set;

b) A is a $\hat{\Phi}_5$-set;

c) A is precompact.

The proposition follows immediately from Proposition 1.4.5. □

Remark. A precompact set is not necessarily a $\hat{\Phi}_4$-set, as the following example shows. Let us endow $\mathbb{N}$ with the coarsest uniformity for which the bounded real functions on $\mathbb{N}$ are uniformly continuous. Then $\mathbb{N}$ is precompact and the sequence $(n)_{n\in\mathbb{N}}$ does not belong to $\Theta_4(\mathbb{N})$. Hence $\mathbb{N}$ is not a $\hat{\Phi}_4$-set (Proposition 1.5.5). A. Weil proved (1938) ([1] Théorème VIII) that any Φ_2-set of a uniform space is precompact.

Proposition 1.5.7. *If a Θ_2-sequence in a topological space possesses a unique adherent point then it converges to this point.*

Let X be a topological space, $x \in X$ and $(x_n)_{n\in\mathbb{N}}$ be a Θ_2-sequence in X for which x is the unique adherent point. Assume $(x_n)_{n\in\mathbb{N}}$ does not converge to x. Then there exists a neighbourhood U of x and a subsequence $(x_{k_n})_{n\in\mathbb{N}}$ of $(x_n)_{n\in\mathbb{N}}$ such that $x_{k_n} \notin U$ for any $n \in \mathbb{N}$. Let y be an adherent point of $(x_{k_n})_{n\in\mathbb{N}}$. Then y is an adherent point of $(x_n)_{n\in\mathbb{N}}$ and therefore $y = x$, which is a contradiction. □

Theorem 1.5.8. *(C. Constantinescu (1973) [2] Corollary 2.4). Any paracompact (and therefore any metrizable) space is Φ_2-compact. Any Φ_2-set of a paracompact space is relatively compact.*

Let X be a paracompact space and let $\mathfrak{F}$ be a Φ_2-ultrafilter on X. Assume $\mathfrak{F}$ does not converge. Then for any $x \in X$ there exists an open neighbourhood U_x of x not belonging to $\mathfrak{F}$. We get $X \setminus U_x \in \mathfrak{F}$. $\{U_x \mid x \in X\}$ is an open covering of X. Since X is paracompact there exists a locally finite covering $\mathfrak{A}$ of X finer than $\{U_x \mid x \in X\}$. Hence, $X \setminus A \in \mathfrak{F}$ for any $A \in \mathfrak{A}$.

Let (I, f) be a Θ_2-net in X such that $f(\mathfrak{G}) = \mathfrak{F}$, where $\mathfrak{G}$ denotes the section filter of I. We construct inductively an increasing sequence $(\iota_n)_{n\in\mathbb{N}}$ in I and a sequence $(A_n)_{n\in\mathbb{N}}$ in $\mathfrak{A}$ such that

$$f(\iota_n) \in A_n \setminus \bigcup_{m<n} A_m$$

for any $n \in \mathbb{N}$. Let $n \in \mathbb{N}$ and assume the sequences were constructed up to $n-1$. Since $\mathfrak{F}$ is an ultrafilter we have $\bigcup_{m<n} A_m \notin \mathfrak{F}$ so there exists $\iota_n \in I$ such that $\iota_n \geq \iota_{n-1}$ and

$$f(\iota_n) \in X \setminus \bigcup_{m<n} A_m .$$

Since $\mathfrak{A}$ is a covering of X there exists $A_n \in \mathfrak{A}$ such that $f(\iota_n) \in A_n$.

Let x be an adherent point of the sequence $(f(\iota_n))_{n\in\mathbb{N}}$. Since $\mathfrak{A}$ is locally finite there exist a neighbourhood U of x and a natural number m such that

$$U \cap (\bigcup_{n \geq m} A_n) = \emptyset .$$

We get $f(\iota_n) \notin U$ for any $n \geq m$, which contradicts the fact that x is an adherent point of the sequence $(f(\iota_n))_{n\in\mathbb{N}}$.

The second assertion follows immediately from the first and from Proposition 1.2.10. □

Proposition 1.5.9. *Let $i \in \{1, 2\}$ $(i \in \{3, 4, 5\})$ and $(X_\iota)_{\iota \in I}$ be a family of Φ_i-compact topological (uniform) spaces. Then $\prod_{\iota \in I} X_\iota$ is Φ_i-compact.*

The assertion follows immediately from the remark that the projection of any Φ_i-filter is a Φ_i-filter. □

Proposition 1.5.10. *Any Φ_5-ultrafilter on a uniform space is a Cauchy filter.*

Let $\mathfrak{F}$ be a Φ_5-ultrafilter on a uniform space X and (I, f) be a Θ_5-net in X such that $f(\mathfrak{G}) = \mathfrak{F}$, where $\mathfrak{G}$ denotes the section filter of I. Let U be an entourage of X. Assume we have $A \times A \not\subset U$ for any $A \in \mathfrak{F}$. Let V be a symmetrical entourage of X such that $V \circ V \subset U$ and ι_{-1} be an arbitrary element of I. We construct inductively an increasing sequence $(\iota_n)_{n \in \mathbb{N}}$ in I such that $(f(\iota_{n-1}), f(\iota)) \notin V$ for any $n \in \mathbb{N}$ and for any $\iota \in I$ with $\iota \geq \iota_n$. Let $n \in \mathbb{N}$ and assume the sequence was constructed up to $n - 1$. Since

$$V(f(\iota_{n-1})) \times V(f(\iota_{n-1})) \subset V \circ V \subset U$$

we get $V(f(\iota_{n-1})) \notin \mathfrak{F}$ and therefore $X \setminus V(f(\iota_{n-1})) \in \mathfrak{F}$. Hence, there exists $\iota_n \in I$ with $\iota_n \geq \iota_{n-1}$ and

$$\{f(\iota) \mid \iota \in I,\ \iota \geq \iota_n\} \subset X \setminus V(f(\iota_{n-1})),$$

which means precisely that $(f(\iota_{n-1}), f(\iota)) \notin V$ for any $\iota \in I$ with $\iota \geq \iota_n$. The sequence $(\iota_n)_{n \in \mathbb{N}}$ is therefore constructed and for any different $m, n \in \mathbb{N}$ we have $(f(\iota_m), f(\iota_n)) \notin V$, which is a contradiction. □

Corollary 1.5.11. *Any complete uniform space is Φ_5-compact.* □

Corollary 1.5.12. *$\hat{\Phi}_5(X) \subset \hat{\Phi}_C(X)$ for any uniform space X.*

The assertion follows immediately from Proposition 1.5.10 and Proposition 1.2.2. □

Remark. The above inclusion may be strict as the following example shows. Let X be an uncountable set. For any countable subset A of X we set

$$U_A := \{(x, x) \mid x \in A\} \cup ((X \setminus A) \times (X \setminus A)).$$

$\{U_A \mid A \text{ countable, } A \subset X\}$ is a fundamental system of entourages of a uniformity on X for which the precompact sets are finite. Hence any ultrafilter of $\Phi_5(X)$ is trivial (Propositions 1.4.2, 1.4.7, and 1.5.6). It follows that any ultrafilter on X finer than the filter

$$\{X \setminus A \mid A \quad \text{countable,} \quad A \subset X\}$$

is a Cauchy ultrafilter, which is not a $\hat{\Phi}_5$-filter.

Proposition 1.5.13. *Let (I, f) be a net in a uniform space X and $\mathfrak{F}$ be the section filter of I. We assume that for any $\iota \in I$ and for any increasing sequence $(\iota_n)_{n \in \mathbb{N}}$ in I for which* *the set* $\{n \in \mathbb{N} \mid \iota_n \geqq \iota\}$ *is empty there exist different* $m, n \in \mathbb{N}$ *with* $f(\iota_m) = f(\iota_n)$. *Then the following assertions are equivalent;*

a) $f(\mathfrak{F}) \in \Phi_5(X)$;
b) $f(\mathfrak{F}) \in \hat{\Phi}_5(X)$;
c) $f(\mathfrak{F}) \in \hat{\Phi}_C(X)$.

a $\Rightarrow$ b is trivial.
b $\Rightarrow$ c follows from Corollary 1.5.12.
c $\Rightarrow$ a. Let $(\iota_n)_{n \in \mathbb{N}}$ be an increasing sequence in I and U be an entourage of X. By Proposition 1.2.4 there exists a finite set $\mathfrak{A}$ of subsets of X such that

$$\bigcup_{A \in \mathfrak{A}} (A \times A) \subset U, \quad \bigcup_{A \in \mathfrak{A}} A \in \mathfrak{F}$$

and therefore there exists $\iota \in I$ with

$$\{f(\kappa) \mid \kappa \geq \iota\} \subset \bigcup_{A \in \mathfrak{A}} A.$$

If the set $\{n \in \mathbb{N} \mid \iota_n \geq \iota\}$ is nonempty we deduce that there exist different $m, n \in \mathbb{N}$ with $(f(\iota_m), f(\iota_n)) \in U$. By the hypothesis this last relation holds automatically if the set $\{n \in \mathbb{N} \mid \iota_n \geq \iota\}$ is empty. Hence, $(f(\iota_n))_{n \in \mathbb{N}} \in \Theta_5(X)$ and therefore $f(\mathfrak{F}) \in \Phi_5(X)$. □

The following Corollary will not be used in the sequel.

Corollary 1.5.14. *The following assertions coincide for any elementary filter* $\mathfrak{F}$ *on a uniform space* X:

a) $\mathfrak{F} \in \Phi_5(X)$;
b) $\mathfrak{F} \in \hat{\Phi}_5(X)$;
c) $\mathfrak{F} \in \hat{\Phi}_C(X)$. □

Proposition 1.5.15. *Let* X *be a topological (uniform) space,* $i \in \{1, 2\}$ $(i \in \{3, 4, 5\})$ *and* $\mathfrak{F}, \mathfrak{G} \in \Phi_i(X)$. *Then* $\mathfrak{F} \cap \mathfrak{G} \in \Phi_i(X)$.
The assertion follows immediately from Proposition 1.4.9. □

Proposition 1.5.16. *Let* X *be a topological (uniform) space,* $i \in \{1, 2\}$ $(i \in \{3, 4, 5\})$ *and* $(x_n)_{n \in \mathbb{N}}$ *be a sequence in* X. *Then* $(x_n)_{n \in \mathbb{N}}$ *is a* Θ_i*-sequence iff* $\{x_n \mid n \in \mathbb{N}\}$ *is a* Φ_i*-set.*
The assertion follows immediately from Proposition 1.4.7. □

Proposition 1.5.17. *Let* X *be a uniform space,* (I, f) *be a net in* X *such that* $(f(\iota_n))_{n \in \mathbb{N}}$ *is a Cauchy sequence in* X *for any increasing sequence* $(\iota_n)_{n \in \mathbb{N}}$ *in* I, *and let* $\mathfrak{F}$ *be the section filter of* I. *Then* $f(\mathfrak{F})$ *is a Cauchy filter.*

Assume the contrary. Then there exists an entourage U of X such that $f(A) \times f(A) \not\subset U$ for any $A \in \mathfrak{F}$. Let V be an entourage of X such that $V \circ \overset{-1}{V} \subset U$

We construct inductively an increasing sequence $(\iota_n)_{n\in\mathbb{N}}$ in I such that

$$(f(\iota_n),\, f(\iota_{n+1}))\notin V$$

for any $n\in\mathbb{N}$. Let $n\in\mathbb{N}$ and assume the sequence was constructed up to n. Then

$$\{\iota\in I\,|\,\iota\geq\iota_n\}\in\mathfrak{F}$$

and therefore there exist $\iota', \iota''\in I$ such that

$$\iota'\ \geq\iota_n,\ \iota''\geq\iota_n,\ (f(\iota'), f(\iota''))\notin U\,.$$

Hence, either $(f(\iota_n), f(\iota'))\notin V$ or $(f(\iota_n), f(\iota''))\notin V$ and we put respectively either $\iota_{n+1}:=\iota'$ or $\iota_{n+1}:=\iota''$. Since $(f(\iota_n))_{n\in\mathbb{N}}$ is a Cauchy sequence this is a contradiction. Hence, $f(\mathfrak{F})$ is a Cauchy filter. □

Definition 1.5.18. *Let X be a topological (uniform) space, A be a subset of X, φ be the inclusion map $A\to X$, and let $i\in\{1,2\}$ $(i\in\{1,2,3,4,5\})$. We say that A is a Φ_i-closed set of X if for any filter $\mathfrak{F}$ of $\Phi_i(A)$ the limit points of $\varphi(\mathfrak{F})$ belong to A. We say A is Φ_i-dense in X if any Φ_i-closed set of X containing A is equal to X.*

We can replace filter by ultrafilter (Proposition 1.4.2) or $\Phi_i(A)$ by $\hat{\Phi}_i(A)$ in the above definition of a Φ_i-closed set.

Any Φ_i-closed set of a Φ_i-compact space is Φ_i-compact with respect to the induced topology (uniformity). Any Φ_i-compact subspace of X is Φ_i-closed if X is Hausdorff.

Definition 1.5.19. *Let X be a topological space, A be a subset of X and φ be the inclusion map $A\to X$. We denote by $\Theta_1(A,X)$ the set of sequences $(x_n)_{n\in\mathbb{N}}$ in A for which $(\varphi(x_n))_{n\in\mathbb{N}}\in\Theta_1(X)$ and set*

$$\Theta_2(A,X):=\Theta_1(A,X)\cup\Theta_2(A)\,.$$

Let $i\in\{1,2\}$. We say A is a strongly Φ_i-closed set of X if for any filter $\mathfrak{F}\in\Phi(\Theta_i(A,X))$ the limit points of $\varphi(\mathfrak{F})$ belong to A. We say that A is loosely Φ_i-dense in X if any strongly Φ_i-closed set of X containing A is equal to X.

Any strongly Φ_i-closed set is Φ_i-closed and any Φ_i-dense set is loosely Φ_i-dense $(i\in\{1,2\})$.

Proposition 1.5.20. *Let X be a uniform space.*

a) Any Φ_4-closed (Φ_3-closed) set of X is strongly Φ_2-closed (strongly Φ_1-closed);
b) any loosely Φ_2-dense (loosely Φ_1-dense) set of X is Φ_4-dense (Φ_3-dense).

We have $\Theta_1(A,X)\subset\Theta_3(A)$ and $\Theta_2(A,X)\subset\Theta_4(A)$ and therefore

$$\Phi(\Theta_1(A,X))\subset\Phi_3(A),\quad \Phi(\Theta_2(A,X))\subset\Phi_4(A)$$

for any $A\subset X$, which proves a). b) follows immediately from a). □

Proposition 1.5.21. *Let X be a topological space and A be a subset of X. If $X \setminus A$ is countable, then any $\Theta_2(X)$-sequence in A belongs to $\Theta_2(A, X)$. If, in addition, the neighbourhood filter of any point of $X \setminus A$ is a $\hat{\Phi}_2(X)$-filter and A is dense in X, then A is loosely Φ_2-dense in X.*

Let $(x_n)_{n \in \mathbb{N}}$ be a $\Theta_2(X)$-sequence in A which does not belong to $\Theta_1(A, X)$. We have to show $(x_n)_{n \in \mathbb{N}} \in \Theta_2(A)$. Let $(y_n)_{n \in \mathbb{N}}$ be a sequence in $X \setminus A$ such that $\{y_n \mid n \in \mathbb{N}\} = X \setminus A$. We construct inductively a sequence $(U_n)_{n \in \mathbb{N}}$ of subsets of X and a decreasing sequence $(M_n)_{n \in \mathbb{N}}$ of infinite subsets of $\mathbb{N}$ such that we have for any $n \in \mathbb{N}$:

a) U_n is a neighbourhood of y_n;

b) $m \in M_n \Rightarrow x_m \notin U_n$.

Let $n \in \mathbb{N}$ and assume the sequences were constructed up to $n-1$. Since $(x_m)_{m \in M_{n-1}}$ ($M_{-1} := \mathbb{N}$) does not converge to y_n there exists a neighbourhood U_n of y_n such that

$$M_n := \{m \in M_{n-1} \mid x_m \notin U_n\}$$

is infinite.

Let $(k_n)_{n \in \mathbb{N}}$ be a strictly increasing sequence in $\mathbb{N}$ such that $k_n \in M_n$ for any $n \in \mathbb{N}$. Then no point of $X \setminus A$ is an adherent point of $(x_{k_n})_{n \in \mathbb{N}}$. Hence, $(x_{k_n})_{n \in \mathbb{N}}$ possesses adherent points in A. We get $(x_n)_{n \in \mathbb{N}} \in \Theta_2(A)$.

Assume now that the neighbourhood filter of any point of $X \setminus A$ is a $\hat{\Phi}_2$-filter. Let B be a strongly Φ_2-closed set of X containing A, $x \in X \setminus A$, $\mathfrak{F}$ be the neighbourhood filter of x and $\mathfrak{G}$ be an ultrafilter on X finer than $\mathfrak{F}$ such that $A \in \mathfrak{G}$. By Proposition 1.4.2 $\mathfrak{G} \in \Phi_2(X)$. Hence, there exists a Θ_2-net (I, f) in X such that $\mathfrak{G} = f(\mathfrak{G}')$, where $\mathfrak{G}'$ denotes the section filter of I. We set

$$J := \{\iota \in I \mid f(\iota) \in B\}$$

and denote by $\mathfrak{H}$ the section filter of J and by g the map

$$J \to B, \quad \iota \mapsto f(\iota).$$

By the above considerations (J, g) is a $\Theta_2(B, X)$-net in B so $g(\mathfrak{H}) \in \Phi(\Theta_2(B, X))$. If φ denotes the inclusion map $B \to X$, then $\varphi(g(\mathfrak{H})) = \mathfrak{G}$ so $\varphi(g(\mathfrak{H}))$ converges to x. Since B is a strongly Φ_2-closed set of X it contains x. Hence, $X = B$ and A is loosely Φ_2-dense in X. □

Corollary 1.5.22. *Let X be a countably compact space and A be a subset of X such that $X \setminus A$ is countable. Then*

a) any sequence in A belongs to $\Theta_2(A, X)$;

b) if A is dense in X then it is loosely Φ_2-dense in X.

The assertions follow immediately from Proposition 1.5.21 and Proposition 1.5.5 c ⇒ a. □

Proposition 1.5.23. *Let X be a topological space, $\mathfrak{A}$ be a nonempty set of Φ_2-sets of X such that $A, A' \cup A'' \in \mathfrak{A}$ for any $A', A'' \in \mathfrak{A}$ and $A \subset A'$, and let a be a point which does not belong to X. We assume $X \notin \mathfrak{A}$ and denote by Y the set $X \cup \{a\}$ endowed with the topology*

$$\{U \subset Y \mid U \cap X \quad \text{is open in } X, \quad (a \in U \;\Rightarrow\; X \backslash U \in \mathfrak{A})\}.$$

Then:

a) X is a subspace of Y;

b) Y is a countably compact space;

c) any loosely Φ_2-dense set of X is a loosely Φ_2-dense set of Y.

a) is obvious.

b) Let $(x_n)_{n \in \mathbb{N}}$ be a sequence in Y. If there exists $A \in \mathfrak{A}$ such that $\{n \in \mathbb{N} \mid x_n \in A\}$ is infinite, then $(x_n)_{n \in \mathbb{N}}$ possesses an adherent point in X and therefore in Y. Otherwise $(x_n)_{n \in \mathbb{N}}$ converges to a. Hence, Y is a countably compact space.

c) Let A be a loosely Φ_2-dense set of X and B be a strongly Φ_2-closed set of Y containing A. Then $B \cap X$ is a strongly Φ_2-closed set of X containing A so $B \cap X = X$. By b) and Corollary 1.5.22 b) $B = Y$. Hence, A is a loosely Φ_2-dense set of Y. □

Corollary 1.5.24. *Any locally compact non-compact space is loosely Φ_2-dense in its Alexandrov compactification.* □

Proposition 1.5.25. *Let X be a topological space, A be a subset of X, φ be the inclusion map $A \to X$ and $\mathfrak{F}$ be a filter on A such that $\varphi(\mathfrak{F})$ converges. If $\mathfrak{F}$ possesses a countable base or if $A \backslash B$ is finite for any $B \in \mathfrak{F}$ then $\mathfrak{F} \in \Phi(\Theta_2(A, X))$.*

Assume first $\mathfrak{F}$ possesses a countable base. Let $(A_n)_{n \in \mathbb{N}}$ be a decreasing sequence in $\mathfrak{F}$ such that $\{A_n \mid n \in \mathbb{N}\}$ is a base of $\mathfrak{F}$. We set

$$I := \{(x, n) \in A \times \mathbb{N} \mid x \in A_n\}$$

and denote by $\leq$ the upper directed order relation on I defined by

$$(x, m) < (y, n) \;:\Leftrightarrow\; m < n$$

for any $(x, m), (y, n) \in I$ and by f the map

$$I \to A, \quad (x, n) \mapsto x.$$

Then (I, f) is a $\Theta_1(A, X)$-net in A and $\mathfrak{F} = f(\mathfrak{G})$, where $\mathfrak{G}$ denotes the section filter of I. Hence, $\mathfrak{F} \in \Phi(\Theta_1(A, X))$.

Assume now $A \backslash B$ is finite for any $B \in \mathfrak{F}$. We set

$$I := \{(x, B) \in A \times \mathfrak{F} \mid x \in B\}$$

and denote by $\leq$ the upper directed preorder relation on I defined by

$$(x, B) \leq (y, C) \;:\Leftrightarrow\; C \subset B$$

for any $(x, B), (y, C) \in I$ and by f the map

$$I \longrightarrow A, \quad (x, B) \mapsto x.$$

Let $(x_n, B_n)_{n \in \mathbb{N}}$ be an increasing sequence in I such that $(x_n)_{n \in \mathbb{N}}$ has no adherent point in A and let $B \in \mathfrak{F}$. Since $A \backslash B$ is finite there exists $n \in \mathbb{N}$ such that $x_m \in B$ for any $m \geq n$. Hence, $(\varphi(x_n))_{n \in \mathbb{N}}$ is convergent. We deduce (I, f) is a $\Theta_2(A, X)$-net in A. Since $\mathfrak{F} = f(\mathfrak{G})$, where $\mathfrak{G}$ denotes the section filter of I, it follows $\mathfrak{F} \in \Phi(\Theta_2(A, X))$. □

Proposition 1.5.26. *Let $(X_\iota)_{\iota \in I}$ be a countable family of topological spaces. Then:*

a) $(\mathfrak{F}_\iota)_{\iota \in I} \in \prod_{\iota \in I} \Phi_1(X_\iota) \Rightarrow \prod_{\iota \in I} \mathfrak{F}_\iota \in \Phi_1(\prod_{\iota \in I} X_\iota)$;

b) $(\mathfrak{F}_\iota)_{\iota \in I} \in \prod_{\iota \in I} \hat{\Phi}_1(X_\iota) \Rightarrow \prod_{\iota \in I} \mathfrak{F}_\iota \in \hat{\Phi}_1(\prod_{\iota \in I} X_\iota)$;

c) if for any $\iota \in I$ the neighbourhood filter of any point of X_ι belongs to $\Phi_1(X_\iota)$ ($\hat{\Phi}_1(X_\iota)$), then the neighbourhood filter of any point of $\prod_{\iota \in I} X_\iota$ belongs to $\Phi_1(\prod_{\iota \in I} X_\iota)$ ($\hat{\Phi}_1(\prod_{\iota \in I} X_\iota)$).

The assertion follows immediately from Proposition 1.4.12. □

The next five results will not be used later.

Proposition 1.5.27. *Let X be a uniform space and $(x_n)_{n \in \mathbb{N}}$ be a sequence in X. Then the following assertions are equivalent:*

a) $(x_n)_{n \in \mathbb{N}}$ is a $\Theta_5(X)$-sequence;

b) for any entourage U of X there exists a subsequence $(x_{k_n})_{n \in \mathbb{N}}$ of $(x_n)_{n \in \mathbb{N}}$ such that

$$\{x_{k_n} \mid n \in \mathbb{N}\} \times \{x_{k_n} \mid n \in \mathbb{N}\} \subset U.$$

a ⇒ b. Assume there exists an entourage U of X such that the assertion of b) does not hold. Let V be an entourage of X such that $V \circ \overset{-1}{V} \subset U$. Then for any $x \in X$ the set

$$\{n \in \mathbb{N} \mid (x, x_n) \in V\}$$

is finite. Hence, there exists a subsequence $(x_{k_n})_{n \in \mathbb{N}}$ of $(x_n)_{n \in \mathbb{N}}$ such that $(x_{k_m}, x_{k_n}) \notin V$ for any different $m, n \in \mathbb{N}$ and this is a contradiction.

b ⇒ a is trivial. □

Proposition 1.5.28. *Let $(X_\iota)_{\iota \in I}$ be a countable family of uniform spaces and $i \in \{3, 5\}$. Then:*

a) $(\mathfrak{F}_\iota)_{\iota \in I} \in \prod_{\iota \in I} \Phi_i(X_\iota) \Rightarrow \prod_{\iota \in I} \mathfrak{F}_\iota \in \Phi_i(\prod_{\iota \in I} X_\iota)$;

b) $(\mathfrak{F}_\iota)_{\iota \in I} \in \prod_{\iota \in I} \hat{\Phi}_i(X_\iota) \Rightarrow \prod_{\iota \in I} \mathfrak{F}_\iota \in \hat{\Phi}_i(\prod_{\iota \in I} X_\iota)$;

c) if for any $\iota \in I$ the neighbourhood filter of any point of X_ι belongs to $\Phi_i(X_\iota)$ $(\hat{\Phi}_i(X_\iota))$, then the neighbourhood filter of any point of $\prod_{\iota \in I} X_\iota$ belongs to $\Phi_i(\prod_{\iota \in I} X_\iota)$ $(\hat{\Phi}_i(\prod_{\iota \in I} X_\iota))$.

The assertions follow immediately from Proposition 1.4.12 for $i = 3$. Now let $i = 5$.

a) Without loss of generality we may assume $I = \mathbb{N}$. For any $n \in \mathbb{N}$ let (I_n, f_n) be a Θ_5-net in X_n such that $f_n(\mathfrak{G}_n) = \mathfrak{F}_n$, where $\mathfrak{G}_n$ denotes the section filter of I_n, and such that $I_n \cap X_n = \emptyset$. We set

$$J := \bigcup_{n \in \mathbb{N}} (\prod_{m \leq n} I_m) \times (\prod_{m > n} X_m).$$

For any $g \in J$ there exists exactly one $n(g) \in \mathbb{N}$ such that

$$g \in (\prod_{m \leq n(g)} I_m) \times (\prod_{m > n(g)} X_m).$$

We define an upper directed order relation $\leq$ on J as follows. Let $g, h \in J$. We set $g \leq h$ if $g = h$ or if $n(g) < n(h)$ and $g(m) \leq h(m)$ for any $m \leq n(g)$. Let f be the map of J into $\prod_{n \in \mathbb{N}} X_n$ defined in the following manner: for any $g \in J$ we set

$$f(g)_n := f_n(g(n))$$

if $n \leq n(g)$ and

$$f(g)_n := g(n)$$

if $n > n(g)$. Then (J, f) is a net in $\prod_{n \in \mathbb{N}} X_n$ such that $f(\mathfrak{G}) = \prod_{n \in \mathbb{N}} \mathfrak{F}_n$, where $\mathfrak{G}$ denotes the section filter of J. We want to show that (J, f) is a Θ_5-net. Let $(g_n)_{n \in \mathbb{N}}$ be an increasing sequence in J and U be an entourage of $\prod_{n \in \mathbb{N}} X_n$. There exists a sequence $(U_n)_{n \in \mathbb{N}}$ such that U_n is an entourage of X_n for any $n \in \mathbb{N}$, $\{n \in \mathbb{N} \mid U_n \notin X_n\}$ is finite, and $\prod_{n \in \mathbb{N}} U_n \subset U$. If there exist $m, n \in \mathbb{N}$ such that $m < n$ and $g_m = g_n$, then $(f(g_m), f(g_n)) \in U$. In the contrary case there exists $p \in \mathbb{N}$ such that $U_m = X_m$ for any $m \geq n(g_p)$. By applying Proposition 1.5.27 a $\Rightarrow$ b $n(g_p)$-times we may find a subsequence $(g_{k_n})_{n \in \mathbb{N}}$ of $(g_n)_{n \in \mathbb{N}}$ such that

$$\{f(g_{k_n})_m \mid n \in \mathbb{N}\} \times \{f(g_{k_n})_m \mid n \in \mathbb{N}\} \subset U_m$$

for any $m \in \mathbb{N}$. It follows that there exist $n', n'' \in \mathbb{N}$ such that $n' < n''$ and $(f(g_{k_{n'}}), f(g_{k_{n''}})) \in U$. Hence, $(g_n)_{n \in \mathbb{N}}$ is a Θ_5-sequence in $\prod_{n \in \mathbb{N}} X_n$, (J, f) is a Θ_5-net in $\prod_{n \in \mathbb{N}} X_n$, and $\prod_{n \in \mathbb{N}} \mathfrak{F}_n \in \Phi_5(\prod_{n \in \mathbb{N}} X_n)$.

b) follows from a) and Proposition 1.1.10.

c) follows from a) and b). □

Proposition 1.5.29. *Let X be a topological space, let Θ be a set of Θ_1-sequences in X such that for any $(x_n)_{n\in\mathbb{N}} \in \Theta$ and for any limit point x of $(x_n)_{n\in\mathbb{N}}$ the sequence $(y_n)_{n\in\mathbb{N}}$ belongs to Θ, where $y_{2n} := x_n$, $y_{2n+1} := x$ for any $n \in \mathbb{N}$. Furthermore, let Y, Z be uniform spaces and φ be a map of $X \times Y$ into Z such that $\varphi(\mathfrak{F} \times \mathfrak{G})$ is a Cauchy filter for any convergent elementary filter $\mathfrak{F} \in \Phi(\Theta)$ and for any Cauchy filter $\mathfrak{G} \in \Phi_4(Y)$. Then:*

a) $(\mathfrak{F}, \mathfrak{G}) \in \Phi(\Theta) \times \Phi_4(Y) \Rightarrow \varphi(F \times G) \in \Phi_4(Z)$;

b) $(\mathfrak{F}, \mathfrak{G}) \in \hat{\Phi}(\Theta) \times \hat{\Phi}_4(Y) \Rightarrow \varphi(\mathfrak{F} \times \mathfrak{G}) \in \hat{\Phi}_4(Z)$;

c) *$\varphi(A \times B)$ is a Φ_4-set for any $\Phi(\Theta)$-set A of X and for any Φ_4-set B of Y.*

a) Let $(x_n)_{n\in\mathbb{N}}$ be a Θ-sequence in X and $(y_n)_{n\in\mathbb{N}}$ be a Θ_4-sequence in Y. Let $(x_{k_n})_{n\in\mathbb{N}}$ be a convergent subsequence of $(x_n)_{n\in\mathbb{N}}$, $\mathfrak{F}'$ be the elementary filter on X generated by $(x_{k_n})_{n\in\mathbb{N}}$ and x be a limit point of $(x_{k_n})_{n\in\mathbb{N}}$. Then $\mathfrak{F}'$ is a convergent elementary $\Phi(\Theta)$-filter on X. If $(y_{k_n})_{n\in\mathbb{N}}$ possesses a Cauchy subsequence, then $(\varphi(x_{k_n}, y_{k_n}))_{n\in\mathbb{N}}$ possesses a Cauchy subsequence. Assume $(y_{k_n})_{n\in\mathbb{N}}$ does not possess Cauchy subsequences. Then it possesses an adherent point y. Let $\mathfrak{G}'$ be a filter on Y converging to y and finer than the elementary filter on Y generated by $(y_{k_n})_{n\in\mathbb{N}}$. Then $\mathfrak{G}' \in \Phi_4(Y)$ (Proposition 1.4.2). We set

$$\mathfrak{F}'' := \{A \in \mathfrak{F}' \mid x \in A\}, \quad \mathfrak{G}'' := \{B \in \mathfrak{G}' \mid y \in B\}.$$

Then $\mathfrak{F}''$ is a convergent elementary filter of $\Phi(\Theta)$ and $\mathfrak{G}''$ is a convergent filter of $\Phi_4(Y)$ (Proposition 1.5.15). Hence, $\varphi(\mathfrak{F}'' \times \mathfrak{G}'')$ is a Cauchy filter on Z. Since $\varphi(x, y)$ is an adherent point of $\varphi(\mathfrak{F}'' \times \mathfrak{G}'')$ it is a limit point of it and therefore a limit point of $\varphi(\mathfrak{F}' \times \mathfrak{G}')$. We deduce that $\varphi(x, y)$ is an adherent point of $(\varphi(x_{k_n}, y_{k_n}))_{n\in\mathbb{N}}$ and therefore of $(\varphi(x_n, y_n))_{n\in\mathbb{N}}$. Hence, $(\varphi(x_n, y_n))_{n\in\mathbb{N}}$ is a Θ_4-sequence in Z. From this result we easily deduce $\varphi(\mathfrak{F} \times \mathfrak{G}) \in \Phi_4(Z)$ for any $(\mathfrak{F}, \mathfrak{G}) \in \Phi(\Theta) \times \Theta_4(Y)$.

b) follows from a) and Proposition 1.1.11.

c) Let $\mathfrak{H}$ be a filter on Z containing $\varphi(A \times B)$. There exists a filter $\mathfrak{H}_0$ on $X \times Y$ containing $A \times B$ such that $\varphi(\mathfrak{H}_0) = \mathfrak{H}$. Let $\mathfrak{F}$, $\mathfrak{G}$ be the projections of $\mathfrak{H}_0$ on X and Y respectively. Then $A \in \mathfrak{F}$, $B \in \mathfrak{G}$ and therefore $\mathfrak{F} \in \Phi(\Theta)$, $\mathfrak{G} \in \Phi_4(Y)$. By a) $\varphi(\mathfrak{F} \times \mathfrak{G}) \in \Phi_4(Z)$. Since $\mathfrak{H}_0$ is finer than $\mathfrak{F} \times \mathfrak{G}$ we get $\mathfrak{H} \in \Phi_4(Z)$ (Proposition 1.4.2). Hence, $A \times B$ is a Φ_4-set. □

Corollary 1.5.30. *We have for any uniform spaces X, Y:*

a) $(\mathfrak{F}, \mathfrak{G}) \in \Phi_1(X) \times \Phi_4(Y) \Rightarrow \mathfrak{F} \times \mathfrak{G} \in \Phi_4(X \times Y)$;

b) $(\mathfrak{F}, \mathfrak{G}) \in \hat{\Phi}_1(X) \times \hat{\Phi}_4(Y) \Rightarrow \mathfrak{F} \times \mathfrak{G} \in \hat{\Phi}_4(X \times Y)$;

c) *A is a $\Phi_1(X)$-set, B is a $\Phi_4(Y)$-set $\Rightarrow$ $A \times B$ is a $\Phi_4(X \times Y)$ set.* □

Proposition 1.5.31. *Let X be a topological space, $x \in X$ and $\mathfrak{F}$ be the neighbourhood* *filter of x. If* $\mathfrak{F}$ *possesses a countable base, then* $\mathfrak{F} \in \Phi(\Theta) \subset \Phi_1(X)$, *where* Θ *denotes the set of convergent sequences in* X.

Let $(U_n)_{n \in \mathbb{N}}$ be a decreasing sequence in $\mathfrak{F}$ such that $\{U_n | n \in \mathbb{N}\}$ is a basis of $\mathfrak{F}$. We denote by I the set

$$\{(x, n) \in X \times \mathbb{N} \,|\, x \in U_n\}$$

endowed with the upper directed order relation $\leq$ defined by

$$(x, m) < (y, n) \;:\Leftrightarrow\; m < n$$

and by f the map

$$I \to X, \quad (x, n) \mapsto x.$$

Then (I, f) is a Θ-net in X and $f(\mathfrak{G}) = \mathfrak{F}$, where $\mathfrak{G}$ denotes the section filter of I. Hence, $\mathfrak{F} \in \Phi(\Theta) \subset \Phi_1(X)$. □

§ 1.6 δ-stable filters

Definition 1.6.1. *A filter* $\mathfrak{F}$ *is called* <u>δ-stable</u> *if the intersection of any sequence in* $\mathfrak{F}$ *belongs to* $\mathfrak{F}$. *A uniform space is called* <u>δ-complete</u> *if any* δ*-stable Cauchy filter converges.*

Any complete uniform space is δ-complete.

Lemma 1.6.2. *Let* X *be a uniform space possessing a sequence* $(U_n)_{n \in \mathbb{N}}$ *of entourages such that* $\bigcap_{n \in \mathbb{N}} U_n$ *is the diagonal of* X. *Then* X *is* δ*-complete.*

Let $\mathfrak{F}$ be a δ-stable Cauchy filter on X. For any $n \in \mathbb{N}$ there exists $A_n \in \mathfrak{F}$ with $A_n \times A_n \subset U_n$. Then

$$A := \bigcap_{n \in \mathbb{N}} A_n \in \mathfrak{F}.$$

From

$$A \times A \subset \bigcap_{n \in \mathbb{N}} (A_n \times A_n) \subset \bigcap_{n \in \mathbb{N}} U_n$$

we deduce that A is a one-point set. Hence, $\mathfrak{F}$ is convergent. □

Proposition 1.6.3. *Let* X *be a topological space,* $\mathfrak{F} \in \hat{\Phi}_2(X)$ *and* $\mathfrak{G}$ *be the filter on* X *generated by the filter base*

$$\{\bigcap_{n \in \mathbb{N}} \bar{A}_n \,|\, (A_n)_{n \in \mathbb{N}} \text{ sequence in } \mathfrak{F}\}.$$

Then $\mathfrak{G}$ *is* δ*-stable and the adherent points of* $\mathfrak{F}$ *and* $\mathfrak{G}$ *coincide.*

Let $\mathfrak{F}'$ be an ultrafilter on X finer than $\mathfrak{F}$. Then $\mathfrak{F}' \in \Phi_2(X)$ (Proposition 1.4.2) so there exists a Θ_2-net (I, f) in X such that $\mathfrak{F}' = f(\mathfrak{G}')$, where $\mathfrak{G}'$ denotes the section filter of I. Let $(A_n)_{n\in\mathbb{N}}$ be a sequence in $\mathfrak{F}$. For any $n\in\mathbb{N}$ there exists $\iota_n\in I$ such that

$$\{f(\iota)\mid\iota\in I,\ \iota\geq\iota_n\}\subset A_n.$$

There exists an increasing sequence $(\kappa_n)_{n\in\mathbb{N}}$ in I with $\kappa_n\geq\iota_n$ for any $n\in\mathbb{N}$. Let x be an adherent point of the sequence $(f(\kappa_n))_{n\in\mathbb{N}}$. Then $x\in\bigcap_{n\in\mathbb{N}}\bar{A}_n$ and therefore $\bigcap_{n\in\mathbb{N}}\bar{A}_n\neq\emptyset$. This shows that

$$\{\bigcap_{n\in\mathbb{N}}\bar{A}_n\mid(A_n)_{n\in\mathbb{N}}\text{ sequence in }\mathfrak{F}\}$$

is a filter base. The other assertions are obvious. □

Proposition 1.6.4. *Any δ-complete uniform space is Φ_2-compact. In particular any topological group whose one-point sets are G_δ-sets is Φ_2-compact.*

Let X be a δ-complete uniform space and $\mathfrak{F}$ be a $\Phi_2(X)$-ultrafilter. By Proposition 1.5.10 $\mathfrak{F}$ is a Cauchy filter. By Proposition 1.6.3

$$\{\bigcap_{n\in\mathbb{N}}\bar{A}_n\mid(A_n)_{n\in\mathbb{N}}\text{ sequence in }\mathfrak{F}\}$$

is a filter base and the filter $\mathfrak{G}$ on X generated by it is δ-stable and its adherent points are adherent points of $\mathfrak{F}$. It is obvious that $\mathfrak{G}$ is a Cauchy filter and therefore convergent. Hence, $\mathfrak{F}$ converges and X is Φ_2-compact (Proposition 1.4.2).

Let G be a topological group whose one point sets are G_δ-sets. Then there exists a sequence of entourages of G for the left uniformity whose intersection is the diagonal of G. By Lemma 1.6.2 G is δ-complete and by the above result it is Φ_2-compact. □

§ 1.7 Mioritic spaces

Definition 1.7.1. *A* <u>mioritic space</u> *is a topological space X for which $\Phi_1(X)=\Phi_2(X)$.*

Proposition 1.7.2. *Let X be a topological space. The following assertions are equivalent:*

a) X is a mioritic space;

b) any $\Theta_2(X)$-sequence belongs to $\Theta_1(X)$;

c) any $\Theta_2(X)$-sequence is a $\Theta_1(X)$-sequence;

d) any $\Theta_2(X)$-net is a $\Theta_1(X)$-net;

e) $\hat{\Phi}_1(X)=\hat{\Phi}_2(X)$;

f) any $\Phi_2(X)$-set is a $\Phi_1(X)$-set.

The relations a ⇔ c ⇔ d ⇔ e ⇔ f follow from Proposition 1.4.8. b ⇔ c is trivial. □

Proposition 1.7.3. *Let X be a uniform space. The following assertions are equivalent:*

a) X is a mioritic space;

b) any $\Theta_4(X)$-sequence belongs to $\Theta_3(X)$;

c) any $\Theta_4(X)$-sequence is a $\Theta_3(X)$-sequence;

d) $\Phi_3(X) = \Phi_4(X)$;

e) $\hat{\Phi}_3(X) = \hat{\Phi}_4(X)$;

f) any $\Phi_4(X)$-set is a $\Phi_3(X)$-set.

Any subspace of a mioritic uniform space is mioritic.

a $\Rightarrow$ b. Let $(x_n)_{n\in\mathbb{N}}$ be a $\Theta_4(X)$-sequence. If it is a $\Theta_2(X)$-sequence then by Proposition 1.7.2 a $\Rightarrow$ b it belongs to $\Theta_1(X)$ and therefore to $\Theta_3(X)$. If $(x_n)_{n\in\mathbb{N}}$ is not a $\Theta_2(X)$-sequence, then it possesses a subsequence, which does not belong to $\Theta_2(X)$. This subsequence belongs then to $\Theta_3(X)$ and therefore $(x_n)_{n\in\mathbb{N}} \in \Theta_3(X)$.

b $\Rightarrow$ a. Let $(x_n)_{n\in\mathbb{N}}$ be a $\Theta_2(X)$-sequence. By b) $(x_n)_{n\in\mathbb{N}} \in \Theta_3(X)$. Hence, $(x_n)_{n\in\mathbb{N}}$ possesses a Cauchy subsequence $(x_{k_n})_{n\in\mathbb{N}}$. Since $(x_{k_n})_{n\in\mathbb{N}} \in \Theta_2(X)$ it possesses adherent points. Hence, $(x_{k_n})_{n\in\mathbb{N}}$ converges and $(x_n)_{n\in\mathbb{N}} \in \Theta_1(X)$.

b $\Leftrightarrow$ c is trivial.

c $\Leftrightarrow$ d $\Leftrightarrow$ e $\Leftrightarrow$ f follows from Proposition 1.4.8.

The last assertion follows from b $\Leftrightarrow$ a. □

Definition 1.7.4. *Let A be a set of a topological space. A is called* <u>countably compact</u> *if any sequence in A possesses an adherent point in A. A is called* <u>sequentially compact</u> *if any sequence in A possesses a subsequence converging to a point in A.*

Any quasicompact set and any sequentially compact set is countably compact. None of the other four possible implications holds. Any countably compact set is a Φ_2-set (Proposition 1.5.5 c $\Rightarrow$ a) but there exist Φ_2-sets which are not countably compact. Any countably compact set of a mioritic space is sequentially compact. A set of a topological space is countably compact iff it is a countably compact space with respect to the induced topology.

Proposition 1.7.5. *Let X be a uniform (topological) space such that the neighbourhood filter of any point of X is a $\hat{\Phi}_3$-filter ($\hat{\Phi}_1$-filter). Then X is a mioritic space.*

Let us assume first that X is a topological space. Let $(x_n)_{n\in\mathbb{N}}$ be a $\Theta_2(X)$-sequence and x be an adherent point of it. There exists a filter $\mathfrak{F}$ on X finer than the elementary filter of $(x_n)_{n\in\mathbb{N}}$ and converging to x. Then $\mathfrak{F} \in \hat{\Phi}_1(X)$ and by Proposition 1.4.4 there exists a subsequence of $(x_n)_{n\in\mathbb{N}}$ belonging to $\Theta_1(X)$. Hence, $(x_n)_{n\in\mathbb{N}} \in \Theta_1(X)$. By Proposition 1.7.2 b $\Rightarrow$ a X is a mioritic space.

Let now X be a uniform space and let $(x_n)_{n\in\mathbb{N}}$ be a $\Theta_4(X)$-sequence. Assume $(x_n)_{n\in\mathbb{N}} \notin \Theta_3(X)$. Then $(x_n)_{n\in\mathbb{N}}$ possesses an adherent point x and there exists a filter $\mathfrak{F}$ on X finer than the elementary filter of $(x_n)_{n\in\mathbb{N}}$ and converging to x. Then $\mathfrak{F} \in \hat{\Phi}_3(X)$ and by Proposition 1.4.4 there exists a subsequence of $(x_n)_{n\in\mathbb{N}}$ belonging to $\Theta_3(X)$. Hence, $(x_n)_{n\in\mathbb{N}} \in \Theta_3(X)$ and the assertion follows from Proposition 1.7.3 b $\Rightarrow$ a. □

Proposition 1.7.6. *Let X be a mioritic space and Y be a subspace of X. We assume that for any $(x, y) \in X^2$ for which there exists a neighbourhood of y not containing x there exists a closed neighbourhood of y not containing x. The Y is a mioritic space.*

Denote by φ the inclusion map $Y \to X$ and let $(x_n)_{n \in \mathbb{N}}$ be a $\Theta_2(Y)$-sequence. Then $(\varphi(x_n))_{n \in \mathbb{N}}$ is a $\Theta_2(X)$-sequence so there exists a subsequence $(x_{k_n})_{n \in \mathbb{N}}$ of $(x_n)_{n \in \mathbb{N}}$ such that $(\varphi(x_{k_n}))_{n \in \mathbb{N}}$ converges to a point $x \in X$. Let y be an adherent point of $(x_{k_n})_{n \in \mathbb{N}}$ in Y. Assume $(x_{k_n})_{n \in \mathbb{N}}$ does not converge to y. Then y possesses a neighbourhood in X not containing x. By the hypothesis y possesses a closed neighbourhood F not containing x. Since $(\varphi(x_{k_n}))_{n \in \mathbb{N}}$ converges to x there exists $n_0 \in \mathbb{N}$ such that $x_{k_n} \notin F$ for any $n \in \mathbb{N}$ with $n \geq n_0$, which is a contradiction. Hence, $(x_n)_{n \in \mathbb{N}} \in \Theta_1(Y)$. By Proposition 1.7.2 b $\Rightarrow$ a, Y is a mioritic space. □

The next two criteria for a topological space to be mioritic will not be used in the sequel.

Proposition 1.7.7. *Let X be a topological space such that any point possesses a fundamental system of closed neighbourhoods. If any point $x \in X$ possesses a neighbourhood Y such that Y endowed with the topology induced by X is a mioritic space, then X is a mioritic space.*

Let $(x_n)_{n \in \mathbb{N}}$ be a $\Theta_2(X)$-sequence and x be an adherent point of it. By the hypothesis x possesses a neighbourhood Y such that Y endowed with the topology induced by X is a mioritic space. Let F be a closed neighbourhood of x contained in Y. By Proposition 1.7.6 F endowed with the topology induced by Y is a mioritic space. Let $(x_{k_n})_{n \in \mathbb{N}}$ be a subsequence of $(x_n)_{n \in \mathbb{N}}$ such that $x_{k_n} \in F$ for any $n \in \mathbb{N}$. Then $(x_{k_n})_{n \in \mathbb{N}}$ is a $\Theta_2(F)$-sequence. By Proposition 1.7.2 a $\Rightarrow$ b, it belongs to $\Theta_1(F)$ and therefore to $\Theta_1(X)$. By Proposition 1.7.2 b $\Rightarrow$ a, X is a mioritic space. □

Proposition 1.7.8. *The product of a countable family of mioritic spaces is a mioritic space.*

Let $(X_\iota)_{\iota \in I}$ be a countable family of mioritic spaces and $\mathfrak{F} \in \Phi_2(\prod_{\iota \in I} X_\iota)$. For any $\iota \in I$ let $\mathfrak{F}_\iota$ denote the projection of $\mathfrak{F}$ on X_ι. Then $\mathfrak{F}_\iota \in \Phi_2(X_\iota) = \Phi_1(X_\iota)$. By Proposition 1.5.26 a) $\prod_{\iota \in I} F_\iota \in \Phi_1(\prod_{\iota \in I} X_\iota)$. Since $\mathfrak{F}$ is finer than $\prod_{\iota \in I} \mathfrak{F}_\iota$ we deduce $\mathfrak{F} \in \Phi_1(\prod_{\iota \in I} X_\iota)$ (Proposition 1.4.2) and therefore $\prod_{\iota \in I} X_\iota$ is a mioritic space. □

Proposition 1.7.9. *Let X be a uniform (topological) mioritic space, Y be a subspace of X and $\mathfrak{F}$ be a Φ_4-filter (Φ_2-filter) on X such that any set of $\mathfrak{F}$ meets Y. Then the filter induced by $\mathfrak{F}$ on Y belongs to $\Phi_3(Y)$ (to $\Phi(\Theta_1(Y, X))$).*

By Proposition 1.7.3 a $\Rightarrow$ d $\mathfrak{F} \in \Phi_3(X)$ (by the definition of the mioritic space $\mathfrak{F} \in \Phi_1(X)$). Let $\mathfrak{G}$ be the filter induced by $\mathfrak{F}$ on Y and let φ be the inclusion map $Y \to X$. Then $\mathfrak{F} \subset \varphi(\mathfrak{G})$ and therefore (Proposition 1.4.2) $\varphi(\mathfrak{G}) \in \Phi_3(X)$ ($\varphi(\mathfrak{G}) \in \Phi_1(X)$). We get immediately $\mathfrak{G} \in \Phi_3(Y)$ ($\mathfrak{G} \in \Phi(\Theta_1(Y, X))$). □

Proposition 1.7.10. *Let X be a uniform (topological) mioritic space, A be a subset of X, $x \in X$, and let $\mathfrak{F} \in \Phi_4(X)$ ($\mathfrak{F} \in \Phi_2(X)$) such that any set of $\mathfrak{F}$ meets A and such that $\mathfrak{F}$ converges to x. Then x belongs to any Φ_3-closed (strongly Φ_1-closed) set of X containing A.*

Let B be a Φ_3-closed (strongly Φ_1-closed) set of X containing A and let $\mathfrak{G}$ be the filter induced by $\mathfrak{F}$ on B. By Proposition 1.7.9 $\mathfrak{G} \in \Phi_3(B)$ ($(\mathfrak{G} \in \Phi(\Theta_1(B, X)))$) so $x \in B$. □

Corollary 1.7.11. *Let X be a uniform (topological) mioritic space and A be a Φ_4-set (Φ_2-set) of X. Then any Φ_3-closed (strongly Φ_1-closed) set of X containing A contains $\bar{A}$. If A is dense, then it is Φ_3-dense (loosely Φ_1-dense).* □

Proposition 1.7.12. *Let X be a uniform (topological) space such that any of its points possesses a closed neighbourhood which is a Φ_3-set (Φ_1-set). Then any dense set of X is Φ_3-dense (loosely Φ_1-dense).*

Let A be a dense set of X and B be a Φ_3-closed (strongly Φ_1-closed) set of X containing A. Let $x \in X$ and F be a closed neighbourhood of x which is a Φ_3-set (Φ_1-set). Further, let U be an open neighbourhood of x contained in F. Then $\bar{U}$ is a Φ_3-set (Φ_1-set) of itself. By Proposition 1.7.5 $\bar{U}$ is a mioritic space. $A \cap \bar{U}$ is a dense set of $\bar{U}$. By Corollary 1.7.11 $A \cap \bar{U}$ is a Φ_3-dense (loosely Φ_1-dense) set of $\bar{U}$. Since $B \cap \bar{U}$ is a Φ_3-closed (strongly Φ_1-closed) set of $\bar{U}$ containing $A \cap \bar{U}$ we get $B \cap \bar{U} = \bar{U}$ and therefore $x \in B$. Hence, $B = X$ and A is a Φ_3-dense (loosely Φ_1-dense) set of X. □

Proposition 1.7.13. *Let X be a uniform (topological) space and Y be a Φ_3-closed (Φ_1-closed) set of X. If Y endowed with the topology induced by X is a mioritic space, then Y is a Φ_4-closed (Φ_2-closed) set of X.*

Let φ be the inclusion map $Y \to X$, $\mathfrak{F} \in \Phi_4(Y)$ ($\mathfrak{F} \in \Phi_2(Y)$) and x be a limit point of $\varphi(\mathfrak{F})$. Then $\mathfrak{F} \in \Phi_3(Y)$ (Proposition 1.7.3) ($\mathfrak{F} \in \Phi_1(Y)$) and $x \in Y$. Hence, Y is Φ_4-closed (Φ_2-closed). □

§ 1.8 Φ_i-continuous and uniformly Φ_i-continuous maps

Throughout this section we denote by i one of the numbers 1, 2, 3, 4, 5.

Definition 1.8.1. *Let X, Y be uniform spaces. A uniformly $\Phi_i(X)$-continuous map of X into Y will be called* uniformly Φ_i-continuous.

Definition 1.8.2. *Let X be a uniform (topological) space, Y be a topological space and $i \in \{1, 2, 3, 4, 5\}$ ($i \in \{1, 2\}$). A $\Phi_i(X)$-continuous map of X into Y will be called* Φ_i-continuous.

Any Φ_1-continuous map is sequentially continuous.

Proposition 1.8.3. *Any uniformly Φ_i-continuous map is Φ_i-continuous.*

The assertion follows immediately from Corollary 1.3.5. □

Theorem 1.8.4. *Let X, Y be uniform spaces and φ be a map of X into Y. We consider the following assertions:*

a) φ *is uniformly* Φ_i*-continuous;*

b) $\mathfrak{F} \in \Phi_C(X) \cap \Phi_i(X) \Rightarrow \varphi(\mathfrak{F}) \in \Phi_C(Y)$;

c) $\mathfrak{F} \in \Phi_C(X) \cap \hat{\Phi}_i(X) \Rightarrow \varphi(\mathfrak{F}) \in \Phi_C(Y) \cap \hat{\Phi}_i(Y)$;

d) $\mathfrak{F} \in \Phi_C(X) \cap \Phi_i(X) \Rightarrow \varphi(\mathfrak{F}) \in \Phi_C(Y) \cap \Phi_i(Y)$;

e) $(\varphi(x_n))_{n \in \mathbb{N}}$ *is a* $\Theta_i(Y)$*-sequence for any* $\Theta_i(X)$*-sequence* $(x_n)_{n \in \mathbb{N}}$;

f) $(I, \varphi \circ f)$ *is a* $\Theta_i(Y)$*-net for any* $\Theta_i(X)$*-net* (I, f);

g) $\mathfrak{F} \in \Phi_i(X) \Rightarrow \varphi(\mathfrak{F}) \in \Phi_i(Y)$;

h) $\varphi(A)$ *is a* $\Phi_i(Y)$*-set for any* $\Phi_i(X)$*-set* A;

i) $\mathfrak{F} \in \hat{\Phi}_i(X) \Rightarrow \varphi(\mathfrak{F}) \in \hat{\Phi}_i(Y)$;

j) $\varphi | A$ *is uniformly continuous and* $\varphi(A)$ *is a* $\Phi_i(Y)$*-set for any* $\Phi_i(X)$*-set* A.

Then we have the following implications:

$a \Leftrightarrow b \Leftrightarrow c \Leftrightarrow d \Rightarrow j \Rightarrow e \Leftrightarrow f \Leftrightarrow g \Leftrightarrow h \Leftrightarrow i.$

The implications c ⇒ a ⇒ b, d ⇒ b, and j ⇒ h are trivial.

e ⇔ f ⇔ g ⇔ h ⇔ i follows from Proposition 1.4.8.

b ⇒ a. By Proposition 1.4.2 any filter finer than a filter of $\Phi_i(X)$ belongs to $\Phi_i(X)$. By Proposition 1.5.15 $\mathfrak{F} \cap \mathfrak{G} \in \Phi_i(X)$ for any $\mathfrak{F}, \mathfrak{G} \in \Phi_i(X)$ and a) follows from Proposition 1.3.9.

a ⇒ j. By Proposition 1.3.11 the restriction of φ to $\bar{A}$ is uniformly continuous. Hence, $\varphi(A)$ is a $\Phi_i(Y)$-set (Propositions 1.5.4, 1.5.5, 1.5.6).

a ⇒ c & d. By the above results a ⇒ g & i and the implications a & g ⇒ d, a & i ⇒ c are trivial. □

Remark. Let ω_1 be the first uncountable ordinal number, X be the set of ordinal numbers $\leq \omega_1$ endowed with the usual compact topology, and φ be the map of X into $\mathbb{R}$ equal to 1 at ω_1 and equal to 0 elsewhere. Then $\varphi(A)$ is a Φ_i-set for any $A \subset X$. but φ is not uniformly continuous. Hence, the implication e ⇒ j in the above theorem does not hold in general.

Corollary 1.8.5. *The composition of two uniformly Φ_i-continuous maps is uniformly Φ_i-continuous.*

The assertion follows immediately from Theorem 1.8.4 a ⇒ c. □

The next corollary will not be used in the sequel.

Corollary 1.8.6. *The image of a precompact set with respect to a uniformly Φ_5-continuous map is precompact.*

By Proposition 1.5.6 the precompact sets are exactly the Φ_5-sets and the assertion follows from Theorem 1.8.4 a $\Rightarrow$ h. $\square$

Remark. The image of a precompact set with respect to a Φ_4-uniformly continuous map is not always precompact as the following example shows. Let $\tilde{\mathbb{N}}$ be the set $\mathbb{N}$ endowed with the uniformity induced by the Stone-Čech compactification of $\mathbb{N}$. Then $\tilde{\mathbb{N}}$ is a precompact space and the identity map $\tilde{\mathbb{N}} \to \mathbb{N}$ is uniformly Φ_4-continuous, but $\mathbb{N}$ is not precompact. In particular we deduce that there exist precompact sets which are not Φ_4-sets and uniformly Φ_4-continuous maps on precompact spaces which are not uniformly continuous.

Corollary 1.8.7. *Let X', X'' be two uniform spaces defined on the same set X such that the identity map $X' \to X''$ is sequentially continuous and the identity map $X'' \to X'$ is uniformly Φ_3-continuous. If X' is sequentially complete, then so is X''.*

Let $(x_n)_{n\in\mathbb{N}}$ be a Cauchy sequence in X''. By Theorem 1.8.4 a $\Rightarrow$ e $(x_n)_{n\in\mathbb{N}}$ is a $\Theta_3(X')$-sequence. Hence, it possesses a Cauchy and therefore a convergent subsequence $(x_{k_n})_{n\in\mathbb{N}}$ in X'. Since the identity map $X' \to X''$ is sequentially continuous $(x_{k_n})_{n\in\mathbb{N}}$ converges in X''. It follows that $(x_n)_{n\in\mathbb{N}}$ converges in X''. Hence, X'' is sequentially complete. $\square$

Proposition 1.8.8. *Let X', X'' be uniform spaces defined on the same set such that the identity maps $X' \to X''$, $X'' \to X'$ are uniformly Φ_i-continuous. If A is a Φ_i-set of X', then the closures of A in X' and X'' coincide and X', X'' induce the same uniformity on this closure of A.*

The assertion follows immediately from Corollary 1.3.12 and Proposition 1.8.4 a $\Rightarrow$ g. $\square$

Proposition 1.8.9. *Let X, Y be topological spaces, $i \in \{1, 2\}$ and φ be a Φ_i-continuous map of X into Y. Then:*

a) $(\varphi(x_n))_{n\in\mathbb{N}}$ *is a* $\Theta_i(Y)$*-sequence for any* $\Theta_i(X)$*-sequence* $(x_n)_{n\in\mathbb{N}}$;

b) $(I, \varphi \circ f)$ *is a* $\Theta_i(Y)$*-net for any* $\Theta_i(X)$*-net* (I, f);

c) $\mathfrak{F} \in \Phi_i(X) \Rightarrow \varphi(\mathfrak{F}) \in \Phi_i(Y)$;

d) $\mathfrak{F} \in \hat{\Phi}_i(X) \Rightarrow \varphi(\mathfrak{F}) \in \hat{\Phi}_i(Y)$;

e) $\varphi(A)$ *is a* $\Phi_i(Y)$*-set and* $\varphi|A$ *is continuous for any* $\Phi_i(X)$*-set* A*; if any point of* Y *possesses a fundamental system of closed neighbourhoods then* $\varphi|\bar{A}$ *is continuous.*

It is easy to prove a). The other assertions follow from a), Proposition 1.3.7, and Proposition 1.4.2. $\square$

Remark. The map $\mathbb{R} \to \mathbb{R}$, equal to 0 at 0 and equal to 1 elsewhere possesses the above properties a) to d), but it is not Φ_1-continuous.

Corollary 1.8.10. *Let X, Y, Z be topological spaces, $i \in \{1, 2\}$ and $\varphi : X \to Y$, $\psi : Y \to Z$ be Φ_i-continuous maps. Then $\psi \circ \varphi$ is Φ_i-continuous.* □

Proposition 1.8.11. *Let X', X'' be uniform (topological) spaces defined on the same set X such that the identity maps $X' \to X''$, $X'' \to X'$, are uniformly Φ_i-continuous (Φ_i-continuous with $i \in \{1, 2\}$). Then X' is Φ_i-compact iff X'' is Φ_i-compact.*

Assume X' is Φ_i-compact and let $\mathfrak{F}$ be an ultrafilter of $\Phi_i(X'')$. By Theorem 1.8.4 (Proposition 1.8.9) $\mathfrak{F}$ is an ultrafilter of $\Phi_i(X')$. Hence, $\mathfrak{F}$ converges in X'. By Proposition 1.8.3 it converges in X''. Hence, X'' is Φ_i-compact. □

Proposition 1.8.12. *Let X', X'' be two topological spaces defined on the same set such that X' is Hausdorff and the identity map $X'' \to X'$ is Φ_2-continuous. If a Θ_2-sequence in X'' converges to a point in X', then it converges to this point in X''.*

Let $(x_n)_{n \in \mathbb{N}}$ be a Θ_2-sequence in X'' converging in X' to a point x, and y be an adherent point in X'' of $(x_n)_{n \in \mathbb{N}}$. Since the identity map $X'' \to X'$ is Φ_2-continuous y is an adherent point of $(x_n)_{n \in \mathbb{N}}$ in X' and therefore $y = x$. Hence, x is the unique adherent point in X'' of $(x_n)_{n \in \mathbb{N}}$. By Proposition 1.5.7 $(x_n)_{n \in \mathbb{N}}$ converges in X'' to x. □

Proposition 1.8.13. *Let X, Y be uniform (topological) spaces, φ be a map of X into Y and $(\varphi_\iota)_{\iota \in I}$ be a family of maps of Y into uniform (topological) spaces such that the uniformity (topology) of Y is the initial uniformity (topology) with respect to this family. If $\varphi_\iota \circ \varphi$ is uniformly Φ_i-continuous (Φ_i-continuous with $i \in \{1, 2\}$) for any $\iota \in I$, then φ is uniformly Φ_i-continuous (Φ_i-continuous).*

We first prove the topological case. Let $\mathfrak{F} \in \Phi_i(X)$ and x be a limit point of $\mathfrak{F}$. Then $\varphi_\iota \circ \varphi(\mathfrak{F})$ converges to $\varphi_\iota \circ \varphi(x)$ for any $\iota \in I$. We deduce that $\varphi(\mathfrak{F})$ converges to $\varphi(x)$. Hence, φ is Φ_i-continuous. Assume now that X, Y are uniform spaces and that $\mathfrak{F} \in \hat{\Phi}_i(X) \cap \Phi_C(X)$. Then $\varphi_\iota \circ \varphi(\mathfrak{F})$ is a Cauchy filter for any $\iota \in I$. Hence, $\varphi(\mathfrak{F})$ is a Cauchy filter and φ is uniformly Φ_i-continuous. □

Corollary 1.8.14. *Let X, Y be uniform (topological) spaces and φ be a uniformly Φ_i-continuous (Φ_i-continuous with $i \in \{1, 2\}$) map of X into Y. Further, let X' and Y' be subspaces of X and Y respectively such that $\varphi(X') \subset Y'$. Then the map $X' \to Y'$, $x \mapsto \varphi(x)$ is uniformly Φ_i-continuous (Φ_i-continuous).*

Let us denote by φ' the map $X' \to Y'$, $x \mapsto \varphi(x)$ and by f, g the inclusion maps $X' \to X$, $Y' \to Y$ respectively. Then $g \circ \varphi' = \varphi \circ f$ and therefore $g \circ \varphi'$ is uniformly Φ_i-continuous (Φ_i-continuous) (Corollary 1.8.5 (Corollary 1.8.10)). By Proposition 1.8.13 φ' is uniformly Φ_i-continuous (Φ_i-continuous). □

Corollary 1.8.15. *Let X', X'' be uniform (topological) spaces defined on the same set X, Y be a subset of X and Y', Y'' be the set Y endowed with the uniformity (topology) induced by X' and X'' respectively. If the identity map $X' \to X''$ is uniformly Φ_i-continuous (Φ_i-continuous with $i \in \{1, 2\}$), then so is the identity map $Y' \to Y''$.* □

Corollary 1.8.16. *Let $(X_\iota)_{\iota\in I}$, $(Y_\iota)_{\iota\in I}$ be families of uniform (topological) spaces and for each $\iota\in I$ let φ_ι be a uniformly Φ_i-continuous (Φ_i-continuous with $i\in\{1,2\}$) map of X_ι into Y_ι. Then the map*

$$\prod_{\iota\in I} X_\iota \to \prod_{\iota\in I} Y_\iota,\quad (x_\iota)_{\iota\in I} \mapsto (\varphi_\iota(x_\iota))_{\iota\in I}$$

is uniformly Φ_i-continuous (Φ_i-continuous).

For any $\iota\in I$ denote by p_ι (by q_ι) the ι-projection of $\prod_{\iota\in I} X_\iota$ (of $\prod_{\iota\in I} Y_\iota$) and let φ be the map defined above. Then $q\circ\varphi = \varphi_\iota\circ p_\iota$ and so (Corollary 1.8.5 (Corollary 1.8.10)) $q_\iota\circ\varphi$ is uniformly Φ_i-continuous (Φ_i-continuous) for any $\iota\in I$. By Proposition 1.8.13 φ is uniformly Φ_i-continuous (Φ_i-continuous). □

Proposition 1.8.17. *Let X, Y be uniform spaces, φ be a map $X\to Y$, A be a Φ_i-closed set of Y and ψ be the map*

$$\overset{-1}{\varphi}(A) \to A,\ x\mapsto \varphi(x).$$

We assume that one of the following assertions holds:

a) φ is uniformly Φ_i-continuous;

b) ψ is uniformly Φ_i-continuous and there exists a fundamental system $\mathfrak{U}$ of entourages of Y such that $\overset{-1}{\widehat{\varphi\times\varphi}}(U)$ is a closed set of $X\times X$ for any $U\in\mathfrak{U}$.

Then $\overset{-1}{\varphi}(A)$ is a Φ_i-closed set of X.

Let us denote by f and g the inclusion maps $\overset{-1}{\varphi}(A)\to X$ and $A\to Y$ respectively. Let $\mathfrak{F}$ be a Φ_i-filter on $\overset{-1}{\varphi}(A)$ and x be a limit point of $f(\mathfrak{F})$. We have to show that $x\in\overset{-1}{\varphi}(A)$, i.e. $\varphi(x)\in A$.

a) $f(\mathfrak{F})$ is a Φ_i-filter. Since φ is Φ_i-continuous (Proposition 1.8.3) $\varphi(x)$ is a limit point of $\varphi(f(\mathfrak{F}))$. By Corollary 1.8.14 ψ is uniformly Φ_i-continuous so by Theorem 1.8.4 a $\Rightarrow$ g $\psi(\mathfrak{F})$ is a Φ_i-filter on A. We have $g(\psi(\mathfrak{F})) = \varphi(f(\mathfrak{F}))$. Hence, $g(\psi(\mathfrak{F}))$ converges to $\varphi(x)$ so, A being a Φ_i-closed set of Y, $\varphi(x)\in A$.

b) $\mathfrak{F}$ is obviously a Cauchy Φ_i-filter on $\overset{-1}{\varphi}(A)$ so by Theorem 1.8.4 a $\Rightarrow$ d $\psi(\mathfrak{F})$ is a Cauchy Φ_i-filter on A. Let $U\in\mathfrak{U}$. There exists $B\in\mathfrak{F}$ such that $\psi(B)\times\psi(B)\subset U$. We get $B\times B\subset \overset{-1}{\widehat{\varphi\times\varphi}}(U)$. Let $y\in B$ and $\mathfrak{G}$ be the filter on $X\times X$ generated by the filter base $\{C\times\{y\}\mid C\in\mathfrak{F}\}$. Then $\overset{-1}{\widehat{\varphi\times\varphi}}(U)\in\mathfrak{G}$ and $\mathfrak{G}$ converges to (x,y). Since $\overset{-1}{\widehat{\varphi\times\varphi}}(U)$ is a closed set of $X\times X$ we get $(x,y)\in\overset{-1}{\widehat{\varphi\times\varphi}}(U)$. Hence, $(\varphi(x),\varphi(y))\in U$ and $\varphi(y)\in U(\varphi(x))$. y being arbitrary we deduce $\varphi(B)\subset U(\varphi(x))$. Thus $\varphi(f(\mathfrak{F}))$ converges to $\varphi(x)$. We have $g(\psi(\mathfrak{F})) = \varphi(f(\mathfrak{F}))$. Hence, $g(\psi(\mathfrak{F}))$ converges to $\varphi(x)$ and therefore, A being a Φ_i-closed set of Y, $\varphi(x)\in A$. □

Proposition 1.8.18. *Let X, Y be topological spaces, $i\in\{1,2\}$, φ be a Φ_i-continuous map of X into Y and A be a strongly Φ_i-closed set of Y. Then $\overset{-1}{\varphi}(A)$ is a strongly Φ_i-closed set of X.*

Let us denote by ψ the map

$$\overset{-1}{\varphi}(A) \to A\,, \quad x \mapsto \varphi(x)$$

and by f, g the inclusion maps $\overset{-1}{\varphi}(A) \to X$ and $A \to Y$ respectively. Let $\mathfrak{F} \in \Phi(\Theta_i(\overset{-1}{\varphi}(A), X))$ and x be a limit point of $f(\mathfrak{F})$. Let further (I, h) be a $\Theta_i(\overset{-1}{\varphi}(A), X)$-net in $\overset{-1}{\varphi}(A)$ with $h(\mathfrak{G}) = \mathfrak{F}$, where $\mathfrak{G}$ denotes the section filter of I. Then $(I, \psi \circ h)$ is a $\Theta_i(A, Y)$-net (Proposition 1.8.9b) and Corollary 1.8.14) and therefore $\psi(\mathfrak{F}) \in \Phi(\Theta_i(A, Y))$. Since $g(\psi(\mathfrak{F})) = \varphi(f(\mathfrak{F}))$ and $\varphi(f(\mathfrak{F}))$ converges to $\varphi(x)$ we get $\varphi(x) \in A$ and therefore $x \in \overset{-1}{\varphi}(A)$. Hence, $\overset{-1}{\varphi}(A)$ is a strongly Φ_i-closed set of X. □

Proposition 1.8.19. *Let X, Y be uniform or topological spaces, $i, j \in \{1, 2, 3, 4, 5\}$, Z be a uniform space and φ be a map of $X \times Y$ into Z such that $\varphi(\mathfrak{F} \times \mathfrak{G})$ is a Cauchy filter for any Cauchy (convergent) $\hat{\Phi}_i$-filter $\mathfrak{F}$ on X and for any Cauchy (convergent) $\hat{\Phi}_j$-filter $\mathfrak{G}$ on Y. Further let A, B be a Φ_i-set of X and a Φ_j-set of Y respectively. Then the restriction of φ to the closure of $A \times B$ is uniformly continuous (continuous).*

Let us denote by ψ', ψ'' the projections of $X \times Y$ on X and Y respectively and let Φ be the set of filters $\mathfrak{F}$ on $X \times Y$ for which $\psi'(\mathfrak{F}) \in \Phi_i(X)$ and $\psi''(\mathfrak{F}) \in \Phi_j(Y)$. By Corollary 1.1.5 and Proposition 1.4.2 $\psi'(\mathfrak{F}) \in \hat{\Phi}_i(X)$ and $\psi''(\mathfrak{F}) \in \hat{\Phi}_j(Y)$ for any $\mathfrak{F} \in \hat{\Phi}$. $A \times B$ is a Φ-set.

By Corollary 1.1.6 and Proposition 1.3.4 φ is Φ-continuous. By Proposition 1.3.7 its restriction to the closure of $A \times B$ is continuous.

Assume now that both X and Y are uniform spaces. Then φ is uniformly Φ-continuous. By Proposition 1.3.11 the restriction of φ to the closure of $A \times B$ is uniformly continuous. □

Proposition 1.8.20. *Let X, Y be uniform (topological) spaces, $i \in \{1, 2, 3, 4, 5\}$ ($i \in \{1, 2\}$), φ be a map of X into Y, $(Y_\iota)_{\iota \in I}$ be a family of subspaces of Y such that for any countable subset A of Y there exists $\iota \in I$ with $A \subset Y_\iota$ and, for each $\iota \in I$ let φ_ι be a map of Y into Y_ι such that $\varphi_\iota \mid Y_\iota$ is the identity map and such that $\varphi_\iota \circ \varphi$ is uniformly Φ_i-continuous (Φ_i-continuous). Then $\varphi(A)$ is a Φ_i-set of Y for any Φ_i-set A of X.*

Let A be a Φ_i-set of X and $(y_n)_{n \in \mathbb{N}}$ be a sequence in $\varphi(A)$. By the hypothesis there exists $\iota \in I$ such that $\{y_n \mid n \in \mathbb{N}\} \subset Y_\iota$. Let $(x_n)_{n \in \mathbb{N}}$ be a sequence in A such that $\varphi(x_n) = y_n$ for any $n \in \mathbb{N}$. By Proposition 1.4.5 a $\Rightarrow$ c $(x_n)_{n \in \mathbb{N}}$ is a Θ_i-sequence in X. Since $\varphi_\iota \circ \varphi$ is uniformly Φ_i-continuous (Φ_i-continuous) and since $\varphi_\iota \circ \varphi(x_n) = y_n$ for any $n \in \mathbb{N}$, $(y_n)_{n \in \mathbb{N}}$ is a Θ_i-sequence in Y_ι (Theorem 1.8.4 a $\Rightarrow$ e, (Proposition 1.8.9a))) and therefore a Θ_i-sequence in Y. By Proposition 1.4.5 c $\Rightarrow$ a $\varphi(A)$ is a Φ_i-set of Y. □

Proposition 1.8.21. *Let X', X'' be two uniform spaces defined on the same set X such that X' is a Hausdorff mioritic space, the identity map $X'' \to X'$ is uniformly continuous, and X'' is metrizable. Let Y be a dense set of X'' and Y', Y'' be the uniform spaces* *obtained by endowing Y with the uniformities induced by X' and X'' respectively. If the identity map $Y' \to Y''$ is uniformly Φ_3-continuous, then the identity map $X' \to X''$ is uniformly Φ_4-continuous.*

Let $\mathfrak{F}$ be a Cauchy Φ_4-filter on X' and (I, f) be a Θ_4-net in X' such that $f(\mathfrak{G}) = \mathfrak{F}$, where $\mathfrak{G}$ denotes the section filter of I. Further, let d be a metric on X generating the uniformity of X''. We set

$$L := \{(\iota, n, x) \in I \times \mathbb{N} \times Y \mid d(x, f(\iota)) < \frac{1}{n+1}\},$$

$$g : L \to Y, \ (\iota, n, x) \mapsto x$$

and endow L with the upper directed order relation $\leq$ defined by

$$(\iota, n, x) < (\iota', n', x') :\Leftrightarrow (\iota \leq \iota', n < n').$$

We want to show (L, g) is a Θ_3-net in Y'. Let $(\iota_m, n_m, x_m)_{m \in \mathbb{N}}$ be an increasing sequence in L. Then $(\iota_m)_{m \in \mathbb{N}}$ is an increasing sequence in I and $(f(\iota_m))_{m \in \mathbb{N}}$ is a Θ_4-sequence in X'. By Proposition 1.7.3 a $\Rightarrow$ c it is a Θ_3-sequence in X' and it contains therefore a Cauchy subsequence in X'. Then $(x_m)_{m \in \mathbb{N}}$ possesses a Cauchy subsequence in Y'. Hence, (L, g) is a Θ_3-net in Y'. It is easy to see that $g(\mathfrak{H})$ is a Cauchy filter on Y', where $\mathfrak{H}$ denotes the section filter of L. Hence, $g(\mathfrak{H})$ is a Cauchy filter on Y''. We deduce $\mathfrak{F}$ is a Cauchy filter on X''. By Theorem 1.8.4 b $\Rightarrow$ a the identity map $X' \to X''$ is uniformly Φ_4-continuous. □

The next result will not be used in the sequel.

Proposition 1.8.22. *Let X be a mioritic Hausdorff topological space and Y be a subset of X endowed with a topology such that the inclusion map $\varphi : Y \to X$ is Φ_2-continuous. Then Y is a mioritic space. In particular, any subspace of X is mioritic.*

Let $(x_n)_{n \in \mathbb{N}}$ be a Θ_2-sequence in Y. By Proposition 1.8.9a) $(\varphi(x_n))_{n \in \mathbb{N}}$ is a Θ_2-sequence in X and therefore by Proposition 1.7.2 a $\Rightarrow$ b there exists a subsequence $(x_{k_n})_{n \in \mathbb{N}}$ of $(x_n)_{n \in \mathbb{N}}$ such that $(\varphi(x_{k_n}))_{n \in \mathbb{N}}$ converges to a point x in X. Let y be an arbitrary adherent point of $(x_{k_n})_{n \in \mathbb{N}}$ in Y. Then $\varphi(y) = x$. Hence, $(x_{k_n})_{n \in \mathbb{N}}$ possesses a unique adherent point and therefore it is a convergent sequence in Y (Proposition 1.5.7). By Proposition 1.7.2 b $\Rightarrow$ a Y is a mioritic space. □

Chapter 2: Spaces of functions

§ 2.1 Uniformities on spaces of functions

This section is dedicated to the definition of some uniformities on the spaces of functions and to some elementary properties of these uniformities which are needed later.

Definition 2.1.1. *Let X be a set and Y be a uniform space. For any subset A of X, for any filter $\mathfrak{F}$ on X, and for any entourage U of Y we set*

$$\underline{\mathscr{W}(A, U; X, Y)} := \underline{\mathscr{W}(A, U)} := \{(f, g) \in Y^X \times Y^X \mid x \in A \Rightarrow (f(x), g(x)) \in U\},$$

$$\underline{\mathscr{W}(\mathfrak{F}, U; X, Y)} := \underline{\mathscr{W}(\mathfrak{F}, U)} := \{(f, g) \in Y^X \times Y^X \mid \{x \in X \mid (f(x), g(x)) \in U\} \in \mathfrak{F}\}.$$

Definition 2.1.2. *Let X be a set, $\mathfrak{A}$ be a set of subsets of X, Φ be a set of filters on X, Y be a uniform space, $\mathfrak{U}$ be the uniformity of Y and $\mathscr{F}$ be a subset of Y^X. The set of finite intersections of the sets of the form $\mathscr{W}(A, U) \cap (\mathscr{F} \times \mathscr{F})$, where (A, U) runs through $\mathfrak{A} \times \mathfrak{U}$ (of the sets of the form $\mathscr{W}(\mathfrak{F}, U) \cap (\mathscr{F} \times \mathscr{F})$, where $(\mathfrak{F}, U)$ runs through $\Phi \times \mathfrak{U}$) is a fundamental system of entourages of a uniformity on $\mathscr{F}$, which will be called the* uniformity of $\mathfrak{A}$-convergence (uniformity of Φ-convergence) on $\mathscr{F}$. *Its corresponding topology will be called the* topology of $\mathfrak{A}$-convergence (topology of Φ-convergence) on $\mathscr{F}$. *(In the above definition we have assumed $\Phi \neq \emptyset$ and $\mathfrak{A} \neq \emptyset$; otherwise we take the coarsest uniformity on $\mathscr{F}$).*

If we set $\mathfrak{F}(A) := \{B \mid A \subset B \subset X\}$ for any $A \in \mathfrak{A}$, then $\Phi(\mathfrak{A}) := \{\mathfrak{F}(A) \mid A \in \mathfrak{A}\}$ is a set of filters on X and the uniformity (topology) of $\mathfrak{A}$-convergence is nothing else but the uniformity (topology) of $\Phi(\mathfrak{A})$-convergence.

Definition 2.1.3. *Let X be a set, Y be a uniform space, $\mathscr{F}$ be a subset of Y^X, $\mathfrak{A}$ be a set of subsets of X and Φ be a set of filters on X. We denote by $\underline{\mathscr{F}_{\mathfrak{A}}}$ and $\underline{\mathscr{F}_{\Phi}}$ the set $\mathscr{F}$ endowed with the uniformity of $\mathfrak{A}$-convergence and Φ-convergence respectively. In particular, $\underline{\mathscr{F}_{\{X\}}}$ will stand for $\mathscr{F}$ endowed with the uniformity of uniform convergence and $\underline{Y^X_{\mathfrak{A}}}$, $\underline{Y^X_{\Phi}}$ will stand for the set of all maps of X into Y endowed with the uniformity of $\mathfrak{A}$-convergence and Φ-convergence respectively.*

Definition 2.1.4. *Let X be a set, A be a subset of X, Y be a uniform (topological) space and $\mathscr{F}$ be a subset of Y^X. We denote by $\underline{\mathscr{F}_A}$ the set $\mathscr{F}$ endowed with the uniformity (topology) of pointwise convergence in A. If $A = X$ we drop the index. This means that any set of maps of X into Y will be considered to be endowed with the uniformity (topology) of pointwise convergence if no other uniformity (topology) is mentioned.*

In particular, $\underline{Y}_A^X$ *and* $\underline{Y}^X$ *mean the set of all maps of* X *into* Y *endowed with the uniformity (topology) of pointwise convergence in* A *or in* X *respectively.*

Let X be a set, A be a subset of X, Y be a uniform space and $\mathscr{F}$ be a subset of Y^X. Then $\mathscr{F}_A = \mathscr{F}_{\{\{x\} | x \in A\}}$.

Proposition 2.1.5. *Let* X *be a topological space,* Φ *be a set of filters on* X, Y *be a uniform space and* $\mathscr{F}$ *be a set of continuous maps of* X *into* Y. *For any filter* $\mathfrak{F}$ *on* X *we set*

$$\mathfrak{F}' := \{A \mid \exists B \in \mathfrak{F},\ \bar{B} \subset A \subset X\},$$

$$\mathfrak{F}'' := \{A \mid \exists B \in \mathfrak{F},\ A \text{ is a neighbourhood of } B\}.$$

If we put

$$\Phi' := \{\mathfrak{F}' \mid \mathfrak{F} \in \Phi\}, \quad \Phi'' := \{\mathfrak{F}'' \mid \mathfrak{F} \in \Phi\},$$

then $\mathscr{F}_\Phi = \mathscr{F}_{\Phi'} = \mathscr{F}_{\Phi''}$.

Let $\mathfrak{F} \in \Phi$ and U be an entourage of Y. There exist a closed entourage V of Y such that $V \subset U$ and an open entourage W of Y such that $W \subset U$. Let $f, g \in \mathscr{F}$. Then

$$\{x \in X \mid (f(x), g(x)) \in V\}$$

is a closed set and

$$\{x \in X \mid (f(x), g(x)) \in W\}$$

is an open set. Hence,

$$\mathscr{W}(\mathfrak{F}, V) = \mathscr{W}(\mathfrak{F}', V) \subset \mathscr{W}(\mathfrak{F}', U)$$

$$\mathscr{W}(\mathfrak{F}, W) = \mathscr{W}(\mathfrak{F}'', W) \subset \mathscr{W}(\mathfrak{F}'', U)$$

and we deduce that the uniformities of Φ'-convergence and Φ''-convergence on $\mathscr{F}$ are coarser than the uniformity of Φ-convergence on $\mathscr{F}$. The converse assertion is trivial and therefore they coincide. □

Proposition 2.1.6. *Let* X, Y *be uniform spaces,* $\Phi := \hat{\Phi}_C(X)$ *and* $\mathscr{F}$ *be a uniformly equicontinuous set of maps of* X *into* Y. *Then* $\mathscr{F} = \mathscr{F}_\Phi$.

It is obvious that the uniformity of Φ-convergence is finer than the uniformity of pointwise convergence. Let $\mathfrak{F} \in \Phi$ and U be an arbitrary entourage of Y. Let further V be an entourage of Y such that $\overset{-1}{V} \circ V \circ V \subset U$. Since $\mathscr{F}$ is uniformly equicontinuous there exists an entourage W of X such that $(f(x), f(y)) \in V$ for any $f \in \mathscr{F}$ and for any $(x, y) \in W$. Let $\mathfrak{A}$ be a finite set of nonempty subsets of X such that

$$\bigcup_{A \in \mathfrak{A}} (A \times A) \subset W, \quad \bigcup_{A \in \mathfrak{A}} A \in \mathfrak{F}$$

(Proposition 1.2.4). Let x_A be a point of A for any $A \in \mathfrak{A}$. Let $f, g \in \mathscr{F}$ such that $(f(x_A), g(x_A)) \in V$ for any $A \in \mathfrak{A}$. Let $x \in \bigcup_{A \in \mathfrak{A}} A$. Then there exists $A \in \mathfrak{A}$ such that

$x \in A$. We get $(x, x_A) \in W$ and therefore

$$(f(x), f(x_A)) \in V, \quad (g(x), g(x_A)) \in V.$$

Hence,

$$(f(x), g(x)) \in \overset{-1}{V} \circ V \circ V \subset U.$$

Since x is arbitrary we get $(f, g) \in \mathscr{W}(\mathfrak{F}, U)$ and therefore

$$\bigcap_{A \in \mathfrak{A}} (\mathscr{W}(\{x_A\}, V) \cap (\mathscr{F} \times \mathscr{F})) \subset \mathscr{W}(\mathfrak{F}, U) \cap (\mathscr{F} \times \mathscr{F}).$$

This shows that the uniformity of pointwise convergence on $\mathscr{F}$ is finer than the uniformity of Φ-convergence on $\mathscr{F}$. □

Proposition 2.1.7. *Let X be a set, Y be a uniform space, Φ be a set of filters on X, and $\mathscr{F}$ be the set of $f \in Y^X$ such that $f(\mathfrak{F})$ is a Cauchy filter for any $\mathfrak{F} \in \Phi$. Then $\mathscr{F}$ is a closed set of Y_Φ^X.*

Let f be an adherent point of $\mathscr{F}$ in Y_Φ^X. Let $\mathfrak{F} \in \Phi$ and U be an arbitrary entourage of Y. Further, let V be an entourage of Y such that $\overset{-1}{V} \circ V \circ V \subset U$. There exists $g \in \mathscr{F}$ with $(f, g) \in \mathscr{W}(\mathfrak{F}, V)$ i.e.

$$A := \{x \in X \mid (f(x), g(x)) \in V\} \in \mathfrak{F}.$$

Since $g(\mathfrak{F})$ is a Cauchy filter there exists $B \in \mathfrak{F}$ with $g(B) \times g(B) \subset V$. We get

$$f(A \cap B) \times f(A \cap B) \subset \overset{-1}{V} \circ V \circ V \subset U.$$

Hence, $f(\mathfrak{F})$ is a Cauchy filter, $f \in \mathscr{F}$, and $\mathscr{F}$ is a closed set of Y_Φ^X. □

Proposition 2.1.8. *(Osgood (1897) [1] Theorem IV). Let X be a topological space, Y be a pseudometrizable uniform space and $\mathscr{F}$ be the set of continuous maps of X into Y. Then any Cauchy sequence in $\mathscr{F}$ is equicontinuous at each point of the complement of a meagre set of X.*

Let $(f_n)_{n \in \mathbb{N}}$ be a Cauchy sequence in $\mathscr{F}$ and d be a pseudometric on Y defining its uniformity. We set for any $m, p \in \mathbb{N}$

$$F_{mp} := \{x \in X \mid \ n \geq m \Rightarrow d(f_m(x), f_n(x)) \leq \frac{1}{p+1}\}$$

Since the maps $f_n (n \in \mathbb{N})$ are continuous, the sets F_{mp} $(m, p \in \mathbb{N})$ are closed. $(f_n)_{n \in \mathbb{N}}$ being a Cauchy sequence in $\mathscr{F}$ we have

$$\bigcup_{m \in \mathbb{N}} F_{mp} = X$$

for any $p \in \mathbb{N}$. The set

$$A := \bigcap_{p \in \mathbb{N}} \bigcup_{m \in \mathbb{N}} \overset{\circ}{F}_{mp}$$

is the complement of a meagre set of X.

Let $x \in A$ and U be an entourage of Y. Then there exists $p \in \mathbb{N}$ such that

$$\{(y, z) \in Y \times Y \mid d(y, z) < \frac{3}{p+1}\} \subset U.$$

In addition, there exists $m \in \mathbb{N}$ such that $x \in \mathring{F}_{m,p}$. Let B be a neighbourhood of x such that $d(f_n(x), f_n(y)) < \dfrac{1}{p+1}$ for any $n \leq m$ and for any $y \in B$. We get

$$d(f_n(x), f_n(y)) \leq d(f_n(x), f_m(x)) + d(f_m(x), f_m(y)) + d(f_m(y), f_n(y)) < \frac{3}{p+1}$$

for any $n \geq m$ and for any $y \in B \cap \mathring{F}_{mp}$. Hence, $(f_n(x), f_n(y)) \in U$ for any $n \in \mathbb{N}$ and for any $y \in B \cap \mathring{F}_{mp}$. This shows that $(f_n)_{n \in \mathbb{N}}$ is equicontinuous at x. □

Definition 2.1.9. *Let X, Y be uniform spaces and $\mathscr{F}$ be a subset of Y^X. A filter $\mathfrak{F}$ on $\mathscr{F}$ is called* uniformly equicontinuous *if for any entourage V of Y there exist an entourage U of X and a set $\mathscr{F}' \in \mathfrak{F}$ such that $(f(x), f(y)) \in V$ for any $(x, y) \in U$ and any $f \in \mathscr{F}'$.*

A subset $\mathscr{G}$ of Y^X is uniformly equicontinuous iff the filter $\{\mathscr{H} \mid \mathscr{G} \subset \mathscr{H} \subset Y^X\}$ is uniformly equicontinuous. Any filter finer than a uniformly equicontinuous filter is uniformly equicontinuous.

Proposition 2.1.10. *Let X, Y be uniform spaces, $\mathscr{F}$ be a set of uniformly continuous maps of X into Y and $\mathfrak{F}$ be a Cauchy filter on $\mathscr{F}_{\{X\}}$. Then $\mathfrak{F}$ is uniformly equicontinuous.*

Let V be an arbitrary entourage of Y and W be an entourage of Y such that $W \circ W \circ W \subset V$. Since $\mathfrak{F}$ is a Cauchy filter on $\mathscr{F}_{\{X\}}$ there exists $\mathscr{G} \in \mathfrak{F}$ such that $(f(x), g(x)) \in W$ for any $f, g \in \mathscr{G}$ and any $x \in X$. Let $f \in \mathscr{G}$. Since f is uniformly continuous there exists an entourage U of X such that $(f(x), f(y)) \in W$ for any $(x, y) \in U$. We get

$$(g(x), g(y)) \in W \circ W \circ W \subset V$$

for any $g \in \mathscr{G}$ and for any $(x, y) \in U$. □

Proposition 2.1.11. *Let X, Y be uniform spaces and $\mathfrak{F}$ be a uniformly equicontinuous Cauchy filter of Y^X. If X is precompact, then for any entourage V of Y there exist an entourage U of X and an $\mathscr{F} \in \mathfrak{F}$ such that $(f(x), g(y)) \in V$ for any $(x, y) \in U$ and for any $f, g \in \mathscr{F}$. In particular, $\mathfrak{F}$ is a Cauchy filter on $Y^X_{\{X\}}$. If $\mathfrak{F}$ converges in Y^X to an f, then it converges to f in $Y^X_{\{X\}}$.*

Let W be an entourage of Y such that $W \circ W \circ \overset{-1}{W} \subset V$. Since $\mathfrak{F}$ is uniformly equicontinuous there exist an entourage U of X and $\mathscr{G} \in \mathfrak{F}$ such that $(g(x), g(y)) \in W$ for any $g \in \mathscr{G}$ and any $(x, y) \in U$. Since X is precompact there exists a finite subset A of X such that $X = \bigcup_{z \in A} U(z)$. Since $\mathfrak{F}$ is a Cauchy filter on Y^X there exists an $\mathscr{H} \in \mathfrak{F}$ such that $(g(z), h(z)) \in W$ for any $z \in A$ and any $g, h \in \mathscr{H}$. We get

$$(g(x), h(y)) \in W \circ W \circ \overset{-1}{W} \subset V$$

for any $(x, y) \in U$ and any $g, h \in \mathscr{G} \cap \mathscr{H}$. In particular, $\mathfrak{F}$ is a Cauchy filter on $Y^X_{\{X\}}$. If $\mathfrak{F}$ converges in Y^X to an f, then we have $(g(x), f(x)) \in \bar{V}$ for any $x \in X$ and for any $g \in \mathscr{G} \cap \mathscr{H}$, so $\mathfrak{F}$ converges to f in $Y^X_{\{X\}}$. □

Proposition 2.1.12. *Let X, Y be uniform spaces, $\mathscr{F}$ be a subset of Y^X and Φ be the set of uniformly equicontinuous filters on $\mathscr{F}$. Then $\hat{\Phi} = \Phi$. In particular, a filter $\mathfrak{F}$ on $\mathscr{F}$ is uniformly equicontinuous if any ultrafilter on $\mathscr{F}$ finer than it is uniformly equicontinuous.*

Let $\mathfrak{F} \in \hat{\Phi}$ and let V be an entourage of Y. For any $\mathfrak{G} \in \Phi$ there exist an entourage $U(\mathfrak{G})$ of X and a set $\mathscr{F}(\mathfrak{G}) \in \mathfrak{G}$ such that $(f(x), f(y)) \in V$ for any $(x, y) \in U(\mathfrak{G})$ and any $f \in \mathscr{F}(\mathfrak{G})$. Then there exists a finite subset Ψ of Φ such that

$$\bigcup_{\mathfrak{G} \in \Psi} \mathscr{F}(\mathfrak{G}) \in \mathfrak{F}.$$

We get $(f(x), f(y)) \in V$ for any $f \in \bigcup_{\mathfrak{G} \in \Psi} \mathscr{F}(\mathfrak{G})$ and any $(x, y) \in \bigcap_{\mathfrak{G} \in \Psi} U(\mathfrak{G})$. Hence, $\mathfrak{F} \in \Phi$. The last assertion follows from Corollary 1.1.5. □

Proposition 2.1.13. *Let X, Y be uniform spaces, $\mathfrak{F}$ be a uniformly equicontinuous filter on Y^X and $\mathscr{F}$ be the set of adherent points of $\mathfrak{F}$ in Y^X. Then $\mathscr{F}$ is uniformly equicontinuous.*

Let V be a closed entourage of Y. There exist an entourage U of X and a $\mathscr{G} \in \mathfrak{F}$ such that $(g(x), g(y)) \in V$ for any $(x, y) \in U$ and for any $g \in \mathscr{G}$. Let $f \in \mathscr{F}$. Since f is adherent to $\mathscr{G}$ in Y^X and since V is closed we get $(f(x), f(y)) \in V$ for any $(x, y) \in U$. Hence, $\mathscr{F}$ is uniformly equicontinuous. □

Proposition 2.1.14. *Let X be a set, Y be a uniform space, $i \in \{1, 2, 3, 4, 5\}$ and $\mathscr{F}$ be a Φ_i-closed set of Y^X. If Y is Φ_i-compact (sequentially complete and $i \geq 3$) then so is $\mathscr{F}$.*

Let φ be the inclusion map $\mathscr{F} \to Y^X$.

Assume that Y is Φ_i-compact. By Proposition 1.5.9 Y^X is Φ_i-compact. Let $\mathfrak{F}$ be an ultrafilter on $\mathscr{F}$ belonging to $\Phi_i(\mathscr{F})$. Then $\varphi(\mathfrak{F}) \in \Phi_i(Y^X)$ so $\varphi(\mathfrak{F})$ converges. Since $\mathscr{F}$ is a Φ_i-closed set of Y^X, $\mathfrak{F}$ converges. Hence, $\mathscr{F}$ is Φ_i-compact.

Assume that Y is sequentially complete and that $i \geq 3$. Let $(f_n)_{n \in \mathbb{N}}$ be a Cauchy sequence in $\mathscr{F}$. Then $(f_n(x))_{n \in \mathbb{N}}$ is a Cauchy sequence and therefore a convergent sequence for any $x \in X$. We set

$$f: X \to Y, \quad x \mapsto \lim_{n \to \infty} f_n(x).$$

Then $(\varphi(f_n))_{n \in \mathbb{N}}$ is a Θ_3-sequence in Y^X converging to f. Since $\mathscr{F}$ is a Φ_3-closed set of Y^X, we get $f \in \mathscr{F}$. Hence, $\mathscr{F}$ is sequentially complete. □

Proposition 2.1.15. *Let X be a set, A be a subset of X, let $\mathfrak{S}$ be a set of subsets of A, Y be a uniform space, $i \in \{1, 2, 3, 4, 5\}$, $\mathscr{F}$ be a Φ_i-closed set of $Y^X_{\mathfrak{S}}$ such that the iden-*

tity map $\mathscr{F}_A \to \mathscr{F}_{\mathfrak{S}}$ *is uniformly* Φ_i*-continuous, and* $\mathscr{F}'$ *be the set* $\mathscr{F}$ *endowed with a uniformity. We have:*

a) $\mathscr{F}$ *is a* Φ_i*-closed set of* Y^X*;*

b) if the identity maps $\mathscr{F}_{\mathfrak{S}} \to \mathscr{F}' \to \mathscr{F}$ *are uniformly* Φ_i*-continuous and if* Y *is* Φ_i*-compact, then* $\mathscr{F}'$ *is* Φ_i*-compact;*

c) if the identity map $\mathscr{F}_{\mathfrak{S}} \to \mathscr{F}'$ *is sequentially continuous, if the identity map* $\mathscr{F}' \to \mathscr{F}$ *is uniformly* Φ_3*-continuous, if* Y *is sequentially complete, and if* $i \geq 3$*, then* $\mathscr{F}'$ *is sequentially complete.*

a) Let U be a closed entourage of Y and let $B \in \mathfrak{S}$. Then $\mathscr{W}(B, U)$ is a closed set of $Y^X \times Y^X$ and therefore by Proposition 1.8.17 $\mathscr{F}$ is a Φ_i-closed of Y^X.

b) By a) and Proposition 2.1.14 $\mathscr{F}$ is Φ_i-compact. By Corollary 1.8.5 the identity map $\mathscr{F} \to \mathscr{F}'$ is uniformly Φ_i-continuous so, by Proposition 1.8.11, $\mathscr{F}'$ is Φ_i-compact.

c) By a) and Proposition 2.1.14 $\mathscr{F}$ is sequentially complete. By Proposition 1.8.3 the identity map $\mathscr{F} \to \mathscr{F}'$ is sequentially continuous. By Corollary 1.8.7 $\mathscr{F}'$ is sequentially complete. □

Proposition 2.1.16. *Let* X, Y *be topological spaces such that any point of* Y *possesses a fundamental system of closed neighbourhoods, let* Θ *be a set of* Θ_2*-sequences in* X *and* $\mathscr{F}$ *be the set of* $\Phi(\Theta)$*-continuous maps of* X *into* Y*. Then* $\mathscr{F}$ *is a* Φ_2*-closed set of* Y^X*. If* Y *is* Φ_2*-compact, then so is* $\mathscr{F}$*.*

Denote by φ the inclusion map $\mathscr{F} \to Y^X$, let $f \in Y^X$, and let $\mathfrak{F} \in \Phi_2(\mathscr{F})$ such that $\varphi(\mathfrak{F})$ converges to f. We want to show that $f \in \mathscr{F}$. Assume the contrary. Then there exist $\mathfrak{G} \in \Phi(\Theta)$ and a limit point x of $\mathfrak{G}$ such that $f(\mathfrak{G})$ does not converge to $f(x)$. Let U be a closed neighbourhood of $f(x)$ such that $f(A) \setminus U \neq \emptyset$ for any $A \in \mathfrak{G}$ and let V be an open neighbourhood of $f(x)$ whose closure lies in the interior of U.

Let (I, g) be a Θ_2-net in $\mathscr{F}$ such that $g(\mathfrak{F}') = \mathfrak{F}$, where $\mathfrak{F}'$ denotes the section filter of I, and let (J, h) be a Θ-net in X such that $h(\mathfrak{G}') = \mathfrak{G}$, where $\mathfrak{G}'$ denotes the section filter of J. We construct inductively an increasing sequence $(\iota_n)_{n \in \mathbb{N}}$ in I and an increasing sequence $(\lambda_n)_{n \in \mathbb{N}}$ in J such that we have for any $n \in \mathbb{N}$:

a) $f(h(\lambda_n)) \in Y \setminus U$;

b) $m \in \mathbb{N},\ m < n \Rightarrow g_{\iota_n}(h(\lambda_m)) \in Y \setminus U$;

c) $\lambda \in J,\ \lambda \geq \lambda_n \Rightarrow g_{\iota_n}(h(\lambda)) \in V$.

Let $n \in \mathbb{N}$ and assume that the sequences were constructed up to $n-1$. Since $f(h(\lambda_m))$ belongs to the open set $Y \setminus U$ for any $m \leq n-1$ and since $\varphi(\mathfrak{F})$ converges to f there exists $\iota_n \in I$ such that $\iota_n \geq \iota_{n-1}$, b) is fulfilled, and $g_{\iota_n}(x) \in V$. Since V is open and $g_{\iota_n}(\mathfrak{G})$ converges to $g_{\iota_n}(x)$ there exists $\lambda_n \in J$ so that $\lambda_n \geq \lambda_{n-1}$ and c) is fulfilled. By the hypothesis of the proof we may choose λ_n in such a way that a) is also fulfilled. This finishes the inductive construction.

Let x_0 be an adherent point of the sequence $(h(\lambda_n))_{n \in \mathbb{N}}$ and f_0 be an adherent function in $\mathscr{F}$ of the sequence $(g_{\iota_n})_{n\in\mathbb{N}}$. By b) we get $f_0(h(\lambda_n)) \in Y\backslash U$ and by c) we get $g_{\iota_n}(x_0) \in \bar{V}$ for any $n \in \mathbb{N}$. From the first relation we get $f_0(x_0) \in \overline{Y\backslash U}$ and from the second one $f_0(x_0) \in \bar{V}$, which is a contradiction since $\bar{V} \cap \overline{Y\backslash U} = \emptyset$. □

Remark. Assume X countably compact. If we take $\Theta = \Theta_2$, then by Proposition 1.5.5 c ⇒ b and by Proposition 1.3.7 $\mathscr{F}$ is exactly the set $\mathscr{C}$ of continuous maps of X into Y. Hence, the closure in Y^X of any Φ_2-set of $\mathscr{C}$ is contained in $\mathscr{C}$. This result was proved by A. Grothendieck (1952) ([1] Théorème 2, 2°)).

Proposition 2.1.17. *Let X be a set, $i \in \{1, 2, 3, 4, 5\}$, Y_1, Y_2 be uniform spaces (topological spaces and $i \in \{1, 2\}$) defined on the same set Y such that the identity map $Y_1 \to Y_2$ is uniformly Φ_i-continuous (Φ_i-continuous), and let $\mathscr{F}$ be a subset of Y^X. If $\mathscr{F}_1$ and $\mathscr{F}_2$ denote the set $\mathscr{F}$ endowed with the uniformity (topology) induced by Y_1^X and Y_2^X respectively, then the identity map $\mathscr{F}_1 \to \mathscr{F}_2$ is uniformly Φ_i-continuous (Φ_i-continuous).*

By Corollary 1.8.16 the identity map $Y_1^X \to Y_2^X$ is uniformly Φ_i-continuous (Φ_i-continuous) and the assertion follows from Corollary 1.8.15. □

The next two results will not be used later.

Proposition 2.1.18. *Let X be a set, Y be a uniform space, $i \in \{1, 2, 3, 4, 5\}$, $\mathscr{F}$ be a subset of Y^X and A, B be subsets of X such that for any Cauchy Φ_i-filter $\mathfrak{F}$ on $\mathscr{F}_A$, for any $y \in B$, and for any entourage U of Y there exist $x \in A$ and $\mathscr{G} \in \mathfrak{F}$ such that $(f(x), f(y)) \in U$ for any $f \in \mathscr{G}$. Then the identity map $\mathscr{F}_A \to \mathscr{F}_B$ is uniformly Φ_i-continuous.*

Let $\mathfrak{F}$ be a Cauchy Φ_i-filter on $\mathscr{F}_A$. Let $y \in B$, U be an entourage of Y and V be an entourage Y such that $V \circ V \circ \overset{-1}{V} \subset U$. By the hypothesis there exist $x \in A$ and $\mathscr{G} \in \mathfrak{F}$ such that $(f(x), f(y)) \in V$ for any $f \in \mathscr{G}$. There exists further $\mathscr{H} \in \mathfrak{F}$ such that $(f(x), g(x)) \in V$ for any $f, g \in \mathscr{H}$. We get

$$(f(y), g(y)) \in V \circ V \circ \overset{-1}{V} \subset U$$

for any $f, g \in \mathscr{G} \cap \mathscr{H}$. Hence, $\mathfrak{F}$ is a Cauchy filter on $\mathscr{F}_B$ and the identity map $\mathscr{F}_A \to \mathscr{F}_B$ is uniformly Φ_i-continuous (Theorem 1.8.4 b ⇒ a). □

Proposition 2.1.19. *Let X be a topological space, Y be a uniform space, $\mathscr{C}$ be the set of continuous maps of X into Y, $\mathscr{G}$ be a subset of Y^X, $\mathscr{F} := \mathscr{C} \cap \mathscr{G}$, $i \in \{1, 2, 3, 4, 5\}$, $\Phi := \Phi_i(\mathscr{F})$ and φ be the map*

$$\mathscr{F} \times X \to Y, \quad (f, x) \mapsto f(x).$$

If the map

$$X \to Y_\Phi^{\mathscr{F}}, \quad x \mapsto \varphi(\cdot, x)$$

is continuous, then $\mathscr{F}$ is a Φ_i-closed set of $\mathscr{G}$.

Let ψ be the inclusion map $\mathscr{F} \to \mathscr{G}$ and $\mathfrak{F} \in \Phi_i(\mathscr{F})$ such that $\psi(\mathfrak{F})$ converges to an $f \in \mathscr{G}$. We have to show that $f \in \mathscr{C}$.

Let $x \in X$ and V be a closed entourage of Y. By the hypothesis there exists a neighbourhood U of x such that

$$\{g \in \mathscr{F} \mid (g(x), g(y)) \in V\} \in \mathfrak{F}$$

for any $y \in U$. We get $(f(x), f(y)) \in V$ for any $y \in U$. Hence, f is continuous. □

Proposition 2.1.20. *Let X, Y be uniform spaces, $\mathscr{F}$ be a subset of Y^X, $i \in \{1, 2, 3, 4, 5\}$, $\Psi := \hat{\Phi}_i(\mathscr{F})$ and φ, ψ be the maps*

$$\mathscr{F} \times X \to Y, \quad (f, x) \mapsto f(x),$$
$$X \to Y_{\Psi}^{\mathscr{F}}, \quad x \mapsto \varphi(\cdot, x)$$

respectively. Let us consider the following assertions:

a) any $\hat{\Phi}_i$-filter on $\mathscr{F}$ is uniformly equicontinuous;

b) ψ is uniformly continuous;

c) ψ is continuous;

d) any Φ_i-set of $\mathscr{F}$ is uniformly equicontinuous;

e) any Φ_i-set of $\mathscr{F}$ is equicontinuous;

f) for any Φ_i-set $\mathscr{G}$ of $\mathscr{F}$ the restriction of φ to $\mathscr{G} \times X$ is continuous.

Then

$$\begin{array}{ccccc} \Rightarrow & b & \Rightarrow & d & \\ & \Downarrow & & \Downarrow & \\ & c & \Rightarrow & e & \Rightarrow f. \end{array}$$

a ⇒ b. Let V be an entourage of Y and $\mathfrak{F} \in \Psi$. Then there exist $\mathscr{G} \in \mathfrak{F}$ and an entourage U of X such that $(f(x), f(y)) \in V$ for any $f \in \mathscr{G}$ and any $(x, y) \in U$. Hence,

$$(\psi(x), \psi(y)) \in \mathscr{W}(\mathfrak{F}, V)$$

for any $(x, y) \in U$ so ψ is uniformly continuous.

b ⇒ c and d ⇒ e are trivial.

b ⇒ d. Let $\mathscr{G}$ be a Φ_i-set of $\mathscr{F}$ and V be an entourage of Y. There exists an entourage U of X such that

$$(\psi(x), \psi(y)) \in \mathscr{W}(\mathscr{G}, V)$$

for any $(x, y) \in U$. We get $(f(x), f(y)) \in V$ for any $f \in \mathscr{G}$ and any $(x, y) \in U$. Hence, $\mathscr{G}$ is uniformly equicontinuous.

c ⇒ e. Let $\mathscr{G}$ be a Φ_i-set of $\mathscr{F}$, $x \in X$ and V be an entourage of Y. There exists a neighbourhood U of x such that

$$(\psi(x), \psi(y)) \in \mathscr{W}(\mathscr{G}, V)$$

for any $y \in U$. We get $(f(x), f(y)) \in V$ for any $f \in \mathscr{G}$ and any $y \in U$. Hence, $\mathscr{G}$ is equicontinuous.

e ⇒ f. Let $(f, x) \in \mathscr{G} \times X$ and V be an entourage of Y. Let W be another entourage of Y such that $W \circ W \subset V$. There exists a neighbourhood U of x such that $(g(x), g(y)) \in W$ for any $g \in \mathscr{G}$ and any $y \in U$. We get

$$(f(x), g(y)) \in W \circ W \subset V$$

for any $(g, y) \in \mathscr{G} \times X$ with $(f(x), g(x)) \in W$ and $y \in U$. □

Proposition 2.1.21. *Let X, Y be uniform spaces and $\mathfrak{F}$ be a uniformly equicontinuous Cauchy filter on Y^X. Then for any Cauchy filter $\mathfrak{G}$ on X the image of $\mathfrak{F} \times \mathfrak{G}$ with respect to the map*

$$Y^X \times X \longrightarrow Y, \quad (f, x) \longmapsto f(x)$$

is a Cauchy filter.

Let V be an entourage of Y. There exist $\mathscr{F} \in \mathfrak{F}$ and an entourage U of X such that $(f(x), f(y)) \in V$ for any $f \in \mathscr{F}$ and any $(x, y) \in U$. Since $\mathfrak{G}$ is a Cauchy filter on X there exists $A \in \mathfrak{G}$ with $A \times A \subset U$. We get $(f(x), f(y)) \in V$ for any $f \in \mathscr{F}$ and any $x, y \in A$.

Since $\mathfrak{F}$ is a Cauchy filter on Y^X the assertion follows from Lemma 2.2.2. □

Proposition 2.1.22. *Let X, Y be uniform spaces, A be a subset of X, $i \in \{1, 2, 3, 4, 5\}$, $\mathscr{F}$ be a set of uniformly continuous maps of X into Y such that the identity map $\mathscr{F}_A \to \mathscr{F}_{\{X\}}$ is uniformly Φ_i-continuous, $\Phi := \hat{\Phi}_i(\mathscr{F}_A)$ and φ, ψ be the maps*

$$\mathscr{F}_A \times X \longrightarrow Y, \quad (f, x) \longmapsto f(x),$$

$$X \longrightarrow Y_{\Phi}^{\mathscr{F}}, \quad x \longmapsto \varphi(\cdot, x)$$

respectively. Then:

a) any $\hat{\Phi}_i$-filter in $\mathscr{F}_A$ is uniformly equicontinuous;

b) $\varphi(\mathfrak{F} \times \mathfrak{G})$ is a Cauchy filter for any Cauchy $\hat{\Phi}_i$-filter $\mathfrak{F}$ on $\mathscr{F}_A$ and any Cauchy filter $\mathfrak{G}$ on X;

c) φ is uniformly Φ_i-continuous;

d) ψ is uniformly continuous;

e) the restriction of φ to $\bar{\mathscr{G}}_A \times X$ is uniformly continuous for any Φ_i-set $\mathscr{G}$ of $\mathscr{F}_A$, where $\bar{\mathscr{G}}$ denotes the closure of $\mathscr{G}$ in $\mathscr{F}_A$.

a) Let $\mathfrak{F}$ be a Φ_i-ultrafilter on $\mathscr{F}_A$. By Proposition 1.5.10 $\mathfrak{F}$ is a Cauchy filter on $\mathscr{F}_A$ and therefore a Cauchy filter on $\mathscr{F}_{\{X\}}$. By Proposition 2.1.10 $\mathfrak{F}$ is uniformly equicontinuous. By Proposition 2.1.12 (and Proposition 1.4.2) any $\hat{\Phi}_i$-filter on $\mathscr{F}_A$ is uniformly equicontinuous.

b) follows immediately from a) and Proposition 2.1.21.

c) follows immediately from b).

d) follows from a) and from Proposition 2.1.20 a ⇒ b.

e) Let $\mathscr{G}$ be a Φ_i-set of $\mathscr{F}_A$. Then $\mathscr{G}$ is a Φ_i-set (Theorem 1.8.4 a ⇒ h) and therefore a precompact set of $\mathscr{F}_{\{X\}}$ (Proposition 1.5.6 a ⇒ c). By Proposition 1.8.8 $\bar{\mathscr{G}}$

is the closure of $\mathscr{G}$ in $\mathscr{F}_{\{X\}}$ and $\overline{\mathscr{G}}_A = \overline{\mathscr{G}}_{\{X\}}$. Hence, $\mathscr{G}$ is a precompact set of $\mathscr{F}_{\{X\}}$ and the restriction of φ to $\overline{\mathscr{G}}_A \times X$ is uniformly continuous. □

Proposition 2.1.23. *Let $(Y_\iota)_{\iota\in I}$ be a family of sets, $(x_\iota)_{\iota\in I} \in \prod_{\iota\in I} Y_\iota$, $\mathfrak{P}_f(I)$ be the set of finite subsets of I ordered by the inclusion relation and $\mathfrak{F}$ be the section filter of $\mathfrak{P}_f(I)$. For any $(J, y) \in \mathfrak{P}(I) \times \prod_{\iota\in I} Y_\iota$ we denote by y^J the element of $\prod_{\iota\in I} Y_\iota$ defined by*

$$(y^J)_\iota := \begin{cases} y_\iota & \text{for } \iota \in J \\ x_\iota & \text{for } \iota \in I \backslash J\,. \end{cases}$$

Further, let X be a set, Y be a topological space defined on a subset of $\prod_{\iota\in I} Y_\iota$, $\mathscr{F}$ be a strongly Φ_1-closed set of Y^X and $f \in Y^X$ such that:

a) $y^J \in Y$ for any $y \in Y$ and for any countable $J \subset I$;

b) y is a limit point of the image of $\mathfrak{F}$ under the map

$$\mathfrak{P}_f(I) \to Y, \quad J \mapsto y^J$$

for any $y \in Y$;

c) the map

$$X \to Y, \quad x \mapsto f(x)^J$$

belongs to $\mathscr{F}$ for any $J \in \mathfrak{P}_f(I)$.
Then $f \in \mathscr{F}$.

For any $J \in \mathfrak{P}_f(I)$ denote by g_J the map

$$X \to Y,\ x \mapsto f(x)^J,$$

by g the map

$$\mathfrak{P}_f(I) \to \mathscr{F}, \quad J \mapsto g_J,$$

and by φ the inclusion map $\mathscr{F} \to Y^X$. Then $g(\mathfrak{F}) \in \Phi(\Theta_1(\mathscr{F}, Y^X))$ and $\varphi(g(\mathfrak{F}))$ converges to f. Since $\mathscr{F}$ is strongly Φ_1-closed we get $f \in \mathscr{F}$. □

Definition 2.1.24. *For any additive group G (i.e. the group law of G is commutative and is denoted by $+$) we denote by $\underline{0}$ its zero element and for any subsets A, B of G and any $x \in G$ we set*

$$\underline{A+B} := \{y+z \mid (y,z) \in A \times B\}, \quad \underline{0A} := \{0\},$$

$$\underline{-A} := \{-y \mid y \in A\},$$

$$\underline{A+x} := \underline{x+A} := \{x\} + A\,.$$

We define nA inductively for any $n \in \mathbb{N}$ by setting

$$\underline{(n+1)A} := A + nA$$

and put

$$\underline{-nA} := n(-A).$$

If R is a ring and G is a left R-module, then we set

$$\underline{AB} := \{\alpha y \mid (\alpha, y) \in A \times B\},$$

$$\underline{\alpha B} := \{\alpha\} B, \quad \underline{Ax} := A\{x\}$$

for any $A \subset R$, $B \subset G$, $\alpha \in R$, $x \in G$. *We identify the additive groups and the* $\mathbb{Z}$*-modules as well as the topological additive groups and the topological* $\mathbb{Z}$*-modules.*

If G is a left R-module, if $A \subset G$, and if $n \in R \cap \mathbb{Z}$, then the set nA has two different meanings depending on whether we consider n an element of R or of $\mathbb{Z}$. We hope that no confusion will arise from this awkward situation. If $R = \mathbb{R}$ and if A is convex then the two meanings coincide.

Definition 2.1.25. *Let R be a topological ring and G be a topological left R-module. A filter* $\mathfrak{F}$ *on G is called* R-bounded *or simply* bounded *if for any 0-neighbourhood U in G there exist* $A \in \mathfrak{F}$ *and a 0-neighbourhood V in R such that* $VA \subset U$. *A subset A of G is called* R-bounded *or simply* bounded *if the filter* $\{B \mid A \subset B \subset G\}$ *on G is R-bounded.*

Proposition 2.1.26. *Let R be a topological ring and G be a topological left R-module. Then any filter of* $\hat{\Phi}_C(G)$ *is R-bounded.*

Let $\mathfrak{F} \in \hat{\Phi}_C(G)$ and U be an arbitrary 0-neighbourhood in G. Let U' be a 0-neighbourhood in G such that $U' + U' \subset U$, U'' be a 0-neighbourhood in G and V be a 0-neighbourhood in R such that $VU'' \subset U'$. By Proposition 1.2.4 there exists a finite set $\mathfrak{A}$ of nonempty subsets of G such that

$$\bigcup_{A \in \mathfrak{A}} (A - A) \subset U'', \quad B := \bigcup_{A \in \mathfrak{A}} A \in \mathfrak{F}.$$

For any $A \in \mathfrak{A}$ let $x_A \in A$. Let W be a 0-neighbourhood in R such that $Wx_A \subset U'$ for any $A \in \mathfrak{A}$. Let $\alpha \in V \bigcap W$ and let $x \in B$. Then there exists $A \in \mathfrak{A}$ such that $x \in A$ and we get

$$\alpha x \in \alpha x_A + \alpha U'' \subset U' + U' \subset U$$

and therefore $(V \cap W) B \subset U$. □

Proposition 2.1.27. *Let R be a topological ring, G be a topological left R-module, X be a set,* Φ *be a set of filters on X and* $\mathscr{F}$ *be the set of maps* $f: X \to G$ *such that* $f(\mathfrak{F})$ *is an R-bounded filter on G (this happens if* $f(\mathfrak{F}) \in \hat{\Phi}_C(G)$*) for any* $\mathfrak{F} \in \Phi$. *Then* $\mathscr{F}$ *is an R-submodule of* G^X *and* $\mathscr{F}_\Phi$ *is a topological left R-module.*

It is easy to see that $\mathscr{F}$ is an R-submodule of G^X. By Proposition 2.1.26 any $\hat{\Phi}_C(G)$-filter is R-bounded.

Let $\mathfrak{F} \in \Phi$. For any 0-neighbourhood U in G we denote by U' the set

$\{f \in \mathscr{F} \mid \overset{-1}{f}(U) \in \mathfrak{F}\}$. If $\mathfrak{U}$ is a fundamental system of 0-neighbourhoods in G, then $\{U' \mid U \in \mathfrak{U}\}$ is a fundamental system of 0-neighbourhoods in $\mathscr{F}_{\{\mathfrak{F}\}}$. We have

$$U' + V' \subset (U + V)', \quad (-U)' = -U'$$

for any 0-neighbourhood U, V in G. Since the topology of $\mathscr{F}_\Phi$ is obviously translation invariant we deduce from the above considerations that it is a group topology on $\mathscr{F}$.

We now want to show that the topology of $\mathscr{F}_\Phi$ is a topology of R-module. Let $f \in \mathscr{F}$, $\mathfrak{F} \in \Phi$ and U be a 0-neighbourhood in G. Since $f(\mathfrak{F})$ is R-bounded there exist $A \in \mathfrak{F}$ and a 0-neighbourhood V in R such that $Vf(A) \subset U$. This shows that the map

$$R \to \mathscr{F}_\Phi, \quad \alpha \mapsto \alpha f$$

is continuous at $0 \in R$.

Let now $\alpha \in R$, $\mathfrak{F} \in \Phi$ and U be a 0-neighbourhood in G. There exists a 0-neighbourhood V in G such that $\alpha V \subset U$. Let $f \in \mathscr{F}$ such that $\overset{-1}{f}(V) \in \mathfrak{F}$. Then $\overset{-1}{\widehat{\alpha f}}(U) \in \mathfrak{F}$. Hence, the map

$$\mathscr{F}_\Phi \to \mathscr{F}_\Phi, \quad f \mapsto \alpha f$$

is continuous at $0 \in \mathscr{F}$.

Let $\mathfrak{F} \in \Phi$ and U be a 0-neighbourhood in G. There exist a 0-neighbourhood V in G and a 0-neighbourhood W in R such that $WV \subset U$. Let $f \in \mathscr{F}$ such that $\overset{-1}{f}(V) \in \mathfrak{F}$ and let $\alpha \in W$. We get $\overset{-1}{\widehat{\alpha f}}(U) \in \mathfrak{F}$. Hence, the map

$$R \times \mathscr{F}_\Phi \to \mathscr{F}_\Phi, \quad (\alpha, f) \mapsto \alpha f$$

is continuous at $(0, 0) \in R \times \mathscr{F}$. We have proved that $\mathscr{F}_\Phi$ is a topological left R-module. $\square$

Proposition 2.1.28. *Let E be a locally convex space, $\mathscr{P}$ be a set of semi-norms on E generating its topology, X be a set, Φ be a set of filters on X and $\mathscr{F}$ be the set of maps f of X into E such that $f(\mathfrak{F})$ is a bounded filter of E (this happens if $f(\mathfrak{F}) \in \hat{\Phi}_C(E)$) for any $\mathfrak{F} \in \Phi$. For any $(p, \mathfrak{F}) \in \mathscr{P} \times \Phi$ we denote by $p_\mathfrak{F}$ the map*

$$\mathscr{F} \to \mathbb{R}, \quad f \mapsto \inf_{A \in \mathfrak{F}} \sup_{x \in A} p(f(x)).$$

Then:

a) $\mathscr{F}$ is a vector subspace of E^X;

b) $p_\mathfrak{F}$ is a semi-norm on $\mathscr{F}$ for any $(p, \mathfrak{F}) \in \mathscr{P} \times \Phi$;

c) $\mathscr{F}_\Phi$ is a locally convex space and its topology is generated by the set $\{p_\mathfrak{F} \mid (p, \mathfrak{F}) \in \mathscr{P} \times \Phi\}$ of semi-norms.

a) follows from Proposition 2.1.27.

b) Let $f, g \in \mathscr{F}$ and α be a scalar. It is obvious that $p_\mathfrak{F}(\alpha f) = |\alpha| p_\mathfrak{F}(f)$. We have

$$p_{\mathfrak{F}}(f+g) \leq \sup_{x \in A} p((f+g)(x)) \leq \sup_{x \in A} (p(f(x)) + p(g(x))) \leq$$
$$\leq \sup_{x \in A} p(f(x)) + \sup_{x \in A} p(g(x))$$

for any $A \in \mathfrak{F}$. Since $\mathfrak{F}$ is lower directed with respect to the inclusion relation we get

$$p_{\mathfrak{F}}(f+g) \leq p_{\mathfrak{F}}(f) + p_{\mathfrak{F}}(g).$$

c) Let $(p, \mathfrak{F}) \in \mathscr{P} \times \Phi$ and $\mathscr{U}$ be the set

$$\{f \in \mathscr{F} \mid \{x \in X \mid p(f(x)) < 1\} \in \mathfrak{F}\}.$$

We have

$$p_{\mathfrak{F}}(f) < 1 \Rightarrow f \in \mathscr{U} \Rightarrow p_{\mathfrak{F}}(f) \leq 1$$

for any $f \in \mathscr{F}$. Hence, the set $\{p_{\mathfrak{F}} \mid (p, \mathfrak{F}) \in \mathscr{P} \times \Phi\}$ of semi-norms on $\mathscr{F}$ generates the topology of $\mathscr{F}_\Phi$, which is therefore locally convex. □

Corollary 2.1.29. *Let R be a topological ring, G be a topological left R-module, X be a uniform (topological) space,* $i \in \{1, 2, 3, 4, 5\}$ $(i \in \{1, 2\})$, Φ *be a subset of* $\hat{\Phi}_i(X)$ *and* $\mathscr{F}$ *be the set of uniformly* Φ_i*-continuous (the set of* Φ_i*-continuous) maps of X into G. Then* $\mathscr{F}$ *is an R-submodule of* G^X *and* $\mathscr{F}_\Phi$ *is a topological left R-module. If G is locally convex, then* $\mathscr{F}_\Phi$ *is locally convex.*

Let $f, g \in \mathscr{F}$ and $\alpha \in R$. Since the maps

$$G \times G \to G, \quad (x, y) \mapsto x + y,$$
$$G \to G, \quad x \mapsto \alpha x$$

are uniformly continuous we get $f+g$, $\alpha f \in \mathscr{F}$. Hence, $\mathscr{F}$ is an R-submodule of G^X.

Let $(f, \mathfrak{F}) \in \mathscr{F} \times \Phi$. By Theorem 1.8.4 a $\Rightarrow$ i (by Proposition 1.8.9d)), $f(\mathfrak{F}) \in \hat{\Phi}_i(G)$ so (Corollary 1.5.12) $f(\mathfrak{F}) \in \hat{\Phi}_C(G)$. The assertions now follow from Propositions 2.1.27 and 2.1.28. □

Proposition 2.1.30. *Let E be a locally convex space and E′ be its dual. Then:*

a) a filter $\mathfrak{F}$ *on E′ is uniformly equicontinuous iff there exists a 0-neighbourhood U in E such that* $U^\circ \in \mathfrak{F}$*;*

b) if E is a Fréchet space, then any $\hat{\Phi}_5$*-filter on* E'_E *is uniformly equicontinuous,* E'_E *is* Φ_5*-compact, and* $E = E_\Phi$*, where* $\Phi := \hat{\Phi}_5(E'_E)$.

a) First, assume that $\mathfrak{F}$ is uniformly equicontinuous. Then there exist a 0-neighbourhood U in E and an A' in $\mathfrak{F}$ such that $|x'(x)| < 1$ for any $x \in U$ and any $x' \in A'$. Hence, $A' \subset U^\circ$ and $U^\circ \in \mathfrak{F}$.

The converse implication is trivial.

b) Let $\mathfrak{F}$ be a Φ_5-filter on E'_E and (I, f) be a Θ_5-net in E'_E such that $f(\mathfrak{G}) = \mathfrak{F}$, where $\mathfrak{G}$ denotes the section filter of I. Furthermore, let $(U_n)_{n \in \mathbb{N}}$ be a sequence of subsets of E forming a fundamental system of 0-neighbourhoods in E. Assume

$U_n^\circ \notin \mathfrak{F}$ for any $n \in \mathbb{N}$. Then we may construct an increasing sequence $(\iota_n)_{n\in\mathbb{N}}$ in I such that $f(\iota_n) \notin U_n^\circ$ for any $n \in \mathbb{N}$. $\{f(\iota_n) \mid n \in \mathbb{N}\}$ being a bounded set of E'_E and E being barrelled, $\{f(\iota_n) \mid n \in \mathbb{N}\}$ is equicontinuous and this is a contradiction. Hence, there exists $n \in \mathbb{N}$ such that $U_n^\circ \in \mathfrak{F}$ and by a) $\mathfrak{F}$ is uniformly equicontinuous. By Proposition 2.1.12 any $\hat{\Phi}_5$-filter on E'_E is uniformly equicontiuous.

The other assertions follow immediately from the first one. □

The next two results will not be used in the sequel.

Theorem 2.1.31. *Let X be a topological space, for any $x \in X$ let $\mathfrak{B}_x$ be the neighbourhood filter of x, let Φ be the set $\{\mathfrak{B}_x \mid x \in X\}$, and Y be a uniform space. Then the set of continuous maps of X into Y is a closed set of Y_Φ^X.*

Let $\mathscr{C}$ be the set of continuous maps of X into Y and f be an adherent point of $\mathscr{C}$ in Y_Φ^X. Further, let $x \in X$, U be an arbitrary entourage of Y and V be an entourage of Y such that $\overset{-1}{V} \circ V \circ V \subset U$. There exists $g \in \mathscr{C}$ such that

$$A := \{y \in X \mid (f(y), g(y)) \in V\} \in \mathfrak{B}_x .$$

Furthermore, there exists $B \in \mathfrak{B}_x$ such that

$$\{(g(y), g(x)) \mid y \in B\} \subset V .$$

We get

$$\{(f(y), f(x)) \mid y \in A \cap B\} \subset \overset{-1}{V} \circ V \circ V \subset U .$$

Hence, f is continuous at x. Since x is arbitrary $f \in \mathscr{C}$ and $\mathscr{C}$ is a closed set of Y_Φ^X. □

Theorem 2.1.32. *Let X be a topological space, Θ be a set of Θ_2-sequences in X, Φ be $\hat{\Phi}(\Theta)$ and $\overline{\mathscr{F}}$ ($\underline{\mathscr{F}}$) be the set of real functions f on X such that for any Θ-sequence $(x_n)_{n\in\mathbb{N}}$ in X for any adherent point x of this sequence, and for any ultrafilter $\mathfrak{F}$ on X converging to x and finer than the elementary filter of the sequence $(x_n)_{n\in\mathbb{N}}$*

$$f(x) \geq \lim f(\mathfrak{F}) \; (f(x) \leq \lim f(\mathfrak{F})) .$$

In addition, let $(f_n)_{n\in\mathbb{N}}$ be an increasing sequence in $\underline{\mathscr{F}}$ such that the map

$$f: X \to \mathbb{R}, \quad x \mapsto \lim_{n\to\infty} f_n(x)$$

exists and belongs to $\overline{\mathscr{F}}$. Then $(f_n)_{n\in\mathbb{N}}$ converges to f in $\mathbb{R}_\Phi^X$.

Let $\mathfrak{G} \in \Phi(\Theta)$ and (I, g) be a Θ-net in X such that $g(\mathfrak{H}) = \mathfrak{G}$ where $\mathfrak{H}$ denotes the section filter of I. Let ε be a strictly positive real number. We want to show there exists $(n, A) \in \mathbb{N} \times \mathfrak{G}$ such that $f(x) - f_n(x) < \varepsilon$ for any $x \in A$. Assume the contrary. Then we may construct an increasing sequence $(\iota_n)_{n\in\mathbb{N}}$ in I inductively such that

$$f(g(\iota_n)) - f_n(g(\iota_n)) \geq \varepsilon$$

for any $n \in \mathbb{N}$. Let x be an adherent point of $(g(\iota_n))_{n\in\mathbb{N}}$ and $\mathfrak{F}$ be an ultrafilter on X

converging to x, and which is finer than the elementary filter of $(g(\iota_n))_{n\in\mathbb{N}}$. Further, let $m\in\mathbb{N}$ such that $f(x)-f_m(x)<\varepsilon$. We have

$$\lim f(\mathfrak{F})-\lim f_m(\mathfrak{F})\leq f(x)-f_m(x)<\varepsilon.$$

Hence, there exists $A\in\mathfrak{F}$ such that $f(y)-f_m(y)<\varepsilon$ for any $y\in A$. We deduce that there exists $n\in\mathbb{N}$ such that $n\geq m$ and

$$f(g(\iota_n))-f_m(g(\iota_n))<\varepsilon.$$

Hence, it follows

$$\varepsilon\leq f(g(\iota_n))-f_n(g(\iota_n))\leq f(g(\iota_n))-f_m(g(\iota_n))<\varepsilon,$$

which is a contradiction.

Now let $\mathfrak{G}_0\in\Phi$ and ε be a strictly positive real number. By the above considerations there exists $(n(\mathfrak{G}),A(\mathfrak{G}))\in\mathbb{N}\times\mathfrak{G}$ for any $\mathfrak{G}\in\Phi(\Theta)$ such that $f(x)-f_{n(\mathfrak{G})}(x)<\varepsilon$ for any $x\in A(\mathfrak{G})$. Let Ψ be a finite subset of $\Phi(\Theta)$ such that $\bigcup_{\mathfrak{G}\in\Psi}A(\mathfrak{G})\in\mathfrak{G}_0$. Then $f(x)-f_n(x)<\varepsilon$ for any $x\in\bigcup_{\mathfrak{G}\in\Psi}A(\mathfrak{G})$ and any $n\geq\sup_{\mathfrak{G}\in\Psi}n(\mathfrak{G})$. Hence, $(f_n)_{n\in\mathbb{N}}$ converges to f in $\mathbb{R}^X_\Phi$. □

Remark. Assume that f and the functions f_n $(n\in\mathbb{N})$ are continuous. Then $f\in\mathscr{F}$ and $\{f_n\mid n\in\mathbb{N}\}\subset\mathscr{F}$ so by the above theorem $(f_n)_{n\in\mathbb{N}}$ converges to f uniformly on any quasicompact set of X. This result is called Dini's theorem. It was proved by U. Dini (1878) for a compact interval X of $\mathbb{R}$ ([1] 99.).

In fact Dini's theorem holds not only for increasing sequences but even for upper directed families. In case of Theorem 2.1.32 the same is not true even if the functions are continuous as the following example shows. Let $\beta\mathbb{N}$ be the Stone-Čech compactification of $\mathbb{N}$ and let $x\in\beta\mathbb{N}\setminus\mathbb{N}$. We denote by X the subspace $\beta\mathbb{N}\setminus\{x\}$ of $\beta\mathbb{N}$, by $\mathfrak{M}$ the set of subsets M of $\mathbb{N}$ such that x does not belong to the closure of M in $\beta\mathbb{N}$, and for any $M\in\mathfrak{M}$ we denote by f_M the continuous real function on X extending $1^{\mathbb{N}}_M$. Then X is countably compact, $(f_M)_{M\in\mathfrak{M}}$ is an upper directed family of continuous real functions on X, whose supremum in $\mathbb{R}^X$ is the function 1_X, but $(f_M)_{M\in\mathfrak{M}}$ does not converge uniformly to 1_X.

§ 2.2 Uniformities on spaces of functions defined by sets of sequences

Throughout this section we denote by X a set, by Θ a set of sequences in X, by $\mathfrak{S}$ a set of subsets of X, such that for any Θ-sequence $(x_n)_{n\in\mathbb{N}}$ there exists $S\in\mathfrak{S}$ for which the set $\{n\in\mathbb{N}\mid x_n\in S\}$ is infinite, by A a subset of X containing $\bigcup_{S\in\mathfrak{S}}S$, by Φ the set $\hat{\Phi}(\Theta)$, and by Y a uniform space.

As $\mathfrak{S}$ one may take the set of Φ-sets (Proposition 1.4.7) or the set

$$\{\{x_n \mid n \in \mathbb{N}\} \mid (x_n)_{n\in\mathbb{N}} \text{ is a } \theta\text{-sequence}\}.$$

Proposition 2.2.1. *Let Θ' be the set of sequences $(f_n, x_n)_{n\in\mathbb{N}}$ in $Y^X \times X$ such that for any entourage U of Y there exist $m, n \in \mathbb{N}$ with $m < n$ and $(f_m(x_m), f_n(x_m)) \in U$. Then for any $(\mathfrak{F}, \mathfrak{G}) \in \Phi_5(Y^X_{\mathfrak{S}}) \times \Phi(\Theta)$ we have $\mathfrak{F} \times \mathfrak{G} \in \Phi(\Theta')$.*

Let $(f_n)_{n\in\mathbb{N}}$ be a $\Theta_5(Y^X_{\mathfrak{S}})$-sequence and $(x_n)_{n\in\mathbb{N}}$ be a Θ-sequence. Then there exists $S \in \mathfrak{S}$ such that $\{p \in \mathbb{N} \mid x_p \in S\}$ is infinite. For any entourage U of Y there exist $m, n \in \{p \in \mathbb{N} \mid x_p \in S\}$ with $m < n$ and $(f_m, f_n) \in \mathscr{W}(S, U)$. We get $(f_m(x_m), f_n(x_m)) \in U$. Hence, $(f_n, x_n)_{n\in\mathbb{N}} \in \Theta'$.

Let (I, f) be a $\Theta_5(Y^X_{\mathfrak{S}})$-net in $Y^X_{\mathfrak{S}}$ such that $f(\mathfrak{F}') = \mathfrak{F}$, where $\mathfrak{F}'$ denotes the section filter of I, and let (J, g) be a Θ-net in X such that $g(\mathfrak{G}') = \mathfrak{G}$, where $\mathfrak{G}'$ denotes the section filter of J. We endow $I \times J$ with the product preorder relation and denote by h the map

$$I \times J \to Y^X \times X, \quad (\iota, \kappa) \mapsto (f_\iota, g(\kappa))$$

By the above considerations $(I \times J, h)$ is a Θ'-net. Since $\mathfrak{F}' \times \mathfrak{G}'$ is the section filter of $I \times J$ and $h(\mathfrak{F}' \times \mathfrak{G}') = \mathfrak{F} \times \mathfrak{G}$ we get $\mathfrak{F} \times \mathfrak{G} \in \Phi(\Theta')$. □

Lemma 2.2.2. *Let X, Y be sets, $\mathfrak{F}$, $\mathfrak{G}$ be filters on X and Y respectively, Z be a uniform space, and φ be a map of $X \times Y$ into Z such that:*

a) for any entourage U of Z there exist $A \in \mathfrak{F}$ and $B \in \mathfrak{G}$ such that $(\varphi(x, y), \varphi(x', y)) \in U$ for any $x, x' \in A$ and $y \in B$:

b) $\{x \in X \mid \varphi(x, \mathfrak{G})$ is a Cauchy filter on $Z\} \in \mathfrak{F}$.

Then $\varphi(\mathfrak{F} \times \mathfrak{G})$ is a Cauchy filter.

Let U be an arbitrary entourage of Z and let V be another entourage of Z such that $V \circ V \circ V \subset U$. By the hypothesis there exist $A \in \mathfrak{F}$ and $B \in \mathfrak{G}$ such that $(\varphi(x, y), \varphi(x', y)) \in V$ for any $x, x' \in A$ and $y \in B$. We set

$$A' := \{x \in X \mid \varphi(x, \mathfrak{G}) \text{ is a Cauchy filter on } Z\}.$$

Let $x_0 \in A \cap A'$. Since $\varphi(x_0, \mathfrak{G})$ is a Cauchy filter on Z there exists $B' \in \mathfrak{G}$ such that $(\varphi(x_0, y), \varphi(x_0, y')) \in V$ for any $y, y' \in B'$. We get

$$(\varphi(x, y), \varphi(x', y')) \in V \circ V \circ V \subset U$$

for any $x, x' \in A \cap A'$ and $y, y' \in B \cap B'$. Hence,

$$\varphi((A \cap A') \times (B \cap B')) \times \varphi((A \cap A') \times (B \cap B')) \subset U$$

and therefore $\varphi(\mathfrak{F} \times \mathfrak{G})$ is a Cauchy filter. □

Theorem 2.2.3. *Let $\mathfrak{F}$ be a Cauchy filter on Y^X_A which belongs to $\hat{\Phi}_5(Y^X_{\mathfrak{S}})$, let $\mathfrak{G} \in \Phi$ and φ be the map*

$$Y^X \times X \to Y, \quad (f, x) \mapsto f(x).$$

a) For any entourage U of Y there exist $\mathscr{F} \in \mathfrak{F}$ and $B \in \mathfrak{G}$ such that $(f(x), g(x)) \in U$ for any $f, g \in \mathscr{F}$ and for any $x \in B$. In particular, $\mathfrak{F}$ is a Cauchy filter on Y^X_Φ.

b) If

$$\{f \in Y^X \mid f(\mathfrak{G}) \text{ is a Cauchy filter}\} \in \mathfrak{F},$$

then $\varphi(\mathfrak{F} \times \mathfrak{G})$ is a Cauchy filter.

c) If

$$\{f \in Y^X \mid (x_n)_{n \in \mathbb{N}} \ \Theta\text{-sequence} \Rightarrow (f(x_n))_{n \in \mathbb{N}} \in \Theta_5(Y)\} \in \mathfrak{F}$$

then $\varphi(\mathfrak{F} \times \mathfrak{G}) \in \hat{\Phi}_5(Y)$.

Let Θ' be the set of sequences $(f_n, x_n)_{n \in \mathbb{N}}$ in $Y^X \times X$ such that for any entourage U of Y there exist $m, n \in \mathbb{N}$ with $m < n$ and $(f_m(x_m), f_n(x_m)) \in U$. By Proposition 2.2.1 $\mathfrak{F}' \times \mathfrak{G}' \in \Phi(\Theta')$ for any $(\mathfrak{F}', \mathfrak{G}') \in \Phi_5(Y^X_\mathfrak{S}) \times \Phi(\Theta)$.

a) By Proposition 1.1.10 $\mathfrak{F} \times \mathfrak{G} \in \hat{\Phi}(\Theta')$. $\varphi(\mathfrak{F}, x)$ is a Cauchy filter for any $x \in A$. By Proposition 1.4.6 $A \in \mathfrak{G}$ and therefore, by Proposition 1.4.11, for any entourage U of Y there exists $(\mathscr{F}, B) \in \mathfrak{F} \times \mathfrak{G}$ such that $(f(x), g(x)) \in U$ for any $f, g \in \mathscr{F}$ and any $x \in B$.

b) follows from a) and Lemma 2.2.2.

c) We set

$$\mathscr{G} := \{f \in Y^X \mid (x_n)_{n \in \mathbb{N}} \ \Theta\text{-sequence} \Rightarrow (f(x_n))_{n \in \mathbb{N}} \in \Theta_5(Y)\}$$

and denote by Θ'' the set of sequences $(f_n, x_n)_{n \in \mathbb{N}}$ in $Y^X \times X$ such that $(f_n)_{n \in \mathbb{N}} \in \Theta_5(Y^X_\mathfrak{S})$, $(x_n)_{n \in \mathbb{N}} \in \Theta$, $\{f_n \mid n \in \mathbb{N}\} \subset \mathscr{G}$. Let $(f_n, x_n)_{n \in \mathbb{N}}$ be a Θ''-sequence, $\mathfrak{N}$ be an ultrafilter on $\mathbb{N}$ finer than the section filter of $\mathbb{N}$, and g, h be the maps

$$\mathbb{N} \to Y^X, \quad n \mapsto f_n$$

$$\mathbb{N} \to X, \quad n \mapsto x_n$$

respectively. Then $g(\mathfrak{N})$ is a $\Phi_5(Y^X_\mathfrak{S})$-ultrafilter (Proposition 1.4.2) and therefore a Cauchy filter on $Y^X_\mathfrak{S}$ (Proposition 1.5.10), $h(\mathfrak{N}) \in \Phi$, and

$$\{f \in Y^X \mid f(h(\mathfrak{N})) \text{ is a Cauchy filter}\} \in g(\mathfrak{N}).$$

By b) $\varphi(g(\mathfrak{N}) \times h(\mathfrak{N}))$ is a Cauchy filter. Hence, for any entourage U of Y there exist different $m, n \in \mathbb{N}$ with $(f_m(x_m), f_n(x_n)) \in U$ so $(f_n(x_n))_{n \in \mathbb{N}} \in \Theta_5(Y)$. By Proposition 1.4.8 a $\Rightarrow$ c $\varphi(\mathfrak{H}) \in \Phi_5(Y)$ for any $\mathfrak{H} \in \Phi(\Theta'')$.

Let $\mathfrak{H}$ be an ultrafilter on Y finer than $\varphi(\mathfrak{F} \times \mathfrak{G})$. Then there exists an ultrafilter $\mathfrak{H}'$ on $Y^X \times X$ finer than $\mathfrak{F} \times \mathfrak{G}$ such that $\varphi(\mathfrak{H}') = \mathfrak{H}$. Let $\mathfrak{F}', \mathfrak{G}'$ be the projections of $\mathfrak{H}'$ on Y^X and X respectively. Then $\mathfrak{F} \subset \mathfrak{F}'$, $\mathfrak{G} \subset \mathfrak{G}'$ so $(\mathfrak{F}', \mathfrak{G}') \in \Phi_5(Y^X_\mathfrak{S}) \times \Phi(\Theta)$ (Proposition 1.4.2). Let (I, g) be a Θ_5-net in $Y^X_\mathfrak{S}$ such that $g(\mathfrak{F}'') = \mathfrak{F}'$, where $\mathfrak{F}''$ denotes the section filter of I, and let (J, h) be a Θ-net in X such that $h(\mathfrak{G}'') = \mathfrak{G}'$, where $\mathfrak{G}''$ denotes the section filter of J. By the hypothesis there exists $\iota_0 \in I$ such that $\{g_\iota \mid \iota \in I, \iota \geq \iota_0\} \subset \mathscr{G}$. We set $I_0 := \{\iota \in I \mid \iota \geq \iota_0\}$ and denote by g_0 the restriction of g to I_0 and by $\mathfrak{F}_0''$ the section filter of I_0. Then (I_0, g_0) is a Θ_5-net in $Y^X_{\mathfrak{S}}$, $g_0(\mathfrak{F}_0'') = \mathfrak{F}'$ so $\mathfrak{F}' \times \mathfrak{G}' \in \Phi(\Theta'')$. Hence $\mathfrak{H}' \in \Phi(\Theta'')$ (Proposition 1.4.2) and by the above considerations $\varphi(\mathfrak{H}') \in \Phi_5(Y)$. Since $\mathfrak{H}$ is arbitrary we get $\varphi(\mathfrak{F} \times \mathfrak{G}) \in \hat{\Phi}_5(Y)$ (Proposition 1.4.2). □

Corollary 2.2.4. *The identity map $Y^X_{\mathfrak{S}} \to Y^X_{\Phi}$ is uniformly Φ_5-continuous.*

The assertion follows immediately from Theorem 2.2.3 a). □

Corollary 2.2.5. *Let $i \in \{1, 2, 3, 4, 5\}$ and $\mathscr{F}$ be a subset of Y^X such that the identity map $\mathscr{F}_A \to \mathscr{F}_{\mathfrak{S}}$ is uniformly Φ_i-continuous and such that $\mathscr{F}$ is a Φ_i-closed set of Y^X_{Φ}. Then $\mathscr{F}$ is a Φ_i-closed set of Y^X.*

By Corollary 2.2.4 the identity map $Y^X_{\mathfrak{S}} \to Y^X_{\Phi}$ is uniformly Φ_i-continuous and therefore by Proposition 1.8.17 $\mathscr{F}$ is a Φ_i-closed set of $Y^X_{\mathfrak{S}}$. By Proposition 2.1.15 a) $\mathscr{F}$ is a Φ_i-closed set of Y^X. □

Corollary 2.2.6. *Let $f \in Y^X$ and $\mathfrak{F}$ be a filter of $\hat{\Phi}_5(Y^X_{\mathfrak{S}})$ converging to f in Y^X_A. Then for any $\mathfrak{G} \in \Phi$ and any entourage U of Y there exists $\mathscr{F} \in \mathfrak{F}$ and $B \in \mathfrak{G}$ such that $(f(x), g(x)) \in U$ for any $g \in \mathscr{F}$ and any $x \in B$. In particular, $\mathfrak{F}$ converges to f in Y^X_{Φ}.*

Without loss of generality we may assume U closed. By Theorem 2.2.3 a) there exist $\mathscr{F} \in \mathfrak{F}$ and $C \in \mathfrak{G}$ such that $(g(x), h(x)) \in U$ for any $g, h \in \mathscr{F}$ and any $x \in C$. By Proposition 1.4.6 $A \in \mathfrak{G}$ and therefore $B := A \cap C \in \mathfrak{G}$. Since f is an adherent point of $\mathscr{F}$ in Y^X_A and U is closed we get $(f(x), g(x)) \in U$ for any $g \in \mathscr{F}$ and any $x \in B$. □

Lemma 2.2.7. *Let X, Y, Z be sets, Φ be a set of filters on X, $\mathfrak{G}$ be a filter on Y, φ be a map of $X \times Y$ into Z, and let U be a subset of $Z \times Z$ such that for any $\mathfrak{F} \in \Phi$ there exist $A \in \mathfrak{F}$ and $B \in \mathfrak{G}$ with $(\varphi(x, y), \varphi(x, y')) \in U$ for any $x \in A$ and $y, y' \in B$. Then for any $\mathfrak{F}_0 \in \hat{\Phi}$ there exist $A \in \mathfrak{F}_0$ and $B \in \mathfrak{G}$ with $(\varphi(x, y), \varphi(x, y')) \in U$ for any $x \in A$ and $y, y' \in B$.*

By the hypothesis there exist maps $M : \Phi \to \mathfrak{P}(X)$, $P : \Phi \to \mathfrak{G}$ such that:

a) $M(\mathfrak{F}) \in \mathfrak{F}$ for any $\mathfrak{F} \in \Phi$;

b) for any $\mathfrak{F} \in \Phi$ and for any $x \in M(\mathfrak{F})$ and $y, y' \in P(\mathfrak{F})$ we have $(\varphi(x, y), \varphi(x, y')) \in U$.

Since $\mathfrak{F}_0 \in \hat{\Phi}$ there exists a finite subset Ψ of Φ such that

$$A := \bigcup_{\mathfrak{F} \in \Psi} M(\mathfrak{F}) \in \mathfrak{F}_0 .$$

We set

$$B := \bigcap_{\mathfrak{F} \in \Psi} P(\mathfrak{F}) \in \mathfrak{G} .$$

Then $(\varphi(x, y), \varphi(x, y')) \in U$ for any $x \in A$ and $y, y' \in B$. □

Proposition 2.2.8. *Let $i \in \{1, 2, 3, 4, 5\}$, $\mathscr{F}$ be a subset of Y^X such that the identity* *map* $\mathscr{F} \to \mathscr{F}_{\mathfrak{S}}$ *is uniformly* Φ_i*-continuous,* $\mathfrak{F} \in \hat{\Phi}_i(\mathscr{F})$ *and* $\mathfrak{G} \in \Phi$ *such that*

$$\{f \in \mathscr{F} \mid f(\mathfrak{G}) \text{ is a Cauchy filter}\} \in \mathfrak{F}.$$

Then for any entourage U of Y there exist $\mathscr{F}_0 \in \mathfrak{F}$ *and* $B \in \mathfrak{G}$ *such that* $(f(x), f(y)) \in U$ *for any* $f \in \mathscr{F}_0$ *and any* $x, y \in B$.

We set $\Psi = \{\mathfrak{H} \in \hat{\Phi}_i(\mathscr{F}) \cap \Phi_C(\mathscr{F}) \mid \mathfrak{F} \subset \mathfrak{H}\}$. By Theorems 1.8.4 a $\Rightarrow$ c and 2.2.3 b) for any $\mathfrak{H} \in \Psi$ there exist $\mathscr{G}(\mathfrak{H}) \in \mathfrak{H}$ and $B(\mathfrak{H}) \in \mathfrak{G}$ such that $(f(x), f(y)) \in U$ for any $f \in \mathscr{G}(\mathfrak{H})$ and for any $x, y \in B(\mathfrak{H})$. Since $\mathfrak{F} \in \hat{\Phi}$ (Corollary 1.5.12) the assertion follows from Lemma 2.2.7. □

Proposition 2.2.9. *Let X be a uniform (topological) space,* $\mathscr{F}$ *be a set of uniformly* $\Phi(\Theta)$*-continuous* ($\Phi(\Theta)$*-continuous) maps of X into Y,* $\Psi := \hat{\Phi}_5(\mathscr{F}_{\mathfrak{S}})$*, and* φ, ψ *be the maps*

$$\mathscr{F}_{\mathfrak{S}} \times X \to Y, \quad (f, x) \mapsto f(x),$$

$$X \to Y^{\mathscr{F}}_{\Psi}, \quad x \mapsto \varphi(\cdot, x)$$

respectively. Then:

a) for any $\mathfrak{F} \in \Psi \cap \Phi_C(\mathscr{F}_A)$ *and any* $\mathfrak{G} \in \Phi \cap \Phi_C(X)$ $(\mathfrak{G} \in \Phi \cap \Phi_c(X))$, $\varphi(\mathfrak{F} \times \mathfrak{G})$ *is a Cauchy filter;*

b) ψ *is uniformly* Φ*-continuous (if any constant sequence in X belongs to* Θ *then* ψ *is* Φ*-continuous);*

c) if $\Theta = \Theta_i(X)$ *for an* $i \in \{1, 2, 3, 4, 5\}$ $(i \in \{1, 2\})$ *then* φ *is uniformly* Φ_i*-continuous (*Φ_i*-continuous).*

a) By Proposition 1.3.6 and Proposition 1.4.2 $f(\mathfrak{G})$ is a Cauchy filter for any $f \in \mathscr{F}$. Let φ_0 be the inclusion map $\mathscr{F}_{\mathfrak{S}} \to Y^X_{\mathfrak{S}}$. Then $\varphi_0(\mathfrak{F}) \in \hat{\Phi}_5(Y^X_{\mathfrak{S}}) \cap \Phi_C(Y^X_A)$ and

$$\{f \in Y^X \mid f(\mathfrak{G}) \text{ is a Cauchy filter}\} \in \varphi_0(\mathfrak{F}).$$

By Theorem 2.2.3 b) $\varphi(\mathfrak{F} \times \mathfrak{G})$ is a Cauchy filter.

b) Let $\mathfrak{G} \in \Phi \cap \Phi_C(X)$ $(\mathfrak{G} \in \Phi \cap \Phi_c(X))$ and $\mathfrak{F} \in \Psi$. Let Ψ_0 be the set of ultrafilters on $\mathscr{F}$ finer than $\mathfrak{F}$. Then $\mathfrak{F}' \in \Psi \cap \Phi_C(\mathscr{F}_{\mathfrak{S}})$ (Propositions 1.4.2, 1.5.10) so by a), $\varphi(\mathfrak{F}' \times \mathfrak{G})$ is a Cauchy filter for any $\mathfrak{F}' \in \Psi_0$. Let U be an entourage of Y. For any $\mathfrak{F}' \in \Psi_0$ there exist $\mathscr{F}' \in \mathfrak{F}'$ and $B \in \mathfrak{G}$ with $(f(x), f(y)) \in U$ for any $f \in \mathscr{F}'$ and any $x, y \in B$. Since $\mathfrak{F} \in \hat{\Psi}_0$ (Proposition 1.1.3) there exist by Lemma 2.2.7 $\mathscr{F}' \in \mathfrak{F}$ and $B \in \mathfrak{G}$ with $(f(x), f(y)) \in U$ for any $f \in \mathscr{F}'$ and any $x, y \in B$. Hence,

$$(\psi(x), \psi(y)) \in \mathscr{W}(U, \mathfrak{F}; \mathscr{F}, Y).$$

Since U and $\mathfrak{F}$ are arbitrary $\psi(\mathfrak{G})$ is a Cauchy filter. Hence, ψ is uniformly Φ-continuous (Φ-continuous (Proposition 1.3.4)).

c) Let $\mathfrak{H} \in \hat{\Phi}_i(\mathscr{F}_{\mathfrak{S}} \times X) \cap \Phi_C(\mathscr{F}_{\mathfrak{S}} \times X)$ $(\mathfrak{H} \in \hat{\Phi}_i(\mathscr{F}_{\mathfrak{S}} \times X) \cap \Phi_c(\mathscr{F}_{\mathfrak{S}} \times X))$, and let $\mathfrak{F}$, $\mathfrak{G}$ be its projections on $\mathscr{F}$ and X respectively. By a) $\varphi(\mathfrak{F} \times \mathfrak{G})$ and a fortiori $\varphi(\mathfrak{H})$ are Cauchy filters. Hence, φ is uniformly Φ_i-continuous (Φ_i-continuous (Proposition 1.3.4)). □

Proposition 2.2.10. *Let $\mathscr{F}$ be a subset of Y^X and $i \in \{1,2,3,4,5\}$. Then the following assertions are equivalent:*

a) the identity map $\mathscr{F}_A \to \mathscr{F}_\Phi$ is uniformly Φ_i-continuous;
b) $\Phi_i(\mathscr{F}_A) \subset \Phi_i(\mathscr{F}_\Phi)$;
c) $\Phi_i(\mathscr{F}_A) \subset \hat{\Phi}_C(\mathscr{F}_\Phi)$;
d) any Φ_i-set of $\mathscr{F}_A$ is a precompact set of $\mathscr{F}_\Phi$.

If $\mathfrak{A}$ denotes the set of Φ-sets of X and $\mathfrak{B}$ the set

$$\{\{x_n \mid n \in \mathbb{N}\} \mid (x_n)_{n\in\mathbb{N}} \text{ is a } \Theta\text{-sequence}\}$$

and if any sequence in X belongs to Θ if it possesses a subsequence belonging to Θ, then all these assertions are equivalent to the assertions obtained from them by replacing $\mathscr{F}_\Phi$ with $\mathscr{F}_\mathfrak{A}$ or $\mathscr{F}_\mathfrak{B}$.

Let us denote by a($\mathfrak{A}$), b($\mathfrak{A}$), c($\mathfrak{A}$), d($\mathfrak{A}$), a($\mathfrak{B}$), b($\mathfrak{B}$), c($\mathfrak{B}$), and d($\mathfrak{B}$) the assertions obtained from the assertions a), b), c), and d) by replacing Φ with $\mathfrak{A}$ and $\mathfrak{B}$ respectively.

a $\Rightarrow$ b, a($\mathfrak{A}$) $\Rightarrow$ b($\mathfrak{A}$), a($\mathfrak{B}$) $\Rightarrow$ b($\mathfrak{B}$) follow from Theorem 1.8.4 a $\Rightarrow$ g.

b $\Rightarrow$ c, b($\mathfrak{A}$) $\Rightarrow$ c($\mathfrak{A}$), b($\mathfrak{B}$) $\Rightarrow$ c($\mathfrak{B}$) follow from Corollary 1.5.12.

c $\Rightarrow$ d, c($\mathfrak{A}$) $\Rightarrow$ d($\mathfrak{A}$), c($\mathfrak{B}$) $\Rightarrow$ d($\mathfrak{B}$) follow from Corollary 1.2.3 and Proposition 1.4.2.

a $\Rightarrow$ a($\mathfrak{A}$) $\Rightarrow$ a($\mathfrak{B}$), d $\Rightarrow$ d($\mathfrak{A}$) $\Rightarrow$ d($\mathfrak{B}$) follow from the fact that the identity maps $\mathscr{F}_\Phi \to \mathscr{F}_\mathfrak{A}$ and $\mathscr{F}_\mathfrak{A} \to \mathscr{F}_\mathfrak{B}$ are uniformly continuous (Propositions 1.4.2 and 1.4.7).

d($\mathfrak{B}$) $\Rightarrow$ a. Let $\mathscr{G}$ be a $\Phi_i(\mathscr{F}_A)$-set. Then $\mathscr{G}$ is a precompact set of $\mathscr{F}_\mathfrak{B}$ and by Proposition 1.5.6 it is a $\Phi_5(\mathscr{F}_\mathfrak{B})$-set. By Proposition 1.4.8 d $\Rightarrow$ e $\hat{\Phi}_i(\mathscr{F}_A) \subset \hat{\Phi}_5(\mathscr{F}_\mathfrak{B})$.

Let $\mathfrak{F} \in \Phi_C(\mathscr{F}_A) \cap \hat{\Phi}_i(\mathscr{F}_A)$. By the above considerations $\mathfrak{F} \in \hat{\Phi}_5(\mathscr{F}_\mathfrak{B})$. By Theorem 2.2.3 $\mathfrak{F}$ is a Cauchy filter on $\mathscr{F}_\Phi$. Hence, the identity map $\mathscr{F}_A \to \mathscr{F}_\Phi$ is uniformly Φ_i-continuous. □

Corollary 2.2.11. *Let $\mathscr{F}$ be a subset of Y^X, $i \in \{1,2,3,4,5\}$, $\mathfrak{Z}$ be a set of closed subspaces of Y, and for any $Z \in \mathfrak{Z}$ let $\mathscr{F}(Z)$ be the set of $f \in \mathscr{F}$ with $f(X) \subset Z$. We assume:*

a) for any Θ_i-sequence $(f_n)_{n\in\mathbb{N}}$ in $\mathscr{F}$ there exists $Z \in \mathfrak{Z}$ such that $\{f_n \mid n \in \mathbb{N}\} \subset \mathscr{F}(Z)$;
b) the identity map

$$\mathscr{F}(Z) \to \mathscr{F}(Z)_\Phi$$

is uniformly Φ_i-continuous for any $Z \in \mathfrak{Z}$.

Then the identity map $\mathscr{F} \to \mathscr{F}_\Phi$ is uniformly Φ_i-continuous.

Let $(f_n)_{n\in\mathbb{N}}$ be a Θ_i-sequence in $\mathscr{F}$. By a) there exists $Z \in \mathfrak{Z}$ such that $\{f_n \mid n \in \mathbb{N}\} \subset \mathscr{F}(Z)$. Since Z is closed, $(f_n)_{n\in\mathbb{N}}$ is a Θ_i-sequence in $\mathscr{F}(Z)$. By b) and Theorem 1.8.4 a $\Rightarrow$ e, $(f_n)_{n\in\mathbb{N}}$ is a Θ_i-sequence in $\mathscr{F}(Z)_\Phi$. The uniformity of $\mathscr{F}(Z)_\Phi$ being the uniformity induced by $\mathscr{F}_\Phi$, $(f_n)_{n\in\mathbb{N}}$ is a Θ_i-sequence in $\mathscr{F}_\Phi$. We get

$\Phi_i(\mathscr{F}) \subset \Phi_i(\mathscr{F}_\Phi)$ and by Proposition 2.2.10 b $\Rightarrow$ a the identity map $\mathscr{F} \to \mathscr{F}_\Phi$ is uniformly Φ_i-continuous. □

The next four results will not be ned later.

Proposition 2.2.12. *Let $\mathscr{F}$ be a subset of Y^X, $i \in \{1,2,3,4,5\}$, $\Psi := \hat{\Phi}_i(\mathscr{F})$ and φ, ψ be the maps*

$$\mathscr{F} \times X \to Y, \quad (f,x) \mapsto f(x),$$

$$X \to Y_\Psi^{\mathscr{F}}, \quad x \mapsto \varphi(\cdot, x)$$

respectively. We assume X is a uniform space and $\Phi \subset \hat{\Phi}_5(X)$. Then:

a) if any Φ_i-set of $\mathscr{F}$ is uniformly equicontinuous, then the identity map $\mathscr{F} \to \mathscr{F}_\Phi$ is uniformly Φ_i-continuous;

b) if the neighbourhood filter of any point of X is a Φ-filter, if any $f \in \mathscr{F}$ is uniformly Φ-continuous, and if the identity map $\mathscr{F} \to \mathscr{F}_\Phi$ is uniformly Φ_i-continuous, then ψ is continuous.

a) Let $\mathfrak{A}$ be the set of precompact sets of X and $\mathscr{G}$ be a Φ_i-set of $\mathscr{F}$. Since $\mathscr{G}$ is uniformly equicontinuous we have $\mathscr{G} = \mathscr{G}_{\mathfrak{A}}$. Hence, $\mathscr{G}$ is a precompact set of $\mathscr{F}_{\mathfrak{A}}$ (Proposition 1.5.6 a $\Rightarrow$ c). By Proposition 2.2.10 d $\Rightarrow$ a and Proposition 1.5.6 c $\Rightarrow$ a the identity map $\mathscr{F} \to \mathscr{F}_\Phi$ is uniformly Φ_i-continuous.

b) By Theorem 1.8.4 a $\Rightarrow$ i $\Psi = \hat{\Phi}_i(\mathscr{F}_\Phi)$. By Proposition 2.2.9 b) ψ is uniformly Φ-continuous. By Proposition 1.3.3 ψ is continuous. □

Proposition 2.2.13. *Let Z be a uniform space, $i \in \{1,2,3,4,5\}$, $\mathscr{F}$ be the set of uniformly Φ_i-continuous maps of Y into Z, $\Psi := \hat{\Phi}_i(Y)$, $\mathfrak{F}$ be a Cauchy filter on Z_Ψ^Y belonging to $\hat{\Phi}_5(Z_\Psi^Y)$ such that $\mathscr{F} \in \mathfrak{F}$, and let $\mathfrak{G}$ be a Cauchy filter on Y^X belonging to $\hat{\Phi}_i(Y^X)$. Then the image of $\mathfrak{F} \times \mathfrak{G}$ through the map*

$$Z^Y \times Y^X \to Z^X, \quad (f,g) \mapsto f \circ g$$

is a Cauchy filter.

Let $x \in X$ and $\mathfrak{H}$ be the image of $\mathfrak{G}$ through the map

$$Y^X \to Y, \quad g \mapsto g(x).$$

Then $\mathfrak{H}$ is a Cauchy filter belonging to $\hat{\Phi}_i(Y)$. Hence, $f(\mathfrak{H})$ is a Cauchy filter for any $f \in \mathscr{F}$. By Theorem 2.2.3 b) the image of $\mathfrak{F} \times \mathfrak{H}$ under the map

$$Z^Y \times Y \to Z, \quad (f,y) \mapsto f(y)$$

is a Cauchy filter. Hence, the image of $\mathfrak{F} \times \mathfrak{G}$ with respect to the map

$$Z^Y \times Y^X \to Z, \quad (f,g) \mapsto f \circ g(x)$$

is a Cauchy filter. The assertion follows from the fact that x is arbitrary. □

Corollary 2.2.14. *Let Z be a uniform space, $i \in \{1, 2, 3, 4, 5\}$, $\mathscr{F}$ be a set of uniformly Φ_i-continuous maps of Y into Z, $\Psi := \hat{\Phi}_i(Y)$, $\mathscr{F}_0$ be the set $\mathscr{F}$ endowed with a uniformity such that the identity map $\mathscr{F}_0 \to \mathscr{F}_\Psi$ is uniformly Φ_i-continuous, and $\mathscr{G}$ be a subspace of Y^X. Then the map*

$$\mathscr{F}_0 \times \mathscr{G} \to Z^X, \quad (f, g) \mapsto f \circ g$$

is uniformly Φ_i-continuous.

Let $\mathfrak{H} \in \Phi_C(\mathscr{F}_0 \times \mathscr{G}) \cap \hat{\Phi}_i(\mathscr{F}_0 \times \mathscr{G})$ and $\mathfrak{F}$, $\mathfrak{G}$ be its projections on $\mathscr{F}_0$ and $\mathscr{G}$ respectively. Then $\mathfrak{F} \in \Phi_C(\mathscr{F}_0) \cap \hat{\Phi}_i(\mathscr{F}_0)$, $\mathfrak{G} \in \Phi_C(\mathscr{G}) \cap \hat{\Phi}_i(\mathscr{G})$ so $\mathfrak{F} \in \Phi_C(\mathscr{F}_\Psi) \cap \hat{\Phi}_i(\mathscr{F}_\Psi)$ (Theorem 1.8.4 a $\Rightarrow$ c). Let us denote by $\mathfrak{F}'$, $\mathfrak{G}'$ the images of $\mathfrak{F}$ and $\mathfrak{G}$ with respect to the inclusion maps $\mathscr{F} \to Z^Y$, $\mathscr{G} \to Y^X$ respectively. Then $\mathfrak{F}' \in \Phi_C(Z^Y_\Psi) \cap \hat{\Phi}_i(Z^Y_\Psi)$, $\mathfrak{G}' \in \Phi_C(Y^X) \cap \hat{\Phi}_i(Y^X)$, and $\mathscr{F} \in \mathfrak{F}'$. Proposition 2.2.13 the image of $\mathfrak{F}' \times \mathfrak{G}'$ with respect to the map

$$Z^Y \times Y^X \to Z^X, \quad (f, g) \mapsto f \circ g$$

is a Cauchy filter and this is nothing else but the image of $\mathfrak{F} \times \mathfrak{G}$ with respect to the map

$$\mathscr{F} \times \mathscr{G} \to Z^X, \quad (f, g) \mapsto f \circ g\,.$$

Since $\mathfrak{H}$ is finer than $\mathfrak{F} \times \mathfrak{G}$ we deduce that the image of $\mathfrak{H}$ under the above map is a Cauchy filter. The assertion follows now from the fact that $\mathfrak{H}$ is arbitrary. □

Corollary 2.2.15. *Let $i \in \{1, 2, 3, 4, 5\}$, $\mathscr{F}$ be a set of uniformly Φ_i-continuous maps of Y into itself such that $f \circ g \in \mathscr{F}$ for any $f, g \in \mathscr{F}$, $\Psi := \hat{\Phi}_i(Y)$ and $\mathscr{F}_0$ be the set $\mathscr{F}$ endowed with a uniformity such that identity maps $\mathscr{F} \to \mathscr{F}_0$, $\mathscr{F}_0 \to \mathscr{F}_\Psi$ are uniformly Φ_i-continuous. Then the map*

$$\mathscr{F}_0 \times \mathscr{F}_0 \to \mathscr{F}_0\,, \quad (f, g) \mapsto f \circ g$$

is uniformly Φ_i-continuous.

By Corollary 2.2.14 the map

$$\mathscr{F}_0 \times \mathscr{F} \to Y^X, \quad (f, g) \mapsto f \circ g$$

is uniformly Φ_i-continuous and therefore the map

$$\mathscr{F}_0 \times \mathscr{F} \to \mathscr{F}, \quad (f, g) \mapsto f \circ g$$

is uniformly Φ_i-continuous (Corollary 1.8.14). Since the identity map $\mathscr{F}_\Psi \to \mathscr{F}$ is uniformly continuous, the identity map $\mathscr{F}_0 \to \mathscr{F}$ is uniformly Φ_i-continuous. Hence, the identity map $\mathscr{F}_0 \times \mathscr{F}_0 \to \mathscr{F}_0 \times \mathscr{F}$ is uniformly Φ_i-continuous (Corollary 1.8.16). By the above considerations and Corollary 1.8.5 the map

$$\mathscr{F}_0 \times \mathscr{F}_0 \to \mathscr{F}_0\,, \quad (f, g) \mapsto f \circ g$$

is uniformly Φ_i-continuous. □

§ 2.3 Šmulian spaces

Definition 2.3.1. *A* Šmulian space *is a topological space X possessing the following property: if (I, f) is a Θ_2-net in X, if $\mathfrak{F}$ is the section filter of I, and if $f(\mathfrak{F})$ is an ultrafilter on X, then for any limit point x of $f(\mathfrak{F})$ there exists an increasing sequence $(\iota_n)_{n\in\mathbb{N}}$ in I such that $(f(\iota_n))_{n\in\mathbb{N}}$ converges to x.*

Any subspace of a Šmulian space is a Šmulian space.

Proposition 2.3.2. *Let X be a Šmulian space, (I, f) be a Θ_2-net in X, $\mathfrak{F}$ be the section filter of I, $(I_n)_{n\in\mathbb{N}}$ be a sequence in $\mathfrak{F}$ and x be an adherent point of $f(\mathfrak{F})$. Then there exists an increasing sequence $(\iota_n)_{n\in\mathbb{N}}$ in I such that $(f(\iota_n))_{n\in\mathbb{N}}$ converges to x and $\iota_n \in I_n$ for any $n \in \mathbb{N}$.*

Let A_0 be the intersection of all neighbourhoods of x. We may assume there exists $A \in \mathfrak{F}$ such that $A \cap \overset{-1}{f}(A_0) = \emptyset$ since otherwise the assertion is trivial. Let $\mathfrak{G}$ be an ultrafilter on I finer than $\mathfrak{F}$ such that $f(\mathfrak{G})$ converges to x. We set

$$J := \{(B, n, \iota) \mid B \in \mathfrak{G},\ n \in \mathbb{N},\ \iota \in A \cap B \cap (\bigcap_{m\leq n} I_m)\}$$

and denote by $\leq$ the upper directed order relation of J defined by

$$(B, n, \iota) \leq (B', n', \iota') :\Leftrightarrow ((B, n, \iota) = (B', n', \iota') \quad \text{or} \quad B' \subset B,\ n < n',\ \iota \leq \iota'),$$

by $\mathfrak{H}$ the section filter of J, and by g the map

$$J \to I, \quad (B, n, \iota) \mapsto \iota.$$

Then $(J, f\circ g)$ is a Θ_2-net in X and $g(\mathfrak{H}) = \mathfrak{G}$. Since X is a Šmulian space there exists an increasing sequence $(\lambda_n)_{n\in\mathbb{N}}$ in J such that $(f\circ g(\lambda_n))_{n\in\mathbb{N}}$ converges to x. We have $f\circ g(\lambda_n) \notin A_0$ for any $n \in \mathbb{N}$ and therefore $(\lambda_n)_{n\in\mathbb{N}}$ possesses a strictly increasing subsequence $(\lambda_{k_n})_{n\in\mathbb{N}}$. Then $(g(\lambda_{k_n}))_{n\in\mathbb{N}}$ is an increasing sequence in I such that $(f(g(\lambda_{k_n})))_{n\in\mathbb{N}}$ converges to x and $g(\lambda_{k_n}) \in I_n$ for any $n \in \mathbb{N}$. □

Proposition 2.3.3. *Any Šmulian space is a mioritic space.*

Let X be a Šmulian space and $(x_n)_{n\in\mathbb{N}}$ be a $\Theta_2(X)$-sequence. Let x be an adherent point of $(x_n)_{n\in\mathbb{N}}$. By Proposition 2.3.2 there exists a subsequence of $(x_n)_{n\in\mathbb{N}}$ converging to x. Hence, $(x_n)_{n\in\mathbb{N}} \in \Theta_1(X)$ and therefore X is a mioritic space (Proposition 1.7.2 b $\Rightarrow$ a). □

Proposition 2.3.4. *Any subspace of a Φ_2-compact Šmulian Hausdorff space is a Φ_2-compact Šmulian space.*

Let X be a Φ_2-compact Šmulian Hausdorff space, Y be a subspace of X and φ be the inclusion map $Y \to X$. Further, let $\mathfrak{F}$ be a Φ_2-ultrafilter on Y and (I, f) be a Θ_2-net in Y such that $f(\mathfrak{G}) = \mathfrak{F}$, where $\mathfrak{G}$ denotes the section filter of I. Since X is Φ_2-compact $\varphi(\mathfrak{F})$ possesses a limit point x. Since X is a Smulian space there exists an increasing sequence $(\iota_n)_{n\in\mathbb{N}}$ in I such that $(\varphi\circ f(\iota_n))_{n\in\mathbb{N}}$ converges to x. Let y be an adherent point of $(f(\iota_n))_{n\in\mathbb{N}}$ in Y. Since X is Hausdorff we get $x = y \in Y$. Hence, $\mathfrak{F}$ possesses an adherent point in Y so Y is Φ_2-compact.

Y is obviously a Šmulian space. □

Definition 2.3.5. *Let A be a subset of a topological space X and φ be the inclusion map $A \to X$. The set A is called* sequentially dense *if for any net (I, f) in A and for any adherent point x of $\varphi(f(\mathfrak{F}))$, where $\mathfrak{F}$ denotes the section filter of I, there exists an increasing sequence $(\iota_n)_{n\in\mathbb{N}}$ in I such that $(\varphi(f(\iota_n)))_{n\in\mathbb{N}}$ converges to x.*

If A is sequentially dense then for any $x \in \bar{A}$ there is a sequence in A converging to x. Any countably compact sequentially dense set is sequentially compact.

Theorem 2.3.6. *a) Any Φ_2-set of a Šmulian space is sequentially dense.*

b) Any countably compact set of a Φ_2-compact Šmulian Hausdorff space is compact and sequentially compact.

a) Let A be a Φ_2-set of a Šmulian space X, φ be the inclusion map $A \to X$, (I, f) be a net in A, $\mathfrak{F}$ be the section filter of I and x be an adherent point of $\varphi(f(\mathfrak{F}))$. Then $(I, \varphi \circ f)$ is a Θ_2-net in X (Proposition 1.5.5a $\Rightarrow$ c) and the assertion follows from Proposition 2.3.2.

b) Let A be a countably compact set of a Φ_2-compact Šmulian Hausdorff space X. By a) it is sequentially dense so it is sequentially compact. By Proposition 2.3.4 A is a Φ_2-compact Šmulian space. Let $\mathfrak{F}$ be an ultrafilter on A. Since A is a Φ_2-set of A, $\mathfrak{F}$ is a Φ_2-filter and therefore converges. Hence, A is compact. □

Proposition 2.3.7. *Let X', X'' be two topological spaces defined on the same set such that X' is Šmulian and Hausdorff and such that the identity map $X'' \to X'$ is Φ_2-continuous. Then X'' is Šmulian. If, in addition, X' is Φ_2-compact, then so is X''.*

Let us denote by φ the identity map $X'' \to X'$. Let (I, f) be a Θ_2-net in X'', $\mathfrak{F}$ be the section filter of I and x be a limit point of $f(\mathfrak{F})$. Then $(I, \varphi \circ f)$ is a Θ_2-net in X' (Proposition 1.8.9 b)) and $\varphi(x)$ is a limit point of $\varphi \circ f(\mathfrak{F})$. Since X' is a Šmulian space there exists an increasing sequence $(\iota_n)_{n\in\mathbb{N}}$ in I such that $(\varphi \circ f(\iota_n))_{n\in\mathbb{N}}$ converges to $\varphi(x)$. By Proposition 1.8.12 $(f(\iota_n))_{n\in\mathbb{N}}$ converges to x. Hence, X'' is a Šmulian space.

In addition, assume now that X' is Φ_2-compact and let ψ be the identity map $X' \to X''$. Let $\mathfrak{F}$ be a $\Phi_2(X'')$-ultrafilter. Then $\varphi(\mathfrak{F})$ is a $\Phi_2(X')$-ultrafilter (Proposition 1.8.9 c)) so it converges in X' to an x. We want to show that $\mathfrak{F}$ converges to $\psi(x)$.

Assume the contrary. Then there exists a neighbourhood U of $\psi(x)$ such that $X''\backslash U \in \mathfrak{F}$. Let (I, f) be a $\Theta_2(X'')$-net such that $f(\mathfrak{G}) = \mathfrak{F}$, where $\mathfrak{G}$ denotes the section filter of I. Then $\overset{-1}{f}(X''\backslash U) \in \mathfrak{G}$ and $(I, \varphi \circ f)$ is a $\Theta_2(X')$-net (Proposition 1.8.9 b)) such that $\varphi \circ f(\mathfrak{G}) = \varphi(\mathfrak{F})$. Since X' is a Šmulian space there exists an increasing sequence $(\iota_n)_{n\in\mathbb{N}}$ in I such that $(\varphi \circ f(\iota_n))_{n\in\mathbb{N}}$ converges to x and $\iota_n \in \overset{-1}{f}(X''\backslash U)$ for any $n \in \mathbb{N}$ (Proposition 2.3.2). By Proposition 1.8.12 $(f(\iota_n))_{n\in\mathbb{N}}$ converges to $\psi(x)$, which contradicts the relation $f(\iota_n) \in X''\backslash U$ for any $n \in \mathbb{N}$. □

Proposition 2.3.8. *Let X be a topological space, $x \in X$, (I, f) be a Θ_2-net in X, $\mathfrak{F}$ be the section filter of I and $(F_n)_{n\in\mathbb{N}}$ be a sequence in $f(\mathfrak{F})$ of closed sets of X such that*

$$\bigcap_{n\in\mathbb{N}} F_n = \{x\}.$$

Then $f(\mathfrak{F})$ converges to x and there exists an increasing sequence $(\iota_n)_{n\in\mathbb{N}}$ in I such that $(f(\iota_n))_{n\in\mathbb{N}}$ converges to x.

Let $(\iota_n)_{n\in\mathbb{N}}$ be an increasing sequence in I such that

$$f(\iota_n) \in \bigcap_{m\leq n} F_m$$

for any $n \in \mathbb{N}$. We want to show that $(f(\iota_n))_{n\in\mathbb{N}}$ converges to x. Let y be an adherent point of $(f(\iota_n))_{n\in\mathbb{N}}$. Then $y \in \bigcap_{n\in\mathbb{N}} F_n = \{x\}$ and therefore x is the unique adherent point of $(f(\iota_n))_{n\in\mathbb{N}}$. By Proposition 1.5.7 $(f(\iota_n))_{n\in\mathbb{N}}$ converges to x.

Assume $f(\mathfrak{F})$ does not converge to x. Then there exists a neighbourhood U of x such that

$$\bigcap_{m\leq n} F_m \not\subset U$$

for any $n \in \mathbb{N}$. We may construct inductively an increasing sequence $(\iota_n)_{n\in\mathbb{N}}$ in I such that

$$f(\iota_n) \in \bigcap_{m\leq n} F_m \backslash U$$

for any $n \in \mathbb{N}$. By the above considerations, $(f(\iota_n))_{n\in\mathbb{N}}$ converges to x, which is the expected contradiction. □

Proposition 2.3.9. *Let X be a topological space such that for any $x \in X$ there exists a countable family of closed neighbourhoods of x whose intersection is $\{x\}$. Then X is a Šmulian space. In particular, any uniformizable topological space whose one point sets are G_δ-sets is a Šmulian space.*

Let (I, f) be a Θ_2-net in X, $\mathfrak{F}$ be the section filter of I and x be a limit point of $f(\mathfrak{F})$. By the hypothesis there exists a sequence $(F_n)_{n\in\mathbb{N}}$ of closed neighbourhoods of x such that $\bigcap_{n\in\mathbb{N}} F_n = \{x\}$. Then $F_n \in f(\mathfrak{F})$ for any $n \in \mathbb{N}$. By Proposition 2.3.8 there exists an increasing sequence $(\iota_n)_{n\in\mathbb{N}}$ in I such that $(f(\iota_n))_{n\in\mathbb{N}}$ converges to x. Hence, X is a Šmulian space. □

Corollary 2.3.10. *Let X be a topological space possessing a coarser metrizable topology. Then:*

a) X is a Φ_2-compact Šmulian space;

b) any countably compact set of X is metrizable and compact;

c) if, in addition, X is regular, then $\bar{A}$ is metrizable and compact for any Φ_2-set A of X.

a) follows immediately from Proposition 2.3.7, Proposition 2.3.9 and Theorem 1.5.8.

b) Let A be a countably compact set of X. By a) and by Theorem 2.3.6 b) A is compact. Since there exists a coarser metrizable topology on A, A is metrizable.

c) By a) and Proposition 1.2.10 $\bar{A}$ is compact. $\bar{A}$ is metrizable since it possesses a coarser metrizable topology. □

Remark. b) and c) were proved by A. Grothendieck (1952) ([1] Théorème 3) under the additional hypothesis that X is completely regular. a) was proved by C. Constantinescu (1973) ([2] Corollary 3.19 and Proposition 3.1).

Lemma 2.3.11. *Let X, Y be topological spaces such that Y is Hausdorff and such that any compact subspace of Y is metrizable, let $\mathfrak{K}$ be a countable set of quasicompact sets of X such that $\bigcup_{K\in\mathfrak{K}} K = X$, and let $\mathscr{F}$ be a separable subspace of Y^X such that $f|K$ is continuous for any $(f, K)\in\mathscr{F}\times\mathfrak{K}$. Then there exists a countable subset A of X such that $\mathscr{F}_A$ is Hausdorff.*

Let $\mathscr{G}$ be a countable dense set of $\mathscr{F}$. We denote for any $x\in X$ by φ_x the map

$$\mathscr{G}\to Y,\quad f\mapsto f(x)$$

and for any $K\in\mathfrak{K}$ by ψ_K and ψ'_K the maps

$$K\to Y^{\mathscr{G}},\quad x\mapsto\varphi_x,$$

$$K\to\psi_K(K),\quad x\mapsto\varphi_x$$

respectively.

Let $(f, K)\in\mathscr{F}\times\mathfrak{K}$ and $x, y\in K$ with $f(x)\neq f(y)$. Then there exist disjoint neighbourhoods U, V of $f(x)$ and $f(y)$ respectively. Since f is an adherent point of $\mathscr{G}$ in $\mathscr{F}$ there exists $g\in\mathscr{G}$ with $g(x)\in U$ and $g(y)\in V$. We get $g(x)\neq g(y)$ so $\psi_K(x)\neq\psi_K(y)$. Hence, there exists a map $f_K:\psi_K(K)\to Y$ such that $f|K = f_K\circ\psi'_K$. Since $f_K\circ\psi'_K$ and ψ'_K are continuous and since K and $\psi_K(K)$ are quasicompact and compact respectively we deduce that f_K is continuous.

Let $K\in\mathfrak{K}$. Since $\psi_K(K)$ is a compact set of $Y^{\mathscr{G}}$ and since $\mathscr{G}$ is countable we deduce from the hypothesis above Y that $\psi_K(K)$ is separable. There exists, therefore, a countable subset A_K of K such that $\psi'_K(A_K)$ is a dense set of $\psi_K(K)$.

We set $A := \bigcup_{K\in\mathfrak{K}} A_K$. Let $f, g\in\mathscr{F}$ with $f\neq g$. There exists $K\in\mathfrak{K}$ with $f|K\neq g|K$ so $f_K\neq g_K$. Since f_K and g_K are continuous and $\psi'_K(A_K)$ dense there exists $x\in A_K$ with $f(x)\neq g(x)$. Hence, $\mathscr{F}_A$ is Hausdorff. □

Proposition 2.3.12. *Let X, Y be topological spaces, $(L_n)_{n\in\mathbb{N}}$ be an increasing sequence of quasicompact sets of X whose union is X and $\mathscr{F}$ be the set of maps f of X into Y such that $f|L_n$ is continuous for any $n\in\mathbb{N}$. If there exists a sequence $(U_n)_{n\in\mathbb{N}}$ of open sets of $Y\times Y$ such that*

$$\bigcap_{n\in\mathbb{N}} U_n = \bigcap_{n\in\mathbb{N}} \bar{U}_n = \{(y, y)\,|\,y\in Y\},$$

then $\mathscr{F}$ is a Šmulian space. If, in addition, Y is Φ_2-compact and any point of Y possesses a fundamental system of closed neighbourhoods, then $\mathscr{F}$ is Φ_2-compact.

We may assume that $(U_n)_{n\in\mathbb{N}}$ is decreasing.

Let (I, f) be a Φ_2-net in $\mathscr{F}$ such that $f(\mathfrak{F})$ converges to a $g\in\mathscr{F}$, where $\mathfrak{F}$ denotes the section filter of I. We construct inductively a sequence $(I_n)_{n\in\mathbb{N}}$ of finite subsets of I such that we have for any $n\in\mathbb{N}$:

a) $\iota\in\bigcup_{m<n} I_m,\ \lambda\in I_n \Rightarrow \iota\le\lambda$;

b) for any family $(x_m)_{m\le n}$ in L_n there exists $\iota\in I_n$ such that

$$(g(x_m),\ f_\iota(x_m))\in U_n$$

for any $m\le n$.

Let $n\in\mathbb{N}$ and assume the sequence was constructed up to $n-1$. Let $\lambda\in I$ be an upper bound of $\bigcup_{m<n} I_m$ and $x := (x_m)_{m\le n}$ be a point of L_n^{n+1}. Since g is adherent to the set

$$\{f_\iota \mid \iota\in I,\ \iota\ge\lambda\}$$

there exists $\iota_x\in I$ with $\iota_x\ge\lambda$ and

$$(g(x_m),\ f_{\iota_x}(x_m))\in U_n$$

for any $m\le n$. The restrictions of g and f_{ι_x} to L_n, being continuous,

$$U'_x := \{(y_m)_{m\le n}\in L_n^{n+1} \mid m\le n \Rightarrow (g(y_m), f_{\iota_x}(y_m))\in U_n\}$$

is an open set of L_n^{n+1}, which obviously contains x. Since L_n^{n+1} is quasicompact, there exists a finite subset A of L_n^{n+1} such that

$$L_n^{n+1} = \bigcup_{x\in A} U'_x.$$

We set

$$I_n := \{\iota_x \mid x\in A\}.$$

It is obvious that I_n is a finite subset of I satisfying the above condition a). In order to check b) let $(x_m)_{m\le n}$ be a family in L_n. Then there exists $y\in A$ such that $(x_m)_{m\le n}\in U'_Y$ so we have

$$(g(x_m),\ f_{\iota_y}(x_m))\in U_n$$

for any $m\le n$.

We set $J := \bigcup_{n\in\mathbb{N}} I_n$. J is an upper directed countable subset of I. We denote by $\mathfrak{G}$ the section filter of J and set $f' := f\mid J$. We want to show that g is an adherent point of $f'(\mathfrak{G})$. Let $\iota\in J$, $(x_\lambda)_{\lambda\in L}$ be a finite family in X and, for any $\lambda\in L$, let V_λ be a neighbourhood of $g(x_\lambda)$. There exists $n_0\in\mathbb{N}$ such that:

1) card $L\le n_0+1$;

2) $\lambda \in L \Rightarrow x_\lambda \in L_{n_0}$;

3) $\iota \in \bigcup_{n<n_0} I_m$.

By b) there exists for any $n \geq n_0$ an $\iota_n \in I_n$ such that

$$(g(x_\lambda), f_{\iota_n}(x_\lambda)) \in U_n$$

for any $\lambda \in L$. We want to show that there exists $n \geq n_0$ such that $f_{\iota_n}(x_\lambda) \in V_\lambda$ for any $\lambda \in L$. Assume the contrary. Then there exist $\lambda \in L$ and a strictly increasing sequence $(k(n))_{n\in\mathbb{N}}$ in $\mathbb{N}$ such that $f_{\iota_{k(n)}}(x_\lambda) \notin V_\lambda$ for any $n \in \mathbb{N}$. By a) $(\iota_{k(n)})_{n\in\mathbb{N}}$ is increasing in I. Since (I, f) is a Θ_2-net in $\mathscr{F}$ there exists $h \in \mathscr{F}$ which is adherent to the sequence $(f_{\iota_{k(n)}})_{n\in\mathbb{N}}$. Then $h(x_\lambda)$ is adherent to the sequence $(f_{\iota_{k(n)}}(x_\lambda))_{n\in\mathbb{N}}$. From $f_{\iota_{k(n)}}(x_\lambda) \notin V_\lambda$ we deduce that $h(x_\lambda) \neq g(x_\lambda)$. From

$$(g(x_\lambda), f_{\iota_{k(n)}}(x_\lambda)) \in U_{k(n)}$$

for any $n \in \mathbb{N}$ we get

$$(g(x_\lambda), h(x_\lambda)) \in \bigcap_{n\in\mathbb{N}} \bar{U}_{k(n)}$$

so $g(x_\lambda) = h(x_\lambda)$, which is the expected contradiction. Hence, g is adherent to $f'(\mathfrak{G})$.

Let K be a compact set of Y. From the hypothesis about Y it follows that $\{(y, y) \mid y \in K\}$ is a G_δ-set of $K \times K$. Hence, K is metrizable.

We denote by $\mathscr{G}$ the closure of

$$\{f_\iota \mid \iota \in J\} \cup \{g\}$$

in $\mathscr{F}$. By Lemma 2.3.11 there exists a countable subset A of X such that $\mathscr{G}_A$ is Hausdorff. Hence, by the hypothesis about Y any point h of $\mathscr{G}_A$ possesses a countable set of closed neighbourhoods whose intersection is $\{h\}$. By Proposition 2.3.9 $\mathscr{G}_A$ is a Šmulian space. By Proposition 2.3.7 $\mathscr{G}$ is a Šmulian space. Let f'' be the map

$$J \to \mathscr{G}, \quad \iota \mapsto f_\iota.$$

Then (J, f'') is a Θ_2-net in $\mathscr{G}$ and g is an adherent point of $f''(\mathfrak{G})$. Hence, there exists an increasing sequence $(\iota_n)_{n\in\mathbb{N}}$ in J such that $(f''_{\iota_n})_{n\in\mathbb{N}}$ converges to g. We have proved that $\mathscr{F}$ is a Šmulian space.

Now assume Y has the additional properties. Let Θ be the set of sequences $(x_n)_{n\in\mathbb{N}}$ in X for which there exists $m \in \mathbb{N}$ with $\{x_n \mid n \in \mathbb{N}\} \subset K_m$. By Propositions 1.4.2, 1.4.3, and 1.3.7 $\mathscr{F}$ is the set of $\Phi(\Theta)$-continuous maps of X into Y. By Proposition 2.1.16 $\mathscr{F}$ is Φ_2-compact. □

Proposition 2.3.13. *Let X be a topological space which is a σ-quasicompact set, Y be a topological space possessing a coarser metrizable topology and $\mathscr{F}$ be a set of maps of X into Y such that the restriction of any $f \in \mathscr{F}$ to any quasicompact set of X is continuous. Then $\mathscr{F}$ is a Φ_2-compact Šmulian space.*

Let $(K_n)_{n\in\mathbb{N}}$ be a sequence of quasicompact sets of X whose union is X, Z be the space Y endowed with a coarser metrizable topology, $\mathscr{H}$ be the set of maps f of X into Z such that $f|K_n$ is continuous for any $n\in\mathbb{N}$ and let $\mathscr{G}$ be the set $\mathscr{H}$ endowed with the topology induced by Y^X. By Proposition 2.3.12 and Corollary 2.3.10 a) $\mathscr{H}$ is a Φ_2-compact Šmulian space. By Proposition 2.3.7 $\mathscr{G}$ is a Φ_2-compact Šmulian space. By Proposition 2.3.4 $\mathscr{F}$ is a Φ_2-compact Šmulian space. □

Theorem 2.3.14. *(C. Constantinescu (1973) [2] Corollary 5.16). Let X be a topological space possessing a dense σ-quasicompact set, let Y be a topological space possessing a coarser metrizable topology, and let $\mathscr{F}$ be a set of continuous maps of X into Y. Then $\mathscr{F}$ is a Φ_2-compact Šmulian mioritic space.*

Let $(K_n)_{n\in\mathbb{N}}$ be a sequence of quasicompact sets of X whose union Z is dense in X and let $\mathscr{G}$ be the set $\{f|Z \mid f\in\mathscr{F}\}$. By Proposition 2.3.13 $\mathscr{G}$ is a Φ_2-compact Šmulian space. The map

$$\mathscr{F}_Z \to \mathscr{G}, \quad f \mapsto f|Z$$

being a homeomorphism $\mathscr{F}_Z$ is a Φ_2-compact Šmulian space. By Proposition 2.3.7 $\mathscr{F}$ is a Φ_2-compact Šmulian space and by Proposition 2.3.3 it is a mioritic space. □

Remark. Let $\mathscr{G}$ be a Φ_2-set of $\mathscr{F}$ and let f be an adherent point of $\mathscr{G}$ in Y^X. By the above theorem and by Theorem 2.3.6 a) f belongs to $\mathscr{F}$ and there exists a sequence in $\mathscr{G}$ converging to f. This result was proved by J.D. Pryce (1971) ([1] Theorem 3.2).

Proposition 2.3.15. *Let X be a topological space possessing a sequence of quasicompact sets whose union is dense in X, Θ be a set of sequences in X, $\Phi := \hat{\Phi}(\Theta)$, Y be a uniform space whose topology possesses a coarser metrizable topology and $\mathscr{F}$ be a set of continuous maps of X into Y such that any Cauchy sequence in $\mathscr{F}$ is a Cauchy sequence in $\mathscr{F}_\Phi$. Then the identity map $\mathscr{F} \to \mathscr{F}_\Phi$ is uniformly Φ_4-continuous.*

By Theorem 2.3.14 $\mathscr{F}$ is a mioritic space. Let $(f_n)_{n\in\mathbb{N}}$ be a $\Theta_4(\mathscr{F})$-sequence. By Proposition 1.7.3 a ⇒ c $(f_n)_{n\in\mathbb{N}}$ is a $\Theta_3(\mathscr{F})$-sequence. By the hypothesis $(f_n)_{n\in\mathbb{N}}$ is a $\Theta_3(\mathscr{F}_\Phi)$-sequence. By Theorem 1.8.4 e ⇒ g, $\Phi_4(\mathscr{F}) \subset \Phi_4(\mathscr{F}_\Phi)$ and therefore by Proposition 2.2.10 b ⇒ a the identity map $\mathscr{F} \to \mathscr{F}_\Phi$ is uniformly Φ_4-continuous. □

Proposition 2.3.16. *Let X, Y be topological spaces, A be a σ-quasicompact set of X, g be the inclusion map $\bar{A} \to X$, d be a pseudometric on Y such that the map*

$$Y \to [0, \infty], \quad y' \mapsto d(y, y')$$

is continuous for any $y\in Y$ and Z be the canonical metric space and $h: Y \to Z$ be the canonical map associated to (Y, d). Further, let $\mathscr{F}$ be a set of continuous maps of X into Y, φ be the inclusion map $\mathscr{F} \to Y^X$ and $\mathfrak{F}$ be a Φ_2-filter on $\mathscr{F}$ such that $\varphi(\mathfrak{F})$ converges to an $f\in Y^X$. Then there exists $f'\in\mathscr{F}$ such that

$$h \circ f \circ g = h \circ f' \circ g .$$

Let (I, ι) be a Θ_2-net in $\mathscr{F}$ such that $\iota(\mathfrak{G}) = \mathfrak{F}$, where $\mathfrak{G}$ denotes the section filter of I. We set

$$\mathscr{G} := \{h \circ f'' \circ g \mid f'' \in \mathscr{F}\}$$

and denote by φ' the inclusion map $\mathscr{G} \to Z^{\bar{A}}$ and by ψ, ψ' the maps

$$\mathscr{F} \to \mathscr{G}, \quad f'' \mapsto h \circ f'' \circ g,$$
$$Y^X \to Z^{\bar{A}}, \quad f'' \mapsto h \circ f'' \circ g$$

respectively. Then $\psi' \circ \varphi = \varphi' \circ \psi$, $(I, \psi \circ \iota)$ is a Θ_2-net in $\mathscr{G}$, and $\varphi'(\psi(\iota(\mathfrak{G})))$ converges to $\psi'(f)$. By Theorem 2.3.14 $\mathscr{G}$ is a Φ_2-compact Šmulian space. Hence, $\psi \circ \iota(\mathfrak{G})$ possesses an adherent point f_0 in $\mathscr{G}$ and there exists an increasing sequence $(\iota_n)_{n \in \mathbb{N}}$ in I such that $(\psi \circ \iota(\iota_n))_{n \in \mathbb{N}}$ converges to f_0 (Proposition 2.3.2). Let f' be an adherent point in $\mathscr{F}$ of the sequence $(\iota(\iota_n))_{n \in \mathbb{N}}$. Then $\psi(f')$ is an adherent point of $(\psi \circ \iota(\iota_n))_{n \in \mathbb{N}}$ and therefore $\psi(f') = f_0$. Since $\varphi'(\psi(f')) = \varphi'(f_0)$ is an adherent point of $\varphi'(\psi(\iota(\mathfrak{G})))$ we get

$$\psi'(\varphi(f')) = \varphi'(\psi(f')) = \psi'(f).$$

Hence,

$$h \circ f \circ g = h \circ f' \circ g. \quad \square$$

Proposition 2.3.17. *Let X, Y be topological spaces, $\mathfrak{A}$ be a set of σ-quasicompact sets of X, for each $A \in \mathfrak{A}$ let g_A be the inclusion map $\bar{A} \to X$, let $\mathscr{D}$ be a set of pseudometrics on Y such that the map*

$$Y \to [0, \infty], \quad y' \mapsto d(y, y')$$

is continuous for any $(y, d) \in Y \times \mathscr{D}$ and for any $d \in \mathscr{D}$ let Y_d be the canonical metric space and $h_d : Y \to Y_d$ be the canonical map associated with (Y, d). Further, let $\mathscr{F}$ be a set of continuous maps of X into Y and $\mathscr{G}$ be a Φ_2-closed set of Y^X containing $\mathscr{F}$ such that any $f \in \mathscr{G}$ belongs to $\mathscr{F}$ if for any $(A, d) \in \mathfrak{A} \times \mathscr{D}$ there exists $f' \in \mathscr{F}$ such that

$$h_d \circ f \circ g_A = h_d \circ f' \circ g_A.$$

We have

a) $\mathscr{F}$ is a Φ_2-closed set of Y^X;

b) if Y is Φ_2-compact, then $\mathscr{F}$ is Φ_2-compact;

c) if Y is Φ_2-compact and $\mathscr{F}$ is countably compact, then $\mathscr{F}$ is quasicompact.

d) if Y is regular and Φ_2-compact and if $\mathscr{F}$ is a Φ_2-set, then $\bar{\mathscr{F}}$ is compact.

a) Let us denote by φ the inclusion map $\mathscr{F} \to \mathscr{G}$. Let $\mathfrak{F}$ be a Φ_2-filter on $\mathscr{F}$ such that $\varphi(\mathfrak{F})$ converges to an $f \in \mathscr{G}$. We have to show that $f \in \mathscr{F}$.

Let $(A, d) \in \mathfrak{A} \times \mathscr{D}$. By Proposition 2.3.16 there exists $f' \in \mathscr{F}$ such that

$$h_d \circ f \circ g_A = h_d \circ f' \circ g_A.$$

Since (A, d) is arbitrary we get $f \in \mathscr{F}$.

b) follows from a) and Proposition 1.5.9.

c) By b) $\mathscr{F}$ is Φ_2-compact. By Proposition 1.5.5 c $\Rightarrow$ a $\mathscr{F}$ is a Φ_2-set of itself and therefore $\mathscr{F}$ is quasicompact.

d) Y^X is regular and Φ_2-compact (Proposition 1.5.9) and the assertion follows from Proposition 1.2.10. □

Corollary 2.3.18. *Let E be a Hausdorff locally convex space and let E' be its dual endowed with the topology of pointwise convergence. Assume that any linear form x' on E is continuous if for any weakly σ-compact vector subspace F of E there exists $y' \in E'$ so that $x' = y'$ on $\bar{F}$. Then E' is Φ_2-compact.*

Any weakly σ-compact set of E is enclosed by a weakly σ-compact vector subspace of E (Lemma 2.3.24) and the assertion follows from Proposition 2.3.17 b) and Theorem 1.5.8. □

Corollary 2.3.19. *Let E be a Hausdorff locally convex space and E' be its dual endowed with the topology of pointwise convergence. Assume that any linear form x^* on E' is continuous if for any σ-compact vector subspace F' of E' there exists $x \in E$ with $x = x^*$ on $\bar{F}'$. Then E endowed with its weak topology is Φ_2-compact. In particular, any Hausdorff complete locally convex space is weakly Φ_2-compact.*

The first assertion follows immediately from the preceding result by inverting E and E'. The second one follows from the first and from Grothendieck's criterium for completeness (H. H. Schaefer [1] Corollary 2 of Theorem IV 6.2). □

Remark. In particular, any weakly Φ_2-set of E is weakly relatively compact. This result was proved by J. D. Pryce (1971) ([1] Theorem 4.4). The same result was proved by J. Dieudonné and L. Schwartz (1950) ([1] Proposition 18) in the case when E is a Fréchet space or E an LF-space. W. F. Eberlein proved (1947) ([1] Theorem), that any weakly closed weakly Φ_2-set of a Banach space is weakly compact. The last assertion of the corollary was proved by C. Constantinescu (1973) ([2] Corollary 6.16).

Lemma 2.3.20. *Let E be a locally convex space whose one point sets are G_δ-sets and E' be its dual endowed with the topology of pointwise convergence. Then E' possesses a dense σ-compact set.*

Let $(U_n)_{n \in \mathbb{N}}$ be a sequence of convex closed neighbourhoods of the origin of E whose intersection is equal to $\{0\}$. If for any $n \in \mathbb{N}$, U_n^0 denotes the polar set of U_n with respect to the duality $\langle E, E' \rangle$, then $(U_n^0)_{n \in \mathbb{N}}$ is a sequence of compact sets of E'. Let $x \in E$ be equal to 0 on $\bigcup_{n \in \mathbb{N}} U_n^0$. Then $x \in U_n$ for any $n \in \mathbb{N}$ and therefore $x = 0$. This shows that $\bigcup_{n \in \mathbb{N}} U_n^0$ is dense in E'. □

The next result will not be used later.

Corollary 2.3.21. *Let E, F be Hausdorff locally convex spaces such that the one-point sets of F are G_δ-sets, F_σ be the vector space F endowed with its weak topology and $\mathscr{L}$ be the set of continuous linear maps of E into F_σ endowed with the topology induced by F_σ^E. Then $\mathscr{L}$ is Φ_2-compact.*

Let E', F' be the duals of E and F respectively endowed with the topologies of pointwise convergence and let $\mathscr{L}'$ be the set of continuous linear maps of F' into E' endowed with the topology of pointwise convergence. For any $u \in \mathscr{L}$ let $u' \in \mathscr{L}'$ be the adjoint of u. The map

$$\mathscr{L} \to \mathscr{L}', \quad u \mapsto u',$$

being a homeomorphism, it is sufficient to prove that $\mathscr{L}'$ is Φ_2-compact.

By Lemma 2.3.20 F' possesses a dense σ-compact set. Let f be a map of F' into E' such that for any $x \in E$ there exists $u' \in \mathscr{L}'$ with $x \circ f = x \circ u'$. Then $f \in \mathscr{L}'$ and the assertion follows from Proposition 2.3.17 b). □

Theorem 2.3.22. *Let E be a locally convex space such that its onepoint sets are G_δ-sets. Then E endowed with any topology finer than its weak topology is a Φ_2-compact Šmulian mioritic space. In particular, any weakly Φ_2-set of E is weakly relatively compact and sequentially dense and any weakly countably compact set of E is weakly compact and weakly sequentially compact.*

Let E' be the dual of E. By Lemma 2.3.20 E'_E possesses a dense σ-compact set. By Theorem 2.3.14 E endowed with its weak topology is a Φ_2-compact Šmulian space. By the Propositions 2.3.7 and 2.3.3 E endowed with any topology finer than the weak topology is a Φ_2-compact Šmulian mioritic space. The last assertion follows from Proposition 1.2.10 and Theorem 2.3.6. □

Remark. V. Šmulian proved (1940) ([1] Satz 5) that any sequence contained in a weakly Φ_2-set of a Banach space possesses a weakly convergent subsequence. J. Dieudonné and L. Schwartz proved (1950) ([1] Proposition 17) that any Fréchet space is a Šmulian space. A. Grothendieck showed (1952) ([1] Proposition 6) that under the hypothesis of Theorem 2.3.22 E endowed with its weak topology is a mioritic space. The above result was proved by C. Constantinescu (1973) [(2] Corollary 6.3.).

Proposition 2.3.23. *Let E be a locally convex space possessing a weakly σ-compact weakly dense set and let E' be its dual endowed with the topology of pointwise convergence. Then any equicontinuous set of E' is a Φ_1-set of E'.*

Let A' be an equicontinuous set of E'. By Proposition 1.5.5 c $\Rightarrow$ a A' is a Φ_2-set of E'. By Theorem 2.3.14 E' is a mioritic space and therefore by Proposition 1.7.2 a $\Rightarrow$ e A' is a Φ_1-set of E'. □

Lemma 2.3.24. *Let E be a Hausdorff locally convex space. Any σ-compact set of E is enclosed in a σ-compact vector subspace of E.*

Let $(K_n)_{n\in\mathbb{N}}$ be a sequence of compact sets of E. We set

$$L_n := [-n, n]^n \times \Big(\bigcup_{m=1}^{n} K_m\Big)^n,$$

$$\varphi_n : L_n \to E, \quad ((\alpha_m)_{1\le m\le n}, (x_m)_{1\le m\le n}) \mapsto \sum_{m=1}^{n} \alpha_m x_m$$

for any $n\in\mathbb{N}$. Then for any $n\in\mathbb{N}$, L_n is compact, φ_n is continuous, and consequently $\varphi_n(L_n)$ is compact. Hence, $\bigcup_{n\in\mathbb{N}} \varphi_n(L_n)$ is σ-compact, and it is easy to see, it is even a vector subspace of E containing $\bigcup_{n\in\mathbb{N}} K_n$. □

§ 2.4 Constructions with spaces of functions

Proposition 2.4.1. *Let X be a set, Y be a uniform space, and let $(X_\lambda, g_\lambda, \Phi_\lambda)_{\lambda\in L}$, $(Y_\mu, h_\mu)_{\mu\in M}$ be families such that:*

a) X_λ is a set, $g_\lambda \in X^{X_\lambda}$, and Φ_λ is a set of filters on X_λ for any $\lambda\in L$:

b) Y_μ is a uniform space and $h_\mu \in (Y_\mu)^Y$ for any $\mu\in M$;

c) the uniformity of Y is the initial uniformity on Y with respect to the family $(h_\mu)_{\mu\in M}$.

We set

$$\Phi := \{g_\lambda(\mathfrak{F}) \mid \lambda\in L, \mathfrak{F}\in\Phi_\lambda\}$$

and, for each $(\lambda, \mu)\in L\times M$, denote by $\varphi_{\lambda\mu}$ the map

$$Y^X \to (Y_\mu)^{X_\lambda}_{\Phi_\lambda}, \quad f \mapsto h_\mu \circ f \circ g_\lambda .$$

Then the uniformity of Φ-convergence on Y^X is the initial uniformity on Y^X with respect to the family $(\varphi_{\lambda\mu})_{(\lambda,\mu)\in L\times M}$.

Let $(\lambda, \mu)\in L\times M$, $\mathfrak{F}\in\Phi_\lambda$ and U be an entourage of Y_μ. Then $V := \overset{-1}{\widehat{h_\mu \times h_\mu}}(U)$ is an entourage of Y. Let $f', f''\in Y^X$. The following assertions are equivalent:

$$(f', f'')\in\mathscr{W}(g_\lambda(\mathfrak{F}), V),$$
$$\{x\in X \mid (f'(x), f''(x))\in V\}\in g_\lambda(\mathfrak{F}),$$
$$\{x\in X_\lambda \mid (h_\mu\circ f'\circ g_\lambda(x), h_\mu\circ f''\circ g_\lambda(x))\in U\}\in\mathfrak{F},$$
$$(\varphi_{\lambda\mu}(f'), \varphi_{\lambda\mu}(f''))\in\mathscr{W}(\mathfrak{F}, U).$$

This shows that the uniformity of Φ-convergence on Y^X is the initial uniformity on Y^X with respect to the family $(\varphi_{\lambda\mu})_{(\lambda,\mu)\in L\times M}$. □

Proposition 2.4.2. *Let X be a set, Y be a uniform space, $\mathscr{F}$ be a subset of Y^X, $i\in\{1, 2, 3, 4, 5\}$ and $(X_\lambda, g_\lambda, \Phi_\lambda)_{\lambda\in L}$, $(Y_\mu, h_\mu)_{\mu\in M}$ and $(\mathscr{F}_{\lambda\mu})_{(\lambda,\mu)\in L\times M}$ be families such* *that:*

a) X_λ *is a set,* $g_\lambda \in X^{X_\lambda}$, *and* Φ_λ *is a set of filters on* X_λ *for any* $\lambda \in L$;

b) Y_μ *is a uniform space and* $h_\mu \in (Y_\mu)^Y$ *for any* $\mu \in M$;

c) the uniformity of Y *is the initial uniformity on* Y *with respect to the family* $(h_\mu)_{\mu \in M}$;

d) $\mathscr{F}_{\lambda\mu} \subset (Y_\mu)^{X_\lambda}$ *and the identity map* $\mathscr{F}_{\lambda\mu} \to (\mathscr{F}_{\lambda\mu})_{\Phi_\lambda}$ *is uniformly* Φ_i*-continuous for any* $(\lambda, \mu) \in L \times M$;

e) $h_\mu \circ f \circ g_\lambda \in \mathscr{F}_{\lambda\mu}$ *for any* $(\lambda, \mu, f) \in L \times M \times \mathscr{F}$.

We set

$$\Phi := \{g_\lambda(\mathfrak{F}) \mid \lambda \in L, \mathfrak{F} \in \Phi_\lambda\}, \quad A := \bigcup_{\lambda \in L} g_\lambda(X_\lambda).$$

Then the identity map $\mathscr{F}_A \to \mathscr{F}_\Phi$ *is uniformly* Φ_i*-continuous.*

For each $(\lambda, \mu) \in L \times M$ we denote by $\varphi_{\lambda\mu}$, $\varphi'_{\lambda\mu}$, and $\psi_{\lambda\mu}$ the maps

$$\mathscr{F}_A \to \mathscr{F}_{\lambda\mu}, \quad f \mapsto h_\mu \circ f \circ g_\lambda,$$
$$\mathscr{F}_\Phi \to (\mathscr{F}_{\lambda\mu})_{\Phi_\lambda}, \quad f \mapsto h_\mu \circ f \circ g_\lambda,$$
$$\mathscr{F}_{\lambda\mu} \to (\mathscr{F}_{\lambda\mu})_{\Phi_\lambda}, \quad f \mapsto f$$

respectively. If ψ denotes the identity map $\mathscr{F}_A \to \mathscr{F}_\Phi$, then

$$\psi_{\lambda\mu} \circ \varphi_{\lambda\mu} = \varphi'_{\lambda\mu} \circ \psi$$

for any $(\lambda, \mu) \in L \times M$. Since $\psi_{\lambda\mu}$ is uniformly Φ_i-continuous and since $\varphi_{\lambda\mu}$ is uniformly continuous it follows that $\varphi'_{\lambda\mu} \circ \psi$ is uniformly Φ_i-continuous for any $(\lambda, \mu) \in L \times M$. By Proposition 2.4.1 the uniformity of $\mathscr{F}_\Phi$ is the initial uniformity on $\mathscr{F}$ with respect to the family $(\varphi'_{\lambda\mu})_{(\lambda, \mu) \in L \times M}$. Hence, ψ is uniformly Φ_i-continuous (Proposition 1.8.13). □

Theorem 2.4.3. *Let* X *be a set,* Y *be a uniform space and let* $(X_\lambda, g_\lambda, \Phi_\lambda)_{\lambda \in L}$, $(Y_\mu, h_\mu)_{\mu \in M}$, *and* $(\mathscr{F}_{\lambda\mu})_{(\lambda, \mu) \in L \times M}$ *be families such that:*

a) X_λ *is a uniform space possessing a dense* σ*-quasicompact set,* $g_\lambda \in X^{X_\lambda}$, *and* $\Phi_\lambda \subset \hat{\Phi}_5(X_\lambda)$ *for any* $\lambda \in L$;

b) Y_μ *is a pseudo-metrizable uniform space and* $h_\mu \in (Y_\mu)^Y$ *for any* $\mu \in M$;

c) the uniformity of Y *is the initial uniformity on* Y *with respect to the family* $(h_\mu)_{\mu \in M}$;

d) $\mathscr{F}_{\lambda\mu}$ *is a set of continuous maps of* X_λ *into* Y_μ *and any Cauchy sequence in* $\mathscr{F}_{\lambda\mu}$ *is uniformly equicontinuous if it is equicontinuous at any point of the complement of a meagre set of* X_λ *for any* $(\lambda, \mu) \in L \times M$.

We set

$$\Phi := \{g_\lambda(\mathfrak{F}) \mid \lambda \in L,\ \mathfrak{F} \in \Phi_\lambda\}, \quad A := \bigcup_{\lambda \in L} g_\lambda(X_\lambda),$$

$$\mathscr{F} := \{f \in Y^X \mid (\lambda, \mu) \in L \times M \Rightarrow h_\mu \circ f \circ g_\lambda \in \mathscr{F}_{\lambda\mu}\}$$

Then the identity map $\mathscr{F}_A \to \mathscr{F}_\Phi$ *is uniformly* Φ_4*-continuous.*

For any $\mu \in M$ let Y'_μ be the canonical metric space associated with Y_μ, φ_μ be the canonical map $Y_\mu \to Y'_\mu$ associated with Y_μ, and h'_μ be the map $\varphi_\mu \circ h_\mu$. We set

$$\mathscr{F}'_{\lambda\mu} := \{\varphi_\mu \circ f \mid f \in \mathscr{F}_{\lambda\mu}\}$$

for any $(\lambda, \mu) \in L \times M$. Then for any $(\lambda, \mu) \in L \times M$ any Cauchy sequence in $\mathscr{F}'_{\lambda\mu}$ is uniformly equicontinuous if it is equicontinuous at any point of the complement of a meagre set of X_λ and

$$\mathscr{F} = \{f \in Y^X \mid (\lambda, \mu) \in L \times M \Rightarrow h'_\mu \circ f \circ g_\lambda \in \mathscr{F}'_{\lambda\mu}\}.$$

Let $(\lambda, \mu) \in L \times M$ and $(f_n)_{n \in \mathbb{N}}$ be a Cauchy sequence in $\mathscr{F}'_{\lambda\mu}$. By Proposition 2.1.8 $(f_n)_{n \in \mathbb{N}}$ is equicontinuous at any point of the complement of a meagre set of X_λ. From the hypothesis we deduce that $(f_n)_{n \in \mathbb{N}}$ is uniformly equicontinuous. Hence, $(f_n)_{n \in \mathbb{N}}$ is a Cauchy sequence in $(\mathscr{F}'_{\lambda\mu})_{\hat{\Phi}_\lambda}$ (Proposition 2.1.6 and Corollary 1.5.12). By Proposition 2.3.15 the identity map $\mathscr{F}'_{\lambda\mu} \to (\mathscr{F}'_{\lambda\mu})_{\Phi_\lambda}$ is uniformly Φ_4-continuous. By Proposition 2.4.2 the identity map $\mathscr{F}_A \to \mathscr{F}_\Phi$ is uniformly Φ_4-continuous. □

§ 2.5 Spaces of parametrized functions

Throughout this section X shall denote a set, Y a uniform space, $\mathscr{F}$ a subset of Y^X, T a uniform (topological) space, and i a number belonging to $\{1, 2, 3, 4, 5\}$ (to $\{1, 2\}$).

Proposition 2.5.1. *Let Φ be a set of filters on T and let f be a map of $X \times T$ into Y such that $f(\cdot, t) \in \mathscr{F}$ for any $t \in T$. We denote by g the map*

$$T \to \mathscr{F}, \quad t \mapsto f(\cdot, t)$$

and by Ψ the set $\{g(\mathfrak{F}) \mid \mathfrak{F} \in \Phi\}$. Then the map

$$Y_\Psi^{\mathscr{F}} \to Y_\Phi^T, \quad h \mapsto h \circ g$$

is uniformly continuous.

Let U be an entourage of Y, $\mathfrak{F} \in \Phi$ and $(h, h') \in \mathscr{W}(g(\mathfrak{F}), U; \mathscr{F}, Y)$. Then

$$\mathscr{G} := \{f' \in \mathscr{F} \mid (h(f'), h'(f')) \in U\} \in g(\mathfrak{F})$$

so

$$\{t \in T \mid (h \circ g(t), h' \circ g(t)) \in U\} = \overset{-1}{g}(\mathfrak{G}) \in \mathfrak{F},$$
$$(h \circ g, h' \circ g) \in \mathscr{W}(\mathfrak{F}, U; T, Y).$$

Since $g(\mathfrak{F}) \in \Psi$, the map

$$Y^{\mathscr{F}}_{\Psi} \to Y^{T}_{\Phi}, \quad h \mapsto h \circ g$$

is uniformly continuous. □

Proposition 2.5.2. *Let f be a map of $X \times T$ into Y such that $f(\cdot, t) \in \mathscr{F}$ for any $t \in T$ and such that $f(x, \cdot)$ is uniformly Φ_i-continuous (Φ_i-continuous) for any $x \in X$ and let g be the map*

$$T \to \mathscr{F}, \quad t \mapsto f(\cdot, t).$$

We set $\Phi := \hat{\Phi}_i(T)$ and $\Psi := \hat{\Phi}_i(\mathscr{F})$. Then g is uniformly Φ_i-continuous (Φ_i-continuous) and the map

$$Y^{\mathscr{F}}_{\Psi} \to Y^{T}_{\Phi}, \quad h \mapsto h \circ g$$

is uniformly continuous.

For any $x \in X$ let φ_x denote the map

$$\mathscr{F} \to Y, \quad f' \mapsto f'(x).$$

Then the uniformity of $\mathscr{F}$ is the initial uniformity with respect to the family $(\varphi_x)_{x \in X}$. By Proposition 1.8.13 g is uniformly Φ_i-continuous (Φ_i-continuous). By Theorem 1.8.4 a ⇒ i (Proposition 1.8.9 d)) $g(\mathfrak{F}) \in \Psi$ for any $\mathfrak{F} \in \Phi$. By Proposition 2.5.1 the map

$$Y^{\mathscr{F}}_{\Psi} \to Y^{T}_{\Phi}, \quad h \mapsto h \circ g$$

is uniformly continuous. □

Corollary 2.5.3. *Let f be a map of $X \times T$ into Y such that $f(x, \cdot)$ is uniformly Φ_i-continuous (Φ_i-continuous) for any $x \in X$. If $\mathscr{F}$ is a Φ_i-closed (strongly Φ_i-closed) set of Y^X, then $\{t \in T \mid f(\cdot, t) \in \mathscr{F}\}$ is a Φ_i-closed (strongly Φ_i-closed) set of T.*

By Proposition 2.5.2 the map

$$T \to Y^X, \quad t \mapsto f(\cdot, t)$$

is uniformly Φ_i-continuous (Φ_i-continuous) and the assertions follow from Proposition 1.8.17 and Proposition 1.8.18. □

Theorem 2.5.4. *Let X be a uniform (topological) space, Θ be a set of sequences in X (containing any constant sequence in X), $\mathfrak{S}$ be a set of subsets of X such that for any Θ-sequence $(x_n)_{n \in \mathbb{N}}$ there exists $S \in \mathfrak{S}$ for which the set $\{x_n \in \mathbb{N} \mid x_n \in S\}$ is infinite, let $\Phi := \hat{\Phi}_i(T)$ and f be a map of $X \times T$ into Y such that $f(\cdot, t) \in \mathscr{F}$ for any $t \in T$ and* *such that $f(x,\cdot)$ is uniformly Φ_i-continuous (Φ_i-continuous) for any $x \in X$. If the identity map $\mathscr{F} \to \mathscr{F}_{\mathfrak{S}}$ is uniformly Φ_i-continuous (Φ_i-continuous) and if any $g \in \mathscr{F}$ is uniformly $\Phi(\Theta)$-continuous ($\Phi(\Theta)$-continuous), then:*

a) the map

$$X \to Y_\Phi^T, \quad x \mapsto f(x,\cdot)$$

is uniformly $\Phi(\Theta)$-continuous ($\Phi(\Theta)$-continuous);

b) if Ψ denotes the set of filters on $X \times T$ the projections of which belong to $\Phi(\Theta)$ and $\Phi_i(T)$ respectively, then f is uniformly Ψ-continuous (Ψ-continuous):

c) if $\Theta \supset \Theta_i(X)$, then f is uniformly Φ_i-continuous (Φ_i-continuous).

a) We set $\Psi := \hat{\Phi}_i(\mathscr{F})$ and denote by φ and g the maps

$$\mathscr{F} \times X \to Y, \quad (h, x) \mapsto h(x),$$

$$T \to \mathscr{F}, \quad t \mapsto f(\cdot, t)$$

respectively. By Proposition 2.5.2 g is uniformly Φ_i-continuous (Φ_i-continuous) and the map

$$Y_\Psi^{\mathscr{F}} \to Y_\Phi^T, \quad h \to h \circ g$$

is uniformly continuous. By Theorem 1.8.4 a $\Rightarrow$ i (Proposition 1.8.9 d)) $\Psi \subset \hat{\Phi}_5(\mathscr{F}_{\mathfrak{S}})$ and therefore by Proposition 2.2.9 b) the map

$$X \to Y_\Psi^{\mathscr{F}}, \quad x \mapsto \varphi(\cdot, x)$$

is uniformly $\Phi(\Theta)$-continuous ($\Phi(\Theta)$-continuous). Hence, the map

$$X \to Y_\Phi^T, \quad x \mapsto f(x,\cdot)$$

is uniformly $\Phi(\Theta)$-continuous ($\Phi(\Theta)$-continuous).

b) Let $\mathfrak{H} \in \hat{\Psi} \cap \Phi_c(X \times T)$ ($\mathfrak{H} \in \hat{\Psi} \cap \Phi_C(X \times T)$) and $\mathfrak{F}$, $\mathfrak{G}$ be its projections on T an X respectively. Then $\mathfrak{F} \in \hat{\Phi}_i(T) \cap \Phi_C(T)$ ($\mathfrak{F} \in \hat{\Phi}_i(T) \cap \Phi_c(T)$) and $\mathfrak{G} \in \hat{\Phi}(\Theta) \cap \Phi_C(X)$ ($\mathfrak{G} \in \hat{\Phi}(\Theta) \cap \Phi_c(X)$) (Propositions 1.4.2, 1.1.3, 1.1.10). Since g and the identity map $\mathscr{F} \to \mathscr{F}_{\mathfrak{S}}$ are uniformly Φ_i-continuous (Φ_i-continuous) we get $g(\mathfrak{F}) \in \hat{\Phi}_i(\mathscr{F}_{\mathfrak{S}}) \cap \Phi_C(\mathscr{F}_{\mathfrak{S}})$, ($g(\mathfrak{F}) \in \hat{\Phi}_i(\mathscr{F}_{\mathfrak{S}}) \cap \Phi_c(\mathscr{F}_{\mathfrak{S}})$) (Corollary 1.8.5 and Theorem 1.8.4 a $\Rightarrow$ c (Corollary 1.8.10 and Proposition 1.8.9 d))). By Proposition 2.2.9 a) $f(\mathfrak{G} \times \mathfrak{F}) = \varphi(g(\mathfrak{F}) \times \mathfrak{G})$ is a Cauchy filter. Hence, $f(\mathfrak{H})$ is a Cauchy filter and therefore f is uniformly Ψ-continuous (Ψ-continuous) (Proposition 1.3.4).

c) Since $\Phi_i(X \times T) \subset \Psi$ the assertion follows immediately from b). □

The next result will not be used later.

Proposition 2.5.5. *Assume Y is Hausdorff and $\mathscr{F}$ is Φ_i-closed in Y^X. Let f be a map of $X \times T$ into Y such that $f(x,\cdot)$ is uniformly Φ_i-continuous (Φ_i-continuous) for any $x \in X$ and such that $f(\cdot, t) \in \mathscr{F}$ for any $t \in T$ and let $\mathfrak{F} \in \hat{\Phi}_i(T)$. If $f(x, \mathfrak{F})$ converges for any $x \in X$, then the map*

$$X \to Y, \quad x \mapsto \lim f(x, \mathfrak{F})$$

belongs to $\mathscr{F}$. *The filter* $f(x, \mathfrak{F})$ *converges for any* $x \in X$ *if one of the following assertions holds: a) Y is* Φ_i*-compact and* $\mathfrak{F}$ *is an ultrafilter; b) Y is* Φ_i*-compact, T is a uniform space and* $\mathfrak{F}$ *is a Cauchy filter; c) Y is sequentially complete, T is a uniform space and* $\mathfrak{F}$ *is a Cauchy filter with a countable base.*

By Proposition 2.5.2 the map

$$T \to \mathscr{F}, \quad t \mapsto f(\cdot, t)$$

is uniformly Φ_i-continuous (Φ_i-continuous) and therefore the image of $\mathfrak{F}$ in $\mathscr{F}$ with respect to this map belongs to $\hat{\Phi}_i(\mathscr{F})$ (Theorem 1.8.4 a ⇒ i (Proposition 1.8.9d))). Since $\mathscr{F}$ is a Φ_i-closed set of Y^X the map

$$X \to Y, \quad x \mapsto \lim f(x, \mathfrak{F})$$

belongs to $\mathscr{F}$. The last assertion follows from Theorem 1.8.4 a ⇒ i (Proposition 1.8.9 d)) and Proposition 1.4.2. □

Chapter 3: Spaces of supersummable families

§ 3.1 The set $\mathcal{G}(I, G)$

The sum of a family in a topological additive group only may be defined usefully if the topology is Hausdorff. Nevertheless, many properties concerning the spaces of supersummable families still hold without this property. We present them in this section without the Hausdorff hypothesis.

Definition 3.1.1. *For any set I we denote by $\mathfrak{P}_f(I)$ the set of finite subsets of I and by $\mathfrak{P}_c(I)$ the set of countable subsets of I. $\mathfrak{P}(I)$, $\mathfrak{P}_c(I)$, and $\mathfrak{P}_f(I)$ will be always ordered by the inclusion relation. If we identify the subsets of I with their characteristic functions, then $\mathfrak{P}(I)$ is identified with $\{0,1\}^I$. We shall endow $\mathfrak{P}(I)$ with the corresponding product uniformity of $\{0,1\}^I$ and $\mathfrak{P}_f(I)$, $\mathfrak{P}_c(I)$ with the induced uniformities.*

By the Tychonoff theorem $\mathfrak{P}(I)$ is compact. A subset U of $\mathfrak{P}(I)$ is open iff for any $J \in U$ there exists a finite subset K of I such that

$$\{L \in \mathfrak{P}(I) \mid L \cap K = J \cap K\} \subset U .$$

Proposition 3.1.2. *For any set I the set $\mathfrak{P}_f(I)$ is Φ_3-dense and loosely Φ_1-dense in $\mathfrak{P}(I)$.*

Let $J \subset I$, $\mathfrak{F}$ be the section filter of $\mathfrak{P}_f(J)$ and φ, ψ be the inclusion maps $\mathfrak{P}_f(J) \to \mathfrak{P}_f(I)$ and $\mathfrak{P}_f(I) \to \mathfrak{P}(I)$ respectively. Then $(\mathfrak{P}_f(J), \varphi)$ is a $\Theta_1(\mathfrak{P}_f(I), \mathfrak{P}(I))$-net in $\mathfrak{P}_f(I)$ and therefore $\varphi(\mathfrak{F})$ is a $\Phi(\Theta_1(\mathfrak{P}_f(I), \mathfrak{P}(I)))$-filter on $\mathfrak{P}_f(I)$. Since $\psi(\varphi(\mathfrak{F}))$ converges to J any strongly Φ_1-closed set of $\mathfrak{P}(I)$ containing $\mathfrak{P}_f(I)$ contains J. Since J is arbitrary, $\mathfrak{P}_f(I)$ is loosely Φ_1-dense in $\mathfrak{P}(I)$. By Proposition 1.5.20 b) $\mathfrak{P}_f(I)$ is Φ_3-dense in $\mathfrak{P}(I)$. □

Proposition 3.1.3. *Let I be a set. Any Φ_1-continuous map of $\mathfrak{P}(I)$ into a topological space X is continuous if any point of X possesses a fundamental system of closed neighbourhoods.*

For any $J \subset I$, let $\mathfrak{F}(J)$ be the filter on $\mathfrak{P}(I)$ generated by the section filter of $\mathfrak{P}_f(J)$. Then $\mathfrak{F}(J)$ is a Φ_1-filter on $\mathfrak{P}(I)$ converging to J.

Let φ be a Φ_1-continuous map of $\mathfrak{P}(I)$ into X and U be a closed neighbourhood of $\varphi(I)$. Since $\varphi(\mathfrak{F}(I))$ converges to $\varphi(I)$ there exists $K \in \mathfrak{P}_f(I)$ such that $\varphi(J) \in U$ for any $J \in \mathfrak{P}_f(I)$ so that $J \supset K$. Let J_0 be a subset of I containing K. Then $\varphi(\mathfrak{F}(J_0))$ converges to $\varphi(J_0)$. Since U is closed we get $\varphi(J_0) \in U$ by the above result. Hence, φ is continuous at I.

Since for any point of $\mathfrak{P}(I)$ there exists a homeomorphism of $\mathfrak{P}(I)$ into itself mapping this point into I, φ is continuous at any point of $\mathfrak{P}(I)$. □

Definition 3.1.4. *For any set I and for any additive group G we denote by $\underline{\mathscr{H}(I,G)}$ the set of maps g of $\mathfrak{P}(I)$ into G such that*

$$g(J\cup K) = g(J)+g(K)$$

for any disjoint subsets J, K of I. If G is a group endowed with a topology, then we denote by $\underline{\mathscr{G}(I,G)}$ the set of continuous maps of $\mathscr{H}(I,G)$.

If φ is a group homomorphism of G into H, then for any set I and for any $g\in\mathscr{H}(I,G)$ the map $\varphi\circ g$ belongs to $\mathscr{H}(I,H)$.

Proposition 3.1.5. *Let G, H be topological additive groups, u be a Φ_1-continuous (sequentially continuous) group homomorphism of G into H and I be a set (a countable set). Then $u\circ f\in\mathscr{G}(I,H)$ for any $f\in\mathscr{G}(I,G)$.*

If I is countable then the assertion follows from the fact that $\mathfrak{P}(I)$ is metrizable. If I is not countable then the assertion follows from Proposition 3.1.3. □

Theorem 3.1.6. *Let G be a topological additive group and I be a set. If a subset of $\mathscr{H}(I,G)$ is equicontinuous at a point of $\mathfrak{P}(I)$, then it is uniformly equicontinuous.*

Let $\mathscr{F}$ be a subset of $\mathscr{H}(I,G)$ which is equicontinuous at a point J_0 of $\mathfrak{P}(I)$ and let U be a 0-neighbourhood in G. Let V be another 0-neighbourhood in G such that $V-V\subset U$. Since $\mathscr{F}$ is equicontinuous at J_0 there exists $K\in\mathfrak{P}_f(I)$ such that $f(J)-f(J_0)\in V$ for any $f\in\mathscr{F}$ and any $J\subset I$ so that $J\cap K = J_0\cap K$. Let $J', J''\in\mathfrak{P}(I)$ such that $J'\cap K = J''\cap K$. We have

$$\begin{aligned}&f(J')-f(J'') =\\ &= f(J'\cap K)+f(J'\backslash K)+f(J_0\cap K)-f(J''\cap K)-f(J''\backslash K)-f(J_0\cap K) =\\ &= f((J'\backslash K)\cup(J_0\cap K))-f(J_0)-f((J''\backslash K)\cup(J_0\cap K))+f(J_0)\in V-V\subset U\end{aligned}$$

for any $f\in\mathscr{F}$. Hence, $\mathscr{F}$ is uniformly equicontinuous. □

Lemma 3.1.7. *Let X be a topological space, Y be a metrizable uniform space and φ be a map of X into Y such that, for any entourage U of Y, there exists a countable closed covering $\mathfrak{M}$ of X such that $\varphi(M)\times\varphi(M)\subset U$ for any $M\in\mathfrak{M}$. Then the set of points where φ is discontinuous is meagre.*

Let $\mathfrak{U}$ be a countable fundamental system of entourages of Y and for any $U\in\mathfrak{U}$ let $\mathfrak{M}(U)$ be a countable closed covering of X such that $\varphi(M)\times\varphi(M)\subset U$ for any $M\in\mathfrak{M}(U)$. Then the set

$$A := \bigcap_{U\in\mathfrak{U}}\ \bigcup_{M\in\mathfrak{M}(U)} \overset{\circ}{M}$$

is the complement of a meagre set.

Let $x\in A$ and V be a neighbourhood of $\varphi(x)$. Then there exists $U\in\mathfrak{U}$ such that $U(\varphi(x))\subset V$. Since $x\in A$ there exists $M\in\mathfrak{M}(U)$ with $x\in\overset{\circ}{M}$. Hence, M is a neighbourhood of x. We get for any $y\in M$

$$(\varphi(x), \varphi(y)) \in \varphi(M) \times \varphi(M) \subset U, \quad \varphi(y) \in U(\varphi(x)) \subset V.$$

We have proved that φ is continuous at x. □

Lemma 3.1.8. *Let Y be a uniform space and $(\varphi_\iota)_{\iota \in I}$ be a family of maps of Y into metrizable uniform spaces such that the uniformity of Y is the initial uniformity with respect to the family $(\varphi_\iota)_{\iota \in I}$. Further, let X be a Baire space and φ be a map of X into Y such that: a) for any $\iota \in I$ the set of points where $\varphi_\iota \circ \varphi$ is not continuous either is empty or it is not a meagre set; b) for any entourage U of Y there exists a countable closed covering $\mathfrak{M}$ of X such that we have $\varphi(M) \times \varphi(M) \subset U$ for any $M \in \mathfrak{M}$. Then φ is continuous.*

Let $\iota \in I$ and let U_ι be an entourage of Y_ι. Then $\overset{-1}{\widehat{\varphi_\iota \times \varphi_\iota}}(U_\iota)$ is an entourage of Y and by b) there exists a countable closed covering $\mathfrak{M}$ of X such that

$$\varphi(M) \times \varphi(M) \subset \overset{-1}{\widehat{\varphi_\iota \times \varphi_\iota}}(U_\iota)$$

for any $M \in \mathfrak{M}$. We deduce

$$\varphi_\iota \circ \varphi(M) \times \varphi_\iota \circ \varphi(M) \subset U_\iota$$

for any $M \in \mathfrak{M}$. By Lemma 3.1.7 the set of points of X at which $\varphi_\iota \circ \varphi$ is discontinuous is meagre. By a) $\varphi_\iota \circ \varphi$ is continuous.

Since ι is arbitrary and since the topology of Y is the initial topology with respect to the family $(\varphi_\iota)_{\iota \in I}$ we deduce that φ is continuous. □

Proposition 3.1.9. *Let G be a topological additive group, I be a set and $f \in \mathscr{H}(I, G)$. If for any 0-neighbourhood U in G there exists a countable closed covering $\mathfrak{M}$ of $\mathfrak{P}(I)$ such that $f(J) - f(K) \in U$ for any $J, K \in M$ and any $M \in \mathfrak{M}$, then f is continuous.*

By Lemma 4.1.20 there exists a family of group homomorphisms of G into metrizable topological additive groups such that the topology of G is the initial topology with respect to this family. Now we apply Lemma 3.1.8 by replacing Y by G, X by $\mathfrak{P}(I)$, and φ by f. Condition a) of Lemma 3.1.8 is satisfied by Theorem 3.1.6. Condition b) of that Lemma is obviously satisfied. Hence, f is continuous. □

Remark. The Lemmata 3.1.7, 3.1.8 and the above proposition still hold (together with the same proof) if we replace the assumption that the sets of $\mathfrak{M}$ are closed by the weaker one that $\bar{M} \setminus M$ is meagre for any $M \in \mathfrak{M}$.

Proposition 3.1.10. *Let G be a topological additive group, A be a strongly Φ_1-closed set of G and $\mathfrak{F} \in \Phi_1(G)$ such that there exists $B \in \mathfrak{F}$ so that $B - B \subset A$. Then $x + A \in \mathfrak{F}$ for any limit point x of $\mathfrak{F}$. If, in addition, $\mathfrak{F}$ has a countable base, then we may replace the hypothesis "A is strongly Φ_1-closed" by "A is sequentially closed".*

Let $y \in B$. We have $B \subset y - A$. Since $\mathfrak{F}$ induces a Φ_1-filter on $y - A$ and since $y - A$ is Φ_1-closed we get $x \in y - A$. We deduce $y \in x + A$ and therefore $B \subset x + A$ and $x + A \in \mathfrak{F}$.

The last assertion can be proved similarly. □

Definition 3.1.11. *Let X be a uniform space. A filter $\mathfrak{F}$ on X is called* semi-separable *if for any entourage U of X there exists a countable set $\mathfrak{A}$ of subsets of X such that*

$$\bigcup_{A\in\mathfrak{A}} (A\times A)\subset U\,, \quad \bigcup_{A\in\mathfrak{A}} A\in\mathfrak{F}\,.$$

A subset B of X is called semi-separable *if the filter $\{C\subset X\,|\,B\subset C\}$ on X is semi-separable.*

A subset B of a uniform space X is semi-separable iff for any entourage U of X there exists a countable set $\mathfrak{A}$ of subsets of X such that

$$\bigcup_{A\in\mathfrak{A}} (A\times A)\subset U\,, \quad B\subset \bigcup_{A\in\mathfrak{A}} A\,.$$

The intersection (union) of a countable family of semi-separable filters (sets) is semi-separable.

If subsets of locally convex spaces are under consideration, the above notion was introduced by H.G. Garnir, M. de Wilde, and J. Schmets (1968) ([1] page 67). (They called those sets "séparable par semi-norm".)

Proposition 3.1.12. *Let G, H be topological additive groups, I be a set (a countable set), $f\in\mathcal{G}(I,G)$ and u be a group homomorphism of G into H for which there exists a fundamental system $\mathfrak{U}$ of 0-neighbourhoods in H such that $\overset{-1}{u}(U)$ is a strongly Φ_1-closed (a sequentially closed) set of G for any $U\in\mathfrak{U}$. Then the following assertions are equivalent;*

a) $u\circ f\in\mathcal{G}(I,H)$;

b) $u\circ f(\mathfrak{P}(I))$ is compact;

c) $u\circ f(\mathfrak{P}(I))$ is semi-separable.

In particular, if $u(G)$ is semi-separable, then $u\circ f\in\mathcal{G}(I,H)$ for any $f\in\mathcal{G}(I,G)$.

a $\Rightarrow$ b $\Rightarrow$ c is trivial.

c $\Rightarrow$ a. First, assume that I is countable. Let $J\in\mathfrak{P}(I)$ and $U\in\mathfrak{U}$. We have

$$\overset{-1}{u}(u\circ f(J)+U)=f(J)+\overset{-1}{u}(U)\,.$$

Since $\overset{-1}{u}(U)$ is sequentially closed and since f is continuous, the set $\overset{-1}{\widehat{u\circ f}}(u\circ f(J)+U)$ is sequentially closed. The space $\mathfrak{P}(I)$ is metrizable, so that set is even closed. By c) and by Proposition 3.1.9 $u\circ f$ is continuous.

Now let I be arbitrary. By the above proof the restriction of $u\circ f$ to $\mathfrak{P}(J)$ is continuous for any $J\in\mathfrak{P}_c(I)$. Let $\mathfrak{F}$ be the section filter of $\mathfrak{P}_f(I)$. For any increasing sequence $(I_n)_{n\in\mathbb{N}}$ in $\mathfrak{P}_f(I)$ the sequence $(u\circ f(I_n))_{n\in\mathbb{N}}$ converges. Hence, $(\mathfrak{P}_f(I),u\circ f)$ is a Θ_1-net in H and by Proposition 1.5.17 $u\circ f(\mathfrak{F})$ is a Cauchy filter on H. Let $U\in\mathfrak{U}$. There exists $A\in\mathfrak{F}$ so that

$$u(f(A))-u(f(A))\subset U$$

and therefore $f(A)-f(A)\subset\overset{-1}{u}(U)$. Since $\overset{-1}{u}(U)$ is strongly Φ_1-closed and since $f(\mathfrak{F})$

is a Φ_1-filter which converges to $f(I)$ we deduce by Proposition 3.1.10 that

$$f(I) + \overset{-1}{u}(U) \in f(\mathfrak{F}).$$

Hence, $u(f(I)) + U \in u \circ f(\mathfrak{F})$. Since the set U is arbitrary, $u \circ f(\mathfrak{F})$ converges to $u(f(I))$ so $u \circ f$ is continuous at I. By Theorem 3.1.6 $u \circ f$ is continuous. □

§ 3.2 Structures on $\mathcal{G}(I, G)$

Throughout this section R shall denote a topological ring, G, H two topological left R-modules, I a set, and $\mathfrak{J}$ a set of subsets of I such that

a) $\mathfrak{P}_f(I) \subset \mathfrak{J}$;

b) $K \subset J \in \mathfrak{J} \Rightarrow K \in \mathfrak{J}$;

c) for any disjoint sequence $(J_n)_{n \in \mathbb{N}}$ *in* $\mathfrak{P}_f(I)$ *there exists an infinite subset M of* $\mathbb{N}$ *such that* $\bigcup_{n \in M} J_n \in \mathfrak{J}$.

We may choose $\mathfrak{P}_c(I)$ as $\mathfrak{J}$. Lemmata 3.2.8 and 3.2.10 present other examples of $\mathfrak{J}$.

Any topological additive group will be considered being endowed with its canonical structure of topological left $\mathbb{Z}$-module.

Proposition 3.2.1. *The identity map*

$$\mathcal{G}(I, G) \to \mathcal{G}(I, G)_{\{\mathfrak{P}(I)\}}$$

is uniformly Φ_4*-continuous.*

By Lemma 4.1.20 there exists a family $(h_\mu)_{\mu \in M}$ of group homomorphisms of G into metrizable topological additive groups such that the topology of G is the initial topology with respect to that family. Let $\mu \in M$, let G_μ be the target of h_μ, and let $(f_n)_{n \in \mathbb{N}}$ be a Cauchy sequence in $\mathcal{G}(I, G_\mu)$. By Theorem 3.1.6 this sequence is uniformly equicontinuous if it is equicontinuous at a point of $\mathfrak{P}(I)$. $\mathfrak{P}(I)$ is a compact space and therefore a Baire space. Hence by Theorem 2.4.3 the identity map

$$\mathcal{G}(I, G) \to \mathcal{G}(I, G)_{\{\mathfrak{P}(I)\}}$$

is uniformly Φ_4-continuous. □

Remark. The identity map

$$\mathcal{G}(I, G) \to \mathcal{G}(I, G)_{\{\mathfrak{P}(I)\}}$$

is not uniformly Φ_5-continuous in general. Indeed if G is precompact $\mathcal{G}(I, G)$ is precompact too, whilst $\mathcal{G}(I, G)_{\{\mathfrak{P}(I)\}}$ is not precompact if I is infinite and the topology of G is not trivial.

Proposition 3.2.2. *Let X be a set, $(I_\lambda)_{\lambda\in L}$ be a family of sets, for each $\lambda \in L$ let g_λ be a map of $\mathfrak{P}(I_\lambda)$ into X, let Θ be the set of sequences $(x_n)_{n\in\mathbb{N}}$ in X for which there exists $\lambda\in L$ such that*

$$\{n\in\mathbb{N} \mid x_n \in g_\lambda(\mathfrak{P}(I_\lambda))\}$$

is infinite, Φ be the set $\hat{\Phi}(\Theta)$, A be the set $\bigcup_{\lambda\in L} g_\lambda(\mathfrak{P}(I_\lambda))$, $\mathscr{F}$ be the set of maps $f: X \to G$ such that $f\circ g_\lambda \in \mathscr{G}(I_\lambda, G)$ for any $\lambda\in L$ and $\mathscr{F}'$ be the set $\mathscr{F}$ endowed with a uniformity such that the identity maps $\mathscr{F}_\Phi \to \mathscr{F}' \to \mathscr{F}$ are uniformly Φ_4-continuous. Then:

a) the identity map $\mathscr{F}_A \to \mathscr{F}_\Phi$ is uniformly Φ_4-continuous;

b) if G is Hausdorff, then $\mathscr{F}$ is a Φ_4-closed set of G^X and of G^X_A;

c) if G is Hausdorff and Φ_i-compact for an $i\in\{1, 2, 3, 4\}$ (e.g. G complete), then $\mathscr{F}'$ is Φ_i-compact:

d) if G is Hausdorff and sequentially complete, then $\mathscr{F}'$ is sequentially complete;

e) if $\{0\}$ is a G_δ-set of G, then $\mathscr{F}'$ is Φ_2-compact and if, in addition G is sequentially complete, then $\mathscr{F}'$ is Φ_4-compact;

f) $f(\mathfrak{F})$ is a Φ_2-filter (Φ_1-filter if I_λ is countable for any $\lambda\in L$) for any $f\in\mathscr{F}$ and for any $\mathfrak{F}\in\Phi(\Theta)$:

g) $f(\mathfrak{F})\in\hat{\Phi}_2(G)$ ($\hat{\Phi}_1(G)$ if I_λ is countable for any $\lambda\in L$) for any $f\in\mathscr{F}$ and for any $\mathfrak{F}\in\Phi$:

h) let $\mathfrak{F}\in\hat{\Phi}_4(\mathscr{F})$, (K, h) be a net in X such that K possesses no greatest element and such that for any strictly increasing sequence $(\kappa_n)_{n\in\mathbb{N}}$ in K there exist $\lambda\in L$ and a sequence $(J_n)_{n\in\mathbb{N}}$ in $\mathfrak{P}(I_\lambda)$ converging to $\emptyset$ such that $(g_\lambda(J_n))_{n\in\mathbb{N}}$ is a subsequence of $(h(\kappa_n))_{n\in\mathbb{N}}$, let $\mathfrak{G}$ be the section filter of K, and φ be the map

$$\mathscr{F}\times K \to G, \quad (f,\kappa)\mapsto f(h(\kappa));$$

then $\varphi(\mathfrak{F}\times\mathfrak{G})$ converges to 0;

i) $\mathscr{F}$ is an R-submodule of G^X;

j) $\mathscr{F}_A$, $\mathscr{F}$ and $\mathscr{F}_\Phi$ are topological left R-modules, locally convex if G is locally convex.

a) By Proposition 3.2.1 the identity map

$$\mathscr{G}(I_\lambda, G) \to \mathscr{G}(I_\lambda, G)_{\{\mathfrak{P}(I_\lambda)\}}$$

is uniformly Φ_4-continuous for every $\lambda\in L$ so by Proposition 2.4.2 the identity map $\mathscr{F}_A \to \mathscr{F}_\mathfrak{S}$ is uniformly Φ_4-continuous, where

$$\mathfrak{S} := \{g_\lambda(\mathfrak{P}(I_\lambda)) \mid \lambda\in L\}.$$

By Corollary 2.2.4 the identity map $\mathscr{F}_\mathfrak{S} \to \mathscr{F}_\Phi$ is uniformly Φ_5-continuous so by Corollary 1.8.5 the identity map $\mathscr{F}_A \to \mathscr{F}_\Phi$ is uniformly Φ_4-continuous.

b), c), d). $\mathscr{F}$ is a closed set of $G^X_\mathfrak{S}$ and the assertions follow from a), Proposition 2.1.15, and Corollary 1.5.11.

e) follows from c), d), and Proposition 1.6.4.

f) Let (K, φ) be a Θ-net in X such that $\varphi(\mathfrak{F}') = \mathfrak{F}$, where $\mathfrak{F}'$ denotes the section filter of K. Further, let $(\kappa_n)_{n\in\mathbb{N}}$ be an increasing sequence in K. There exist an infinite subset M of $\mathbb{N}$ and a $\lambda \in L$ such that $\varphi(\kappa_n) \in g_\lambda(\mathfrak{P}(I_\lambda))$ for any $n \in M$. Let J_n be a subset of I_λ such that $g_\lambda(J_n) = \varphi(\kappa_n)$ for any $n \in M$. Since $\mathfrak{P}(I_\lambda)$ is compact (and metrizable if I_λ is countable) and since $f \circ g_\lambda$ is continuous $(f \circ g_\lambda(J_n))_{n\in\mathbb{N}}$ possesses an adherent point (possesses a convergent subsequence if I_λ is countable). We find that $(f \circ \varphi(\kappa_n))_{n\in\mathbb{N}}$ possesses an adherent point (possesses a convergent subsequence if I_λ is countable). Hence, $f(\mathfrak{F})$ is a Φ_2-filter (a Φ_1-filter if I_λ is countable for any $\lambda \in L$).

g) follows from f) and Proposition 1.1.9.

h) We may assume $\mathfrak{F} \in \Phi_4(\mathscr{F})$. Let (M, f) be a Θ_4-net in $\mathscr{F}$ such that $f(\mathfrak{F}') = \mathfrak{F}$, where $\mathfrak{F}'$ denotes the section filter of M. Assume the assertion does not hold. Then there exist a neighbourhood U of 0 in G, an increasing sequence $(\mu_n)_{n\in\mathbb{N}}$ in M and a strictly increasing sequence $(\kappa_n)_{n\in\mathbb{N}}$ in K such that $f_{\mu_n}(h(\kappa_n)) \notin U$ for any $n \in \mathbb{N}$. There exist further $\lambda \in L$ and a sequence $(J_n)_{n\in\mathbb{N}}$ in $\mathfrak{P}(I_\lambda)$ converging to $\emptyset$ such that $(g_\lambda(J_n))_{n\in\mathbb{N}}$ is a subsequence of $(h(\kappa_n))_{n\in\mathbb{N}}$. By Proposition 1.4.7 $\{f_{\mu_n} \mid n \in \mathbb{N}\}$ is *a* Φ_4-set of $\mathscr{F}$ and therefore $\{f_{\mu_n} \circ g_\lambda \mid n \in \mathbb{N}\}$ is a Φ_4-set of $\mathscr{G}(I_\lambda, G)$. By Proposition 3.2.1 and Theorem 1.8.4 a $\Rightarrow$ h $\{f_{\mu_n} \circ g_\lambda \mid n \in \mathbb{N}\}$ is a Φ_4-set and therefore (Proposition 1.5.6 a $\Rightarrow$ c) a precompact set of $\mathscr{G}(I_\lambda, G)_{\{\mathfrak{P}(I_\lambda)\}}$. In particular, $\{f_{\mu_n} \circ g_\lambda \mid n \in \mathbb{N}\}$ is an equicontinuous set of $\mathscr{G}(I_\lambda, G)$ and therefore $(f_{\mu_n} \circ g_\lambda(J_m))_{m\in\mathbb{N}}$ converges to 0 uniformly in $n \in \mathbb{N}$, which is a contradiction.

i) Let $f, g \in \mathscr{F}$, $\alpha, \beta \in R$, $\lambda \in L$ and J, K be disjoint subsets of I_λ. We have

$$\begin{aligned}(\alpha f + \beta g) \circ g_\lambda(J \cup K) &= \alpha f \circ g_\lambda(J \cup K) + \beta g \circ g_\lambda(J \cup K) = \\ &= \alpha f \circ g_\lambda(J) + \alpha f \circ g_\lambda(K) + \beta g \circ g_\lambda(J) + \beta g \circ g_\lambda(K) = \\ &= (\alpha f + \beta g) \circ g_\lambda(J) + (\alpha f + \beta g) \circ g_\lambda(K).\end{aligned}$$

Hence, $\alpha f + \beta g \in \mathscr{F}$ and $\mathscr{F}$ is an R-submodule of G^X.

j) It is obvious that $\mathscr{F}_A$ and $\mathscr{F}$ are topological R-modules. Let $f \in \mathscr{F}$ and $\mathfrak{F} \in \Phi$. By g) $f(\mathfrak{F}) \in \hat{\Phi}_2(G)$ so $f(\mathfrak{F}) \in \hat{\Phi}_C(G)$ by Corollary 1.5.12. By Propositions 2.1.27 and 2.1.28 $\mathscr{F}_\Phi$ is a topological R-module, specifically locally convex if G is locally convex. □

Theorem 3.2.3. *a)* $\mathscr{G}(I, G)$ *is an R-submodule of* $G^{\mathfrak{P}(I)}$;

b) $\mathscr{G}(I, G)_{\mathfrak{J}}$, $\mathscr{G}(I, G)$, *and* $\mathscr{G}(I, G)_{\{\mathfrak{P}(I)\}}$ *are topological left R-modules, specifically locally convex if G is locally convex;*

c) the identity map

$$\mathscr{G}(I, G)_{\mathfrak{J}} \to \mathscr{G}(I, G)_{\{\mathfrak{P}(I)\}}$$

is uniformly Φ_4*-continuous.*

a) and b) are obvious.

c) Let $\mathscr{F}$ be the set of maps f of $\mathfrak{P}(I)$ into G such that $f \mid \mathfrak{P}(J) \in \mathscr{G}(J, G)$ for any $J \in \mathfrak{J}$. We denote by Θ the set of sequences $(A_n)_{n \in \mathbb{N}}$ in $\mathfrak{P}_f(I)$ such that $A_m \cap A_n = \emptyset$ or $A_m = A_n$ for any $m, n \in \mathbb{N}$, we denote by L the set

$$\{(J, K) \in \mathfrak{P}_f(I) \times \mathfrak{P}_f(I) \mid J \cap K = \emptyset\},$$

by $\leq$ the upper directed order relation on L defined by

$$(J, K) \leq (J', K') :\Leftrightarrow ((J, K) = (J', K') \quad \text{or} \quad J \cup K \subset J'),$$

by $\mathfrak{G}$ the section filter of L and by φ the map

$$L \to \mathfrak{P}(I), \quad (J, K) \mapsto K.$$

Then (L, φ) is a Θ-net and so $\varphi(\mathfrak{G}) \in \Phi(\Theta)$.

By the hypothesis about $\mathfrak{J}$ and by Proposition 3.2.2 a) the identity map $\mathscr{F}_{\mathfrak{J}} \to \mathscr{F}_{\Phi(\Theta)}$ is uniformly Φ_4-continuous. We deduce by Corollary 1.8.15 that the identity map

$$\mathscr{G}(I, G)_{\mathfrak{J}} \to \mathscr{G}(I, G)_{\Phi(\Theta)}$$

is uniformly Φ_4-continuous. Let $\mathfrak{F}$ be a Cauchy $\hat{\Phi}_4$-filter on $\mathscr{G}(I, G)_{\mathfrak{J}}$. By Theorem 1.8.4 a $\Rightarrow$ c $\mathfrak{F}$ is a Cauchy $\hat{\Phi}_4$-filter on $\mathscr{G}(I, G)_{\Phi(\Theta)}$.

Let U be an arbitrary 0-neighbourhood in G and let V be a closed 0-neighbourhood in G such that $2V \subset U$. By Theorem 2.2.3 a) there exist $\mathscr{F}' \in \mathfrak{F}$ and $J \in \mathfrak{P}_f(I)$ such that

$$f(K) - g(K) \in V$$

for any $f, g \in \mathscr{F}'$ and for any $K \in \mathfrak{P}_f(I \setminus J)$. Since V is closed and the functions of $\mathscr{G}(I, G)$ continuous we get

$$f(K) - g(K) \in V$$

for any $f, g \in \mathscr{F}'$ and for any $K \in \mathfrak{P}(I \setminus J)$. There exists $\mathscr{F}'' \in \mathfrak{F}$ such that

$$f(K) - g(K) \in V$$

for any $K \subset J$. We get

$$f(K) - g(K) = (f(K \cap J) - g(K \cap J)) + (f(K \setminus J) - g(K \setminus J)) \in 2V \subset U$$

for any $f, g \in \mathscr{F}' \cap \mathscr{F}'' \in \mathfrak{F}$ and for any $K \subset I$. Hence, $\mathfrak{F}$ is a Cauchy filter on $\mathscr{G}(I, G)_{\{\mathfrak{P}(I)\}}$ and the identity map

$$\mathscr{G}(I, G)_{\mathfrak{J}} \to \mathscr{G}(I, G)_{\{\mathfrak{P}(I)\}}$$

is uniformly Φ_4-continuous. □

Corollary 3.2.4. *Let $\mathscr{G}(I, G)'$ be the set $\mathscr{G}(I, G)$ endowed with a uniformity such that the identity maps*

$$\mathscr{G}(I, G)_{\{\mathfrak{P}(I)\}} \to \mathscr{G}(I, G)' \to \mathscr{G}(I, G)_{\mathfrak{J}}$$

are uniformly Φ_4-continuous and assume G Hausdorff. Then:

a) $\mathscr{G}(I,G)$ *is a* Φ_4*-closed set of* $G^{\mathfrak{P}(I)}$;

b) *if* G *is* Φ_i*-compact for an* $i \in \{1,2,3,4\}$ *then so is* $\mathscr{G}(I,G)'$;

c) *if* G *is sequentially complete then so is* $\mathscr{G}(I,G)'$.

By Theorem 3.2.3 the identity map

$$\mathscr{G}(I,G)_{\mathfrak{J}} \to \mathscr{G}(I,G)_{\{\mathfrak{P}(I)\}}$$

is uniformly Φ_4-continuous. Since $\mathscr{G}(I,G)$ is a closed set of $G^{\mathfrak{P}(I)}_{\{\mathfrak{P}(I)\}}$ the assertions follow from Proposition 2.1.15 and Corollary 1.8.5. □

Theorem 3.2.5. *We denote by* Φ *the set of* $\hat{\Phi}_4$*-filters on* $\mathscr{G}(I,G)$ *and by* φ, ψ *the maps*

$$\mathscr{G}(I,G)_{\mathfrak{J}} \times \mathfrak{P}(I) \to G, \quad (f,J) \mapsto f(J),$$

$$\mathfrak{P}(I) \to G_{\Phi}^{\mathscr{G}(I,G)}, \quad J \mapsto \varphi(\cdot, J)$$

respectively. Then:

a) *any* $\hat{\Phi}_4$*-filter on* $\mathscr{G}(I,G)_{\mathfrak{J}}$ *is uniformly equicontinuous;*

b) $\varphi(\mathfrak{F} \times \mathfrak{G})$ *is a Cauchy filter for any Cauchy* $\hat{\Phi}_4$*-filter* $\mathfrak{F}$ *on* $\mathscr{G}(I,G)_{\mathfrak{J}}$ *and for any Cauchy filter* $\mathfrak{G}$ *on* $\mathfrak{P}(I)$;

c) φ *is uniformly* Φ_4*-continuous;*

d) ψ *is uniformly continuous;*

e) *let* $\mathscr{F}$ *be a* Φ_4*-set of* $\mathscr{G}(I,G)_{\mathfrak{J}}$ *and* $\bar{\mathscr{F}}$ *be its closure in* $\mathscr{G}(I,G)_{\mathfrak{J}}$; *then*

e$_1$) $\bar{\mathscr{F}}$ *is the closure of* $\mathscr{F}$ *in* $\mathscr{G}(I,G)_{\{\mathfrak{P}(I)\}}$ *and*

$$\bar{\mathscr{F}}_{\mathfrak{J}} = \bar{\mathscr{F}} = \bar{\mathscr{F}}_{\{\mathfrak{P}(I)\}};$$

e$_2$) *the restriction of* φ *to* $\bar{\mathscr{F}}_{\mathfrak{J}} \times \mathfrak{P}(I)$ *is uniformly continuous.*

By Theorem 3.2.3 the identity map

$$\mathscr{G}(I,G)_{\mathfrak{J}} \to \mathscr{G}(I,G)_{\{\mathfrak{P}(I)\}}$$

is uniformly Φ_4-continuous.

a), b), c), d), e$_2$) follow from Proposition 2.1.22.

e$_1$) follows from Proposition 1.8.8. □

Corollary 3.2.6. *Let* G *be Hausdorff,* $\mathfrak{F}$ *be a* $\hat{\Phi}_4$*-filter on* $\mathscr{G}(I,G)_{\mathfrak{J}}$ *and* $\mathfrak{K}$ *be the set of* $J \in \mathfrak{P}(I)$ *for which* $\lim_{f,\mathfrak{F}} f(J)$ *exists. Then the map*

$$g: \mathfrak{K} \to G, \quad J \mapsto \lim_{f,\mathfrak{F}} f(J)$$

is uniformly continuous with respect to the uniformity on $\mathfrak{K}$ *induced by* $\mathfrak{P}(I)$, $\mathfrak{F}$ *converges to* g *uniformly on* $\mathfrak{K}$ *and, for any 0-neighbourhood* U *in* G, *there exist* $\mathscr{F} \in \mathfrak{F}$ *and a finite subset* K *of* I *such that*

$$f(J) - g(J') \in U$$

for any $f \in \mathscr{F}$, *any* $J \in \mathfrak{P}(I)$ *and any* $J' \in \mathfrak{K}$ *with* $J \cap K = J' \cap K$. *If* G *is* Φ_1*-compact and* $\mathfrak{K} \supset \mathfrak{P}_c(I)$ *or if* G *is* Φ_4*-compact and* $\mathfrak{K} \supset \mathfrak{J}$, *then* $\mathfrak{K} = \mathfrak{P}(I)$ *and* $g \in \mathscr{G}(I, G)$.

By Theorem 3.2.5 a) $\mathfrak{F}$ is uniformly equicontinuous.

Let φ be the map

$$\mathscr{G}(I, G) \to G^{\mathfrak{K}}, \quad f \mapsto f|\mathfrak{K}.$$

Then $\varphi(\mathfrak{F})$ is uniformly equicontinuous. By Proposition 2.1.13 g is uniformly continuous. The other assertions follow from Proposition 2.1.11.

Assume now that G Φ_1-compact and $\mathfrak{K} \supset \mathfrak{P}_c(I)$. Since $\mathfrak{P}_c(I)$ is Φ_1-dense in $\mathfrak{P}(I)$ and since g is uniformly continuous, there exists a uniformly continuous map h of $\mathfrak{P}(I)$ into G whose restriction to $\mathfrak{K}$ is equal to g. By the above result it follows immediately that $\mathfrak{F}$ converges to h in $G^{\mathfrak{P}(I)}$. Hence, $\mathfrak{K} = \mathfrak{P}(I)$ and $g \in \mathscr{G}(I, G)$.

Assume G is Φ_4-compact and $\mathfrak{K} \supset \mathfrak{J}$. Then $\mathfrak{F}$ is a Cauchy filter on $\mathscr{G}(I, G)_{\mathfrak{J}}$ so by Theorem 3.2.3 it is a Cauchy filter on $\mathscr{G}(I, G)$. By Corollary 3.2.4 b) it is a convergent filter on $\mathscr{G}(I, G)$. □

Proposition 3.2.7. *Let* $i \in \{1, 2, 3, 4\}$ *and* u *be a uniformly* Φ_i*-continuous group homomorphism of* G *into* H. *Then the map*

$$\mathscr{G}(I, G)_{\mathfrak{J}} \to \mathscr{G}(I, H)_{\{\mathfrak{P}(I)\}}, \quad f \mapsto u \circ f$$

is uniformly Φ_i*-continuous. If* u *is continuous, then the maps*

$$\mathscr{G}(I, G)_{\mathfrak{J}} \to \mathscr{G}(I, H)_{\mathfrak{J}}, \quad f \mapsto u \circ f,$$

$$\mathscr{G}(I, G)_{\{\mathfrak{P}(I)\}} \to \mathscr{G}(I, H)_{\{\mathfrak{P}(I)\}}, \quad f \mapsto u \circ f$$

are uniformly continuous.

The map

$$\mathscr{G}(I, G)_{\mathfrak{J}} \to H, \quad f \mapsto u \circ f(J)$$

is uniformly Φ_i-continuous for any $J \in \mathfrak{J}$. It follows that the map (Proposition 3.1.5)

$$\mathscr{G}(I, G)_{\mathfrak{J}} \to \mathscr{G}(I, H)_{\mathfrak{J}}, \quad f \mapsto u \circ f$$

is uniformly Φ_i-continous too (Proposition 1.8.13). By Theorem 3.2.3 the identity map

$$\mathscr{G}(I, H)_{\mathfrak{J}} \to \mathscr{G}(I, H)_{\{\mathfrak{P}(I)\}}$$

is uniformly Φ_4-continuous. Hence the map

$$\mathscr{G}(I, G)_{\mathfrak{J}} \to \mathscr{G}(I, H)_{\{\mathfrak{P}(I)\}}, \quad f \mapsto u \circ f$$

is uniformly Φ_i-continuous (Corollary 1.8.5).

The last assertion is obvious. □

The results of this section depend on $\mathfrak{J}$ and they improve if we take $\mathfrak{J}$ smaller. The aim of the next lemmata is to give examples for $\mathfrak{J}$.

Lemma 3.2.8. *Let* $(\mathfrak{I}_\lambda)_{\lambda \in L}$ *be a countable family of subsets of* $\mathfrak{P}(I)$ *such that we have for any* $\lambda \in L$

a) $\mathfrak{P}_f(I) \subset \mathfrak{I}_\lambda$;

b) $K \subset J \in \mathfrak{I}_\lambda \Rightarrow K \in \mathfrak{I}_\lambda$;

c) *for any disjoint sequence* $(J_n)_{n \in \mathbb{N}}$ *in* $\mathfrak{P}_f(I)$ *there exists an infinite subset M of* $\mathbb{N}$ *such that* $\bigcup_{n \in M} J_n \in \mathfrak{I}_\lambda$;

d) $J, K \in \mathfrak{I}_\lambda, \ J \cap K = \emptyset \Rightarrow J \cup K \in \mathfrak{I}_\lambda$.

Then $\bigcap_{\lambda \in L} \mathfrak{I}_\lambda$ *possesses the above properties* a), b), c), d) *too.*

a), b) and d) are trivial. c) follows from d) by the diagonal method. $\quad\square$

Lemma 3.2.9. *(W. Sierpiński (1928)* [1] *page 239). There exists a set* $\mathfrak{M}$ *of infinite subsets of* $\mathbb{N}$ *having the power of continuum and such that* $M' \cap M''$ *is finite for any different* $M', M'' \in \mathfrak{M}$.

We set

$$\varphi: \mathfrak{P}(\mathbb{N}) \times \mathbb{N} \longrightarrow \mathbb{N}, \ (A, n) \mapsto \sum_{\substack{k \in A \\ k \leq n}} 2^k,$$

$$\psi: \mathfrak{P}(\mathbb{N}) \longrightarrow \mathfrak{P}(\mathbb{N}), \ A \mapsto \{\varphi(A, n) \mid n \in A\}.$$

Let $A, B \in \mathfrak{P}(\mathbb{N})$ and let $m, n \in \mathbb{N}$, $m \leq n$. Then $\varphi(A, m) = \varphi(B, n)$ implies

$$\{k \in A \mid k \leq m\} = \{k \in B \mid k \leq m\}.$$

Hence $\psi(A) \cap \psi(B)$ is finite for all different $A, B \in \mathfrak{P}(\mathbb{N})$. The set $M := \psi(\mathfrak{P}(\mathbb{N}) \backslash \mathfrak{P}_f(\mathbb{N}))$ possesses the required properties. $\quad\square$

Lemma 3.2.10. *Let* Φ *be a set of free ultrafilters on* I *such that the power of* $\{\mathfrak{F} \in \Phi \mid K \in \mathfrak{F}\}$ *is strictly smaller than the power of continuum for any* $K \in \mathfrak{P}_c(I)$ *and let* $\mathfrak{I}_0$ *be the set*

$$\mathfrak{P}_c(I) \backslash \bigcup_{\mathfrak{F} \in \Phi} \mathfrak{F}.$$

Then:

a) $\mathfrak{P}_f(I) \subset \mathfrak{I}_0$;

b) $K \subset J \in \mathfrak{I}_0 \Rightarrow K \in \mathfrak{I}_0$;

c) *for any disjoint sequence* $(J_n)_{n \in \mathbb{N}}$ *in* $\mathfrak{I}_0$ *there exists an infinite subset M of* $\mathbb{N}$ *such that* $\bigcup_{n \in M} J_n \in \mathfrak{I}_0$.

a) and b) are trivial. Let $(J_n)_{n \in \mathbb{N}}$ be a disjoint sequence in $\mathfrak{I}_0$. By Lemma 3.2.9 there exists a set $\mathfrak{M}$ of infinite subsets of $\mathbb{N}$ having the power of continuum and such that $M' \cap M''$ is finite for any different $M', M'' \in \mathfrak{M}$. Then

$$(\{\mathfrak{F}\in\Phi \mid \bigcup_{n\in M} J_n\in\mathfrak{F}\})_{M\in\mathfrak{M}}$$

is a disjoint family of subsets of Φ whose union has a power strictly smaller than that of continuum. Hence, there exists $M\in\mathfrak{M}$ for which $\{\mathfrak{F}\in\Phi \mid \bigcup_{n\in M} J_n\in\mathfrak{F}\}$ is empty and therefore $\bigcup_{n\in M} J_n\in\mathfrak{J}_0$. □

§ 3.3 Spaces of supersummable families

Throughout this section R shall denote a topological ring, G, H two Hausdorff topological left R-modules, I a set and $\mathfrak{J}$ a set of subsets of I such that:

a) $\mathfrak{P}_f(I)\subset\mathfrak{J}$;

b) $K\subset J\in\mathfrak{J} \Rightarrow K\in\mathfrak{J}$;

c) for any disjoint sequence $(J_n)_{n\in\mathbb{N}}$ *in* $\mathfrak{P}_f(I)$ *there exists an infinite subset M of* $\mathbb{N}$ *such that* $\bigcup_{n\in M} J_n\in\mathfrak{J}$.

Any topological additive group will be considered being endowed with its canonical structure of topological left $\mathbb{Z}$-module.

Definition 3.3.1. *Let* $\mathfrak{F}(I)$ *be the section filter of* $\mathfrak{P}_f(I)$ *and* $(x_\iota)_{\iota\in I}$ *be a family in G. For any* $J\in\mathfrak{P}_f(I)$ *we set*

$$x(J):=\sum_{\iota\in J} x_\iota ,$$

where the sum of families indexed by finite sets is defined in a natural way by induction (equal to 0 if J is empty). The family $(x_\iota)_{\iota\in I}$ *is called* <u>summable</u> *if the filter* $x(\mathfrak{F}(I))$ *is convergent. In that case we set*

$$\underline{\sum_{\iota\in I} x_\iota} := \lim x(\mathfrak{F}(I))$$

and call this element the <u>sum of the family $(x_\iota)_{\iota\in I}$</u>. *We say that* $(x_\iota)_{\iota\in I}$ *is* <u>supersummable</u> *if the family* $(x_\iota)_{\iota\in J}$ *is summable for any* $J\subset I$.

Remark. The summability for arbitrary families (in some special fields) was introduced by E. H. Moore (1939) ([1] Definition 3.8.1).

Definition 3.3.2. *We denote by* <u>$\ell(I,G)$</u> *the set of supersummable families in G indexed by I. We shall identify the family* $(x_\iota)_{\iota\in I}$ *in G with the map*

$$I\to G,\quad \iota\mapsto x_\iota$$

and consequently $\ell(I,G)$ *will be identified with the set of maps* f *of* I *into* G *such that* $(f(\iota))_{\iota\in I}$ *is supersummable. It is easy to see that* $\ell(I,G)$ *is an R-submodule of* G^I. *We call* coarse topology of $\ell(I,G)$ *the coarsest topology of* $\ell(I,G)$ *for which the maps*

$$\ell(I,G)\to G,\quad f\mapsto\sum_{\iota\in J}f(\iota)$$

are continuous for all $J\in\mathfrak{J}$. *By the* fine topology of $\ell(I,G)$ *we shall mean the coarsest topology on* $\ell(I,G)$ *for which the set of all these maps, while* J *goes through* $\mathfrak{P}(I)$, *is equicontinuous. We denote by* $\ell_c(I,G)$ *(by* $\ell_f(I,G)$*) the left R-module* $\ell(I,G)$ *endowed with the coarse topology (with the fine topology).* $\ell_c(I,G)$ *and* $\ell_f(I,G)$ *are topological left R-modules (Theorem 3.3.5), locally convex if G is locally convex. Their uniformities will be called* coarse uniformity *and* fine uniformity *respectively.*

The fine topology of $\ell(\mathbb{N},E)$ was considered by G. Birkhoff (1935) ([1] page 361) for E being a Banach space.

The coarse topology and the coarse uniformity of $\ell(I,G)$ depend on $\mathfrak{J}$ which was fixed in this section. Subsequently we will use them without specifying $\mathfrak{J}$; then the meaning will be "for any $\mathfrak{J}$ satisfying the conditions a), b), c)" that is from the beginning of § 3.3 (e.g. for $\mathfrak{J}=\mathfrak{P}_c(I)$).

The sets of the form

$$\bigcap_{K\in\mathfrak{K}}\{f\in\ell(I,G)\mid\sum_{\iota\in K}f(\iota)\in U\}$$

while $\mathfrak{K}$ goes through the set of finite subsets of $\mathfrak{J}$ and U through a fundamental system of 0-neighbourhoods in G, constitute a fundamental system of 0-neighbourhoods in $\ell_c(I,G)$. The sets of the form

$$\bigcap_{J\in\mathfrak{P}(I)}\{f\in\ell(I,G)\mid\sum_{\iota\in J}f(\iota)\in U\}$$

where U runs through a fundamental system of 0-neighbourhoods in G, constitute a fundamental system of 0-neighbourhoods for the fine topology of $\ell(I,G)$. The fine topology is finer than the coarse topology and finer than the topology of uniform convergence ($\ell(I,G)\subset G^I$). The coarse topology is finer than the topology of pointwise convergence. We remark that $\ell(I,\mathbb{R})$ is nothing but the set of summable families of real numbers indexed by I, denoted usually by $\ell^1(I)$, and that the fine topology of $\ell(I,\mathbb{R})$ is the norm topology of $\ell^1(I)$ (Proposition 3.5.6 e)). The coarse topology on $\ell(I,\mathbb{R})$ is the weak topology on $\ell^1(I)$ associated to the duality $<\ell^1(I),\ell^\infty_*(I)>$, where $\ell^\infty_*(I)$ denotes the vector space of families of real numbers, indexed by I, taking a finite number of values only, and equal to 0 outside of a finite union of sets of $\mathfrak{J}$. Since $\ell^\infty_*(I)\subset\ell^\infty(I)$ the coarse topology is coarser than the weak topology of $\ell^1(I)$.

Proposition 3.3.3. *Let* $(x_\iota)_{\iota\in I}$ *be a family in G the countable subfamilies of which are summable. If G is* Φ_1*-compact then* $(x_\iota)_{\iota\in I}$ *is supersummable.*

Let J be a subset of I, $\mathfrak{F}$ be the section filter of $\mathfrak{P}_f(J)$ and φ be the map

$$\mathfrak{P}_f(J)\to G,\quad K\mapsto\sum_{\iota\in K}x_\iota\,.$$

Then for any increasing sequence $(J_n)_{n\in\mathbb{N}}$ in $\mathfrak{P}_f(J)$ the sequence $(\varphi(J_n))_{n\in\mathbb{N}}$ converges. By Proposition 1.5.17 $\varphi(\mathfrak{F})$ is a Cauchy Φ_1-filter on G. Since G is Φ_1-compact it converges. Hence, $(x_\iota)_{\iota\in J}$ is summable. □

Proposition 3.3.4. *Assume H is a strongly Φ_1-closed subgroup of G. Any family in H which is supersummable in G is supersummable in H.*

Let $(x_\iota)_{\iota\in I}$ be a family in H which is supersummable in G. Let $J\subset I$, $\mathfrak{F}$ be the section filter of $\mathfrak{P}_f(J)$, φ be the map

$$\mathfrak{P}_f(J)\to H, K\mapsto \sum_{\iota\in K} x_\iota ,$$

and ψ be the inclusion map $H\to G$. Then $\varphi(\mathfrak{F})$ is a $\Phi(\Theta_1(H,G))$-filter in H and $\psi(\varphi(\mathfrak{F}))$ converges. Since H is strongly Φ_1-closed $\varphi(\mathfrak{F})$ converges and therefore $(x_\iota)_{\iota\in J}$ is summable in H. Hence, $(x_\iota)_{\iota\in I}$ is supersummable in H. □

Theorem 3.3.5. *For any $f\in\ell(I,G)$ we denote by $\bar{f}$ the map*

$$\mathfrak{P}(I)\to G,\quad J\mapsto \sum_{\iota\in J} f(\iota).$$

Then $\bar{f}\in\mathcal{G}(I,G)$ for any $f\in\ell(I,G)$ and the maps

$$\ell_c(I,G)\to\mathcal{G}(I,G)_{\mathfrak{Z}},\quad f\mapsto\bar{f}$$

$$\ell_f(I,G)\to\mathcal{G}(I,G)_{\{\mathfrak{P}(I)\}},\quad f\mapsto\bar{f}$$

are isomorphisms of topological R-modules.

It is obvious that $\bar{f}\in\mathcal{H}(I,G)$ and that $\bar{f}$ is continuous at the point $I\in\mathfrak{P}(I)$. By Theorem 3.1.6 $\bar{f}\in\mathcal{G}(I,G)$. Since the map

$$\ell(I,G)\to\mathcal{G}(I,G),\quad f\mapsto\bar{f}$$

is obviously bijective, the corresponding maps are isomorphisms of topological R-modules. □

Corollary 3.3.6. *Let $(x_\iota)_{\iota\in I}\in\ell(I,G)$. Then the set $\{\sum_{\iota\in J} x_\iota \mid J\subset I\}$ is compact and for any 0-neighbourhood U in G there exists a finite subset K of I such that*

$$\sum_{\iota\in J'} x_\iota-\sum_{\iota\in J''} x_\iota\in U,\quad \sum_{\iota\in J} x_\iota\in U$$

for any subsets J, J', J'' of I with $J'\cap K=J''\cap K$ and $J\subset I\backslash K$. The set of maps $\{\mathfrak{P}(I)\to G, J\mapsto \sum_{\iota\in J\cap K} x_\iota \mid K\subset I\}$ is uniformly equicontinuous.

The first assertion follows immediately from Theorem 3.3.5. In oder to prove the second, we remark that the set of maps $\{\mathfrak{P}(I)\to G, J\mapsto \sum_{\iota\in J\cap K} x_\iota \mid K\subset I\}$ is obviously equicontinuous at $\emptyset$. By the Theorems 3.3.5 and 3.1.6 this set is uniformly equicontinuous. □

Corollary 3.3.7. *The identity map*

$$\ell_c(I,G) \to \ell_f(I,G)$$

is uniformly Φ_4-continuous.

The assertion follows from Theorem 3.3.5 and Theorem 3.2.3c). □

Remark. Let $\ell_*^\infty(I)$ be the vector space of real functions on I taking a finite number of values only and vanishing outside of a finite union of sets of $\mathfrak{J}$. By the above result (taking $G = \mathbb{R}$) any Cauchy sequence in $\ell^1(I)$ with respect to the weak topology $\sigma(\ell^1(I), \ell_*^\infty(I))$ associated to the duality $< \ell^1(I), \ell_*^\infty(I) >$ is a Cauchy sequence in $\ell^1(I)$ for the norm topology (Proposition 3.5.6e)). In particular, the weak and the strong Cauchy sequences in $\ell^1(I)$ coincide. This result was proved by J. Schur (1920) ([1] III page 82) for $I = \mathbb{N}$. H. Hahn showed (1922) ([1] V c) that for sequences in $\ell^1(\mathbb{N})$ the weak convergence and the convergence with respect to $\sigma(\ell^1(\mathbb{N}), \ell_*^\infty(\mathbb{N}))$ coincide.

Corollary 3.3.8. *Let $\mathfrak{F}$ be a $\hat{\Phi}_4$-filter on $\ell_c(I,G)$ such that $\lim\limits_{x,\mathfrak{F}} x_\iota$ exists for any $\iota \in I$ and let $\mathfrak{R}$ be a set of subsets of I such that $\lim\limits_{x,\mathfrak{F}} \sum\limits_{\iota\in J} x_\iota$ exists and $\mathfrak{P}(J) \subset \mathfrak{R}$ for any $J \in \mathfrak{R}$. Then*

a) $(\lim\limits_{x,\mathfrak{F}} x_\iota)_{\iota\in J}$ is summable and

$$\sum_{\iota\in J} \lim_{x,\mathfrak{F}} x_\iota = \lim_{x,\mathfrak{F}} \sum_{\iota\in J} x_\iota$$

for any $J \in \mathfrak{R}$;

b) for any 0-neighbourhood U in G there exist $\mathscr{F} \in \mathfrak{F}$ and a finite subset K of I such that

$$\sum_{\iota\in J} x_\iota - \sum_{\iota\in J'} \lim_{x,\mathfrak{F}} x_\iota \in U$$

for any $x \in \mathscr{F}$, any $J \subset I$ and for any $J' \in \mathfrak{R}$ with $J \cap K = J' \cap K$;

c) if G is Φ_1-compact (Φ_4-compact) and $\mathfrak{R} \subset \mathfrak{P}_c(I)$ ($\mathfrak{R} \supset \mathfrak{J}$) then $\lim\limits_{x,\mathfrak{F}} \sum\limits_{\iota\in J} x_\iota$ exists for any $J \subset I$ and $(\lim\limits_{x,\mathfrak{F}} x_\iota)_{\iota\in I}$ is supersummable.

The assertions follow immediately from Theorem 3.3.5 and Corollary 3.2.6. □

Corollary 3.3.9. *Let $(x_{\iota n})_{(\iota,n)\in I\times\mathbb{N}}$ be a family in G such that $(x_{\iota n})_{\iota\in J}$ is summable for any $J \subset I$ and any $n \in \mathbb{N}$, such that $(x_{\iota n})_{n\in\mathbb{N}}$ converges for any $\iota \in I$, and such that $(\sum\limits_{\iota\in J} x_{\iota n})_{n\in\mathbb{N}}$ is a Cauchy sequence for every $J \in \mathfrak{J}$ and let $\mathfrak{R}$ be a set of subsets of I such that $(\sum\limits_{\iota\in J} x_{\iota n})_{n\in\mathbb{N}}$ converges and $\mathfrak{P}(J) \subset \mathfrak{R}$ for any $J \in \mathfrak{R}$. Then* *a)* $(\lim_{n\to\infty} x_{\iota n})_{\iota \in J}$ *is summable and*

$$\sum_{\iota \in J} \lim_{n\to\infty} x_{\iota n} = \lim_{n\to\infty} \sum_{\iota \in J} x_{\iota n}$$

for any $J \in \mathfrak{K}$;

b) for any 0-neighbourhood U *in* G *there exist* $n_0 \in \mathbb{N}$ *and a finite subset* K *of* I *such that*

$$\sum_{\iota \in J} x_{\iota n} - \sum_{\iota \in J'} \lim_{n\to\infty} x_{\iota n} \in U$$

for any $n \in \mathbb{N}$ *with* $n \geq n_0$, *any* $J \subset I$ *and any* $J' \in \mathfrak{K}$ *with* $J \cap K = J' \cap K$. □

Remark. The above corollary was proved by A. P. Robertson (1972) ([2] Lemma and Corollary 1) for $I = \mathbb{N}$ and $\mathfrak{K} = \mathfrak{P}(\mathbb{N})$.

Corollary 3.3.10. *We denote by* Φ *the set of* $\hat{\Phi}_4$*-filters on* $\ell_c(I, G)$, *by* $\mathscr{F}$ *the set of uniformly continuous maps of* $\ell_f(I, G)$ *into* G, *and by* φ *the map*

$$\ell_c(I, G) \times \mathfrak{P}(I) \to G, \quad (x, J) \mapsto \sum_{\iota \in J} x_\iota .$$

Then:

a) $\varphi(\mathfrak{F} \times \mathfrak{G})$ *is a Cauchy filter for any Cauchy* $\hat{\Phi}_4$*-filter* $\mathfrak{F}$ *on* $\ell_c(I, G)$ *and for any Cauchy filter* $\mathfrak{G}$ *on* $\mathfrak{P}(I)$;

b) φ *is uniformly* Φ_4*-continuous;*

c) $\mathscr{F}$ *is an R-submodule of* $G^{\ell(I,G)}$, $\mathscr{F}_\Phi$ *is a Hausdorff topological left R-module (locally convex if G is locally convex),* $\varphi(\cdot, \{\iota\}) \in \mathscr{F}$ *for any* $\iota \in I$, *and the map*

$$I \to \mathscr{F}_\Phi, \quad \iota \mapsto \varphi(\cdot, \{\iota\})$$

belongs to $\ell(I, \mathscr{F}_\Phi)$.

The assertions follow immediately from Theorem 3.3.5, Proposition 2.1.29 and Theorem 3.2.5 b), c), d). □

Corollary 3.3.11. *Let* $\mathfrak{F}$ *be a* $\hat{\Phi}_4$*-filter on* $\ell_c(I, G)$ *and* U *be a 0-neighbourhood in* G. *Then there exist* $\mathscr{F} \in \mathfrak{F}$ *and a finite subset* K *of* I *such that:*

a) $\sum_{\iota \in J'} x_\iota - \sum_{\iota \in J''} x_\iota \in U, \quad \sum_{\iota \in J} x_\iota \in U$

for any $(x_\iota)_{\iota \in I} \in \mathscr{F}$ *and any* $J, J', J'' \in \mathfrak{P}(I)$ *with* $J' \cap K = J'' \cap K$ *and* $J \subset I \backslash K$;

b) if, in addition, $\mathfrak{F}$ *is a Cauchy filter on* $\ell_c(I, G)$, *then*

$$\sum_{\iota \in J} x_\iota - \sum_{\iota \in J'} y_\iota \in U$$

for any $(x_\iota)_{\iota \in I}, (y_\iota)_{\iota \in I} \in \mathscr{F}$ *and any* J, J' *with* $J \cap K = J' \cap K$.

The corollary follows immediately from Theorem 3.3.5 by Theorem 3.2.5 a), b). □

Corollary 3.3.12. *Let* $i \in \{1, 2, 3, 4\}$ *and* $\ell(I, G)'$ *be the set* $\ell(I, G)$ *endowed with a uniformity finer than the coarse uniformity and coarser than the fine uniformity. Then:*

a) if G is Φ_i*-compact then* $\ell(I, G)'$ *is* Φ_i*-compact;*

b) if G is sequentially complete then $\ell(I, G)'$ *is sequentially complete.*

The Corollary follows from Theorem 3.3.5 and Corollary 3.2.4. □

Theorem 3.3.13. *Let* $\mathscr{F}$ *be a* Φ_4*-set of* $\ell_c(I, G)$, $\bar{\mathscr{F}}$ *be its closure in* $\ell_c(I, G)$ *and* $\mathscr{G}$ *be the R-submodule of* $G^{\bar{\mathscr{F}}}$ *of uniformly continuous maps. Then;*

a) $\bar{\mathscr{F}}$ *is the closure of* $\mathscr{F}$ *in* $\ell_f(I, G)$;

b) $\ell_c(I, G)$ *and* $\ell_f(I, G)$ *induce the same uniformity on* $\bar{\mathscr{F}}$:

c) the map

$$\bar{\mathscr{F}} \times \mathfrak{P}(I) \to G, \quad ((x_\iota)_{\iota \in I}, J) \mapsto \sum_{\iota \in J} x_\iota$$

is uniformly continuous;

d) the set $\{\sum_{\iota \in J} x_\iota \mid J \subset I, (x_\iota)_{\iota \in I} \in \bar{\mathscr{F}}\}$ *is precompact and it is compact if* $\bar{\mathscr{F}}$ *is compact;*

e) if G is Φ_4*-compact (e.g. complete), then* $\bar{\mathscr{F}}$ *is compact;*

f) $\mathscr{G}_{\{\bar{\mathscr{F}}\}}$ *is a Hausdorff topological left R-module, for any* $\iota \in I$ *the map*

$$g_\iota : \bar{\mathscr{F}} \to G, \quad f \mapsto f(\iota)$$

belongs to $\mathscr{G}$, $(g_\iota)_{\iota \in I} \in \ell(I, \mathscr{G}_{\{\bar{\mathscr{F}}\}})$, *and* $\sum_{\iota \in J} g_\iota$ *is the map*

$$\bar{\mathscr{F}} \to G, \quad f \mapsto \sum_{\iota \in J} f(\iota)$$

for any $J \subset I$;

g) for any 0-neighbourhood U in G there exists a finite subset K of I such that

$$\sum_{\iota \in J'} x_\iota - \sum_{\iota \in J''} x_\iota \in U, \quad \sum_{\iota \in J} x_\iota \in U$$

for any $(x_\iota)_{\iota \in I} \in \bar{\mathscr{F}}$ *and any* $J, J', J'' \in \mathfrak{P}(I)$ *with* $J \subset I \backslash K$ *and* $J' \cap K = J'' \cap K$.

a), b), c) follow from Theorem 3.3.5 and Theorem 3.2.5 e).

d) follows from c) and Proposition 1.5.6 a ⇒ c.

e) By Corollary 3.3.12 a) $\ell_c(I, G)$ is Φ_4-compact and the assertion follows from Proposition 1.2.10 (and Corollary 1.5.11).

f) By Proposition 2.1.5 $\mathscr{G}_{\{\bar{\mathscr{F}}\}} = \mathscr{G}_{\{\mathscr{F}\}}$. By Proposition 2.1.29 $\mathscr{G}_{\{\mathscr{F}\}}$ is a topological left R-module which obviously is Hausdorff. The other assertions follow from c).

g) follows from c), from Theorem 3.3.5 and from Theorem 3.2.5 a). □

Corollary 3.3.14. *Let $(x_{\iota n})_{(\iota, n) \in I \times \mathbb{N}}$ be a family in G such that:*

1) for any $n \in \mathbb{N}$ and for any $J \subset I$ the family $(x_{\iota n})_{\iota \in J}$ is summable;

2) for any $J \in \mathfrak{J}$, $(\sum_{\iota \in J} x_{\iota n})_{n \in \mathbb{N}}$ is a Cauchy sequence in G.

Then:

a) the set $\{\sum_{\iota \in J} x_{\iota n} \mid J \subset I, n \in \mathbb{N}\}$ is precompact and it is relatively compact if G is Φ_4-compact;

b) for any 0-neighbourhood U in G there exist $n_0 \in \mathbb{N}$ and a finite subset K of I such that

$$\sum_{\iota \in J} x_{\iota m} - \sum_{\iota \in J'} x_{\iota n} \in U,$$

$$\sum_{\iota \in J} x_{\iota p} - \sum_{\iota \in J'} x_{\iota p} \in U, \quad \sum_{\iota \in J''} x_{\iota p} \in U$$

for any $m, n, p \in \mathbb{N}$ with $m \geq n_0, n \geq n_0$, and any $J, J', J'' \in \mathfrak{P}(I)$ with $J \cap K = J' \cap K$ and $J'' \subset I \backslash K$.

a) follows immediately from Proposition 1.4.7 and Theorem 3.3.13 d) and e).

b) follows immediately from Proposition 1.4.7, Theorem 3.3.13 g) and Corollary 3.3.11 b). □

Remark. The last assertion of b) was proved by P. Antosik (1973) ([1] Corollary 2 (a)) for I countable.

Proposition 3.3.15. *Let $(x_\iota)_{\iota \in I}$ be a family in G such that for any infinite subset J of I there exists an infinite subset K of J such that $(x_\iota)_{\iota \in K}$ is summable. If $\{0\}$ is a G_δ-set of G, then $\{\iota \in I \mid x_\iota \neq 0\}$ is countable.*

Let $(U_n)_{n \in \mathbb{N}}$ be a sequence of open sets of G with

$$\bigcap_{n \in \mathbb{N}} U_n = \{0\}.$$

Then for any $n \in \mathbb{N}$ the set $\{\iota \in I \mid x_\iota \notin U_n\}$ is finite and therefore

$$\{\iota \in I \mid x_\iota \neq 0\} = \bigcup_{n \in \mathbb{N}} \{\iota \in I \mid x_\iota \notin U_n\}$$

is countable. □

Remark. L. Kluvanek proved (1961) ([1] page 181 Remark 2) that if G is a metrizable topological vector space and if $(x_\iota)_{\iota \in I}$ is a family in G whose countable subfamilies are summable, then

$$\{\iota \in I \mid x_\iota \neq 0\}$$

is countable. He also gave an example (page 181 Example 1) of a topological vector space with this property which is not metrizable.

Theorem 3.3.16. *Let $\mathscr{F}$ be a subset of $\ell(I, G)$, $\bar{\mathscr{F}}$ be its closure in $\ell_c(I, G)$ and $\ell(I, G)'$ be the set $\ell(I, G)$ endowed with a topology such that the identity map*

$$\ell(I, G)' \to \ell_c(I, G)$$

is Φ_2-continuous. If $\{0\}$ is a G_δ-set of G, then:

a) there exists a metrizable uniformity on $\ell(I, G)$ coarser than the fine uniformity;

b) $\ell(I, G)'$ is a Φ_2-compact Šmulian space;

c) if $\mathscr{F}$ is a Φ_4-set of $\ell_c(I, G)$, then the set

$$\{\iota \in I \mid \exists f \in \bar{\mathscr{F}},\ f(\iota) \neq 0\}$$

is countable;

d) if $\mathscr{F}$ is a Φ_2-set of $\ell_c(I, G)$, then $\ell_c(I, G)$ and $\ell_f(I, G)$ induce the same compact metrizable uniformity on $\bar{\mathscr{F}}$;

e) any countably compact set of $\ell_c(I, G)$ is a metrizable compact set of $\ell_c(I, G)$ and of $\ell_f(I, G)$.

a) follows from the fact that there exists a countable family of entourages of $\ell_f(I, G)$ whose intersection is the diagonal of $\ell(I, G)$.

b) By a) and Corollary 2.3.10a) $\ell_f(I, G)$ is a Φ_2-compact Šmulian space. By Corollary 3.3.7 the identity map

$$\ell_c(I, G) \to \ell_f(I, G)$$

is uniformly Φ_4-continuous. By Proposition 1.8.3 and Corollary 1.8.10 the identity map

$$\ell(I, G)' \to \ell_f(I, G)$$

is Φ_2-continuous. By Proposition 2.3.7 $\ell(I, G)'$ is a Φ_2-compact Šmulian space.

c) follows from Theorem 3.3.13f) and Proposition 3.3.15.

d) By Theorem 3.3.13b) $\ell_c(I, G)$ and $\ell_f(I, G)$ induce the same uniformity on $\bar{\mathscr{F}}$ and by a) and Corollary 2.3.10c) this uniformity is metrizable and compact.

e) follows immediately from d). □

Remark. There exists a locally convex space E for which $\{0\}$ is a G_δ-set which possesses a Φ_3-set $\mathscr{F}$ such that the uniformity induced by E on $\mathscr{F}$ is not metrizable. We take as E the real vector space of bounded sequences of real numbers (i.e. ℓ^∞) endowed with the topology generated by the semi-norms

$$E \to \mathbb{R}_+,\quad (\alpha_n)_{n\in\mathbb{N}} \mapsto \lim_{n,\mathfrak{F}} |\alpha_n|,$$

where $\mathfrak{F}$ runs through the set of all ultrafilters on $\mathbb{N}$. E is a locally convex space for which $\{0\}$ is a G_δ-set. Let $\mathfrak{M}$ be an uncountable set of infinite subsets of $\mathbb{N}$ such that $M' \cap M''$ is finite for any different $M', M'' \in \mathfrak{M}$ (Lemma 3.2.9). We denote, for any $M \in \mathfrak{M}$, by 1_M the characteristic function of M. Then $\mathscr{F} := \{1_M \mid M \in \mathfrak{M}\}$ is a Φ_3-set of E. Since $\{1_M\}$ is an open set of $\mathscr{F}$ for any $M \in \mathfrak{M}$, $\mathscr{F}$ has no countable base. Being a precompact set of E (Proposition 1.5.6), its uniformity is not metrizable. This example shows that d) no longer holds if we replace Φ_2-sets by Φ_3-sets.

Corollary 3.3.17. *Let $i \in \{1, 2, 3, 4\}$ and u be a uniformly Φ_i-continuous group homomorphism of G into H. Then:*

a) for any $(x_\iota)_{\iota \in I} \in \ell(I, G)$ we have $(u(x_\iota))_{\iota \in I} \in \ell(I, H)$ and

$$u\Big(\sum_{\iota \in J} x_\iota\Big) = \sum_{\iota \in J} u(x_\iota)$$

for any $J \subset I$.

b) the map

$$\ell_c(I, G) \rightarrow \ell_f(I, H), \quad (x_\iota)_{\iota \in I} \mapsto (u(x_\iota))_{\iota \in I}$$

is uniformly Φ_i-continuous;

c) if u is continuous, then the maps

$$\ell_c(I, G) \rightarrow \ell_c(I, H), \quad (x_\iota)_{\iota \in I} \mapsto (u(x_\iota))_{\iota \in I},$$
$$\ell_f(I, G) \rightarrow \ell_f(I, H), \quad (x_\iota)_{\iota \in I} \mapsto (u(x_\iota))_{\iota \in I}$$

are uniformly continuous.

The assertions follow immediately from Theorem 3.3.5 and Proposition 3.2.7. □

Definition 3.3.18. *We denote by $\underline{\mathscr{P}(G, H)}$ the set of homomorphisms of R-modules $u: G \rightarrow H$ such that for any supersummable family $(x_\lambda)_{\lambda \in L}$ in G $(u(x_\lambda))_{\lambda \in L}$ is summable and*

$$\sum_{\lambda \in L} u(x_\lambda) = u\Big(\sum_{\lambda \in L} x_\lambda\Big).$$

Further, we denote by $\underline{\mathscr{P}^c(G, H)}$ the set of homomorphisms of R-modules $u: G \rightarrow H$ such that for any countable supersummable family $(x_\lambda)_{\lambda \in L}$ in G, $(u(x_\lambda))_{\lambda \in L}$ is summable and

$$\sum_{\lambda \in L} u(x_\lambda) = u\Big(\sum_{\lambda \in L} x_\lambda\Big).$$

The maps of $\mathscr{P}^c(G, H)$ were considered by C. Bennet and N.J. Kalton (1972) ([1] page 819) for G, H Hausdorff locally convex spaces (they called them subcontinuous), and by P. Dierolf (1975) ([1] Definition 8.6) for G, H Hausdorff topological commutative groups (he called them Σ-continuous).

We have $\mathscr{P}(G, H) \subset \mathscr{P}^c(G, H)$ (in general $\mathscr{P}(G, H) \neq \mathscr{P}^c(G, H)$). By Theorem 3.3.5 any homomorphism of R-modules $u: G \rightarrow H$ belongs to $\mathscr{P}(G, H)$ (to

$\mathscr{P}^c(G, H)$) iff for any set (for any countable set) L and for any $f \in \mathscr{G}(L, G)$ we have $u \circ f \in \mathscr{G}(L, H)$.

If H is a Hausdorff topological group having the same underlying group as G such that the identity map $H \to G$ is continuous, then the identity map $G \to H$ belongs to $\mathscr{P}(G, H)$ iff $\ell(J, G) = \ell(J, H)$ for any set J.

Proposition 3.3.19. *Let G, H be topological real vector spaces and G', H' be the underlying topological groups of G and H respectively. Then any map of $\mathscr{P}^c(G', H')$ is linear.*

Let $u \in \mathscr{P}^c(G', H')$ and $x \in G$. We have $u(nx) = nu(x)$ for any $n \in \mathbb{Z}$. We get

$$nu\left(\frac{m}{n}x\right) = u(mx) = mu(x)$$

and therefore

$$u\left(\frac{m}{n}x\right) = \frac{m}{n}u(x)$$

for any $m, n \in \mathbb{Z}$ with $n \neq 0$. Let $\alpha \in \mathbb{R}$. There exists a sequence $(\alpha_n)_{n \in \mathbb{N}} \in \mathbb{Q}$ which is summable in $\mathbb{R}$ and whose sum is α. Then $(\alpha_n x)_{n \in \mathbb{N}}$ is supersummable in G and its sum is αx. We get by the above considerations

$$u(\alpha x) = \sum_{n \in \mathbb{N}} u(\alpha_n x) = \sum_{n \in \mathbb{N}} \alpha_n u(x) = \alpha u(x). \quad \square$$

Proposition 3.3.20. *Any Φ_1-continuous (sequentially continuous) homomorphism of R-modules of G into H belongs to $\mathscr{P}(G, H)$ (to $\mathscr{P}^c(G, H)$).*

The assertions follow immediately from Proposition 3.1.5 and Theorem 3.3.5. □

Proposition 3.3.21. *Let X be a set, $(I_\lambda)_{\lambda \in L}$ be a family of sets, for each $\lambda \in L$ let g_λ be a map of $\mathfrak{P}(I_\lambda)$ into X such that $\bigcup_{\lambda \in L} g_\lambda(\mathfrak{P}(I_\lambda)) = X$, let Θ be the set of sequences $(x_n)_{n \in \mathbb{N}}$ in X for which there exists $\lambda \in L$ such that*

$$\{n \in \mathbb{N} \mid x_n \in g_\lambda(\mathfrak{P}(I_\lambda))\}$$

is infinite, Φ be the set $\hat{\Phi}(\Theta)$ and $\mathscr{F}$ be the set of maps $f: X \to G$ such that $f \circ g_\lambda \in \mathscr{G}(I_\lambda, G)$ for any $\lambda \in L$. Then:

a) $\mathscr{F}$ is a Φ_4-closed R-submodule of G^X;

b) if G is Φ_i-compact for an $i \in \{1, 2, 3, 4\}$ (e.g. G complete) or sequentially complete, then so is $\mathscr{F}$;

c) $\mathscr{F}$ and $\mathscr{F}_\Phi$ are Hausdorff topological left R-modules, locally convex if G is locally convex;

d) $\ell(I, \mathscr{F}) = \ell(I, \mathscr{F}_\Phi)$ *and the identity map*

$$\ell_c(I, \mathscr{F}) \to \ell_f(I, \mathscr{F}_\Phi)$$

is uniformly Φ_4*-continuous;*

e) any family in $\mathscr{F}$ *which is supersummable in* G^X *is supersummable in* $\mathscr{F}_\Phi$.

a) follows from Proposition 3.2.2 b), i).

b) follows from Proposition 3.2.2 c), d).

c) follows from Proposition 3.2.2 j).

d) By Proposition 3.2.2 a) the identity map $\mathscr{F} \to \mathscr{F}_\Phi$ is uniformly Φ_4-continuous and the assertions follow Corollary 3.3.17 a), b).

e) By a) $\mathscr{F}$ is a Φ_4-closed set and therefore (Proposition 1.5.20 a)) a strongly Φ_2-closed set of G^X. By Proposition 3.3.4 any family in $\mathscr{F}$ which is supersummable in G^X is supersummable in $\mathscr{F}$ and by d) it is even supersummable in $\mathscr{F}_\Phi$. □

Theorem 3.3.22. *Let* Θ *be the set of sequences* $(x_n)_{n\in\mathbb{N}}$ *in* G *for which there exists a supersummable (a countable supersummable) family* $(y_\lambda)_{\lambda\in L}$ *in* G *such that*

$$\{n \in \mathbb{N} \mid x_n \in \{\sum_{\lambda\in K} y_\lambda \mid K \subset L\}\}$$

is an infinite set and let Φ *be the set* $\hat{\Phi}(\Theta)$. *We have*

a) $\mathscr{P}(G,H)$ *and* $\mathscr{P}^c(G,H)$ *are* Φ_4*-closed subgroups of* H^G*; if* R *is commutative then they are* R*-submodules of* H^G*;*

b) the identity map

$$\mathscr{P}(G,H) \to \mathscr{P}(G,H)_\Phi, \quad (\mathscr{P}^c(G,H) \to \mathscr{P}^c(G,H)_\Phi)$$

is uniformly Φ_4*-continuous;*

c) if H *is* Φ_i*-compact for an* $i \in \{1,2,3,4\}$ *(e.g. complete) or sequentially complete, then so are* $\mathscr{P}(G,H)$ *and* $\mathscr{P}^c(G,H)$*;*

d) if $\{0\}$ *is a* G_δ*-set of* H*, then* $\mathscr{P}(G,H)$ *and* $\mathscr{P}^c(G,H)$ *are* Φ_2*-compact; if, in addition,* H *is sequentially complete, then* $\mathscr{P}(G,H)$ *and* $\mathscr{P}^c(G,H)$ *are* Φ_4*-compact;*

e) $f(\mathfrak{F})$ *is a* Φ_2*-filter (*Φ_1*-filter) for any* $f \in \mathscr{P}(G,H)$ *(for any* $f \in \mathscr{P}^c(G,H)$*) and for any* $\mathfrak{F} \in \Phi(\Theta)$*;*

f) $f(\mathfrak{F})$ *is a* $\hat{\Phi}_2$*-filter (*$\hat{\Phi}_1$*-filter) for any* $f \in \mathscr{P}(G,H)$ *(for any* $f \in \mathscr{P}^c(G,H)$*) and for any* $\mathfrak{F} \in \hat{\Phi}(\Theta)$*;*

g) let $\mathfrak{F} \in \hat{\Phi}_4(\mathscr{P}(G,H))$, (K,h) *be a net in* G *such that* K *possesses no greatest element and such that for any strictly increasing sequence* $(\kappa_n)_{n\in\mathbb{N}}$ *in* K *there exist a supersummable (a countable supersummable) family* $(y_\lambda)_{\lambda\in L}$ *in* G *and a sequence* $(L_n)_{n\in\mathbb{N}}$ *in* $\mathfrak{P}(L)$ *converging to* $\emptyset$ *such that* $(\sum_{\lambda\in L_n} y_\lambda)_{n\in\mathbb{N}}$ *is a subsequence of* $(h(\kappa_n))_{n\in\mathbb{N}}$*, let* $\mathfrak{G}$ *be the section filter of* K *and* φ *be the map*

$$\mathscr{P}(G,H) \times K \to G, \quad (f,\kappa) \mapsto f(h(\kappa));$$

then $\varphi(\mathfrak{F} \times \mathfrak{G})$ *converges to 0;*

h) $\mathscr{P}(G,H)$, $\mathscr{P}(G,H)_\Phi$, ($\mathscr{P}^c(G,H)$, $\mathscr{P}^c(G,H)_\Phi$) *are Hausdorff topological groups: if R is commutative then they are topological R-modules; if H is locally convex, then they are locally convex;*

i) $\ell(I,\mathscr{P}(G,H)) = \ell(I,\mathscr{P}(G,H)_\Phi)$

$$(\ell(I,\mathscr{P}^c(G,H)) = \ell(I,\mathscr{P}^c(G,H)_\Phi))$$

and the identity maps

$$\ell_c(I,\mathscr{P}(G,H)) \to \ell_f(I,\mathscr{P}(G,H)_\Phi)$$

$$(\ell_c(I,\mathscr{P}^c(G,H)) \to \ell_f(I,\mathscr{P}^c(G,H)_\Phi))$$

are uniformly Φ_4*-continuous;*

j) any family in $\mathscr{P}(G,H)$ *(in* $\mathscr{P}^c(G,H)$*) which is supersummable in* H^G *is supersummable in* $\mathscr{P}(G,H)_\Phi$ *(in* $\mathscr{P}^c(G,H)_\Phi$*).*

Let G', H' be the underlying topological groups of G and H respectively and let $\mathscr{F}$ be the set of homomorphisms of R-modules of G into H. Then $\mathscr{F}$ is a closed set of H^G and

$$\mathscr{P}(G,H) = \mathscr{P}(G',H')\cap\mathscr{F}, \quad \mathscr{P}^c(G,H) = \mathscr{P}^c(G',H')\cap\mathscr{F}.$$

Let us denote by Δ the class of supersummable (of countable supersummable) families in G. Without loss of generality we may assume that Δ is a set. For any $y := (y_\lambda)_{\lambda\in L(y)} \in \Delta$ we denote by g_y the map

$$\mathfrak{P}(L(y)) \to G, \quad J \mapsto \sum_{\lambda\in J} y_\lambda .$$

By Theorem 3.3.5 a map $u: G\to H$ belongs to $\mathscr{P}(G',H')$ (to $\mathscr{P}^c(G',H')$) iff $u\circ g_y \in \mathscr{G}(L(y),H)$ for any $y\in\Delta$ and the assertions of the theorem follow from Propositions 3.2.2 and 3.3.21 if we replace G by G' and H by H' in these assertions. Now the assertions for G and H follow with the aid of the above remarks and of Corollary 1.8.15. □

Proposition 3.3.23. *Let u be a homomorphism of R-modules of G into H for which there exists a fundamental system* $\mathfrak{U}$ *of 0-neighbourhoods in H such that* $\overset{-1}{u}(U)$ *is strongly* Φ_1*-closed for any* $U\in\mathfrak{U}$ *and let* $(x_\iota)_{\iota\in I}\in\ell(I,G)$*. Then the following assertions are equivalent;*

a) $(u(x_\iota))_{\iota\in I}$ *is supersummable und* $u(\sum_{\iota\in J} x_\iota) = \sum_{\iota\in J} u(x_\iota)$ *for any* $J\subset I$*;*

b) $\{u(\sum_{\iota\in J} x_\iota) \mid J\subset I\}$ *is compact;*

c) $\{u(\sum_{\iota\in J} x_\iota) \mid J\subset I\}$ *is semi-separable.*

In particular, if $u(G)$ *is semi-separable, then* $u\in\mathscr{P}(G,H)$*. If I is countable we may replace the strongly* Φ_1*-closed condition by sequentially closed.*

The assertion follows immediately from Theorem 3.3.5 and Proposition 3.1.12. □

Proposition 3.3.24. *Assume G and H have the same underlying group and there exists a fundamental system of 0-neighbourhoods in H which are strongly Φ_1-closed in G. Then the identity map $G \to H$ is $\Phi_1(G) \cap \Phi_5(H)$-continuous.*

Let $\mathfrak{F}$ be a convergent $\Phi_1(G) \cap \Phi_5(H)$-ultrafilter on G and let x be its limit point in G. Furthermore, let u be the identity map $G \to H$ and let U be a 0-neighbourhood in H which is strongly Φ_1-closed in G. By Proposition 1.5.10 $\mathfrak{F}$ is a Cauchy filter on H and therefore there exists $A \in \mathfrak{F}$ with $A - A \subset U$. By Proposition 3.1.10 we have $x + U \in \mathfrak{F}$. Since U is arbitrary $\mathfrak{F}$ converges to x in H. □

Corollary 3.3.25. *Assume G and H have the same underlying R-module and there exists a fundamental system of 0-neighbourhoods in H which are strongly Φ_1-closed in G and let $(x_\iota)_{\iota \in I} \in \ell(I, G)$. Then the following assertions are equivalent;*

a) $(x_\iota)_{\iota \in I}$ is supersummable in H and the sums in G and H of its subfamilies coincide;

b) $\{\sum_{\iota \in J} x_\iota \mid J \subset I\}$ is a compact set of H;

c) $\{\sum_{\iota \in J} x_\iota \mid J \subset I\}$ is a semi-separable set of H;

d) any countable subfamily of $(x_\iota)_{\iota \in I}$ is summable in H and its sums in G and H coincide.

In particular, if H is semi-separable, then the identity map $G \to H$ belongs to $\mathscr{P}(G, H)$. If I is countable we may replace the strongly Φ_1-closed condition by sequentially closed.

The implications a ⇔ b ⇔ c as well as the last two assertions follow immediately from Proposition 3.3.23. The implication a ⇒ d is trivial.

d ⇒ a. Let $J \subset I$, $\mathfrak{F}$ be the section filter of $\mathfrak{P}_c(J)$ and f be the map

$$\mathfrak{P}_c(J) \to H, \quad K \mapsto \sum_{\iota \in K} x_\iota .$$

Then $(\mathfrak{P}_c(J), f)$ is a Θ_1-net in G and in H so $f(\mathfrak{F})$ is a Φ_1-filter on G and on H which converges to the sum of $(x_\iota)_{\iota \in J}$ in G. By Proposition 3.3.24 $f(\mathfrak{F})$ converges to this sum in H. □

Remark. This result was proved by N.J. Kalton (1971) ([1] Theorem 7) for H separable, I countable, with the assumptions sequentially closed and strongly Φ_1-closed replaced by closed. L. Drewnowski (1973) improved this result by replacing in the last hypothesis closed by sequentially closed and replacing H separable by $\{\sum_{\iota \in J} x_\iota \mid J \subset I\}$ is a separable set of H ([4] Theorem 1). In a later paper (1975) ([6] Theorem 1) Drewnowski replaced separable by semi-separable.

Lemma 3.3.26. *(J.D. Weston (1957)* [1] *Theorem 8). Let $\mathfrak{T}_1$, $\mathfrak{T}_2$ be two Hausdorff topologies defined on the same set X such that:*

a) $\mathfrak{T}_1 \subset \mathfrak{T}_2$;

b) $\mathfrak{T}_2$ is metrizable by a complete metric:

c) for any $x \in X$ and for any $\mathfrak{T}_2$-neighbourhood U of x, the $\mathfrak{T}_1$-closure of U is a $\mathfrak{T}_1$-neighbourhood of x.

Then $\mathfrak{T}_1 = \mathfrak{T}_2$.

For any subset A of X we denote by $\bar{A}$ the $\mathfrak{T}_1$-closure of A. Let d be a complete metric on X generating $\mathfrak{T}_2$. For any $x \in X$ and $\varepsilon > 0$ we set

$$U(x, \varepsilon) := \{y \in X \mid d(x, y) < \varepsilon\}.$$

Let $x_0 \in X$ and let $\varepsilon > 0$. We want to prove

$$\overline{U\left(x_0, \frac{\varepsilon}{2}\right)} \subset U(x_0, 2\varepsilon).$$

Let $y_0 \in \overline{U\left(x_0, \frac{\varepsilon}{2}\right)}$. We construct inductively two sequences $(x_n)_{n \in \mathbb{N}}$, $(y_n)_{n \in \mathbb{N}}$ such that we have for any $n \in \mathbb{N}$, $n \neq 0$:

1) $d(x_n x_{n-1}) < \frac{\varepsilon}{2^n}$;

2) $d(y_n, y_{n-1}) < \frac{\varepsilon}{2^n}$;

3) $y_n \in \overline{U\left(x_n, \frac{\varepsilon}{2^{n+1}}\right)}$.

Let $n \in \mathbb{N}$ with $n \neq 0$ and assume the sequences were constructed up to $n-1$. By c) $\overline{U\left(y_{n-1}, \frac{\varepsilon}{2^n}\right)}$ is a $\mathfrak{T}_1$-neighbourhood of y_{n-1} so by 3) there exists

$$x_n \in \overline{U\left(y_{n-1}, \frac{\varepsilon}{2^n}\right)} \cap U\left(x_{n-1}, \frac{\varepsilon}{2^n}\right).$$

Again by c) $\overline{U\left(x_n, \frac{\varepsilon}{2^{n+1}}\right)}$ is a $\mathfrak{T}_1$-neighbourhood of x_n so there exists

$$y_n \in \overline{U\left(x_n, \frac{\varepsilon}{2^{n+1}}\right)} \cap U\left(y_{n-1}, \frac{\varepsilon}{2^n}\right).$$

$(x_n)_{n \in \mathbb{N}}$ and $(y_n)_{n \in \mathbb{N}}$ are Cauchy sequences and therefore convergent sequences with respect to d. We set

$$x := \lim_{n \to \infty} x_n, \quad y := \lim_{n \to \infty} y_n.$$

We want to prove $x = y$. Assume the contrary. Then there exists a $\mathfrak{T}_1$-neighbourhood U of x such that $y \notin \bar{U}$. By a) there exists $\eta > 0$ such that

$$U(x, \eta) \subset U.$$

For a sufficiently large $n \in \mathbb{N}$ we have

$$d(x_n, x) < \frac{\eta}{2}, \quad \frac{\varepsilon}{2^n} < \frac{\eta}{2}$$

so

$$U\left(x_n, \frac{\varepsilon}{2^n}\right) \subset U(x, \eta) \subset U.$$

By 3) we get $y_n \in \bar{U}$, which is a contradiction. Hence, $x = y$. We have

$$d(x, x_0) = \lim_{n \to \infty} d(x_n, x_0) < \varepsilon,$$

$$d(y, y_0) = \lim_{n \to \infty} d(y_n, y_0) < \varepsilon$$

so $d(x_0, y_0) < 2\varepsilon$ and $y_0 \in U(x_0, 2\varepsilon)$. Since y_0 is arbitrary we have proved

$$\overline{U\left(x_0, \frac{\varepsilon}{2}\right)} \subset U(x_0, 2\varepsilon)$$

so by c) $\mathfrak{T}_1 \supset \mathfrak{T}_2$. □

Theorem 3.3.27. *Assume G and H have the same underlying group G_0 and the identity map $H \to G$ is continuous and let $(x_\iota)_{\iota \in I} \in \ell(I, G)$. If $\{\sum_{\iota \in J} x_\iota \mid J \subset I\}$ is a semi-separable set of H and if it is contained in a subset A of G_0 such that the topology induced on A by H is metrizable by a complete metric, then $(x_\iota)_{\iota \in I}$ is supersummable in H.*

Let $\mathfrak{T}$ be the finest group topology on G_0 coarser than the topology of H and such that the map

$$\mathfrak{P}(I) \to G_0, \quad J \mapsto \sum_{\iota \in J} x_\iota$$

is continuous with respect to $\mathfrak{T}$. By Theorem 3.3.5 $\mathfrak{T}$ is finer than the topology of G and $(x_\iota)_{\iota \in I}$ is supersummable with respect to $\mathfrak{T}$. Let $\mathfrak{U}$ be the neighbourhood filter of 0 in H and, for any $U \in \mathfrak{U}$, denote by $\bar{U}$ its closure with respect to $\mathfrak{T}$. Then $\{\bar{U} \mid U \in \mathfrak{U}\}$ is a fundamental system of neighbourhoods of 0 for a group topology $\mathfrak{T}'$ on G which is finer than $\mathfrak{T}$ and coarser than the topology of H. Hence $\{\sum_{\iota \in J} x_\iota \mid J \subset I\}$ is semi-separable with respect to $\mathfrak{T}'$. By Corollary 3.3.25 $(x_\iota)_{\iota \in I}$ is supersummable with respect to $\mathfrak{T}'$. We deduce $\mathfrak{T} = \mathfrak{T}'$. It follows that for any $U \in \mathfrak{U}$ the set $\bar{U}$ is a neighbourhood of 0 with respect to $\mathfrak{T}$. By Lemma 3.3.26 $\mathfrak{T}$ and H induce the same topology on A. Hence, the map

$$\mathfrak{P}(I) \to H, \quad J \mapsto \sum_{\iota \in J} x_\iota$$

is continuous. We deduce $\mathfrak{T}$ is the topology of H. □

Remarks. 1) This result was proved by N.J. Kalton (1971) ([1] Theorem 3) for I countable, G and H separable, and H metrizable and complete. It was improved by L. Drewnowski (1973) ([4] Theorem 2) for I countable, $\{\sum_{\iota \in J} x_\iota \mid J \subset I\}$ separable in H, and the topology induced on $\{\sum_{\iota \in J} x_\iota \mid J \subset I\}$ by H metrizable and complete. The above proof follows very closely the proof of the last quoted theorem.

2) We cannot drop the semi-separable hypothesis in the above theorem as the example $G := \mathbb{R}$, $H :=$ ($\mathbb{R}$ endowed with the discrete topology) shows.

Proposition 3.3.28. *Let u be a homomorphism of R-modules of G into H. If G is metrizable and complete then the following assertions are equivalent:*

a) u is continuous;

b) $u \in \mathscr{P}(G, H)$;

c) $u \in \mathscr{P}^c(G, H)$;

d) $\lim_{n \to \infty} u(x_n) = 0$ for any summable sequence $(x_n)_{n \in \mathbb{N}}$ in G.

a ⇒ b ⇒ c ⇒ d is trivial.

d ⇒ a. Let $(x_n)_{n \in \mathbb{N}}$ be a sequence in G converging to 0. We have to show that $(u(x_n))_{n \in \mathbb{N}}$ converges to 0 in H. Assume the contrary. Then there exists a 0-neighbourhood V in H such that $\{n \in \mathbb{N} \mid u(x_n) \notin V\}$ is an infinite set. Let d be an invariant metric on G generating its uniformity. There exists a subsequence $(x_{k_n})_{n \in \mathbb{N}}$ of $(x_n)_{n \in \mathbb{N}}$ such that

$$u(x_{k_n}) \notin V, \quad d(x_{k_n}, 0) < \frac{1}{2^n}$$

for any $n \in \mathbb{N}$. Then $(\sum_{n=0}^{p} x_{k_n})_{p \in \mathbb{N}}$ is a Cauchy sequence. Since G is complete $(x_{k_n})_{n \in \mathbb{N}}$ is summable and by d) $\lim_{n \to \infty} u(x_{k_n}) = 0$, which is the expected contradiction. □

Corollary 3.3.29. *If G is metrizable and complete then the topology of G is the finest Hausdorff group topology on the underlying group of G for which the summable sequences (families) in G are summable and have the same sum.* □

Remarks. 1) This result was proved by P. Dierolf (1975) ([1] Korollar 11.9 (a)).

2) Let $(u_\iota)_{\iota \in I}$ be a family of group homomorphisms of metrizable complete topological commutative groups into G such that the topology of G is the final topology with respect to this family. P. Dierolf proved (1979) ([3] Proposition 3.17) that in this case the above corollary holds too.

The next result will not be used in the sequel.

Proposition 3.3.30. *Let $(x_n)_{n \in \mathbb{N}}$ be a sequence in G such that $\lim_{n \to \infty} \sum_{\substack{m \in M \\ m \leq n}} x_m$ exists for* *any* $M \subset \mathbb{N}$. *Then* $(x_n)_{n \in \mathbb{N}}$ *is supersummable and the above convergence is uniform in* M.

We set $x_{mn} := x_m$ for all $m, n \in \mathbb{N}$ with $m \leq n$ and $x_{mn} := 0$ for all $m, n \in \mathbb{N}$ with $m > n$. Then $(x_{mn})_{m \in \mathbb{N}} \in \ell^1(\mathbb{N}, G)$ for every $n \in \mathbb{N}$ and $(\sum_{m \in M} x_{mn})_{n \in \mathbb{N}}$ converges for every $M \subset \mathbb{N}$ and the assertions follow from Corollary 3.3.9. □

Proposition 3.3.31. *Let* $(x_\iota)_{\iota \in I}$ *be a family in G and* $(I_\lambda)_{\lambda \in L}$ *be a disjoint family of sets such that:*

a) $\bigcup_{\lambda \in L} I_\lambda = I$;

b) $(x_\iota)_{\iota \in I_\lambda}$ *is supersummable for any* $\lambda \in L$;

c) $(\sum_{\iota \in J \cap I_\lambda} x_\iota)_{\lambda \in L}$ *is summable for any* $J \subset I$.

Then $(x_\iota)_{\iota \in I}$ *is supersummable and* $\sum_{\iota \in J} x_\iota = \sum_{\lambda \in L} \sum_{\iota \in J \cap J_\lambda} x_\iota$ *for any* $J \subset I$.

We set

$$x_{K\iota} := \begin{cases} x_\iota & \text{if } \iota \in \bigcup_{\lambda \in K} I_\lambda \\ 0 & \text{if } \iota \notin \bigcup_{\lambda \in K} I_\lambda \end{cases}$$

for all $(K, \iota) \in \mathfrak{P}_f(L) \times I$ and denote by $\mathfrak{F}$ the section filter of $\mathfrak{P}_f(L)$ and by φ the map

$$\mathfrak{P}_f(L) \to \ell(I, G), \ K \mapsto (x_{K\iota})_{\iota \in I}.$$

Then $\varphi(\mathfrak{F})$ is a Φ_3-filter on $\ell_c(I, G)$ such that

$$\lim_{K, \mathscr{F}} \sum_{\iota \in J} (\varphi(K))(\iota) = \sum_{\lambda \in L} (\sum_{\iota \in J \cap I_\lambda} x_\iota)$$

for every $J \subset I$. By Corollary 3.3.8 a) $(x_\iota)_{\iota \in I}$ is supersummable and

$$\sum_{\iota \in J} x_\iota = \sum_{\lambda \in L} \sum_{\iota \in J \cap J_\lambda} x_\iota$$

for every $J \subset I$. □

Theorem 3.3.32. *Let R be Hausdorff and* $(g_\lambda)_{\lambda \in L}$ *be a supersummable family in* R^I *such that* $\{\sum_{\lambda \in M} g_\lambda(\iota) \mid M \subset L\}$ *possesses an invertible element for any* $\iota \in I$. *We set*

$$\mathscr{G} := \prod_{\iota \in I} \{\sum_{\lambda \in M} g_\lambda(\iota) \mid M \subset L\},$$

$$\mathscr{G}' := \{g \in \mathscr{G} \mid \{\iota \in I \mid g(\iota) \neq 0\} \quad \textit{countable}\},$$

$$\mathscr{F} := \{f \in G^I \mid g \in \mathscr{G} \Rightarrow (g(\iota) f(\iota))_{\iota \in I} \in \ell(I, G)\},$$

and

$$<g, f> := \sum_{\iota \in I} g(\iota) f(\iota)$$

for any $(g,f)\in\mathcal{G}\times\mathcal{F}$ *and denote by* $\mathcal{F}'$ *(by* $\mathcal{F}''$*) the set* $\mathcal{F}$ *endowed with the coarsest uniformity for which* $\langle g,\cdot\rangle$ *is uniformly continuous for any* $g\in\mathcal{G}'$ *(for which the set* $\{\langle g,\cdot\rangle\,|\,g\in\mathcal{G}\}$ is *uniformly equicontinuous). We have:*

a) the map

$$\mathcal{G}\to G,\quad g\mapsto\langle g,f\rangle$$

is continuous for any $f\in\mathcal{F}$ *(in particular,* $\{\langle g,f\rangle\,|\,g\in\mathcal{G}\}$ *is compact for any* $f\in\mathcal{F}$*);*

b) the identity map $\mathcal{F}'\to\mathcal{F}''$ *is uniformly* Φ_4*-continuous;*

c) if we embed $\mathcal{F}$ *in* $G^{\mathcal{G}}$ *via the map*

$$\mathcal{F}\to G^{\mathcal{G}},\quad f\mapsto\langle\cdot,f\rangle,$$

then $\mathcal{F}$ *is a* Φ_4*-closed set of* $G^{\mathcal{G}}$*;*

d) if $(f_n)_{n\in\mathbb{N}}$ *is a sequence in* $\mathcal{F}$ *such that* $(\langle g,f_n\rangle)_{n\in\mathbb{N}}$ *converges for any* $g\in\mathcal{G}$*, then* $(f_n)_{n\in\mathbb{N}}$ *converges in* $\mathcal{F}_I$ *to an* f*,*

$$\langle g,f\rangle=\lim_{n\to\infty}\langle g,f_n\rangle$$

for any $g\in\mathcal{G}$*, and for any 0-neighbourhood* U *in* G *there exist* $n_0\in\mathbb{N}$ *and a finite subset* J *of* I *such that*

$$\langle g',f_m\rangle-\langle g'',f\rangle\in U,\quad\langle g,f_n\rangle\in U$$

for any $m,n\in\mathbb{N}$ *with* $m\geq n_0$*, and for any* $g,\ g',\ g''\in\mathcal{G}$ *with* $g'|J=g''|J$ *and* $g|J=0$.

For any $\iota\in I$ let L_ι be a subset of L such that $\sum_{\lambda\in L_\iota} g_\lambda(\iota)$ is an invertible element of R. We set

$$\mathcal{H}:=\{h\in\mathcal{G}(I\times L,G)\,|\,\iota\in I,\ M\subset L\Rightarrow$$
$$h(\{\iota\}\times M)=(\sum_{\lambda\in M}g_\lambda(\iota))(\sum_{\lambda\in L_\iota}g_\lambda(\iota))^{-1}h(\{\iota\}\times L_\iota)\},$$

$$A_\iota:=\{\lambda\in L\,|\,(\iota,\lambda)\in A\}$$

for any $A\subset I\times L$ and any $\iota\in I$,

$$\mathfrak{A}:=\{A\subset I\times L\,|\,\{\iota\in I\,|\,A_\iota\neq\emptyset\}\ \text{ countable}\},$$

$$\varphi:\mathfrak{P}(I\times L)\to\mathcal{G},\quad A\mapsto(\sum_{\lambda\in A_\iota}g_\lambda(\iota))_{\iota\in I},$$

$$f':\mathfrak{P}(I\times L)\to G,\quad A\mapsto\langle\varphi(A),f\rangle$$

for any $f\in\mathcal{F}$.

Let $f\in\mathcal{F}$. We want to show $f'\in\mathcal{H}$. First, we prove $f'\in\mathcal{G}(I\times L,G)$. For this purpose, we apply Proposition 3.3.31 replacing I by $I\times L$, I_λ by $\{\lambda\}\times L$ for any $\lambda\in I$, and $x_{(\iota,\lambda)}$ by $g_\lambda(\iota)f(\iota)$ for any $(\iota,\lambda)\in I\times L$. Hence, $(g_\lambda(\iota)f(\iota))_{(\iota,\lambda)\in I\times L}$ is supersummable and by Theorem 3.3.5 $f'\in\mathcal{G}(I\times L,G)$.

Let $\iota \in I$ and $M \subset L$. We have

$$f'(\{\iota\} \times L_\iota) = (\sum_{\lambda \in L_\iota} g_\lambda(\iota)) f(\iota),$$

$$f'(\{\iota\} \times M) = (\sum_{\lambda \in M} g_\lambda(\iota)) f(\iota) = (\sum_{\lambda \in M} g_\lambda(\iota))(\sum_{\lambda \in L_\iota} g_\lambda(\iota))^{-1} f'(\{\iota\} \times L_\iota).$$

Hence, $f' \in \mathscr{H}$.

Now let $h \in \mathscr{H}$. We set

$$f: I \to G, \quad \iota \mapsto (\sum_{\lambda \in L_\iota} g_\lambda(\iota))^{-1} h(\{\iota\} \times L_\iota).$$

Let $M \subset L$. We have

$$\sum_{\lambda \in M} g_\lambda(\iota) f(\iota) = (\sum_{\lambda \in M} g_\lambda(\iota))(\sum_{\lambda \in L_\iota} g_\lambda(\iota))^{-1} \mathrm{h}(\{\iota\} \times L_\iota) = h(\{\iota\} \times M)$$

for any $\iota \in I$. By Theorem 3.3.5 $f \in \mathscr{F}$ and $f' = h$. Hence, the map

$$\mathscr{F} \to \mathscr{H}, \quad f \mapsto f'$$

is bijective.

a) Let $f \in \mathscr{F}$ and ψ be the map

$$\mathscr{G} \to G, \quad g \mapsto \langle g, f \rangle.$$

Then $\psi \circ \varphi = f' \in \mathscr{H}$ so $\psi \circ \varphi$ is continuous. Since $\mathfrak{P}(I \times L)$ is compact and φ is continuous and surjective, ψ is continuous.

b) We identify $\mathscr{F}$ with $\mathscr{H}$ via the bijection

$$\mathscr{F} \to \mathscr{H}, \quad f \mapsto f'.$$

By this identification $\mathscr{F}' = \mathscr{H}_{\mathfrak{A}}$ and $\mathscr{F}'' = \mathscr{H}_{\{\mathfrak{P}(I \times L)\}}$. We have

$$\mathfrak{P}_f(I \times L) \subset \mathfrak{A},$$

$$A \subset B \in \mathfrak{A} \Rightarrow A \in \mathfrak{A},$$

and for any disjoint sequence $(A_n)_{n \in \mathbb{N}}$ in $\mathfrak{P}_f(I \times L)$ there exists an infinite subset M of $\mathbb{N}$ such that $\bigcup_{n \in \mathbb{N}} A_n \in \mathfrak{A}$. By Theorem 3.2.3 c) and Corollary 1.8.15 the identity map $\mathscr{H}_{\mathfrak{A}} \to \mathscr{H}_{\{\mathfrak{P}(I \times L)\}}$ is uniformly Φ_4-continuous. By the above identification the identity map $\mathscr{F}' \to \mathscr{F}''$ is uniformly Φ_4-continuous.

c) By Corollary 3.2.4 a) $\mathscr{G}(I \times L, G)$ is a Φ_4-closed set of $G^{\mathfrak{P}(I \times L)}$. Since $\mathscr{H}$ is a closed set of $\mathscr{G}(I \times L, G)$ with respect to the topology induced by $G^{\mathfrak{P}(I \times L)}$ we deduce $\mathscr{H}$ is a Φ_4-closed set of $G^{\mathfrak{P}(I \times L)}$. Denote by ψ the map

$$\mathscr{F} \to \mathscr{H}, \quad f \mapsto f$$

and by ψ' and ψ'' the inclusion maps $\mathscr{F} \to G^{\mathscr{G}}$ and $\mathscr{H} \to G^{\mathfrak{P}(I \times L)}$ respectively. Let $\mathfrak{F}$ be a Φ_4-filter on $\mathscr{F}$ such that $\psi'(\mathfrak{F})$ converges to an $f \in G^{\mathscr{G}}$. Then $\psi(\mathfrak{F})$ is a Φ_4-filter on $\mathscr{H}$ such that $\psi''(\psi(\mathfrak{F}))$ converges to $f \circ \varphi$. Since $\mathscr{H}$ is a Φ_4-closed set of $G^{\mathfrak{P}(I \times L)}$ we get $f \circ \varphi \in \mathscr{H}$. Hence, $f \in \mathscr{F}$ and $\mathscr{F}$ is a Φ_4-closed set of $G^{\mathscr{G}}$.

d) By c) and the fact that $\{\sum_{\lambda \in M} g_\lambda(\iota) \mid M \subset L\}$ possesses an invertible element for any $\iota \in I$ we deduce that $(f_n)_{n\in\mathbb{N}}$ converges to an f in $\mathscr{F}_I$. Then $(f'_n)_{n\in\mathbb{N}}$ converges to f' in $\mathscr{G}(I \times L, G)$ and by Corollary 3.2.6 there exist $n_0 \in \mathbb{N}$ and a finite subset A of $I \times L$ such that

$$f'_n(B) - f'(C) \in U$$

for any $n \in \mathbb{N}$ with $n \geq n_0$ and any $B, C \in \mathfrak{P}(I \times L)$ with $A \cap B = A \cap C$. We get

$$\langle g', f_n \rangle - \langle g'', f \rangle \in U$$

for any $n \in \mathbb{N}$ with $n \geq n_0$ and any $g', g'' \in \mathscr{G}$ with $g' \mid J' = g'' \mid J'$, where $J' := \{\iota \in I \mid A_\iota \neq \emptyset\}$. By a) there exists $J'' \in \mathfrak{P}_f(I)$ such that $\langle g, f_m \rangle \in U$ for any $g \in \mathscr{G}$ with $g \mid J'' = 0$ and for any $m \in \mathbb{N}$ in case $m \leq n_0$. The set $J := J' \cup J''$ possesses the required properties. □

Corollary 3.3.33. *Let G be a Hausdorff topological vector space and let $(\alpha_\iota)_{\iota \in I}$ be a family of strictly positive real numbers. We set*

$$\mathscr{G} := \{g \in \mathbb{R}^I (\textit{or } \mathbb{C}^I) \mid \iota \in I \Rightarrow |g(\iota)| \leq \alpha_\iota\},$$

$$\mathscr{G}' := \{g \in \mathscr{G} \mid \{\iota \in I \mid g(\iota) \neq 0\} \quad \textit{countable}\},$$

$$\mathscr{F} := \{f \in G^I \mid g \in \mathscr{G} \Rightarrow (g(\iota) f(\iota))_{\iota \in I} \in \ell(I, G)\},$$

and

$$\langle g, f \rangle := \sum_{\iota \in I} g(\iota) f(\iota)$$

for any $(g, f) \in \mathscr{G} \times \mathscr{F}$ and denote by $\mathscr{F}'$ (by $\mathscr{F}''$) the set $\mathscr{F}$ endowed with the coarsest uniformity for which $\langle g, \cdot \rangle$ is uniformly continuous for any $g \in \mathscr{G}'$ (for which the set $\{\langle g, \cdot \rangle \mid g \in \mathscr{G}\}$ is uniformly equicontinuous). We have:

a) the map

$$\mathscr{G} \to G, \quad g \mapsto \langle g, f \rangle$$

is continuous for any $f \in \mathscr{F}$ (in particular, $\{\langle g, f \rangle \mid g \in \mathscr{G}\}$ is compact for any $f \in \mathscr{F}$);

b) the identity map $\mathscr{F}' \to \mathscr{F}''$ is uniformly Φ_4-continuous;

c) if we embed $\mathscr{F}$ in $G^{\mathscr{G}}$ via the map

$$\mathscr{F} \to G^{\mathscr{G}}, \quad f \mapsto \langle \cdot, f \rangle,$$

then $\mathscr{F}$ is a Φ_4-closed set of $G^{\mathscr{G}}$;

d) if $(f_n)_{n\in\mathbb{N}}$ is a sequence in $\mathscr{F}$ such that $(\langle g, f_n \rangle)_{n\in\mathbb{N}}$ converges for any $g \in \mathscr{G}$, then $(f_n(\iota))_{n\in\mathbb{N}}$ converges for any $\iota \in I$, the map

$$f : I \to G, \quad \iota \mapsto \lim_{n\to\infty} f_n(\iota)$$

belongs to $\mathscr{F}$,

$$\sum_{\iota \in I} g(\iota) f(\iota) = \lim_{n \to \infty} \sum_{\iota \in I} g(\iota) f_n(\iota)$$

for any $g \in \mathcal{G}$, and for any 0-neighbourhood U in G there exists $n_0 \in \mathbb{N}$ and a finite subset J of I such that

$$\sum_{\iota \in I} g'(\iota) f_n(\iota) - \sum_{\iota \in I} g''(\iota) f(\iota) \in U \, ,$$

$$\sum_{\iota \in I} g(\iota) f_m(\iota) \in U$$

for any $m, n \in \mathbb{N}$ so that $n \geq n_0$ and for any $g, g', g'' \in \mathcal{G}$ with $g' \mid J = g'' \mid J$ and $g \mid J = 0$.

The real case follows immediately from Theorem 3.3.32 since

$$[-\alpha, \alpha] = \left\{ \sum_{n \in M} \frac{n}{|n|} \frac{\alpha}{2^{|n|}} \mid M \subset \mathbb{Z} \setminus \{0\} \right\}$$

for any positive real number α. Now let us discuss the complex case.

We have $(g(\iota) f(\iota))_{\iota \in I} \in \ell(I, G)$ for any $f \in \mathcal{F}$ and for any $g \in \mathbb{C}^I$ such that $|g(\iota)| \leq 2\alpha_\iota$ for any $\iota \in I$. We set

$$K_\iota := \left\{ \sum_{n \in M'} \frac{n}{|n|} \frac{\alpha_\iota}{2^{|n|}} + i \sum_{n \in M''} \frac{n}{|n|} \frac{\alpha_\iota}{2^{|n|}} \mid M', M'' \in \mathfrak{P}(\mathbb{Z} \setminus \{0\}) \right\}$$

for any $\iota \in I$. Then for every $\iota \in I$

$$\{z \in \mathbb{C} \mid |z| \leq \alpha_\iota\} \subset K_\iota \subset \{z \in \mathbb{C} \mid |z| \leq 2\alpha_\iota\}$$

so $(g(\iota) f(\iota))_{\iota \in I} \in \ell(I, G)$ for any $g \in \prod_{\iota \in I} K_\iota$. We now apply Theorem 3.3.32 with respect to the family $(K_\iota)_{\iota \in I}$ and restrict the results for the family $(\{z \in \mathbb{C} \mid |z| \leq \alpha_\iota\})_{\iota \in I}$. □

Remark. Let K be the set of sequences $(\alpha_n)_{n \in \mathbb{N}}$ in $\mathbb{R}$ (in $\mathbb{C}$) such that $|\alpha_n| \leq 1$ for any $n \in \mathbb{N}$. Furthermore, let $(x_{mn})_{(m,n) \in \mathbb{N} \times \mathbb{N}}$ be a family in G such that $(\alpha_n x_{mn})_{n \in \mathbb{N}}$ is summable for any $(\alpha_n)_{n \in \mathbb{N}} \in K$ and for any $m \in \mathbb{N}$ and such that $\lim_{m \to \infty} \sum_{n \in \mathbb{N}} \alpha_n x_{mn}$ exists for any $(\alpha_n)_{n \in \mathbb{N}} \in K$. If we set

$$x_n := \lim_{m \to \infty} x_{mn}$$

for any $n \in \mathbb{N}$, then by d) $(\alpha_n x_n)_{n \in \mathbb{N}}$ is summable for any $(\alpha_n)_{n \in \mathbb{N}} \in K$,

$$\lim_{m \to \infty} \sum_{n \in \mathbb{N}} \alpha_n x_{mn} = \sum_{n \in \mathbb{N}} \alpha_n x_n$$

uniformly on K, and the series $\sum_{n \in \mathbb{N}} \alpha_n x_{mn}$ converge uniformly on K. This result was proved by C. Swartz (to appear) ([2] Theorem 3 and Corollary 4).

Corollary 3.3.34. *Let G be a Hausdorff topological vector space and $\mathscr{K}$ be a nonempty set of real functions on I such that $k(\iota) > 0$ for any $(k, \iota) \in \mathscr{K} \times I$. We set*

$$\mathscr{G} := \{g \in \mathbb{R}^I \text{ (or } \mathbb{C}^I) \mid \exists k \in \mathscr{K},\ (\iota \in I \Rightarrow |g(\iota)| \le k(\iota))\}$$
$$\mathscr{F} := \{f \in G^I \mid g \in \mathscr{G} \Rightarrow (g(\iota) f(\iota))_{\iota \in I} \in \ell(I, G)\}.$$

If $(f_n)_{n \in \mathbb{N}}$ is a sequence in $\mathscr{F}$ such that $(\sum_{\iota \in I} g(\iota) f_n(\iota))_{n \in \mathbb{N}}$ converges for any $g \in \mathscr{G}$, then $(f_n(\iota))_{n \in \mathbb{N}}$ converges for any $\iota \in I$, the map

$$f : I \to G, \ \iota \mapsto \lim_{n \to \infty} f_n(\iota)$$

belongs to $\mathscr{F}$, and

$$\sum_{\iota \in I} g(\iota) f(\iota) = \lim_{n \to \infty} g(\iota) f_n(\iota)$$

for any $g \in \mathscr{G}$.

The assertion follows immediately from Corollary 3.3.33 d). □

Remark. Let c_0 be the set of sequences in $\mathbb{R}$ (in $\mathbb{C}$) which converge to 0, $\mathscr{F}$ be the set

$$\{f \in G^{\mathbb{N}} \mid (\alpha_n)_{n \in \mathbb{N}} \in c_0 \Rightarrow (\alpha_n f(n))_{n \in \mathbb{N}} \text{ is supersummable in } G\},$$

and $(f_n)_{n \in \mathbb{N}}$ be a sequence in $\mathscr{F}$ such that $\sum_{m \in \mathbb{N}} \alpha_m f_n(m)$ converges for any $(\alpha_m)_{m \in \mathbb{N}} \in c_0$. Then by the above corollary,

$$f : \mathbb{N} \to G, \quad m \mapsto \lim_{n \to \infty} f_n(m)$$

belongs to $\mathscr{F}$ and

$$\sum_{m \in \mathbb{N}} \alpha_m f(m) = \lim_{n \to \infty} \sum_{m \in \mathbb{N}} \alpha_m f_n(m)$$

for any $(\alpha_m)_{m \in \mathbb{N}} \in c_0$. This result was proved by C. Swartz (to appear) ([2] Proposition 5).

§ 3.4 Spaces of supersummable families of functions

Throughout this section G shall denote a Hausdorff topological additive group and I shall denote a set.

Theorem 3.4.1. *Let T be a uniform (topological) space, $i \in \{1, 2, 3, 4\}$ ($i \in \{1, 2\}$), Φ be a subset of $\hat{\Phi}_i(T)$, T_0 be a Φ_i-dense (loosely Φ_i-dense) set of T and $\mathscr{F}$ be a subgroup of G^T such that any element of $\mathscr{F}$ is uniformly Φ_i-continuous (Φ_i-continuous). If the identity map $\mathscr{F}_\Phi \to \mathscr{F}_{T_0}$ is uniformly continuous, then:*

a) $\mathcal{F}_{T_0}$, $\mathcal{F}$, $\mathcal{F}_\Phi$ *are Hausdorff topological groups;*

b) $\ell(I, \mathcal{F}_{T_0}) = \ell(I, \mathcal{F}) = \ell(I, \mathcal{F}_\Phi)$;

c) for any $(f_\iota)_{\iota \in I} \in \ell(I, \mathcal{F})$, *for any Cauchy* $\hat{\Phi}_i$*-filter (for any* Φ_i*-ultrafilter or convergent* $\hat{\Phi}_i$*-filter)* $\mathfrak{G}$ *on* T, *and for any 0-neighbourhood* U *in* G *there exist* $A \in \mathfrak{G}$ *and* $K \in \mathfrak{P}_f(I)$ *such that*

$$\sum_{\iota \in J'} f_\iota(t') - \sum_{\iota \in J''} f_\iota(t'') \in U$$

for any $t', t'' \in A$ *and any* $J', J'' \in \mathfrak{P}(I)$ *with* $J' \cap K = J'' \cap K$;

d) for any $(f_\iota)_{\iota \in I} \in \ell(I, \mathcal{F})$, *any* Φ_i*-set* A *of* T, *and any 0-neighbourhood* U *in* G *there exists* $K \in \mathfrak{P}_f(I)$ *such that*

$$\sum_{\iota \in J'} f_\iota(t) - \sum_{\iota \in J''} f_\iota(t) \in U$$

for any $t \in \bar{A}$ *and any* $J', J'' \in \mathfrak{P}(I)$ *with* $J' \cap K = J'' \cap K$;

e) if the identity map $\mathcal{F} \to \mathcal{F}_\Phi$ $(\mathcal{F}_{T_0} \to \mathcal{F})$ *is uniformly* Φ_j*-continuous for a* $j \in \{1, 2, 3, 4\}$, *then the identity map*

$$\ell_c(I, \mathcal{F}) \to \ell_f(I, \mathcal{F}_\Phi) \quad (\ell_c(I, \mathcal{F}_{T_0}) \to \ell_f(I, \mathcal{F}))$$

is uniformly Φ_j*-continuous.*

By Theorem 3.2.3 and Corollary 3.2.4 a) the identity map

$$\mathcal{G}(I, G) \to \mathcal{G}(I, G)_{\{\mathfrak{P}(I)\}}$$

is uniformly Φ_4-continuous and $\mathcal{G}(I, G)$ is a Φ_4-closed set of $G^{\mathfrak{P}(I)}$. By Proposition 1.5.20 a) $\mathcal{G}(I, G)$ is a strongly Φ_2-closed set of $G^{\mathfrak{P}(I)}$.

a) By the Propositions 1.8.17 a) and 1.8.18 $\mathcal{F}_{T_0}$ is Hausdorff and a fortiori $\mathcal{F}$, $\mathcal{F}_\Phi$ are Hausdorff. By Corollary 2.1.29 $\mathcal{F}_{T_0}$, $\mathcal{F}$, and $\mathcal{F}_\Phi$ are topological groups.

b) Let $f \in \ell(I, \mathcal{F}_{T_0})$ and $\bar{f}$ be the map

$$\mathfrak{P}(I) \to \mathcal{F}_{T_0}, \quad J \mapsto \sum_{\iota \in J} f_\iota .$$

By a) and Theorem 3.3.5 $\bar{f} \in \mathcal{G}(I, \mathcal{F}_{T_0})$. We denote by g the map

$$\mathfrak{P}(I) \times T \to G, \quad (J, t) \mapsto (\bar{f}(J))(t).$$

We have $g(J, \cdot) \in \mathcal{F}$ for any $J \subset I$ and $g(\cdot, t) \in \mathcal{G}(I, G)$ for any $t \in T_0$. By Corollary 2.5.3 $g(\cdot, t) \in \mathcal{G}(I, G)$ for any $t \in T$. By Theorem 2.5.4 a) the map

$$\mathfrak{P}(I) \to G^T_\Phi, \quad J \mapsto g(J, \cdot)$$

is uniformly Φ_5-continuous and therefore uniformly continuous (Propositions 1.3.11 and 1.5.6). Hence, the map

$$\mathfrak{P}(I) \to \mathcal{F}_\Phi, \quad J \mapsto \sum_{\iota \in J} f_\iota$$

is continuous. By a) and Theorem 3.3.5 $f \in \ell(I, \mathcal{F}_\Phi)$.

c) Denote by φ the map

$$T \to \ell_c(I, G), \quad t \mapsto (f_\iota(t))_{\iota \in I}.$$

Here, the coarse uniformity of $\ell(I, G)$ is defined with respect to $\mathfrak{P}(I)$. By Proposition 2.5.2 (and Theorem 3.3.5) φ is uniformly Φ_i-continuous (is Φ_i-continuous) so by Theorem 1.8.4 a $\Rightarrow$ c (by Proposition 1.8.9d) and Proposition 1.5.10) $\varphi(\mathfrak{F})$ is a Cauchy $\hat{\Phi}_i$-filter on $\ell_c(I, G)$. The assertion now follows from Corollary 3.3.11 b).

d) follows immediately from c) and Proposition 1.8.19 (and Theorem 3.3.5).

e) By Proposition 2.1.17 the identity map

$$\ell_c(I, \mathscr{F}) \to \ell_c(I, \mathscr{F}_\Phi) \ (\ell_c(I, \mathscr{F}_{T_0}) \to \ell_c(I, \mathscr{F}))$$

is uniformly Φ_j-continuous and the assertion follows from Corollary 3.3.7 and Corollary 1.8.5. □

Remarks. 1) It is not possible to replace the hypothesis "T_0 is Φ_i- dense (loosely Φ_i-dense)" with the weaker hypothesis "T_0 is dense", even if T is compact and the elements of $\mathscr{F}$ are continuous. This may be seen by taking the Stone-Čech compactification of $\mathbb{N}$ as T and $\mathbb{N}$ as T_0.

2) Let T be an additive topological group and $\mathscr{F}$ be the group of continuous group homomorphisms of T into G. By Theorem 3.4.1 b) any supersummable sequence in $\mathscr{F}$ is supersummable in $\mathscr{F}_{\mathfrak{S}}$, where $\mathfrak{S}$ denotes the set of Φ_2-sets. This result was proved by C. Swartz (1978) ([1] Theorem 4).

Corollary 3.4.2. *Let T be a set, $\mathfrak{F}$ be a filter with a countable base on T (the filter $\{A \subset T \mid T \backslash A$ finite$\}$) and $(f_\iota)_{\iota \in I}$ be a family of maps of T into G such that $(f_\iota(t))_{\iota \in J}$ is summable for any $t \in T$ and $\lim_{t, \mathfrak{F}} \sum_{\iota \in I} f_\iota(t)$ exists for any $J \subset I$. Then*

a) $(\lim_{\mathfrak{F}} f_\iota)_{\iota \in J}$ is summable and

$$\sum_{\iota \in J} (\lim_{\mathfrak{F}} f_\iota) = \lim_{t, \mathfrak{F}} \sum_{\iota \in J} f_\iota(t)$$

for any $J \subset I$;

b) for any 0-neighbourhood U in G there exists $A \in \mathfrak{F}$ such that

$$\sum_{\iota \in J} f_\iota(t) \in \Big(\sum_{\iota \in J} (\lim_{\mathfrak{F}} f_\iota)\Big) + U$$

for any $J \subset I$ and for any $t \in A$;

c) if $T \backslash A$ is finite for any $A \in \mathfrak{F}$ then for any 0-neighbourhood U in G there exists a finite subset K of I such that

$$\sum_{\iota \in J'} f_\iota(t) - \sum_{\iota \in J''} f_\iota(t) \in U$$

for any $t \in T$ and any $J', J'' \in \mathfrak{P}(I)$ with $J' \cap K = J'' \cap K$.

Let t_0 be a point which does not belong to T and let us endow $T \cup \{t_0\}$ with the topology

$$\{V \subset T \cup \{t_0\} \mid t_0 \in V \Rightarrow V \setminus \{t_0\} \in \mathfrak{F}\}.$$

For any $\iota \in I$ we denote by f'_ι the continuous map of $T \cup \{t_0\}$ into G equal to f_ι on T and equal to $\lim_{\mathfrak{F}} f_\iota$ at t_0. By Proposition 1.5.25 $\mathfrak{F} \in \Phi(\Theta_2(T, T \cup \{t_0\}))$. If φ denotes the inclusion map $T \to T \cup \{t_0\}$, then $\varphi(\mathfrak{F})$ converges to t_0 so T is loosely Φ_2-dense in $T \cup \{t_0\}$. a) and b) follow from Theorem 3.4.1 b) and c) respectively. If $T \setminus A$ is finite for any $A \in \mathfrak{F}$, then $T \cup \{t_0\}$ is compact and c) follows from Theorem 3.4.1 d). □

The following special case of Corollary 3.4.2 was proved by A. P. Robertson (1972) ([2] Lemma and Corollary 1).

Corollary 3.4.3. *Let T be a subset of a metrizable topological space, t_0 be a point in the closure of T and $(f_n)_{n \in \mathbb{N}}$ be a sequence of maps of T into G such that for any $M \subset \mathbb{N}$ and any $t \in T$ the family $(f_n(t))_{n \in M}$ is summable and*

$$\lim_{T \ni t \to t_0} \sum_{n \in M} f_n(t)$$

exists. Then $(\lim_{T \ni t \to t_0} f_n(t))_{n \in M}$ is summable and

$$\sum_{n \in M} (\lim_{T \ni t \to t_0} f_n(t)) = \lim_{T \ni t \to t_0} \sum_{n \in M} f_n(t)$$

for any $M \subset \mathbb{N}$. For any 0-neighbourhood V in G there exists a neighbourhood U of t_0 such that

$$\sum_{n \in M} f_n(t) - \lim_{T \ni t \to t_0} \sum_{n \in M} f_n(t) \in V$$

for any $M \subset \mathbb{N}$ and any $t \in U$. □

Corollary 3.4.4. *Let T be a countably compact topological space, T_0 be a loosely Φ_2-dense set of T and $\mathscr{F}$ be the set of continuous maps of T into G. Then:*

a) $\mathscr{F}$ is a subgroup of G^T and $\mathscr{F}_{T_0}$, $\mathscr{F}_{\{T\}}$ are Hausdorff topological groups;

b) $\ell(I, \mathscr{F}_{T_0}) = \ell(I, \mathscr{F}_{\{T\}})$;

c) for any $f \in \ell(I, \mathscr{F}_{T_0})$, any ultrafilter or convergent filter $\mathfrak{F}$ on T and any 0-neighbourhood U in G there exist $A \in \mathfrak{F}$ and a finite subset K of I such that

$$\sum_{\iota \in J'} f_\iota(t') - \sum_{\iota \in J''} f_\iota(t'') \in U$$

for any $t', t'' \in A$ and for any $J', J'' \in \mathfrak{P}(I)$ with $J' \cap K = J'' \cap K$;

d) for any $f \in \ell(I, \mathscr{F}_{T_0})$ and for any 0-neighbourhood U in G there exists a finite subset K of I such that

$$\sum_{\iota \in J'} f_\iota(t) - \sum_{\iota \in J''} f_\iota(t) \in U$$

for any $t \in T$ *and for any* $J', J'' \in \mathfrak{P}(I)$ *with* $J' \cap K = J'' \cap K$.

If T *is a mioritic space (e.g. sequentially compact) or if* $T \setminus T_0$ *is countable, then we may take as* T_0 *any dense set of* T.

By the hypothesis T is a Φ_2-set and a), b), c), d) follow immediately from Theorem 3.4.1 a), b), c), and d) respectively. If T is a mioritic space then by Corollary 1.7.11 any dense set of T is loosely Φ_2-dense. If $T \setminus T_0$ is countable and T_0 is dense in T by Corollary 1.5.22 T_0 is loosely Φ_2-dense. □

Remark. The assertion b) was proved by E. Thomas (1970) ([3] Théorème II 4) for G a normed space or a normed space endowed with its weak topology, for T compact and $T_0 = T$ or for T metrizable and compact and T_0 dense. It was extended by J. Labuda (1972) to G a Hausdorff topological additive group, T sequentially compact, and $T_0 = T$ ([1] Théorème 2.3) or T compact and metrizable and T_0 dense ([1] Théorème 2.4) or T compact, $T_0 = T$, and I countable ([1] Théorème 2.5). The assertion d) was proved by A. P. Robertson (1972) ([2] Corollary 2) for G Hausdorff topological additive group, T compact, $T_0 = T$, and $J'' = \emptyset$.

Corollary 3.4.5. *Let* T *be a well ordered set endowed with the topology*

$$\{U \subset T \mid \xi \in U,\ \xi \neq \inf T \Rightarrow \exists \eta \in T,\ \eta < \xi,\ \{\zeta \in T \mid \eta < \zeta \leq \xi\} \subset U\},$$

T_0 *be the subset of* T

$$T_0 := \{\xi \in T \mid \sup\{\eta \in T \mid \eta < \xi\} < \xi\} \cup \{\inf T\},$$

and $\mathscr{F}$ *be the set of continuous maps of* T *into* G. *Then:*

a) $\ell(I, \mathscr{F}_{T_0}) = \ell(I, \mathscr{F})$;

b) if any increasing sequence in T *possesses a supremum (e.g.* T *is the first uncountable ordinal number), then* $\ell(I, \mathscr{F}_{T_0}) = \ell(I, \mathscr{F}_{\{T\}})$.

a) T_0 is loosely Φ_1-dense and the assertion follows from Theorem 3.4.1 b).

b) T is countably compact and the assertion follows from Corollary 3.4.4 b). □

Corollary 3.4.6. *Let* L *be a set and* $\mathscr{F}$ *be the set of continuous maps of* $\mathfrak{P}(L)$ *into* G. *Then:*

$$\ell(I, \mathscr{F}_{\mathfrak{P}_f(L)}) = \ell(I, \mathscr{F}_{\{\mathfrak{P}(L)\}})$$

By Proposition 3.1.2 $\mathfrak{P}_f(L)$ is loosely Φ_1-dense in $\mathfrak{P}(L)$ and the assertion follows from Corollary 3.4.4 b). □

Corollary 3.4.7. *Let* T *be a topological space,* $\mathfrak{A}$ *be the set of* Φ_2*-sets of* T *and* T_0 *be a loosely* Φ_2*-dense set of* T. *We assume* $T \notin \mathfrak{A}$ *and denote by* $\mathfrak{F}$ *the filter*

$\{T\backslash A\,|\,A\in\mathfrak{A}\}$ on T and by $\mathscr{F}$ the group of continuous maps f of T into G which converge with respect to $\mathfrak{F}$. Then:

a) $\ell\,(I,F_{T_0})=\ell\,(I,\mathscr{F}_{\{T\}})$;

b) for any $f\in\ell\,(I,\mathscr{F}_{T_0})$ and any 0-neighbourhood U in G there exists a finite subset K of I such that

$$\sum_{\iota\in J'}f_\iota(t)-\sum_{\iota\in J''}f_\iota(t)\in U$$

for any $t\in T$ and any $J',J''\in\mathfrak{P}(I)$ with $J'\cap K=J''\cap K$.

Let a be a point not belonging to T and T^* be the set $T\cup\{a\}$ endowed with the topology

$$\{U\subset T^*\,|\,U\cap T\quad\text{open in }T,\quad(a\in U\;\Rightarrow\;T\backslash U\in\mathfrak{A})\}\,.$$

By Proposition 1.5.23 T is a subspace of T^*, T^* is countably compact, and T_0 is a loosely Φ_2-dense set of T^*. Let $\mathscr{C}$ be the group of continuous maps of T^* into G and, for any $f\in\mathscr{F}$, let $\tilde{f}$ be the element of $\mathscr{C}$ whose restriction to T equals f (the existence of $\tilde{f}$ is assured by the hypothesis about $\mathscr{F}$). Then the maps

$$\mathscr{F}_{T_0}\to\mathscr{C}_{T_0},\quad f\mapsto\tilde{f},$$
$$\mathscr{F}_{\{T\}}\to\mathscr{C}_{\{T^*\}},\quad f\mapsto\tilde{f}$$

are isomorphisms of topological groups and the assertions follow from Corollary 3.4.4 b) and d).   □

Proposition 3.4.8. *Let T be a topological space, T_0 be a loosely Φ_2-dense set of T, $\mathfrak{A}$ be an upper directed covering of T with Φ_2-sets and, for each $A\in\mathfrak{A}$, let $\mathscr{F}(A)$ be the set of continuous maps of T into G vanishing outside of A. We set $\mathscr{F}:=\bigcup\limits_{A\in\mathfrak{A}}\mathscr{F}(A)$ and denote by $\tilde{\mathscr{F}}$ the group $\mathscr{F}$ endowed with a Hausdorff group topology inducing on $\mathscr{F}(A)$ a topology coarser than the topology of uniform convergence for any $A\in\mathfrak{A}$. Then any $(f_\iota)_{\iota\in I}\in\ell\,(I,\mathscr{F}_{T_0})$ is supersummable in $\tilde{\mathscr{F}}$ if there exists $A\in\mathfrak{A}$ such that $f_\iota(t)=0$ for any*

$$(\iota,t)\in I\times(T\backslash A)\,.$$

$\mathscr{F}_{T_0}$ is obviously a Hausdorff topological group. $\mathscr{F}_{\mathfrak{A}}$ is a Hausdorff topological group and by Theorem 3.4.1 b) $(f_\iota)_{\iota\in I}\in\ell\,(I,\mathscr{F}_{\mathfrak{A}})$. Hence, $(f_\iota)_{\iota\in I}\in\ell\,(I,\mathscr{F}(A)_{\{A\}})$ and therefore $(f_\iota)_{\iota\in I}$ is supersummable in $\tilde{\mathscr{F}}$.   □

Theorem 3.4.9. *Let T be a locally compact paracompact space, T_0 be a loosely Φ_2-dense set of T, $\mathfrak{R}$ be the set of compact sets of T and, for each $K\in\mathfrak{R}$, let $\mathscr{F}(K)$ be the set of continuous maps of T into G vanishing outside of K. We set $\mathscr{F}:=\bigcup\limits_{K\in\mathfrak{R}}\mathscr{F}(K)$ and denote by $\tilde{\mathscr{F}}$ the group $\mathscr{F}$ endowed with a Hausdorff group topology inducing on $\mathscr{F}(K)$ a topology coarser than the topology of uniform convergence for any $K\in\mathfrak{R}$.*

Then, for any $(f_\iota)_{\iota\in I}\in\ell(I,\mathscr{F}_{T_0})$, *there exists* $K\in\mathfrak{K}$ *such that* $f_\iota(t)=0$ *for any* $(\iota,t)\in I\times(T\backslash K)$ *and* $(f_\iota)_{\iota\in I}$ *is supersummable in* $\tilde{\mathscr{F}}$.

The second assertion follows immediately from the first and from Proposition 3.4.8. Accordingly we only have to prove the first assertion. By Theorem 3.4.1 b) we may assume $(f_\iota)_{\iota\in I}\in\ell(I,\mathscr{F})$. We set

$$A:=\bigcup_{\iota\in I}\{t\in T\,|\,f_\iota(t)\neq 0\}.$$

We want to show A is a Φ_2-set of T. Let $(t_n)_{n\in\mathbb{N}}$ be a sequence in A. By Proposition 1.5.5 c $\Rightarrow$ a we have to show that $(t_n)_{n\in\mathbb{N}}$ possesses an adherent point. Assume the contrary. Let $(\iota_n)_{n\in\mathbb{N}}$ be a sequence in I such that $f_{\iota_n}(t_n)\neq 0$ for any $n\in\mathbb{N}$. Further, let $(U_n)_{n\in\mathbb{N}}$ be a sequence of closed 0-neighbourhoods in G such that $f_{\iota_n}(t_n)\notin U_n$ for any $n\in\mathbb{N}$. We construct a strictly increasing map $\varphi:\mathbb{N}\to\mathbb{N}$ inductively such that

$$\sum_{p=0}^{n} f_{\iota_{\varphi(p)}}(t_{\varphi(m)})\notin U_m$$

for any $m,n\in\mathbb{N}$ so that $m<n$. Let $n\in\mathbb{N}$ and assume that φ was constructed up to $n-1$. Since $(t_p)_{p\in\mathbb{N}}$ does not possess adherent points there exists $p_0\in\mathbb{N}$ such that $f_{\iota_{\varphi(m)}}(t_p)=0$ for any $m,p\in\mathbb{N}$ so that $m<n$ and $p\geq p_0$. Since $(f_{\iota_p}(t_{\varphi(m)}))_{p\in\mathbb{N}}$ converges to 0 for any $m\in\mathbb{N}$, $m<n$, there exists $\varphi(n)\in\mathbb{N}$ such that $\varphi(n)\geq p_0$, $\varphi(n)>\varphi(n-1)$ if $n\neq 0$, and

$$\sum_{p=0}^{n} f_{\iota_{\varphi(p)}}(t_{\varphi(m)})\notin U_m$$

for any $m\in\mathbb{N}$, where $m<n$. We have

$$\sum_{p=0}^{n} f_{\iota_{\varphi(p)}}(t_{\varphi(n)})=f_{\iota_{\varphi(n)}}(t_{\varphi(n)})\notin U_n.$$

This concludes the inductive construction. We set

$$f:=\sum_{n\in\mathbb{N}} f_{\iota_{\varphi(n)}}.$$

There exists $K\in\mathfrak{K}$ such that $f=0$ on $T\backslash K$. Since $f(t_{\varphi(n)})\neq 0$ for any $n\in\mathbb{N}$ we deduce that $\{t_{\varphi(n)}\,|\,n\in\mathbb{N}\}\subset K$, which contradicts the hypothesis that $(t_n)_{n\in\mathbb{N}}$ has no adherent points. Hence, A is a Φ_2-set of T.

By Theorem 1.5.8 T is Φ_2-compact and A is a relatively compact set of T. □

§ 3.5 Supersummable families in special spaces

Theorem 3.5.1. *Let G be a Hausdorff topological additive group, I, J be sets and $\ell_p(J,G)$ be the group $\ell(J,G)$ endowed with the topology of pointwise convergence (i.e. the topology induced by G^J). Then:*

a) $\ell_p(J,G)$ is a Hausdorff topological group and

$$\ell(I,\ell_p(J,G)) = \ell(I,\ell_c(J,G)) = \ell(I,\ell_f(J,G));$$

b) the identity map $\ell_p(I,G) \to \ell_f(I,G)$ belongs to

$$\mathscr{P}(\ell_p(I,G),\ \ell_f(I,G));$$

c) the map

$$\ell_p(I,G) \to G, \quad (x_\iota)_{\iota\in I} \mapsto \sum_{\iota\in J} x_\iota$$

belongs to $\mathscr{P}(\ell_p(I,G),G)$ for any $J\subset I$;

d) the identity map

$$\ell_c(I,\ell_c(J,G)) \to \ell_f(I,\ell_f(J,G))$$

is uniformly Φ_4-continuous.

a) The first assertion is obvious. By Theorem 3.3.5 we may replace $\ell_p(J,G)$ and $\ell_f(J,G)$ by $\mathscr{G}(J,G)_{\mathfrak{P}_f(J)}$ and $\mathscr{G}(J,G)_{\{\mathfrak{P}(J)\}}$ respectively and the assertion follows from Corollary 3.4.6.

b) follows from a).

c) follows from b).

d) By Corollary 3.3.7 the identity map

$$\ell_c(J,G) \to \ell_f(J,G)$$

is uniformly Φ_4-continuous and the assertion follows from Corollary 3.3.17 b). □

Remark. b) was proved by E. Thomas (1970) ([3] Théorème II 4) for $G=\mathbb{R}$. The generalization to an arbitrary G was announced by C. Constantinescu (1976) ([3] Corollaire 8).

Theorem 3.5.2. *(C. Constantinescu (1976) [4] Théorème 7). Let G be a Hausdorff topological additive group and $(x_{\iota\lambda})_{(\iota,\lambda)\in I\times L}$ be a family in G such that; 1) the family $(x_{\iota\lambda})_{\iota\in I}$ is supersummable for any $\lambda\in L$; 2) the family $(\sum_{\iota\in J} x_{\iota\lambda})_{\lambda\in L}$ is supersummable for any $J\subset I$. Then:*

a) $((x_{\iota\lambda})_{\lambda\in L})_{\iota\in I}\in\ell(I,\ell_f(L,G))$,
$((x_{\iota\lambda})_{\iota\in I})_{\lambda\in L}\in\ell(L,\ell_f(I,G))$;

b) the family $(\sum_{\lambda \in K} x_{\iota\lambda})_{\iota \in I}$ *is supersummable for any* $K \subset L$ *and*

$$\sum_{\iota \in J} (\sum_{\lambda \in K} x_{\iota\lambda}) = \sum_{\lambda \in K} (\sum_{\iota \in J} x_{\iota\lambda})$$

any $J \subset I$;

c) the map

$$\mathfrak{P}(I) \times \mathfrak{P}(L) \to G, \quad (J, K) \mapsto \sum_{\iota \in J} (\sum_{\lambda \in K} x_{\iota\lambda})$$

is uniformly continuous.

If G is Φ_1*-compact then we may assume 2) for countable J only.*

Let us denote for any $J \subset I$ by f_J the map

$$\mathfrak{P}(L) \to G, \quad K \mapsto \sum_{\lambda \in K} (\sum_{\iota \in J} x_{\iota\lambda})$$

We set $f_\iota := f_{\{\iota\}}$ for any $\iota \in I$. By Theorem 3.3.5 $f_J \in \mathscr{G}(L, G)$ for any $J \subset I$.

a), b) The family $(f_\iota(K))_{\iota \in J}$ is summable and

$$\sum_{\iota \in J} f_\iota(K) = \sum_{\iota \in J} (\sum_{\lambda \in K} x_{\iota\lambda}) = \sum_{\lambda \in K} (\sum_{\iota \in J} x_{\iota\lambda}) = f_J(K)$$

for any $J \subset I$ and any $K \in \mathfrak{P}_f(L)$. Hence,

$$(f_\iota)_{\iota \in I} \in \ell(I, \mathscr{G}(L, G)_{\mathfrak{P}_f(L)})$$

and

$$\sum_{\iota \in J} f_\iota = f_J$$

for any $J \subset I$. By Corollary 3.4.6

$$(f_\iota)_{\iota \in I} \in \ell(I, \mathscr{G}(L, G)_{\{\mathfrak{P}(L)\}}).$$

Hence, $(\sum_{\lambda \in K} x_{\iota\lambda})_{\iota \in J}$ is summable and

$$\sum_{\iota \in J} (\sum_{\lambda \in K} x_{\iota\lambda}) = \sum_{\iota \in J} f_\iota(K)) = f_J(K) = \sum_{\lambda \in K} (\sum_{\iota \in J} x_{\iota\lambda})$$

for any $J \subset I$ and any $K \subset L$. By the above relation and Theorem 3.3.5 we also get

$$((x_{\iota\lambda})_{\lambda \in L})_{\iota \in I} \in \ell(I, \ell_f(L, G))$$

and, by symmetry,

$$((x_{\iota\lambda})_{\iota \in I})_{\lambda \in L} \in \ell(L, \ell_f(I, G)).$$

c) By Theorem 3.3.5 the map

$$\mathfrak{P}(I) \to \mathscr{G}(L, G)_{\{\mathfrak{P}(L)\}}, \quad J \mapsto f_J$$

is continuous. Hence, the map

$$\mathfrak{P}(I)\times\mathfrak{P}(L) \to G, \quad (J,K)\mapsto f_J(K)$$

is continuous (N. Bourbaki [1] X § 1 Proposition 9) and therefore uniformly continuous.

Assume now that G is Φ_1-compact and that 2) is fulfilled for countable J only. By the above considerations $(f_\iota)_{\iota\in J}$ is summable in $\mathscr{G}(L,G)_{\{\mathfrak{P}(L)\}}$ for any countable J. Since $\mathscr{G}(L,G)_{\{\mathfrak{P}(L)\}}$ is Φ_1-compact $(f_\iota)_{\iota\in I}$ is supersummable (Proposition 3.3.3). □

Remark. This result is obviously related to Fubini's theorem but the hypothesis is weaker since it is not assumed that $(x_{\iota\lambda})_{(\iota,\lambda)}\in I\times L$ is summable. This last assertion does not follow from the hypothesis even in the case of real numbers. More exactly, there exists a family $(\alpha_{mn})_{(m,n)\in\mathbb{N}\times\mathbb{N}}$ of real numbers such that:

1) $(\alpha_{mp})_{m\in\mathbb{N}}, (\alpha_{pn})_{n\in\mathbb{N}}$ are summable for any $p\in\mathbb{N}$;

2) $(\sum_{m\in A}\alpha_{mp})_{p\in\mathbb{N}}$, $(\sum_{n\in B}\alpha_{pn})_{p\in\mathbb{N}}$ are summable and

$$\sum_{m\in A}(\sum_{n\in B}\alpha_{mn}) = \sum_{n\in B}(\sum_{m\in A}\alpha_{mn})$$

for any $A, B\in\mathfrak{P}(N)$;

3) $(\alpha_{mn})_{(m,n)\in\mathbb{N}\times\mathbb{N}}$ is not summable.

In order to construct such a family we need some notation. For any $m\in\mathbb{N}$ denote by $(k_{mn})_{n\in\mathbb{N}}$ the sequence in $\{0,1\}$ such that

$$m = \sum_{n\in\mathbb{N}} k_{mn}2^n.$$

We set for any $p\in\mathbb{N}$

$$j(p) = \sum_{q=1}^{p} 2^{2q-1}$$

and for any $m\in\mathbb{N}$ with $j(p)\le m < j(p+1)$

$$\alpha_{mn} := \begin{cases} \dfrac{2k_{m-j(p),\,n-p^2}-1}{(p+1)(2p+1)2^{2p+1}} & \text{if } p^2\le n<(p+1)^2 \\ 0 & \text{if } n<p^2 \text{ or } n\ge (p+1)^2. \end{cases}$$

The hypothesis 1) is trivially fulfilled since for any $p\in\mathbb{N}$ the sets $\{m\in\mathbb{N}\mid \alpha_{mp}\neq 0\}$, $\{n\in\mathbb{N}\mid\alpha_{pn}\neq 0\}$ are finite. 3) follows from

$$\sum_{m,n\in\mathbb{N}}|\alpha_{mn}| = \sum_{p\in\mathbb{N}}\sum_{n=p^2}^{(p+1)^2-1}\sum_{m\in\mathbb{N}}|\alpha_{mn}| = \sum_{p\in\mathbb{N}}\frac{1}{p+1} = \infty$$

In order to prove 2), let A be a subset of $\mathbb{N}$ and $p\in\mathbb{N}\setminus\{0\}$. We have

$$\sum_{n=p^2}^{(p+1)^2-1} |\sum_{m\in A} \alpha_{mn}| < \frac{1}{(p+1)(2p+1)2^{2p+1}} \sum_{r=0}^{p} \binom{2p+1}{r}(2p+1-2r) =$$

$$= \frac{1}{(p+1)2^{2p+1}} \sum_{r=0}^{p} \binom{2p+1}{r} - \frac{2}{(p+1)2^{2p+1}} \sum_{r=1}^{p} \binom{2p}{r-1} =$$

$$= \frac{2^{2p} - \left(2^{2p} - \binom{2p}{p}\right)}{(p+1)2^{2p+1}} = \frac{\binom{2p}{p}}{(p+1)2^{2p+1}} = \frac{(2p)!}{(p+1)2^{2p+1}(p!)^2} \sim$$

$$= \left(\frac{2p}{e}\right)^{2p} \frac{\sqrt{4\pi p}}{4p\,2^{2p}\pi p} \left(\frac{e}{p}\right)^{2p} = \frac{1}{2\sqrt{\pi p^3}}$$

by Stirling's formula. Hence, $(\sum_{m\in A} \alpha_{mn})_{n\in\mathbb{N}}$ is summable and the assertion follows from Theorem 3.5.2.

Corollary 3.5.3. *Let G, H be Hausdorff topological additive groups, I be a set, $(x_\iota)_{\iota\in I} \in \ell(I, G)$ and $(u_\lambda)_{\lambda\in L}$ be a family in $\mathscr{P}(G, H)$ such that $(u_\lambda(\sum_{\iota\in J} x_\iota))_{\lambda\in L}$ is supersummable for any $J \subset I$. Then:*

a) $(\sum_{\lambda\in K} u_\lambda(x_\iota))_{\iota\in I}$ is supersummable for any $K \subset L$ and

$$\sum_{\iota\in J}(\sum_{\lambda\in K} u_\lambda(x_\iota)) = \sum_{\lambda\in K} u_\lambda(\sum_{\iota\in J} x_\iota);$$

for any $J \subset I$;

b) for any 0-neighbourhood U in H there exist a finite subset I_0 of I and a finite subset L_0 of L such that

$$\sum_{\lambda\in L'} u_\lambda(\sum_{\iota\in I'} x_\iota) - \sum_{\lambda\in L''} u_\lambda(\sum_{\iota\in I''} x_\iota) \in U$$

for any $I', I'' \in \mathfrak{P}(I)$ with $I' \cap I_0 = I'' \cap I_0$ and any $L', L'' \in \mathfrak{P}(L)$ with $L' \cap L_0 = L'' \cap L_0$.

Let $\lambda \in L$. Since $u_\lambda \in \mathscr{P}(G, H)$ the family $(u_\lambda(x_\iota))_{\iota\in I}$ is supersummable and

$$\sum_{\iota\in J} u_\lambda(x_\iota) = u_\lambda(\sum_{\iota\in J} x_\iota)$$

for any $J \subset I$.

The assertions follow immediately from this remark and Theorem 3.5.2. □

Definition 3.5.4. *Let G be an additive group. A* <u>*quasi-norm*</u> *on G is a map $p : G \to \mathbb{R}_+$ such that:*

a) $p(0) = 0$;

b) $x, y \in G \Rightarrow p(x+y) \le p(x) + p(y)$;

c) $x \in G \Rightarrow p(-x) = p(x)$.

Let $\mathcal{P}$ be a set of quasi-norms on G. The sets of the form $\{x \in G \mid p(x) < \varepsilon\}$, where p runs through $\mathcal{P}$ and ε runs through the set of strictly positive real numbers, form a fundamental system of 0-neighbourhoods for a group topology on G. We call this topology the topology generated by $\mathcal{P}$.

If G is a topological additive group, then its topology is generated by the set of continuous quasi-norms on G.

Definition 3.5.5. *Let G be an additive group, $\mathcal{P}$ be a set of quasi-norms on G, I be a set and α be a strictly positive real number. We denote by $\bar{\ell}^{\alpha}(I, G, \mathcal{P})$ the set of maps f of I into G such that $((p \circ f(\iota))^{\alpha})_{\iota \in I}$ is summable for any $p \in \mathcal{P}$. $\bar{\ell}^{\alpha}(I, G, \mathcal{P})$ is a subgroup of G^I. The finite intersections of the sets of the form*

$$\{f \in \bar{\ell}^{\alpha}(I, G, \mathcal{P}) \mid \sum_{\iota \in I} (p \circ f(\iota))^{\alpha} < \varepsilon\},$$

where p runs through $\mathcal{P}$ and ε runs through the set of strictly positive real numbers, form a fundamental system of 0-neighbourhoods of a group topology on $\bar{\ell}^{\alpha}(I, G, \mathcal{P})$. We denote by $\ell^{\alpha}(I, G, \mathcal{P})$ the group $\bar{\ell}^{\alpha}(I, G, \mathcal{P})$ endowed with this topology. If G is a locally convex space, then we set $\ell^{\alpha}(I, G) := \ell^{\alpha}(I, G, \mathcal{P})$, where $\mathcal{P}$ denotes the set of continuous semi-norms on G. We set $\ell^{\alpha}(I) := \ell^{\alpha}(I, \mathbb{R})$ for any set I and $\ell^{\alpha} := \ell^{\alpha}(\mathbb{N})$.

Proposition 3.5.6. *Let E be a Hausdorff complete locally convex space and I be a set. We have:*

a) $(f(\iota))_{\iota \in I}$ is supersummable for any $f \in \bar{\ell}^{\alpha}(I, E)$ and for any $\alpha \leq 1$;

b) $\bar{\ell}^1(I, E)$ is a vector subspace of the vector space E^I, $\ell^1(I, E)$ is a Hausdorff complete locally convex space, and the map

$$\ell^1(I, E) \to \ell_f(I, E), \quad f \mapsto (f(\iota))_{\iota \in I}$$

is injective, linear and continuous;

c) if E is a normed space then $\ell^1(I, E)$ is cannically normed;

d) the map

$$\ell^1(I, \mathbb{R}^L) \to \ell_f(I, \mathbb{R}^L), \quad f \mapsto (f(\iota))_{\iota \in I}$$

is an isomorphism of locally convex spaces for any set L;

e) if E is finite dimensional, then the map of b) is an isomorphism of locally convex spaces.

Let $\mathcal{P}$ be the set of continuous semi-norms on E.

a) Let $\mathfrak{F}$ be the section filter of $\mathfrak{P}_f(I)$, $p \in \mathcal{P}$ and ε be a strictly positive real number. There exists $J \in \mathfrak{P}_f(I)$ such that

$$\sum_{\iota \in I \setminus J} p(f(\iota))^{\alpha} \leq \inf(\varepsilon, 1).$$

Then

$$\sum_{\iota \in I \setminus J} p(f(\iota)) < \varepsilon$$

so the image of $\mathfrak{F}$ with respect to the map

$$\mathfrak{P}_f(I) \to E, \quad K \mapsto \sum_{\iota \in K} f(\iota)$$

is a Cauchy filter. Since E is complete $(f(\iota))_{\iota \in I}$ is supersummable.

b) It is obvious that $\overline{\ell}^1(I, E)$ is a vector subspace of the vector space E^I and that $\ell^1(I, E)$ is a Hausdorff locally convex space. Let $\mathfrak{F}$ be a Cauchy filter on $\ell^1(I, E)$ and for any $\iota \in I$ let π_ι be the ι-projection of $\ell^1(I, E)$. Then $\pi_\iota(\mathfrak{F})$ is a Cauchy and therefore a convergent filter for any $\iota \in I$. We set

$$f_0 : I \to E, \quad \iota \mapsto \lim \pi_\iota(\mathfrak{F}).$$

Let $p \in \mathscr{P}$ and ε be a strictly positive real number. There exists $\mathscr{F} \in \mathfrak{F}$ such that

$$\sum_{\iota \in I} p(f(\iota) - g(\iota)) < \varepsilon$$

for any $f, g \in \mathscr{F}$. We get

$$\sum_{\iota \in J} p(f(\iota) - f_0(\iota)) \le \varepsilon$$

for any $J \in \mathfrak{P}_f(I)$ and $f \in \mathscr{F}$ so

$$\sum_{\iota \in I} p(f(\iota) - f_0(\iota)) \le \varepsilon.$$

Hence, $f_0 \in \ell^1(I, E)$, $\mathfrak{F}$ converges to f_0 and $\ell^1(I, E)$ is complete.

Let us denote by φ the map

$$\ell^1(I, E) \to \ell_f(I, E), \quad f \mapsto (f(\iota))_{\iota \in I}.$$

$\ell_f(I, E)$ is a locally convex space. It is obvious that φ is linear and injective. Let U be a 0-neighbourhood in E. There exist $p \in \mathscr{P}$ and a strictly positive real number α such that

$$\{x \in E \mid p(x) < \alpha\} \subset U.$$

We set

$$\mathscr{U} := \{f \in \ell^1(I, E) \mid \sum_{\iota \in I} p(f(\iota)) < \alpha\}.$$

Then $\mathscr{U}$ is a 0-neighbourhood in $\ell^1(I, E)$ and

$$p(\sum_{\iota \in J} f(\iota)) \le \sum_{\iota \in J} p(f(\iota)) < \alpha$$

for any $f \in \mathscr{U}$ and for any $J \in \mathfrak{P}(I)$. We get

$$\{\sum_{\iota \in J} f(\iota) \mid f \in \mathscr{U}, J \subset I\} \subset U$$

so φ is continuous.

c) is obvious.

d) We denote by φ the map

$$\ell^1(I, \mathbb{R}^L) \to \ell_f(I, \mathbb{R}^L), \quad f \mapsto (f(\iota))_{\iota \in I}.$$

We prove first that φ is bijective. Let $(x_\iota)_{\iota \in I} \in \ell(I, \mathbb{R}^L)$. We set

$$f: I \to \mathbb{R}^L, \quad \iota \mapsto x_\iota.$$

Let $\lambda \in L$ and $I(\lambda)$ be the set $\{\iota \in I \mid x_\iota(\lambda) > 0\}$. Then

$$\sum_{\iota \in I} |f(\iota)_\lambda| = \sum_{\iota \in I(\lambda)} x_\iota(\lambda) - \sum_{\iota \in I \setminus I(\lambda)} x_\iota(\lambda).$$

Hence, $f \in \ell^1(I, \mathbb{R}^L)$. Since $\varphi(f) = (x_\iota)_{\iota \in I}$, φ is surjective. By b) φ is bijective.

We prove now that φ is a homeomorphism. By b) it is continuous. Let p be a continuous semi-norm on $\mathbb{R}^L$ and ε be a strictly positive real number. There exist $K \in \mathfrak{P}_f(L)$ and a strictly positive real number α such that

$$U := \{(\alpha_\lambda)_{\lambda \in L} \in \mathbb{R}^L \mid \sum_{\lambda \in K} |\alpha_\lambda| < \alpha\} \subset \{x \in \mathbb{R}^L \mid p(x) < \varepsilon\}.$$

We set

$$\mathcal{U} := \{(x_\iota)_{\iota \in I} \in \ell(I, \mathbb{R}^L) \mid \{\sum_{\iota \in J} x_\iota \mid J \subset I\} \subset U\}.$$

Then $\mathcal{U}$ is a 0-neighbourhood in $\ell_f(I, \mathbb{R}^L)$ and

$$\sum_{\iota \in I} p(f(\iota)) \leq \frac{\varepsilon}{\alpha} \sum_{\iota \in I} \sum_{\lambda \in K} |f(\iota)_\lambda| \leq \frac{\varepsilon}{\alpha} 2n\alpha = 2n\varepsilon$$

for any $f \in \overset{-1}{\varphi}(\mathcal{U})$, where n is the cardinal number of K. Hence, $\overset{-1}{\varphi}$ is continuous so φ is a homeomorphism.

e) follows immediately from d). □

Remarks. By the theorem of Dvoretzky-Rogers (1950) ([1] Theorem 1) if E is a Banach space for which the map of b) is surjective, then E is finite dimensional.

Definition 3.5.7. *Let G be an additive group, $\mathcal{P}$ be a set of quasi-norms on G and I be a set. If I is finite, then we set $\bar{c}_0(I, G, \mathcal{P}) := G^I$. If I is infinite, then we denote by $\mathfrak{F}$ the filter*

$$\{A \subset I \mid I \setminus A \quad \text{is finite}\}$$

on I and set

$$\bar{c}_0(I, G, \mathcal{P}) := \{f \in G^I \mid \forall p \in \mathcal{P} \Rightarrow \lim_{\mathfrak{F}} p \circ f = 0\}.$$

$\bar{c}_0(I, G, \mathcal{P})$ is a subgroup of G^I. The finite intersections of the sets of the form

$$\{f \in \bar{c}_0(I, G, \mathcal{P}) \mid \sup_{\iota \in I} p(f(\iota)) < \varepsilon\},$$

where p runs through $\mathcal{P}$ and ε runs through the set of strictly positive real numbers, *form a fundamental system of 0-neighbourhoods of a group topology on* $\bar{c}_0(I, G, \mathscr{P})$. *We denote by* $\underline{c_0(I, G, \mathscr{P})}$ *the group* $\bar{c}_0(I, G, \mathscr{P})$ *endowed with this topology. If G is a locally convex space then we set* $\underline{c_0(I, G)} := c_0(I, G, \mathscr{P})$, *where* $\mathscr{P}$ *denotes the set of continuous semi-norms on G. We set* $\underline{c_0(I)} := c_0(I, \mathbb{R})$ *for any set I and* $\underline{c_0} := c_0(\mathbb{N})$.

Theorem 3.5.8. *Let G be a Hausdorff topological additive group,* $\mathscr{P}$ *be a set of quasi-norms on G generating its topology, I be a set and* α *be a strictly positive real number. Any family in* $\ell^{\alpha}(I, G, \mathscr{P})$ *(in* $c_0(I, G, \mathscr{P})$*) is supersummable if it is supersummable with respect to the topology of pointwise convergence (i.e. with respect to the topology induced by* G^I*).*

Let us denote by H the group $\ell^{\alpha}(I, G, \mathscr{P})$ endowed with the topology of pointwise convergence and $\mathfrak{F}$ be the section filter of $\mathfrak{P}_f(I)$. For any $K \in \mathfrak{P}_f(I)$ and for any $f \in \ell^{\alpha}(I, G, \mathscr{P})$ denote by f_K the map of I into G equal to f on K and equal to 0 on $I \backslash K$. For any $K \in \mathfrak{P}_f(I)$ denote by u_K the map

$$H \to \ell^{\alpha}(I, G, \mathscr{P}), \quad f \mapsto f_K,$$

by u the map

$$\mathfrak{P}_f(I) \to \mathscr{P}(H, \ell^{\alpha}(I, G, \mathscr{P})), \quad K \mapsto u_K$$

and by φ the inclusion map

$$\mathscr{P}(H, \ell^{\alpha}(I, G, \mathscr{P})) \to \ell^{\alpha}(I, G, \mathscr{P})^H.$$

Then $(\mathfrak{P}_f(I), u)$ is a Θ_1-net so $u(\mathfrak{F})$ is a Φ_1-filter on $\mathscr{P}(H, \ell^{\alpha}(I, G, \mathscr{P}))$. Since $\varphi(u(\mathfrak{F}))$ converges to the identity map $u_0 : H \to \ell^{\alpha}(I, G, \mathscr{P})$ we deduce by Theorem 3.3.22 a) that

$$u_0 \in \mathscr{P}(H, \ell^{\alpha}(I, G, \mathscr{P})),$$

which is equivalent to the assertion that any supersummable family in H is supersummable in $\ell^{\alpha}(I, G, \mathscr{P})$.

The proof for $c_0(I, G, \mathscr{P})$ is similar. □

Remarks. 1) This result was proved by E. Thomas (1970) for G a normed space and $\alpha \geq 1$ ([3] Théorème II 4) and for $G = \mathbb{C}$ and α arbitrary (1968) (1970) ([2] page A 7 and [3] Proposition II 8). It was extended to a locally convex space and an arbitrary α by C. Constantinescu (1976) ([3] Théorème 7) and to a quasi-normed additive group by C. Swartz (1978) ([1] Theorem 13).

2) It is easy to see that the above result does not hold for the Banach space ℓ^{∞} of bounded sequences of real numbers.

Proposition 3.5.9. *Let I be a set,* α *be a strictly positive real number, G, H be topological additive groups such that H is Hausdorff and* $\mathscr{P}$ *be a set of quasi-norms on H generating its topology such that the set of continuous group homomorphisms of G into* $\ell^{\alpha}(I, H, \mathscr{P})$ *is strongly* Φ_1*-closed in* $\ell^{\alpha}(I, H, \mathscr{P})^G$. *For any* $(\iota, y) \in I \times H^I$ *we denote* *by* y^ι *the element of* $\ell^\alpha(I, H, \mathscr{P})$ *defined by*

$$(y^\iota)_\lambda := \begin{cases} y_\iota & \text{if } \lambda = \iota \\ 0 & \text{if } \lambda \neq \iota \end{cases} \qquad (\lambda \in I).$$

Then any group homomorphism f *of* G *into* $\ell^\alpha(I, H, \mathscr{P})$ *is continuous if the map*

$$G \to \ell^\alpha(I, H, \mathscr{P}), \quad x \mapsto f(x)^\iota$$

is continuous for any $\iota \in I$. *The assertion still holds if we replace* $\ell^\alpha(I, H, \mathscr{P})$ *by* $c_0(I, H, \mathscr{P})$.

Let $\mathfrak{F}$ be the section filter of $\mathfrak{P}_f(I)$. For any $(J, y) \in \mathfrak{P}(I) \times H^I$ denote by y^J the map

$$I \to H, \quad \iota \mapsto \begin{cases} y_\iota & \text{if } \iota \in J \\ 0 & \text{if } \iota \in I \setminus J. \end{cases}$$

We have $y^J \in \ell^\alpha(I, H, \mathscr{P})$ for any $(J, y) \in \mathfrak{P}(I) \times \ell^\alpha(I, H, \mathscr{P})$ and

$$\lim_{J, \mathfrak{F}} y^J = y$$

for any $y \in \ell^\alpha(I, H, \mathscr{P})$, the limit being taken in $\ell^\alpha(I, H, \mathscr{P})$.

Let f be a group homomorphism of G into $\ell^\alpha(I, H, \mathscr{P})$ such that the map

$$G \to \ell^\alpha(I, H, \mathscr{P}), \quad x \mapsto f(x)^\iota$$

is continuous for any $\iota \in I$. Then the map

$$G \to \ell^\alpha(I, H, \mathscr{P}), \quad x \mapsto f(x)^J$$

is continuous for any $J \in \mathfrak{P}_f(I)$. By Proposition 2.1.23 f is continuous. □

Theorem 3.5.10. *Let* E *be a Hausdorff locally convex space,* Δ *be a set of supersummable families in* E, E^* *be the algebraic dual of* E, F *be the set of* $y \in E^*$ *for which* $(y(x_\iota))_{\iota \in J}$ *is summable with the sum* $y(\sum_{\iota \in J} x_\iota)$ *for any* $(x_\iota)_{\iota \in J} \in \Delta$ *(endowed with the topology of pointwise convergence) and* Φ *be the set* $\hat{\Phi}_4(F)$. *For any* $x \in E$ *we denote by* $\tilde{x}$ *the map*

$$F \to \mathbb{R}, \quad y \mapsto y(x)$$

and identify E *with a subset of* $\mathbb{R}^F$ *via the map*

$$E \to \mathbb{R}^F, \quad x \mapsto \tilde{x}.$$

Then:

a) F *is a* Φ_4*-closed subspace of* E^* *and therefore a* Φ_4*-compact space;*

b) E_Φ *is a locally convex space and any family of* Δ *is supersummable in* E_Φ;

c) the topology of Φ*-convergence on* E *is the Mackey topology on* E *associated* *with the duality* $\langle E, F\rangle$ *and it is the finest Hausdorff locally convex topology on E for which the families of* Δ *are supersummable and have the same sum as in E;*

d) the convex circled closed hull of any Φ_4*-set of F is compact;*

e) if Δ *is the set of all supersummable (of all countable supersummable) families in E, then* $F = \mathscr{P}(E, \mathbb{R})$ $(F = \mathscr{P}^c(E, \mathbb{R}))$.

a) and b). It is obvious that F is a subspace of E^*. By Corollary 2.1.29 E_Φ is a locally convex space. Let T be the intersection of all Φ_4-closed sets of E^* containing F. Then T is a Φ_4-closed set of E^* and F is a Φ_4-dense set of T. By Theorem 3.4.1 b) we get $F = T$ and b) follows from Theorem 3.4.1 b).

c) Let $\mathfrak{T}$ be a Hausdorff locally convex topology on E such that any family of Δ is supersummable and has the same sum as in E. Further, let G be the set of elements of E^* continuous with respect to $\mathfrak{T}$. Then $G \subset F$ so $\mathfrak{T}$ is coarser than the Mackey topology on E of the duality $\langle E, F\rangle$. By b) and by the above considerations, the topology of Φ-convergence on E is coarser than the Mackey topology on E of the duality $\langle E, F\rangle$. The converse is trivial.

d) Let A be a Φ_4-set of F. By c) A is equicontinuous with respect to the Mackey topology on E of the duality $\langle E, F\rangle$ and therefore the convex circled closed hull of A in F is compact.

e) follows from Proposition 3.3.19. □

Remarks. 1) Let Δ be the set of supersummable sequences in E and let $\mathfrak{A}$ be the set of compact sets of F. By b) any sequence of Δ is supersummable in $E_\mathfrak{A}$ and by d) the convex circled closed hull of any set of $\mathfrak{A}$ in F is compact. This result was proved by I. Tweddle (1970) ([3] Proposition 1). In the same proposition I. Tweddle proves that the Mackey topology on E of the duality $\langle E, F\rangle$ is the finest Hausdorff locally convex topology on E for which the sequences of Δ are supersummable and have the same sum.

By a) F_E is sequentially complete; this result was proved by P. Dierolf (1979) ([3] Proposition 3.18 (d)). In the same paper P. Dierolf proved (Theorem 3.1) that any Hausdorff topological vector space (Hausdorff topological additive group) E possesses a finest vector space topology (group topology) $\mathfrak{T}$ such that the supersummable sequences in E are supersummable with respect to $\mathfrak{T}$.

2) Let X be a set and let E be the vector space of bounded real functions on X endowed with the topology of pointwise convergence. If there exists an atomfree strictly positive real measure μ on $\mathfrak{P}(X)$, then the map

$$E \to \mathbb{R}, \quad f \mapsto \int f d\mu$$

belongs to $\mathscr{P}^c(E, \mathbb{R}) \setminus \mathscr{P}(E, \mathbb{R})$.

Corollary 3.5.11. *Let E be a Hausdorff locally convex space, E′ be its dual (endowed with the topology of pointwise convergence) and* Φ *be the set* $\hat{\Phi}_4(E')$. *Then any supersummable family in E is supersummable in* E_Φ. □

Remark. Let $\mathfrak{A}$ be a set of Φ_4-sets on E'. By the above result any supersummable sequence in E is supersummable in $E_{\mathfrak{A}}$. This result was proved by J. Tweddle (1970) ([2] Corollary of Theorem 7) for $\mathfrak{A}$ the set of compact sets of E' and by P. Dierolff (1975) ([1] (4.5) Korollar (c) and (e)) (1977) ([2] Corollary 2.4 c) and e)) for $\mathfrak{A}$ the set of Φ_2-sets of E' and the set of Φ_3-sets of E'.

Corollary 3.5.12. *(Orlicz-Pettis). Let E be a Hausdorff locally convex space. Any weakly supersummable family in E is supersummable.* □

Remark. This result was proved by B.J. Pettis (1938) ([1] Theorem 2.32) (and formulated without proof by S. Banach (1932) ([1] page 240 (3))) for weakly supersummable sequences in Banach spaces. The above formulation was given by A. Grothendieck (1953) ([2] page 166). If we assume E weakly sequentially complete, then any weakly summable sequence in E is weakly supersummable and therefore supersummable. This result was proved by W. Orlicz (1929) ([1] Satz 2) for E a Banach space.

Proposition 3.5.13. *Let E be a Hausdorff complete locally convex space,* $\mathfrak{F} \in \hat{\Phi}_C(E)$ *and K be the adherence of* $\mathfrak{F}$. *For any* $A \subset E$ *denote by* $\Gamma(A)$ *the closed circled convex hull of A. Then* $\Gamma(K)$ *is compact, nonempty and equal to* $\bigcap_{A \in \mathfrak{F}} \Gamma(A)$.

Assume there exists $x \in \bigcap_{A \in \mathfrak{F}} \Gamma(A) \setminus \Gamma(K)$. Then there exist a continuous linear form x' on E and an $\alpha \in \mathbb{R}$ with

$$\sup_{y \in \Gamma(K)} |x'(y)| < \alpha < x'(x).$$

For any $A \in \mathfrak{F}$ we have $A \cap \{y \in E \mid |x'(y)| > \alpha\} \neq \emptyset$. Hence, there exists an ultrafilter $\mathfrak{G}$ on E finer than $\mathfrak{F}$ and containing $\{y \in E \mid |x'(y)| > \alpha\}$. By Proposition 1.2.2 $\mathfrak{G}$ converges to a point $y \in E$ and we obtain the contradictory relation $y \in K$ and $|x'(y)| \geq \alpha$. Hence, $\bigcap_{\mathscr{A} \in \mathfrak{F}} \Gamma(A) \subset \Gamma(K)$. The converse inclusion is trivial. By Proposition 1.2.7 a $\Rightarrow$ b K is compact and nonempty. By Krein's theorem (H. H. Schaefer [1] IV 11.5) $\Gamma(K)$ is compact. □

Proposition 3.5.14. *Let E, F be Hausdorff locally convex spaces, F be complete and* $\mathscr{L}(E, F)$ *be the vector space of linear maps of E into F endowed with the topology of pointwise convergence. For any* $\mathscr{A} \subset \mathscr{L}(E, F)$ *denote by* $\Gamma(\mathscr{A})$ *the closed circled convex hull of* $\mathscr{A}$ *in* $\mathscr{L}(E, F)$. *Then:*

a) $\mathscr{P}(E, F)$ *and* $\mathscr{P}^c(E, F)$ *are vector subspaces of* $\mathscr{L}(E, F)$ *and* Φ_4*-compact;*

b) for any $\hat{\Phi}_4$*-filter* $\mathfrak{F}$ *on* $\mathscr{P}(E, F)$ *(on* $\mathscr{P}^c(E, F)$*) the set* $\bigcap_{\mathscr{A} \in \mathfrak{F}} \Gamma(\mathscr{A})$ *is compact, nonempty, contained in* $\mathscr{P}(E, F)$ *(in* $\mathscr{P}^c(E, F)$*) and equal to* $\Gamma(\mathscr{B})$, *where* $\mathscr{B}$ *denotes the adherence of* $\mathfrak{F}$ *in* $\mathscr{P}(E, F)$ *(in* $\mathscr{P}^c(E, F)$*);*

c) the following assertions are equivalent for any $\mathscr{A} \subset \mathscr{P}(E, F)$ ($\mathscr{A} \subset \mathscr{P}^c(E, F)$): 1) $\mathscr{A}$ is a Φ_4-set; 2) $\mathscr{A}$ is relatively compact; 3) $\Gamma(\mathscr{A})$ is compact and contained in $\mathscr{P}(E, F)$ (in $\mathscr{P}^c(E, F)$).

a) By Theorem 3.3.21 a) (and Theorem 3.3.5) $\mathscr{P}(E, F)$ and $\mathscr{P}^c(E, F)$ are vector subspaces of $\mathscr{L}(E, F)$ and by Theorem 3.3.21 b) they are Φ_4-compact.

b) Let $\Phi := \hat{\Phi}_4(\mathscr{P}(E, F))$ ($\Phi := \hat{\Phi}_4(\mathscr{P}^c(E, F))$ and let $\mathscr{C}$ be the vector space of continuous maps of $\mathscr{P}(E, F)$ (of $\mathscr{P}^c(E, F)$) into F. We denote for any $x \in E$ by $\hat{x}$ the map

$$\mathscr{P}(E, F) \to F, \quad f \mapsto f(x)$$
$$(\mathscr{P}^c(E, F) \to F, \quad f \mapsto f(x)).$$

Let $(x_\iota)_{\iota \in I} \in \ell(I, E)$ (I countable in the $\mathscr{P}^c(E, F)$ case). Then $(\hat{x}_\iota)_{\iota \in I} \in \ell(I, \mathscr{C})$ and by Theorem 3.4.1 b) $(\hat{x}_\iota)_{\iota \in I} \in \ell(I, \mathscr{C}_\Phi)$. Let U be a closed circled convex 0-neighbourhood in F and let $f \in \bigcap_{\mathscr{A} \in \mathfrak{F}} \Gamma(\mathscr{A})$. Since $\mathfrak{F} \in \Phi$ there exists a finite subset K of I such that

$$\{g \in \mathscr{P}(E, F) \mid g(\sum_{\iota \in J} x_\iota) \in \tfrac{1}{2} U\} \in \mathfrak{F}$$
$$(\{g \in \mathscr{P}^c(E, F) \mid g(\sum_{\iota \in J} x_\iota) \in \tfrac{1}{2} U\} \in \mathfrak{F})$$

for any $J \subset I \backslash K$. U being closed circled and convex we get $f(\sum_{\iota \in J} x_\iota) \in \frac{1}{2} U$ for any $J \subset I \backslash K$. We get

$$\sum_{\iota \in J} f(x_\iota) = f(\sum_{\iota \in J} x_\iota) = f(\sum_{\iota \in I} x_\iota) - f(\sum_{\iota \in I \backslash K} x_\iota) + f(\sum_{\iota \in J \backslash K} x_\iota) \in f(\sum_{\iota \in I} x_\iota) + U$$

for any finite subset J of I containing K. Since U is arbitrary we deduce that

$$\sum_{\iota \in I} f(x_\iota) = f(\sum_{\iota \in I} x_\iota).$$

Hence, $f \in \mathscr{P}(E,F)$ ($f \in \mathscr{P}^c(E, F)$) and $\bigcap_{\mathscr{A} \in \mathfrak{F}} \Gamma(\mathscr{A}) \subset \mathscr{P}(E, F)$ ($\bigcap_{\mathscr{A} \in \mathfrak{F}} \Gamma(\mathscr{A}) \subset \mathscr{P}^c(E, F)$)

By Corollary 1.5.12 and Proposition 3.5.13 $\Gamma(\mathscr{B})$ is compact nonempty and equal to $\bigcap_{\mathscr{A} \in \mathfrak{F}} \Gamma(\mathscr{A})$.

c) $1 \Rightarrow 3$ follows from b); $3 \Rightarrow 2 \Rightarrow 1$ are trivial. □

Proposition 3.5.15. *Let E be a B_r-complete locally convex space (H. H. Schaefer [1] IV 8), E' be its dual (endowed with the topology of pointwise convergence) and A' be a subset of E' containing a dense vector subspace of E' and such that the limit point of any convergent Φ_2-filter $\mathfrak{F}$ on E' belongs to A' if $A' \in \mathfrak{F}$. Then $A' = E'$.*

Let F' be a dense vector subspace of E' contained in A' and $\mathfrak{B}$ be the set of subsets B' of E' containing F' with the property that the limit point of any convergent Φ_2-filter $\mathfrak{F}$ on E' belongs to B' if $B' \in \mathfrak{F}$. We set $B'_0 := \bigcap_{B' \in \mathfrak{B}} B'$. Then $B'_0 \in \mathfrak{B}$. Let $x' \in F'$. Then $x' + B'_0 \in \mathfrak{B}$ so $B'_0 \subset x' + B'_0$. Since x' is arbitrary we get $F' + B'_0 = B'_0$.

Let $y' \in B'_0$ and $B' \in \mathfrak{B}$. Then, by the above inclusion, $y' + F' \subset B'$. Hence, $-y' + B' \in \mathfrak{B}$ so $B'_0 \subset -y' + B'$, $B'_0 + y' \subset B'$. Since y' and B' are arbitrary we get $B'_0 + B'_0 \subset B'_0$. We have $\alpha B'_0 \subset B'_0$ for any $\alpha \in \mathbb{R}$ so B'_0 is a subspace of E'.

Let C' be an equicontinuous set of E' and let x' be an adherent point of $B'_0 \cap C'$ in C'. Let $\mathfrak{F}$ be the filter on E' generated by the filter base

$$\{B'_0 \cap C' \cap U' \mid U' \quad \text{neighbourhood of } x' \text{ in } E'\}.$$

Since C' is a relatively compact set of E', $\mathfrak{F}$ is a Φ_2-filter on E'. Since $B'_0 \in \mathfrak{B}$ we get $x' \in B'_0$. Hence, $B'_0 \cap C'$ is a closed set of C'. Since E is B_r-complete we deduce that $B'_0 = E'$ so $A' = E'$. □

Theorem 3.5.16. *Let E be a B_r-complete locally convex space, E' be its dual (endowed with the topology of pointwise convergence) and $\Phi := \hat{\Phi}_4(E')$. For any $x \in E$ we denote by $\tilde{x}$ the map*

$$F \longrightarrow \mathbb{R}, \quad y \mapsto y(x)$$

and identity E with a subset of $\mathbb{R}^F$ via the map

$$E \longrightarrow \mathbb{R}^F, \quad x \mapsto \tilde{x}.$$

Further let A' be a dense subset of E' such that any countable subset of E is contained in a closed subspace F of $E_{A'}$ with the property that $F_{A'}$ possesses a dense σ-compact set. Then any supersummable family in $E_{A'}$ is supersummable in E_Φ.

We may assume that A' is a vector subspace of E'. Let $(x_\iota)_{\iota \in I}$ be a supersummable family in $E_{A'}$ and J be a countable subset of I. By the hypothesis there exists a closed vector subspace F of $E_{A'}$ containing $\{x \mid \iota \in J\}$ such that $F_{E'}$ possesses a dense σ-compact set. We deduce that

$$\{\sum_{\iota \in K} x_\iota \mid K \subset J\} \subset F,$$

where the sums are defined in $E_{A'}$.

Let us denote by F' the set of continuous linear forms on F endowed with the topology of pointwise convergence and by φ the map

$$E' \longrightarrow F', \quad x' \mapsto x' \mid F.$$

We want to show that $\varphi(A')$ is loosely Φ_2-dense in F'. Let B' be a strongly Φ_2-closed set of F' containing $\varphi(A')$, $\mathfrak{F}$ be a convergent Φ_2-filter on E' with $\overset{-1}{\varphi}(B') \in \mathfrak{F}$ and x' be the limit point of $\mathfrak{F}$. Then $\varphi(\mathfrak{F})$ is a Φ_2-filter on F' converging to $\varphi(x')$ and $B' \in \varphi(\mathfrak{F})$. By Theorem 2.3.14 F' is a mioritic space and by Proposition 1.7.10 $\varphi(x') \in B'$. Hence, $x' \in \overset{-1}{\varphi}(B')$. Since $A' \subset \overset{-1}{\varphi}(B')$ we get, by Proposition 3.5.15, $\overset{-1}{\varphi}(B') = E'$ so $B' = F'$. This shows that $\varphi(A')$ is loosely Φ_2-dense in F'.

Let $\mathscr{F}$ be the vector space of continuous real functions on F' and for any $x \in F$ let $\hat{x}$ be the map

$$F' \longrightarrow \mathbb{R}, \quad x' \mapsto x'(x).$$

Then $\hat{x} \in \mathscr{F}$ for any $x \in F$ and $(\hat{x}_\iota)_{\iota \in J}$ is supersummable in $\mathscr{F}_{\varphi(A')}$. By Theorem 3.4.1 b) it is summable in $\mathscr{F}$. We get $(x_\iota)_{\iota \in J}$ is weakly summable in E. By Corollary 3.5.12 $(x_\iota)_{\iota \in J}$ is summable in E.

Let ψ be the map

$$\mathfrak{P}_c(I) \to E, \quad J \mapsto \sum_{\iota \in J} x_\iota$$

and let $\mathfrak{F}$ be the section filter of $\mathfrak{P}_c(I)$. Then $(\psi(J_n))_{n \in \mathbb{N}}$ is a Cauchy sequence for any increasing sequence $(J_n)_{n \in \mathbb{N}}$ in $\mathfrak{P}_c(I)$ and therefore by Proposition 1.5.17 $\psi(\mathfrak{F})$ is a Cauchy filter. Since E is complete (H.H. Schaefer [1] IV 8.1) $\psi(\mathfrak{F})$ converges and therefore $(x_\iota)_{\iota \in I}$ is summable. By Corollary 3.5.11 $(x_\iota)_{\iota \in I}$ is supersummable in E_Φ. □

Remark. This theorem improves two earlier results of E. Thomas (1968), (1970), (1974) ([2] Théorème 1, [3] Théorème II 3, and [4] Theorem 0.3) where 1) a countable dense set replaces the above σ-compact set; 2) the summability is proved only for the initial topology of E instead of E_Φ; 3) the B_r-completeness is replaced by the stronger hypothesis that E is a Banach space in the first two papers and a Fréchet space in the last paper.

Theorem 3.5.17. *(A.P. Robertson (1970) [1] Proposition 3). Let G be a Hausdorff topological additive group possessing no nontrivial compact subgroup and let $(x_\iota)_{\iota \in I}$ be a family in G. Then $(x_\iota)_{\iota \in I}$ is supersummable iff $\{\sum_{\iota \in J} x_\iota \mid J \in \mathfrak{P}_f(I)\}$ is a relatively compact set of G.*

If $(x_\iota)_{\iota \in I}$ is supersummable, then $\{\sum_{\iota \in J} x_\iota \mid J \subset I\}$ is compact (Corollary 3.3.6) so $\{\sum_{\iota \in J} x_\iota \mid J \in \mathfrak{P}_f(I)\}$ is relatively compact.

Now let K be a compact set of G containing $\{\sum_{\iota \in J} x_\iota \mid J \in \mathfrak{P}_f(I)\}$. We have to show that $(x_\iota)_{\iota \in I}$ is summable. Assume the contrary. Then there exist a 0-neighbourhood U in G and a disjoint sequence $(I_n)_{n \in \mathbb{N}}$ in $\mathfrak{P}_f(I)$ such that

$$y_n := \sum_{\iota \in I_n} x_\iota \notin U$$

for any $n \in \mathbb{N}$. Let y be an adherent point of $(y_n)_{n \in \mathbb{N}}$ in K and let $m \in \mathbb{N}$. We want to show that $my \in K$. Let V be an arbitrary 0-neighbourhood in G and W be a 0-neighbourhood in G such that $mW \subset -V$. Then $y + W$ contains an infinity of y_n $(n \in \mathbb{N})$. Let z be the sum of m of these. Then

$$z \in m\,(y + W) \subset my - V, \quad my \in z + V \subset K + V.$$

Since V is arbitrary, $my \in K$. Hence, the subgroup of G generated by y is contained in $K \cup (-K)$ so it is compact. This group is non-trivial since $y \notin \overset{\circ}{U}$, which is the expected contradiction. □

Theorem 3.5.18. *(C. Constantinescu (1976) [4] Théorème 6). Let $n \in \mathbb{N}$, $n \neq 0$, X be an open set of $\mathbb{R}^n$, $p \in (\mathbb{N} \cup \{\infty\})^n$, $\mathscr{C}$ be the vector space of real functions on X for which the derivatives of order $< p$ exist and are continuous and $\mathfrak{T}$ be the topology on $\mathscr{C}$ defined by the semi-norms*

$$\mathscr{C} \to \mathbb{R}_+, \quad f \mapsto \sup_{x \in K} |D^r f(x)|,$$

where K runs through the set of compact subsets of X, r runs through the set of indices $<p$, and D^r denotes the derivative of order r. Then any supersummable family in $\mathscr{C}$ with respect to the topology of pointwise convergence in a dense set of X is supersummable with respect to $\mathfrak{T}$.

Let $(f_\iota)_{\iota \in I}$ be a supersummable family in $\mathscr{C}$ with respect to the topology of pointwise convergence in a dense set A of X. By Theorem 3.4.1 b) (replacing G by $\mathbb{R}$, T by the uniform space X, i by 4, T_0 by A, and $\mathscr{F}$ by $\mathscr{C}$) $(f_\iota)_{\iota \in I}$ is supersummable with respect to the topology on $\mathscr{C}$ of compact convergence. Let $\mathscr{M}$ be the set of real valued measures on X with compact carrier. We have

$$\int (\sum_{\iota \in J} f_\iota) d\mu = \sum_{\iota \in J} \int f_\iota d\mu$$

for any $\mu \in \mathscr{M}$ and for any $J \subset I$. Let $x \in X$ and let $r \in \mathbb{N}^n$, $r < p$. There exists a sequence $(\mu_m)_{m \in \mathbb{N}}$ in $\mathscr{M}$ such that

$$D^r f(x) = \lim_{m \to \infty} \int f d\mu_m$$

for any $f \in \mathscr{C}$. By Corollary 3.3.9 a) $(D^r f_\iota(x))_{\iota \in I}$ is summable in $\mathbb{R}$ and

$$(D^r(\sum_{\iota \in I} f_\iota))(x) = \sum_{\iota \in J} D^r f_\iota(x)$$

for any $J \subset I$. We conclude using Theorem 3.4.1 b) once more. □

Theorem 3.5.19. *Let T be a topological space, $\mathscr{G}$ be an upper directed set of upper semi-continuous positive real functions on T such that*

$$0 < \sup_{g \in \mathscr{G}} g(t) \leq \infty$$

for any $t \in T$, E be a Hausdorff topological vector space, $\mathfrak{U}$ be a fundamental system of 0-neighbourhoods in E and $\mathscr{F}$ be the set of continuous maps f of T into E such that

$$\{t \in T \mid g(t) f(t) \notin U\}$$

is a Φ_2-set of T for any $(g, U) \in \mathscr{G} \times \mathfrak{U}$. We set

$$\mathscr{U}(g, U) := \{f \in \mathscr{F} \mid t \in T \Rightarrow g(t) f(t) \in U\}$$

for each $(g, U) \in \mathscr{G} \times \mathfrak{U}$. Then:

a) $\mathscr{F}$ is a vector subspace of the vector space E^T containing any continuous map of T into E vanishing outside a Φ_2-set of T and

$$\{\mathscr{U}(g,U) \mid (g,U) \in \mathscr{G} \times \mathfrak{U}\}$$

is a fundamental system of 0-neighbourhoods in $\mathscr{F}$ for a Hausdorff vector space topology $\mathfrak{T}$ on $\mathscr{F}$, locally convex if E is locally convex;

b) any supersummable family in $\mathscr{F}$ with respect to the topology of pointwise convergence in a loosely Φ_2-dense set of T is summable with respect to $\mathfrak{T}$;

c) if $\mathscr{G}$ is the set of all upper semi-continuous positive real functions g on T such that $\{t \in T \mid g(t) \geq \varepsilon\}$ is a Φ_2-set of T for any $\varepsilon > 0$, then $\mathscr{F}$ is the set of continuous bounded maps of T into E ($f \in T^E$ is bounded iff $f(T)$ is a bounded set of E).

a) is obvious.

b) For any $f \in \mathscr{F}$ denote by $\tilde{f}$ the map

$$T \times \mathbb{R} \to E, \quad (t,\alpha) \mapsto \alpha f(t)$$

and let us endow $T \times \mathbb{R}$ with the coarsest topology for which all maps $\tilde{f}$ with $f \in \mathscr{F}$ are continuous. For any $g \in \mathscr{G}$ we set

$$A(g) := \{(t,\alpha) \in T \times \mathbb{R} \mid |\alpha| \leq g(t)\}.$$

Let $g \in \mathscr{G}$ and $(t_n, \alpha_n)_{n \in \mathbb{N}}$ be a sequence in $A(g)$. Assume first that $(t_n)_{n \in \mathbb{N}}$ possesses an adherent point $t \in T$. Let $\mathfrak{F}$ be an ultrafilter on $\mathbb{N}$ finer than the section filter of $\mathbb{N}$ such that

$$\lim_{n,\mathfrak{F}} t_n = t.$$

Since g is upper semi-continuous

$$\alpha := \lim_{n,\mathfrak{F}} \alpha_n \in \mathbb{R}$$

exists and $(t,\alpha) \in A(g)$. Since the elements of $\mathscr{F}$ are continuous we deduce that

$$\lim_{n,\mathfrak{F}} (t_n, \alpha_n) = (t,\alpha)$$

in $T \times \mathbb{R}$. Hence, the sequence $(t_n, \alpha_n)_{n \in \mathbb{N}}$ possesses an adherent point in $A(g)$. Assume now that $(t_n)_{n \in \mathbb{N}}$ does not possess adherent points in T. Let $f \in \mathscr{F}$ and let U be a circled closed 0-neighbourhood in E. By the definition of $\mathscr{F}$

$$\{t \in T \mid g(t) f(t) \notin U\}$$

is a Φ_2-set of T. Hence, $g(t_n) f(t_n) \in U$ for sufficiently large numbers $n \in \mathbb{N}$. Since U is arbitrary $(\alpha_n f(t_n))_{n \in \mathbb{N}}$ converges to 0. Hence $(\tilde{f}(t_n, \alpha_n))_{n \in \mathbb{N}}$ converges to 0 so $(t_n, \alpha_n)_{n \in \mathbb{N}}$ converges to $(t_0, 0) \in A(g)$. We have proved that $A(g)$ is countably compact and therefore a Φ_2-set of $T \times \mathbb{R}$ (Proposition 1.5.5 c $\Rightarrow$ a).

Let $(f_\iota)_{\iota \in I}$ be a supersummable family in $\mathscr{F}$ with respect to the topology of pointwise convergence in a loosely Φ_2-dense set of T. By Theorem 3.4.1 b) $(f_\iota)_{\iota \in I}$ is supersummable in $\mathscr{F}$ (with respect to the topology of pointwise convergence). Let $\mathscr{C}$ be the vector space of continuous maps of $T \times \mathbb{R}$ into E. Then $(\tilde{f}_\iota)_{\iota \in I}$ is supersummable in $\mathscr{C}$. By Theorem 3.4.1b) it is supersummable in $\mathscr{C}$ with respect to the topology of uniform convergence on the sets $A(g)$ with $g \in \mathscr{G}$. Let $(g, U) \in \mathscr{G} \times \mathfrak{U}$. There exists a finite subset K of I such that

$$\sum_{\iota \in J} \tilde{f}_\iota(t, \alpha) \in U$$

for any $J \subset I \backslash K$ and for any $(t, \alpha) \in A(g)$. Since $(t, g(t)) \in A(g)$ for any $t \in T$ we get

$$g(t) \sum_{\iota \in J} f_\iota(t) \in U$$

for any $t \in T$ and any $J \subset I \backslash K$. Hence,

$$\sum_{\iota \in J} f_\iota \in \mathscr{U}(g, U)$$

for any $J \subset I \backslash K$. It follows that $(f_\iota)_{\iota \in I}$ is supersummable with respect to $\mathfrak{T}$.

c) Let f be a continuous bounded map of T into E and $(g, U) \in \mathscr{G} \times \mathfrak{U}$. There exists $\varepsilon > 0$ such that $\alpha f(T) \subset U$ for any $\alpha \in \mathbb{R}$ with $|\alpha| < \varepsilon$. Hence,

$$\{t \in T \mid g(t) f(t) \notin U\} \subset \{t \in T \mid g(t) \geq \varepsilon\}$$

so $\{t \in T \mid g(t) f(t) \notin U\}$ is a Φ_2-set of T. We get $f \in \mathscr{F}$.

Let $f \in \mathscr{F}$ and assume $f(T)$ is not bounded. Then there exist $U \in \mathfrak{U}$ and a sequence $(t_n)_{n \in \mathbb{N}}$ in T with $\frac{1}{n+1} f(t_n) \notin U$ for any $n \in \mathbb{N}$ and $t_m \neq t_n$ for any different $m, n \in \mathbb{N}$. Then $(t_n)_{n \in \mathbb{N}}$ has no adherent points in T and the real function g on T, equal to $\frac{1}{n+1}$ at t_n for any $n \in \mathbb{N}$ and equal to 0 elsewhere, belongs to $\mathscr{G}$. From

$$\{t_n \mid n \in \mathbb{N}\} = \{t \in T \mid g(t) f(t) \notin U\}$$

we get $\{t_n \mid n \in \mathbb{N}\}$ is a Φ_2-set, which is a contradiction. □

Remark. In particular, any supersummable family in $\mathscr{F}_T$ is supersummable with regard to $\mathfrak{T}$. This result was announced by C. Constantinescu (1976) [4] Proposition 4.

Theorem 3.5.20. *Let T be a Hausdorff topological space, T_0 be a loosely Φ_2-dense set of T, E be a Hausdorff locally convex space, $\mathscr{P}$ be a set of semi-norms on E generating its topology and $\mathscr{G}$ be a set of positive real functions g on $\mathscr{P} \times T$ such that:*

1) $g(p, \cdot)$ is upper semi-continuous for any $p \in \mathscr{P}$;

2) any $t \in T$ possesses a neighbourhood U such that the set

$$\{p \in \mathscr{P} \mid \sup_{s \in U} g(p, s) > 0\}$$

is finite;

3) for any $(t, x) \in T \times (E \backslash \{0\})$ there exists $(g, p) \in \mathscr{G} \times \mathscr{P}$ so that $g(p, t) p(x) \neq 0$.

We set

$$q_g : E^T \to [0, \infty], \quad f \mapsto \sum_{p \in \mathscr{P}} \sup_{t \in T} g(p, t) p(f(t))$$

for any $g \in \mathscr{G}$ *and denote by* $\mathscr{F}$ *the set of continuous maps of* T *into* E *such that:*

4) $q_g(f) < \infty$ *for any* $g \in \mathscr{G}$;

5) the set

$$\{t \in T \mid g(p, t) p(f(t)) \geq \varepsilon\}$$

is compact for any $(g, p) \in \mathscr{G} \times \mathscr{P}$ *and any* $\varepsilon > 0$.

We set

$$\mathscr{F}(A) := \{f \in \mathscr{F} \mid f = 0 \text{ on } T \setminus A\}$$

for any $A \subset T$ *and denote by* $\mathfrak{A}$ *the set of subsets* A *of* T *such that*

$$\sum_{p \in \mathscr{P}} \sup_{t \in A} g(p, t) < \infty$$

for any $g \in \mathscr{G}$ *and such that any* $f \in \mathscr{F}(A)$ *is bounded. We have:*

a) $\mathscr{F}$ *is a vector subspace of* E^T *containing any continuous map of* T *into* E *with compact carrier;*

b) $q_g | \mathscr{F}$ *is a semi-norm for any* $g \in \mathscr{G}$ *and the topology* $\mathfrak{T}$ *on* $\mathscr{F}$ *generated by these semi-norms (for* g *running through* $\mathscr{G}$*) is Hausdorff;*

c) any supersummable family in $\mathscr{F}_{T_0}$ *is supersummable with respect to* $\mathfrak{T}$;

d) if $\mathscr{G}$ *is the set of all positive real functions* g *on* $\mathscr{P} \times T$ *satisfying 1) and 2), then any* $f \in \mathscr{F}$ *is bounded and* $\{t \in T \mid f(t) \neq 0\}$ *is a* Φ_2*-set of* T;

e) any Φ_2*-set* A *of* T *belongs to* $\mathfrak{A}$ *and*

$$\{p \in \mathscr{P} \mid \sup_{t \in A} g(p, t) > 0\}$$

is finite for any $g \in \mathscr{G}$;

f) $\mathfrak{T}$ *induces a topology on* $\mathscr{F}(A)$ *which is coarser than the topology of uniform convergence for any* Φ_2*-set* A *of* T;

g) if T *is locally compact and* σ*-compact and if* $\mathscr{G}$ *is the set of all positive real functions on* $\mathscr{P} \times T$ *satisfying 1) and 2), then* $\mathscr{F} = \bigcup_{K \in \mathfrak{K}} \mathscr{F}(K)$, *where* $\mathfrak{K}$ *denotes the set of compact sets of* T *and* $\mathfrak{T}$ *is the finest locally convex topology on* $\mathscr{F}$ *inducing on* $\mathscr{F}(K)$ *a coarser topology than the topology of uniform convergence for any* $K \in \mathfrak{K}$. *Moreover, the topology* $\mathfrak{T}$ *is generated by the family of semi-norms*

$$r_g : \mathscr{F} \to \mathbb{R}, \quad f \mapsto \sup_{p \in \mathscr{P}} \sup_{t \in T} g(p, t) p(f(t)),$$

g *running through* $\mathscr{G}$.

a) It is easy to check that $\mathscr{F}$ is a vector subspace of E^T. Let f be a continuous map of T into E with compact carrier K. Take $g \in \mathscr{G}$. By 2) there exists a finite subset

$\mathscr{P}_0$ of $\mathscr{P}$ such that $g(p,t) = 0$ for any $(p,t) \in (\mathscr{P} \backslash \mathscr{P}_0) \times K$. Since $f(K)$ is compact we have

$$\alpha(p) := \sup_{t \in K} p(f(t)) < \infty$$

for any $p \in \mathscr{P}$. By 1)

$$\beta(p) := \sup_{t \in K} g(p,t) < \infty$$

for any $p \in \mathscr{P}$ so

$$q_g(f) = \sum_{p \in \mathscr{P}_0} \sup_{t \in K} g(p,t) p(f(t)) \leq \sum_{p \in \mathscr{P}_0} \beta(p) \alpha(p) < \infty$$

Hence, 4) is fulfilled. 5) is trivial.

b) It is easy to see that $q_g | \mathscr{F}$ is a semi-norm for any $g \in \mathscr{G}$. $\mathfrak{T}$ is Hausdorff by 3).

c) For each $p \in \mathscr{P}$ let us denote by $(E_p, \| \ \|_p)$ the Banach space and by u_p the canonical map $E \to E_p$ associated with p. Further, denote by F the vector subspace of $\prod_{p \in \mathscr{P}} E_p$

$$\{(x_p)_{p \in \mathscr{P}} \in \prod_{p \in \mathscr{P}} E_p \mid \sum_{p \in \mathscr{P}} \|x_p\|_p < \infty\}$$

endowed with the norm

$$F \to \mathbb{R}, \quad (x_p)_{p \in \mathscr{P}} \mapsto \sum_{p \in \mathscr{P}} \|x_p\|_p$$

and, for each $g \in \mathscr{G}$, denote by A_g the set

$$\{(v,k) \in T^{\mathscr{P}} \times \mathbb{R}^{\mathscr{P}} \mid p \in \mathscr{P} \Rightarrow |k(p)| \leq g(p, v(p))\}.$$

We set

$$S := \bigcup_{g \in \mathscr{G}} A_g$$

and, for any $f \in \mathscr{F}$, denote by $\tilde{f}$ the map

$$S \to F, \quad (v,k) \mapsto (u_p(k(p) f(v(p))))_{p \in \mathscr{P}},$$

which is well defined by virtue of 4). We endow S with the initial topology with respect to the family $(\tilde{f})_{f \in \mathscr{F}}$.

Let $g \in \mathscr{G}$. We want to show that A_g is a quasicompact set of S. Let $\mathfrak{F}$ be an ultrafilter on A_g and for any $p \in \mathscr{P}$ let φ_p and ψ_p be the maps

$$A_g \to T, \quad (v,k) \mapsto v(p),$$
$$A_g \to \mathbb{R}, \quad (v,k) \mapsto k(p)$$

respectively. We denote by $\mathscr{P}'$ the set of $p \in \mathscr{P}$ for which $\varphi_p(\mathfrak{F})$ converges and set

$$v_0 : \mathscr{P} \to T, \quad p \mapsto \begin{cases} \lim \varphi_p(\mathfrak{F}) & \text{for } p \in \mathscr{P}' \\ t_0 & \text{for } p \in \mathscr{P} \backslash \mathscr{P}', \end{cases}$$

where t_0 is an arbitrary but fixed point of T. By 1)

$$|\lim \psi_p(\mathfrak{F})| \leq g(p, v_0(p))$$

for any $p \in \mathscr{P}'$. We set

$$k_0 : \mathscr{P} \to \mathbb{R}, \quad p \mapsto \begin{cases} \lim \psi_p(\mathfrak{F}) & \text{for} \quad p \in \mathscr{P}' \\ 0 & \text{for} \quad p \in \mathscr{P} \setminus \mathscr{P}'. \end{cases}$$

We have $(v_0, k_0) \in A_g$. Let $f \in \mathscr{F}$. By 5)

$$\lim_{(v,k),\mathfrak{F}} p(k(p) f(v(p))) = 0$$

for any $p \in \mathscr{P} \setminus \mathscr{P}'$ so

$$\lim_{(v,k),\mathfrak{F}} p(k(p) f(v(p)) - k_0(p) f(v_0(p))) = 0$$

for any $p \in \mathscr{P}$. Let ε be a strictly positive real number. By 4) there exists a finite subset $\mathscr{P}''$ of $\mathscr{P}$ such that

$$\sum_{p \in \mathscr{P} \setminus \mathscr{P}''} \sup_{t \in T} g(p, t) p(f(t)) < \varepsilon .$$

We get

$$\sum_{p \in \mathscr{P} \setminus \mathscr{P}''} p(k(p) f(v(p)) - k_0(p) f(v_0(p)) \leq$$

$$\leq \sum_{p \in \mathscr{P} \setminus \mathscr{P}''} p(k(p) f(v(p))) + \sum_{p \in \mathscr{P} \setminus \mathscr{P}''} p(k_0(p) f(v_0(p))) \leq 2\varepsilon$$

for any $(v, k) \in A_g$ so

$$\lim_{(v,k),\mathfrak{F}} \|\tilde{f}(v, k) - \tilde{f}(v_0, k_0)\| =$$

$$= \lim_{(v,k),\mathfrak{F}} \sum_{p \in \mathscr{P}} \|u_p(k(p) f(v(p))) - u_p(k_0(p) f(v_0(p)))\|_p =$$

$$= \sum_{p \in \mathscr{P}''} \lim_{(v,k),\mathfrak{F}} p(k(p) f(v(p)) - k_0(p) f(v_0(p))) +$$

$$+ \lim_{(v,k),\mathfrak{F}} \sum_{p \in \mathscr{P} \setminus \mathscr{P}''} p(k(p) f(v(p)) - k_0(p) f(v_0(p))) \leq 2\varepsilon .$$

Since ε is arbitrary we also find that

$$\lim_{\mathfrak{F}} \tilde{f} = \tilde{f}(v_0, k_0) .$$

Hence, $\mathfrak{F}$ converges to (v_0, k_0) and A_g is quasicompact,

Let $(f_\iota)_{\iota \in I}$ be a supersummable family in $\mathscr{F}_{T_0}$. By Theorem 3.4.1 b) $(f_\iota)_{\iota \in I}$ is supersummable in $\mathscr{F}$. Let $\mathscr{C}$ be the vector space of continuous maps of S into F. Let $(v, k) \in S$ and $J \subset I$. We have

$$\sum_{\iota \in J} \tilde{f}_\iota(v, k)_p = \sum_{\iota \in J} u_p(k(p) f_\iota(v(p))) =$$

$$= u_p(k(p)(\sum_{\iota \in J} f_\iota)(v(p))) = (\widetilde{\sum_{\iota \in J} f_\iota}(v, k))_p$$

for any $p \in \mathscr{P}$. By Theorem 3.5.2 a) (replacing G by F and $(x_{\iota\lambda})_{(\iota,\lambda)\in I\times L}$ by $(\tilde{f}_\iota(v,k)_p)_{(\iota,p)\in I\times\mathscr{P}}$) we deduce that $(\tilde{f}_\iota(v,k))_{\iota\in J}$ is summable for any $J \subset I$ and its sum is $\sum_{\iota\in J} f_\iota(v,k)$. Hence, $(\tilde{f}_\iota)_{\iota\in I}$ is supersummable in $\mathscr{C}$. By Theorem 3.4.1 b) (and Proposition 1.5.5 c $\Rightarrow$ a) $(\tilde{f}_\iota)_{\iota\in I}$ is supersummable in $\mathscr{C}_{\mathfrak{K}}$, where $\mathfrak{K}$ denotes the set of quasicompact sets of S.

Let $g \in \mathscr{G}$ and let ε be a strictly positive real number. There exists a finite subset J of I such that

$$\sum_{p\in\mathscr{P}} \| \sum_{\iota\in L} u_p(k(p) f_\iota(v(p))) \|_p < \varepsilon$$

for any $L \subset I \backslash J$ and for any $(v,k) \in A_g$. Since

$$(v, g(\cdot, v(\cdot))) \in A_g$$

for any $v \in T^{\mathscr{P}}$ we get

$$\begin{aligned} q_g(\sum_{\iota\in L} f_\iota) &= \sum_{p\in\mathscr{P}} \sup_{t\in T} g(p,t) p(\sum_{\iota\in L} f_\iota(t)) = \\ &= \sum_{p\in\mathscr{P}} \sup_{t\in T} \| \sum_{\iota\in L} u_p(g(p,t) f_\iota(t)) \|_p = \\ &= \sup_{v\in T^{\mathscr{P}}} \sum_{p\in\mathscr{P}} \| \sum_{\iota\in L} u_p(g(p,v(p)) f_\iota(v(p))) \|_p \leq \varepsilon \end{aligned}$$

for any $L \subset I \backslash J$. It follows immediately that $(f_\iota)_{\iota\in I}$ is supersummable with respect to $\mathfrak{T}$.

d) Let $f \in \mathscr{F}$. Assume $\{t \in T \mid f(t) \neq 0\}$ is not a Φ_2-set of T. There exists then a sequence $(t_n)_{n\in\mathbb{N}}$ in T with no adherent point such that $t_m \neq t_n$ for any different $m, n \in \mathbb{N}$ and such that $f(t_n) \neq 0$ for any $n \in \mathbb{N}$. There exists further a sequence $(p_n)_{n\in\mathbb{N}}$ in $\mathscr{P}$ such that $p_n(f(t_n)) \neq 0$ for any $n \in \mathbb{N}$. The function g on $\mathscr{P} \times T$ equal to $\dfrac{n}{p_n(f(t_n))}$ at (p_n, t_n) for any $n \in \mathbb{N}$ and equal to 0 elsewhere belongs to $\mathscr{G}$ and we get the contradictory relation $q_g(f) = \infty$. Hence, $\{t \in T \mid f(t) \neq 0\}$ is a Φ_2-set of T so f is bounded.

e) Let A be a Φ_2-set of T. Assume there exists $g \in \mathscr{G}$ such that the set

$$\{p \in \mathscr{P} \mid \sup_{t\in A} g(p,t) > 0\}$$

is infinite. Then there exist a sequence $(p_n)_{n\in\mathbb{N}}$ in $\mathscr{P}$ and a sequence $(t_n)_{n\in\mathbb{N}}$ in A such that $p_m \neq p_n$ for any different $m, n \in \mathbb{N}$ and such that $g(p_n, t_n) > 0$ for any $n \in \mathbb{N}$. Let t be an adherent point in T of the sequence $(t_n)_{n\in\mathbb{N}}$. By 2) there exists a neighbourhood U of t such that

$$\{p \in \mathscr{P} \mid \sup_{s\in U} g(p,s) > 0\}$$

is finite so the set $\{n \in \mathbb{N} \mid t_n \in U\}$ is finite, which is a contradiction. Hence,

$$\{p \in \mathscr{P} \mid \sup_{t\in A} g(p,t) > 0\}$$

is finite and

$$\sum_{p \in \mathscr{P}} \sup_{t \in A} g(p,t) < \infty$$

for any $g \in \mathscr{G}$.

f) Let $g \in \mathscr{G}$. We set

$$\mathscr{P}_0 := \{p \in \mathscr{P} \mid \sup_{t \in A} g(p,t) > 0\}.$$

By e) $\mathscr{P}_0$ is finite. We get

$$q_g(f) \leq \Big(\sum_{p \in \mathscr{P}_0} \sup_{t \in A} g(p,t)\Big) \sup_{(p,t) \in \mathscr{P}_0 \times T} p(f(t))$$

for any $f \in \mathscr{F}(A)$, which proves the assertion.

g) The relation $\mathscr{F} = \bigcup_{K \in \mathfrak{K}} \mathscr{F}(K)$ follows immediately from d), e) and Theorem 1.5.8. Let $\mathfrak{S}$ be a locally convex topology on $\mathscr{F}$ inducing on $\mathscr{F}(K)$ a topology which is coarser than the topology of uniform convergence for any $K \in \mathfrak{K}$ and let $\mathscr{U}$ be a convex 0-neighbourhood in $\mathscr{F}$ with respect to $\mathfrak{S}$. Let $(U_n)_{n \in \mathbb{N}}$ be a covering of T with relatively compact open sets such that $\bar{U}_n \subset U_{n+1}$ for any $n \in \mathbb{N}$. Then there exist a decreasing sequence $(\alpha_n)_{n \in \mathbb{N}}$ of strictly positive real numbers and an increasing sequence $(\mathscr{P}_n)_{n \in \mathbb{N}}$ of finite subsets of $\mathscr{P}$ such that

$$n \in \mathbb{N},\ f \in \mathscr{F}(\bar{U}_n),\quad \sup_{(p,t) \in \mathscr{P}_n \times T} p(f(t)) \leq \alpha_n \Rightarrow f \in \mathscr{U}.$$

We denote by g the real function on $\mathscr{P} \times T$ equal to $\dfrac{2^{n+2}}{\alpha_n}$ on $\mathscr{P}_n \times (U_n \setminus U_{n-1})$ for any $n \in \mathbb{N}$, $(U_{-1} := \emptyset)$, and equal to 0 elsewhere. Then $g \in \mathscr{G}$. Let $f \in \mathscr{F}$ such that $r_g(f) \leq 1$. We set

$$\varphi_n : T \to \mathbb{R},\quad t \mapsto \sup_{p \in \mathscr{P}_n} p(f(t)),$$

$$f_n : T \to E,\quad t \mapsto \begin{cases} f(t) & \text{if } \varphi_n(t) \leq \dfrac{\alpha_n}{2^{n+2}} \\[2ex] \dfrac{\alpha_n}{2^{n+2}\varphi_n(t)} f(t) & \text{if } \varphi_n(t) > \dfrac{\alpha_n}{2^{n+2}} \end{cases}$$

for any $n \in \mathbb{N}$. Let $n \in \mathbb{N} \setminus \{0\}$ and $t \in T$. Then

$$f_{n-1}(t) - f_n(t) = 0 \text{ if } \varphi_n(t) \leq \frac{\alpha_n}{2^{n+2}},$$

$$f_{n-1}(t) - f_n(t) = f(t) - \frac{\alpha_n}{2^{n+2}\varphi_n(t)} f(t),$$

$$\sup_{p \in \mathscr{P}_{n-1}} p(f_{n-1}(t) - f_n(t)) = \left(1 - \frac{\alpha_n}{2^{n+2}\varphi_n(t)}\right) \varphi_{n-1}(t) \leq \frac{\alpha_{n-1}}{2^{n+1}}$$

if $\frac{\alpha_n}{2^{n+2}} < \varphi_n(t)$ and $\varphi_{n-1}(t) \leq \frac{\alpha_{n-1}}{2^{n+1}}$, and

$$f_{n-1}(t) - f_n(t) = \frac{\alpha_{n-1}}{2^{n+1}\varphi_{n-1}(t)} f(t) - \frac{\alpha_n}{2^{n+2}\varphi_n(t)} f(t),$$

$$\sup_{p \in \mathscr{P}_{n-1}} p(f_{n-1}(t) - f_n(t)) = \left(\frac{\alpha_{n-1}}{2^{n+1}\varphi_{n-1}(t)} - \frac{\alpha_n}{2^{n+2}\varphi_n(t)} \right) \varphi_{n-1}(t) \leq \frac{\alpha_{n-1}}{2^{n+1}}$$

if $\frac{\alpha_{n-1}}{2^{n+1}} < \varphi_{n-1}(t)$. Since

$$\frac{2^{n+2}}{\alpha_n} \varphi_n(t) \leq \sup_{p \in \mathscr{P}_n} g(p, t) p(f(t)) \leq 1$$

for any $t \in T \backslash U_{n-1}$ we get $f_{n-1} - f_n \in \mathscr{F}(\bar{U}_{n-1})$ and $f_0 = f$. Moreover,

$$\sup_{(p,t) \in \mathscr{P}_{n-1} \times T} p(f_{n-1}(t) - f_n(t)) \leq \frac{\alpha_{n-1}}{2^{n+1}}$$

so $f_{n-1} - f_n \in \frac{1}{2^{n+1}} \mathscr{U}$. We deduce that

$$\sum_{n=1}^{\infty} (f_{n-1} - f_n) \in \tfrac{1}{2} \mathscr{U}.$$

Since

$$f - \sum_{n=1}^{\infty} (f_{n-1} - f_n) \in \tfrac{1}{2} \mathscr{U}$$

we get $f \in \mathscr{U}$. Hence

$$\{f \in \mathscr{F} \mid r_g(f) \leqq l\} \subset \mathscr{U}.$$

Let $\mathfrak{T}'$ be the topology on $\mathscr{F}$ generated by the family $(r_g)_{g \in \mathscr{G}}$ of semi-norms. By the above considerations $\mathfrak{S} \subset \mathfrak{T}'$. We deduce, by f), $\mathfrak{T} \subset \mathfrak{T}'$. Since the converse inclusion is trivial we get $\mathfrak{T} = \mathfrak{T}'$. The assertions now follow from f). □

Remark. Let us denote by $\tilde{\mathscr{F}}$ the vector space $\mathscr{F}$ endowed with the finest locally convex topology inducing on $\mathscr{F}(K)$ a topology which is coarser than the topology of uniform convergence for any compact set K of T. If T is locally compact and σ-compact, then by g) and c) any supersummable countable family in $\mathscr{F}$ is supersummable in $\tilde{\mathscr{F}}$. This result was proved by I. Labuda (1973) ([3] Corollary 2.2).

Theorem 3.5.21. *Let X be a topological space, $i \in \{1, 2\}$ and $\mathscr{F}$ be the set of Φ_i-continuous real functions on X. Then:*

a) $\mathscr{F}$ is a subgroup of $\mathbb{R}^X$;

b) if $(f_\iota)_{\iota \in I}$ is a supersummable family in $\mathscr{F}$, then $(|f_\iota|)_{\iota \in I}$ is supersummable in $\mathscr{F}$ too.

a) is easy to check.

b) It is sufficient to show that $f := \sum_{\iota \in I} |f_\iota|$ is Φ_i-continuous. Let $\mathfrak{F}$ be a convergent Φ_i-filter on X, x be a limit point of $\mathfrak{F}$ in X and (L, g) be a Θ_i-net in X such that $g(\mathfrak{G}) = \mathfrak{F}$, $\mathfrak{G}$ denoting the section filter of L. We have to show that $\lim f(\mathfrak{F}) = f(x)$. Assume the contrary. Then there exists a strictly positive real number α such that for any $A \in \mathfrak{F}$ there exists $y \in A$ so that $f(y) - f(x) > \alpha$. We construct an increasing sequence $(\lambda_n)_{n \in \mathbb{N}}$ in L and an increasing sequence $(K_n)_{n \in \mathbb{N}}$ in $\mathfrak{P}_f(I)$ inductively such that we have for any $n \in \mathbb{N} \setminus \{0\}$:

1) $\lambda \in L, \quad \lambda \geq \lambda_n \Rightarrow \sum_{\iota \in K_{n-1}} |f_\iota|(g(\lambda)) < f(x) + \frac{\alpha}{4}$;

2) $\sum_{\iota \in K_n} |f_\iota|(g(\lambda_n)) > f(x) + \alpha$.

We choose K_0 and λ_0 arbitrarily. Let $n \in \mathbb{N} \setminus \{0\}$ and assume the sequences were constructed up to $n-1$. Since $\sum_{\iota \in K_{n-1}} |f_\iota|$ is Φ_i-continuous there exists $\lambda' \in L$ such that

$$\sum_{\iota \in K_{n-1}} |f_\iota|(g(\lambda)) < \sum_{\iota \in K_{n-1}} |f_\iota|(x) + \frac{\alpha}{4} < f(x) + \frac{\alpha}{4}$$

for any $\lambda \in L$ with $\lambda \geq \lambda'$. By the hypothesis there exists $\lambda_n \in L$ such that $\lambda_n \geq \lambda_{n-1}$, $\lambda_n \geq \lambda'$, and

$$\sum_{\iota \in I} |f_\iota|(g(\lambda_n)) > f(x) + \alpha.$$

Then there exists $K_n \in \mathfrak{P}_f(I)$ such that $K_n \supset K_{n-1}$ and 2) holds. This finishes the inductive construction.

Let Φ be the set $\Phi_i(X)$ and let K be the set $\bigcup_{n \in \mathbb{N}} K_n$. By Theorem 3.4.1 a), b) $\mathscr{F}_\Phi$ is a Hausdorff topological group and $(f_\iota)_{\iota \in K}$ is a supersummable family in $\mathscr{F}_\Phi$. Hence by Theorem 3.3.5 the map

$$\mathfrak{P}(K) \to \mathscr{F}_\Phi, \quad J \mapsto \sum_{\iota \in J} f_\iota$$

is continuous. By Proposition 1.4.7 $A := \{g(\lambda_n) \mid n \in \mathbb{N}\}$ is a $\Phi_i(X)$-set. We deduce that there exists $m \in \mathbb{N}$ such that $|\sum_{\iota \in J} f_\iota| < \frac{\alpha}{4}$ on A for any $J \subset K \setminus K_m$. We get $\sum_{\iota \in K \setminus K_m} |f_\iota| < \frac{\alpha}{2}$ on A and therefore by 1)

$$\sum_{\iota \in K} |f_\iota|(g(\lambda_n)) < f(x) + \frac{3\alpha}{4}$$

for any $n \in \mathbb{N}$ with $n > m$, which contradicts 2). $\square$

Corollary 3.5.22. *Let X be a topological space such that the neighbourhood filter of any point of X belongs to $\hat{\Phi}_2(X)$ (e.g. a metrizable or a locally compact topological space), let $\mathscr{C}$ be the vector space of continuous real functions on X and $(f_\iota)_{\iota \in I}$ be a supersummable family in $\mathscr{C}$. Then $(|f_\iota|)_{\iota \in I}$ is supersummable in $\mathscr{C}$.*

By the hypothesis about X any Φ_2-continuous function on X is continuous and the assertion follows from Theorem 3.5.21. □

Remark. We cannot drop the supplementary hypothesis about X in the above corollary as the following example shows. We choose the topological space $\ell_c(\mathbb{N}, \mathbb{R})$ as X and for any $n \in \mathbb{N}$ we denote by φ_n the map

$$\ell_c(\mathbb{N}, \mathbb{R}) \to \mathbb{R}, \quad f \mapsto f(n).$$

Then

$$\sum_{n \in M} \varphi_n(f) = \sum_{n \in M} f(n)$$

for any $M \subset \mathbb{N}$ and any $f \in \ell(\mathbb{N}, \mathbb{R})$. Hence, $(\varphi_n)_{n \in \mathbb{N}}$ is a supersummable sequence in $\mathscr{C}$. In order to show that $(|\varphi_n|)_{n \in \mathbb{N}}$ is not supersummable in $\mathscr{C}$, it is sufficient to show that $\sum_{n \in \mathbb{N}} |\varphi_n|$ is not continuous at $0 \in \ell(\mathbb{N}, \mathbb{R})$. Let U be an arbitrary neighbourhood of 0 in $\ell_c(\mathbb{N}, \mathbb{R})$. Then there exist a strictly positive real number ε and a finite partition $(M_\iota)_{\iota \in I}$ of $\mathbb{N}$ such that

$$\{f \in \ell(\mathbb{N}, \mathbb{R}) \mid \forall J \subset I \Rightarrow |\sum_{\iota \in J} \sum_{n \in M_\iota} f(n)| < \varepsilon\} \subset U.$$

Let $\iota_0 \in I$ such that M_{ι_0} possesses at least two points a, b. We denote by f the real function on $\mathbb{N}$ which is equal to 1 at a, equal to -1 at b, and equal to 0 otherwise. Then $f \in U$ and $\sum_{n \in \mathbb{N}} |\varphi_n(f)| = 2$. Since U is arbitrary and $\sum_{n \in \mathbb{N}} |\varphi_n(0)| = 0$ the function $\sum_{n \in \mathbb{N}} |\varphi_n|$ is not continuous.

Chapter 4: Spaces of measures

§ 4.1 Measures and exhaustive maps

Definition 4.1.1. *A* ring of sets *is a nonempty set* $\mathfrak{R}$ *of sets, such that*

$$A, B \in \mathfrak{R} \Rightarrow A \varDelta B := (A \backslash B) \cup (B \backslash A) \in \mathfrak{R}, \quad A \cap B \in \mathfrak{R}.$$

If, in addition, $\bigcap_{n \in \mathbb{N}} A_n \in \mathfrak{R}$ $(\bigcup_{n \in \mathbb{N}} A_n \in \mathfrak{R})$ *for any sequence* $(A_n)_{n \in \mathbb{N}}$ *in* $\mathfrak{R}$, *then* $\mathfrak{R}$ *is called a* δ-ring (σ-ring).

If $\mathfrak{R}$ is a ring of sets, then $\mathfrak{R}$ endowed with the composition laws $\varDelta$ (as addition) and $\cap$ (as multiplication) is a commutative ring. Any σ-ring is a δ-ring. If $\mathfrak{R}$ is a δ-ring and if $(A_n)_{n \in \mathbb{N}}$ is a sequence in $\mathfrak{R}$, then $\bigcup_{n \in \mathbb{N}} A_n$ belongs to $\mathfrak{R}$ if it is contained in a set of $\mathfrak{R}$.

Definition 4.1.2. *A family of sets* $(A_\iota)_{\iota \in I}$ *is called* disjoint *if*

$$\iota, \lambda \in I, \iota \neq \lambda \Rightarrow A_\iota \cap A_\lambda = \emptyset.$$

Definition 4.1.3. *Let* $\mathfrak{R}$ *be a set of sets. We denote by* $\Gamma(\mathfrak{R})$ *the set of disjoint sequences* $(A_n)_{n \in \mathbb{N}}$ *in* $\mathfrak{R}$ *for which*

$$\{\bigcup_{n \in M} A_n \mid M \subset \mathbb{N}\} \subset \mathfrak{R}$$

and set

$$\Sigma(\mathfrak{R}) := \{\{\bigcup_{n \in M} A_n \mid M \in \mathfrak{P}(\mathbb{N})\} \mid (A_n)_{n \in \mathbb{N}} \in \Gamma(\mathfrak{R})\},$$

$$\Sigma_f(\mathfrak{R}) := \{\{\bigcup_{n \in M} A_n \mid M \in \mathfrak{P}_f(\mathbb{N})\} \mid (A_n)_{n \in \mathbb{N}} \in \Gamma(\mathfrak{R})\}.$$

Further, denote by $\Theta(\mathfrak{R})$ $(\Theta_f(\mathfrak{R}))$ *the set of sequences* $(A_n)_{n \in \mathbb{N}}$ *in* $\mathfrak{R}$ *for which there exists* $\mathfrak{A} \in \Sigma(\mathfrak{R})$ $(\mathfrak{A} \in \Sigma_f(\mathfrak{R}))$ *such that* $\{n \in \mathbb{N} \mid A_n \in \mathfrak{A}\}$ *is infinite and by* $\Psi(\mathfrak{R})$ $(\Psi_f(\mathfrak{R}))$ *the set* $\hat{\Phi}(\Theta(\mathfrak{R}))$ $(\hat{\Phi}(\Theta_f(\mathfrak{R})))$.

We have $\Gamma(\mathfrak{R}) \subset \Theta_f(\mathfrak{R}) \subset \Theta(\mathfrak{R})$ and $\Psi_f(\mathfrak{R}) \subset \Psi(\mathfrak{R})$.

Definition 4.1.4. *Let* $\mathfrak{R}$ *be a ring of sets.* $\mathfrak{R}$ *is called a* quasi-σ-ring *if any disjoint sequence in* $\mathfrak{R}$ *possesses a subsequence which belongs to* $\Gamma(\mathfrak{R})$. $\mathfrak{R}$ *is called a* quasi-δ-ring *if any disjoint sequence in* $\mathfrak{R}$ *whose union is contained in a set of* $\mathfrak{R}$ *possesses a subsequence which belongs to* $\Gamma(\mathfrak{R})$.

Any quasi-σ-ring is a quasi-δ-ring. Any σ-ring (δ-ring) is a quasi-σ-ring (quasi-δ-ring).

Proposition 4.1.5. *Let $\mathfrak{R}$ be a set of sets and $(A_n)_{n\in\mathbb{N}}$ be a disjoint sequence in $\mathfrak{R}$. Then the following assertions are equivalent;*

a) there exists a subsequence of $(A_n)_{n\in\mathbb{N}}$ which belongs to $\Gamma(\mathfrak{R})$;

b) $(A_n)_{n\in\mathbb{N}} \in \Theta_f(\mathfrak{R})$;

c) $(A_n)_{n\in\mathbb{N}} \in \Theta(\mathfrak{R})$;

a $\Rightarrow$ b $\Rightarrow$ c is trivial.

c $\Rightarrow$ a. There exists $(B_n)_{n\in\mathbb{N}} \in \Gamma(\mathfrak{R})$ and a subsequence $(A_{k_n})_{n\in\mathbb{N}}$ of $(A_n)_{n\in\mathbb{N}}$ such that

$$\{A_{k_n} \mid n\in\mathbb{N}\} \subset \{\bigcup_{n\in M} B_n \mid M \subset N\}.$$

There exists, in addition, a disjoint sequence $(M_n)_{n\in\mathbb{N}}$ in $\mathfrak{P}(\mathbb{N})$ such that

$$A_{k_n} = \bigcup_{m\in M_n} B_m$$

for any $n\in\mathbb{N}$. Then

$$\bigcup_{n\in M} A_{k_n} = \bigcup_{n\in M}\bigcup_{m\in M_n} B_m = \bigcup_{m\in M'} B_m \in \mathfrak{R}$$

for any $M \subset \mathbb{N}$, where $M' = \bigcup_{n\in M} M_n$. □

Proposition 4.1.6. *Let $\mathfrak{R}$ be a quasi-σ-ring ordered by the inclusion relation and let $\mathfrak{A}$ be an upper directed set of $\mathfrak{R}$. We set*

$$I := \{(A, B) \in \mathfrak{A} \times \mathfrak{A} \mid A \subset B\}$$

and endow I with the upper directed order relation $\leq$ defined by

$$(A, B) \leq (C, D) \ :\Leftrightarrow\ ((A, B) = (C, D) \quad \text{or} \quad B \subset C)$$

for any $(A, B), (C, D) \in I$. If φ denotes the map

$$I \to \mathfrak{R}, \quad (A, B) \mapsto B\backslash A,$$

then (I, φ) is a $\Theta_f(\mathfrak{R})$-net in $\mathfrak{R}$. If $\mathfrak{A}$ is upper bounded or if $\mathfrak{R}$ and I are ordered by the converse order relation, then we may replace the hypothesis "$\mathfrak{R}$ is a quasi-σ-ring" by the weaker one "$\mathfrak{R}$ is a quasi-δ-ring".

Let $((A_n, B_n))_{n\in\mathbb{N}}$ be an increasing sequence in I. Then $(B_n\backslash A_n)_{n\in\mathbb{N}}$ contains a disjoint sequence in $\mathfrak{R}$, or $\{B_n\backslash A_n \mid n\in\mathbb{N}\}$ is finite. Since $\mathfrak{R}$ is a quasi-σ-ring $(B_n\backslash A_n)_{n\in\mathbb{N}} \in \Theta_f(\mathfrak{R})$. Hence, (I, φ) is a $\Theta_f(\mathfrak{R})$-net in $\mathfrak{R}$.

The last assertion is obvious. □

Definition 4.1.7. *Let $\mathfrak{R}$ be a ring of sets, G be a topological additive group and μ be a map of $\mathfrak{R}$ into G. μ is called* additive *if*

$$\mu(A \cup B) = \mu(A) + \mu(B)$$

for any disjoint sets A, B *of* $\mathfrak{R}$. μ *is called* exhaustive *if* $(\mu(A_n))_{n\in\mathbb{N}}$ *converges to* 0 *for any* $(A_n)_{n\in\mathbb{N}} \in \Gamma(\mathfrak{R})$. μ *is called a* measure *if it is additive and if* $\left(\sum_{m=0}^{n} \mu(A_m)\right)_{n\in\mathbb{N}}$ *converges to* $\mu(\bigcup_{n\in\mathbb{N}} A_n)$ *for any* $(A_n)_{n\in\mathbb{N}} \in \Gamma(\mathfrak{R})$. *We denote by* $\mathscr{E}(\mathfrak{R}, G)$ *the set of exhaustive additive maps of* $\mathfrak{R}$ *into* G *and by* $\mathscr{M}(\mathfrak{R}, G)$ *the set of* G*-valued measures on* $\mathfrak{R}$.

In the usual definition of a positive real measure it is only required that $\bigcup_{n\in\mathbb{N}} A_n \in \mathfrak{R}$ (instead of $(A_n)_{n\in\mathbb{N}} \in \Gamma(\mathfrak{R})$), so our definition is more general. We made this change because it is precisely the relation $(A_n)_{n\in\mathbb{N}} \in \Gamma(\mathfrak{R})$ which appears in the proofs. On the other hand, integration theory with respect to positive real measures uses the usual definition (with $\bigcup_{n\in\mathbb{N}} A_n \in \mathfrak{R}$) but in this book the integral will only be used if $\mathfrak{R}$ is a δ-ring, in which case both definitions coincide. The second condition in the definition of measure is usually called σ-additivity. If G is Hausdorff then the σ-additivity implies the additivity.

We have $\mu(\emptyset) = 0$ for any additive map. Any measure is additive and exhaustive i.e.

$$\mathscr{M}(\mathfrak{R}, G) \subset \mathscr{E}(\mathfrak{R}, G).$$

The exhaustive additive maps were introduced by C.E. Rickart (1943) ([1] Definition 2.1) who called them strongly bounded. The reason for this terminology is given in the next propositions.

Proposition 4.1.8. *Let* I *be a set, for any* $\iota \in I$ *let* π_ι *be the* ι*-projection of* $\mathbb{R}^I$, *let* $\mathfrak{R}$ *be a ring of sets and* μ *be an additive map of* $\mathfrak{R}$ *into* $\mathbb{R}^I$ *such that* $\pi_\iota \circ \mu$ *is bounded for any* $\iota \in I$. *Then* μ *is exhaustive.*

Assume μ is not exhaustive. Then there exist $\iota \in I$ and $(A_n)_{n\in\mathbb{N}} \in \Gamma(\mathfrak{R})$ such that $(\pi_\iota \circ \mu(A_n))_{n\in\mathbb{N}}$ does not converge to 0. There exists then a strictly positive real number ε such that at least one of the sets

$$M' := \{n \in \mathbb{N} \mid \pi_\iota \circ \mu(A_n) > \varepsilon\}, \quad M'' := \{n \in \mathbb{N} \mid \pi_\iota \circ \mu(A_n) < -\varepsilon\}.$$

is infinite. Without loss of generality we may assume M' is infinite. We get

$$\pi_\iota \circ \mu(\bigcup_{n\in M} A_n) > m\varepsilon$$

for any finite subset M of M', where m denotes the cardinal number of M, which contradicts the hypothesis. □

Proposition 4.1.9. *Let* $\mathfrak{R}$ *be a ring of sets,* G *be a topological additive group and* μ *be an exhaustive map of* $\mathfrak{R}$ *into* G. *Then* $(\mu(A_n))_{n\in\mathbb{N}}$ *converges to* 0 *for any disjoint* $\Theta(\mathfrak{R})$*-sequence* $(A_n)_{n\in\mathbb{N}}$.

Let U be a 0-neighbourhood in G. Assume

$$M := \{n \in \mathbb{N} \mid \mu(A_n) \notin U\}$$

is infinite. By Proposition 4.1.5 c $\Rightarrow$ a there exists a strictly increasing sequence $(k_n)_{n\in\mathbb{N}}$ in M such that $(A_{k_n})_{n\in\mathbb{N}} \in \Gamma(\mathfrak{R})$ so $(\mu(A_{k_n}))_{n\in\mathbb{N}}$ converges to 0, which is a contradiction. Hence, $\{n \in \mathbb{N} \mid \mu(A_n) \notin U\}$ is finite and $(\mu(A_n))_{n\in\mathbb{N}}$ converges to 0. □

Proposition 4.1.10. *Let $\mathfrak{R}$ be a ring of sets, $(A_\iota)_{\iota\in I}$ be a disjoint family in $\mathfrak{R}$ such that for any disjoint sequence $(I_n)_{n\in\mathbb{N}}$ in $\mathfrak{P}_f(I)$ there exists an infinite subset M of $\mathbb{N}$ such that $\bigcup_{n\in M'} \bigcup_{\iota\in I_n} A_\iota \in \mathfrak{R}$ for any $M' \subset M$, let φ be the map*

$$\mathfrak{P}_f(I) \to \mathfrak{R}, \quad J \mapsto \bigcup_{\iota\in J} A_\iota,$$

G be a topological additive group and $\mu \in \mathscr{E}(\mathfrak{R}, G)$. We denote by $\mathfrak{F}$ the section filter of $\mathfrak{P}_f(I)$ and by $\mathfrak{G}$ the neighbourhood filter of $\emptyset$ in $\mathfrak{P}_f(I)$. We have:

a) $\varphi(\mathfrak{G}) \in \Psi_f(\mathfrak{R})$;

b) $\mu(\varphi(\mathfrak{F}))$ is a Cauchy filter and $\mu(\varphi\,\mathfrak{G}))$ converges to 0;

c) if G is Hausdorff and sequentially complete, then any countable subfamily of $(\mu(A_\iota))_{\iota\in I}$ is summable;

d) if G is Hausdorff and Φ_1-compact, then $(\mu(A_\iota))_{\iota\in I}$ is supersummable.

a) We set

$$\Delta := \{(A, B) \in \mathfrak{P}_f(I) \times \mathfrak{P}_f(I) \mid A \cap B = \emptyset\},$$

$$\psi : \Delta \to \mathfrak{R}, \quad (A, B) \mapsto \bigcup_{\iota\in A} A_\iota$$

and denote by $\leq$ the upper directed order relation on Δ defined by

$$(A, B) \leq (A', B') :\Leftrightarrow ((A, B) = (A', B') \quad \text{or} \quad A \cup B \subset B').$$

Let $(A_n, B_n)_{n\in\mathbb{N}}$ be an increasing sequence in Δ. By Proposition 4.1.5 a $\Rightarrow$ b $(\psi(A_n, B_n))_{n\in\mathbb{N}} \in \Theta_f(\mathfrak{R})$ and therefore (Δ, ψ) is a $\Theta_f(\mathfrak{R})$-net in $\mathfrak{R}$. Since $\varphi(\mathfrak{G}) = \psi(\mathfrak{H})$, where $\mathfrak{H}$ denotes the section filter of Δ, $\varphi(\mathfrak{G})$ belongs to $\Psi_f(\mathfrak{R})$.

b) Assume the contrary. Then there exist a 0-neighbourhood U in G and a disjoint sequence $(I_n)_{n\in\mathbb{N}}$ in $\mathfrak{P}_f(I)$ such that $\mu(\bigcup_{\iota\in I_n} A_\iota) \notin U$ for any $n \in \mathbb{N}$. By Proposition 4.1.5 a $\Rightarrow$ c $(\bigcup_{\iota\in I_n} A_\iota)_{n\in\mathbb{N}} \in \Theta(\mathfrak{R})$ so by Proposition 4.1.9 $(\mu(\bigcup_{\iota\in I_n} A_\iota))_{n\in\mathbb{N}}$ converges to 0, which is a contradiction.

c) Assume first that I is countable. Then $\mathfrak{F}$ possesses a countable base so, by b), $\mu(\varphi(\mathfrak{F}))$ converges. Hence, $(\mu(A_\iota))_{\iota\in I}$ is summable.

Now let I be arbitrary and J be a countable subset of I. Replacing I by J in the above considerations we deduce that $(\mu(A_\iota))_{\iota\in J}$ is summable.

d) follows from c) and Proposition 3.3.3. □

Corollary 4.1.11. *Let I be a set and μ be an additive exhaustive map on $\mathfrak{P}(I)$. Then the restriction of μ on $\mathfrak{P}_f(I)$ is uniformly continuous.*

Let U be a 0-neighbourhood in the target of μ. By Proposition 4.1.10 b) there exists a finite subset K of I such that

$$\{\mu(J) \mid J \in \mathfrak{P}_f(I), \quad J \cap K = \emptyset\} \subset U.$$

We get

$$\mu(J') - \mu(J'') \in U - U$$

for any $J', J'' \in \mathfrak{P}_f(I)$ with $J' \cap K = J'' \cap K$. Hence, the restriction of μ to $\mathfrak{P}_f(I)$ is uniformly continuous. □

Proposition 4.1.12. *Let $\mathfrak{R}$ be a quasi-σ-ring (quasi-δ-ring), $(A_\iota)_{\iota \in I}$ be a disjoint family of sets (whose union is contained in a set of $\mathfrak{R}$), G be a topological additive group, U be a 0-neighbourhood in G, and let $\mu \in \mathscr{E}(\mathfrak{R}, G)$. Then there exists a finite subset J of I such that $\mu(A) \in U$ for any $A \in \mathfrak{R}$ with $A \subset \bigcup_{\iota \in I \setminus J} A_\iota$ for which $\{\iota \in I \mid A \cap A_\iota \neq \emptyset\}$ is a finite set.*

Assume the contrary. Then we may construct inductively a disjoint sequence $(J_n)_{n \in \mathbb{N}}$ in $\mathfrak{P}_f(I)$ and a sequence $(B_n)_{n \in \mathbb{N}}$ in $\mathfrak{R}$ such that

$$B_n \subset \bigcup_{\iota \in J_n} A_\iota, \quad \mu(B_n) \notin U$$

for any $n \in \mathbb{N}$. By the hypothesis there exists a subsequence $(B_{k_n})_{n \in \mathbb{N}}$ of $(B_n)_{n \in \mathbb{N}}$ which belongs to $\Gamma(\mathfrak{R})$ so $(\mu(B_{k_n}))_{n \in \mathbb{N}}$ converges to 0, which is a contradiction. □

Proposition 4.1.13. *Let $\mathfrak{R}$ be a quasi-σ-ring (quasi-δ-ring), $(A_\iota)_{\iota \in I}$ be a countable disjoint family in $\mathfrak{R}$ (whose union is contained in a set of $\mathfrak{R}$), G be a sequentially complete Hausdorff topological additive group and $\mu \in \mathscr{E}(\mathfrak{R}, G)$. Then:*

a) $(\mu(A_\iota))_{\iota \in I}$ is summable;

b) for any 0-neighbourhood U in G there exists a finite subset J of I such that

$$\sum_{\kappa \in K} \mu(B_\kappa) \in U$$

for any countable disjoint family $(B_\kappa)_{\kappa \in K}$ in $\mathfrak{R}$ such that for any $\kappa \in K$ there exists $\iota \in I \setminus J$ with $B_\kappa \subset A_\iota$.

a) follows from Proposition 4.1.10 c).

b) We may assume U closed. By Proposition 4.1.12 there exists a finite subset J of I such that $\mu(A) \in U$ for any $A \in \mathfrak{R}$ with $A \subset \bigcup_{\iota \in I \setminus J} A_\iota$ for which $\{\iota \in I \mid A \cap A_\iota \neq \emptyset\}$ is a finite set. Let $(B_\kappa)_{\kappa \in K}$ be a countable disjoint family in $\mathfrak{R}$ such that for any $\kappa \in K$ there exists $\iota \in I \setminus J$ with $B_\kappa \subset A_\iota$. We get

$$\sum_{\kappa \in K'} \mu(B_\kappa) = \mu\Big(\bigcup_{\kappa \in K'} B_\kappa\Big) \in U$$

for any finite subset K' of K. By a) $(\mu(B_\kappa))_{\kappa \in K}$ is summable so

$$\sum_{\kappa \in K} \mu(B_\kappa) \in U . \quad \square$$

Proposition 4.1.14. *Let $\mathfrak{R}$ be a ring of sets, G be a topological additive group and μ be an additive map of $\mathfrak{R}$ into G. The following assertions are equivalent:*

a) $\mu \in \mathcal{M}(\mathfrak{R}, G)$;

b) the map

$$\mathfrak{P}(\mathbb{N}) \to G, \quad M \mapsto \mu(\bigcup_{n \in M} A_n)$$

belongs to $\mathcal{G}(\mathbb{N}, G)$ for any $(A_n)_{n \in \mathbb{N}} \in \Gamma(\mathfrak{R})$;

c) the map

$$\mathfrak{P}(\mathbb{N}) \to G, \quad M \mapsto \mu(\bigcup_{n \in M} A_n)$$

is continuous for any $(A_n)_{n \in \mathbb{N}} \in \Gamma(\mathfrak{R})$;

d) the map

$$\mathfrak{P}(\mathbb{N}) \to G, \quad M \mapsto \mu(\bigcup_{n \in M} A_n)$$

is continuous at at least one point for any $(A_n)_{n \in \mathbb{N}} \in \Gamma(\mathfrak{R})$.

If G is Hausdorff then the above assertions are equivalent with the following.

e) $(\mu(A_n))_{n \in \mathbb{N}}$ is summable and its sum is $\mu(\bigcup_{n \in \mathbb{N}} A_n)$ for any $(A_n)_{n \in \mathbb{N}} \in \Gamma(\mathfrak{R})$.

In particular, if G is Hausdorff and $(x_\iota)_{\iota \in I}$ supersummable in G, then

$$\mathfrak{P}(I) \to G, \quad J \mapsto \sum_{\iota \in J} x_\iota$$

is a measure.

Let $(A_n)_{n \in \mathbb{N}} \in \Gamma(\mathfrak{R})$ and f be the map

$$\mathfrak{P}(\mathbb{N}) \to G, \quad M \mapsto \mu(\bigcup_{n \in M} A_n).$$

Since μ is additive we have $f \in \mathcal{H}(\mathbb{N}, G)$ so b $\Leftrightarrow$ c. c $\Rightarrow$ d and c $\Rightarrow$ a are trivial. d $\Rightarrow$ c follows from Theorem 3.1.6. From a) it follows easily that f is continuous at $\mathbb{N}$ so a $\Rightarrow$ d.

If G is Hausdorff, then b $\Leftrightarrow$ e follows from Theorem 3.3.5. The last assertion follows immediately from e). $\square$

Proposition 4.1.15. *Let $\mathfrak{R}$ be a δ-ring, G be a topological additive group and μ be an additive map of $\mathfrak{R}$ into G. The following assertions are equivalent:*

a) $\mu \in \mathcal{M}(\mathfrak{R}, G)$;

b) $(\mu(A_n))_{n\in\mathbb{N}}$ *converges to* $\mu(\bigcup_{n\in\mathbb{N}} A_n)$ *for any increasing sequence* $(A_n)_{n\in\mathbb{N}}$ *in* $\mathfrak{R}$ *whose union belongs to* $\mathfrak{R}$;

c) $(\mu(A_n))_{n\in\mathbb{N}}$ *converges to* $\mu(\bigcap_{n\in\mathbb{N}} A_n)$ *for any decreasing sequence* $(A_n)_{n\in\mathbb{N}}$ *in* $\mathfrak{R}$;

d) $(\mu(A_n))_{n\in\mathbb{N}}$ *converges to* 0 *for any decreasing sequence* $(A_n)_{n\in\mathbb{N}}$ *in* $\mathfrak{R}$ *with empty intersection.*

a ⇒ d. $(A_n \backslash A_{n+1})_{n\in\mathbb{N}} \in \Gamma(\mathfrak{R})$ so $(\mu(A_n))_{n\in\mathbb{N}}$ converges to 0.
d ⇒ c ⇒ b ⇒ a are trivial. □

Remark. The implication a ⇒ d does not hold for any quasi-σ-ring.

Proposition 4.1.16. *Let* $\mathfrak{R}$ *be a quasi-*σ*-ring (*σ*-ring) ordered by the inclusion relation,* $\mathfrak{A}$ *be an upper directed set of* $\mathfrak{R}$ *and* $\mathfrak{F}$ *be the filter on* $\mathfrak{R}$ *generated by the section filter of* $\mathfrak{A}$*. Then* $\mu(\mathfrak{F})$ *is a Cauchy* Φ_3*-filter (Cauchy* Φ_1*-filter) for any exhaustive additive map (for any measure)* μ *defined on* $\mathfrak{R}$*. If* $\mathfrak{A}$ *is upper bounded or if* $\mathfrak{R}$ *is ordered by the converse inclusion relation, then we may replace the hypothesis "*$\mathfrak{R}$ *is a quasi-*σ*-ring (*σ*-ring)" by the weaker one "*$\mathfrak{R}$ *is a quasi-*δ*-ring (*δ*-ring)".*

Let G be a topological additive group and $\mu \in \mathscr{E}(\mathfrak{R}, G)$. Further, let $(A_n)_{n\in\mathbb{N}}$ be an increasing sequence in $\mathfrak{A}$. We want to show $(\mu(A_n))_{n\in\mathbb{N}}$ is a Cauchy sequence. Assume the contrary. Then there exist a 0-neighbourhood U in G and a subsequence $(A_{k_n})_{n\in\mathbb{N}}$ of $(A_n)_{n\in\mathbb{N}}$ such that

$$\mu(A_{k_{n+1}}) - \mu(A_{k_n}) \notin U$$

for any $n \in \mathbb{N}$. Since $\mathfrak{R}$ is a quasi-σ-ring $(A_{k_{n+1}} \backslash A_{k_n})_{n\in\mathbb{N}}$ is a $\Theta_f(\mathfrak{R})$-sequence and by Proposition 4.1.9 $(\mu(A_{k_{n+1}} \backslash A_{k_n}))_{n\in\mathbb{N}}$ converges to 0 which is a contradiction. By Proposition 1.5.17 $\mu(\mathfrak{F})$ is a Cauchy Φ_3-filter.

Now suppose G is a topological additive group and let $\mu \in \mathscr{M}(\mathfrak{R}, G)$. Then $(\mu(A_n))_{n\in\mathbb{N}}$ is a convergent sequence for any increasing sequence $(A_n)_{n\in\mathbb{N}}$ in $\mathfrak{R}$. Hence, $\mu(\mathfrak{F})$ is a Φ_1-filter. By the above proof it is a Cauchy filter.

The last assertion is obvious. □

Proposition 4.1.17. *Let* G*,* H *be topological additive groups,* u *be a sequentially continuous group homomorphism of* G *into* H*,* $\mathfrak{R}$ *be a ring of sets and* $\mu \in \mathscr{E}(\mathfrak{R}, G)$ $(\mu \in \mathscr{M}(\mathfrak{R}, G))$*. Then* $u \circ \mu \in \mathscr{E}(\mathfrak{R}, H)$ $(u \circ \mu \in \mathscr{M}(\mathfrak{R}, H))$.

Assume first that $\mu \in \mathscr{E}(\mathfrak{R}, G)$. Let $(A_n)_{n\in\mathbb{N}} \in \Gamma(\mathfrak{R})$. Then $(\mu(A_n))_{n\in\mathbb{N}}$ converges to 0 so $(u \circ \mu(A_n))_{n\in\mathbb{N}}$ converges to 0. Hence, $u \circ \mu \in \mathscr{E}(\mathfrak{R}, H)$.

Now take $\mu \in \mathscr{M}(\mathfrak{R}, G)$ and $(A_n)_{n\in\mathbb{N}} \in \Gamma(\mathfrak{R})$. Then $(\sum_{m=0}^{n} \mu(A_m))_{n\in\mathbb{N}}$ converges to $\mu(\bigcup_{n\in\mathbb{N}} A_n)$ so $(\sum_{m=0}^{n} u \circ \mu(A_m))_{n\in\mathbb{N}}$ converges to $u \circ \mu(\bigcup_{n\in\mathbb{N}} A_n)$. Hence, $u \circ \mu \in \mathscr{M}(\mathfrak{R}, H)$. □

Proposition 4.1.18. *(L. Drewnowski (1972) [2] Lemma). Let* $\mathfrak{R}$ *be a ring of sets,* G

be a metrizable topological additive group, $(\mu_\iota)_{\iota \in I}$ be a countable family in $\mathscr{E}(\mathfrak{R}, G)$ and $(A_n)_{n \in \mathbb{N}}$ be a disjoint sequence in $\mathfrak{R}$ belonging to $\Theta(\mathfrak{R})$. Then there exists an infinite subset M_0 of $\mathbb{N}$ such that $(A_n)_{n \in M_0} \in \Gamma(\mathfrak{R})$ and such that

$$\mathfrak{P}(M_0) \to G, \; M \mapsto \mu_\iota(\bigcup_{n \in M} A_n)$$

is a measure for any $\iota \in I$.

The map

$$\mathfrak{R} \to G^I, \quad A \mapsto (\mu_\iota(A))_{\iota \in I}$$

is additive and exhaustive. Since G^I is metrizable it is sufficient to prove the proposition for a unique additive exhaustive map $\mu : \mathfrak{R} \to G$. Moreover we may assume G complete. By Proposition 4.1.5 c $\Rightarrow$ a there exists an infinite subset N of $\mathbb{N}$ such that $(A_n)_{n \in \mathbb{N}} \in \Gamma(\mathfrak{R})$.

By Proposition 4.1.13 $(\mu(\{n\}))_{n \in \mathbb{N}}$ is supersummable. We set

$$v : \mathfrak{P}(N) \to G, \quad M \mapsto \mu(\bigcup_{n \in M} A_n) - \sum_{n \in M} \mu(A_n).$$

Then v is an additive exhaustive map vanishing on $\mathfrak{P}_f(N)$. By Lemma 3.2.9 there exists an uncountable set $\mathfrak{M}$ of infinite subsets of N such that $M' \cap M''$ is finite for any different $M', M'' \in \mathfrak{M}$. Let U be a 0-neighbourhood in G. We want to show

$$\{M \in \mathfrak{M} \mid \exists M' \subset M, \; v(M') \notin U\}$$

is finite. Assume the contrary. Then there exists an injective map $\varphi: \mathbb{N} \to \mathfrak{M}$ and a sequence $(N_n)_{n \in \mathbb{N}}$ such that $M_n \subset \varphi(n)$ and $v(N_n) \notin U$ for any $n \in \mathbb{N}$. Since $N_n \cap (\bigcup_{m=0}^{n-1} N_m)$ is finite we have

$$v(N_n \setminus \bigcup_{m=0}^{n-1} N_m) = v(N_n) - v(N_n \cap (\bigcup_{m=0}^{n-1} N_m)) = v(N_n) \notin U$$

for any $n \in \mathbb{N}$, and this is a contradiction because $(N_n \setminus \bigcup_{m=0}^{n-1} N_m)_{n \in \mathbb{N}}$ is a disjoint sequence. Hence

$$\{M \in \mathfrak{M} \mid \exists M' \subset M, \; v(M') \notin U\}$$

is finite. Since G is metrizable

$$\{M \in \mathfrak{M} \mid \exists M' \subset M, \; v(M') \neq 0\}$$

is countable. Hence there exists $M_0 \in \mathfrak{M}$ such that $v(M) = 0$ for any $M \subset M_0$. We get

$$\mu(\bigcup_{n \in M} A_n) = \sum_{n \in M} \mu(A_n)$$

for any $M \subset M_0$. By Proposition 4.1.14 e $\Rightarrow$ a the map

$$\mathfrak{P}(M_0) \to G, \quad M \mapsto \mu(\bigcup_{n \in M} A_n)$$

is a measure. □

Lemma 4.1.19. *Let G be a topological additive group and U be a 0-neighbourhood in G. Then there exist a metrizable topological additive group H, a 0-neighbourhood V in H and a continuous surjective group homomorphism* $u: G \to H$ *such that* $\overset{-1}{u}(V) \subset U$.

Let $(U_n)_{n \in \mathbb{N}}$ be a sequence of symmetric 0-neighbourhoods in G such that $2U_0 \subset U$ and $2U_{n+1} \subset U_n$ for any $n \in \mathbb{N}$. Then

$$G_0 := \bigcap_{n \in \mathbb{N}} U_n$$

is a subgroup of G. We denote by H the quotient group G/G_0 and by u the canonical map $G \to H$. Then $\{u(U_n) \mid n \in \mathbb{N}\}$ is a fundamental system of 0-neighbourhoods in H for a metrizable group topology on H. We endow H with this topology. We have

$$\overset{-1}{u}(u(U_0)) \subset U$$

and

$$\overset{-1}{u}(u(U_n)) \supset U_n$$

for any $n \in \mathbb{N}$. Hence, u is continuous and we may take $V := u(U_0)$. □

Lemma 4.1.20. *For any topological additive group G there exists a family of group homomorphisms of G into metrizable topological additive groups such that the topology of G is the initial topology with respect to this family.*

The assertion follows immediately from the above Lemma. □

Theorem 4.1.21. *Let I be a set,* $\mathfrak{R}$ *be a ring of subsets of I containing* $\mathfrak{P}_f(I)$ *and such that any disjoint sequence in* $\mathfrak{P}_f(I)$ *belongs to* $\Theta(\mathfrak{R})$, *and let G be a topological additive group. Then the identity map*

$$\mathscr{E}(\mathfrak{R}, G) \to \mathscr{E}(\mathfrak{R}, G)_{\{\mathfrak{P}_f(I)\}}$$

is uniformly Φ_3*-continuous.*

First assume G metrizable. Let $(\mu_n)_{n \in \mathbb{N}}$ be a Θ_3-sequence in $\mathscr{E}(\mathfrak{R}, G)$ and let $(J_n)_{n \in \mathbb{N}}$ be a disjoint sequence in $\mathfrak{P}_f(I)$. By Proposition 4.1.18 there exists an infinite subset M_0 of $\mathbb{N}$ such that $(J_n)_{n \in M_0} \in \Gamma(\mathfrak{R})$ and such that

$$f_n : \mathfrak{P}(M_0) \to G, \quad M \mapsto \mu_n(\bigcup_{m \in M} J_m)$$

is a measure for any $n \in \mathbb{N}$. By Proposition 4.1.14 a $\Rightarrow$ b $f_n \in \mathscr{G}(M_0, G)$ for any $n \in \mathbb{N}$. By Proposition 3.2.1 the identity map

$$\mathscr{G}(M_0, G)_{\mathfrak{P}(M_0)} \to \mathscr{G}(M_0, G)_{\{\mathfrak{P}(M_0)\}}$$

is uniformly Φ_4-continuous. Since $(f_n)_{n \in \mathbb{N}}$ is a Θ_3-sequence in $\mathscr{G}(M_0, G)_{\mathfrak{P}(M_0)}$ it is also a Θ_3-sequence in $\mathscr{G}(M_0, G)_{\{\mathfrak{P}(M_0)\}}$ (Theorem 1.8.4 a $\Rightarrow$ e). Hence, for any 0-neighbourhood U in G, there exist $m, n \in M_0$ such that $m < n$ and

$$\mu_m(J_m) - \mu_n(J_m) \in U .$$

Let $\mathfrak{F}$ be a Cauchy Φ_3-filter on $\mathscr{E}(\mathfrak{R}, G)$. We want to show that $\mathfrak{F}$ is a Cauchy filter on $\mathscr{E}(\mathfrak{R}, G)_{\{\mathfrak{P}_f(I)\}}$. Assume the contrary. Let (L, μ) be a Θ_3-net in $\mathscr{E}(\mathfrak{R}, G)$ such that $\mu(\mathfrak{G}) = \mathfrak{F}$, where $\mathfrak{G}$ denotes the section filter of L. There exists a 0-neighbourhood U in G such that for any $\lambda \in L$ there exist $\lambda', \lambda'' \in L$ and $J \in \mathfrak{P}_f(I)$ such that $\lambda' \geq \lambda$, $\lambda'' \geq \lambda$, and

$$\mu_{\lambda'}(J) - \mu_{\lambda''}(J) \notin U .$$

Let V be a symmetric 0-neighbourhood in G such that $4V \subset U$. We denote by λ'_{-1} an arbitrary element of L and construct inductively two increasing sequences $(\lambda_n)_{n \in \mathbb{N}}$, $(\lambda'_n)_{n \in \mathbb{N}}$ in L and a disjoint sequence $(J_n)_{n \in \mathbb{N}}$ in $\mathfrak{P}_f(I)$ such that for any $n \in \mathbb{N}$:

a) $\lambda'_{n-1} \leq \lambda_n \leq \lambda'_n$,

b) $\lambda \in L,\ \lambda \geq \lambda'_n \Rightarrow \mu_\lambda(J_n) - \mu_{\lambda_n}(J_n) \notin V$,

c) $\lambda', \lambda'' \in L,\ \lambda' \geq \lambda'_n,\ \lambda'' \geq \lambda'_n,\ J \subset \bigcup_{m=0}^{n} J_m \Rightarrow \mu_{\lambda'}(J) - \mu_{\lambda''}(J) \in V$.

Let $n \in \mathbb{N}$ and assume the sequences were constructed up to $n-1$. There exist $\lambda', \lambda'' \in L$ and $J \in \mathfrak{P}_f(I)$ such that $\lambda' \geq \lambda'_{n-1}$, $\lambda'' \geq \lambda'_{n-1}$, and

$$\mu_{\lambda'}(J) - \mu_{\lambda''}(J) \notin U .$$

We set $J_n := J \setminus \bigcup_{m=0}^{n-1} J_m$. By c)

$$\mu_{\lambda'}\Big(J \cap \Big(\bigcup_{m=0}^{n-1} J_m\Big)\Big) - \mu_{\lambda''}\Big(J \cap \Big(\bigcup_{m=0}^{n-1} J_m\Big)\Big) \in V$$

so

$$\mu_{\lambda'}(J_n) - \mu_{\lambda''}(J_n) \notin 3V .$$

Since $\mathfrak{F}$ is a Cauchy filter on $\mathscr{E}(\mathfrak{R}, G)$ there exists $\lambda'_n \in L$ such that $\lambda'_n \geq \lambda'$, $\lambda'_n \geq \lambda''$, and

$$\mu_{\xi'}(K) - \mu_{\xi''}(K) \in V$$

for any $\xi', \xi'' \in L$ and $K \subset \bigcup_{m=0}^{n} J_m$ with $\xi' \geq \lambda'_n$, $\xi'' \geq \lambda'_n$. Then either

$$\mu_\lambda(J_n) - \mu_{\lambda'}(J_n) \notin V$$

for any $\lambda \in L$ with $\lambda \geq \lambda'_n$ or

$$\mu_\lambda(J_n) - \mu_{\lambda''}(J_n) \notin V$$

for any $\lambda \in L$ with $\lambda \geq \lambda'_n$. We set $\lambda_n := \lambda'$ in the first case and $\lambda_n := \lambda''$ in the second case. This finishes the inductive construction.

By the first part of the proof there exist $m, n \in \mathbb{N}$ with $m < n$ and

$$\mu_{\lambda_m}(J_m) - \mu_{\lambda_n}(J_m) \in V,$$

which contradicts a & b. Hence, $\mathfrak{F}$ is a Cauchy filter on $\mathscr{E}(\mathfrak{R}, G)_{\{\mathfrak{P}_f(I)\}}$. By Theorem 1.8.4 b $\Rightarrow$ a the identity map $\mathscr{E}(\mathfrak{R}, G) \to \mathscr{E}(\mathfrak{R}, G)_{\{\mathfrak{P}_f(I)\}}$ is uniformly Φ_3-continuous.

Let now G be arbitrary. There exists a family $(G_\lambda)_{\lambda \in L}$ of metrizable topological groups and for any $\lambda \in L$ a group homomorphism $\varphi_\lambda : G \to G_\lambda$ such that the topology of G is the initial topology with respect to the family $(\varphi_\lambda)_{\lambda \in L}$ (Lemma 4.1.20). By the above considerations the identity map

$$\mathscr{E}(\mathfrak{R}, G_\lambda) \to \mathscr{E}(\mathfrak{R}, G_\lambda)_{\{\mathfrak{P}_f(I)\}}$$

is uniformly Φ_3-continuous for any $\lambda \in L$. Since $\varphi_\lambda \circ \mu \in \mathscr{E}(\mathfrak{R}, G_\lambda)$ for any $\mu \in \mathscr{E}(\mathfrak{R}, G)$ and any $\lambda \in L$ we deduce, by Proposition 2.4.2, that the identity map

$$\mathscr{E}(\mathfrak{R}, G) \to \mathscr{E}(\mathfrak{R}, G)_{\{\mathfrak{P}_f(I)\}}$$

is uniformly Φ_3-continuous. □

Remark. The following example will show that the identity map

$$\mathscr{E}(\mathfrak{R}, G) \to \mathscr{E}(\mathfrak{R}, G)_{\{\mathfrak{P}_f(I)\}}$$

is not always Φ_2-continuous (so a fortiori it is not uniformly Φ_2-continuous (Proposition 1.8.3)). We take $I := \mathbb{N}$, $\mathfrak{R} := \mathfrak{P}(\mathbb{N})$, and $G := \mathbb{R}$ and, for any $n \in \mathbb{N}$, denote by δ_n the Dirac measure on $\mathfrak{R}$ at the point n. Then $(\delta_n)_{n \in \mathbb{N}}$ is a Θ_2-sequence in $\mathscr{E}(\mathfrak{R}, G)$ but not a Θ_2-sequence in $\mathscr{E}(\mathfrak{R}, G)_{\{\mathfrak{P}_f(\mathbb{N})\}}$. By Proposition 1.8.9a) the above map is not Φ_2-continuous.

Theorem 4.1.22. *Let R be a topological ring, G be a topological left R-module, X be a set, $\mathscr{G}$ be a set of maps of $\mathfrak{P}(\mathbb{N})$ into X such that for any $g \in \mathscr{G}$ and for any disjoint sequence $(A_n)_{n \in \mathbb{N}}$ in $\mathfrak{P}(\mathbb{N})$ the map*

$$\mathfrak{P}(\mathbb{N}) \to X, \quad M \mapsto g(\bigcup_{n \in M} A_n)$$

belongs to $\mathscr{G}$, let A be the set $\bigcup_{g \in \mathscr{G}} g(\mathfrak{P}(\mathbb{N}))$, Θ be the set of sequences $(x_n)_{n \in \mathbb{N}}$ in X for which there exists $g \in \mathscr{G}$ such that the set $\{n \in \mathbb{N} \mid x_n \in g(\mathfrak{P}_f(\mathbb{N}))\}$ (the set $\{n \in \mathbb{N} \mid x_n \in g(\mathfrak{P}(\mathbb{N}))\}$) is infinite, let Φ be the set $\hat{\Phi}(\Theta)$, $\mathscr{F}$ be the set of maps $f : X \to G$ such that $f \circ g \in \mathscr{E}(\mathfrak{P}(\mathbb{N}), G)$ (such that $f \circ g \in \mathscr{M}(\mathfrak{P}(\mathbb{N}), G)$) for any $g \in \mathscr{G}$, and let $i = 3$, ($i = 4$). We have:

a) $\mathscr{F}$ is an R-submodule of G^X;

b) $\mathscr{F}$, $\mathscr{F}_A$, and $\mathscr{F}_\Phi$ are topological left R-modules, locally convex if G is locally convex;

c) the identity map $\mathscr{F}_A \to \mathscr{F}_\Phi$ is uniformly Φ_i-continuous;

d) if G is Hausdorff, then $\mathscr{F}$ is a Φ_4-closed set of G^X;

e) if G is Hausdorff and Φ_j-compact for an $j \in \{1, 2, 3, 4\}$, then so is $\mathscr{F}$;

f) if G is Hausdorff and sequentially complete, then so is $\mathscr{F}$;

g) if $\{0\}$ *is a* G_δ*-set of G, then* $\mathscr{F}$ *is* Φ_2*-compact; if in addition G is sequentially complete then* $\mathscr{F}$ *is* Φ_4*-compact;*

h) $f(\mathfrak{F}) \in \hat{\Phi}_3(G)$ $(f(\mathfrak{F}) \in \hat{\Phi}_1(G))$ *for any* $f \in \mathscr{F}$ *and any* $\mathfrak{F} \in \Phi$;

i) let $\mathfrak{F} \in \hat{\Phi}_i(\mathscr{F}_A)$, (K, h) *be a net in X such that K possesses no greatest element and such that for any strictly increasing sequence* $(\kappa_n)_{n \in \mathbb{N}}$ *in K there exist* $g \in \mathscr{G}$ *and a sequence* $(M_n)_{n \in \mathbb{N}}$ *in* $\mathfrak{P}_f(\mathbb{N})$ *(in* $\mathfrak{P}(\mathbb{N})$*) converging to* $\emptyset$ *such that* $(g(M_n))_{n \in \mathbb{N}}$ *is a subsequence of* $(h(\kappa_n))_{n \in \mathbb{N}}$, *let* $\mathfrak{G}$ *be the section filter of K and* φ *be the map*

$$\mathscr{F} \times K \to G, \quad (f, \kappa) \mapsto f(h(\kappa));$$

then $\varphi(\mathfrak{F} \times \mathfrak{G})$ *converges to* 0.

a) is obvious.

b) Let $f \in \mathscr{F}$ and $\mathfrak{F} \in \Phi(\Theta)$. There exists a Θ-net (I, φ) in X such that $\varphi(\mathfrak{G}) = \mathfrak{F}$, where $\mathfrak{G}$ is the section filter of I. Let $(\iota_n)_{n \in \mathbb{N}}$ be an increasing sequence in I. There exists $g \in \mathscr{G}$ such that the set $\{n \in \mathbb{N} \mid \varphi(\iota_n) \in g(\mathfrak{P}_f(\mathbb{N}))\}$ (such that $\{n \in \mathbb{N} \mid \varphi(\iota_n) \in g(\mathfrak{P}(\mathbb{N}))\}$) is infinite. Hence, the set

$$\{n \in \mathbb{N} \mid f(\varphi(\iota_n)) \in f \circ g(\mathfrak{P}_f(\mathbb{N}))\}$$
$$(\{n \in \mathbb{N} \mid f(\varphi(\iota_n)) \in f \circ g(\mathfrak{P}(\mathbb{N}))\})$$

is infinite. By Theorem 4.1.11 $f \circ g(\mathfrak{P}_f(\mathbb{N}))$ is precompact (by Proposition 4.1.14 a $\Rightarrow$ c $f \circ g(\mathfrak{P}(\mathbb{N}))$ is compact). We deduce that $(I, f \circ \varphi)$ is a Θ_5-net and $f(\mathfrak{F})$ is a Φ_5-filter in G. By Proposition 1.1.9 and Corollary 1.5.12 the image of any Φ-filter with respect to f belongs to $\hat{\Phi}_C(G)$. By the Propositions 2.1.27 and 2.1.28, $\mathscr{F}_\Phi$ is a topological R-module, locally convex if G is locally convex.

c) By Theorem 4.1.21 the identity map

$$\mathscr{E}(\mathfrak{P}(\mathbb{N}), G) \to \mathscr{E}(\mathfrak{P}(\mathbb{N}), G)_{\{\mathfrak{P}_f(\mathbb{N})\}}$$

is uniformly Φ_3-continuous. By Proposition 3.2.1 and Proposition 4.1.14 a $\Leftrightarrow$ b the identity map

$$\mathscr{M}(\mathfrak{P}(\mathbb{N}), G) \to \mathscr{M}(\mathfrak{P}(\mathbb{N}), G)_{\{\mathfrak{P}(\mathbb{N})\}}$$

is uniformly Φ_4-continuous. We set

$$\mathfrak{S} := \{g(\mathfrak{P}_f(\mathbb{N})) \mid g \in \mathscr{G}\}$$
$$(\mathfrak{S} := \{g(\mathfrak{P}(\mathbb{N})) \mid g \in \mathscr{G}\}).$$

By Proposition 2.4.2 the identity map $\mathscr{F}_A \to \mathscr{F}_\mathfrak{S}$ is uniformly Φ_i-continuous. By Corollaries 2.2.4 and 1.8.15 the identity map $\mathscr{F}_\mathfrak{S} \to \mathscr{F}_\Phi$ is uniformly Φ_4-continuous. Hence, (Corollary 1.8.5) the identity map $\mathscr{F}_A \to \mathscr{F}_\Phi$ is uniformly Φ_i-continuous.

d) First, we prove the theorem for measures. $\mathscr{F}$ is a closed set of G^X_Φ and the assertion follows from c) and Corollary 2.2.5.

Now let us discuss the case of exhaustive additive maps. Let φ be the inclusion map $\mathscr{F} \to G^X$, $\mathfrak{F}$ be a Φ_4-filter on $\mathscr{F}$ and $f \in G^X$ such that $\varphi(\mathfrak{F})$ converges to f. We have to show that $f \in \mathscr{F}$. Assume the contrary. There exists $g \in \mathscr{G}$ such that $f \circ g \notin \mathscr{E}(\mathfrak{P}(\mathbb{N}), G)$. Hence, there exist a 0-neighbourhood U in G and a disjoint sequence $(A_n)_{n \in \mathbb{N}}$ in $\mathfrak{P}(\mathbb{N})$ such that $f \circ g(A_n) \notin U$ for any $n \in \mathbb{N}$. Furthermore, by Lemma 4.1.19 there exist a metrizable topological group H, a closed 0-neighbourhood V in H, and a continuous group homomorphism $u: G \to H$ such that $\overset{-1}{u}(V+V) \subset U$. We get $u \circ f \circ g(A_n) \notin V+V$ for any $n \in \mathbb{N}$. Let (I, f') be a Φ_4-net in $\mathscr{F}$ such that $f'(\mathfrak{G}) = \mathfrak{F}$, where $\mathfrak{G}$ denotes the section filter of I. Since $\varphi(f'(\mathfrak{G}))$ converges to f in G^X, we may construct an increasing sequence $(\iota_m)_{m \in \mathbb{N}}$ in I inductively such that $(u \circ f'_{\iota_m} \circ g(A_n))_{m \in \mathbb{N}}$ converges to $u \circ f \circ g(A_n)$ for any $n \in \mathbb{N}$. Assume that $(f'_{\iota_m})_{m \in \mathbb{N}}$ has an adherent point f'' in $\mathscr{F}$. Then $f'' \circ g \in \mathscr{E}(\mathfrak{P}(\mathbb{N}), G)$ hence $(f'' \circ g(A_n))_{n \in \mathbb{N}}$ converges to 0. We get

$$0 = \lim_{n \to \infty} u \circ f'' \circ g(A_n) = \lim_{n \to \infty} u \circ f \circ g(A_n),$$

which is a contradiction. Hence, $(f'_{\iota_m})_{m \in \mathbb{N}}$ has no adherent point in $\mathscr{F}$ so it possesses a Cauchy subsequence $(f'_{\iota_{k(m)}})_{m \in \mathbb{N}}$. Let us denote by g' the map

$$\mathfrak{P}(\mathbb{N}) \to X, \quad M \mapsto g(\bigcup_{n \in M} A_n).$$

By the hypothesis, $g' \in \mathscr{G}$. By Theorem 4.1.21 and Theorem 1.8.4 a $\Rightarrow$ e $(u \circ f'_{\iota_{k(m)}} \circ g')_{m \in \mathbb{N}}$ is a Cauchy sequence in $\mathscr{E}(\mathfrak{P}(\mathbb{N}), H)_{\{\mathfrak{P}_f(\mathbb{N})\}}$. Hence, there exists $m \in \mathbb{N}$ such that

$$u \circ f'_{\iota_{k(m')}} \circ g'(\{n\}) - u \circ f'_{\iota_{k(m)}} \circ g'(\{n\}) \in V$$

for any $m', n \in \mathbb{N}$ with $m' \geq m$. We get

$$u \circ f \circ g(A_n) - u \circ f'_{\iota_{k(m)}} \circ g(A_n) \in V$$

so

$$u \circ f'_{\iota_{k(m)}} \circ g(A_n) \notin V$$

for any $n \in \mathbb{N}$, which is a contradiction. Hence, $f \in \mathscr{F}$ and $\mathscr{F}$ is a Φ_4-closed set of G^X.

e) and f) follow immediately from d) and Proposition 2.1.14.

g) follows from e) and Proposition 1.6.4.

h) Let (I, h) be a Θ-net in X and $(\iota_n)_{n \in \mathbb{N}}$ be an increasing sequence in I. There exists $g \in \mathscr{G}$ such that the set $\{n \in \mathbb{N} \mid h(\iota_n) \in g(\mathfrak{P}_f(\mathbb{N}))\}$ ($\{n \in \mathbb{N} \mid h(t_n) \in g(\mathfrak{P}(\mathbb{N}))\}$) is infinite. There exists further a Cauchy (a convergent) sequence $(M_n)_{n \in \mathbb{N}}$ in $\mathfrak{P}_f(\mathbb{N})$ (in $\mathfrak{P}(\mathbb{N})$) such that $(g(M_n))_{n \in \mathbb{N}}$ is a subsequence of $(h(\iota_n))_{n \in \mathbb{N}}$. By Proposition 4.1.11 (by Proposition 4.1.14 a $\Rightarrow$ c) $(f(g(M_n)))_{n \in \mathbb{N}}$ is a Cauchy (a convergent) sequence. Hence, $(I, f \circ h)$ is a Θ_3-net (a Θ_1-net). The assertion now follows from Proposition 1.1.9.

i) We may assume $\mathfrak{F} \in \Phi_i(\mathscr{F}_A)$. Let (I, f) be a Θ_i-net in $\mathscr{F}_A$ such that $f(\mathfrak{F}') = \mathfrak{F}$, where $\mathfrak{F}'$ denotes the section filter of I. Assume $\varphi(\mathfrak{F}\times\mathfrak{G})$ does not converge to 0. Then there exist a 0-neighbourhood U in G, an increasing sequence $(\iota_n)_{n\in\mathbb{N}}$ in I, and a strictly increasing sequence $(\kappa_n)_{n\in\mathbb{N}}$ in K such that $f_{\iota_n}(h(\kappa_n))\notin U$ for any $n\in\mathbb{N}$. There also exist $g\in\mathscr{G}$ and a sequence $(M_n)_{n\in\mathbb{N}}$ in $\mathfrak{P}_f(\mathbb{N})$ (in $\mathfrak{P}(\mathbb{N})$) converging to $\emptyset$ such that $(g(M_n))_{n\in\mathbb{N}}$ is a subsequence of $(h(\kappa_n))_{n\in\mathbb{N}}$. By Proposition 1.4.7 $\{f_{\iota_n}\,|\,n\in\mathbb{N}\}$ is a Φ_i-set of $\mathscr{F}_A$ so $\{f_{\iota_n}\circ g\,|\,n\in\mathbb{N}\}$ is a Φ_i-set of $\mathscr{E}(\mathfrak{P}(\mathbb{N}),G)$ (of $\mathscr{M}(\mathfrak{P}(\mathbb{N}),G)$). By Theorem 4.1.21 (by Proposition 4.1.14 a $\Rightarrow$ b and Proposition 3.2.1) and Theorem 1.8.4 h) $\{f_{\iota_n}\circ g\,|\,n\in\mathbb{N}\}$ is a Φ_i-set and therefore (Proposition 1.5.6 a $\Rightarrow$ c) a precompact set of $\mathscr{E}(\mathfrak{P}(\mathbb{N}),G)_{\{\mathfrak{P}_f(\mathbb{N})\}}$ (of $\mathscr{G}(\mathbb{N},G)_{\{\mathfrak{P}(\mathbb{N})\}}$). In particular, (with the aid of Proposition 4.1.11 in the $\mathscr{E}$-case) $\{f_{\iota_n}\circ g\,|\,n\in\mathbb{N}\}$ is an equicontinuous set of $G^{\mathfrak{P}_f(\mathbb{N})}$ (of $G^{\mathfrak{P}(\mathbb{N})}$) so $(f_{\iota_n}\circ g(M_m))_{m\in\mathbb{N}}$ converges to 0 uniformly in $n\in\mathbb{N}$, which is a contradiction. □

Corollary 4.1.23. *Let G, H be topological additive groups, G be Hausdorff, Δ be the set of supersummable sequences in G, Θ be the set of sequences $(x_n)_{n\in\mathbb{N}}$ in G for which there exists $(y_n)_{n\in\mathbb{N}}\in\Delta$ such that*

$$\{x_n\,|\,n\in\mathbb{N}\}\cap\{\sum_{n\in M}y_n\,|\,M\in\mathfrak{P}_f(\mathbb{N})\}$$

is infinite, Φ be the set $\hat{\Phi}(\Theta)$ and $\mathscr{F}$ be the set of group homomorphisms f of G into H such that $(f(y_n))_{n\in\mathbb{N}}$ converges to 0 for any $(y_n)_{n\in\mathbb{N}}\in\Delta$. Then:

a) the identity map $\mathscr{F}\to\mathscr{F}_\Phi$ is uniformly Φ_3-continuous;

b) if H is Hausdorff, then $\mathscr{F}$ is a Φ_4-closed set of H^G;

c) if H is Hausdorff and Φ_i-compact for an $i\in\{1,2,3,4\}$, then so is $\mathscr{F}$;

d) if H is Hausdorff and sequentially complete, then so is $\mathscr{F}$;

e) if $\{0\}$ is a G_δ-set of H, then $\mathscr{F}$ is Φ_2-compact and if, in addition, G is sequentially complete, then $\mathscr{F}$ is Φ_4-compact;

f) $f(\mathfrak{F})\in\hat{\Phi}_3(H)$ for any $f\in\mathscr{F}$ and any $\mathfrak{F}\in\Phi$;

g) let $\mathfrak{F}\in\hat{\Phi}_3(\mathscr{F})$, (I,g) be a net in G such that I possesses no greatest element and such that for any strictly increasing sequence $(\iota_n)_{n\in\mathbb{N}}$ in I there exist a supersummable sequence $(y_m)_{m\in\mathbb{N}}$ in G and a sequence $(M_n)_{n\in\mathbb{N}}$ in $\mathfrak{P}_f(\mathbb{N})$ converging to $\emptyset$ such that $(\sum_{m\in M_n}y_m)_{n\in\mathbb{N}}$ is a subsequence of $(g(\iota_n))_{n\in\mathbb{N}}$, let $\mathfrak{G}$ be the section filter of I and φ be the map

$$\mathscr{F}\times I\to H,\quad (f,\iota)\mapsto f(g(\iota));$$

then $\varphi(\mathfrak{F}\times\mathfrak{G})$ converges to 0.

For any $y:=(y_n)_{n\in\mathbb{N}}\in\Delta$ denote by g_y the map

$$\mathfrak{P}(\mathbb{N})\to G,\quad M\mapsto\sum_{n\in M}y_n.$$

A map $f\colon G\to H$ belongs to $\mathscr{F}$ iff $f\circ g_y\in\mathscr{E}(\mathfrak{P}(\mathbb{N}),H)$ for any $y\in\Delta$ and the assertions follow from Theorem 4.1.22. □

Definition 4.1.24. *Let $\mathfrak{R}$ be a ring of sets and $\mathfrak{K}$ be a subset of $\mathfrak{R}$ closed under finite unions (i.e. $\bigcup_{\iota\in I} K_\iota \in \mathfrak{K}$ for any finite family $(K_\iota)_{\iota\in I}$ in $\mathfrak{K}$; in particular $\emptyset\in\mathfrak{K}$). For any $A\in\mathfrak{R}$, denote by $\underline{\mathfrak{F}_{\mathfrak{R}}(A,\mathfrak{K})}$ or simply by $\underline{\mathfrak{F}(A,\mathfrak{K})}$ the filter on $\mathfrak{R}$ generated by the filter base*

$$\{\{B\in\mathfrak{R}\mid K\subset B\subset A\}\mid K\in\mathfrak{K},\ K\subset A\},$$

by $\underline{\mathfrak{G}_{\mathfrak{R}}(A,\mathfrak{K})}$ or simply $\underline{\mathfrak{G}(A,\mathfrak{K})}$ the filter on $\mathfrak{R}$ generated by the filter base

$$\{\{L\in\mathfrak{K}\mid K\subset L\subset A\}\mid K\in\mathfrak{K},\ K\subset A\}$$

and by $\underline{\mathfrak{H}_{\mathfrak{R}}(A,\mathfrak{K})}$ or simply by $\underline{\mathfrak{H}(A,\mathfrak{K})}$ the filter on $\mathfrak{R}$ generated by the filter base

$$\{\{L\in\mathfrak{K}\mid L\subset A\backslash K\}\mid K\in\mathfrak{K},\ K\subset A\}.$$

A map μ of $\mathfrak{R}$ into a topological space is called $\underline{\mathfrak{K}\text{-regular}}$ if $\mu(\mathfrak{F}(A,\mathfrak{K}))$ converges to $\mu(A)$ for any $A\in\mathfrak{R}$. We set

$$\underline{\mathscr{E}(\mathfrak{R},G;\mathfrak{K})}:=\{\mu\in\mathscr{E}(\mathfrak{R},G)\mid \mu \text{ is } \mathfrak{K}\text{-regular}\}$$

$$\underline{\mathscr{M}(\mathfrak{R},G;\mathfrak{K})}:=\{\mu\in\mathscr{M}(\mathfrak{R},G)\mid \mu \text{ is } \mathfrak{K}\text{-regular}\}$$

for any topological additive group G. For any subset $\mathscr{M}$ of $\mathscr{E}(\mathfrak{R},G;\mathfrak{K})$ denote by $\underline{\bar{\mathscr{M}}}$ its closure in $\mathscr{E}(\mathfrak{R},G;\mathfrak{K})$ and by $\underline{\bar{\bar{\mathscr{M}}}}$ its closure in $\mathscr{E}(\mathfrak{R},G;\mathfrak{K})_{\mathfrak{R}}$ (we have $\bar{\mathscr{M}}\subset\bar{\bar{\mathscr{M}}}$).

Any map of $\mathfrak{R}$ into a topological space is $\mathfrak{R}$-regular so

$$\mathscr{E}(\mathfrak{R},G)=\mathscr{E}(\mathfrak{R},G;\mathfrak{R}),\quad \mathscr{M}(\mathfrak{R},G)=\mathscr{M}(\mathfrak{R},G;\mathfrak{R}).$$

Hence, the results for the spaces $\mathscr{E}(\mathfrak{R},G;\mathfrak{K})$, $\mathscr{M}(\mathfrak{R},G;\mathfrak{K})$ include those for the spaces $\mathscr{E}(\mathfrak{R},G)$ and $\mathscr{M}(\mathfrak{R},G)$ respectively. For this reason we prefer to formulate the theorems for the first spaces.

Proposition 4.1.25. *Let $\mathfrak{R}$ be a ring of sets and $\mathfrak{K}$ be a subset of $\mathfrak{R}$ closed under finite unions. Then:*

a) $\{\mathfrak{H}(A,\mathfrak{K})\mid A\in\mathfrak{K}\}\subset\Psi_f(\mathfrak{R})$;

b) if any disjoint sequence in $\mathfrak{K}$, with the property that its union is contained in a set of $\mathfrak{R}$, belongs to $\Theta(\mathfrak{R})$ then

$$\{\mathfrak{H}(A,\mathfrak{K})\mid A\in\mathfrak{R}\}\subset\Psi_f(\mathfrak{R});$$

c) if $\mathfrak{R}$ is a quasi-δ-ring (δ-ring), then

$$\{\mathfrak{H}(A,\mathfrak{K})\mid A\in\mathfrak{R}\}\subset\Psi_f(\mathfrak{R})\ \ (\{\mathfrak{G}(A,\mathfrak{K})\mid A\in\mathfrak{R}\}\subset\Psi_f(\mathfrak{R})).$$

a) is trivial.

b) Let $A\in\mathfrak{R}$, I be the set

$$\{(K,L)\in\mathfrak{K}\times\mathfrak{K}\mid K\subset A,\ L\subset A\backslash K\},$$

$\leq$ be the upper directed order relation on I defined by

$$(K,L) \leq (K',L') :\Leftrightarrow ((K,L) = (K',L') \quad \text{or} \quad K \cup L \subset K')$$

for any (K,L), $(K',L') \in I$, $\mathfrak{F}$ be the section filter of I and φ be the map

$$I \longrightarrow \mathfrak{R}, \quad (K,L) \mapsto L.$$

Then (I, φ) is a $\Theta_f(\mathfrak{R})$-net in $\mathfrak{R}$ (Proposition 4.1.5 c $\Rightarrow$ b) and

$$\mathfrak{H}(A,\mathfrak{K}) = \varphi(\mathfrak{F}) \in \Psi_f(\mathfrak{R}).$$

c) follows immediately from b). □

Proposition 4.1.26. *Let $\mathfrak{R}$ be a ring of sets, $\mathfrak{K}$ be a subset of $\mathfrak{R}$ closed under finite unions, G be a topological additive group and μ be an additive map of $\mathfrak{R}$ into G. Then the following assertions are equivalent:*

a) μ is $\mathfrak{K}$-regular;

b) $\mu(\mathfrak{G}(A,\mathfrak{K}))$ converges to $\mu(A)$ for any $A \in \mathfrak{R}$.

a $\Rightarrow$ b follows from the fact that $\mathfrak{G}(A,\mathfrak{K})$ is finer than $\mathfrak{F}(A,\mathfrak{K})$ for any $A \in \mathfrak{R}$.

b $\Rightarrow$ a. Let $A \in \mathfrak{R}$ and U be an arbitrary 0-neighbourhood in G. Let V be another 0-neighbourhood in G such that $V - V \subset U$. By b) there exists $K \in \mathfrak{K}$ such that $K \subset A$ and

$$\mu(A) - \mu(L) \in V$$

for any $L \in \mathfrak{K}$ with $K \subset L \subset A$. Let $B \in \mathfrak{R}$ such that $K \subset B \subset A$. By b) there exists $L \in \mathfrak{K}$ such that $L \subset B \backslash K$ and

$$\mu(B \backslash K) - \mu(L) \in V.$$

We get

$$\mu(A) - \mu(B) = (\mu(A) - \mu(K \cup L)) - (\mu(B \backslash K) - \mu(L)) \in V - V \subset U.$$

Hence, $\mu(\mathfrak{F}(A,\mathfrak{K}))$ converges to $\mu(A)$ and μ is $\mathfrak{K}$-regular. □

Proposition 4.1.27. *Let $\mathfrak{R}$ be a ring of sets, $\mathfrak{K}$ be a subset of $\mathfrak{R}$ closed under finite unions, $A \in \mathfrak{R}$, G be a topological additive group and μ be an additive map of $\mathfrak{R}$ into G. Then:*

a) if $\mu(\mathfrak{G}(A,\mathfrak{K}))$ is a Cauchy filter, then $\mu(\mathfrak{H}(A,\mathfrak{K}))$ converges to 0;

b) if μ is $\mathfrak{K}$-regular, then $\mu(\mathfrak{H}(A,\mathfrak{K}))$ converges to 0.

a) Let U be a 0-neighbourhood in G. There exists $K \in \mathfrak{K}$ such that $K \subset A$ and

$$\mu(L) - \mu(K) \in U$$

for any $L \in \mathfrak{K}$ with $K \subset L \subset A$. We get

$$\{\mu(L) \mid L \in \mathfrak{K},\ L \subset A \backslash K\} \subset U.$$

Hence, $\mu(\mathfrak{H}(A, \mathfrak{K}))$ converges to 0.

b) follows from a) and Proposition 4.1.26 a $\Rightarrow$ b. □

Proposition 4.1.28. *Let G, H be topological additive groups, u be a group homomorphism of G into H, $\mathfrak{R}$ be a quasi-δ-ring and $\mathfrak{K}$ be a subset of $\mathfrak{R}$ closed under finite unions. We have:*

a) if u is Φ_3-continuous, then

$$\mu \in \mathscr{E}(\mathfrak{R}, G; \mathfrak{K}) \Rightarrow u \circ \mu \in \mathscr{E}(\mathfrak{R}, H; \mathfrak{K}),$$
$$\mu \in \mathscr{M}(\mathfrak{R}, G; \mathfrak{K}) \Rightarrow u \circ \mu \in \mathscr{M}(\mathfrak{R}, H; \mathfrak{K});$$

b) if u is Φ_1-continuous and if $\mathfrak{R}$ is a δ-ring, then

$$\mu \in \mathscr{M}(\mathfrak{R}, G; \mathfrak{K}) \Rightarrow u \circ \mu \in \mathscr{M}(\mathfrak{R}, H; \mathfrak{K}).$$

By Proposition 4.1.17 the hypotheses imply

$$\mu \in \mathscr{E}(\mathfrak{R}, G) \Rightarrow u \circ \mu \in \mathscr{E}(\mathfrak{R}, H),$$
$$\mu \in \mathscr{M}(\mathfrak{R}, G) \Rightarrow u \circ \mu \in \mathscr{M}(\mathfrak{R}, H).$$

Let $A \in \mathfrak{R}$. By Proposition 4.1.26 a $\Rightarrow$ b $\mu(\mathfrak{G}(A, \mathfrak{K}))$ converges to $\mu(A)$ for any $\mu \in \mathscr{E}(\mathfrak{R}, G; \mathfrak{K})$. By Proposition 4.1.16 $\mu(\mathfrak{G}(A, \mathfrak{K}))$ is a Φ_3-filter for any $\mu \in \mathscr{E}(\mathfrak{R}, G; \mathfrak{K})$ and it is a Φ_1-filter for any $\mu \in \mathscr{M}(\mathfrak{R}, G; \mathfrak{K})$ if $\mathfrak{R}$ is a δ-ring. Using these remarks and the given hypotheses, we deduce that $u \circ \mu(\mathfrak{G}(A, \mathfrak{K}))$ converges to $u \circ \mu(A)$. We conclude by Proposition 4.1.26 b $\Rightarrow$ a. □

Theorem 4.1.29. *Let $\mathfrak{R}$ be a ring of sets, $\mathfrak{K}$ be a subset of $\mathfrak{R}$ which is closed under finite unions and G, H be Hausdorff topological groups having the same underlying additive group. Consider the following assertions:*

1) Any supersummable sequence in G is summable in H and has the same sum in G and H (i.e. the identity map $G \longrightarrow H$ belongs to $\mathscr{P}^c(G, H)$).

2) Any supersummable sequence in G converges to 0 in H.

3) G is metrizable or sequentially complete.

4) There exists a fundamental system of 0-neighbourhoods in H which are Φ_3-closed in G, or H is Φ_3-compact and the identity map $H \longrightarrow G$ is Φ_3-continuous.

5) $\mathfrak{R}$ is a quasi-δ-ring.

Then:

a) 1 $\Rightarrow$ $\mathscr{M}(\mathfrak{R}, G) \subset \mathscr{M}(\mathfrak{R}, H)$;

b) 2 & 3 $\Rightarrow$ $\mathscr{E}(\mathfrak{R}, G) \subset \mathscr{E}(\mathfrak{R}, H)$;

c) $$4 \,\&\, 5 \Rightarrow \begin{cases} \mathscr{M}(\mathfrak{R}, G; \mathfrak{K}) \cap \mathscr{M}(\mathfrak{R}, H) \subset \mathscr{M}(\mathfrak{R}, H; \mathfrak{K}), \\ \mathscr{E}(\mathfrak{R}, G; \mathfrak{K}) \cap \mathscr{E}(\mathfrak{R}, H) \subset \mathscr{E}(\mathfrak{R}, H; \mathfrak{K}) \end{cases}$$

a) follows from Proposition 4.1.14 a $\Leftrightarrow$ e.

b) Let $\mu \in \mathscr{E}(\mathfrak{R}, G)$ and $(A_n)_{n\in\mathbb{N}} \in \Gamma(\mathfrak{R})$. First assume that G is metrizable and let U be a 0-neighbourhood in H. Assume

$$M := \{n \in \mathbb{N} \mid \mu(A_n) \notin U\}$$

is infinite. By Proposition 4.1.18 there exists an infinite subset M_0 of M such that $(\mu(A_n))_{n\in M_0}$ is supersummable in G. Hence, $(\mu(A_n))_{n\in M_0}$ converges to 0 in H, which is a contradiction. Hence, M is finite and $(\mu(A_n))_{n\in\mathbb{N}}$ converges to 0. We deduce that $\mu \in \mathscr{E}(\mathfrak{R}, H)$. Now assume that G is sequentially complete. By Proposition 4.1.10 c) $(\mu(A_n))_{n\in\mathbb{N}}$ is supersummable in G and converges therefore to 0 in H. Hence, $\mu \in \mathscr{E}(\mathfrak{R}, H)$.

c) Let $\mu \in \mathscr{M}(\mathfrak{R}, G; \mathfrak{K}) \cap \mathscr{M}(\mathfrak{R}, H)$, $(\mu \in \mathscr{E}(\mathfrak{R}, G; \mathfrak{K}) \cap \mathscr{E}(\mathfrak{R}, H))$ and $A \in \mathfrak{R}$. By Proposition 4.1.16 $\mu(\mathfrak{G}(A, \mathfrak{K}))$ is a Cauchy Φ_3-filter on G and H. By Proposition 4.1.26 a $\Rightarrow$ b $\mu(\mathfrak{G}(A, \mathfrak{K}))$ converges to $\mu(A)$ in G. We want to show that $\mu(\mathfrak{G}(A, \mathfrak{K}))$ converges to $\mu(A)$ in H.

First assume that there exists a fundamental system $\mathfrak{U}$ of 0-neighbourhoods in H which are Φ_3-closed in G. Let $U \in \mathfrak{U}$. There exists $\mathfrak{A} \in \mathfrak{G}(A, \mathfrak{K})$ such that

$$\mu(B) - \mu(C) \in U$$

for any $B, C \in \mathfrak{A}$. Let $B \in \mathfrak{A}$. Then $\mu(B) - U$ is a Φ_3-closed set of G belonging to $\mu(\mathfrak{G}(A, \mathfrak{K}))$. The filter induced by $\mu(\mathfrak{G}(A, \mathfrak{K}))$ on $\mu(B) - U$ is a Φ_3-filter with respect to the uniformity induced by G. Since $\mu(\mathfrak{G}(A, \mathfrak{K}))$ converges to $\mu(A)$ in G and since $\mu(B) - U$ is a Φ_3-closed set of G we get

$$\mu(A) \in \mu(B) - U$$

so

$$\mu(B) \in \mu(A) + U .$$

Hence, $\mu(\mathfrak{A}) \subset \mu(A) + U$. Since U is arbitrary $\mu(\mathfrak{G}(A, \mathfrak{K}))$, converges to $\mu(A)$in H.

Now assume that H is Φ_3-compact and the identity map $H \longrightarrow G$ is Φ_3-continuous. Then $\mu(\mathfrak{G}(A, \mathfrak{K}))$ converges to an element x of H and $x = \mu(A)$.

By Proposition 4.1.26 b $\Rightarrow$ a μ is $\mathfrak{K}$-regular. □

Corollary 4.1.30. *Let $\mathfrak{R}$ be a ring of sets, E be a Hausdorff locally convex space, E' be its dual and Φ be $\hat{\Phi}_4(\mathscr{P}^c(E, \mathbb{R}))$. For any $x \in E$ we denote by $\tilde{x}$ (by x^*) the map*

$$\mathscr{P}^c(E, \mathbb{R}) \longrightarrow \mathbb{R}, \quad f \longmapsto f(x)$$

$$(E' \longrightarrow \mathbb{R}, \quad f \longmapsto f(x))$$

and identify E with a subset of $\mathbb{R}^{\mathscr{P}^c(E,\mathbb{R})}$ (of $\mathbb{R}^{E'}$) via the map

$$E \longrightarrow \mathbb{R}^{\mathscr{P}^c(E,\mathbb{R})}, \quad x \longmapsto \tilde{x}$$

$$(E \longrightarrow \mathbb{R}^{E'}, \quad x \longmapsto x^*).$$

Then:

a) $\mathscr{M}(\mathfrak{R}, E_{E'}) = \mathscr{M}(\mathfrak{R}, E_{\Phi})$;

b) if E is weakly sequentially complete, then $\mathscr{E}(\mathfrak{R}, E_{E'}) = \mathscr{E}(R, E_{\Phi})$.

We have

$$\mathscr{M}(\mathfrak{R}, E_{\Phi}) \subset \mathscr{M}(\mathfrak{R}, E_{E'}), \quad \mathscr{E}(\mathfrak{R}, E_{\Phi}) \subset \mathscr{E}(\mathfrak{R}, E_{E'}).$$

Consequently, only the converse inclusions have to be proved. By Theorem 3.5.10 b), e) and Corollary 3.5.12 the identity map $E_{E'} \to E_{\Phi}$ belongs to $\mathscr{P}^c(E_{E'}, E_{\Phi})$. The assertions now follow from Theorem 4.1.29 a), b). □

Remark. In b) we cannot drop the hypothesis that E is weakly sequentially complete, as the example ℓ^∞ shows. In Theorem 4.6.7 we shall give another sufficient condition in order to obtain an Orlicz-Pettis type result for exhaustive additive maps.

Corollary 4.1.31. *Let $\mathfrak{R}$ be a quasi-δ-ring, $\mathfrak{K}$ be a subset of $\mathfrak{R}$ which is closed under finite unions, E be a Hausdorff locally convex space, E′ be the dual of E and $\mathfrak{A}$ be a covering of E′ with Φ_4-sets of E'_E. We identify E with a subset of $\mathbb{R}^{E'}$ via the map*

$$E \to \mathbb{R}^{E'}, \; x \mapsto \langle x, \cdot \rangle.$$

Then:

a) $\mathscr{M}(\mathfrak{R}, E_{E'}; \mathfrak{K}) = \mathscr{M}(\mathfrak{R}, E_{\mathfrak{A}}; \mathfrak{K}), \quad \mathscr{E}(\mathfrak{R}, E_{\mathfrak{A}}) \cap \mathscr{E}(\mathfrak{R}, E_{E'}; \mathfrak{K}) = \mathscr{E}(\mathfrak{R}, E_{\mathfrak{A}}; \mathfrak{K})$.

b) if E is weakly sequentially complete, then $\mathscr{E}(\mathfrak{R}, E_{E'}; \mathfrak{K}) = \mathscr{E}(\mathfrak{R}, E_{\mathfrak{A}}; \mathfrak{K})$.

a) We have

$$\mathscr{M}(\mathfrak{R}, E_{\mathfrak{A}}; \mathfrak{K}) \subset \mathscr{M}(\mathfrak{R}, E_{E'}; \mathfrak{K}), \quad \mathscr{E}(\mathfrak{R}, E_{\mathfrak{A}}; \mathfrak{K}) \subset \mathscr{E}(\mathfrak{R}, E_{E'}; \mathfrak{K})$$

so only the converse inclusions have to be proved. Let $\mu \in \mathscr{M}(\mathfrak{R}, E_{E'}; \mathfrak{K})$, $(\mu \in \mathscr{E}(\mathfrak{R}, E_{\mathfrak{A}}) \cap \mathscr{E}(\mathfrak{R}, E_{E'}; \mathfrak{K}))$. By Corollary 4.1.30 a) $\mu \in \mathscr{M}(\mathfrak{R}, E_{E'}; \mathfrak{K}) \cap \mathscr{M}(\mathfrak{R}, E_{\mathfrak{A}})$. By Theorem 4.1.29 c) $\mu \in \mathscr{M}(\mathfrak{R}, E_{\mathfrak{A}}; \mathfrak{K})$ $(\mu \in \mathscr{E}(\mathfrak{R}, E_{\mathfrak{A}}; \mathfrak{K}))$.

b) follows immediately from a) and Corollary 4.1.30 b). □

Remark. In Corollary 4.1.31 we cannot replace $E_{\mathfrak{A}}$ by the E_{Φ} that was dealt with in Corollary 4.1.30, as the following example shows. We take as $\mathfrak{R}$ the σ-ring of Borel sets of $[0, 1]$, as $\mathfrak{K}$ the set of finite subsets of $[0, 1]$, and as E the vector space of bounded Borel functions on $[0, 1]$ endowed with the topology of pointwise convergence. The map μ, which assigns to any $A \in \mathfrak{R}$ the characteristic function of A, belongs to $\mathscr{M}(\mathfrak{R}, E_{E'}; \mathfrak{K})$. Let $(f_n)_{n \in \mathbb{N}}$ be a supersummable sequence in $E_{E'}$. We want to show that $\{\sum_{n \in M} f_n \mid M \in \mathfrak{P}_f(\mathbb{N})\}$ is uniformly bounded. Assume the contrary. Then we may even assume that $\{\sum_{n \in M} f_n \mid M \in \mathfrak{P}_f(\mathbb{N})\}$ is not uniformly upper bounded. We construct inductively a sequence $(x_n)_{n \in \mathbb{N}}$ in $[0, 1]$, a disjoint sequence $(I_n)_{n \in \mathbb{N}}$ in

$\mathfrak{P}_f(\mathbb{N})$, and an increasing sequence $(J_n)_{n\in\mathbb{N}}$ in $\mathfrak{P}_f(\mathbb{N})$ such that for any $n\in\mathbb{N}$:

a) $\sum_{m=0}^{n} \sum_{p\in I_m} f_p(x_n) \geq n$,

b) $\bigcup_{m=0}^{n} I_m \subset J_n$,

c) $I_n \cap J_{n-1} = \emptyset$, $(J_{-1} := \emptyset)$,

d) $\sum_{m\in\mathbb{N}\setminus J_n} |f_m(x_n)| < 1$.

Let $n\in\mathbb{N}$ and assume that the sequences were constructed up to $n-1$. We set

$$\alpha := \sum_{p\in J_{n-1}} \sup_{x\in[0,1]} |f_p(x)|.$$

By the hypothesis of the proof there exist $M\in\mathfrak{P}_f(\mathbb{N})$ and $x_n\in[0,1]$ such that

$$\sum_{p\in M} f_p(x_n) > n + 2\alpha.$$

Set $I_n := M\setminus J_{n-1}$. Since $(f_m(x_n))_{m\in\mathbb{N}}$ is summable there exists $J_n\in\mathfrak{P}_f(\mathbb{N})$ such that $J_{n-1}\subset J_n$ and b & d holds. We get (by b) for $n-1$)

$$\sum_{m=0}^{n} \sum_{p\in I_m} f_p(x_n) \geq \sum_{p\in I_n} f_p(x_n) - \sum_{m=0}^{n-1} \Big| \sum_{p\in I_m} f_p(x_n) \Big| \geq$$

$$\geq \sum_{p\in M} f_p(x_n) - 2 \sum_{p\in J_{n-1}} |f_p(x_n)| \geq n + 2\alpha - 2\alpha = n,$$

which proves a). This finishes the inductive construction.

We set $I := \bigcup_{n\in\mathbb{N}} I_n$, $f := \sum_{n\in I} f_n$. Now

$$f(x_n) = \sum_{m=0}^{n} \sum_{p\in I_m} f_p(x_n) + \sum_{m=n+1}^{\infty} \sum_{p\in I_m} f_p(x_n) > n-1,$$

which contradicts the fact that f is bounded.

Set

$$\varphi : E \to \mathbb{R}, \quad f \mapsto \int_0^1 f(x)dx.$$

By the above considerations and Lebesgue's dominated convergence theorem, $\varphi\in\mathscr{P}^c(E_{E'}, \mathbb{R})$. It is obvious, however, that $\varphi\circ\mu$ is not $\mathfrak{K}$-regular.

Lemma 4.1.32. *Let G', G'' be two Hausdorff topological groups defined on the same additive group G such that G'' possesses a countable base and the identity map $G''\to G'$ is continuous. Then there exists a metrizable group topology on G being coarser than the topology of G'.*

Let $\mathfrak{U}$ be the set of closed 0-neighbourhoods of G'. Then $\bigcap_{U\in\mathfrak{U}} U = \{0\}$. Let $\mathfrak{W}$ be a countable base of G''. We set

$$\mathfrak{W}_0 := \{W \in \mathfrak{W} \mid U \in \mathfrak{U},\ U \cap W = \emptyset\}.$$

For any $W \in \mathfrak{W}_0$ let $U_W \in \mathfrak{U}$ such that $U_W \cap W = \emptyset$. Further, set $\mathfrak{V} := \{U_W \mid W \in \mathfrak{W}_0\}$. Then $\mathfrak{V}$ is a countable subset of $\mathfrak{U}$. Let $x \in G \backslash \{0\}$. There exists $U \in \mathfrak{U}$ such that $x \notin U$. Since U is closed in G'' there exists $W \in \mathfrak{W}$ such that $x \in W$ and $U \cap W = \emptyset$. Hence, $W \in \mathfrak{W}_0$, $x \notin U_W$, and $x \notin \bigcap_{V \in \mathfrak{V}} V$. Further $\bigcap_{V \in \mathfrak{V}} V = \{0\}$ and we see that $\{0\}$ is a G_δ-set of G'. The assertion follows. □

Theorem 4.1.33. *Let $\mathfrak{R}$ be a ring of sets, $\mathfrak{K}$ be a subset of $\mathfrak{R}$ which is closed under finite unions and G, H be Hausdorff topological groups having the same underlying additive group such that H is metrizable, separable and complete, and such that the identity map $H \to G$ is continuous. Then*

$$\mathcal{M}(\mathfrak{R}, G) = \mathcal{M}(\mathfrak{R}, H), \quad \mathcal{E}(\mathfrak{R}, G) = \mathcal{E}(\mathfrak{R}, H)$$

and, if $\mathfrak{R}$ is a quasi-δ-ring,

$$\mathcal{M}(\mathfrak{R}, G; \mathfrak{K}) = \mathcal{M}(\mathfrak{R}, H; \mathfrak{K}), \quad \mathcal{E}(\mathfrak{R}, G; \mathfrak{K}) = \mathcal{E}(\mathfrak{R}, H; \mathfrak{K}).$$

By Lemma 4.1.32 there exists a metrizable topological group F defined on the underlying group of G such that the identity map $G \to F$ is continuous. By Theorem 3.3.27 the identity map $F \to H$ belongs to $\mathcal{P}(F, H)$ so by Theorem 4.1.29a), b)

$$\mathcal{M}(\mathfrak{R}, F) \subset \mathcal{M}(\mathfrak{R}, H), \quad \mathcal{E}(\mathfrak{R}, F) \subset \mathcal{E}(\mathfrak{R}, H).$$

We immediately find that

$$\mathcal{M}(\mathfrak{R}, G) = \mathcal{M}(\mathfrak{R}, H), \quad \mathcal{E}(\mathfrak{R}, G) = \mathcal{E}(\mathfrak{R}, H).$$

The last assertion follows immediately from the first and from Theorem 4.1.29c). □

Remark. The above relation $\mathcal{M}(\mathfrak{R}, G) = \mathcal{M}(\mathfrak{R}, H)$ was proved by N.J. Kalton (1971) ([1] Theorem 3) and the relation $\mathcal{E}(\mathfrak{R}, G) = \mathcal{E}(\mathfrak{R}, H)$ by I. Labuda (1973) ([2] Theorem) and by N.J. Kalton (1974) ([2] Theorem 7).

Lemma 4.1.34. (*L. Drewnowski (1975) [6] Lemma*). *Let G be an additive group and $\mathfrak{T}'$, $\mathfrak{T}''$ be group topologies on G such that $\mathfrak{T}''$ is Hausdorff and possesses a countable base and such that there exists a fundamental system of 0-neighbourhoods with respect to $\mathfrak{T}''$, which are closed with respect to $\mathfrak{T}'$. Then there exists a metrizable group topology $\mathfrak{T}$ on G, which is coarser than $\mathfrak{T}'$, such that there exists a fundamental system of 0-neighbourhoods with respect to $\mathfrak{T}''$, which are closed with respect to $\mathfrak{T}$.*

Let $\mathfrak{U}$ be the set of 0-neighbourhoods with respect to $\mathfrak{T}'$ and let $(V_n)_{n \in \mathbb{N}}$ be a fundamental system of 0-neighbourhoods with respect to $\mathfrak{T}''$ which are closed with respect to $\mathfrak{T}'$ and such that $V_{n+1} - V_{n+1} \subset V_n$ for any $n \in \mathbb{N}$. We set

$$G(n, U) := \{x \in G \mid (x + U) \cap V_n = \emptyset\}$$

for any $(n, U) \in \mathbb{N} \times \mathfrak{U}$.

Let $(n, U) \in \mathbb{N} \times \mathfrak{U}$ and $x \in G(n, U)$. Then $x + V_{n+1} \subset G(n+1, U)$ so $x \in \mathring{G}(n+1, U)$, where $\mathring{G}(n+1, U)$ denotes the interior of $G(n+1, U)$ with respect to $\mathfrak{T}''$, and $G(n, U) \subset \mathring{G}(n+1, U)$.

Let $n \in \mathbb{N}$. Since V_n is closed with respect to $\mathfrak{T}'$

$$G \backslash V_n = \bigcup_{U \in \mathfrak{U}} G(n, U) \subset \bigcup_{U \in \mathfrak{U}} \mathring{G}(n+1, U).$$

Since $\mathfrak{T}''$ possesses a countable base there exists a sequence $(U_{n,k})_{k \in \mathbb{N}}$ in $\mathfrak{U}$ such that

$$\bigcup_{k \in \mathbb{N}} \mathring{G}(n+1, U_{n,k}) = \bigcup_{U \in \mathfrak{U}} \mathring{G}(n+1, U) \supset G \backslash V_n.$$

Let $\mathfrak{B}$ be a countable subset of $\mathfrak{U}$ such that $\{U_{n,k} \mid n, k \in \mathbb{N}\} \subset \mathfrak{B}$ and such that for any $V \in \mathfrak{B}$ there exists $W \in \mathfrak{B}$ with $W - W \subset V$. There exists a group topology $\mathfrak{T}$ on G for which $\mathfrak{B}$ is a fundamental system of 0-neighbourhoods. It is obvious that $\mathfrak{T} \subset \mathfrak{T}'$.

Let $n \in \mathbb{N}$ and $x \in G \backslash V_n$. By the above relation there exists $U \in \mathfrak{B}$ such that $(x + U) \cap V_{n+1} = \emptyset$. Hence, the closure of V_{n+1} with respect to $\mathfrak{T}$ is contained in V_n. We deduce that there exists a fundamental system of 0-neighbourhoods with respect to $\mathfrak{T}''$, which are closed with respect to $\mathfrak{T}$, and that $\mathfrak{T}$ is Hausdorff. Hence, $\mathfrak{T}$ is metrizable. □

Theorem 4.1.35. *Let $\mathfrak{R}$ be a ring of sets, $\mathfrak{K}$ be a subset of $\mathfrak{R}$, closed under finite unions and G, H be Hausdorff topological groups having the same underlying additive group such that H is semi-separable. Consider the following assertions.*

1) There exists a fundamental system of 0-neighbourhoods in H which are sequentially closed in G.

2) There exists a fundamental system of 0-neighbourhoods in H which are Φ_3-closed in G.

3) There exists a fundamental system of 0-neighbourhoods in H which are closed in G.

4) $\mathfrak{R}$ is a quasi-δ-ring.

Then:

a) 1 ⇒ $\mathcal{M}(\mathfrak{R}, G) \subset \mathcal{M}(\mathfrak{R}, H)$;

b) 3 ⇒ $\mathcal{E}(\mathfrak{R}, G) \subset \mathcal{E}(\mathfrak{R}, H)$;

c) 2 & 4 ⇒ $\mathcal{M}(\mathfrak{R}, G; \mathfrak{K}) \subset \mathcal{M}(\mathfrak{R}, H; \mathfrak{K})$;

d) 3 & 4 ⇒ $\mathcal{E}(\mathfrak{R}, G; \mathfrak{K}) \subset \mathcal{E}(\mathfrak{R}, H; \mathfrak{K})$.

a) follows from Corollary 3.3.25 and Theorem 4.1.29 a).

b) First assume that H is metrizable. By Lemma 4.1.34 there exists a metrizable topological group F defined on the underlying group of G such that the identity map $G \rightarrow F$ is continuous and such that there exists a fundamental system of 0-

neighbourhoods in H which are closed in F. Obviously $\mathscr{E}(\mathfrak{R}, G) \subset \mathscr{E}(\mathfrak{R}, F)$. By a) and Theorem 4.1.29 b) $\mathscr{E}(\mathfrak{R}, F) \subset \mathscr{E}(\mathfrak{R}, H)$, so $\mathscr{E}(\mathfrak{R}, G) \subset \mathscr{E}(\mathfrak{R}, H)$.

Now let H be arbitrary and V_0 be a 0-neighbourhood in H which is closed in G. We construct inductively a sequence $(V_n)_{n \in \mathbb{N}}$ of 0-neighbourhoods in H, which are closed in G, and such that $V_{n+1} - V_{n+1} \subset V_n$ for any $n \in \mathbb{N}$. Set $M := \bigcap_{n \in \mathbb{N}} V_n$ and denote by F the topological group defined on the underlying group of G for which $\{V_n \mid n \in \mathbb{N}\}$ is a fundamental system of 0-neighbourhoods. M is a closed subgroup of F and of G. Denote by F', G' the quotient topological groups F/M, G/M respectively and by φ the canonical map $F \to F'$. F' is metrizable. Let $n \in \mathbb{N}$ and $x' \in F' \backslash \varphi(V_n)$. There exists $x \in F \backslash V_n$ such that $x' = \varphi(x)$. Since V_n is closed in G there exists a neighbourhood U of x in G such that $U \cap V_n = \emptyset$. Then $\varphi(U)$ is a neighbourhood of x' in G'. Assume $\varphi(U) \cap \varphi(V_{n+1}) \neq \emptyset$. Then there exists $(y, z) \in U \times V_{n+1}$ with $z - y \in V_{n+1}$ and

$$y \in z - V_{n+1} \subset V_{n+1} - V_{n+1} \subset V_n,$$

which is a contradiction. Hence, $\varphi(U) \cap \varphi(V_{n+1}) = \emptyset$. We deduce that the closure of $\varphi(V_{n+1})$ in G' is contained in $\varphi(V_n)$. Since n is arbitrary there exists a fundamental system of 0-neighbourhoods in F' which are closed in G'. Let $\mu \in \mathscr{E}(\mathfrak{R}, G)$. Then $\varphi \circ \mu \in \mathscr{E}(\mathfrak{R}, G')$ and by the above considerations $\varphi \circ \mu \in \mathscr{E}(\mathfrak{R}, F')$. Since V_0 is arbitrary $\mu \in \mathscr{E}(\mathfrak{R}, H)$.

c) follows immediately from a) and Theorem 4.1.29 c).

d) follows immediately from b) and Theorem 4.1.29 c). □

Remarks. 1) a) was proved by N.J. Kalton (1971) ([1] Theorem 7) with 1) replaced by 3) and for H separable. L. Drewnowski improved Kalton's result (1973) by replacing the hypothesis 3) with 1) but keeping H separable ([4] Theorem 1). The above formulation was proved by L. Drewnowski (1975) ([6] Theorem 1). b) was proved by N.J. Kalton (1974) ([2] Theorem 8) for H separable and by L. Drewnowski (1975) ([6] Theorem 2) in the general case.

2) Let $\langle E, F \rangle$ be a separated duality of vector spaces and $\mathfrak{A}$ be a set of bounded sets of F_E covering F. If $E_{\mathfrak{A}}$ is separable, then by a) $\mathscr{M}(\mathfrak{R}, E_F) = \mathscr{M}(\mathfrak{R}, E_{\mathfrak{A}})$. This result was proved by N.J. Kalton (1971) ([1] Theorem 8).

Definition 4.1.36. *For any set I we denote by $\underline{\ell^\infty(I)}$ the vector subspace of $\mathbb{R}^I$ of bounded functions endowed with the norm*

$$\ell^\infty(I) \to \mathbb{R}, \quad f \mapsto \sup_{\iota \in I} |f(\iota)|.$$

We set $\underline{\ell^\infty} := \ell^\infty(\mathbb{N})$ and denote by $\underline{\ell^*}$ the topological vector subspace of ℓ^∞ formed by the elements of ℓ^∞, which take a finite number of values only.

$\ell^\infty(I)$ is a Banach space for any set I.

Lemma 4.1.37. *(L. Drewnowski (1975) [7] Lemme) Let I be an infinite set, for any $\imath \in I$ let $x_\imath$ be the element of $\ell^\infty(I)$ equal to 1 at $\imath$ and equal to 0 otherwise, let $\mathfrak{I}$ be the set of subsets of I having the same cardinality as I, A be a subset of $\ell^\infty(I)$ containing $\{x_\imath | \imath \in I\}$ such that $x - x_\imath \in A$ for any $(x, \imath) \in A \times I$ with $x(\imath) = 1$, and, for any $J \subset I$, let $A(J)$ be the set $\{x \in A| x|(I \backslash J) = 0\}$. Furthermore, let E be a topological vector space and φ be a map $A \to E$ such that $\varphi(A)$ is $\mathbb{R}$-bounded, 0 is not an adherent point of $\{\varphi(x_\imath) | \imath \in I\}$, and $x, y \in A$, $xy = 0$ (that is $\imath \in I \Rightarrow x(\imath) y(\imath) = 0$) implies $x + y \in A$ and $\varphi(x + y) = \varphi(x) + \varphi(y)$. Then:*

a) there exist $J \in \mathfrak{I}$ and a 0-neighbourhood U in E such that for any $K \in \mathfrak{I}$, $K \subset J$, there exists $\imath \in K$ such that $\varphi(x_\imath + x) \notin U$ for any $x \in A(J \backslash K)$;

b) there exist $K \in \mathfrak{I}$ and a 0-neighbourhood U in E such that $\varphi(x) \notin U$ for any $x \in A(K)$ for which $\{\imath \in K | x(\imath) = 1\}$ is not empty.

a) Let V be a 0-neighbourhood in E such that $\varphi(x_\imath) \notin V - V$ for any $\imath \in I$. Since $\varphi(A)$ is $\mathbb{R}$-bounded there exists $n \in \mathbb{N}$ such that $\varphi(A) \subset \{nz | z \in V\}$. There exists a 0-neighbourhood U in E such that $nU \subset \{nz | z \in V\}$, where $nU := \{\sum_{k=1}^{n} x_k | (x_k)_{1 \le k \le n}$ family in $U\}$. We want to show that there exists $J \in \mathfrak{I}$ which possesses the described property with respect to U.

Assume the contrary. We may construct inductively a decreasing sequence $(I_m)_{m \in \mathbb{N}}$ in $\mathfrak{I}$, starting with $I_0 := I$, such that for any $m \in \mathbb{N} \backslash \{0\}$ and for any $\imath \in I_m$ there exists $x \in A(I_{m-1} \backslash I_m)$ such that $\varphi(x_\imath + x) \in U$. Let $\imath \in I_n$. Then for any $m \in \mathbb{N}$, $1 \le m \le n$, there exists $y_m \in A(I_{m-1} \backslash I_m)$ such that $\varphi(x_\imath + y_m) \in U$. We have $\sum_{m=1}^{n} y_m \in A$ and

$$n\varphi(x_\imath) + \varphi\left(\sum_{m=1}^{n} y_m\right) = \sum_{m=1}^{n} \varphi(x_\imath + y_m) \in nU \subset \{nz | z \in V\}.$$

Hence, there exist $z, z' \in V$ such that

$$n\varphi(x_\imath) + \varphi\left(\sum_{m=1}^{n} y_m\right) = nz, \quad \varphi\left(\sum_{m=1}^{n} y_m\right) = nz'.$$

We deduce that $n\varphi(x_\imath) + nz' = nz$, i.e. $\varphi(x_\imath) = z - z' \in V - V$, which is the expected contradiction.

b) Let J and U be as in a) and $(I_\imath)_{\imath \in I}$ be a disjoint family in $\mathfrak{I}$ such that $\bigcup_{\imath \in I} I_\imath = J$. Then for any $\imath \in I$ there exists $\lambda_\imath \in I_\imath$ such that $\varphi(x_{\lambda_\imath} + x) \notin U$ for any $x \in A(J \backslash I_\imath)$. We set $K := \{\lambda_\imath | \imath \in I\}$. Then $K \in \mathfrak{I}$. Let $x \in A(K)$ such that there exists $\imath \in I$ with $x(\lambda_\imath) = 1$. Then $x - x_{\lambda_\imath} \in A(J \backslash I_\imath)$ so $\varphi(x) = \varphi(x_{\lambda_\imath} + (x - x_{\lambda_\imath})) \notin U$. $\square$

Theorem 4.1.38. *(I. Labuda (1975) [5] Corollaire 2 B (ii) $\Leftrightarrow$ (iii)). Let E be a topological vector space. The following assertions are equivalent.*

a) There exists no additive injective map $\mu : \mathfrak{P}(\mathbb{N}) \to E$ such that $\mu(\mathfrak{P}(\mathbb{N}))$ is discrete and $\mathbb{R}$-bounded.

b) Any additive map μ of a ring of sets $\mathfrak{R}$ into E is exhaustive if $\mu(\mathfrak{R})$ is $\mathbb{R}$-bounded.

a $\Rightarrow$ b. Let $\mathfrak{R}$ be a ring of sets and let $\mu : \mathfrak{R} \to E$ be an additive map such that $\mu(\mathfrak{R})$ is $\mathbb{R}$-bounded. Assume that μ is not exhaustive. Then there exists $(A_n)_{n\in\mathbb{N}} \in \Gamma(\mathfrak{R})$ such that 0 is not an adherent point of $\{\mu(A_n) \mid n \in \mathbb{N}\}$. We denote by A the set of elements of ℓ^∞ taking the values -1, 0, and 1 only and set

$$\varphi : A \to E, \quad x \mapsto \mu(\bigcup_{x(n)=1} A_n) - \mu(\bigcup_{x(n)=-1} A_n).$$

By Lemma 4.1.37 b) there exist an infinite subset M of $\mathbb{N}$ and a symmetric 0-neighbourhood U in E such that $\varphi(x) \notin U$ for any $x \in A$ vanishing on $\mathbb{N}\setminus M$ and equal to 1 at a point of M. Let M', M'' be different subsets of M. We denote by x the element of A equal to 0 on $(\mathbb{N}\setminus M) \cup (M' \cap M'')$, equal to 1 on $M'\setminus M''$, and equal to -1 on $M''\setminus M'$. Then $\varphi(x) \notin U$, i.e.

$$\mu(\bigcup_{n\in M'} A_n) - \mu(\bigcup_{n\in M''} A_n) \notin U.$$

Hence, the map

$$\mathfrak{P}(M) \to E, \quad M' \mapsto \mu(\bigcup_{n\in M'} A_n)$$

is bounded, injective, and additive and the image of $\mathfrak{P}(M)$ is discrete, which contradicts a).

b $\Rightarrow$ a is trivial. □

Lemma 4.1.39. *(L. Drewnowski (1975) [7] Théorème). Let I be an infinite set, for any $\iota \in I$ let x_ι be the element of $\ell^\infty(I)$ equal to 1 at ι and equal to 0 otherwise, let F be a vector subspace of $\ell^\infty(I)$ containing $\{x_\iota \mid \iota \in I\}$, for any $J \subset I$ let $F(J)$ be the vector subspace of F formed by the elements of F which vanish on $I\setminus J$ and endowed with the topology induced by $\ell^\infty(I)$, let E be a topological vector space and $u : F \to E$ be a continuous linear map such that 0 is not an adherent point of $\{u(x_\iota) \mid \iota \in I\}$. Then there exists a subset J of I having the same cardinality as I such that the map $F(J) \to u(F(J))$ defined by u is an isomorphism of topological vector spaces.*

We set $A := \{x \in F \mid \|x\| \le 2\}$. Since u is continuous $u(A)$ is $\mathbb{R}$-bounded. By Lemma 4.1.37 b) there exist a subset J of I having the same cardinality as I and a circled 0-neighbourhood U in E such that $u(x) \notin U$ for any $x \in A$ which vanishes on $I\setminus J$ and for which there exists $\iota \in J$ with $x(\iota) = 1$. In order to prove that the map $F(J) \to u(F(J))$ defined by u is an isomorphism of topological vector spaces, it is sufficient to show that $u(x) \notin U$ for any $x \in F(J)$ with $\|x\| > 1$. Assume the contrary. Then there exists $x \in F(J)$ such that $u(x) \in U$ and $\|x\| > 1$. Furthermore, there exists $\iota \in J$ such that

$$|x(\iota)| > \sup\left(1, \frac{\|x\|}{2}\right).$$

We have $\frac{1}{x(\iota)} x \in F(J) \cap A$, $\left(\frac{1}{x(\iota)} x\right)(\iota) = 1$ and $\left|\frac{1}{x(\iota)}\right| < 1$, which yields the contradictory relation

$$U \ni \frac{1}{x(\iota)} u(x) = u\left(\frac{1}{x(\iota)} x\right) \notin U. \quad \square$$

Definition 4.1.40. *Let E, F be topological vector spaces. We say E* <u>contains no copy of F</u> *in order to express that there exists no map $u : F \to E$ such that the map $F \to u(F)$ defined by it is an isomorphism of topological vector spaces.*

Lemma 4.1.41. *(L. Drewnowski (1975) [7] Corollaire). For any $n \in \mathbb{N}$ let x_n be the element of ℓ^∞ equal to 1 at n and equal to 0 otherwise, F be a vector subspace of ℓ^∞ containing $\{x_n \mid n \in \mathbb{N}\}$, for any $M \subset \mathbb{N}$ let $F(M)$ be the vector subspace of F formed by the elements of F which vanish on $\mathbb{N} \backslash M$ and endowed with the topology induced by ℓ^∞, let E be a topological vector space containing no copy of $F(M)$, M being any infinite subset of $\mathbb{N}$, and $u : F \to E$ be a continuous linear map. Then $(u(x_n))_{n \in \mathbb{N}}$ converges to 0.*

Assume the contrary. Then there exist a 0-neighbourhood U in E and an infinite subset M of $\mathbb{N}$ such that $u(x_n) \notin U$ for any $n \in M$. By Lemma 4.1.39 there exists an infinite subset L of M such that the map $F(L) \to u(F(L))$ defined by u is an isomorphism of topological vector spaces. Hence, E contains a copy of $F(L)$, which is a contradiction. $\square$

Lemma 4.1.42. *For any $n \in \mathbb{N}$ let x_n be the element of ℓ^∞ equal to 1 at n and equal to 0 otherwise, F be a topological vector subspace of ℓ^∞ containing $\{x_n \mid n \in \mathbb{N}\}$ such that $\bar{F} = \ell^\infty$ ($\bar{F} = c_0$) and E be a Hausdorff sequentially complete topological vector space containing no copy of ℓ^∞ (of c_0). Then E contains no copy of F.*

Assume the contrary. There exists a continuous linear map $u : F \to E$ such that $(u(x_n))_{n \in \mathbb{N}}$ does not converge to 0. Since E is Hausdorff and sequentially complete there exists a continuous linear map $v : \ell^\infty \to E$ ($v : c_0 \to E$) extending u. Then $(v(x_n))_{n \in \mathbb{N}}$ does not converge to 0 so by Lemma 4.1.41 E contains a copy of ℓ^∞ (of c_0). $\square$

Theorem 4.1.43. *Let E be a topological vector space. Then the following assertions are equivalent.*

a) E contains no copy of ℓ^*.

b) Any additive map μ of a ring of sets $\mathfrak{R}$ into E is exhaustive if the convex hull of $\{\mu(\bigcup_{n \in M} A_n) \mid M \subset \mathbb{N}\}$ is $\mathbb{R}$-bounded for any $(A_n)_{n \in \mathbb{N}} \in \Gamma(\mathfrak{R})$.

c) Any additive map μ of a ring of sets $\mathfrak{R}$ into E is exhaustive if the convex hull of $\mu(\mathfrak{R})$ is $\mathbb{R}$-bounded.

If E is Hausdorff and sequentially complete then the above assertions are equivalent to the following:

d) E contains no copy of ℓ^∞.

a $\Rightarrow$ b. Assume that μ is not exhaustive. Then there exists $(A_n)_{n\in\mathbb{N}} \in \Gamma(\mathfrak{R})$ such that $(\mu(A_n))_{n\in\mathbb{N}}$ does not converge to 0. Let u be the linear map $\ell^* \to E$ defined by $u(1_M) := \mu(\bigcup_{n\in M} A_n)$ for any $M \subset \mathbb{N}$. Since the convex hull of $\{\mu(\bigcup_{n\in M} A_n) \mid M \subset \mathbb{N}\}$ is $\mathbb{R}$-bounded, u is continuous. For any $M \subset \mathbb{N}$ we denote by $\ell^*(M)$ the vector subspace of ℓ^∞ generated by $\{1_{M'} \mid M' \subset M\}$ endowed with the induced topology. By Lemma 4.1.41 there exists an infinite subset M of $\mathbb{N}$ such that the map $\ell^*(M) \to u(\ell^*(M))$ defined by u is an isomorphism of topological vector spaces. Since $\ell^*(M)$ is isomorphic to ℓ^* as topological vector space, the above relation contradicts *a*).

b $\Rightarrow$ c is trivial.

c $\Rightarrow$ a. Let $u : \ell^* \to E$ be a map such that the map $\ell^* \to u(\ell^*)$ defined by it is an isomorphism of topological vector spaces. Set

$$\mu : \mathfrak{P}(\mathbb{N}) \to E, \quad M \mapsto u(1_M).$$

Then μ is additive and the convex hull of $\mu(\mathfrak{P}(\mathbb{N}))$ is $\mathbb{R}$-bounded but μ ist not exhaustive, which contradicts *c*).

d $\Rightarrow$ a follows from Lemma 4.1.42.

a $\Rightarrow$ d is trivial. □

Corollary 4.1.44. *Let E be a locally convex space. The following assertions are equivalent.*

a) E contains no copy of ℓ^*.

b) Any additive map μ of a ring of sets $\mathfrak{R}$ into E is exhaustive if $\{\mu(\bigcup_{n\in M} A_n) \mid M \subset \mathbb{N}\}$ is bounded for any $(A_n)_{n\in\mathbb{N}} \in \Gamma(\mathfrak{R})$.

c) Any bounded additive map of a ring of sets into E is exhaustive.

If E is Hausdorff and sequentially complete then the above assertions are equivalent to the following one:

d) E contains no copy of ℓ^∞. □

Remark. The above equivalence c $\Leftrightarrow$ d was proved by J. Diestel and B. Faires (1974) for the case when E is a Banach space ([1] Theorem 1.1 (i) $\Leftrightarrow$ (ii)) and by I. Labuda (1975) in the general case ([4] (7)).

Theorem 4.1.45. *The following assertions are equivalent for any topological vector space E.*

a) E contains no copy of $\ell^* \cap c_0$.

b) Any sequence $(x_n)_{n\in\mathbb{N}}$ *in E converges to 0 if the convex hull of* $\{\sum_{n\in M} x_n \mid M \in \mathfrak{P}_f(\mathbb{N})\}$ *is* $\mathbb{R}$*-bounded.*

c) Any sequence $(x_n)_{n\in\mathbb{N}}$ *in E is a Cauchy sequence if the convex hull of* $\{\sum_{n\in M} x_n \mid M \in \mathfrak{P}_f(\mathbb{N})\}$ *is* $\mathbb{R}$*-bounded.*

If E is Hausdorff and sequentially complete then the above assertions are equivalent to the following:

d) E contains no copy of c_0.

e) Any sequence $(x_n)_{n\in\mathbb{N}}$ *in E is summable if the convex hull of* $\{\sum_{n\in M} x_n \mid M \in \mathfrak{P}_f(\mathbb{N})\}$ *is* $\mathbb{R}$*-bounded.*

a ⇒ b. We denote by u the linear map $\ell^* \cap c_0 \to E$ defined by $u(1_M) := \sum_{n\in M} x_n$ for any $M \in \mathfrak{P}_f(\mathbb{N})$. By the hypothesis on $(x_n)_{n\in\mathbb{N}}$ u is continuous. By Lemma 4.1.41 $(x_n)_{n\in\mathbb{N}}$ converges to 0.

b ⇒ c is trivial.

c ⇒ a. The property described in c) belongs to any topological vector subspace of E. Since $\ell^* \cap c_0$ does not possess this property E contains no copy of $\ell^* \cap c_0$.

a ⇒ d and e ⇒ b are trivial.

d ⇒ a follows from Lemma 4.1.42.

b ⇒ e. Assume the contrary. Then there exists a disjoint sequence $(M_n)_{n\in\mathbb{N}}$ in $\mathfrak{P}_f(\mathbb{N})$ such that $(\sum_{m\in M_n} x_m)_{n\in\mathbb{N}}$ does not converge to 0. But the convex hull of $\{\sum_{n\in M}\sum_{m\in M_n} x_m \mid M \in \mathfrak{P}_f(\mathbb{N})\}$ is $\mathbb{R}$-bounded, which contradicts b). □

Remark. The above equivalences b ⇔ d ⇔ e were proved by C. Bessaga and A. Pełczyński (1958) ([1] Theorem 5) for a Banach space E. Ju. B. Tumarkin proved the equivalence d ⇔ e (1970) ([1] Theorem 4) for the case when E is locally convex.

Corollary 4.1.46. *Let E be a topological vector space containing no copy of* $\ell^* \cap c_0$. *Then any additive map* μ *of a ring of sets* $\mathfrak{R}$ *into E is exhaustive if the convex hull of* $\{\mu(\bigcup_{n\in M} A_n) \mid M \in \mathfrak{P}_f(\mathbb{N})\}$ *is* $\mathbb{R}$*-bounded. If E is Hausdorff and sequentially complete, then the above hypothesis is equivalent to the statement that E contains no copy of* c_0.

The assertions follow immediately from Theorem 4.1.45 d ⇔ a ⇒ b. □

Remark. The above result was proved by J. Diestel (1973) ([2] Theorem 1.8) when E is a Banach space.

Proposition 4.1.47. *Let* $\mathfrak{R}$ *be a ring of sets,* $\mathfrak{K}$ *be a subset of* $\mathfrak{R}$ *closed under finite unions, G be a topological additive group,* μ *be a* $\mathfrak{K}$*-regular, additive map of* $\mathfrak{R}$ *into G,* $A \in \mathfrak{R}$*, and let* $\mathfrak{T}$ *be a subset of* $\mathfrak{R}$ *such that:*

a) $\{T \in \mathfrak{T} \mid A \subset T\}$ *is lower directed;*

b) for any $K \in \mathfrak{K}$ with $A \cap K = \emptyset$ there exist disjoint sets $S, T \in \mathfrak{T}$ such that $K \subset S$, $A \subset T$.

Then for any 0-neighbourhood U in G there exists $T \in \mathfrak{T}$ such that $A \subset T$ and

$$\{\mu(B) \mid B \in \mathfrak{R},\ B \subset T \backslash A\} \subset U\,.$$

Since $\emptyset \in \mathfrak{K}$ there exists by b) a $T_0 \in \mathfrak{T}$ such that $A \subset T_0$. μ being $\mathfrak{K}$-regular and additive there exists $K \in \mathfrak{K}$ such that $K \subset T_0 \backslash A$ and

$$\{\mu(B) \mid B \in \mathfrak{R},\ B \subset (T_0 \backslash A) \backslash K\} \subset U\,.$$

By b) there exist disjoint sets $T', T'' \in \mathfrak{T}$ such that $A \subset T'$, $K \subset T''$. By a) there exists $T \in \mathfrak{T}$ with

$$A \subset T \subset T_0 \cap T'.$$

Let $B \in \mathfrak{R}$ such that $B \subset T \backslash A$. Then $B \subset (T_0 \backslash A) \backslash K$ so $\mu(B) \in U$. □

Proposition 4.1.48. *Let H be an additive group, let $\mu : \mathfrak{P}(\mathbb{N}) \to H$ be an additive map, let $u : H \to H$ be a group homomorphism such that $u(\mu(\{n\})) = \mu(\{n\})$ for every $n \in \mathbb{N}$, and let $(u_n)_{n \in \mathbb{N}}$ be a sequence of group homomorphisms of H into $\mathbb{R}$ such that $u_n \circ u \circ \mu$ is bounded for every $n \in \mathbb{N}$ and such that*

$$\lim_{n \to \infty} u_n(u(\mu(M)))$$

exists for every $M \subset \mathbb{N}$. Then $(u_n \circ \mu)_{n \in \mathbb{N}}$ converges to

$$\mathfrak{P}(\mathbb{N}) \to \mathbb{R}\,, \quad M \mapsto \lim_{n \to \infty} u_n(u(\mu(M)))$$

in $\mathbb{R}^{\mathfrak{P}(\mathbb{N})}_{\{\mathfrak{P}_f(\mathbb{N})\}}$.

$u_n \circ u \circ \mu$ is additive and bounded and therefore exhaustive (Proposition 4.1.8) for every $n \in \mathbb{N}$. By the hypothesis $(u_n \circ u \circ \mu)_{n \in \mathbb{N}}$ is a Cauchy sequence in $\mathscr{E}(\mathfrak{P}(\mathbb{N}), \mathbb{R})$. By Theorem 4.1.21 we deduce it is a Cauchy sequence in $\mathscr{E}(\mathfrak{P}(\mathbb{N}), \mathbb{R})_{\{\mathfrak{P}_f(\mathbb{N})\}}$ and the assertion follows from the fact that $u_n \circ u \circ \mu$ and $u_n \circ \mu$ coincide on $\mathfrak{P}_f(\mathbb{N})$ for any $n \in \mathbb{N}$. □

Theorem 4.1.49. *Let T be a locally compact space on which there exists a locally finite sequence of open nonempty sets, (such a sequence exists if T is paracompact and non-compact), let $\mathscr{C}_b$ be the vector space of continuous bounded real functions on T endowed with the norm*

$$\mathscr{C}_b \to \mathbb{R}_+\,, \quad f \mapsto \sup_{t \in T} |f(t)|\,,$$

and let $\mathscr{F}$ be the vector subspace of $\mathscr{C}_b$ formed by the functions which have a limit at the Alexandrov point of T. Then there exists no continuous linear map $\mathscr{C}_b \to \mathscr{F}$ keeping invariant each function of $\mathscr{C}_b$ which has a compact carrier.

Let $(U_n)_{n\in\mathbb{N}}$ be a locally finite sequence of open nonempty sets of T. Without loss of generality we may assume each U_n $(n\in\mathbb{N})$ relatively compact and the sequence $(U_n)_{n\in\mathbb{N}}$ disjoint. For each $n\in\mathbb{N}$ let t_n be a point of U_n and let f_n be a continuous real function on T with the carrier contained in U_n such that $|f_n|\leq 1$ and $f_n(t_n)=1$. We set

$$u_n:\mathscr{C}_b\to\mathbb{R},\quad f\mapsto f(t_n)$$

for every $n\in\mathbb{N}$,

$$g_M:T\to\mathbb{R},\quad t\mapsto\sum_{n\in M}f_n(t)$$

for every $M\subset\mathbb{N}$, and

$$\mu:\mathfrak{P}(\mathbb{N})\to\mathscr{C}_b,\quad M\mapsto g_M.$$

Assume there exists a continuous linear map $u:\mathscr{C}_b\to\mathscr{F}$ keeping invariant each function of $\mathscr{C}_b$ with compact carrier. Then

$$u(\mu(\{n\}))=\mu(\{n\})$$

for every $n\in\mathbb{N}$ and $(u_n)_{n\in\mathbb{N}}$ is a sequence of linear forms on $\mathscr{C}_b$ such that $u_n\circ u\circ\mu$ is bounded for every $n\in\mathbb{N}$ and such that

$$\lim_{n\to\infty}u_n(u(\mu(M)))$$

exists for every $M\subset\mathbb{N}$. By Proposition 4.1.48 $(u_n\circ\mu)_{n\in\mathbb{N}}$ is a Cauchy sequence in $\mathbb{R}^{\mathfrak{P}(\mathbb{N})}_{\{\mathfrak{P}_f(\mathbb{N})\}}$ and this contradicts the relation

$$u_m\circ\mu(\{m\})-u_n\circ\mu(\{m\})=1,$$

which holds for every $m,n\in\mathbb{N}$, $m\neq n$. □

Corollary 4.1.50. *Let I be a set endowed with the discrete topology and let $c(I)$ be the vector subspace of $\ell^\infty(I)$ formed by the functions of $\ell^\infty(I)$ which have a limit of the Alexandrov point of I. Then there exists no continuous linear map $\ell^\infty(I)\to c(I)$ keeping invariant every element of $c_0(I)$.* □

Remark. In particular, there exists no projection $\ell^\infty\to c(\mathbb{N})$. This result was proved by R.S. Phillips (1940) ([1] (7.5)). In fact, every continuous linear map $\ell^\infty(I)\to c(I)$ is weakly compact (see Theorem 5.9.40).

Theorem 4.1.51. *Let T be a completely regular space on which there exists a convergent sequence $(t_n)_{n\in\mathbb{N}}$ such that $t_m\neq\lim_{n\to\infty}t_n$ for every $m\in\mathbb{N}$, let $\mathscr{B}$ be the vector space of bounded Borel functions on T endowed with the norm*

$$\mathscr{B}\to\mathbb{R}_+,\quad f\mapsto\sup_{t\in T}|f(t)|,$$

and let $\mathscr{C}_b$ be the vector subspace of $\mathscr{B}$ formed by the continuous functions. Then there exists no continuous linear map $\mathscr{B} \to \mathscr{C}_b$ keeping invariant each function of $\mathscr{C}_b$.

We may assume without loss of generality that $t_m \neq t_n$ for any different $m, n \in \mathbb{N}$. Since T is completely regular we may construct inductively a sequence $(f_n)_{n \in \mathbb{N}}$ in $\mathscr{C}_b$ such that $|f_n| \leq 1$ and $f_n(t_n) = 1$ for every $n \in \mathbb{N}$ and such that the carriers of these functions are pairwise disjoint. We set

$$u_n : \mathscr{B} \to \mathbb{R}, \quad f \mapsto f(t_n)$$

for every $n \in \mathbb{N}$,

$$g_M : T \to \mathbb{R}, \quad t \mapsto \sum_{n \in M} f_n(t)$$

for every $M \subset \mathbb{N}$, and

$$\mu : \mathfrak{P}(\mathbb{N}) \to \mathscr{B}, \quad M \mapsto g_M .$$

Assume there exists a continuous linear map $u : \mathscr{B} \to \mathscr{C}_b$ keeping invariant each function of $\mathscr{C}_b$. Then

$$u(\mu(\{n\})) = \mu(\{n\})$$

for every $n \in \mathbb{N}$ and $(u_n)_{n \in \mathbb{N}}$ is a sequence of linear forms on $\mathscr{B}$ such that $u_n \circ u \circ \mu$ is bounded for every $n \in \mathbb{N}$ and such that

$$\lim_{n \to \infty} u_n(u(\mu(M)))$$

exists for every $M \subset \mathbb{N}$. By Proposition 4.1.48 $(u_n \circ \mu)_{n \in \mathbb{N}}$ is a Cauchy sequence in $\mathbb{R}^{\mathfrak{P}(\mathbb{N})}_{\{\mathfrak{P}_f(\mathbb{N})\}}$ and this contradicts the relation

$$u_m \circ \mu(\{m\}) - u_n \circ \mu(\{m\}) = 1 ,$$

which holds for every $m, n \in \mathbb{N}, m \neq n$. □

Remark. Let T be a completely regular space such that the closure of any open set is open. With the notation of the above theorem there exists a continuous linear map $\mathscr{B} \to \mathscr{C}_b$ keeping invariant each function of $\mathscr{C}_b$ (Ch. D. Aliprantis, O. Burkinshaw [1] Theorem 3.4). It is easy to see directly that there exists no convergent sequence $(t_n)_{n \in \mathbb{N}}$ in T such that $t_m \neq \lim_{n \to \infty} t_n$ for every $m \in \mathbb{N}$.

Theorem 4.1.52. *Let T be a set, let $\mathfrak{R}$ be a δ-ring of subsets of T, let $\mathscr{F}$ be the vector space of bounded $\mathfrak{R}$-measurable real functions on T, (a real function f on T is called $\mathfrak{R}$-measurable if $A \cap \{t \in T \mid f(t) < \alpha\} \in \mathfrak{R}$ for every $A \in \mathfrak{R}$ and for every $\alpha \in \mathbb{R}$), endowed with the norm*

$$\mathscr{F} \to \mathbb{R}_+, \quad f \mapsto \sup_{t \in T} |f(t)| ,$$

and let $\mathscr{F}_0$ *be the closed vector subspace of* $\mathscr{F}$ *generared by the characteristic functions of the sets of* $\mathfrak{R}$. *If there exists a sequence* $(t_n)_{n\in\mathbb{N}}$ *in* T, *such that any set of* $\mathfrak{R}$ *contains only a finite number of* t_n $(n\in\mathbb{N})$, *and if there exists a disjoint sequence* $(A_n)_{n\in\mathbb{N}}$ *in* $\mathfrak{R}$, *such that* $t_n\in A_n$ *for every* $n\in\mathbb{N}$, *then there exists no continuous linear map* $\mathscr{F}\to\mathscr{F}_0$ *keeping invariant each element of* $\mathscr{F}_0$.

Assume there exists a continuous linear map $u:\mathscr{F}\to\mathscr{F}_0$ keeping invariant each element of $\mathscr{F}_0$. We set

$$f_M : T \to \mathbb{R}, \quad t \mapsto \begin{cases} 1 & \text{if } t\in \bigcup_{n\in M} A_n \\ 0 & \text{otherwise} \end{cases}$$

for every $M\subset\mathbb{N}$ and

$$\mu : \mathfrak{P}(\mathbb{N}) \to \mathscr{F}, \quad M \mapsto f_M .$$

Then μ is an additive map such that

$$u(\mu(\{n\})) = \mu(\{n\})$$

for every $n\in\mathbb{N}$. We denote for every $n\in\mathbb{N}$ by u_n the map

$$\mathscr{F} \to \mathbb{R}, \quad f \mapsto f(t_n) .$$

Then $(u_n)_{n\in\mathbb{N}}$ is a sequence of linear forms on $\mathscr{F}$ such that $u_n\circ u\circ\mu$ is bounded for every $n\in\mathbb{N}$ and such that

$$\lim_{n\to\infty} u_n(u(\mu(M))) = 0$$

for every $M\subset\mathbb{N}$. By Proposition 4.1.48 we get the contradictory relation

$$1 = \lim_{n\to\infty} u_n(\mu(\{n\})) = 0 . \quad \square$$

§ 4.2 Spaces of measures and of exhaustive additive maps

Throughout this section $\mathfrak{R}$ *shall denote a ring of sets,* $\mathfrak{K}$ *a subset of* $\mathfrak{R}$ *closed under finite unions,* R *a topological ring, and* G *a topological left* R*-module. We shall identify the topological additive groups with the topological* $\mathbb{Z}$*-modules.*

Proposition 4.2.1. $\mathscr{E}(\mathfrak{R}, G; \mathfrak{K})$ *and* $\mathscr{M}(\mathfrak{R}, G; \mathfrak{K})$ *are* R*-submodules of* $G^{\mathfrak{R}}$.

The proof is straightforward. $\square$

Remark. Henceforth, we consider $\mathscr{E}(\mathfrak{R}, G; \mathfrak{K})$ and $\mathscr{M}(\mathfrak{R}, G; \mathfrak{K})$ endowed with the structure of left R-modules induced by $G^{\mathfrak{R}}$.

Proposition 4.2.2. *Let* $\mathfrak{R}$ *be a quasi-δ-ring endowed with the coarsest uniformity for which every* $\mu \in \mathscr{E}(\mathfrak{R}, G; \mathfrak{K})$ *(every* $\mu \in \mathscr{M}(\mathfrak{R}, G; \mathfrak{K})$*) is uniformly continuous and let* Φ *be a subset of* $\hat{\Phi}_C(\mathfrak{R})$*. Then:*

a) $\mathfrak{K}$ *is a* Φ_3*-dense set of* $\mathfrak{R}$*;*

b) $\Psi_f(\mathfrak{R}) \subset \hat{\Phi}_3(\mathfrak{R}) \subset \hat{\Phi}_C(\mathfrak{R})$ $(\Psi(\mathfrak{R}) \subset \hat{\Phi}_1(\mathfrak{R}) \subset \hat{\Phi}_C(\mathfrak{R}))$*;*

c) $\mathscr{E}(\mathfrak{R}, G; \mathfrak{K})_\Phi$ $(\mathscr{M}(\mathfrak{R}, G; \mathfrak{K})_\Phi)$ *is a topological left R-module;*

d) if G is locally convex, then so are $\mathscr{E}(\mathfrak{R}, G; \mathfrak{K})_\Phi$ *and* $\mathscr{M}(\mathfrak{R}, G; \mathfrak{K})_\Phi$*.*

We set

$$\mathscr{U}(\mathscr{N}, U) := \bigcap_{\mu \in \mathscr{N}} \{(A, B) \in \mathfrak{R}^2 \mid \mu(A) - \mu(B) \in U\}$$

for any 0-neighbourhood U in G and any finite subset $\mathscr{N}$ of $\mathscr{E}(\mathfrak{R}, G; \mathfrak{K})$ (of $\mathscr{M}(\mathfrak{R}, G; \mathfrak{K})$). The sets of the form $\mathscr{U}(\mathscr{N}, U)$, when U runs through the set of 0-neighbourhoods in G and $\mathscr{N}$ through the set of finite subsets of $\mathscr{E}(\mathfrak{R}, G; \mathfrak{K})$ (of $\mathscr{M}(\mathfrak{R}, G; \mathfrak{K})$), generate a uniformity on $\mathfrak{R}$ which is obviously the coarsest uniformity for which every $\mu \in \mathscr{E}(\mathfrak{R}, G; \mathfrak{K})$ (every $\mu \in \mathscr{M}(\mathfrak{R}, G; \mathfrak{K})$) is uniformly continuous.

a) follows from Proposition 4.1.16 and Proposition 4.1.26 $a \Rightarrow b$.

b) Let $(A_n)_{n \in \mathbb{N}} \in \Gamma(\mathfrak{R})$. We set

$$\mathfrak{A} := \{\bigcup_{n \in M} A_n \mid M \in \mathfrak{P}_f(\mathbb{N})\} \quad (\mathfrak{A} := \{\bigcup_{n \in M} A_n \mid M \in \mathfrak{P}(\mathbb{N})\}),$$

$$\varphi : \mathfrak{P}_f(\mathbb{N}) \to \mathfrak{A}, \quad M \mapsto \bigcup_{n \in M} A_n$$

$$(\varphi : \mathfrak{P}(\mathbb{N}) \to \mathfrak{A}, \quad M \mapsto \bigcup_{n \in M} A_n).$$

We endow $\mathfrak{A}$ with the uniformity for which φ is an isomorphism of uniform spaces. Then $\mathfrak{A}$ is a precompact (compact) metrizable space. By Proposition 4.1.11 (Proposition 4.1.14 $a \Rightarrow c$) the uniformity of $\mathfrak{A}$ is finer than the uniformity induced by $\mathfrak{R}$ on $\mathfrak{A}$ so by Proposition 1.5.4 $c \Rightarrow a$ $\mathfrak{A}$ is a Φ_3-set (a Φ_1-set) of $\mathfrak{R}$. Hence, $\Psi_f(\mathfrak{R}) \subset \hat{\Phi}_3(\mathfrak{R})$ $(\Psi(\mathfrak{R}) \subset \hat{\Phi}_1(\mathfrak{R}))$. By Corollary 1.5.12

$$\hat{\Phi}_1(\mathfrak{R}) \subset \hat{\Phi}_3(\mathfrak{R}) \subset \hat{\Phi}_C(\mathfrak{R}).$$

c), d) follow immediately from the Propositions 4.2.1, 2.1.27 and 2.1.28. □

Proposition 4.2.3. *The identity map*

$$\mathscr{E}(\mathfrak{R}, G; \mathfrak{K}) \to \mathscr{E}(\mathfrak{R}, G; \mathfrak{K})_{\Psi_f(\mathfrak{R})}$$

is uniformly Φ_3*-continuous and the identity map*

$$\mathscr{M}(\mathfrak{R}, G; \mathfrak{K}) \to \mathscr{M}(\mathfrak{R}, G; \mathfrak{K})_{\Psi(\mathfrak{R})}$$

is uniformly Φ_4*-continuous.*

The assertion follows from Theorem 4.1.22 c) (and Corollary 1.8.15) by taking $X := \mathfrak{R}$ and $\mathscr{G}$ equal to the set of maps

$$\mathfrak{P}(\mathbb{N}) \to \mathfrak{R}, \quad M \mapsto \bigcup_{n \in M} A_n,$$

where $(A_n)_{n \in \mathbb{N}}$ runs through $\Gamma(\mathfrak{R})$. □

Theorem 4.2.4. *Let $\mathfrak{R}$ be a quasi-δ-ring, G be Hausdorff and I be a set. Then:*

a) $\mathscr{E}(\mathfrak{R}, G; \mathfrak{K})_{\mathfrak{K}}$, $\mathscr{E}(\mathfrak{R}, G; \mathfrak{K})$, and $\mathscr{E}(\mathfrak{R}, G; \mathfrak{K})_{\Psi_f(\mathfrak{R})}$ are Hausdorff topological left R-modules;

b) the sets $\ell(I, \mathscr{E}(\mathfrak{R}, G; \mathfrak{K})_{\mathfrak{K}})$, $\ell(I, \mathscr{E}(\mathfrak{R}, G; \mathfrak{K}))$, and $\ell(I, \mathscr{E}(\mathfrak{R}, G; \mathfrak{K})_{\Psi_f(\mathfrak{R})})$ coincide;

c) the identity map

$$\ell_c(I, \mathscr{E}(\mathfrak{R}, G; \mathfrak{K})) \to \ell_f(I, \mathscr{E}(\mathfrak{R}, G; \mathfrak{K})_{\Psi_f(\mathfrak{R})})$$

is uniformly Φ_3-continuous.
All the above assertions hold if we replace $\mathscr{E}$ by $\mathscr{M}$, Ψ_f by Ψ, and 3 by 4.

Let us endow $\mathfrak{R}$ with the coarsest uniformity for which every $\mu \in \mathscr{E}(\mathfrak{R}, G; \mathfrak{K})$ (every $\mu \in \mathscr{M}(\mathfrak{R}, G; \mathfrak{K})$) is uniformly continuous.

a) follows from Proposition 4.2.2 a), b), c).

b) follows from a), Proposition 4.2.2 a), b), and Theorem 3.4.1 b).

c) follows from Theorem 3.4.1 e) and Proposition 4.2.3. □

Proposition 4.2.5. *Let $\mathfrak{F}$ be a $\hat{\Phi}_3$-filter on $\mathscr{E}(\mathfrak{R}, G; \mathfrak{K})$ (a $\hat{\Phi}_4$-filter on $\mathscr{M}(\mathfrak{R}, G; \mathfrak{K})$), $\mathfrak{G} \in \Psi_f(\mathfrak{R})$ ($\mathfrak{G} \in \Psi(\mathfrak{R})$) such that*

$$\{\mu \in \mathscr{E}(\mathfrak{R}, G; \mathfrak{K}) \mid \mu(\mathfrak{G}) \text{ is a Cauchy filter}\} \in \mathfrak{F}$$

$$(\{\mu \in \mathscr{M}(\mathfrak{R}, G; \mathfrak{K}) \mid \mu(\mathfrak{G}) \text{ is a Cauchy filter}\} \in \mathfrak{F}),$$

and U be a 0-neighbourhood in G. Then there exist $\mathscr{M} \in \mathfrak{F}$, $\mathfrak{A} \in \mathfrak{G}$, and an entourage $\mathscr{U}$ of $\mathscr{E}(\mathfrak{R}, G; \mathfrak{K})$ (of $\mathscr{M}(\mathfrak{R}, G; \mathfrak{K})$) such that

$$\mu(A) - \nu(B) \in U$$

for any $(\mu, \nu) \in (\bar{\mathscr{M}} \times \bar{\mathscr{M}}) \cap \mathscr{U}$ and any $A, B \in \mathfrak{A}$.

Let V be a closed symmetric 0-neighbourhood in G such that $3V \subset U$. By Proposition 4.2.3 and Proposition 2.2.8 there exist $\mathscr{M} \in \mathfrak{F}$ and $\mathfrak{A} \in \mathfrak{G}$ such that

$$\mu(A) - \mu(B) \in V$$

for any $\mu \in \mathscr{M}$ and any $A, B \in \mathfrak{A}$. Since V is closed the same relation holds for any $\mu \in \bar{\mathscr{M}}$ and any $A, B \in \mathfrak{A}$. Let $C \in \mathfrak{A}$ and $\mathscr{U}$ be an entourage of $\mathscr{E}(\mathfrak{R}, G; \mathfrak{K})$ (of $\mathscr{M}(\mathfrak{R}, G; \mathfrak{K})$) such that

$$\mu(C) - \nu(C) \in V$$

for any $(\mu, \nu) \in \mathscr{U}$. We get

$$\mu(A) - \nu(B) = (\mu(A) - \mu(C)) + (\mu(C) - \nu(C)) + (\nu(C) - \nu(B)) \in 3V \subset U$$

for any $(\mu, \nu) \in (\bar{\mathscr{M}} \times \bar{\mathscr{M}}) \cap \mathscr{U}$ and any $A, B \in \mathfrak{A}$. □

Proposition 4.2.6. *Let $\mathfrak{F}$ be a $\hat{\Phi}_3$-filter on $\mathscr{E}(\mathfrak{R}, G; \mathfrak{K})$ (a $\hat{\Phi}_4$-filter on $\mathscr{M}(\mathfrak{R}, G; \mathfrak{K})$), $(A_n)_{n \in \mathbb{N}}$ be a disjoint $\Theta(\mathfrak{R})$-sequence, $\mathfrak{G}$ be the elementary filter on $\mathfrak{R}$ generated by $(A_n)_{n \in \mathbb{N}}$ and φ be the map*

$$\mathscr{E}(\mathfrak{R}, G; \mathfrak{K}) \times \mathfrak{R} \to G, \quad (\mu, A) \mapsto \mu(A)$$

$$(\mathscr{M}(\mathfrak{R}, G; \mathfrak{K}) \times \mathfrak{R} \to G, \quad (\mu, A) \mapsto \mu(A)).$$

Then $\varphi(\mathfrak{F} \times \mathfrak{G})$ converges to 0.

The assertion follows from Theorem 4.1.22i) (and Proposition 4.1.5 b ⇒ a) by taking $X := \mathfrak{R}$ and $\mathscr{G}$ equal to the set of maps

$$\mathfrak{P}(\mathbb{N}) \to \mathfrak{R}, \quad M \mapsto \bigcup_{n \in M} B_n,$$

where $(B_n)_{n \in \mathbb{N}}$ runs through $\Gamma(\mathfrak{R})$. □

Proposition 4.2.7. *Let $\mathfrak{F}$ be a $\hat{\Phi}_3$-filter on $\mathscr{E}(\mathfrak{R}, G; \mathfrak{K})$ (a $\hat{\Phi}_4$-filter on $\mathscr{M}(\mathfrak{R}, G; \mathfrak{K})$), $A \in \mathfrak{R}$ and U be a 0-neighbourhood in G. If $\mathfrak{H}(A, \mathfrak{K}) \in \Psi_f(\mathfrak{R})$ (this happens if $\mathfrak{R}$ is a quasi-δ-ring or if $\mathfrak{K} = \mathfrak{R}$), then there exist $\mathscr{M} \in \mathfrak{F}$, $K \in \mathfrak{K}$, and an entourage $\mathscr{U}$ of $\mathscr{E}(\mathfrak{R}, G; \mathfrak{K})_{\mathfrak{K}}$ such that $K \subset A$ and*

$$\mu(B) - \nu(C) \in U$$

for any $(\mu, \nu) \in (\bar{\bar{\mathscr{M}}} \times \bar{\bar{\mathscr{M}}}) \cap \mathscr{U}$ and any $B, C \in \mathfrak{R}$ with $K \subset B \subset A$, $K \subset C \subset A$.

Let V be a symmetric closed 0-neighbourhood in G such that $3V \subset U$. By Proposition 4.1.27 b) $\mu(\mathfrak{H}(A, \mathfrak{K}))$ converges to 0 for any $\mu \in \mathscr{E}(\mathfrak{R}, G; \mathfrak{K})$ so, by Proposition 4.2.5, there exists $(\mathscr{M}, K) \in \mathfrak{F} \times \mathfrak{K}$ such that $K \subset A$ and

$$\mu(L) - \mu(L') \in V$$

for any $\mu \in \mathscr{M}$ and any $L, L' \in \mathfrak{K}$ with $L \subset A \backslash K$, $L' \subset A \backslash K$. In particular, $\mu(L) \in V$ for any $\mu \in \bar{\bar{\mathscr{M}}}$ and any $L \in \mathfrak{K}$ with $L \subset A \backslash K$.

Let $B \in \mathfrak{R}$ such that $K \subset B \subset A$. Since V is closed we find, by Proposition 4.1.26 a ⇒ b, that

$$\mu(B \backslash K) = \lim \mu(\mathfrak{G}(B \backslash K, \mathfrak{K})) \in V$$

for any $\mu \in \bar{\bar{\mathscr{M}}}$. Let $\mathscr{U}$ be an entourage of $\mathscr{E}(\mathfrak{R}, G; \mathfrak{K})_{\mathfrak{K}}$ such that

$$\mu(K) - \nu(K) \in V$$

for any $(\mu, \nu) \in \mathscr{U}$. Then

$$\mu(B) - \nu(C) = \mu(B \backslash K) - \nu(C \backslash K) + (\mu(K) - \nu(K)) \in 3V \subset U$$

for any $(\mu, \nu) \in (\bar{\bar{\mathscr{M}}} \times \bar{\bar{\mathscr{M}}}) \cap \mathscr{U}$ and any $B, C \in \mathfrak{R}$ with $K \subset B \subset A$, $K \subset C \subset A$.

By Proposition 4.1.25 c) $\mathfrak{H}(A, \mathfrak{K}) \in \Psi_f(\mathfrak{R})$ if $\mathfrak{R}$ is a quasi-δ-ring. □

Corollary 4.2.8. *If $\{\mathfrak{H}(A, \mathfrak{K}) \mid A \in \mathfrak{R}\} \subset \Psi_f(\mathfrak{R})$ (which happens if $\mathfrak{R}$ is a quasi-δ-ring or if $\mathfrak{K} = \mathfrak{R}$), then for any $\hat{\Phi}_3$-filter $\mathfrak{F}$ on $\mathscr{E}(\mathfrak{R}, G; \mathfrak{K})$ ($\hat{\Phi}_4$-filter $\mathfrak{F}$ on $\mathscr{M}(\mathfrak{R}, G; \mathfrak{K})$) and for any entourage $\mathscr{U}$ of $\mathscr{E}(\mathfrak{R}, G; \mathfrak{K})$ (of $\mathscr{M}(\mathfrak{R}, G; \mathfrak{K})$) there exist $\mathscr{M} \in \mathfrak{F}$ and an entourage $\mathscr{V}$ of $\mathscr{E}(\mathfrak{R}, G; \mathfrak{K})_{\mathfrak{K}}$ such that*

$$(\bar{\bar{\mathscr{M}}} \times \bar{\bar{\mathscr{M}}}) \cap \mathscr{V} \subset (\bar{\bar{\mathscr{M}}} \times \bar{\bar{\mathscr{M}}}) \cap \mathscr{U}.$$

Let U be a symmetric 0-neighbourhood in G and $(A_\iota)_{\iota \in I}$ be a finite family in $\mathfrak{R}$ such that

$$\{(\mu, \nu) \in \mathscr{E}(\mathfrak{R}, G; \mathfrak{K})^2 \ \ ((\mu, \nu) \in \mathscr{M}(\mathfrak{R}, G; \mathfrak{K})^2) \mid \ \iota \in I \Rightarrow \mu(A_\iota) - \nu(A_\iota) \in 3U\} \subset \mathscr{U}.$$

By Proposition 4.2.7 there exist $\mathscr{M} \in \mathfrak{F}$ and a family $(K_\iota)_{\iota \in I}$ in $\mathfrak{K}$ such that $K_\iota \subset A_\iota$ for any $\iota \in I$ and such that

$$\mu(K_\iota) - \mu(A_\iota) \in U$$

for any $\mu \in \bar{\bar{\mathscr{M}}}$. We set

$$\mathscr{V} := \{(\mu, \nu) \in \mathscr{E}(\mathfrak{R}, G; \mathfrak{K})^2 \ \ ((\mu, \nu) \in \mathscr{M}(\mathfrak{R}, G; \mathfrak{K})^2) \mid \iota \in I \Rightarrow \mu(K_\iota) - \nu(K_\iota) \in U\}.$$

Then $\mathscr{V}$ is an entourage of $\mathscr{E}(\mathfrak{R}, G; \mathfrak{K})_{\mathfrak{K}}$ (of $\mathscr{M}(\mathfrak{R}, G; \mathfrak{K})_{\mathfrak{K}}$) and

$$\mu(A_\iota) - \nu(A_\iota) = (\mu(A_\iota) - \mu(K_\iota)) + (\mu(K_\iota) - \nu(K_\iota)) + (\nu(K_\iota) - \nu(A_\iota)) \in 3U$$

for any $\iota \in I$ and any $(\mu, \nu) \in (\bar{\bar{\mathscr{M}}} \times \bar{\bar{\mathscr{M}}}) \cap \mathscr{V}$. Hence,

$$(\bar{\bar{\mathscr{M}}} \times \bar{\bar{\mathscr{M}}}) \cap \mathscr{V} \subset (\bar{\bar{\mathscr{M}}} \times \bar{\bar{\mathscr{M}}}) \cap \mathscr{U}.$$

By Proposition 4.1.25 c) $\{\mathfrak{H}(A, \mathfrak{K}) \mid A \in \mathfrak{R}\} \subset \Psi_f(\mathfrak{R})$ if $\mathfrak{R}$ is a quasi-δ-ring. □

Proposition 4.2.9. *Let $\mathfrak{R}$ be a quasi-σ-ring ordered by the inclusion relation, $\mathfrak{A}$ be an upper directed set of $\mathfrak{R}$, $\mathfrak{F}$ be a $\hat{\Phi}_3$-filter on $\mathscr{E}(\mathfrak{R}, G; \mathfrak{K})$ (a $\hat{\Phi}_4$-filter on $\mathscr{M}(\mathfrak{R}, G; \mathfrak{K})$) and U be a 0-neighbourhood in G. Then there exist $\mathscr{M} \in \mathfrak{F}$, $A \in \mathfrak{A}$ and an entourage $\mathscr{U}$ of $\mathscr{E}(\mathfrak{R}, G; \mathfrak{K})_{\mathfrak{K}}$ (of $\mathscr{M}(\mathfrak{R}, G; \mathfrak{K})_{\mathfrak{K}}$) such that*

$$\mu(B) - \nu(C) \in U$$

for any $(\mu, \nu) \in (\bar{\bar{\mathscr{M}}} \times \bar{\bar{\mathscr{M}}}) \cap \mathscr{U}$ and any $B, C \in \mathfrak{A}$ with $A \leq B$, $A \leq C$. If $\mathfrak{A}$ is upper bounded or if $\mathfrak{R}$ is ordered by the converse inclusion relation, then the above assertion holds even if we replace the hypothesis "$\mathfrak{R}$ is a quasi-σ-ring" by the weaker one "$\mathfrak{R}$ is a quasi-δ-ring".

Set

$$\mathfrak{L} := \{K \in \mathfrak{K} \mid \exists A \in \mathfrak{A},\ K \subset A\},$$

$$I := \{(K, L) \in \mathfrak{L} \times \mathfrak{L} \mid K \subset L\}$$

and endow I with the upper directed order relation $\leq$ defined by

$$(K,L) \leq (K',L') :\Leftrightarrow ((K,L) = (K',L') \quad \text{or} \quad L \subset K'\}$$

for any (K,L), $(K',L') \in I$. If φ denotes the map

$$I \to \mathfrak{R}, \quad (K,L) \mapsto L \backslash K$$

and if $\mathfrak{G}$ denotes the section filter of I, then by Proposition 4.1.6 $\varphi(\mathfrak{G}) \in \Psi_f(\mathfrak{R})$. By Proposition 4.1.16 $\mu(\varphi(\mathfrak{G}))$ is a Cauchy filter for any $\mu \in \mathscr{E}(\mathfrak{R}, G; \mathfrak{K})$.

Let V be a closed symmetric 0-neighbourhood in G such that $3V \subset U$. By Proposition 4.2.5 there exists $(\mathscr{M}, M) \in \mathfrak{F} \times \mathfrak{L}$ such that

$$\mu(L \backslash K) - \mu(L' \backslash K') \in V$$

for any $\mu \in \mathscr{M}$ and any $K, L, K', L' \in \mathfrak{L}$ with $M \subset K \subset L$, $M \subset K' \subset L'$. Then

$$\mu(K) - \mu(L) = \mu(K \backslash M) - \mu(L \backslash M) \in V$$

for any $\mu \in \mathscr{M}$ and any $K, L \in \mathfrak{L}$ with $M \subset K \cap L$. Since V is closed

$$\mu(K) - \mu(L) \in V$$

for any $\mu \in \bar{\mathscr{M}}$ and any $K, L \in \mathfrak{L}$ with $M \subset K \cap L$.

Put

$$\mathfrak{B} := \{B \in \mathfrak{R} \mid \exists A \in \mathfrak{A},\ B \subset A\}.$$

Using once again the fact that V is closed we find, by Proposition 4.1.26 a $\Rightarrow$ b, that

$$\mu(B) - \mu(C) = \lim \mu(\mathfrak{G}(B, \mathfrak{K})) - \lim \mu(\mathfrak{G}(C, \mathfrak{K})) \in V$$

for any $\mu \in \bar{\bar{\mathscr{M}}}$ and any $B, C \in \mathfrak{B}$ with $M \subset B \cap C$.

Let $\mathscr{U}$ be an entourage of $\mathscr{E}(\mathfrak{R}, G; \mathfrak{K})_{\mathfrak{K}}$ (of $\mathscr{M}(\mathfrak{R}, G; \mathfrak{K})_{\mathfrak{K}}$) such that

$$\mu(M) - \nu(M) \in V$$

for any $(\mu, \nu) \in \mathscr{U}$. Then

$$\mu(B) - \nu(C) = (\mu(B) - \mu(M)) + (\mu(M) - \nu(M)) + (\nu(M) - \nu(C)) \in 3V \subset U$$

for any $(\mu, \nu) \in (\bar{\bar{\mathscr{M}}} \times \bar{\bar{\mathscr{M}}}) \cap \mathscr{U}$ and any $B, C \in \mathfrak{B}$ with $M \subset B \cap C$. Let $A \in \mathfrak{A}$ such that $M \subset A$. Then

$$\mu(B) - \nu(C) \in U$$

for any $(\mu, \nu) \in (\bar{\bar{\mathscr{M}}} \times \bar{\bar{\mathscr{M}}}) \cap \mathscr{U}$ and any $B, C \in \mathfrak{A}$ with $A \subset B \cap C$.

Assume now that $\mathfrak{R}$ is ordered by the converse inclusion relation and let $A_0 \in \mathfrak{A}$. Set

$$\mathfrak{B} := \{A_0 \backslash A \mid A \in \mathfrak{A}\}.$$

Then $\mathfrak{B}$ is upper directed and upper bounded with respect to the inclusion relation.

Let V be a symmetric 0-neighbourhood in G such that $2V \subset U$. By the above proof there exist $\mathcal{M}' \in \mathfrak{F}$, $A \in \mathfrak{A}$, and an entourage $\mathcal{U}'$ of $\mathscr{E}(\mathfrak{R}, G; \mathfrak{K})_{\mathfrak{K}}$ (of $\mathcal{M}(\mathfrak{R}, G; \mathfrak{K})_{\mathfrak{K}}$) such that $A \subset A_0$ and

$$\mu(A_0 \backslash B) - \nu(A_0 \backslash C) \in V$$

for any $(\mu, \nu) \in (\bar{\bar{\mathcal{M}}}' \times \bar{\bar{\mathcal{M}}}') \cap \mathcal{U}'$ and any $B, C \in \mathfrak{A}$ with $B \cup C \subset A$. Let $\mathcal{U}''$ be an entourage of $\mathscr{E}(\mathfrak{R}, G; \mathfrak{K})$ (of $\mathcal{M}(\mathfrak{R}, G; \mathfrak{K})$) such that

$$\mu(A_0) - \nu(A_0) \in V$$

for any $(\mu, \nu) \in \mathcal{U}''$. By Corollary 4.2.8 there exist an entourage $\mathcal{U}$ of $\mathscr{E}(\mathfrak{R}, G; \mathfrak{K})_{\mathfrak{K}}$ (of $\mathcal{M}(\mathfrak{R}, G; \mathfrak{K})_{\mathfrak{K}}$) and an $\mathcal{M} \in \mathfrak{F}$ such that $\mathcal{M} \subset \mathcal{M}'$, $\mathcal{U} \subset \mathcal{U}'$, and

$$(\bar{\bar{\mathcal{M}}} \times \bar{\bar{\mathcal{M}}}) \cap \mathcal{U} \subset (\bar{\bar{\mathcal{M}}} \times \bar{\bar{\mathcal{M}}}) \cap \mathcal{U}''.$$

Then

$$\mu(B) - \nu(C) = (\mu(A_0) - \nu(A_0)) - (\mu(A_0 \backslash B) - \nu(A_0 \backslash C)) \in 2V \subset U$$

for any $(\mu, \nu) \in (\bar{\bar{\mathcal{M}}} \times \bar{\bar{\mathcal{M}}}) \cap \mathcal{U}$ and any $B, C \in \mathfrak{A}$ with $B \cup C \subset A$. □

Theorem 4.2.10. *Let $\mathfrak{F}$ be a $\hat{\Phi}_3$-filter on $\mathscr{E}(\mathfrak{R}, G; \mathfrak{K})$ (a $\hat{\Phi}_4$-filter on $\mathcal{M}(\mathfrak{R}, G; \mathfrak{K})$), $(A_n)_{n \in \mathbb{N}} \in \Gamma(\mathfrak{R})$ and U be a 0-neighbourhood in G. Then there exist $\mathcal{M} \in \mathfrak{F}$, $M \in \mathfrak{P}_f(\mathbb{N})$ and an entourage $\mathcal{U}$ of $\mathscr{E}(\mathfrak{R}, G; \mathfrak{K})$ (of $\mathcal{M}(\mathfrak{R}, G; \mathfrak{K})$) such that*

$$\mu\Big(\bigcup_{n \in M'} A_n\Big) - \nu\Big(\bigcup_{n \in M''} A_n\Big) \in U$$

for any $(\mu, \nu) \in (\bar{\mathcal{M}} \times \bar{\mathcal{M}}) \cap \mathcal{U}$ and any $M', M'' \in \mathfrak{P}_f(\mathbb{N})$ ($M', M'' \in \mathfrak{P}(\mathbb{N})$) with $M \cap M' = M \cap M''$. If $\{\mathfrak{H}(A, \mathfrak{K}) \mid A \in \mathfrak{R}\} \subset \Psi_f(\mathfrak{R})$ (which occurs if $\mathfrak{K} = \mathfrak{R}$ or if $\mathfrak{R}$ is a quasi-δ-ring), then we may choose an entourage of $\mathscr{E}(\mathfrak{R}, G; \mathfrak{K})_{\mathfrak{K}}$ (of $\mathcal{M}(\mathfrak{R}, G; \mathfrak{K})_{\mathfrak{K}}$) as $\mathcal{U}$.

Let us denote by φ the map

$$\mathfrak{P}(\mathbb{N}) \to \mathfrak{R}, \quad M \mapsto \bigcup_{n \in M} A_n$$

and let $\mathfrak{G}$ be the neighbourhood filter of $\emptyset$ in $\mathfrak{P}_f(\mathbb{N})$ (in $\mathfrak{P}(\mathbb{N})$). We have $\varphi(\mathfrak{G}) \in \Psi_f(\mathfrak{R})$ ($\varphi(\mathfrak{G}) \in \Psi(\mathfrak{R})$) and by Proposition 4.1.10 b) (by Proposition 4.1.14 a ⇒ c) $\mu(\varphi(\mathfrak{G}))$ converges to 0 for any $\mu \in \mathscr{E}(\mathfrak{R}, G; \mathfrak{K})$ ($\mu \in \mathcal{M}(\mathfrak{R}, G; \mathfrak{K})$).

Let V be a symmetric 0-neighbourhood in G such that $3V \subset U$. By Proposition 4.2.5 there exist $\mathcal{M} \in \mathfrak{F}$ and $M \in \mathfrak{P}_f(\mathbb{N})$ such that

$$\mu(\varphi(M')) - \mu(\varphi(M'')) \in V$$

for any $\mu \in \bar{\mathcal{M}}$ and any $M', M'' \in \mathfrak{P}_f(\mathbb{N} \backslash M)$ ($M', M'' \in \mathfrak{P}(\mathbb{N} \backslash M)$). Let $\mathcal{U}$ be an entourage of $\mathscr{E}(\mathfrak{R}, G; \mathfrak{K})$ (of $\mathcal{M}(\mathfrak{R}, G; \mathfrak{K})$) such that

$$\mu(\varphi(M_0)) - \nu(\varphi(M_0)) \in V$$

for any $(\mu, \nu) \in \mathscr{U}$ and any $M_0 \subset M$. Then

$$\mu(\bigcup_{n \in M'} A_n) - \nu(\bigcup_{n \in M''} A_n) =$$

$$= (\mu(\bigcup_{n \in M \cap M'} A_n) - \nu(\bigcup_{n \in M \cap M''} A_n)) + \mu(\bigcup_{n \in M' \setminus M} A_n) - \nu(\bigcup_{n \in M'' \setminus M} A_n) \in 3V \subset U$$

for any $(\mu, \nu) \in (\bar{\mathscr{M}} \times \bar{\mathscr{M}}) \cap \mathscr{U}$ and $M', M'' \in \mathfrak{P}_f(\mathbb{N})$ $(M', M'' \in \mathfrak{P}(\mathbb{N}))$ with $M \cap M' = M \cap M''$.

The last assertion follows from the first one and from Corollary 4.2.8. □

Corollary 4.2.11. *Let $(A_n)_{n \in \mathbb{N}} \in \Gamma(\mathfrak{R})$, $\mathfrak{F}$ be a $\hat{\Phi}_3$-filter on $\mathscr{E}(\mathfrak{R}, G; \mathfrak{K})$ (a $\hat{\Phi}_4$-filter on $\mathscr{M}(\mathfrak{R}, G; \mathfrak{K})$) and U be a 0-neighbourhood in G. Then there exists $(\mathscr{M}, M) \in \mathfrak{F} \times \mathfrak{P}_f(\mathbb{N})$ such that*

$$\mu(\bigcup_{n \in M'} A_n) \in U$$

for any $\mu \in \bar{\mathscr{M}}$ and any $M' \in \mathfrak{P}_f(\mathbb{N} \setminus M)$ $(M' \in \mathfrak{P}(\mathbb{N} \setminus M))$.

The assertion follows immediately from Theorem 4.2.10. □

Corollary 4.2.12. *For any $(A_n)_{n \in \mathbb{N}} \in \Gamma(\mathfrak{R})$ the map*

$$\mathscr{E}(\mathfrak{R}, G; \mathfrak{K}) \times \mathfrak{P}_f(\mathbb{N}) \to G, \quad (\mu, M) \mapsto \mu(\bigcup_{n \in M} A_n),$$

is uniformly Φ_3-continuous and the map

$$\mathscr{M}(\mathfrak{R}, G; \mathfrak{K}) \times \mathfrak{P}(\mathbb{N}) \to G, \quad (\mu, M) \mapsto \mu(\bigcup_{n \in M} A_n)$$

is uniformly Φ_4-continuous.

Let $\mathfrak{F}$ be a Cauchy $\hat{\Phi}_3$-filter on $\mathscr{E}(\mathfrak{R}, G; \mathfrak{K}) \times \mathfrak{P}_f(\mathbb{N})$ (a Cauchy $\hat{\Phi}_4$-filter on $\mathscr{M}(\mathfrak{R}, G; \mathfrak{K}) \times \mathfrak{P}(\mathbb{N})$) and let $\mathfrak{G}$, $\mathfrak{H}$ be its two projections. Then $\mathfrak{G}$ is a Cauchy $\hat{\Phi}_3$-filter on $\mathscr{E}(\mathfrak{R}, G; \mathfrak{K})$ (a Cauchy $\hat{\Phi}_4$-filter on $\mathscr{M}(\mathfrak{R}, G; \mathfrak{K})$) and $\mathfrak{H}$ is a Cauchy filter on $\mathfrak{P}_f(\mathbb{N})$ (on $\mathfrak{P}(\mathbb{N})$). By Theorem 4.2.10 the image of $\mathfrak{G} \times \mathfrak{H}$ with respect to the above maps is a Cauchy filter. Since $\mathfrak{F}$ is finer than $\mathfrak{G} \times \mathfrak{H}$ the same holds for $\mathfrak{F}$. □

Corollary 4.2.13. *Assume G Hausdorff and*

$$\{\mathfrak{H}(A, \mathfrak{K}) \mid A \in \mathfrak{R}\} \subset \Psi_f(\mathfrak{R})$$

(which is true if $\mathfrak{K} = \mathfrak{R}$ or if $\mathfrak{R}$ is a quasi-δ-ring). A subset $\mathscr{M}$ of $\mathscr{E}(\mathfrak{R}, G; \mathfrak{K})$ (of $\mathscr{M}(\mathfrak{R}, G; \mathfrak{K})$) is relatively compact if the following conditions are fulfilled.

a) $\{\mu(A) \mid \mu \in \mathscr{M}\}$ is a relatively compact set of G for any $A \in \mathfrak{R}$.

b) $\mu(\mathfrak{F}(A, \mathfrak{K}))$ converges to $\mu(A)$ uniformly on $\mathscr{M}$ for any $A \in \mathfrak{R}$.

c) $(\mu(A_n))_{n \in \mathbb{N}}$ $((\sum_{m=0}^{n} \mu(A_m))_{n \in \mathbb{N}})$ converges to 0 (to $\mu(\bigcup_{n \in \mathbb{N}} A_n)$) uniformly on $\mathscr{M}$ for any $(A_n)_{n \in \mathbb{N}} \in \Gamma(\mathfrak{R})$.

For a subset of $\mathscr{M}(\mathfrak{R}, G; \mathfrak{K})$ *the above conditions are necessary.*

The necessity follows from Proposition 4.2.7 and Theorem 4.2.10. In order to prove the sufficiency, let $\mathfrak{F}$ be an ultrafilter on $\mathscr{M}$ and v be the map

$$\mathfrak{R} \to G, \quad A \mapsto \lim_{\mu, \mathfrak{F}} \mu(A).$$

By c) $v \in \mathscr{E}(\mathfrak{R}, G)$ ($v \in \mathscr{M}(\mathfrak{R}, G)$) and by b) v is $\mathfrak{K}$-regular. □

Remark. The $\mathscr{M}$-part of the above corollary was proved by R. G. Bartle, N. Dunford and J. Schwartz (1955) ([1] Theorem 1.3) for $\mathfrak{R}$ a σ-algebra $G = \mathbb{R}$, and $\mathfrak{K} = \mathfrak{R}$. This result was extended to G a locally convex space by W. H. Graves and W. Ruess (1980) ([1] Theorem 7 $1 \Leftrightarrow 2$) and to an arbitrary G and an arbitrary $\mathfrak{R}$ by H. Weber ([1] Satz 3.1.4 (1) $\Leftrightarrow$ (3)). The $\mathscr{E}$-part of the above corollary was proved by J. K. Brooks and N. Dinculeanu (1972) ([1] Théorème 4) for G a Banach space and $\mathfrak{K} = \mathfrak{R}$. A recent result of H. Weber ([1] Satz 3.2.4 (1) $\Leftrightarrow$ (2)) gives necessary and sufficient conditions for $\mathscr{M} \subset \mathscr{E}(\mathfrak{R}, G)$ to be relatively compact, for G arbitrary.

Theorem 4.2.14. *Let* $\mathfrak{A}$ *be a subset of* $\mathfrak{R}$ *such that the identity map* $\mathscr{E}(\mathfrak{R}, G; \mathfrak{K})_{\mathfrak{A}} \to \mathscr{E}(\mathfrak{R}, G; \mathfrak{K})$ ($\mathscr{M}(\mathfrak{R}, G; \mathfrak{K})_{\mathfrak{A}} \to \mathscr{M}(\mathfrak{R}, G; \mathfrak{K})$) *is uniformly* Φ_3*-continuous (uniformly* Φ_4*-continuous) (e.g.* $\mathfrak{A} = \mathfrak{R}$*). If* $\{\mathfrak{H}(A, \mathfrak{K}) \mid A \in \mathfrak{R}\} \subset \Psi_f(\mathfrak{R})$ *(which is true if* $\mathfrak{K} = \mathfrak{R}$ *or if* $\mathfrak{R}$ *is a quasi-*δ*-ring) and if* G *is Hausdorff, then:*

a) $\mathscr{E}(\mathfrak{R}, G; \mathfrak{K})$ *is a* Φ_3*-closed set of* $G^{\mathfrak{R}}$*;* $\mathscr{E}(\mathfrak{R}, G)$ *and* $\mathscr{M}(\mathfrak{R}, G; \mathfrak{K})$ *are* Φ_4*-closed sets of* $G^{\mathfrak{R}}$*;*

b) if G *is* Φ_i*-compact for an* $i \in \{1, 3\}$, ($i \in \{1, 2, 3, 4\}$), *then so is* $\mathscr{E}(\mathfrak{R}, G; \mathfrak{K})_{\mathfrak{A}}$, *(then so are* $\mathscr{E}(\mathfrak{R}, G)$ *and* $\mathscr{M}(\mathfrak{R}, G; \mathfrak{K})_{\mathfrak{A}}$*);*

c) if G *is sequentially complete, then so is* $\mathscr{E}(\mathfrak{R}, G; \mathfrak{K})_{\mathfrak{A}}$ ($\mathscr{M}(\mathfrak{R}, G; \mathfrak{K})_{\mathfrak{A}}$)*;*

d) if $\{0\}$ *is a* G_δ*-set of* G, *then* $\mathscr{E}(\mathfrak{R}, G; \mathfrak{K})_{\mathfrak{A}}$ *is* Φ_1*-compact,* ($\mathscr{E}(\mathfrak{R}, G)$ *and* $\mathscr{M}(\mathfrak{R}, G; \mathfrak{K})_{\mathfrak{A}}$ *are* Φ_2*-compact) and if, in addition,* G *is sequentially complete, then* $\mathscr{E}(\mathfrak{R}, G; \mathfrak{K})_{\mathfrak{A}}$ *is* Φ_3*-compact,* ($\mathscr{E}(\mathfrak{R}, G)$ *and* $\mathscr{M}(\mathfrak{R}, G; \mathfrak{K})_{\mathfrak{A}}$ *are* Φ_4*-compact);*

e) $\mathscr{E}(\mathfrak{R}, G; \mathfrak{K})_{\mathfrak{A}}$ *and* $\mathscr{M}(\mathfrak{R}, G; \mathfrak{K})_{\mathfrak{A}}$ *are Hausdorff.*

By Proposition 4.1.25 a), c) $\{\mathfrak{H}(A, \mathfrak{K}) \mid A \in \mathfrak{R}\} \subset \Psi_f(\mathfrak{R})$ if $\mathfrak{K} = \mathfrak{R}$ or if $\mathfrak{R}$ is a quasi-δ-ring.

a) Let $\mathfrak{F}$ be a Φ_3-filter on $\mathscr{E}(\mathfrak{R}, G; \mathfrak{K})$, (a Φ_4-filter on $\mathscr{E}(\mathfrak{R}, G)$ or $\mathscr{M}(\mathfrak{R}, G; \mathfrak{K})$), φ be the inclusion map $\mathscr{E}(\mathfrak{R}, G; \mathfrak{K}) \to G^{\mathfrak{R}}$, ($\mathscr{E}(\mathfrak{R}, G) \to G^{\mathfrak{R}}$ or $\mathscr{M}(\mathfrak{R}, G; \mathfrak{K}) \to G^{\mathfrak{R}}$), and let $v \in G^{\mathfrak{R}}$ be such that $\varphi(\mathfrak{F})$ converges to v in $G^{\mathfrak{R}}$. By Theorem 4.1.22 d) $\mathscr{E}(\mathfrak{R}, G)$ and $\mathscr{M}(\mathfrak{R}, G)$ are Φ_4-closed sets of $G^{\mathfrak{R}}$ so $v \in \mathscr{E}(\mathfrak{R}, G)$ or $v \in \mathscr{M}(\mathfrak{R}, G)$. We now only have to show that v is $\mathfrak{K}$-regular.

Let $A \in \mathfrak{R}$ and let U be a closed 0-neighbourhood in G. By Proposition 4.2.7 there exists $(\mathscr{M}, K) \in \mathfrak{F} \times \mathfrak{K}$ such that $K \subset A$ and

$$\mu(B) - \mu(A) \in U$$

for any $\mu \in \bar{\mathscr{M}}$ and any $B \in \mathfrak{R}$ with $K \subset B \subset A$. We get

$$\nu(B) - \nu(A) \in U$$

for any $B \in \mathfrak{R}$ with $K \subset B \subset A$. Since U is arbitrary $\nu(\mathfrak{F}(A, \mathfrak{K}))$ converges to $\nu(A)$. Hence, ν is $\mathfrak{K}$-regular.

b) follows immediately from a) and the Propositions 2.1.14 and 1.8.11.

c) follows immediately from a) and the Propositions 2.1.14 and 1.8.7.

d) follows from b) and Proposition 1.6.4.

e) Let $\mu, \nu \in \mathscr{E}(\mathfrak{R}, G; \mathfrak{K})$ and $\mathfrak{F}$ be the filter on $\mathscr{E}(\mathfrak{R}, G; \mathfrak{K})$ generated by the filter base $\{\{\mu\}\}$. If $\mu(A) = \nu(A)$ for any $A \in \mathfrak{A}$, then $\mathfrak{F}$ converges to ν in $\mathscr{E}(\mathfrak{R}, G; \mathfrak{K})_{\mathfrak{A}}$. Since $\mathfrak{F}$ is a Φ_1-filter on $\mathscr{E}(\mathfrak{R}, G; \mathfrak{K})_{\mathfrak{A}}$ it converges to ν in $\mathscr{E}(\mathfrak{R}, G; \mathfrak{K})$ (Proposition 1.8.3) so $\mu = \nu$. Hence, $\mathscr{E}(\mathfrak{R}, G; \mathfrak{K})_{\mathfrak{A}}$ and $\mathscr{M}(\mathfrak{R}, G; \mathfrak{K})_{\mathfrak{A}}$ are Hausdorff. □

Remarks. 1) Let $\mathfrak{R}$ be a σ-ring, G be Hausdorff, $(\mu_n)_{n \in \mathbb{N}}$ be a sequence of additive exhaustive maps of $\mathfrak{R}$ into G such that $(\mu_n(A))_{n \in \mathbb{N}}$ converges for any $A \in \mathfrak{R}$ and μ be the map

$$\mathfrak{R} \to G, \quad A \mapsto \lim_{n \to \infty} \mu_n(A).$$

By Proposition 4.2.6 or Corollary 4.2.11 $(\mu_n)_{n \in \mathbb{N}}$ is uniformly exhaustive and by Theorem 4.2.14 a) μ is exhaustive. This result was proved by T. Andô (1961) ([1] Main Theorem) for the case when $G := \mathbb{R}$, by J.K. Brooks and S. Jewitt (1970) ([1] Theorem 2 and Corollary 1.2) for the case when G is a locally convex space, and by L. Drewnowski (1972) ([2] Theorem (BJ)) and I. Labuda (1972) ([1] Théorème 4.3) for an arbitrary G.

If any μ_n $(n \in \mathbb{N})$ is a measure, then by Corollary 4.2.11 $(\mu_n)_{n \in \mathbb{N}}$ is uniformly σ-additive and by Theorem 4.2.14 a) μ is a measure. This last result was proved by O. Nikodym (1931) ([1] II, [3] page 427) for $G = \mathbb{R}$ and it is known as Nikodym's theorem or more precisely Nikodym's convergence theorem. N. Dunford and J.T. Schwartz (1957) ([1] Theorem IV. 10.6) extended it to G a Banach space. P. Gänssler showed (1971) ([1] Proposition 3.8) that μ is $\mathfrak{K}$-regular if the μ_n are $\mathfrak{K}$-regular $(n \in \mathbb{N})$ for $G = \mathbb{R}$ and D. Landers and L. Rogge extended these results to an arbitrary G (1971) ([1] Theorem 8 (2)); they even showed ([1] Theorem 8 (1)) that $(\mu_n)_{n \in \mathbb{N}}$ is uniformly $\mathfrak{K}$-regular (this last result follows from Proposition 4.2.7).

2) In general, $\mathscr{E}(\mathfrak{R}, G; \mathfrak{K})$ is not a Φ_2-closed set of $G^{\mathfrak{R}}$. Indeed, let $\mathfrak{F}$ be an

Theorem 4.2.15. *We denote by Φ the set of $\hat{\Phi}_3$-filters on $\mathscr{E}(\mathfrak{R}, G; \mathfrak{K})$ (of $\hat{\Phi}_4$-filters on $\mathscr{M}(\mathfrak{R}, G; \mathfrak{K})$) and by φ the map*

$$\mathscr{E}(\mathfrak{R}, G; \mathfrak{K}) \times \mathfrak{R} \to G, \quad (\mu, A) \mapsto \mu(A)$$

$$(\mathscr{M}(\mathfrak{R}, G; \mathfrak{K}) \times \mathfrak{R} \to G, \quad (\mu, A) \mapsto \mu(A)).$$

If $\{\mathfrak{H}(A, \mathfrak{K}) \mid A \in \mathfrak{R}\} \subset \Psi_f(\mathfrak{R})$ (which occurs if $\mathfrak{K} = \mathfrak{R}$ or if $\mathfrak{R}$ is a quasi-δ-ring), then the map

$$\mathfrak{R} \to G_{\Phi}^{\mathscr{E}(\mathfrak{R}, G, \mathfrak{K})}, \quad A \mapsto \varphi(\cdot, A)$$
$$(\mathfrak{R} \to G_{\Phi}^{\mathscr{M}(\mathfrak{R}, G, \mathfrak{K})}, \quad A \mapsto \varphi(\cdot, A))$$

is $\mathfrak{K}$-regular, additive and exhaustive (is a $\mathfrak{K}$-regular measure).

By Corollary 4.2.11 (and Proposition 4.1.14 d $\Rightarrow$ a) the map is exhaustive (is a measure) and by Proposition 4.2.7 it is $\mathfrak{K}$-regular. □

Theorem 4.2.16. *Let $\mathscr{M}$ be a Φ_3-set of $\mathscr{E}(\mathfrak{R}, G; \mathfrak{K})$ (a Φ_4-set of $\mathscr{M}(\mathfrak{R}, G; \mathfrak{K})$) and $\mathscr{F}$ be the group of uniformly continuous maps of $\bar{\bar{\mathscr{M}}}_{\mathfrak{K}}$ into G. If $\{\mathfrak{H}(A, \mathfrak{K}) \mid A \in \mathfrak{R}\} \subset \Psi_f(\mathfrak{R})$ (which occurs if $\mathfrak{K} = \mathfrak{R}$ or if $\mathfrak{R}$ is a quasi-δ-ring), then:*

a) $\bar{\bar{\mathscr{M}}}$ is the closure of $\mathscr{M}$ in $\mathscr{E}(\mathfrak{R}, G; \mathfrak{K})$ and in $\mathscr{E}(\mathfrak{R}, G; \mathfrak{K})_{\Psi_f(\mathfrak{R})}$ (in $\mathscr{M}(\mathfrak{R}, G; \mathfrak{K})$ and in $\mathscr{M}(\mathfrak{R}, G; \mathfrak{K})_{\Psi(\mathfrak{R})}$) and $\bar{\bar{\mathscr{M}}}_{\mathfrak{K}} = \bar{\bar{\mathscr{M}}} = \bar{\bar{\mathscr{M}}}_{\Psi_f(\mathfrak{R})}$ ($\bar{\bar{\mathscr{M}}}_{\mathfrak{K}} = \bar{\bar{\mathscr{M}}} = \bar{\bar{\mathscr{M}}}_{\Psi(\mathfrak{R})}$); if G is Φ_3-compact (if G is Φ_4-compact) (e.g. complete), then $\bar{\bar{\mathscr{M}}}$ is quasi-compact;

b) the map

$$\bar{\bar{\mathscr{M}}}_{\mathfrak{K}} \times \mathfrak{P}_f(\mathbb{N}) \to G, \quad (\mu, M) \mapsto \mu(\bigcup_{n \in M} A_n)$$
$$(\bar{\bar{\mathscr{M}}}_{\mathfrak{K}} \times \mathfrak{P}(\mathbb{N}) \to G, \quad (\mu, M) \mapsto \mu(\bigcup_{n \in M} A_n))$$

is uniformly continuous for any $(A_n)_{n \in \mathbb{N}} \in \Gamma(\mathfrak{R})$;

c) for any $A \in \mathfrak{R}$ and any 0-neighbourhood U in G there exist $K \in \mathfrak{K}$ and an entourage $\mathscr{U}$ of $\bar{\bar{\mathscr{M}}}_{\mathfrak{K}}$ such that $K \subset A$ and

$$\mu(B) - \nu(C) \in U$$

for any $(\mu, \nu) \in \mathscr{U}$ and any $B, C \in \mathfrak{R}$ with $K \subset B \subset A$, $K \subset C \subset A$;

d) the map

$$\nu(A) : \bar{\bar{\mathscr{M}}}_{\mathfrak{K}} \to G, \quad \mu \mapsto \mu(A)$$

belongs to $\mathscr{F}$ for any $A \in \mathbb{R}$ and $\nu \in \mathscr{E}(\mathfrak{R}, \mathscr{F}_{\{\bar{\bar{\mathscr{M}}}\}}; \mathfrak{K})$ ($\nu \in \mathscr{M}(\mathfrak{R}, \mathscr{F}_{\{\bar{\bar{\mathscr{M}}}\}}; \mathfrak{K})$).

a) We may assume G Hausdorff since otherwise we may replace it by its canonical associated Hausdorff topological group. If $\mathscr{M}$ is a Φ_4-set of $\mathscr{M}(\mathfrak{R}, G; \mathfrak{K})$ then, by Theorem 4.2.14 a), $\bar{\bar{\mathscr{M}}}$ is its closure in $\mathscr{M}(\mathfrak{R}, G; \mathfrak{K})$ (and not only in $\mathscr{E}(\mathfrak{R}, G; \mathfrak{K})$).

Assume first that G is Φ_3-compact (Φ_4-compact). By Theorem 4.2.14 b) $\mathscr{E}(\mathfrak{R}, G; \mathfrak{K})$ ($\mathscr{M}(\mathfrak{R}, G; \mathfrak{K})$) is Φ_3-compact (Φ_4-compact) so, by Proposition 1.2.10, $\bar{\mathscr{M}}$ is compact. Since $\bar{\bar{\mathscr{M}}}_{\mathfrak{K}}$ is Hausdorff we get $\bar{\bar{\mathscr{M}}}_{\mathfrak{K}} = \bar{\mathscr{M}}$. Hence, $\bar{\mathscr{M}} = \bar{\bar{\mathscr{M}}}$ and $\bar{\bar{\mathscr{M}}}$ is compact.

By Corollary 1.5.11 any complete uniform space is Φ_4-compact.

For the rest of the proof we may assume G complete. By the above proof $\bar{\bar{\mathscr{M}}}$ is the closure of $\mathscr{M}$ in $\mathscr{E}(\mathfrak{R}, G; \mathfrak{K})$ (in $\mathscr{M}(\mathfrak{R}, G; \mathfrak{K})$) and $\bar{\bar{\mathscr{M}}}_{\mathfrak{K}} = \bar{\bar{\mathscr{M}}}$. By Proposition 4.2.3 the identity map

$$\mathscr{E}(\mathfrak{R}, G; \mathfrak{K}) \to \mathscr{E}(\mathfrak{R}, G; \mathfrak{K})_{\Psi_f(\mathfrak{R})}$$
$$(\mathscr{M}(\mathfrak{R}, G; \mathfrak{K}) \to \mathscr{M}(\mathfrak{R}, G; \mathfrak{K})_{\Psi(\mathfrak{R})})$$

is uniformly Φ_3-continuous (uniformly Φ_4-continuous). By Proposition 1.8.8 $\bar{\bar{\mathscr{M}}}$ is the closure of $\mathscr{M}$ in $\mathscr{E}(\mathfrak{R}, G; \mathfrak{K})_{\Psi_f(\mathfrak{R})}$ (in $\mathscr{M}(\mathfrak{R}, G; \mathfrak{K})_{\Psi(\mathfrak{R})}$) and $\bar{\bar{\mathscr{M}}} = \bar{\bar{\mathscr{M}}}_{\Psi_f(\mathfrak{R})}$ ($\bar{\bar{\mathscr{M}}} = \bar{\bar{\mathscr{M}}}_{\Psi(\mathfrak{R})}$).

b) follows immediately from a) and Theorem 4.2.10.

c) follows from Proposition 4.2.7.

d) By b) $\nu(A) \in \mathscr{F}$ for any $A \in \mathfrak{R}$ and the assertion follows from Theorem 4.2.15. □

Theorem 4.2.17. *Let G be Hausdorff, $\mathfrak{R}$ be a quasi-δ-ring and $(\mu_\iota)_{\iota \in I}$ be a family in $\mathscr{E}(\mathfrak{R}, G; \mathfrak{K})$ (in $\mathscr{M}(\mathfrak{R}, G; \mathfrak{K})$) such that $(\mu_\iota(A))_{\iota \in I}$ is supersummable for any $A \in \mathfrak{R}$. Then:*

a) the family $(\mu_\iota)_{\iota \in I}$ is supersummable in $\mathscr{E}(\mathfrak{R}, G; \mathfrak{K})_{\Psi_f(\mathfrak{R})}$ (in $\mathscr{M}(\mathfrak{R}, G; \mathfrak{K})_{\Psi(\mathfrak{R})}$);

b) for any $A \in \mathfrak{R}$, any $(A_n)_{n \in \mathbb{N}} \in \Gamma(\mathfrak{R})$ and any 0-neighbourhood U in G there exist $K \in \mathfrak{K}$, a finite subset J of I and a finite subset M of $\mathbb{N}$ such that $K \subset A$ and

$$\sum_{\iota \in J'} \mu_\iota(B) - \sum_{\iota \in J''} \mu_\iota(C) \in U,$$

$$\sum_{\iota \in J'} \mu_\iota(\bigcup_{n \in M'} A_n) - \sum_{\iota \in J''} \mu_\iota(\bigcup_{n \in M''} A_n) \in U$$

for any $B, C \in \mathfrak{R}$ with $K \subset B \subset A$, $K \subset C \subset A$, any subsets J', J'' of I with $J \cap J' = J \cap J''$ and any $M', M'' \in \mathfrak{P}_f(\mathbb{N})$ ($M', M'' \in \mathfrak{P}(\mathbb{N})$) with $M \cap M' = M \cap M''$.

a) Let $\mathfrak{F}$ be the section filter of $\mathfrak{P}_f(I)$ and φ be the map

$$\mathfrak{P}_f(I) \to \mathscr{E}(\mathfrak{R}, G; \mathfrak{K}), \quad J \mapsto \sum_{\iota \in J} \mu_\iota$$

$$(\mathfrak{P}_f(I) \to \mathscr{M}(\mathfrak{R}, G; \mathfrak{K}), \quad J \mapsto \sum_{\iota \in J} \mu_\iota)$$

Then $(\mathfrak{P}_f(I), \varphi)$ is a Θ_3-net and $\varphi(\mathfrak{F})$ is a Φ_3-filter on $\mathscr{E}(\mathfrak{R}, G; \mathfrak{K})$ (on $\mathscr{M}(\mathfrak{R}, G; \mathfrak{K})$). By Theorem 4.2.14 a) the map

$$\mathfrak{R} \to G, \quad A \mapsto \sum_{\iota \in I} \mu_\iota(A)$$

belongs to $\mathscr{E}(\mathfrak{R}, G; \mathfrak{K})$ (to $\mathscr{M}(\mathfrak{R}, G; \mathfrak{K})$). Hence, $(\mu_\iota)_{\iota \in I}$ is summable and therefore supersummable in $\mathscr{E}(\mathfrak{R}, G; \mathfrak{K})$ (in $\mathscr{M}(\mathfrak{R}, G; \mathfrak{K})$). By Theorem 4.2.4 b) it is supersummable in $\mathscr{E}(\mathfrak{R}, G; \mathfrak{K})_{\Psi_f(\mathfrak{R})}$ (in $\mathscr{M}(\mathfrak{R}, G; \mathfrak{K})_{\Psi(\mathfrak{R})}$).

b) By the above proof $\varphi(\mathfrak{F})$ is a convergent Φ_3-filter on $\mathscr{E}(\mathfrak{R}, G; \mathfrak{K})$ (on $\mathscr{M}(\mathfrak{R}, G; \mathfrak{K})$) and the assertions follow from Proposition 4.2.9 and Theorem 4.2.10. □

Proposition 4.2.18. *Let $\mu \in \mathscr{E}(\mathfrak{R}, G)$ and U be a 0-neighbourhood in G. If $\mathfrak{R}$ is a quasi-σ-ring, then there exists $A \in \mathfrak{R}$ such that $\mu(B) \in U$ for any $B \in \mathfrak{R}$ with $A \cap B = \emptyset$.*

Assume the contrary. Then we may construct a disjoint sequence $(A_n)_{n \in \mathbb{N}}$ in $\mathfrak{R}$ such that $\mu(A_n) \notin U$ for any $n \in \mathbb{N}$ and this contradicts Proposition 4.1.9. □

Theorem 4.2.19. *Assume $\mathfrak{R}$ is a quasi-σ-ring and $\{0\}$ is a G_δ-set of G and let $\mathscr{M}$ be a subset of $\mathscr{E}(\mathfrak{R}, G)$ (of $\mathscr{M}(\mathfrak{R}, G)$) such that there exists a countable family $(\mathfrak{F}_\iota)_{\iota\in I}$ of $\hat{\Phi}_3$-filters of $\mathscr{E}(\mathfrak{R}, G)$ (of $\hat{\Phi}_4$-filters of $\mathscr{M}(\mathfrak{R}, G)$) such that $\mathscr{M} \subset \bigcup_{\iota\in I} \mathscr{M}_\iota$ for any $(\mathscr{M}_\iota)_{\iota\in I} \in \prod_{\iota\in I} \mathfrak{F}_\iota$ (which occurs, e.g., if $\mathscr{M}$ is the union of a countable family of Φ_3-sets of $\mathscr{E}(\mathfrak{R}, G)$ (of Φ_4-sets of $\mathscr{M}(\mathfrak{R}, G)$). Then there exists an increasing sequence $(A_n)_{n\in\mathbb{N}}$ in $\mathfrak{R}$ such that $\mu(A) = 0$ for any $\mu \in \mathscr{M}$ and any $A \in \mathfrak{R}$ with $A \cap (\bigcup_{n\in\mathbb{N}} A_n) = \emptyset$.*

We set $\Phi := \{\mathfrak{F}_\iota \mid \iota \in I\}$. Then $G_\Phi^{\mathscr{M}}$ is a topological additive group for which $\{0\}$ is a G_δ-set. By Theorem 4.2.15 the map

$$\mathfrak{R} \to G_\Phi^{\mathscr{M}}, \quad A \mapsto (\mu(A))_{\mu\in\mathscr{M}}$$

is additive and exhaustive and the assertion follows from Proposition 4.2.18. □

Theorem 4.2.20. *Let E be a Hausdorff locally convex space and E_σ be the vector space underlying E endowed with the weak topology of E. Then $\mathscr{E}(\mathfrak{R}, E)$ is a strongly Φ_2-closed set of $E_\sigma^{\mathfrak{R}}$.*

Let φ be the inclusion map $\mathscr{E}(\mathfrak{F}, E) \to E_\sigma^{\mathfrak{R}}$, $\mathfrak{F} \in \Phi(\Theta_2(\mathscr{E}(\mathfrak{R}, E), E_\sigma^{\mathfrak{R}}))$ and μ be a limit point of $\varphi(\mathfrak{F})$ in $E_\sigma^{\mathfrak{R}}$. We have to prove $\mu \in \mathscr{E}(\mathfrak{R}, E)$. Assume the contrary. Then there exist a convex circled closed 0-neighbourhood U in E and an $(A_n)_{n\in\mathbb{N}} \in \Gamma(\mathfrak{R})$ such that $\mu(A_n) \notin U$ for any $n \in \mathbb{N}$. Let E' be the dual space of E and U^0 be the polar set of U in E, i.e.

$$U^\circ := \{x' \in E' \mid x \in U \Rightarrow |x'(x)| \le 1\}.$$

There exists a sequence $(x'_n)_{n\in\mathbb{N}}$ in U^0 such that $x'_n(\mu(A_n)) > 1$ for any $n \in \mathbb{N}$. Let u be the map

$$E \to \ell^\infty, \quad x \mapsto (x'_n(x))_{n\in\mathbb{N}};$$

u is continuous and linear. Let (I, λ) be a $\Theta_2(\mathscr{E}(\mathfrak{R}, E), E_\sigma^{\mathfrak{R}})$-net in $\mathscr{E}(\mathfrak{R}, E)$ such that $\lambda(\mathfrak{G}) = \mathfrak{F}$, where $\mathfrak{G}$ denotes the section filter of I. We may inductively construct an increasing sequence $(\iota_m)_{m\in\mathbb{N}}$ in I such that

$$\lim_{m\to\infty} x'_n(\lambda_{\iota_m}(A_n)) = x'_n(\mu(A_n)) > 1$$

for any $n \in \mathbb{N}$. Assume $(\lambda_{\iota_m})_{m\in\mathbb{N}}$ possesses an adherent point ϱ in $\mathscr{E}(\mathfrak{R}, E)$. Then $\varrho(A_n) \notin U$ for any $n \in \mathbb{N}$ and this is a contradiction. Hence, by replacing $(\lambda_{\iota_m})_{m\in\mathbb{N}}$ by a subsequence, we may assume $(\varphi(\lambda_{\iota_m}))_{m\in\mathbb{N}}$ converges in $E_\sigma^{\mathfrak{R}}$. We have $u \circ \lambda_{\iota_m} \in \mathscr{E}(\mathfrak{R}, \ell^\infty)$ for any $m \in \mathbb{N}$. By Proposition 4.1.18 there exists an infinite subset M_0 of $\mathbb{N}$ such that

$$\lambda'_m : \mathfrak{P}(M_0) \to \ell^\infty, \quad M \mapsto u \circ \lambda_{\iota_m}(\bigcup_{n\in M} A_n)$$

is a measure for any $m \in \mathbb{N}$. Let ℓ_σ^∞ be the vector space underlying ℓ^∞ endowed with the weak topology of ℓ^∞ and ψ be the inclusion map

$\mathscr{M}(\mathfrak{P}(M_0), \ell_\sigma^\infty) \to (\ell_\sigma^\infty)^{\mathfrak{P}(M_0)}$. Then $(\lambda'_m)_{m\in\mathbb{N}}$ is a sequence in $\mathscr{M}(\mathfrak{P}(M_0), \ell_\sigma^\infty)$ such that $(\psi(\lambda'_m))_{m\in\mathbb{N}}$ converges to a point v in $(\ell_\sigma^\infty)^{\mathfrak{P}(M_0)}$. By Theorem 4.2.14a) $v \in \mathscr{M}(\mathfrak{P}(M_0), \ell_\sigma^\infty)$ and by Corollary 4.1.30a) $v \in \mathscr{M}(\mathfrak{P}(M_0), \ell^\infty)$. We have $\|v(\{n\})\| > 1$ for any $n \in M_0$, which is the expected contradiction. □

Remark. Let $(\mu_n)_{n\in\mathbb{N}}$ be a sequence of additive exhaustive maps of $\mathfrak{R}$ into E such that $(\mu_n(A))_{n\in\mathbb{N}}$ converges weakly for any $A \in \mathfrak{R}$ and μ be the map of $\mathfrak{R}$ into E sending any $A \in \mathfrak{R}$ into the weak limit of $(\mu_n(A))_{n\in\mathbb{N}}$. By the above theorem μ is exhaustive. This result was proved by J. Diestel (1973) ([1] Corollary 5) for the case when E is a Banach space.

§ 4.3 Vitali-Hahn-Saks theorem and Phillips lemma

Throughout this section $\mathfrak{R}$ shall denote a ring of sets, $\mathfrak{S}$ a subset of $\mathfrak{R}$ closed with respect to finite unions, and G a topological additive group.

Proposition 4.3.1. *Let $\mathfrak{R}$ be a quasi-δ-ring, $(\mu_n)_{n\in\mathbb{N}}$ be a Θ_4-sequence in $\mathscr{M}(\mathfrak{R}, G; \mathfrak{S})$ or a Θ_3-sequence in $\mathscr{E}(\mathfrak{R}, G; \mathfrak{S})$, $(A_n)_{n\in\mathbb{N}}$ be a sequence in $\mathfrak{R}$, U be a 0-neighbourhood in G and $(U_n)_{n\in\mathbb{N}}$ be a sequence of 0-neighbourhoods in G such that $2U_{n+1} \subset U_n$ and*

$$\{\mu_0(A) \mid A \in \mathfrak{R},\ A \subset A_{n+1}\} \subset U_{n+1}$$

for any $n \in \mathbb{N}$ and such that there exists $k \in \mathbb{N}$ with

$$\mu_0(A_0) \notin U_0 + U_k.$$

Then there exist $n_0 \in \mathbb{N}$ and $A \in \mathfrak{R}$ such that $A \subset A_0$, $\mu_0(A) \notin U_0$, and $\mu_n(A_n \cap A) \in U$ for any $n \in \mathbb{N}$ with $n \geq n_0$.

Assume the contrary. We construct inductively a strictly increasing sequence $(p_n)_{n\in\mathbb{N}}$ in $\mathbb{N}$ such that $p_0 > k$ and

$$\mu_{p_n}(A_{p_n} \cap (A_0 \setminus \bigcup_{m=0}^{n-1} A_{p_m})) \notin U$$

for any $n \in \mathbb{N}$. Let $n \in \mathbb{N}$ and assume the sequence was constructed up to $n-1$. We have

$$\mu_0((A_0 \cap A_{p_m}) \setminus \bigcup_{q=0}^{m-1} A_{p_q}) \in U_{p_m}$$

for any $m \in \mathbb{N}$ with $m < n$ so

$$\mu_0(A_0 \cap (\bigcup_{m=0}^{n-1} A_{p_m})) \in \sum_{m=0}^{n-1} U_{p_m} \subset U_k.$$

We get

$$\mu_0(A_0 \setminus \bigcup_{m=0}^{n-1} A_{p_m}) \notin U_0 .$$

By the hypothesis of the proof there exists $p_n \in \mathbb{N}$ such that $p_n > p_{n-1}$ ($p_0 > k$ if $n = 0$) and

$$\mu_{p_n}(A_{p_n} \cap (A_0 \setminus \bigcup_{m=0}^{n-1} A_{p_m})) \notin U .$$

This finishes the inductive construction.

$(A_{p_n} \cap (A_0 \setminus \bigcup_{m=0}^{n-1} A_{p_m}))_{n\in\mathbb{N}}$ is a disjoint sequence in $\mathfrak{R}$. Since $\mathfrak{R}$ is a quasi-δ-ring it is a $\Theta(\mathfrak{R})$-sequence. The above relation contradicts Proposition 4.2.6. □

Theorem 4.3.2. *(Vitali-Hahn-Saks). Let $\mathfrak{R}$ be a quasi-σ-ring, $\mathfrak{G}$ be a filter base on $\mathfrak{R}$ such that $A \cap B \in \mathfrak{A}$ for any $\mathfrak{A} \in \mathfrak{G}$ and any $(A, B) \in \mathfrak{A} \times \mathfrak{R}$, $\mathfrak{F}$ be a $\hat{\Phi}_3$-filter on $\mathscr{E}(\mathfrak{R}, G; \mathfrak{K})$ (a $\hat{\Phi}_4$-filter on $\mathscr{M}(\mathfrak{R}, G; \mathfrak{K})$) such that*

$$\{\mu \in \mathscr{E}(\mathfrak{R}, G; \mathfrak{K}) \ (\mu \in \mathscr{M}(\mathfrak{R}, G; \mathfrak{K})) \mid \mu(\mathfrak{G}) \ \text{ converges to } \ 0\} \in \mathfrak{F},$$

and let U be a 0-neighbourhood in G. Then there exists $(\mathscr{M}, \mathfrak{A}) \in \mathfrak{F} \times \mathfrak{G}$ such that $\mu(A) \in U$ for any $\mu \in \bar{\mathscr{M}}$ and any $A \in \mathfrak{A}$. If there exists $(A, \mathfrak{A}) \in \mathfrak{R} \times \mathfrak{G}$ such that $\bigcup_{B\in\mathfrak{A}} B \subset A$, then we may replace the hypothesis "$\mathfrak{R}$ is a quasi-σ-ring" by the weaker one "$\mathfrak{R}$ is a quasi-δ-ring".

We may suppose U is closed.

Assume first that $\mathfrak{F}$ is a Φ_3-filter on $\mathscr{E}(\mathfrak{R}, G; \mathfrak{K})$ (a Φ_4-filter on $\mathscr{M}(\mathfrak{R}, G; \mathfrak{K})$). Let (I, μ) be a Θ_3-net in $\mathscr{E}(\mathfrak{R}, G; \mathfrak{K})$ (a Θ_4-net in $\mathscr{M}(\mathfrak{R}, G; \mathfrak{K})$) such that $\mu(\mathfrak{F}') = \mathfrak{F}$, where $\mathfrak{F}'$ denotes the section filter of I. By the hypothesis there exists $\lambda \in I$ such that $\mu_\iota(\mathfrak{G})$ converges to 0 for any $\iota \in I$ with $\iota \geq \lambda$. Assume the assertion does not hold. Then, for any $\iota \in I$ and any $\mathfrak{A} \in \mathfrak{G}$, there exists $(\iota', A) \in I \times \mathfrak{A}$ such that $\iota' \geq \iota$ and $\mu_{\iota'}(A) \notin U$.

Let $(U_n)_{n\in\mathbb{N}}$ be a sequence of 0-neighbourhoods in G such that $2U_0 \subset U$ and $2U_{n+1} \subset U_n$ for any $n \in \mathbb{N}$. We set $\mathfrak{A}_{-1} := \mathfrak{R}$ if $\mathfrak{R}$ is a quasi-σ-ring or $\mathfrak{A}_{-1}$ a set of $\mathfrak{G}$ such that $\bigcup_{B\in\mathfrak{A}_{-1}} B$ is contained in a set of $\mathfrak{R}$ if $\mathfrak{R}$ is a quasi-δ-ring. We may construct inductively an increasing sequence $(\iota_n)_{n\in\mathbb{N}}$ in I, a decreasing sequence $(\mathfrak{A}_n)_{n\in\mathbb{N}}$ in $\mathfrak{G}$, and a sequence $(A_n)_{n\in\mathbb{N}}$ in $\mathfrak{R}$ such that for any $n \in \mathbb{N}$:

a) $\iota_n \geq \lambda, \quad \mathfrak{A}_n \subset \mathfrak{A}_{-1}$,

b) $A_n \in \mathfrak{A}_{n-1}$,

c) $\mu_{\iota_n}(A_n) \notin U$,

d) $\bigcup_{m=0}^{n} \{\mu_{\iota_m}(A) \mid A \in \mathfrak{A}_n\} \subset U_{n+1}$.

We set $p(-1) := 0$ and construct inductively a strictly increasing map $p : \mathbb{N} \longrightarrow \mathbb{N}$ and a disjoint sequence $(B_n)_{n \in \mathbb{N}}$ in $\mathfrak{R}$ such that we have for any $n \in \mathbb{N}$:

1) $\mu_{\iota_{p(n-1)}}(B_n) \notin U_0$,
2) $m \in \mathbb{N},\ m \geq p(n) \Rightarrow \mu_{\iota_m}(A_m \cap B_n) \in U_{n+2}$.

Let $n \in \mathbb{N}$ and assume the sequences were constructed up to $n-1$. By 2)

$$\mu_{\iota_{p(n-1)}}(A_{p(n-1)} \cap B_q) \in U_{q+2}$$

for any $q \in \mathbb{N}$ with $q \leq n-1$ so

$$\mu_{\iota_{p(n-1)}}(A_{p(n-1)} \cap (\bigcup_{q=0}^{n-1} B_q)) =$$
$$= \sum_{q=0}^{n-1} \mu_{\iota_{p(n-1)}}(A_{p(n-1)} \cap B_q) \in \sum_{q=0}^{n-1} U_{q+2} \subset U_1 .$$

From c)

$$\mu_{\iota_{p(n-1)}}(A_{p(n-1)} \setminus \bigcup_{q=0}^{n-1} B_q) \notin U_0 + U_1 .$$

$(\mu_{\iota_{p(n-1)+m}})_{m \in \mathbb{N}}$ is a Θ_3-sequence in $\mathscr{E}(\mathfrak{R}, G; \mathfrak{K})$ (a Θ_4-sequence in $\mathscr{M}(\mathfrak{R}, G; \mathfrak{K})$) and

$$(A_{p(n-1)+m} \setminus \bigcup_{q=0}^{n-1} B_q)_{m \in \mathbb{N}}$$

is a sequence in $\mathfrak{R}$ such that by b) and d)

$$\{\mu_{\iota_{p(n-1)}}(A) \mid A \in \mathfrak{R},\ A \subset A_{p(n-1)+m+1} \setminus \bigcup_{q=0}^{n-1} B_q\} \subset U_{m+1}$$

for any $m \in \mathbb{N}$. By Proposition 4.3.1 there exist $p(n) \in \mathbb{N}$ and $B_n \in \mathfrak{R}$ such that $p(n) > p(n-1)$,

$$B_n \subset A_{p(n-1)} \setminus \bigcup_{q=0}^{n-1} B_q ,$$

$$\mu_{\iota_{p(n-1)}}(B_n) \notin U_0 ,$$

and

$$\mu_{\iota_m}(A_m \cap B_n) \in U_{n+2}$$

for any $m \in \mathbb{N}$ with $m \geq p(n)$. This finishes the inductive construction.

$(B_n)_{n \in \mathbb{N}}$ is a disjoint sequence. The hypotheses imply that $(B_n)_{n \in \mathbb{N}}$ is a $\Theta(\mathfrak{R})$-sequence and 1) contradicts Proposition 4.2.6.

Let $\mathfrak{F}$ be arbitrary now and Φ be the set of Φ_3-filters $\mathfrak{H}$ on $\mathscr{E}(\mathfrak{R}, G; \mathfrak{K})$ such that

$$\{\mu \in \mathscr{E}(\mathfrak{R}, G; \mathfrak{K}) \mid \mu(\mathfrak{G}) \text{ converges to } 0\} \in \mathfrak{H} .$$

By the above proof there exists a family $(\mathscr{M}_{\mathfrak{H}}, \mathfrak{A}_{\mathfrak{H}})_{\mathfrak{H}\in\Phi}$ in $\mathscr{E}(\mathfrak{R}, G; \mathfrak{K}) \times \mathfrak{G}$ such that $\mathscr{M}_{\mathfrak{H}} \in \mathfrak{H}$ and $\mu(A) \in U$ for any $\mathfrak{H} \in \Phi$, any $\mu \in \bar{\mathscr{M}}_{\mathfrak{H}}$ and any $A \in \mathfrak{A}_{\mathfrak{H}}$. Let Ψ be a finite subset of Φ such that

$$\mathscr{M} := \bigcup_{\mathfrak{H}\in\Psi} \mathscr{M}_{\mathfrak{H}} \in \mathfrak{F}.$$

There exists $\mathfrak{A} \in \mathfrak{G}$ such that $\mathfrak{A} \subset \bigcap_{\mathfrak{H}\in\Psi} \mathfrak{A}_{\mathfrak{H}}$. Then $\mu(A) \in U$ for any $\mu \in \bar{\mathscr{M}}$ and for any $A \in \mathfrak{A}$. A similar proof holds for $\mathscr{M}$ instead of $\mathscr{E}$. □

Corollary 4.3.3. *Let $\mathfrak{R}$ be a quasi-σ-ring and $\mathfrak{G}$ be a filter base on $\mathfrak{R}$ such that $A \cap B \in \mathfrak{A}$ for any $\mathfrak{A} \in \mathfrak{G}$ and any $(A, B) \in \mathfrak{A} \times \mathfrak{R}$. If G is Hausdorff, then*

$$\{\mu \in \mathscr{E}(\mathfrak{R}, G; \mathfrak{K}) \mid \mathit{lim}\, \mu(\mathfrak{G}) = 0\},$$

is a Φ_3-closed set of $G^{\mathfrak{R}}$ and

$$\{\mu \in \mathscr{M}(\mathfrak{R}, G; \mathfrak{K}) \mid \lim \mu(\mathfrak{G}) = 0\}$$

is a Φ_4-closed sets of $G^{\mathfrak{R}}$. If there exists $(A, \mathfrak{A}) \in \mathfrak{R} \times \mathfrak{G}$ such that $\bigcup_{B\in\mathfrak{A}} B \subset A$ then we may replace the hypothesis "R is a quasi-σ-ring" by the weaker one "R is a quasi-δ-ring".

We set

$$\mathscr{F} := \{\mu \in \mathscr{E}(\mathfrak{R}, G; \mathfrak{K}) \quad (\mu \in \mathscr{M}(\mathfrak{R}, G; \mathfrak{K})) \mid \lim \mu(\mathfrak{G}) = 0\}.$$

Let $v \in G^{\mathfrak{R}}$, φ be the inclusion map $\mathscr{F} \to G^{\mathfrak{R}}$ and $\mathfrak{F}$ be a Φ_3-filter (a Φ_4-filter) on $\mathscr{F}$ such that $\varphi(\mathfrak{F})$ converges to v. By Theorem 4.2.14 a) $v \in \mathscr{E}(\mathfrak{R}, G; \mathfrak{K})$ $(v \in \mathscr{M}(\mathfrak{R}, G; \mathfrak{K}))$. Let U be a closed 0-neighbourhood in G. By Theorem 4.3.2 there exists $(\mathscr{M}, \mathfrak{A}) \in \mathfrak{F} \times \mathfrak{G}$ such that $\mu(A) \in U$ for any $\mu \in \bar{\mathscr{M}}$ and any $A \in \mathfrak{A}$. We get $v(\mathfrak{A}) \subset U$. Hence, $v(\mathfrak{G})$ converges to 0, $v \in \mathscr{F}$, and $\mathscr{F}$ is a Φ_3-closed (a Φ_4-closed) set of $G^{\mathfrak{R}}$. □

Remarks. 1) Let $\mathfrak{R}$ be a σ-ring, μ be an increasing map of $\mathfrak{R}$ into $\overline{\mathbb{R}}_+$ (e.g. a positive measure on $\mathfrak{R}$) and $(\mu_n)_{n\in\mathbb{N}}$ be a sequence of exhaustive additive maps of $\mathfrak{R}$ into a topological additive group G such that $(\mu_n(A))_{n\in\mathbb{N}}$ converges for any $A \in \mathfrak{R}$ and μ_n is absolutely continuous with respect to μ (i.e. $\mu(A) \to 0 \Rightarrow \mu_n(A) \to 0$) for any $n \in \mathbb{N}$. By the above results, (Theorem 4.3.2 and Corollary 4.3.3) $(\mu_n)_{n\in\mathbb{N}}$ is uniformly absolutely continuous with respect to μ and the limit function is also absolutely continuous with respect to μ. This result was proved by G. Vitali (1907) ([1] Teorema page 147) and H. Hahn (1922) ([1] XXI) for $\mathfrak{R}$ the σ-ring of Borel sets of a bounded closed interval of $\mathbb{R}$, for μ the Lebesgue measure on $\mathfrak{R}$, and for $(\mu_n)_{n\in\mathbb{N}}$ a sequence of real valued measures on $\mathfrak{R}$. It was extended by S. Saks (1933) ([1] Theorem 5) to an arbitrary σ-ring $\mathfrak{R}$, to μ a positive measure on $\mathfrak{R}$, and to $(\mu_n)_{n\in\mathbb{N}}$ a sequence of real valued measures on $\mathfrak{R}$ (i.e. $G = \mathbb{R}$). N. Dunford and J.T. Schwarz (1957) ([1] Theorem III 7.2) extended this case to G a locally convex space. T. Andô proved the above result (1961) ([1] Main Theorem) for $(\mu_n)_{n\in\mathbb{N}}$ a sequence of additive exhaustive maps of $\mathfrak{R}$ into $\mathbb{R}$, the filter base $\mathfrak{G}$ being generated by a band of the vector lattice of the additive exhaustive maps of $\mathfrak{R}$ into $\mathbb{R}$. J.K. Brooks and R.S. Jewett proved the same result (1970) ([1] Theorem 3) for $(\mu_n)_{n\in\mathbb{N}}$ a sequence of additive exhaustive maps of $\mathfrak{R}$ into a locally convex space G and μ an additive map of $\mathfrak{R}$ into $\overline{\mathbb{R}}_+$, (this paper suggested the above proof). It was extended to an arbitrary G and a more general filter on $\mathfrak{R}$ than the one generated by μ by L. Drewnowski (1972) ([2] Theorem VHS) and by I. Labuda (1972) ([1] Proposition 4.4).

Another generalization of the Vitali-Hahn-Saks Theorem, which points in another direction, was given by N.S. Gusel'nikov (1976) ([1] Theorem 3 of § 2).

2) The above proof uses only Proposition 4.2.6 so we can replace the Θ_3-sequences or the Θ_4-sequences with equi-exhaustive sequences. L. Drewnowski mentioned this remark (1972) ([1] Corollary 6.2) as did J. Brooks (1973) ([1] Theorem 1).

Proposition 4.3.4. *Let $\mathfrak{R}$ be a quasi-δ-ring, $A\in\mathfrak{R}$, Δ be the set of countable subsets $\mathfrak{B}$ of $\mathfrak{R}$ such that*

1) $B', B''\in\mathfrak{B},\ B'\neq B'' \Rightarrow B'\cap B''=\emptyset$,

2) $\bigcup_{B\in\mathfrak{B}} B = A$,

$\leq$ be the upper directed order relation on Δ defined by

$$\mathfrak{B}\leq\mathfrak{C} :\Leftrightarrow (C\in\mathfrak{C} \Rightarrow \exists B\in\mathfrak{B},\ C\subset B),$$

and $\mathfrak{F}$ be the section filter of Δ. Further, let G be a sequentially complete Hausdorff topological additive group and $\mu\in\mathscr{E}(\mathfrak{R}, G)$. Then:

a) $(\mu(B))_{B\in\mathfrak{B}}$ *is summable for any* $\mathfrak{B}\in\Delta$;

b) $(\sum_{B\in\mathfrak{B}_n}\mu(B))_{n\in\mathbb{N}}$ *is a Cauchy sequence for any increasing sequence* $(\mathfrak{B}_n)_{n\in\mathbb{N}}$ *in* Δ;

c) $\mu_0(\mathfrak{F})$ *is a Cauchy Φ_1-filter, where μ_0 denotes the map*

$$\Delta\to G,\quad \mathfrak{B}\mapsto\sum_{B\in\mathfrak{B}}\mu(B)$$

a) follows from Proposition 4.1.13 a).

b) Assume the contrary. Then there exist a 0-neighbourhood U in G and an increasing sequence $(\mathfrak{B}_n)_{n\in\mathbb{N}}$ in Δ such that

$$\sum_{B\in\mathfrak{B}_n}\mu(B)-\sum_{B\in\mathfrak{B}_{n+1}}\mu(B)\notin U$$

for any $n\in\mathbb{N}$. Let $(U_n)_{n\in\mathbb{N}}$ be a sequence of symmetric 0-neighbourhoods in G such that $5U_0\subset U$ and $2U_{n+1}\subset U_n$ for any $n\in\mathbb{N}$. By Proposition 4.1.13 b) there exists, for each $n\in\mathbb{N}$, a finite subset $\mathfrak{C}_n$ of $\mathfrak{B}_n$ such that

$$\sum_{B\in\mathfrak{B}}\mu(B)\in U_n$$

for any $m\in\mathbb{N}$ and any

$$\mathfrak{B} \subset \{B \in \mathfrak{B}_{m+n} \mid B \subset \bigcup_{C \in \mathfrak{B}_n \setminus \mathfrak{C}_n} C\}.$$

We set

$$\mathfrak{D}_n := \{B \in \mathfrak{C}_n \mid B \cap (\bigcup_{m=0}^{n-1} (\bigcup_{C \in \mathfrak{B}_m \setminus \mathfrak{C}_m} C)) = \emptyset\},$$

$$A_n := \bigcup_{B \in \mathfrak{D}_n} B \setminus \bigcup_{B \in \mathfrak{D}_{n+1}} B$$

for any $n \in \mathbb{N}$.

Let $n \in \mathbb{N}$. We set

$$\mathfrak{E}_m := \{B \in \mathfrak{C}_n \setminus \mathfrak{D}_n \mid B \subset \bigcup_{C \in \mathfrak{B}_m \setminus \mathfrak{C}_m} C\}, \quad \mathfrak{F}_m := \mathfrak{E}_m \setminus \bigcup_{p=0}^{m-1} \mathfrak{E}_p$$

for any $m \in \mathbb{N}$, $m < n$. Then

$$\sum_{B \in \mathfrak{B}_n} \mu(B) - \mu(\bigcup_{B \in \mathfrak{D}_n} B) = \sum_{B \in \mathfrak{B}_n \setminus \mathfrak{D}_n} \mu(B) =$$

$$= \sum_{B \in \mathfrak{B}_n \setminus \mathfrak{C}_n} \mu(B) + \sum_{m=0}^{n-1} \sum_{B \in \mathfrak{F}_m} \mu(B) \in U_n + \sum_{m=0}^{n-1} U_m \subset 2U_0.$$

Let $B \in \mathfrak{D}_{n+1}$. For any $m \in \mathbb{N}$, $m \leq n$, there exists $B_m \in \mathfrak{B}_m$ with $B \subset B_m$. We have $B_m \in \mathfrak{C}_m$ and $B_n \subset B_m$ for any $m \in \mathbb{N}$ with $m \leq n$. It follows that $B_n \in \mathfrak{D}_n$. Hence,

$$\bigcup_{B \in \mathfrak{D}_{n+1}} B \subset \bigcup_{B \in \mathfrak{D}_n} B.$$

We deduce that

$$\sum_{B \in \mathfrak{B}_n} \mu(B) - \sum_{B \in \mathfrak{B}_{n+1}} \mu(B) =$$

$$= (\sum_{B \in \mathfrak{B}_n} \mu(B) - \mu(\bigcup_{B \in \mathfrak{D}_n} B)) + \mu(A_n) + (\mu(\bigcup_{B \in \mathfrak{D}_{n+1}} B) - \sum_{B \in \mathfrak{B}_{n+1}} \mu(B)) \in$$

$$\in \mu(A_n) + 4U_0$$

so $\mu(A_n) \notin U_0$.

$(A_n)_{n \in \mathbb{N}}$ is a disjoint sequence in $\mathfrak{R}$ and therefore a $\Theta(\mathfrak{R})$-sequence (Proposition 4.1.5 a $\Rightarrow$ c). By Proposition 4.1.9 $(\mu(A_n))_{n \in \mathbb{N}}$ converges to 0, which contradicts the above relation.

c) follows immediately from b) and Proposition 1.5.17. □

Remark. a) was proved by C. E. Rickart (1943) ([1] Theorem 2.3.).

Theorem 4.3.5. *Let $\mathfrak{R}$ be a quasi-δ-ring and let G be Hausdorff, sequentially complete and Φ_1-compact (e.g. complete). For any $A \in \mathfrak{R}$, let $\Delta(A)$ be the set of countable sets of pairwise disjoint sets of $\mathfrak{R}$ the union of which is A, let $\leq$ be the upper directed order relation on $\Delta(A)$ defined by*

$$\mathfrak{B} \leq \mathfrak{C} :\Leftrightarrow (C \in \mathfrak{C} \Rightarrow \exists B \in \mathfrak{B},\ C \subset B),$$

and $\mathfrak{F}_A$ be the section filter of $\Delta(A)$. Then:

a) $\mu_A(\mathfrak{F}_A)$ converges for any $\mu \in \mathscr{E}(\mathfrak{R}, G; \mathfrak{K})$ and any $A \in \mathfrak{R}$, where μ_A denotes the map $\Delta(A) \to G$, $\mathfrak{B} \mapsto \sum_{B \in \mathfrak{B}} \mu(B)$; we denote its limit by $\bar{\mu}(A)$;

b) $\bar{\mu} \in \mathscr{M}(\mathfrak{R}, G; \mathfrak{K})$ for any $\mu \in \mathscr{E}(\mathfrak{R}, G; \mathfrak{K})$;

c) let $A \in \mathfrak{R}$ and $\mu \in \mathscr{E}(\mathfrak{R}, G; \mathfrak{K})$; if any countable subset of A belongs to $\mathfrak{R}$, then $(\mu(\{x\}))_{x \in A}$ is summable; if, in addition, any $K \in \mathfrak{K}$ is countable if it is contained in A, then

$$\bar{\mu}(A) = \sum_{x \in A} \mu(\{x\})\,;$$

d) if any set of $\mathfrak{K}$ is countable and if any countable subset of a set of $\mathfrak{R}$ belongs to $\mathfrak{R}$, then the map

$$\mathscr{E}(\mathfrak{R}, G; \mathfrak{K}) \longrightarrow \mathscr{M}(\mathfrak{R}, G; \mathfrak{K})_{\Psi(\mathfrak{R})}, \quad \mu \mapsto \bar{\mu}$$

is a uniformly Φ_3-continuous group homomorphism (homomorphism of R-modules if R is a ring and G is an R-module).

We remark first that by Corollary 1.5.11 G is Φ_1-compact if it is complete.

a) follows from Proposition 4.3.4 c).

b) Let $\mu \in \mathscr{E}(\mathfrak{R}, G; \mathfrak{K})$, $(A_n)_{n \in \mathbb{N}} \in \Gamma(\mathfrak{R})$ and $(U_n)_{n \in \mathbb{N}}$ be a sequence of 0-neighbourhoods in G such that $2U_{n+1} \subset U_n$ for any $n \in \mathbb{N}$. For any $n \in \mathbb{N}$ there exists $\mathfrak{B}_n \in \Delta(A_n)$ such that

$$\sum_{B \in \mathfrak{B}} \mu(B) - \bar{\mu}(A_n) \in U_{n+2}$$

for any $\mathfrak{B} \in \Delta(A_n)$ with $\mathfrak{B}_n \leq \mathfrak{B}$. We have

$$\bigcup_{n \in \mathbb{N}} \mathfrak{B}_n \in \Delta(\bigcup_{n \in \mathbb{N}} A_n).$$

Let $\mathfrak{B} \in \Delta(\bigcup_{n \in \mathbb{N}} A_n)$ such that $\bigcup_{n \in \mathbb{N}} \mathfrak{B}_n \leq \mathfrak{B}$ and such that $\sum_{B \in \mathfrak{B}} \mu(B) - \bar{\mu}(\bigcup_{n \in \mathbb{N}} A_n) \in U_0$.

We set

$$\mathfrak{C}_n := \{B \in \mathfrak{B} \mid B \subset A_n\}$$

for any $n \in \mathbb{N}$. Then $\mathfrak{C}_n \in \Delta(A_n)$, $\mathfrak{B}_n \leq \mathfrak{C}_n$ so

$$\sum_{B \in \mathfrak{C}_n} \mu(B) - \bar{\mu}(A_n) \in U_{n+2}$$

for any $n \in \mathbb{N}$. There exists $p \in \mathbb{N}$ such that

$$\sum_{B \in \mathfrak{B}} \mu(B) - \sum_{n=0}^{m} \sum_{B \in \mathfrak{C}_n} \mu(B) \in U_1$$

for any $m \in \mathbb{N}$, $m \geq p$. We get

$$\sum_{B \in \mathfrak{B}} \mu(B) - \sum_{n=0}^{m} \bar{\mu}(A_n) =$$

$$= (\sum_{B \in \mathfrak{B}} \mu(B) - \sum_{n=0}^{m} \sum_{B \in \mathfrak{C}_n} \mu(B)) + \sum_{n=0}^{m} (\sum_{B \in \mathfrak{C}_n} \mu(B) - \bar{\mu}(A_n)) \in$$

$$\in U_1 + \sum_{n=0}^{m} U_{n+2} \subset U_0,$$

hence,

$$\bar{\mu}(\bigcup_{n \in \mathbb{N}} A_n) - \sum_{n=0}^{m} \bar{\mu}(A_n) \in 2\bar{U}_0$$

for any $m \in \mathbb{N}$, $m \geq p$. Since U_0 is arbitrary we get

$$\bar{\mu}(\bigcup_{n \in \mathbb{N}} A_n) = \lim_{n \to \infty} \sum_{m=0}^{n} \bar{\mu}(A_m).$$

$\bar{\mu}$ is therefore a measure.

Let $A \in \mathfrak{R}$, U be a 0-neighbourhood in G and V be a symmetric 0-neighbourhood in G such that $4V \subset U$. There exist $K \in \mathfrak{K}$ such that $K \subset A$ and

$$\mu(B) - \mu(A) \in V$$

for any $B \in \mathfrak{R}$ with $K \subset B \subset A$. We want to show that

$$\bar{\mu}(B) - \bar{\mu}(A) \in U$$

for any $B \in \mathfrak{R}$ with $K \subset B \subset A$. Let $B \in \mathfrak{R}$ such that $K \subset B \subset A$. There exist $\mathfrak{B} \in \Delta(A)$ and $\mathfrak{B}' \in \Delta(B)$ such that

$$\sum_{C \in \mathfrak{C}} \mu(C) - \bar{\mu}(A) \in V, \quad \sum_{C \in \mathfrak{C}'} \mu(C) - \bar{\mu}(B) \in V$$

for any $\mathfrak{C} \in \Delta(A)$ and $\mathfrak{C}' \in \Delta(B)$ with $\mathfrak{B} \leq \mathfrak{C}$, $B' \leq \mathfrak{C}'$. We have $\mathfrak{B}' \cup \{A \backslash B\} \in \Delta(A)$ so there exists $\mathfrak{C} \in \Delta(A)$ such that $\mathfrak{B} \leq \mathfrak{C}$ and $\mathfrak{B}' \cup \{A \backslash B\} \leq \mathfrak{C}$. We set

$$\mathfrak{C}' := \{C \in \mathfrak{C} \mid C \subset B\}.$$

Then $\mathfrak{C}' \in \Delta(B)$ and $\mathfrak{B}' \leq \mathfrak{C}'$. Let $\mathfrak{C}''$ be a finite subset of $\mathfrak{C} \backslash \mathfrak{C}'$. Then $A \backslash (\bigcup_{C \in \mathfrak{C}''} C) \in \mathfrak{R}$ and

$$K \subset A \backslash \bigcup_{C \in \mathfrak{C}''} C \subset A$$

so

$$\sum_{C \in \mathfrak{C}''} \mu(C) = \mu(\bigcup_{C \in \mathfrak{C}''} C) = \mu(A) - \mu(A \backslash \bigcup_{C \in \mathfrak{C}''} C) \in V.$$

Since $\mathfrak{C}''$ is arbitrary,

$$\sum_{C \in \mathfrak{C} \setminus \mathfrak{C}'} \mu(C) \in \bar{V} \subset 2V .$$

Further

$$\bar{\mu}(B) - \bar{\mu}(A) = (\bar{\mu}(B) - \sum_{C \in \mathfrak{C}'} \mu(C)) - \sum_{C \in \mathfrak{C} \setminus \mathfrak{C}'} \mu(C) + (\sum_{C \in \mathfrak{C}} \mu(C) - \bar{\mu}(A)) \in$$

$$\in V + 2V + V \subset U .$$

Since U and A are arbitrary, μ is $\mathfrak{K}$-regular.

c) $(\mu(\{x\}))_{x \in B}$ is summable for any countable subset B of A so by Proposition 3.3.3 $(\mu(\{x\}))_{x \in A}$ is summable. Assume now that any set of $\mathfrak{K}$ contained in A is countable. Then $\{\{x\} \mid x \in K\}$ is the greatest element of $\Delta(K)$ and

$$\bar{\mu}(K) = \sum_{x \in K} \mu(\{x\})$$

for any $K \in \mathfrak{K}$ with $K \subset A$. By b) $\bar{\mu} \in \mathscr{M}(\mathfrak{R}, G; \mathfrak{K})$. Let U be an arbitrary 0-neighbourhood in G and V be a symmetric closed 0-neighbourhood in G such that $2V \subset U$. There exists $K \in \mathfrak{K}$ such that $K \subset A$ and

$$\bar{\mu}(B) - \bar{\mu}(A) \in V$$

for any $B \in \mathfrak{R}$ with $K \subset B \subset A$. There exists a finite subset L of A such that

$$\sum_{x \in B} \mu(\{x\}) - \sum_{x \in A} \mu(\{x\}) \in V$$

for any finite set B with $L \subset B \subset A$. There exists $K_0 \in \mathfrak{K}$ such that $K_0 \subset L$ and $\mu(\{x\}) = 0$ for any $x \in L \setminus K_0$. Let B be a finite subset of $K \cup K_0$ containing K_0. Then

$$L \subset B \cup (L \setminus K_0) \subset A$$

so

$$\sum_{x \in B} \mu(\{x\}) = \sum_{x \in B \cup (L \setminus K_0)} \mu(\{x\}) \in \sum_{x \in A} \mu(\{x\}) + V .$$

Since V is closed

$$\sum_{x \in K \cup K_0} \mu(\{x\}) \in \sum_{x \in A} \mu(\{x\}) + V .$$

and, since $K \cup K_0 \in \mathfrak{K}$,

$$\sum_{x \in K \cup K_0} \mu(\{x\}) = \bar{\mu}(K \cup K_0) \in \bar{\mu}(A) + V .$$

Hence,

$$\sum_{x \in A} \mu(\{x\}) - \bar{\mu}(A) \in 2V \subset U .$$

U being arbitrary

$$\bar{\mu}(A) = \sum_{x \in A} \mu(\{x\}) .$$

d) The map

$$\mathscr{E}(\mathfrak{R}, G; \mathfrak{K}) \longrightarrow \mathscr{M}(\mathfrak{R}, G; \mathfrak{K}), \quad \mu \mapsto \bar{\mu}$$

is obviously a group homomorphism (a homomorphism of R-modules if R is a ring and G is an R-module).

We now want to show that the map

$$\mathscr{E}(\mathfrak{R}, G; \mathfrak{K})_{\Psi_f(\mathfrak{R})} \longrightarrow \mathscr{M}(\mathfrak{R}, G; \mathfrak{K}), \quad \mu \mapsto \bar{\mu}$$

is uniformly continuous. Let $A \in \mathfrak{R}$ and U be a closed 0-neighbourhood in G. Let $\mathfrak{A}$ be the set of finite subsets of A ordered by the inclusion relation and $\mathfrak{F}$ be the filter on $\mathfrak{R}$ generated by the section filter of $\mathfrak{A}$. Let $(A_n)_{n \in \mathbb{N}}$ be an increasing sequence in $\mathfrak{A}$. Then there exists $(B_n)_{n \in \mathbb{N}} \in \Gamma(\mathfrak{R})$ such that each $(B_n)_{n \in \mathbb{N}}$ possesses at most one point and $\bigcup_{n \in \mathbb{N}} A_n = \bigcup_{n \in \mathbb{N}} B_n$. We deduce that $(A_n)_{n \in \mathbb{N}} \in \Theta_f(\mathfrak{R})$. Hence, $\mathfrak{F} \in \Psi_f(\mathfrak{R})$. Let $\mu \in \mathscr{E}(\mathfrak{R}, G; \mathfrak{K})$ such that

$$\{B \in \mathfrak{R} \mid \mu(B) \in U\} \in \mathfrak{F}.$$

Then there exists a finite subset B of A such that

$$\sum_{x \in C} \mu(\{x\}) \in U$$

for any finite subset C of A containing B. Since U is closed we get by c)

$$\bar{\mu}(A) = \sum_{x \in A} \mu(\{x\}) \in U.$$

This shows that the above map is uniformly continuous.

By Proposition 4.2.3 the identity maps

$$\mathscr{E}(\mathfrak{R}, G; \mathfrak{K}) \longrightarrow \mathscr{E}(\mathfrak{R}, G; \mathfrak{K})_{\Psi_f(\mathfrak{R})}$$
$$\mathscr{M}(\mathfrak{R}, G; \mathfrak{K}) \longrightarrow \mathscr{M}(\mathfrak{R}, G; \mathfrak{K})_{\Psi(\mathfrak{R})}$$

are uniformly Φ_3-continuous. By the above result and Corollary 1.8.5 the map

$$\mathscr{E}(\mathfrak{R}, G; \mathfrak{K}) \longrightarrow \mathscr{M}(\mathfrak{R}, G; \mathfrak{K})_{\Psi(\mathfrak{R})}$$

is uniformly Φ_3-continuous. □

Corollary 4.3.6. *Let G be Hausdorff, sequentially complete and Φ_1-compact (e.g. complete) and I be a set. Then:*

a) the map

$$\mathscr{E}(\mathfrak{P}(I), G)_{\mathfrak{P}_c(I)} \longrightarrow \mathscr{E}(\mathfrak{P}_c(I), G), \quad \mu \mapsto \mu \mid \mathfrak{P}_c(I)$$

is uniformly continuous;

b) $(\mu(\{\iota\}))_{\iota \in I} \in \ell(I, G)$ for any $\mu \in \mathscr{E}(\mathfrak{P}_c(I), G)$;

c) the map

$$\mathscr{E}(\mathfrak{P}_c(I), G) \to \ell_f(I, G), \quad \mu \mapsto (\mu(\{\iota\}))_{\iota \in I}$$

is uniformly Φ_3-continuous.
If I is countable then we may drop the hypothesis "G is Φ_1-compact".

a) is trivial.
b) follows from Theorem 4.3.5 c) and Proposition 3.3.3.
c) The map

$$\mathscr{M}(\mathfrak{P}_c(I), G) \to \ell_c(I, G), \quad \mu \mapsto (\mu(\{\iota\}))_{\iota \in I}$$

(by b) this map is well defined) and the identity map

$$\mathscr{M}(\mathfrak{P}_c(I), G)_{\Psi(\mathfrak{P}_c(I))} \to \mathscr{M}(\mathfrak{P}_c(I), G)$$

are uniformly continuous. Hence, by Theorem 4.3.5 c), d) the map

$$\mathscr{E}(\mathfrak{P}_c(I), G) \to \ell_c(I, G), \quad \mu \mapsto (\mu(\{\iota\}))_{\iota \in I}$$

is uniformly Φ_3-continuous. The identity map

$$\ell_c(I, G) \to \ell_f(I, G),$$

being uniformly Φ_4-continuous (Corollary 3.3.7), the assertion follows from Corollary 1.8.5.
The last assertion can be proved by taking the completion of G. □

Corollary 4.3.7. *(Phillips). Let I be a set and E be a finite dimensional Banach space. We have:*

a) $(\mu(\{\iota\}))_{\iota \in I} \in \ell^1(I, E)$ (i.e. $\sum_{\iota \in I} \|\mu(\{\iota\})\| < \infty$) for any $\mu \in \mathscr{E}(\mathfrak{P}_c(I), E)$;
b) the map

$$\mathscr{E}(\mathfrak{P}_c(I), E) \to \ell^1(I, E), \quad \mu \mapsto (\mu(\{\iota\}))_{\iota \in I}$$

is uniformly Φ_3-continuous;

c) if $\mathscr{M}$ is a Φ_3-set of $\mathscr{E}(\mathfrak{P}_c(I), E)$, then $\bar{\mathscr{M}}$ is compact and the map

$$\bar{\mathscr{M}} \to \ell^1(I, E), \quad \mu \to (\mu(\{\iota\}))_{\iota \in I}$$

is uniformly continuous.

a & b follow from Corollary 4.3.6 b), c) and Proposition 3.5.6 e).
c) By Theorem 4.2.16 a) $\bar{\mathscr{M}}$ is compact. By b) and Theorem 1.8.4 a ⇒ j the map

$$\bar{\mathscr{M}} \to \ell^1(I, E), \quad \mu \to (\mu(\{\iota\}))_{\iota \in I}$$

is uniformly continuous. □

Remark. Assume G Hausdorff and sequentially complete and let $(\mu_n)_{n\in\mathbb{N}}$ be a sequence of additive exhaustive maps of $\mathfrak{P}(\mathbb{N})$ into G such that $(\mu_n(M))_{n\in\mathbb{N}}$ converges to 0 for any $M \subset \mathbb{N}$. By Corollary 4.3.6 c) $(\sum_{m\in M} \mu_n(\{m\}))_{n\in\mathbb{N}}$ converges to 0 uniformly with respect to $M \subset \mathbb{N}$. This result was proved by J.K. Brooks and R.S. Jewett (1970) ([1] Theorem 1) for G a locally convex space. If G is a finite dimensional normed space, then, by Corollary 4.3.7,

$$\lim_{n\to\infty} \sum_{m\in\mathbb{N}} \|\mu_n(\{m\})\| = 0\,;$$

moreover, in this case, any bounded additive map is exhaustive (Proposition 4.1.8). This last result was proved by R.S. Phillips (1940) ([1] Lemma 3.3) for $G = \mathbb{R}$.

Corollary 4.3.8. *Let $\mathfrak{R}$ be a quasi-δ-ring and G', G'' be Hausdorff topological additive groups having the same underlying group such that the identity map $G' \to G''$ belongs to $\mathscr{P}^c(G', G'')$ and G' is sequentially complete. Then*

$$\mathscr{E}(\mathfrak{R}, G'; \mathfrak{K}) \cap \mathscr{M}(\mathfrak{R}, G'') \subset \mathscr{M}(\mathfrak{R}, G'; \mathfrak{K})\,.$$

Let $\mu \in \mathscr{E}(\mathfrak{R}, G'; \mathfrak{K}) \cap \mathscr{M}(\mathfrak{R}, G'')$. For any $A \in \mathfrak{R}$ let $\Delta(A)$ be the set of countable sets of pairwise disjoint sets of $\mathfrak{R}$ the union of which is A, and let μ_A be the map

$$\Delta(A) \to G', \quad \mathfrak{B} \mapsto \sum_{B\in\mathfrak{B}} \mu(B)\,.$$

(Proposition 4.3.4 a)). Since the identity map $G' \to G''$ belongs to $\mathscr{P}^c(G', G'')$

$$\mu_A(\mathfrak{B}) = \sum_{B\in\mathfrak{B}} \mu(B) = \mu(A)$$

for any $A \in \mathfrak{R}$ and any $\mathfrak{B} \in \Delta(A)$. By Theorem 4.3.5 b)

$$\mu \in \mathscr{M}(\mathfrak{R}, G'; \mathfrak{K})\,. \quad \square$$

§ 4.4 Weak topologies on spaces of measures

Throughout this section $\mathfrak{R}$ shall denote a δ-ring, $\mathfrak{K}$ a subset of $\mathfrak{R}$ closed under finite unions, and G a Hausdorff topological additive group.

The aim of this section is to prove some results which will be used in the study of the spaces of measures on topological spaces.

Proposition 4.4.1. *Let (I, μ) be a net in $\mathscr{M}(\mathfrak{R}, G; \mathfrak{K})$ such that $\mu(\mathfrak{F})$ is a Cauchy filter on $\mathscr{M}(\mathfrak{R}, G; \mathfrak{K})_{\mathfrak{K}}$ but not a Cauchy filter on $\mathscr{M}(\mathfrak{R}, G; \mathfrak{K})$, where $\mathfrak{F}$ denotes the* *section filter of I. Then there exist a 0-neighbourhood U in G, an increasing sequence* $(\iota_n)_{n\in\mathbb{N}}$ *in I, and a disjoint sequence* $(K_n)_{n\in\mathbb{N}}$ *in* $\mathfrak{K}$ *such that* $\bigcup_{n\in\mathbb{N}} K_n \in \mathfrak{R}$ *and*

$$\mu_{\iota_n}(K_m) - \mu_{\iota_m}(K_m) \notin U$$

for any $m, n \in \mathbb{N}$ *with* $m < n$.

Since $\mu(\mathfrak{F})$ is not a Cauchy filter on $\mathscr{M}(\mathfrak{R}, G; \mathfrak{K})$ there exist $A \in \mathfrak{R}$ and a 0-neighbourhood V in G such that for any $J \in \mathfrak{F}$ there exist $\iota', \iota'' \in J$ with

$$\mu_{\iota'}(A) - \mu_{\iota''}(A) \notin V.$$

Let U be a 0-neighbourhood in G such that $U = -U$ and

$$6U \subset V.$$

We construct inductively an increasing sequence $(\iota_n)_{n\in\mathbb{N}}$ in I, a decreasing sequence $(I_n)_{n\in\mathbb{N}}$ in $\mathfrak{F}$, and a disjoint sequence $(K_n)_{n\in\mathbb{N}}$ in $\mathfrak{K}$ such that for any $n \in \mathbb{N}$:

a) $K_n \subset A$;

b) $\iota_n \in I_{n-1}$ $(I_{-1} := I)$;

c) $\iota \in I_n \Rightarrow \iota \geq \iota_n,\ \mu_\iota(K_n) - \mu_{\iota_n}(K_n) \notin U$;

d) $\iota', \iota'' \in I_n \Rightarrow \mu_{\iota'}(\bigcup_{m=0}^{n} K_m) - \mu_{\iota''}(\bigcup_{m=0}^{n} K_m) \in U$.

Let $n \in \mathbb{N}$ and assume the sequences were constructed up to $n-1$. By the above remark there exist $\iota', \iota'' \in I_{n-1}$ with

$$\mu_{\iota'}(A) - \mu_{\iota''}(A) \notin V.$$

By a) and d)

$$\mu_{\iota'}(A \setminus \bigcup_{m=0}^{n-1} K_m) - \mu_{\iota''}(A \setminus \bigcup_{m=0}^{n-1} K_m) \notin 5U.$$

Since $\mu_{\iota'}, \mu_{\iota''}$ are $\mathfrak{K}$-regular there exists $K_n \in \mathfrak{K}$ with

$$K_n \subset A \setminus \bigcup_{m=0}^{n-1} K_m,$$

$$\mu_{\iota'}(K_n) - \mu_{\iota'}(A \setminus \bigcup_{m=0}^{n-1} K_m) \in U,$$

$$\mu_{\iota''}(K_n) - \mu_{\iota''}(A \setminus \bigcup_{m=0}^{n-1} K_m) \in U.$$

Since $\mu(\mathfrak{F})$ is a Cauchy filter on $\mathscr{M}(\mathfrak{R}, G; \mathfrak{K})_{\mathfrak{K}}$ there exists $I_n \in \mathfrak{F}$ such that

$$I_n \subset \{\iota \in I_{n-1} \mid \iota \geq \iota', \iota \geq \iota''\}$$

and

$$\mu_\iota(K_n) - \mu_\lambda(K_n) \in U\,, \quad \mu_\iota(\bigcup_{m=0}^{n} K_m) - \mu_\lambda(\bigcup_{m=0}^{n} K_m) \in U$$

for any $\iota, \lambda \in I_n$. We deduce that either

$$\mu_{\iota'}(K_n) - \mu_\iota(K_n) \notin U$$

for any $\iota \in I_n$ or

$$\mu_{\iota''}(K_n) - \mu_\iota(K_n) \notin U$$

for any $\iota \in I_n$. In the first case, we set $\iota_n := \iota'$ and in the second case, $\iota_n := \iota''$.
U and the sequences $(\iota_n)_{n\in\mathbb{N}}$ and $(K_n)_{n\in\mathbb{N}}$ fulfill the conditions of the proposition. □

Proposition 4.4.2. *Let $i \in \{1, 2, 3, 4, 5\}$ and $\mathfrak{A}$ be a subset of $\mathfrak{R}$ containing $\mathfrak{K}$. If the identity map*

$$\mathcal{M}(\mathfrak{R}, G; \mathfrak{K})_{\mathfrak{A}} \to \mathcal{M}(\mathfrak{R}, G; \mathfrak{K})$$

is not uniformly Φ_i-continuous, then there exist a 0-neighbourhood U in G, a Θ_i-sequence $(\mu_n)_{n\in\mathbb{N}}$ in $\mathcal{M}(\mathfrak{R}, G; \mathfrak{K})_{\mathfrak{A}}$, and a disjoint sequence $(K_n)_{n\in\mathbb{N}}$ in $\mathfrak{K}$ such that $\bigcup_{n\in\mathbb{N}} K_n \in \mathfrak{R}$ and

$$\mu_n(K_m) - \mu_m(K_m) \notin U$$

for any $m, n \in \mathbb{N}$ with $m < n$.

By Theorem 1.8.4 b $\Rightarrow$ a there exists a Cauchy Φ_i-filter $\mathfrak{F}$ on $\mathcal{M}(\mathfrak{R}, G; \mathfrak{K})_{\mathfrak{A}}$ which is not a Cauchy filter on $\mathcal{M}(\mathfrak{R}, G; \mathfrak{K})$. Let (I, μ) be a Θ_i-net in $\mathcal{M}(\mathfrak{R}, G; \mathfrak{K})_{\mathfrak{A}}$ such that $\mu(\mathfrak{G}) = \mathfrak{F}$, where $\mathfrak{G}$ denotes the section filter of I. By Proposition 4.4.1 there exist a 0-neighbourhood U in G, an increasing sequence $(\iota_n)_{n\in\mathbb{N}}$ in I, and a disjoint sequence $(K_n)_{n\in\mathbb{N}}$ in $\mathfrak{K}$ such that $\bigcup_{n\in\mathbb{N}} K_n \in \mathfrak{R}$ and

$$\mu_{\iota_n}(K_m) - \mu_{\iota_m}(K_m) \notin U$$

for any $m, n \in \mathbb{N}$ with $m < n$. $(\mu_{\iota_n})_{n\in\mathbb{N}}$ is a Θ_i-sequence in $\mathcal{M}(\mathfrak{R}, G; \mathfrak{K})_{\mathfrak{A}}$. □

Proposition 4.4.3. *Let $\mathfrak{A}$ be a subset of $\mathfrak{R}$, φ be a map $\mathfrak{P}(\mathbb{N}) \to \mathfrak{A}$, $i \in \{3, 4\}$ and $(\mu_n)_{n\in\mathbb{N}}$ be a Θ_i-sequence in $\mathcal{M}(\mathfrak{R}, G; \mathfrak{K})_{\mathfrak{A}}$. We assume that one of the following conditions is fulfilled.*

a) $\mu \circ \varphi \in \mathcal{G}(\mathbb{N}, G)$ for any $\mu \in \mathcal{M}(\mathfrak{R}, G; \mathfrak{K})$, (this condition is fulfilled if there exists a disjoint sequence $(A_n)_{n\in\mathbb{N}}$ in $\mathfrak{A}$ such that $\varphi(M) = \bigcup_{n\in M} A_n$ for any $M \subset \mathbb{N}$).

b) $i = 3$ and $\mu_n \circ \varphi \in \mathcal{G}(\mathbb{N}, G)$ for any $n \in \mathbb{N}$.
Then:

$$\lim_{n\to\infty} \mu_n(\varphi(\{n\})) = 0$$

and for any 0-neighbourhood U in G there exist $m, n \in \mathbb{N}$ such that $m < n$ and

$$\mu_m(\varphi(M)) - \mu_n(\varphi(M)) \in U$$

for any $M \subset \mathbb{N}$.

Assume there exists a disjoint sequence $(A_n)_{n \in \mathbb{N}}$ in $\mathfrak{A}$ such that $\varphi(M) = \bigcup_{n \in M} A_n$ for any $M \subset \mathbb{N}$. By Proposition 4.1.14 a $\Leftrightarrow$ b $\mu \circ \varphi \in \mathscr{G}(\mathbb{N}, G)$ for any $\mu \in \mathscr{M}(\mathfrak{R}, G)$. $(\mu_n \circ \varphi)_{n \in \mathbb{N}}$ is a Θ_i-sequence in $\mathscr{G}(\mathbb{N}, G)$. By Proposition 3.2.1 and Theorem 1.8.4 a $\Rightarrow$ e $(\mu_n \circ \varphi)_{n \in \mathbb{N}}$ is a Θ_i-sequence in $\mathscr{G}(\mathbb{N}, G)_{\{\mathfrak{P}(\mathbb{N})\}}$. Hence, there exist $m, n \in \mathbb{N}$ such that $m < n$ and

$$\mu_m(\varphi(M)) - \mu_n(\varphi(M)) \in U$$

for any $M \in \mathfrak{P}(\mathbb{N})$. By Theorem 3.2.5 a)

$$\lim_{n \to \infty} \mu_n(\varphi(\{n\})) = 0. \quad \square$$

Proposition 4.4.4. *If the union of any disjoint sequence in* $\mathfrak{R}$ *belongs to* $\mathfrak{R}$ *if it belongs to* $\mathfrak{R}$, *then the identity map*

$$\mathscr{M}(\mathfrak{R}, G; \mathfrak{R})_{\mathfrak{R}} \rightarrow \mathscr{M}(\mathfrak{R}, G; \mathfrak{R})$$

is uniformly Φ_4*-continuous.*

Assume the contrary. By Proposition 4.4.2 there exist a 0-neighbourhood U in G, a Θ_4-sequence $(\mu_n)_{n \in \mathbb{N}}$ in $\mathscr{M}(\mathfrak{R}, G; \mathfrak{R})_{\mathfrak{R}}$, and a disjoint sequence $(K_n)_{n \in \mathbb{N}}$ in $\mathfrak{R}$ such that $\bigcup_{n \in \mathbb{N}} K_n \in \mathfrak{R}$ and

$$\mu_n(K_m) - \mu_m(K_m) \notin U$$

for any $m, n \in \mathbb{N}$ with $m < n$. By the hypothesis, $\bigcup_{n \in M} K_n \in \mathfrak{R}$ for any $M \subset \mathbb{N}$ and the above relation contradicts Proposition 4.4.3. $\square$

Definition 4.4.5. *We say that a subset* $\mathfrak{T}$ *of* $\mathfrak{R}$ <u>separates</u> $\mathfrak{R}$ *if for any disjoint sets* K', K'' *of* $\mathfrak{R}$ *there exist disjoint sets* T', T'' *of* $\mathfrak{T}$ *such that* $K' \subset T'$, $K'' \subset T''$.

Proposition 4.4.6. *Let* $\mathfrak{L}$ *and* $\mathfrak{T}$ *be subsets of* $\mathfrak{R}$, φ *be a map defined on* $\mathfrak{T}$ *and* Θ *be a set of sequences in* $\mathscr{M}(\mathfrak{R}, G; \mathfrak{R})$ *such that*

a) for any $K \in \mathfrak{R}$ *and for any* T', T'' *of* $\mathfrak{T}$ *with* $K \subset T' \cap T''$ *there exist* $T \in \mathfrak{T}$ *and* $S \in \varphi(T)$ *such that*

$$K \subset T \subset S \subset T' \cap T'' ;$$

b) for any $L \in \mathfrak{L}$ *the set* $\{T \in \mathfrak{T} \mid L \subset T\}$ *is lower directed;*

c) for any $K \in \mathfrak{R}$ *and* $L \in \mathfrak{L}$ *with* $K \cap L = \emptyset$ *there exist disjoint sets* T', T'' *of* $\mathfrak{T}$ *with* $K \subset T', L \subset T''$;

d) $\mathfrak{T}$ *separates* $\mathfrak{R}$;

e) for any 0-neighbourhood U in G, any Θ-sequence $(\mu_n)_{n\in\mathbb{N}}$ in $\mathcal{M}(\mathfrak{R}, G; \mathfrak{K})$ and any sequence $(T_n)_{n\in\mathbb{N}}$ in $\mathfrak{T}$ whose union is contained in a set of $\mathfrak{T}$ and for which there exists a disjoint sequence $(S_n)_{n\in\mathbb{N}}$ of sets such that $T_n \subset S_n \in \varphi(T_n)$ for any $n \in \mathbb{N}$, there exists $n \in \mathbb{N}$ such that $\mu_n(T_n) \in U$.

Then for any $\hat{\Phi}(\Theta)$-filter $\mathfrak{F}$ on $\mathcal{M}(\mathfrak{R}, G; \mathfrak{K})$, any $L \in \mathfrak{L}$, and for any 0-neighbourhood U in G there exist $T \in \mathfrak{T}$ and $\mathcal{M} \in \mathfrak{F}$ such that $L \subset T$ and

$$\mu(A) - \mu(L) \in U$$

for any $\mu \in \mathcal{M}$ and any $A \in \mathfrak{R}$ with $L \subset A \subset T$.

Assume first that $\mathfrak{F}$ is a $\Phi(\Theta)$-filter on $\mathcal{M}(\mathfrak{R}, G; \mathfrak{K})$ and let (I, ν) be a Θ-net in $\mathcal{M}(\mathfrak{R}, G; \mathfrak{K})$ such that $\nu(\mathfrak{G}) = \mathfrak{F}$, where $\mathfrak{G}$ denotes the section filter of I. Assume the assertion does not hold for $\mathfrak{F}$. Then there exist $L \in \mathfrak{L}$ and a 0-neighbourhood U in G such that for any $T \in \mathfrak{T}$ with $L \subset T$ and for any $\iota \in I$ there exist $\lambda \in I$ and $A \in \mathfrak{R}$ such that $\lambda \geq \iota$, $L \subset A \subset T$, and

$$\nu_\lambda(A) - \nu_\lambda(L) \notin U .$$

Since $\emptyset \in \mathfrak{K}$ there exist by c) a $T'_{-1} \in \mathfrak{T}$ containing L. We denote by ι_{-1} an arbitrary element of I. Let V be a 0-neighbourhood in G such that

$$V + V - V \subset U .$$

We construct inductively an increasing sequence $(\iota_n)_{n\in\mathbb{N}}$ in I, a sequence $(S_n)_{n\in\mathbb{N}}$ of sets, and two sequences $(T_n)_{n\in\mathbb{N}}$, $(T'_n)_{n\in\mathbb{N}}$ in $\mathfrak{T}$ such that for any $n \in \mathbb{N}$:

1) $L \subset T'_n \subset T'_{n-1}$;
2) $T_n \subset S_n \subset T'_{n-1} \backslash T'_n$;
3) $S_n \in \varphi(T_n)$;
4) $\nu_{\iota_n}(T_n) \notin V$.

Let $n \in \mathbb{N}$ and assume the sequences were constructed up to $n-1$. By the above hypothesis there exist $\iota_n \in I$ and $A \in \mathfrak{R}$ such that $\iota_n \geq \iota_{n-1}$, $L \subset A \subset T'_{n-1}$ and

$$\nu_{\iota_n}(A) - \nu_{\iota_n}(L) \notin U .$$

Since ν_{ι_n} is $\mathfrak{K}$-regular there exists $K \in \mathfrak{K}$ such that $K \subset A \backslash L$ and

$$\nu_{\iota_n}(A \backslash L) - \nu_{\iota_n}(K) \in V .$$

Since K and L are disjoint there exist by c) disjoint sets T, $T' \in \mathfrak{T}$ with $K \subset T$, $L \subset T'$. By a), d) and Proposition 4.1.47 there exists $T'' \in \mathfrak{T}$ such that $K \subset T''$ and

$$\nu_{\iota_n}(B) - \nu_{\iota_n}(K) \in V$$

for any $B \in \mathfrak{R}$ with $K \subset B \subset T''$. By 1), a), and b) there exist T_n, $T'_n \in \mathfrak{T}$ and $S_n \in \varphi(T_n)$ such that

$$L \subset T'_n \subset T'_{n-1} \cap T', \quad K \subset T_n \subset S_n \subset T \cap T'' \cap T'_{n-1} .$$

We have

$$\nu_{\iota_n}(A \backslash L) = \nu_{\iota_n}(A) - \nu_{\iota_n}(L) \notin U\,,$$
$$\nu_{\iota_n}(T_n) - \nu_{\iota_n}(K) \in V\,.$$

From

$$\nu_{\iota_n}(A \backslash L) = (\nu_{\iota_n}(A \backslash L) - \nu_{\iota_n}(K)) - (\nu_{\iota_n}(T_n) - \nu_{\iota_n}(K)) + \nu_{\iota_n}(T_n)$$

we get $\nu_{\iota_n}(T_n) \notin V$. This proves the existence of the sequences $(\iota_n)_{n\in\mathbb{N}}$, $(S_n)_{n\in\mathbb{N}}$, $(T_n)_{n\in\mathbb{N}}$ and $(T'_n)_{n\in\mathbb{N}}$ with the properties 1), 2), 3), and 4).

$(\nu_{\iota_n})_{n\in\mathbb{N}}$ is a Θ-sequence and by 2) $(S_n)_{n\in\mathbb{N}}$ is a disjoint sequence of sets. $(T_n)_{n\in\mathbb{N}}$ is a sequence in $\mathfrak{T}$. By 1) and 2) its union is contained in a set of $\mathfrak{T}$ and by 2) and 3) $T_n \subset S_n \in \varphi(T_n)$ for any $n \in \mathbb{N}$. Hence, 4) contradicts e).

Now let $\mathfrak{F}$ be arbitrary. By the above proof, for any $\Phi(\Theta)$-filter $\mathfrak{G}$ on $\mathscr{M}(\mathfrak{R}, G; \mathfrak{K})$, there exist $T(\mathfrak{G}) \in \mathfrak{T}$ and $\mathscr{M}(\mathfrak{G}) \in \mathfrak{G}$ such that $L \subset T(\mathfrak{G})$ and

$$\mu(A) - \mu(L) \in U$$

for any $\mu \in \mathscr{M}(\mathfrak{G})$ and any $A \in \mathfrak{R}$ with $L \subset A \subset T(\mathfrak{G})$. Let Ψ be a finite subset of $\Phi(\Theta)$ such that

$$\mathscr{M} := \bigcup_{\mathfrak{G}\in\Psi} \mathscr{M}(\mathfrak{G}) \in \mathfrak{F}\,.$$

By b) there exists $T \in \mathfrak{T}$ with

$$L \subset T \subset \bigcap_{\mathfrak{G}\in\Psi} T(\mathfrak{G})\,.$$

We get

$$\mu(A) - \mu(L) \in U$$

for any $\mu \in \mathscr{M}$ and any $A \in \mathfrak{R}$ with $L \subset A \subset T$. □

Proposition 4.4.7. *Let $\mathfrak{L}$ and $\mathfrak{T}$ be subsets of $\mathfrak{R}$ and φ be a map defined on $\mathfrak{T}$ such that the following hold:*

1) for any $K \in \mathfrak{K}$ and any $T', T'' \in \mathfrak{T}$ with $K \subset T' \cap T''$ there exist $T \in \mathfrak{T}$ and $S \in \varphi(T)$ such that

$$K \subset T \subset S \subset T' \cap T''\,;$$

2) for any $L \in \mathfrak{L}$ the set $\{T \in \mathfrak{T} \mid L \subset T\}$ is lower directed;

3) for any $K \in \mathfrak{K}$ and $L \in \mathfrak{L}$ with $K \cap L = \emptyset$ there exist disjoint sets $T', T'' \in \mathfrak{T}$ with $K \subset T', L \subset T''$;

4) $\mathfrak{T}$ separates $\mathfrak{K}$;

5) for any Θ_3-sequence $(\mu_n)_{n\in\mathbb{N}}$ in $\mathscr{M}(\mathfrak{R}, G; \mathfrak{K})_{\mathfrak{T}}$, any continuous group homomorphism u of G into a metrizable topological additive group H, and any sequence

$(T_n)_{n\in\mathbb{N}}$ in $\mathfrak{T}$ whose union is contained in a set of $\mathfrak{T}$ and for which there exists a disjoint sequence $(S_n)_{n\in\mathbb{N}}$ of sets such that $T_n \subset S_n \in \varphi(T_n)$ for any $n\in\mathbb{N}$, there exist an infinite subset M of $\mathbb{N}$ and a map $\psi : \mathfrak{P}(M) \to \mathfrak{T}$ such that $u \circ \mu_n \circ \psi \in \mathscr{G}(M, H)$ and $\psi(\{n\}) = T_n$ for any $n \in M$.

Then the following also hold:

a) for any $\hat{\Phi}_3$-filter $\mathfrak{F}$ on $\mathscr{M}(\mathfrak{R}, G; \mathfrak{K})_{\mathfrak{T}}$, any $L \in \mathfrak{L}$ and any 0-neighbourhood U in G there exist $T \in \mathfrak{T}$ and $\mathscr{M} \in \mathfrak{F}$ such that $L \subset T$ and

$$\mu(A) - \mu(L) \in U$$

for any $\mu \in \mathscr{M}$ and any $A \in \mathfrak{R}$ with $L \subset A \subset T$;

b) the identity map

$$\mathscr{M}(\mathfrak{R}, G; \mathfrak{K})_{\mathfrak{T}} \to \mathscr{M}(\mathfrak{R}, G; \mathfrak{K})_{\mathfrak{T} \cup \mathfrak{L}}$$

is uniformly Φ_3-continuous.

a) Let U be a 0-neighbourhood in G, $(\mu_n)_{n\in\mathbb{N}}$ be a Θ_3-sequence in $\mathscr{M}(\mathfrak{R}, G; \mathfrak{K})_{\mathfrak{T}}$ and $(T_n)_{n\in\mathbb{N}}$ be a sequence in $\mathfrak{T}$ whose union is contained in a set of $\mathfrak{T}$ and for which there exists a disjoint sequence $(S_n)_{n\in\mathbb{N}}$ of sets such that $T_n \subset S_n \in \varphi(T_n)$ for any $n \in \mathbb{N}$. There exist a continuous group homomorphism u of G into a metrizable topological additive group H and a 0-neighbourhood V in H such that $\overset{-1}{u}(V) \subset U$. (Lemma 4.1.19). By 5) there exist an infinite subset M of $\mathbb{N}$ and a map $\psi : \mathfrak{P}(M) \to \mathfrak{T}$ such that $u \circ \mu_n \circ \psi \in \mathscr{G}(M, H)$ and $\psi(\{n\}) = T_n$ for any $n \in M$. By Proposition 4.4.3 there exists $n \in M$ such that $u \circ \mu_n(T_n) \in V$ and therefore $\mu_n(T_n) \in U$. The assertion follows now immediately from Proposition 4.4.6.

b) Let $\mathfrak{F}$ be a Cauchy $\hat{\Phi}_3$-filter on $\mathscr{M}(\mathfrak{R}, G; \mathfrak{K})_{\mathfrak{T}}$, $L \in \mathfrak{L}$, U be a 0-neighbourhood in G and V be a 0-neighbourhood in G such that $V + V - V \subset U$. By a) there exist $T \in \mathfrak{T}$ and $\mathscr{M} \in \mathfrak{F}$ such that $L \subset T$ and

$$\mu(T) - \mu(L) \in V$$

for any $\mu \in \mathscr{M}$. Let $\mathscr{M}' \in \mathfrak{F}$ be such that

$$\mu(T) - \nu(T) \in V$$

for any $\mu, \nu \in \mathscr{M}'$. Then

$$\mu(L) - \nu(L) \in -V + V + V \subset U$$

for any $\mu, \nu \in \mathscr{M} \cap \mathscr{M}'$. Hence, $\mathfrak{F}$ is a Cauchy filter on $\mathscr{M}(\mathfrak{R}, G; \mathfrak{K})_{\mathfrak{T} \cup \mathfrak{L}}$ and the identity map

$$\mathscr{M}(\mathfrak{R}, G; \mathfrak{K})_{\mathfrak{T}} \to \mathscr{M}(\mathfrak{R}, G; \mathfrak{K})_{\mathfrak{T} \cup \mathfrak{L}}$$

is uniformly Φ_3-continuous. □

Proposition 4.4.8. *Let $\mathfrak{T}$ be a subset of $\mathfrak{R}$ such that:*

1) the union of any disjoint sequence in $\mathfrak{T}$ belongs to $\mathfrak{T}$ if it is contained in a set of $\mathfrak{T}$;

2) for any $K \in \mathfrak{K}$ the set $\{T \in \mathfrak{T} \mid K \subset T\}$ is lower directed;

3) $\mathfrak{T}$ separates $\mathfrak{K}$.

Then:

a) for any $\hat{\Phi}_4$-filter $\mathfrak{F}$ on $\mathscr{M}(\mathfrak{R}, G; \mathfrak{K})_{\mathfrak{T}}$, any $K \in \mathfrak{K}$ and any 0-neighbourhood U in G, there exist $T \in \mathfrak{T}$ and $\mathscr{M} \in \mathfrak{F}$ such that $K \subset T$ and

$$\mu(A) - \mu(K) \in U$$

for any $\mu \in \mathscr{M}$ and any $A \in \mathfrak{R}$ with $K \subset A \subset T$;

b) the identity map

$$\mathscr{M}(\mathfrak{R}, G; \mathfrak{K})_{\mathfrak{T}} \to \mathscr{M}(\mathfrak{R}, G; \mathfrak{K})_{\mathfrak{T} \cup \mathfrak{K}}$$

is uniformly Φ_4-continuous.

a) Let $(\mu_n)_{n \in \mathbb{N}}$ be a Θ_4-sequence in $\mathscr{M}(\mathfrak{R}, G; \mathfrak{K})_{\mathfrak{T}}$ and $(T_n)_{n \in \mathbb{N}}$ be a disjoint sequence in $\mathfrak{T}$ whose union is contained in a set of $\mathfrak{T}$. By 1) $\bigcup_{n \in M} T_m \in \mathfrak{T}$ for any $M \subset \mathbb{N}$. We set

$$\psi : \mathfrak{P}(\mathbb{N}) \to \mathfrak{T}, \quad M \to \bigcup_{n \in M} T_n .$$

By Proposition 4.4.3

$$\lim_{n \to \infty} \mu_n(T_n) = 0$$

and the assertion follows from Proposition 4.4.6 (by replacing $\mathfrak{L}$, φ and Θ with $\mathfrak{K}$, $T \mapsto \{T\}$ and $\Theta_4(\mathscr{M}(\mathfrak{R}, G; \mathfrak{K})_{\mathfrak{T}})$ respectively).

b) Let $\mathfrak{F}$ be a Cauchy $\hat{\Phi}_4$-filter on $\mathscr{M}(\mathfrak{R}, G; \mathfrak{K})_{\mathfrak{T}}$, $K \in \mathfrak{K}$, U be a 0-neighbourhood in G and V be a 0-neighbourhood in G such that $V + V - V \subset U$. By a) there exist $T \in \mathfrak{T}$ and $\mathscr{M} \in \mathfrak{F}$ such that $K \subset T$ and

$$\mu(T) - \mu(K) \in V$$

for any $\mu \in \mathscr{M}$. Let $\mathscr{M}' \in \mathfrak{F}$ be such that

$$\mu(T) - \nu(T) \in V$$

for any $\mu, \nu \in \mathscr{M}'$. Then

$$\mu(K) - \nu(K) \in -V + V + V \subset U$$

for any $\mu, \nu \in \mathscr{M} \cap \mathscr{M}'$. Hence, $\mathfrak{F}$ is a Cauchy filter on $\mathscr{M}(\mathfrak{R}, G; \mathfrak{K})_{\mathfrak{T} \cup \mathfrak{K}}$ so the identity map

$$\mathscr{M}(\mathfrak{R}, G; \mathfrak{K})_{\mathfrak{T}} \rightarrow \mathscr{M}(\mathfrak{R}, G; \mathfrak{K})_{\mathfrak{T} \cup \mathfrak{K}}$$

is uniformly Φ_4-continuous. □

Proposition 4.4.9. *Let $\mathscr{M}$ be a subset of $\mathscr{M}(\mathfrak{R}, G)$, $\mathfrak{T}$ be a subset of $\mathfrak{R}$ such that the union of any sequence in $\mathfrak{T}$ belongs to $\mathfrak{T}$ if it is contained in a set of $\mathfrak{T}$ and $\mathfrak{B}$ be the set of $B \in \mathfrak{R}$ such that:*

a) $\{T \in \mathfrak{T} \mid B \subset T\}$ is lower directed;
b) for any 0-neighbourhood U in G there exists $T \in \mathfrak{T}$ such that $B \subset T$ and

$$\mu(A) - \mu(B) \in U$$

for any $\mu \in \mathscr{M}$ and any $A \in \mathfrak{R}$ with $B \subset A \subset T$.

Then the union of any sequence in $\mathfrak{B}$ belongs to $\mathfrak{B}$ if it is contained in a set of $\mathfrak{T}$.

Let $(B_n)_{n \in \mathbb{N}}$ be a sequence in $\mathfrak{B}$, $B := \bigcup_{n \in \mathbb{N}} B_n$ and $T', T'' \in \mathfrak{T}$ with $B \subset T' \cap T''$. By a) there exists for any $n \in \mathbb{N}$ a $T_n \in \mathfrak{T}$ such that $B_n \subset T_n \subset T' \cap T''$. Then $\bigcup_{n \in \mathbb{N}} T_n \in \mathfrak{T}$ and

$$B \subset \bigcup_{n \in \mathbb{N}} T_n \subset T' \cap T''.$$

Hence, $\{T \in \mathfrak{T} \mid B \subset T\}$ is lower directed.

Let now $T \in \mathfrak{T}$ with $B \subset T$, U be a closed 0-neighbourhood in G and let $(U_n)_{n \in \mathbb{N}}$ be a sequence of 0-neighbourhoods in G such that $U_0 + U_0 \subset U$ and $U_{n+1} + U_{n+1} \subset U_n$ for any $n \in \mathbb{N}$. By a) and b) there exists, for any $n \in \mathbb{N}$, a $T_n \in \mathfrak{T}$ such that $B_n \subset T_n \subset T$ and

$$\mu(A) - \mu(B_n) \in U_n$$

for any $\mu \in \mathscr{M}$ and any $A \in \mathfrak{R}$ with $B_n \subset A \subset T_n$. Then $\bigcup_{n \in \mathbb{N}} T_n \in \mathfrak{T}$. Let $A \in \mathfrak{R}$ with $B \subset A \subset \bigcup_{n \in \mathbb{N}} T_n$. We set

$$A_n := (A \backslash B) \cap T_n \backslash \bigcup_{m=0}^{n-1} T_m.$$

for any $n \in \mathbb{N}$. Then $A_n \subset T_n \backslash B_n$ so

$$\mu(A_n) = \mu(A_n \cup B_n) - \mu(B_n) \in U_n$$

for any $\mu \in \mathscr{M}$ and any $n \in \mathbb{N}$. Since $(A_n)_{n \in \mathbb{N}}$ is pairwise disjoint and

$$A \backslash B = \bigcup_{n \in \mathbb{N}} A_n$$

we get

$$\mu(A) - \mu(B) = \mu(\bigcup_{n \in \mathbb{N}} A_n) = \sum_{n \in \mathbb{N}} \mu(A_n) \in U$$

for any $\mu \in \mathscr{M}$. Hence, $B \in \mathfrak{B}$. □

Proposition 4.4.10. *Let $\mathfrak{T}$ be a subset of $\mathfrak{R}$ such that:*

a) the union of any sequence in $\mathfrak{T}$ belongs to $\mathfrak{T}$ if it is contained in a set of $\mathfrak{T}$;

b) for any $K \in \mathfrak{K}$ the set $\{T \in \mathfrak{T} \mid K \subset T\}$ is lower directed;

c) $\mathfrak{T}$ separates $\mathfrak{K}$.

Further, let $(K_n)_{n \in \mathbb{N}}$ be a sequence in $\mathfrak{K}$ whose union is contained in a set of $\mathfrak{T}$ and let

$$\mathfrak{A} := \{ \bigcup_{n \in M} K_n \mid M \subset \mathbb{N} \},$$

Then any Θ_4-sequence in $\mathscr{M}(\mathfrak{R}, G; \mathfrak{K})_{\mathfrak{T}}$ is a Θ_4-sequence in $\mathscr{M}(\mathfrak{R}, G; \mathfrak{K})_{\mathfrak{A}}$.

Let $(\mu_n)_{n \in \mathbb{N}}$ be a Cauchy sequence in $\mathscr{M}(\mathfrak{R}, G; \mathfrak{K})_{\mathfrak{T}}$. We want to show that $(\mu_n)_{n \in \mathbb{N}}$ is a Cauchy sequence in $\mathscr{M}(\mathfrak{R}, G; \mathfrak{K})_{\mathfrak{A}}$. Let $\mathfrak{B}$ be the set of $B \in \mathfrak{R}$ such that:

1) $\{T \in \mathfrak{T} \mid B \subset T\}$ is lower directed;

2) for any 0-neighbourhood U in G there exists $T \in \mathfrak{T}$ with $B \subset T$ and

$$\mu_n(A) - \mu_n(B) \in U$$

for any $n \in \mathbb{N}$ and any $A \in \mathfrak{R}$ with $B \subset A \subset T$. By Proposition 1.4.7 $\{\mu_n \mid n \in \mathbb{N}\}$ is a Φ_4-set of $\mathscr{M}(\mathfrak{R}, G; \mathfrak{K})_{\mathfrak{T}}$. By Proposition 4.4.8 a) $\mathfrak{K} \subset \mathfrak{B}$. By Proposition 4.4.9 $\mathfrak{A} \subset \mathfrak{B}$. Let $(B_\iota)_{\iota \in I}$ be a finite family in $\mathfrak{B}$ and U be a 0-neighbourhood in G. Further, let V be a 0-neighbourhood in G such that

$$V + V - V \subset U.$$

There exists for any $\iota \in I$ a $T_\iota \in \mathfrak{T}$ such that

$$\mu_n(T_\iota) - \mu_n(B_\iota) \in V$$

for any $n \in \mathbb{N}$. There exists $m \in \mathbb{N}$ such that

$$\mu_n(T_\iota) - \mu_m(T_\iota) \in V$$

for any $\iota \in I$ and any $n \in \mathbb{N}$ with $n \geq m$. We get

$$\mu_n(B_\iota) - \mu_m(B_\iota) \in -V + V + V \subset U$$

for any $\iota \in I$ and any $n \in \mathbb{N}$ with $n \geq m$. Hence, $(\mu_n)_{n \in \mathbb{N}}$ is a Cauchy sequence in $\mathscr{M}(\mathfrak{R}, G; \mathfrak{K})_{\mathfrak{A}}$.

Let $(\mu_n)_{n \in \mathbb{N}}$ be a Θ_4-sequence in $\mathscr{M}(\mathfrak{R}, G; \mathfrak{K})_{\mathfrak{T}}$ and μ be an adherent point of $(\mu_n)_{n \in \mathbb{N}}$ in $\mathscr{M}(\mathfrak{R}, G; \mathfrak{K})_{\mathfrak{T}}$. We want to show that μ is an adherent point of $(\mu_n)_{n \in \mathbb{N}}$ in $\mathscr{M}(\mathfrak{R}, G; \mathfrak{K})_{\mathfrak{A}}$. We set $\mathscr{M} := \{\mu_n \mid n \in \mathbb{N}\} \cup \{\mu\}$. Let $\mathfrak{B}$ be the set of $B \in \mathfrak{R}$ such that:

1) $\{T \in \mathfrak{T} \mid B \subset T\}$ is lower directed;

2) for any 0-neighbourhood U in G there exists $T \in \mathfrak{T}$ with $B \subset T$ and

$$\nu(A) - \nu(B) \in U$$

for any $\nu \in \mathcal{M}$ and any $A \in \mathfrak{R}$ with $B \subset A \subset T$. By Proposition 1.4.7 $\{\mu_n \mid n \in \mathbb{N}\}$ is a Φ_4-set of $\mathcal{M}(\mathfrak{R}, G; \mathfrak{K})_{\mathfrak{T}}$. By Proposition 4.4.8 a) $\mathfrak{K} \subset \mathfrak{B}$. By Proposition 4.4.9 $\mathfrak{A} \subset \mathfrak{B}$. Let $(B_\iota)_{\iota \in I}$ be a finite family in $\mathfrak{B}$ and U be a 0-neighbourhood in G such that

$$-V + V + V \subset U .$$

There exists for any $\iota \in I$ a $T_\iota \in \mathfrak{T}$ such that

$$\nu(T_\iota) - \nu(B_\iota) \in V$$

for any $\nu \in \mathcal{M}$. Let $m \in \mathbb{N}$. There exists $n \in \mathbb{N}$ with $n \geq m$ and

$$\mu_n(T_\iota) - \mu(T_\iota) \in V$$

for any $\iota \in I$. We get

$$\mu_n(B_\iota) - \mu(B_\iota) \in -V + V + V \subset U .$$

Hence, μ is an adherent point of $(\mu_n)_{n \in \mathbb{N}}$ in $\mathcal{M}(\mathfrak{R}, G; \mathfrak{K})_{\mathfrak{A}}$.

Putting together the above result we deduce that any Θ_4-sequence in $\mathcal{M}(\mathfrak{R}, G; \mathfrak{K})_{\mathfrak{T}}$ is a Θ_4-sequence in $\mathcal{M}(\mathfrak{R}, G; \mathfrak{K})_{\mathfrak{A}}$. □

Proposition 4.4.11. *Let $\mathfrak{A}$ be a subset of $\mathfrak{R}$ closed under finite unions (finite intersections) and $\mathfrak{B}$ be the ring of sets generated by $\mathfrak{A}$. Then*

$$\mathcal{M}(\mathfrak{R}, G; \mathfrak{K})_{\mathfrak{A}} = \mathcal{M}(\mathfrak{R}, G; \mathfrak{K})_{\mathfrak{B}} .$$

Assume first that $\mathfrak{A}$ is closed under finite intersections. We may assume $\emptyset \in \mathfrak{A}$. For any $n \in \mathbb{N}$ set

$$\mathfrak{A}_n := \{A_0 \setminus \bigcup_{m=1}^{n} A_m \mid (A_m)_{0 \leq m \leq n} \text{ family in } \mathfrak{A}\} .$$

Then $\mathfrak{A}_0 = \mathfrak{A}$ and $\mathfrak{A}_n \subset \mathfrak{A}_{n+1}$ for any $n \in \mathbb{N}$. Let $n \in \mathbb{N}$, $(A_m)_{0 \leq m \leq n+1}$ be an arbitrary family in $\mathfrak{A}$ and

$$\mu, \nu \in \mathcal{M}(\mathfrak{R}, G; \mathfrak{K}) .$$

Then

$$A_0 \setminus \bigcup_{m=1}^{n} A_m \in \mathfrak{A}_n , \quad A_0 \in \mathfrak{A}_n$$

$$A_0 \setminus \bigcup_{m=1}^{n} (A_m \cap A_{n+1}) \in \mathfrak{A}_n , \quad A_0 \cap A_{n+1} \in \mathfrak{A}_n .$$

From

$$\mu(A_0\setminus\bigcup_{m=1}^{n+1}A_m)-\nu(A_0\setminus\bigcup_{m=1}^{n+1}A_m)=(\mu(A_0\setminus\bigcup_{m=1}^{n}A_m)-\nu(A_0\setminus\bigcup_{m=1}^{n}A_m))-$$
$$-(\mu(A_0\cap A_{n+1})-\nu(A_0\cap A_{n+1}))$$
$$+(\mu(A_0)-\nu(A_0))+$$
$$-(\mu(A_0\setminus\bigcup_{m=1}^{n}(A_m\cap A_{n+1}))-\nu(A_0\setminus\bigcup_{m=1}^{n}(A_m\cap A_{n+1})))$$

we get

$$\mathscr{M}(\mathfrak{R},G;\mathfrak{K})_{\mathfrak{A}_n}=\mathscr{M}(\mathfrak{R},G;\mathfrak{K})_{\mathfrak{A}_{n+1}}.$$

By induction we deduce that

$$\mathscr{M}(\mathfrak{R},G;\mathfrak{K})_{\mathfrak{A}}=\mathscr{M}(R,G;\mathfrak{K})_{\mathfrak{A}_n}$$

for any $n\in\mathbb{N}$.

Let $n\in\mathbb{N}$ and $(A_m)_{0\le m\le n}$ be families in $\mathfrak{A}$. Then

$$(A_0\setminus\bigcup_{m=1}^{n}A_m)\cap(B_0\setminus\bigcup_{m=1}^{n}B_m)=A_0\cap B_0\setminus\bigcup_{m=1}^{n}(A_m\cup B_m),$$

$$(A_0\setminus\bigcup_{m=1}^{n}A_m)\setminus(B_0\setminus\bigcup_{m=1}^{n}B_m)=$$

$$=(A_0\setminus B_0\cup(\bigcup_{m=1}^{n}A_m))\cup(\bigcup_{m=1}^{n}((A_0\cap B_0\cap B_m)\setminus(\bigcup_{p=1}^{m-1}B_p)\cup(\bigcup_{m=1}^{n}A_m))).$$

Hence, the intersection and the difference of any two sets of $\bigcup_{n\in\mathbb{N}}\mathfrak{A}_n$ may be written as a union of a finite disjoint family in $\bigcup_{n\in\mathbb{N}}\mathfrak{A}_n$. It follows that any set of $\mathfrak{B}$ is the union of a finite disjoint family in $\bigcup_{n\in\mathbb{N}}\mathfrak{A}_n$. We get

$$\mathscr{M}(\mathfrak{R},G;\mathfrak{K})_{\mathfrak{A}}=\mathscr{M}(\mathfrak{R},G;\mathfrak{K})_{\mathfrak{B}}.$$

Assume now that $\mathfrak{A}$ is closed under finite unions. For any $A\in\mathfrak{A}$ set

$$\mathfrak{A}(A):=\{A\setminus B\mid B\in\mathfrak{A},\ B\subset A\}.$$

Let $A\in\mathfrak{A}$. It is obvious that the identity map

$$\mathscr{M}(\mathfrak{R},G;\mathfrak{K})_{\mathfrak{A}}\to\mathscr{M}(\mathfrak{R},G;\mathfrak{K})_{\mathfrak{A}(A)}$$

is uniformly continuous and that $\mathfrak{A}(A)$ is closed under finite intersections. If $\mathfrak{B}(A)$ denotes the ring of sets generated by $\mathfrak{A}(A)$ then, by the above proof, the identity map

$$\mathscr{M}(\mathfrak{R},G;\mathfrak{K})_{\mathfrak{A}}\to\mathscr{M}(\mathfrak{R},G;\mathfrak{K})_{\mathfrak{B}(A)}$$

is uniformly continuous. $\mathfrak{B}(A)$ is the ring of sets generated by $\{B\in\mathfrak{A}\mid B\subset A\}$.

It is easy to see that

$$\mathfrak{B} = \bigcup_{A \in \mathfrak{A}} \mathfrak{B}(A).$$

From this it follows immediately that the identity map

$$\mathscr{M}(\mathfrak{R}, G; \mathfrak{K})_{\mathfrak{A}} \to \mathscr{M}(\mathfrak{R}, G; \mathfrak{K})_{\mathfrak{B}}$$

is uniformly continuous. Hence,

$$\mathscr{M}(\mathfrak{R}, G; \mathfrak{K})_{\mathfrak{A}} = \mathscr{M}(\mathfrak{R}, G; \mathfrak{K})_{\mathfrak{B}}. \quad \square$$

Proposition 4.4.12. *Let $\mathfrak{A}$ be a subset of $\mathfrak{R}$ closed under finite unions (finite intersections) and $\mathscr{M}$ be a Φ_4-set of $\mathscr{M}(\mathfrak{R}, G; \mathfrak{K})$. Then there exists a δ-ring $\mathfrak{B}$ containing $\mathfrak{A}$ and contained in $\mathfrak{R}$ such that the union of any sequence in $\mathfrak{B}$ belongs to $\mathfrak{B}$ if it belongs to $\mathfrak{R}$ and such that $\bar{\mathscr{M}}_{\mathfrak{A}} = \bar{\mathscr{M}}_{\mathfrak{B}}$.*

Let $\mathfrak{B}_0$ be the set of $A \in \mathfrak{R}$ for which the map

$$\bar{\mathscr{M}}_{\mathfrak{A}} \to G, \quad \mu \mapsto \mu(A)$$

is uniformly continuous. Then $\bar{\mathscr{M}}_{\mathfrak{A}} = \bar{\mathscr{M}}_{\mathfrak{B}_0}$. Let $(A_n)_{n \in \mathbb{N}}$ be a decreasing sequence in $\mathfrak{B}_0$ (increasing sequence in $\mathfrak{B}_0$ with $\bigcup_{n \in \mathbb{N}} A_n \in \mathfrak{R}$). Set

$$A := \bigcap_{n \in \mathbb{N}} A_n \qquad \left(A := \bigcup_{n \in \mathbb{N}} A_n\right).$$

Then there exists $\mathfrak{C} \in \Sigma(\mathfrak{R})$ such that $\{A_n \mid n \in \mathbb{N}\} \cup \{A\} \subset \mathfrak{C}$. Let U be an arbitrary 0-neighbourhood in G and V be a 0-neighbourhood in G such that

$$V + V - V \subset U.$$

By Theorem 4.2.16 b) there exists $n \in \mathbb{N}$ such that

$$\mu(A_n) - \mu(A) \in V$$

for any $\mu \in \bar{\mathscr{M}}$. We get

$$\mu(A) - \nu(A) \in -V + V + V \subset U$$

for any $\mu, \nu \in \bar{\mathscr{M}}$ with

$$\mu(A_n) - \nu(A_n) \in V.$$

Hence, $A \in \mathfrak{B}_0$.

By Proposition 4.4.11 $\mathfrak{B}_0$ contains the ring of sets $\mathfrak{A}_0$ generated by $\mathfrak{A}$. Let $\mathfrak{B}$ be the smallest δ-ring containing $\mathfrak{A}_0$ such that the union of any increasing sequence in $\mathfrak{B}$ belongs to $\mathfrak{B}$ if it belongs to $\mathfrak{R}$ and let $\mathfrak{C}$ be the smallest subset of $\mathfrak{R}$ containing $\mathfrak{A}_0$ such that the intersection of any decreasing sequence in $\mathfrak{C}$ belongs to $\mathfrak{C}$ and such that the union of any increasing sequence in $\mathfrak{C}$ belongs to $\mathfrak{C}$ if it belongs to $\mathfrak{R}$. We have $\mathfrak{A}_0 \subset \mathfrak{C} \subset \mathfrak{B}$. Set

$$\Delta(A) := \{B \in \mathfrak{C} \mid A \cup B,\ A \backslash B,\ B \backslash A \in \mathfrak{C}\}$$

for any $A \in \mathfrak{C}$ and let $A \in \mathfrak{C}$. Then the intersection of any decreasing sequence in $\Delta(A)$ belongs to $\Delta(A)$ and the union of any increasing sequence in $\Delta(A)$ belongs to $\Delta(A)$ if it belongs to $\mathfrak{R}$. If $A \in \mathfrak{A}_0$, then $\mathfrak{A}_0 \subset \Delta(A)$ so $\Delta(A) = \mathfrak{C}$. Hence, $\mathfrak{A}_0 \subset \Delta(A)$ so $\Delta(A) = \mathfrak{C}$ for any $A \in \mathfrak{C}$. Hence, $\mathfrak{C}$ is a ring of sets and we deduce that $\mathfrak{C} = \mathfrak{B}$. By the above considerations $\mathfrak{C} \subset \mathfrak{B}_0$. Hence, $\mathfrak{B} \subset \mathfrak{B}_0$ and so $\bar{\mathscr{M}}_{\mathfrak{A}} = \bar{\mathscr{M}}_{\mathfrak{B}}$. □

§ 4.5 Spaces of measures on topological spaces

Throughout this section G denotes a Hausdorff topological additive group.

Some results of this section were published by the author in (1981) [7].

Definition 4.5.1. *Let X be a topological space. A subset of X is called a* G_δ-set of X *if it is the intersection of a countable family of open sets of X. A subset of X is called an* F_σ-set of X *(*K_σ-set of X*) if it is the union of a countable family of closed (compact) sets of X. An open set U of X is called* exact *if there exists a continuous real function f on X such that*

$$U = \{x \in X \mid f(x) \neq 0\}.$$

An open set U of X is called regular *if $U = \overset{\circ}{\bar{U}}$.*

Any exact open set is an F_σ-set.

Theorem 4.5.2. *Let $\mathfrak{R}$ be a δ-ring, $\mathfrak{K}$ be a subset of $\mathfrak{R}$ closed under finite unions and $\mathfrak{T}$ be a subset of $\mathfrak{R}$ such that:*

a) the union of any sequence in $\mathfrak{T}$ belongs to $\mathfrak{T}$ if it is contained in a set of $\mathfrak{T}$;

b) for any $K \in \mathfrak{K}$ the set $\{T \in \mathfrak{T} \mid K \subset T\}$ is lower directed;

c) $\mathfrak{T}$ separates $\mathfrak{K}$;

d) the union of any disjoint sequence in $\mathfrak{K}$ is contained in a set of $\mathfrak{T}$ if it belongs to $\mathfrak{R}$.

Then the identity map

$$\mathscr{M}(\mathfrak{R}, G; \mathfrak{K})_{\mathfrak{T}} \to \mathscr{M}(\mathfrak{R}, G; \mathfrak{K})$$

is uniformly Φ_4-continuous.

Assume the contrary. Then by Proposition 4.4.8 b) the identity map

$$\mathcal{M}(\mathfrak{R}, G; \mathfrak{K})_{\mathfrak{K}\cup\mathfrak{T}} \to \mathcal{M}(\mathfrak{R}, G; \mathfrak{K})$$

is not uniformly Φ_4-continuous (Corollary 1.8.5). By Proposition 4.4.2 there exist a 0-neighbourhood U in G, a Θ_4-sequence $(\mu_n)_{n\in\mathbb{N}}$ in $\mathcal{M}(\mathfrak{R}, G; \mathfrak{K})_{\mathfrak{K}\cup\mathfrak{T}}$, and a disjoint sequence $(K_n)_{n\in\mathbb{N}}$ in $\mathfrak{K}$ such that $\bigcup_{n\in\mathbb{N}} K_n \in \mathfrak{R}$ and

$$\mu_n(K_m) - \mu_m(K_m) \notin U .$$

for any $m, n \in \mathbb{N}$ with $m < n$. We set

$$\mathfrak{A} := \{\bigcup_{n\in M} K_n \mid M \subset \mathbb{N}\} .$$

By Proposition 4.4.10 $(\mu_n)_{n\in\mathbb{N}}$ is a Θ_4-sequence in $\mathcal{M}(\mathfrak{R}, G; \mathfrak{K})_{\mathfrak{A}}$. By Proposition 4.4.3 there exist $m, n \in \mathbb{N}$ with $m < n$ and

$$\mu_m(A) - \mu_n(A) \in U$$

for any $A \in \mathfrak{A}$, which is the expected contradiction. □

Remark. D. Landers and L. Rogge showed (1972) ([2] Theorem 1) that, under the above hypothesis, the identity map

$$\mathcal{M}(\mathfrak{R}, G; \mathfrak{K})_{\mathfrak{T}} \to \mathcal{M}(\mathfrak{R}, G; \mathfrak{K})$$

preserves the Cauchy sequences.

Corollary 4.5.3. *Let X be a Hausdorff space, $\mathfrak{T}$ be a base of X closed under finite unions such that the union of any sequence in $\mathfrak{T}$ belongs to $\mathfrak{T}$ if it is contained in a set of $\mathfrak{T}$, $\mathfrak{R}$ be a δ-ring of subsets of X containing $\mathfrak{T}$ and $\mathfrak{K}$ be a set of compact sets of X belonging to $\mathfrak{R}$ and closed under finite unions such that the union of any disjoint sequence in $\mathfrak{K}$ is contained in a set of $\mathfrak{T}$ if it belongs to $\mathfrak{R}$. Then the identity map*

$$\mathcal{M}(\mathfrak{R}, G; \mathfrak{K})_{\mathfrak{T}} \to \mathcal{M}(\mathfrak{R}, G; \mathfrak{K})$$

is uniformly Φ_4-continuous.

If X is completely regular, we may take as $\mathfrak{T}$ the set of exact open sets. If X is locally compact, we may take as $\mathfrak{T}$ the set of exact open K_σ-sets. □

Remark. J. Dieudonné proved (1951) ([1] Proposition 8) that the identity map $\mathcal{M}(\mathfrak{R}, \mathbb{R}; \mathfrak{K})_{\mathfrak{T}} \to \mathcal{M}(\mathfrak{R}, \mathbb{R}; \mathfrak{K})$ preserves the convergent sequences, where X is compact metrizable, $\mathfrak{R}$ is the σ-algebra of Borel sets of X, $\mathfrak{T}$ is the set of all open sets of X, $\mathfrak{K}$ is the set of compact sets of X, and G is $\mathbb{R}$. This was extended by A. Grothendieck (1953) ([2] page 150) to the case when X is locally compact and by J. Pfanzagl (1971) ([1] Theorem) to the case when X is Hausdorff and $\mathfrak{T}$ is a base of X closed under countable unions. D. Landers and L. Rogge extended the above result of J. Pfanzagl to an arbitrary G (1972) ([2] Corollary 10).

Corollary 4.5.4. *Let X be a locally compact space, $\mathfrak{T}$ be the set of exact open relatively compact sets of X, $\mathfrak{K}$ be the set of compact sets of X and $\mathfrak{R}$ be a δ-ring of relatively compact sets of X containing $\mathfrak{K}$. Then the identity map*

$$\mathscr{M}(\mathfrak{R}, G; \mathfrak{K})_{\mathfrak{T}} \to \mathscr{M}(\mathfrak{R}, G; \mathfrak{K})$$

is uniformly Φ_4-continuous. □

Corollary 4.5.5. *Let X be a normal space, $\mathfrak{T}$ be the set of exact open sets of X, $\mathfrak{R}$ be a δ-ring of subsets of X containing $\mathfrak{T}$ and $\mathfrak{K}$ be the set of closed sets belonging to $\mathfrak{R}$. Then the identity map*

$$\mathscr{M}(\mathfrak{R}, G; \mathfrak{K})_{\mathfrak{T}} \to \mathscr{M}(\mathfrak{R}, G; \mathfrak{K})$$

is uniformly Φ_4-continuous. □

Remark. D. Landers and L. Rogge proved (1972) ([2] Corollary 7) that, under the above assumptions, the identity map

$$\mathscr{M}(\mathfrak{R}, G; \mathfrak{K})_{\mathfrak{T}} \to \mathscr{M}(\mathfrak{R}, G; \mathfrak{K})$$

preserves the Cauchy sequences.

Proposition 4.5.6. *Let $\mathfrak{R}$ be a δ-ring, $\mathfrak{K}$ be a subset of $\mathfrak{R}$ closed under finite unions and $\mathfrak{L}$ be a subset of $\mathfrak{R}$ closed under finite unions and countable intersections such that:*

a) any set of $\mathfrak{R}$ is contained in a set of $\mathfrak{L}$:

b) for any $K', K'' \in \mathfrak{K}$ and $L \in \mathfrak{L}$ with $K' \cup K'' \subset L$ and $K' \cap K'' = \emptyset$ there exist $L', L'' \in \mathfrak{L}$ with

$$K' \subset L' \subset L \setminus K'', \quad K'' \subset L'' \subset L \setminus K', \quad L' \cup L'' = L.$$

Then the identity map

$$\mathscr{M}(\mathfrak{R}, G; \mathfrak{K})_{\mathfrak{L}} \to \mathscr{M}(\mathfrak{R}, G; \mathfrak{K})$$

is uniformly Φ_4-continuous.

We denote by λ the identity map

$$\mathscr{M}(\mathfrak{R}, G; \mathfrak{K})_{\mathfrak{L}} \to \mathscr{M}(\mathfrak{R}, G; \mathfrak{K})$$

and, for any $L \in \mathfrak{L}$, set

$$\mathfrak{R}(L) := \{A \in \mathfrak{R} \mid A \subset L\}, \quad \mathfrak{K}(L) := \mathfrak{R}(L) \cap \mathfrak{K},$$

$$\mathfrak{L}(L) := \mathfrak{R}(L) \cap \mathfrak{L}, \quad \mathfrak{T}(L) := \{L \setminus M \mid M \in \mathfrak{L}\},$$

$$\varphi_L : \mathscr{M}(\mathfrak{R}, G; \mathfrak{K})_{\mathfrak{L}} \to \mathscr{M}(\mathfrak{R}(L), G; \mathfrak{K}(L))_{\mathfrak{L}(L)}, \quad \mu \mapsto \mu \mid \mathfrak{R}(L),$$

$$\psi_L : \mathscr{M}(\mathfrak{R}, G; \mathfrak{K}) \to \mathscr{M}(\mathfrak{R}(L), G; \mathfrak{K}(L)), \quad \mu \mapsto \mu \mid \mathfrak{R}(L),$$

$$\lambda_L : \mathscr{M}(\mathfrak{R}(L), G; \mathfrak{K}(L))_{\mathfrak{L}(L)} \to \mathscr{M}(\mathfrak{R}(L), G; \mathfrak{K}(L)), \quad \mu \mapsto \mu.$$

Let $L \in \mathfrak{L}$. The identity map

$$\mathcal{M}(\mathfrak{R}(L), G; \mathfrak{K}(L))_{\mathfrak{L}(L)} \to \mathcal{M}(\mathfrak{R}(L), G; \mathfrak{K}(L))_{\mathfrak{T}(L)}$$

is uniformly continuous and, therefore, by b) and Theorem 4.5.2 λ_L, is uniformly Φ_4-continuous. Since $\psi_L \circ \lambda = \lambda_L \circ \varphi_L$ it follows that $\psi_L \circ \lambda$ is uniformly Φ_4-continuous. By a) and Proposition 1.8.13 λ is uniformly Φ_4-continuous. □

Corollary 4.5.7. *Let X be a Hausdorff space, $\mathfrak{L}$ be the set of compact G_δ-sets of X, $\mathfrak{R}$ be a δ-ring of subsets of X containing $\mathfrak{L}$ and $\mathfrak{K}$ be the set of compact sets of X belonging to $\mathfrak{R}$. If any set of $\mathfrak{R}$ is contained in a set of $\mathfrak{L}$ then the identity map*

$$\mathcal{M}(\mathfrak{R}, G; \mathfrak{K})_{\mathfrak{L}} \to \mathcal{M}(\mathfrak{R}, G; \mathfrak{K})$$

is uniformly Φ_4-continuous. □

Corollary 4.5.8. *Let X be a locally compact space, $\mathfrak{L}$ be the set of compact G_δ-sets of X, $\mathfrak{R}$ be a δ-ring of relatively compact sets of X containing $\mathfrak{L}$ and $\mathfrak{K}$ be the set of compact sets of X belonging to $\mathfrak{R}$. Then the identity map*

$$\mathcal{M}(\mathfrak{R}, G; \mathfrak{K})_{\mathfrak{L}} \to \mathcal{M}(\mathfrak{R}, G; \mathfrak{K})$$

is uniformly Φ_4-continuous. □

Corollary 4.5.9. *Let X be a completely regular space, $\mathfrak{L}$ be the set of closed G_δ-sets of X, $\mathfrak{R}$ be a δ-ring of subsets of X containing $\mathfrak{L}$ and $\mathfrak{K}$ be the set of compact sets of X belonging to $\mathfrak{R}$. Then the identity map*

$$\mathcal{M}(\mathfrak{R}, G; \mathfrak{K})_{\mathfrak{L}} \to \mathcal{M}(\mathfrak{R}, G; \mathfrak{K})$$

is uniformly Φ_4-contiuous. □

Corollary 4.5.10. *Let X be a normal space, $\mathfrak{L}$ be the set of closed G_δ-sets of X, $\mathfrak{R}$ be a δ-ring of subsets of X containing $\mathfrak{L}$ and $\mathfrak{K}$ be the set of closed sets of X belonging to $\mathfrak{R}$. Then the identity map*

$$\mathcal{M}(\mathfrak{R}, G; \mathfrak{K})_{\mathfrak{L}} \to \mathcal{M}(\mathfrak{R}, G; \mathfrak{K})$$

is uniformly Φ_4-continuous. □

Corollary 4.5.11. *Let X be a Hausdorff space, $\mathfrak{K}$ be the set of compact sets of X and $\mathfrak{R}$ be a δ-ring of relatively compact sets of X containing $\mathfrak{K}$. Then the identity map*

$$\mathcal{M}(\mathfrak{R}, G; \mathfrak{K})_{\mathfrak{K}} \to \mathcal{M}(\mathfrak{R}, G; \mathfrak{K})$$

is uniformly Φ_4-continuous. □

Remark. We cannot drop the hypothesis that the sets of $\mathfrak{R}$ are relatively compact.

Indeed, if X is $\mathbb{N}$ endowed with the discrete topology, G is $\mathbb{R}$, and ε_n denotes the Dirac measure at n for any $n \in \mathbb{N}$, then $(\varepsilon_n)_{n \in \mathbb{N}}$ converges to 0 in $\mathscr{M}(\mathfrak{R}, G; \mathfrak{K})_{\mathfrak{K}}$ but not in $\mathscr{M}(\mathfrak{R}. G; \mathfrak{K})$, if $\mathfrak{R}$ contains a set which is not relatively compact.

Proposition 4.5.12. *Let X be a topological space, $\mathfrak{R}$ be a ring of subsets of X, $(\mu_\iota)_{\iota \in I}$ be a countable family of G-valued measures on $\mathfrak{R}$ and $(T_n)_{n \in \mathbb{N}}$ be a disjoint sequence of open sets of X. We set*

$$A(M) := \overline{\bigcup_{n \in M} T_n}^{\circ} \setminus \bigcup_{n \in M} T_n$$

for any $M \subset \mathbb{N}$ and assume the following.

a) $\bigcup_{n \in M} T_n$ is regular for any finite subset M of $\mathbb{N}$.

b) $A(M) \in \mathfrak{R}$ for any $M \subset \mathbb{N}$.

c) For any sequence $(M_n)_{n \in \mathbb{N}}$ in $\mathfrak{P}(\mathbb{N})$ such that $(A(M_n))_{n \in \mathbb{N}}$ is a disjoint sequence, there exists an infinite subset J of $\mathbb{N}$ with $\{\bigcup_{n \in J'} A(M_n) \mid J' \subset J\} \subset \mathfrak{R}$.

d) $\{0\}$ is a G_δ-set of G.

Then there exists an infinite subset M_0 of $\mathbb{N}$ such that $\mu_\iota(A(M)) = 0$ for any $M \subset M_0$ and any $\iota \in I$.

G^I endowed with the product topology is a Hausdorff topological additive group for which $\{0\}$ is a G_δ-set and

$$\mu : \mathfrak{R} \to G^I, \quad B \mapsto (\mu_\iota(B))_{\iota \in I}$$

is an G^I-valued measure on $\mathfrak{R}$ such that

$$\mu(B) = 0 \Leftrightarrow (\iota \in I \Rightarrow \mu_\iota(B) = 0)$$

for any $B \in \mathfrak{R}$. Hence, it is sufficient to prove the above proposition only for a G-valued measure μ.

By Lemma 3.2.9 there exists an uncountable set $\mathfrak{M}$ of infinite subsets of $\mathbb{N}$ such that $M' \cap M''$ is finite for any different $M', M'' \in \mathfrak{M}$. By a)

$$A(M') \cap A(M'') = \emptyset$$

for any different $M', M'' \in \mathfrak{M}$. Assume for any $M \in \mathfrak{M}$ there exists $M' \subset M$ with

$$\mu(A(M')) \neq 0.$$

Then $(A(M'))_{M \in \mathfrak{M}}$ is an uncountable disjoint family in $\mathfrak{R}$ and by c) for any sequence $(M_n)_{n \in \mathbb{N}}$ in $\mathfrak{M}$ there exists an infinite subset J of $\mathbb{N}$ such that

$$\{\bigcup_{n \in J'} A(M_n') \mid J' \subset J\} \subset \mathfrak{R}$$

By Proposition 3.3.15 the set

$$\{M \in \mathfrak{M} \mid \mu(A(M')) \neq 0\}$$

is countable and this is a contradiction. Hence, there exists $M_0 \in \mathfrak{M}$ with $\mu(A(M)) = 0$ for any $M \subset M_0$. □

Theorem 4.5.13. *Let X be a topological space, $\mathfrak{R}$ be a δ-ring of subsets of X, $\mathfrak{T}$ be a set of regular open sets of X belonging to $\mathfrak{R}$, $\mathfrak{K}$ be a subset of $\mathfrak{R}$ closed under finite unions and $\mathfrak{L}$ be a subset of $\mathfrak{R}$ such that the following hold.*

a) For any $K \in \mathfrak{K}$ and for any $T', T'' \in \mathfrak{T}$ with $K \subset T' \cap T''$ there exists $T \in \mathfrak{T}$ such that

$$K \subset T \subset \bar{T} \subset T' \cap T''.$$

b) $\overline{\bigcup_{n \in \mathbb{N}} T_n}^{\circ} \in \mathfrak{T}$ for any sequence $(T_n)_{n \in \mathbb{N}}$ in $\mathfrak{T}$ the union of which is contained in a set of $\mathfrak{T}$ and for which there exists a disjoint sequence $(S_n)_{n \in \mathbb{N}}$ of open sets of X such that $\bar{T}_n \subset S_n$ for any $n \in \mathbb{N}$.

c) For any $L \in \mathfrak{L}$ the set $\{T \in \mathfrak{T} \mid L \subset T\}$ is lower directed.

d) For any $K \in \mathfrak{K}$ and $L \in \mathfrak{L}$ with $K \cap L = \emptyset$ there exist disjoint sets $T', T'' \in \mathfrak{T}$ with $K \subset T', L \subset T''$.

e) $\mathfrak{T}$ separates $\mathfrak{K}$ and $\emptyset \in \mathfrak{T}$.

Then the identity map

$$\mathscr{M}(\mathfrak{R}, G; \mathfrak{K})_{\mathfrak{T}} \to \mathscr{M}(\mathfrak{R}, G; \mathfrak{K})_{\mathfrak{T} \cup \mathfrak{L}}$$

is uniformly Φ_3-continuous.

Let φ be the map

$$\mathfrak{T} \to \mathfrak{P}(\mathfrak{P}(X)), \quad T \mapsto \{S \mid \bar{T} \subset S, \quad S \text{ open set of } X\}.$$

Let $(\mu_n)_{n \in \mathbb{N}}$ be a sequence of G-valued measures on $\mathfrak{R}$, u be a continuous group homomorphism of G into a metrizable topological additive group H and $(T_n)_{n \in \mathbb{N}}$ be a sequence in $\mathfrak{T}$ whose union is contained in a set of $\mathfrak{T}$ and for which there exists a disjoint sequence $(S_n)_{n \in \mathbb{N}}$ of sets such that $S_n \in \varphi(T_n)$ for any $n \in \mathbb{N}$. By Proposition 4.1.28, $(u \circ \mu_n)_{n \in \mathbb{N}}$ is a sequence of H-valued measures on $\mathfrak{R}$. Then $\bigcup_{n \in M} T_n$ is a regular open set for any finite subset M of $\mathbb{N}$ and by b) and e)

$$\overline{\bigcup_{m \in M} T_m}^{\circ} \in \mathfrak{T}, \quad \overline{\bigcup_{m \in M} T_m}^{\circ} \setminus \bigcup_{m \in M} T_m \in \mathfrak{R}$$

for any $M \subset \mathbb{N}$. Hence, by Proposition 4.5.12 there exists an infinite subset M_0 of $\mathbb{N}$ such that

$$u \circ \mu_n(\overline{\bigcup_{m \in M} T_m}^{\circ} \setminus \bigcup_{m \in M} T_m) = 0$$

for any $n \in \mathbb{N}$ and any $M \subset M_0$. We denote by ψ and ψ' the maps

$$\mathfrak{P}(M_0) \to \mathfrak{T}, \quad M \mapsto \overset{\circ}{\overline{\bigcup_{m \in M} T_m}}$$

and

$$\mathfrak{P}(M_0) \to \mathfrak{R}, \quad M \mapsto \bigcup_{m \in M} T_m$$

respectively. Let $n \in \mathbb{N}$. By the above result,

$$u \circ \mu_n \circ \psi = u \circ \mu_n \circ \psi' .$$

By Proposition 4.1.14 a $\Rightarrow$ b, $u \circ \mu_n \circ \psi' \in \mathscr{G}(M_0, H)$ so $u \circ \mu_n \circ \psi \in \mathscr{G}(M_0, H)$. By a), c), d), e), and Proposition 4.4.7 b) the identity map

$$\mathscr{M}(\mathfrak{R}, G; \mathfrak{K})_{\mathfrak{T}} \to \mathscr{M}(\mathfrak{R}, G; \mathfrak{K})_{\mathfrak{T} \cup \mathfrak{L}}$$

is uniformly Φ_3-continuous. □

Corollary 4.5.14. *Let X be a regular space, $\mathfrak{T}$ be the set of regular open sets of X, $\mathfrak{R}$ be a δ-ring of subsets of X containing $\mathfrak{T}$ and $\mathfrak{K}$ be the set of compact sets of X belonging to $\mathfrak{R}$. Then the identity map*

$$\mathscr{M}(\mathfrak{R}, G; \mathfrak{K})_{\mathfrak{T}} \to \mathscr{M}(\mathfrak{R}, G; \mathfrak{K})$$

is uniformly Φ_3-continuous.

Let $K \in \mathfrak{K}$ and $T', T'' \in \mathfrak{T}$ with $K \subset T' \cap T''$. Since X is regular and K compact there exists an open set V of X such that

$$K \subset V \subset \bar{V} \subset T' \cap T'' .$$

We set $T := \overset{\circ}{\bar{V}}$. Then $T \in \mathfrak{T}$ and

$$K \subset T \subset \bar{T} \subset T' \cap T'' .$$

Let $\mathfrak{L}$ be the set of closed sets of X belonging to $\mathfrak{R}$, $K \in \mathfrak{K}$ and $L \in \mathfrak{L}$ with $K \cap L = \emptyset$. Since K is compact and X is regular there exists an open set U of X such that $K \subset U$, $\bar{U} \cap L = \emptyset$. We set

$$T' := \overset{\circ}{\bar{U}}, \quad T'' := X \setminus \bar{U} .$$

Then T', T'' are disjoint sets of $\mathfrak{T}$ with $K \subset T'$, $L \subset T''$.

If we remark that $\mathfrak{T}$ is closed under finite intersections then the above results and Theorem 4.5.13 show that the identity map

$$\mathscr{M}(\mathfrak{R}, G; \mathfrak{K})_{\mathfrak{T}} \to \mathscr{M}(\mathfrak{R}, G; \mathfrak{K})_{\mathfrak{T} \cup \mathfrak{L}}$$

is uniformly Φ_3-continuous. By Proposition 4.5.6 and Corollary 1.8.5 the identity map

$$\mathcal{M}(\mathfrak{R}, G; \mathfrak{K})_{\mathfrak{T}} \to \mathcal{M}(\mathfrak{R}, G; \mathfrak{K})$$

is uniformly Φ_3-continuous. □

Remark. In particular, any Cauchy sequence in $\mathcal{M}(\mathfrak{R}, G; \mathfrak{K})_{\mathfrak{T}}$ is a Cauchy sequence in $\mathcal{M}(\mathfrak{R}, G; \mathfrak{K})$. This result was proved for $G = \mathbb{R}$ by B.B. Wells Jr. (1969) ([1] Theorem 4) for X compact, by P. Gänssler (1971) ([1] Theorem 4.10) for X normal, and by P. Gänssler (1971) ([2] Theorem 3.1) for X regular. S.S. Khurana extended (1976) ([1] Theorem 4) the last quoted result of P. Gänssler to an arbitrary G.

Corollary 4.5.15. *Let X be a normal space, $\mathfrak{T}$ be the set of regular open sets of X, $\mathfrak{R}$ be a δ-ring of subsets of X containing $\mathfrak{T}$ and $\mathfrak{K}$ be the set of closed sets of X belonging to $\mathfrak{R}$. Then the identity map*

$$\mathcal{M}(\mathfrak{R}, G; \mathfrak{K})_{\mathfrak{T}} \to \mathcal{M}(\mathfrak{R}, G; \mathfrak{K})$$

is uniformly Φ_3-continuous.

Let $K \in \mathfrak{K}$ and let $T', T'' \in \mathfrak{T}$ with $K \subset T' \cap T''$. Since X is normal there exists an open set V of X such that

$$K \subset V \subset \bar{V} \subset T' \cap T''.$$

We set $T := \mathring{\bar{V}}$. Then $T \in \mathfrak{T}$ and

$$K \subset T \subset \bar{T} \subset T' \cap T''.$$

Let K', K'' be disjoint sets of $\mathfrak{K}$. Since X is normal there exist open disjoint sets U', U'' of X such that $K' \subset U'$, $K'' \subset U''$. If we set $T' := \mathring{\bar{U}}'$, and $T'' := \mathring{\bar{U}}''$, then T', T'' are disjoint sets of $\mathfrak{T}$ such that $K' \subset T'$ and $K'' \subset T''$.

By Theorem 4.5.13 the identity map

$$\mathcal{M}(\mathfrak{R}, G; \mathfrak{K})_{\mathfrak{T}} \to \mathcal{M}(\mathfrak{R}, G; \mathfrak{K})_{\mathfrak{K}}$$

is uniformly Φ_3-continuous. By Proposition 4.5.6 and Corollary 1.8.5 the identity map

$$\mathcal{M}(\mathfrak{R}, G; \mathfrak{K})_{\mathfrak{T}} \to \mathcal{M}(\mathfrak{R}, G; \mathfrak{K})$$

is uniformly Φ_3-continuous. □

Remark. L. Rogge proved (1973) ([1] Theorem 1) that the identity map

$$\mathcal{M}(\mathfrak{R}, G; \mathfrak{K})_{\mathfrak{T}} \to \mathcal{M}(\mathfrak{R}, G; \mathfrak{K})$$

preserves the Cauchy sequences.

Corollary 4.5.16. *Let X be a locally compact space, $\mathfrak{T}$ be the set of relatively compact regular open sets of X, $\mathfrak{R}$ be a δ-ring of relatively compact sets of X containing $\mathfrak{T}$ and $\mathfrak{K}$ be the set of compact sets of X belonging to $\mathfrak{R}$. Then the identity map*

$$\mathscr{M}(\mathfrak{R}, G; \mathfrak{K})_{\mathfrak{T}} \to \mathscr{M}(\mathfrak{R}, G; \mathfrak{K})$$

is uniformly Φ_3-continuous.

Let $K \in \mathfrak{K}$ and $T', T'' \in \mathfrak{T}$ with $K \subset T' \cap T''$. There exists an open set V of X such that

$$K \subset V \subset \bar{V} \subset T' \cap T''.$$

We set $T := \bar{V}$. Then $T \in \mathfrak{T}$ and

$$K \subset T \subset \bar{T} \subset T' \cap T''.$$

Let K', K'' be disjoint sets of $\mathfrak{K}$. There exist relatively compact open disjoint sets U', U'' of X such that $K' \subset U'$ and $K'' \subset U''$. We set $T' := \overset{\circ}{\overline{U'}}$ and $T'' := \overset{\circ}{\overline{U''}}$. Then T', T'' are disjoint sets of $\mathfrak{T}$ such that $K' \subset T'$ and $K'' \subset T''$.

By Theorem 4.5.13 the identity map

$$\mathscr{M}(\mathfrak{R}, G; \mathfrak{K})_{\mathfrak{T}} \to \mathscr{M}(\mathfrak{R}, G; \mathfrak{K})_{\mathfrak{K}}$$

is uniformly Φ_3-continuous. By Proposition 4.5.6 and Corollary 1.8.5 the identity map

$$\mathscr{M}(\mathfrak{R}, G; \mathfrak{K})_{\mathfrak{T}} \to \mathscr{M}(\mathfrak{R}, G; \mathfrak{K})$$

is uniformly Φ_3-continuous. □

Corollary 4.5.17. *Let Y be a completely regular space such that the closure of any exact open set of Y is open, X be a subspace of Y, $\mathfrak{T}$ be the set of closed open sets of X, $\mathfrak{R}$ be a δ-ring of subsets of X containing $\mathfrak{T}$ and $\mathfrak{K}$ be the set of compact sets of X belonging to $\mathfrak{R}$. Then the identity map*

$$\mathscr{M}(\mathfrak{R}, G; \mathfrak{K})_{\mathfrak{T}} \to \mathscr{M}(\mathfrak{R}, G; \mathfrak{K})$$

is uniformly Φ_3-continuous.

By $\mathfrak{T}'$ we denote the set of closed open sets of Y, by $\mathfrak{R}'$ the δ-ring generated by $\mathfrak{R} \cup \mathfrak{T}'$, by $\mathfrak{K}'$ the set of compact sets of Y belonging to $\mathfrak{R}'$, and by $\mathfrak{L}'$ the set of closed sets of Y belonging to $\mathfrak{R}'$. By Theorem 4.5.13 the identity map

$$\mathscr{M}(\mathfrak{R}', G; \mathfrak{K}')_{\mathfrak{T}'} \to \mathscr{M}(\mathfrak{R}', G; \mathfrak{K}')_{\mathfrak{T}' \cup \mathfrak{L}'}$$

is uniformly Φ_3-continuous so, by Proposition 4.5.6 and Corollary 1.8.5, the identity map

$$\mathscr{M}(\mathfrak{R}', G; \mathfrak{K}')_{\mathfrak{T}'} \to \mathscr{M}(\mathfrak{R}', G; \mathfrak{K}')$$

is uniformly Φ_3-continuous.

By μ' we denote the map

$$\mathfrak{R}' \to G, \quad A' \mapsto \mu(A' \cap X)$$

for any $\mu \in \mathscr{M}(\mathfrak{R}, G; \mathfrak{K})$. It is easy to see that $\mu' \in \mathscr{M}(\mathfrak{R}', G; \mathfrak{K}')$ for any $\mu \in \mathscr{M}(\mathfrak{R}, G; \mathfrak{K})$ and that the map

$$\mathscr{M}(\mathfrak{R}, G; \mathfrak{K}) \to \mathscr{M}(\mathfrak{R}', G; \mathfrak{K}'), \quad \mu \mapsto \mu'$$

is injective. We identify $\mathscr{M}(\mathfrak{R}, G; \mathfrak{K})$ with a subset of $\mathscr{M}(\mathfrak{R}', G; \mathfrak{K}')$ via this map. By the above result and by Corollary 1.8.15 the identity map

$$\mathscr{M}(\mathfrak{R}, G; \mathfrak{K})_{\mathfrak{T}'} \to \mathscr{M}(\mathfrak{R}, G; \mathfrak{K})_{\mathfrak{R}'}$$

is uniformly Φ_3-continuous. Since the identity maps

$$\mathscr{M}(\mathfrak{R}, G; \mathfrak{K})_{\mathfrak{T}} \to \mathscr{M}(\mathfrak{R}, G; \mathfrak{K})_{\mathfrak{T}'},$$

$$\mathscr{M}(\mathfrak{R}, G; \mathfrak{K})_{\mathfrak{R}'} \to \mathscr{M}(\mathfrak{R}, G; \mathfrak{K})_{\mathfrak{R}}$$

are uniformly continuous the identity map

$$\mathscr{M}(\mathfrak{R}, G; \mathfrak{K})_{\mathfrak{T}} \to \mathscr{M}(\mathfrak{R}, G; \mathfrak{K})_{\mathfrak{R}}$$

is uniformly Φ_3-continuous. □

Corollary 4.5.18. *Let Y be a locally compact space such that the closure of any relatively compact exact open set of Y is open, X be a closed subspace of Y, $\mathfrak{T}$ be the set of compact open sets of X, $\mathfrak{R}$ be a δ-ring of relatively compact sets of X containing $\mathfrak{T}$ and $\mathfrak{K}$ be the set of compact sets of X belonging to $\mathfrak{R}$. Then the identity map*

$$\mathscr{M}(\mathfrak{R}, G; \mathfrak{K})_{\mathfrak{T}} \to \mathscr{M}(\mathfrak{R}, G; \mathfrak{K})$$

is uniformly Φ_3-continuous.

By $\mathfrak{T}'$ we denote the set of compact open sets of Y, by $\mathfrak{R}'$ the δ-ring generated by $\mathfrak{R} \cup \mathfrak{T}'$, and by $\mathfrak{K}'$ the set of compact sets of Y belonging to $\mathfrak{R}'$. By Theorem 4.5.13 the identity map

$$\mathscr{M}(\mathfrak{R}', G; \mathfrak{K}')_{\mathfrak{T}'} \to \mathscr{M}(\mathfrak{R}', G; \mathfrak{K}')_{\mathfrak{T}' \cup \mathfrak{K}'}$$

is uniformly Φ_3-continuous and therefore by Proposition 4.5.6 and Corollary 1.8.5 the identity map

$$\mathscr{M}(\mathfrak{R}', G; \mathfrak{K}')_{\mathfrak{T}'} \to \mathscr{M}(\mathfrak{R}', G; \mathfrak{K}')$$

is uniformly Φ_3-continuous.

By μ' we denote the map

$$\mathfrak{R}' \to G, \quad A' \mapsto \mu(A' \cap X)$$

for any $\mu \in \mathscr{M}(\mathfrak{R}, G; \mathfrak{K})$. It is easy to see that $\mu' \in \mathscr{M}(\mathfrak{R}', G; \mathfrak{K}')$ for any $\mu \in \mathscr{M}(\mathfrak{R}, G; \mathfrak{K})$ and that the map

$$\mathscr{M}(\mathfrak{R}, G; \mathfrak{K}) \to \mathscr{M}(\mathfrak{R}', G; \mathfrak{K}'), \quad \mu \mapsto \mu'$$

is injective. We identify $\mathcal{M}(\mathfrak{R}, G; \mathfrak{K})$ with a subset of $\mathcal{M}(\mathfrak{R}', G; \mathfrak{K}')$ via this map. By the above result and by Corollary 1.8.15 the identity map

$$\mathcal{M}(\mathfrak{R}, G; \mathfrak{K})_{\mathfrak{T}'} \to \mathcal{M}(\mathfrak{R}, G; \mathfrak{K})_{\mathfrak{R}'}$$

is uniformly Φ_3-continuous. Since the identity maps

$$\mathcal{M}(\mathfrak{R}, G; \mathfrak{K})_{\mathfrak{T}} \to \mathcal{M}(\mathfrak{R}, G; \mathfrak{K})_{\mathfrak{T}'},$$

$$\mathcal{M}(\mathfrak{R}, G; \mathfrak{K})_{\mathfrak{R}'} \to \mathcal{M}(\mathfrak{R}, G; \mathfrak{K})_{\mathfrak{R}}$$

are uniformly continuous the identity map

$$\mathcal{M}(\mathfrak{R}, G; \mathfrak{K})_{\mathfrak{T}} \to \mathcal{M}(\mathfrak{R}, G; \mathfrak{K})_{\mathfrak{R}}$$

is uniformly Φ_3-continuous. □

Proposition 4.5.19. *Let X be a Hausdorff space, $\mathfrak{K}$ be the set of compact sets of X, $\mathfrak{R}$ be a δ-ring of subsets of X containing $\mathfrak{K}$, $K \in \mathfrak{K}$ and W be an open set belonging to $\mathfrak{R}$ such that $K \backslash W$ possesses a countable fundamental system of neighbourhoods. Furthermore, let $\mathfrak{F}$ be a $\hat{\Phi}_4$-filter on $\mathcal{M}(\mathfrak{R}, G; \mathfrak{K})$ and V be a 0-neighbourhood in G. Then there exist $\mathcal{M} \in \mathfrak{F}$ and a neighbourhood U of K such that $\mu(A) \in V$ for any $\mu \in \mathcal{M}$ and any $A \in \mathfrak{R}$ with $A \subset U \backslash K$.*

We may assume V closed. Let $(U_n)_{n \in \mathbb{N}}$ be a decreasing sequence such that $\{U_n \mid n \in \mathbb{N}\}$ is a fundamental system of neighbourhoods of $K \backslash W$.

Assume first that $\mathfrak{F}$ is a Φ_4-filter on $\mathcal{M}(\mathfrak{R}, G; \mathfrak{K})$ and let (I, μ) be a Θ_4-net in $\mathcal{M}(\mathfrak{R}, G; \mathfrak{K})$ such that $\mu(\mathfrak{G}) = \mathfrak{F}$, where $\mathfrak{G}$ denotes the section filter of I. Assume the assertion does not hold. We construct inductively a disjoint sequence $(K_n)_{n \in \mathbb{N}}$ in $\mathfrak{K}$ and an increasing sequence $(\iota_n)_{n \in \mathbb{N}}$ in I such that for any $n \in \mathbb{N}$:

a) $K_n \subset (U_n \cup W) \backslash K$;

b) $\mu_{\iota_n}(K_n) \notin V$.

Let $n \in \mathbb{N}$ and assume the sequences were constructed up to $n-1$. Then

$$(U_n \cup W) \backslash \bigcup_{m=0}^{n-1} K_m$$

is a neighbourhood of K and $\{\mu_\iota \mid \iota \in I, \iota \geq \iota_{n-1}\} \in \mathfrak{F}$. By the above hypothesis there exist $\iota_n \in I$ and $A \in \mathfrak{R}$ such that

$$\iota_n \geq \iota_{n-1}, \quad A \subset (U_n \cup W) \backslash (K \cup (\bigcup_{m=0}^{n-1} K_m)), \quad \mu_{\iota_n}(A) \notin V.$$

Since V is closed and μ_{ι_n} $\mathfrak{K}$-regular there exists $K_n \in \mathfrak{K}$ with $K_n \subset A$ and $\mu_{\iota_n}(K_n) \notin V$. This proves the existence of the sequences $(K_n)_{n \in \mathbb{N}}$, $(\iota_n)_{n \in \mathbb{N}}$ with the described properties.

We set

$$L := \overline{\bigcup_{n \in \mathbb{N}} (K_n \backslash W)}.$$

Let $n \in \mathbb{N}$. We want to show that $L \backslash U_n \subset \bigcup_{m=0}^{n-1} (K_m \backslash W)$. Let $x \in L \backslash U_n$. Since X is Hausdorff and $K \backslash W$ compact there exist two disjoint neighbourhoods of x and $K \backslash W$ respectively. Hence, there exists $p \in \mathbb{N}$ such that $p > n$ and $x \notin \bar{U}_p$. We have

$$L = \overline{\bigcup_{m=p}^{\infty} (K_m \backslash W)} \cup \left(\bigcup_{m=0}^{p-1} (K_m \backslash W)\right) \subset \bar{U}_p \cup \left(\bigcup_{m=0}^{p-1} (K_m \backslash W)\right)$$

so $x \in \bigcup_{m=0}^{p-1} (K_m \backslash W)$. Since

$$\bigcup_{m=0}^{p-1} (K_m \backslash W) \subset U_n \cup \left(\bigcup_{m=0}^{n-1} (K_m \backslash W)\right)$$

we get $x \in \bigcup_{m=0}^{n-1} (K_m \backslash W)$, which proves the desired inclusion.

Now we want to show that L is compact. Let $\mathfrak{S}$ be a covering of L with open sets of X. Then $\mathfrak{S} \cup \{X \backslash L\}$ is an open covering of X. Since K is a compact set of X there exists a finite subset $\mathfrak{T}$ of $\mathfrak{S}$ such that

$$K \subset \left(\bigcup_{T \in \mathfrak{T}} T\right) \cup (X \backslash L).$$

Further, there exists $n \in \mathbb{N}$ so that

$$U_n \subset \left(\bigcup_{T \in \mathfrak{T}} T\right) \cup (X \backslash L).$$

Since $\bigcup_{m=0}^{n-1} K_m$ is compact there exists another finite subset $\mathfrak{T}'$ of $\mathfrak{S}$ such that

$$\bigcup_{m=0}^{n-1} K_m \subset \left(\bigcup_{T \in \mathfrak{T}'} T\right) \cup (X \backslash L).$$

We get

$$L \subset U_n \cup \left(\bigcup_{m=0}^{n-1} K_m\right) \subset \bigcup_{T \in \mathfrak{T} \cup \mathfrak{T}'} T.$$

Hence, L is compact.

Since $\bigcup_{n \in \mathbb{N}} (K_n \cap W) \in \mathfrak{R}$, we deduce that $\bigcup_{n \in \mathbb{N}} K_n \in \mathfrak{R}$ and $(K_n)_{n \in \mathbb{N}} \in \Gamma(\mathfrak{R})$. Since $(\mu_{\iota_n})_{n \in \mathbb{N}}$ is a Θ_4-sequence in $\mathscr{M}(\mathfrak{R}, G; \mathfrak{K})$ there exists $n \in \mathbb{N}$ such that $\mu_{\iota_n}(K_n) \in V$ on account of Corollary 4.2.11 and that contradicts b).

Now let $\mathfrak{F}$ be a $\hat{\Phi}_4$-filter on $\mathscr{M}(\mathfrak{R}, G; \mathfrak{K})$. By the above proof for any Φ_4-filter $\mathfrak{G}$ on $\mathscr{M}(\mathfrak{R}, G; \mathfrak{K})$ there exist $\mathscr{M}(\mathfrak{G}) \in \mathfrak{G}$ and a neighbourhood $U(\mathfrak{G})$ of K such that $\mu(A) \in V$ for any $\mu \in \mathscr{M}(\mathfrak{G})$ and any $A \in \mathfrak{R}$ with $A \subset U(\mathfrak{G}) \backslash K$. Let Φ be a finite set of Φ_4-filters on $\mathscr{M}(\mathfrak{R}, G; \mathfrak{K})$ such that

$$\mathscr{M} := \bigcup_{\mathfrak{G} \in \Phi} \mathscr{M}(\mathfrak{G}) \in \mathfrak{F}.$$

Then

$$U := \bigcap_{\mathfrak{G} \in \Phi} U(\mathfrak{G})$$

is a neighbourhood of K and we have $\mu(A) \in V$ for any $\mu \in \mathcal{M}$ and any $A \in \mathfrak{R}$ with $A \subset U \backslash K$. □

Proposition 4.5.20. *Let X be a Hausdorff space, $\mathfrak{K}$ be the set of compact sets of X, $\mathfrak{R}$ be a δ-ring of subsets of X containing $\mathfrak{K}$ and $K \in \mathfrak{K}$. Further, let V be a 0-neighbourhood in G and $\mathfrak{F}$ be a $\hat{\Phi}_4$-filter on $\mathcal{M}(\mathfrak{R}, G; \mathfrak{K})_{\mathfrak{K}}$. If X is metrizable, or locally compact, or if X possesses a countable base, then there exist $\mathcal{M} \in \mathfrak{F}$ and a neighbourhood U of K such that $\mu(A) \in V$ for any $\mu \in \bar{\mathcal{M}}$ and any $A \in \mathfrak{R}$ with $A \subset U \backslash K$.*

Let $\mathfrak{S}$ be the set of relatively compact sets of $\mathfrak{R}$. Then $\mathfrak{S}$ is a δ-ring of relatively compact sets of X containing $\mathfrak{K}$. We have $\mu | \mathfrak{S} \in \mathcal{M}(\mathfrak{S}, G; \mathfrak{K})$ for any $\mu \in \mathcal{M}(\mathfrak{R}, G; \mathfrak{K})$. Denote by φ the map

$$\mathcal{M}(\mathfrak{R}, G; \mathfrak{K}) \to \mathcal{M}(\mathfrak{S}, G; \mathfrak{K}), \quad \mu \mapsto \mu | \mathfrak{S}.$$

Since the map

$$\mathcal{M}(\mathfrak{R}, G; \mathfrak{K})_{\mathfrak{K}} \to \mathcal{M}(\mathfrak{S}, G; \mathfrak{K})_{\mathfrak{K}}, \quad \mu \mapsto \mu | \mathfrak{S},$$

is uniformly continuous $\varphi(\mathfrak{F})$ is a $\hat{\Phi}_4$-filter on $\mathcal{M}(\mathfrak{S}, G; \mathfrak{K})_{\mathfrak{K}}$. By Corollary 4.5.11 the identity map

$$\mathcal{M}(\mathfrak{S}, G; \mathfrak{K})_{\mathfrak{K}} \to \mathcal{M}(\mathfrak{S}, G; \mathfrak{K})$$

is uniformly Φ_4-continuous. Hence, $\varphi(\mathfrak{F})$ is a $\hat{\Phi}_4$-filter on $\mathcal{M}(\mathfrak{S}, G; \mathfrak{K})$ (Theorem 1.8.4 a $\Rightarrow$ i).

We may assume V closed. By Proposition 4.5.19 there exist $\mathcal{M} \in \mathfrak{F}$ and a neighbourhood U of K such that $\mu(A) \in V$ for any $\mu \in \mathcal{M}$ and any $A \in \mathfrak{S}$ with $A \subset U \backslash K$. Since any $\mu \in \mathcal{M}$ is $\mathfrak{K}$-regular we may replace $\mathfrak{S}$ by $\mathfrak{R}$ and $\mathcal{M}$ by $\bar{\mathcal{M}}$ in the above assertion. □

Corollary 4.5.21. *Let X be a Hausdorff space, $\mathfrak{K}$ be the set of compact sets of X, $\mathfrak{R}$ be a δ-ring of subsets of X containing $\mathfrak{K}$ and $K \in \mathfrak{K}$. Further, let V be a 0-neighbourhood in G and $\mathcal{M}$ be a Φ_4-set of $\mathcal{M}(\mathfrak{R}, G; \mathfrak{K})_{\mathfrak{K}}$. If X is metrizable, or locally compact, or possesses a countable base, then there exists a neighbourhood U of K such that $\mu(A) \in V$ for any $\mu \in \bar{\mathcal{M}}$ and any $A \in \mathfrak{R}$ with $A \subset U \backslash K$.* □

Proposition 4.5.22. *Let X be a Hausdorff space with a countable base, $\mathfrak{K}$ be the set of compact sets of X, $\mathfrak{R}$ be a δ-ring of subsets of X containing $\mathfrak{K}$ and $\mathcal{M}$ be a Φ_4-set of $\mathcal{M}(\mathfrak{R}, G; \mathfrak{K})_{\mathfrak{K}}$. Further, let U be a 0-neighbourhood in G and $\mathfrak{L}$ be a set of pairwise disjoint compact sets of X such that for any $L \in \mathfrak{L}$ there exists $\mu \in \bar{\mathcal{M}}$ with $\mu(L) \notin U$. Then $\mathfrak{L}$ is countable.*

Let $\mathfrak{V}$ be a countable base of X closed under finite unions. By Corollary 4.5.21 there exists, for any $L \in \mathfrak{L}$, a $V_L \in \mathfrak{V}$ such that $L \subset V_L$ and $\mu(A) \in U$ for any $\mu \in \bar{\mathcal{M}}$ and any $A \in \mathfrak{R}$ with $A \subset V_L \backslash L$. We see immediately that $V_L \neq V_{L'}$ for any different $L, L' \in \mathfrak{L}$. Hence, $\mathfrak{L}$ is countable. □

Proposition 4.5.23. *Let X be a Hausdorff space with a countable base, $\mathfrak{K}$ be the set of compact sets of X, $\mathfrak{R}$ be a δ-ring of subsets of X containing $\mathfrak{K}$, $\mathcal{M}$ be a Φ_4-set of $\mathcal{M}(\mathfrak{R}, G; \mathfrak{K})_{\mathfrak{K}}$ and $(A_\iota)_{\iota \in I}$ be a disjoint family in $\mathfrak{R}$ such that for any $\iota \in I$ there exists $\mu \in \mathcal{M}$ with $\mu(A_\iota) \neq 0$. If $\{0\}$ is a G_δ-set of G, then I is countable.*

Let $(U_n)_{n \in \mathbb{N}}$ be a sequence of open sets of G such that $\bigcap_{n \in \mathbb{N}} U_n = \{0\}$. For any $\iota \in I$ there exist $K_\iota \in \mathfrak{K}$ and $\mu_\iota \in \bar{\mathcal{M}}$ *with* $\mu_\iota(K_\iota) \neq 0$ and $K_\iota \subset A_\iota$. By Proposition 4.5.22 the set

$$\{\iota \in I \mid \mu_\iota(K_\iota) \neq U_n\}$$

is countable for any $n \in \mathbb{N}$. From

$$I = \bigcup_{n \in \mathbb{N}} \{\iota \in I \mid \mu_\iota(K_\iota) \notin U_n\}$$

it follows that I is countable. □

Proposition 4.5.24. *Let X be a Hausdorff space with a countable base, $\mathfrak{K}$ be the set of compact sets of X, $\mathfrak{R}$ be a δ-ring of subsets of X containing $\mathfrak{K}$, $\mathcal{M}$ be a Φ_4-set of $\mathcal{M}(\mathfrak{R}, G; \mathfrak{K})_{\mathfrak{K}}$. If $\{0\}$ is a G_δ-set of G, then there exists a disjoint sequence $(K_n)_{n \in \mathbb{N}}$ of compact sets of X such that*

$$\mu(A \backslash \bigcup_{n \in \mathbb{N}} K_n) = 0$$

for any $(\mu, A) \in \bar{\mathcal{M}} \times \mathfrak{R}$.

Let Ω be the set of sets $\mathfrak{L}$ of pairwise disjoint compact sets of X such that for any $L \in \mathfrak{L}$ there exists $\mu \in \bar{\mathcal{M}}$ with $\mu(L) \neq 0$. Ω is inductively ordered by the inclusion relation so it possesses a maximal element $\mathfrak{L}$. By Corollary 4.5.23 $\mathfrak{L}$ is countable. Let $(\mu, A) \in \bar{\mathcal{M}} \times \mathfrak{R}$ and assume

$$\mu(A \backslash \bigcup_{L \in \mathfrak{L}} L) \neq 0 .$$

Since μ is $\mathfrak{K}$-regular there exists $K \in \mathfrak{K}$ such that

$$K \subset A \backslash \bigcup_{L \in \mathfrak{L}} L, \quad \mu(K) \neq 0 .$$

But then $\mathfrak{L} \cup \{K\} \in \Omega$, which contradicts the maximality of $\mathfrak{L}$. Hence,

$$\mu(A \backslash \bigcup_{L \in \mathfrak{L}} L) = 0 .$$

for any $(\mu, A) \in \bar{\mathcal{M}} \times \mathfrak{R}$. □

Theorem 4.5.25. *Let X be a Hausdorff space with a countable base, $\mathfrak{K}$ be the set of compact sets of X, $\mathfrak{R}$ be a δ-ring of relatively compact sets of X containing $\mathfrak{K}$ and $\mathscr{M}$ be a Φ_4-set of $\mathscr{M}(\mathfrak{R}, G; \mathfrak{K})_{\mathfrak{K}}$. Then the following hold:*

a) $\mathscr{M}$ is a Φ_4-set of $\mathscr{M}(\mathfrak{R}, G; \mathfrak{K})$, $\bar{\mathscr{M}}$ is the closure of $\mathscr{M}$ in $\bar{\mathscr{M}}(\mathfrak{R}, G; \mathfrak{K})_{\mathfrak{K}}$, and $\bar{\mathscr{M}} = \bar{\mathscr{M}}_{\mathfrak{K}}$;

b) if G is complete, then $\bar{\mathscr{M}}$ is compact;

c) if G is metrizable, then $\bar{\mathscr{M}}$ is metrizable;

d) if $\{0\}$ is a G_δ-set of G and if $\mathscr{M}$ is a Φ_2-set of $\mathscr{M}(\mathfrak{R}, G; \mathfrak{K})_{\mathfrak{K}}$, then $\bar{\mathscr{M}}$ is metrizable and compact; if, in addition, $\mathscr{M}$ is countably compact, then $\mathscr{M} = \bar{\mathscr{M}}$.

a) By Corollary 4.5.11 the identity map

$$\mathscr{M}(\mathfrak{R}, G; \mathfrak{K})_{\mathfrak{K}} \to \mathscr{M}(\mathfrak{R}, G; \mathfrak{K})$$

is uniformly Φ_4-continuous and the assertions follow from Proposition 1.8.8.

b) follows from a) and Theorem 4.2.16 a).

c) Assume first X compact. Let $\mathfrak{T}$ be a countable base of X closed under finite unions and containing X. Then any δ-ring containing $\mathfrak{T}$ contains $\mathfrak{K}$. By a) and Proposition 4.4.12 $\bar{\mathscr{M}} = \bar{\mathscr{M}}_{\mathfrak{T}}$. Hence, $\bar{\mathscr{M}}$ is metrizable.

Now let X be arbitrary. By Proposition 4.5.24 there exists an increasing sequence $(K_n)_{n\in\mathbb{N}}$ of compact sets of X such that

$$\mu(A \setminus \bigcup_{n\in\mathbb{N}} K_n) = 0$$

for any $(\mu, A) \in \bar{\mathscr{M}} \times \mathfrak{R}$. We set

$$\mathfrak{R}_n := \{A \in \mathfrak{R} \mid A \subset K_n\}, \quad \mathfrak{K}_n := \mathfrak{R}_n \cap \mathfrak{K},$$
$$\mathscr{M}_n := \{\mu \mid \mathfrak{R}_n \mid \mu \in \mathscr{M}\}$$

for any $n \in \mathbb{N}$ and

$$\mathfrak{S} := \bigcup_{n\in\mathbb{N}} \mathfrak{R}_n.$$

For any $n \in \mathbb{N}$ the map

$$\mathscr{M}(\mathfrak{R}, G; \mathfrak{K}) \to \mathscr{M}(\mathfrak{R}_n, G; \mathfrak{K}_n), \quad \mu \mapsto \mu \mid \mathfrak{R}_n$$

is uniformly continuous so $\mathscr{M}_n$ is a Φ_4-set of $\mathscr{M}(\mathfrak{R}_n, G; \mathfrak{K}_n)$ and $\{\mu \mid \mathfrak{R}_n \mid \mu \in \bar{\mathscr{M}}\} \subset \bar{\mathscr{M}}_n$. By the above considerations $\bar{\mathscr{M}}_n$ *is* metrizable for any $n \in \mathbb{N}$. Since the topology of $\bar{\mathscr{M}}_{\mathfrak{S}}$ is the initial topology on $\bar{\mathscr{M}}$ with respect to the sequence of maps

$$\bar{\mathscr{M}} \to \bar{\mathscr{M}}_n, \quad \mu \mapsto \mu \mid \mathfrak{R}_n \qquad (n \in \mathbb{N})$$

$\bar{\mathscr{M}}_{\mathfrak{S}}$ is also metrizable. Since $\mathfrak{S}$ is closed under finite unions there exists, by Proposition 4.4.12, a δ-ring $\mathfrak{B}$ containing $\mathfrak{S}$ and contained in $\mathfrak{R}$ such that the union of any sequence in $\mathfrak{B}$ belongs to $\mathfrak{B}$ if it belongs to $\mathfrak{R}$ and such that $\bar{\mathscr{M}}_{\mathfrak{S}} = \bar{\mathscr{M}}_{\mathfrak{B}}$. Let

$A \in \mathfrak{R}$. Then $A \cap (\bigcup_{n \in \mathbb{N}} K_n) \in \mathfrak{B}$ and

$$\mu(A) = \mu(A \cap (\bigcup_{n \in \mathbb{N}} K_n))$$

for any $\mu \in \bar{\mathscr{M}}$. So $\bar{\mathscr{M}}_{\mathfrak{B}} = \bar{\mathscr{M}}$ and $\bar{\mathscr{M}}$ is metrizable.

d) G possesses a coarser metrizable group topology. By c) $\bar{\mathscr{M}}$ possesses a coarser metrizable topology and the assertions follow from Corollary 2.3.10 b), c). □

Proposition 4.5.26. *Let X be a Hausdorff space, $\mathfrak{K}$ be the set of compact sets of X, $\mathfrak{T}$ be the set of open relatively compact σ-compact sets of X and $\mathfrak{R}$ be a δ-ring of subsets of X containing $\mathfrak{K}$. Then for any $\mu \in \mathscr{M}(\mathfrak{R}, G)$ the following assertions are equivalent:*

a) $\mu \in \mathscr{M}(\mathfrak{R}, G; \mathfrak{T})$;

b) $\mu \in \mathscr{M}(\mathfrak{R}, G; \mathfrak{K})$ and $\mu(K) = 0$ for any nowhere dense compact set K of X;

c) $\mu \in \mathscr{M}(\mathfrak{R}, G; \mathfrak{K})$ and $\mu(B) = 0$ for any $B \in \mathfrak{R}$ for which there exists $A \in \mathfrak{R}$ with $B \subset \bar{A} \setminus \mathring{A}$;

d) $\mu \in \mathscr{M}(\mathfrak{R}, G; \mathfrak{K})$ and, for any $A \in \mathfrak{R}$, $\mu(A) = \mu(\mathring{A})$ if $\mathring{A} \in \mathfrak{R}$ and $\mu(A) = \mu(\bar{A})$ if $\bar{A} \in \mathfrak{R}$.

a ⇒ b. It is obvious that $\mu(K) = 0$ for any nowhere dense set K of $\mathfrak{K}$. Let $U \in \mathfrak{T}$ and $(K_n)_{n \in \mathbb{N}}$ be an increasing sequence in $\mathfrak{K}$ whose union is U. We want to show that $\mu(\mathfrak{G}(U, \mathfrak{K}))$ converges to $\mu(U)$. Assume the contrary. Then there exists a neighbourhood V of $\mu(U)$ such that for any $K \in \mathfrak{K}$ with $K \subset U$ there exists $L \in \mathfrak{K}$ such that $K \subset L \subset U$ and $\mu(L) \notin V$. We may construct inductively an increasing sequence $(L_n)_{n \in \mathbb{N}}$ in $\mathfrak{K}$ such that $K_n \subset L_n \subset U$ and $\mu(L_n) \notin V$ for any $n \in \mathbb{N}$. Then $\bigcup_{n \in \mathbb{N}} L_n = U$ so

$$\mu(U) = \lim_{n \to \infty} \mu(L_n),$$

which is a contradiction.

Let $A \in \mathfrak{R}$. We want to show that $\mu(\mathfrak{G}(A, \mathfrak{K}))$ converges to $\mu(A)$. Let V be an arbitrary 0-neighbourhood in G and W be a 0-neighbourhood in G such that

$$W + W + W - W - W \subset V.$$

By a) there exists $U \in \mathfrak{T}$ such that $U \subset A$ and

$$\mu(B) \in \mu(A) + W$$

for any $B \in \mathfrak{R}$ with $U \subset B \subset A$. By the above proof there exists $K \in \mathfrak{K}$ such that $K \subset U$ and

$$\mu(L) \in \mu(U) + W$$

for any $L \in \mathfrak{K}$ with $K \subset L \subset U$. Let $L \in \mathfrak{K}$ be such that $K \subset L \subset A$ and let $(K_n)_{n \in \mathbb{N}}$

be an increasing sequence in $\mathfrak{K}$ such that $K_0 = K$ and $\bigcup_{n\in\mathbb{N}} K_n = U$. Then

$$\mu(L\backslash U) = \mu(L\cup U) - \mu(U) \in W - W.$$

There exists $n\in\mathbb{N}$ such that

$$\mu(L\cap K_n) \in \mu(L\cap U) + W.$$

Since

$$\mu(L\cap K_n) \in \mu(U) + W \subset \mu(A) + W + W$$

we get

$$\mu(L) = \mu(L\cap U) + \mu(L\backslash U) \in \mu(A) + W + W - W + W - W \subset \mu(A) + U.$$

Hence, $\mu(\mathfrak{G}(A,\mathfrak{K}))$ converges to $\mu(A)$ and by Proposition 4.1.26 b $\Rightarrow$ a $\mu\in\mathcal{M}(\mathfrak{R}, G; \mathfrak{K})$.

b $\Rightarrow$ a. Let $A\in\mathfrak{R}$, V be a 0-neighbourhood in G and W be a 0-neighbourhood in G such that $W + W \subset V$. By b) there exists $K\in\mathfrak{K}$ such that $K\subset A$ and

$$\mu(B) - \mu(A) \in W$$

for any $B\in\mathfrak{R}$ with $K\subset B\subset A$. Again by b), $\mu(B) = 0$ for any $B\in\mathfrak{R}$ contained in $K\backslash\mathring{K}$. There exists $L\in\mathfrak{K}$ such that $L\subset\mathring{K}$ and

$$\mu(B) - \mu(\mathring{K}) \in W$$

for any $B\in\mathfrak{R}$ with $L\subset B\subset\mathring{K}$. Let $U\in\mathfrak{T}$ be such that $L\subset U\subset\mathring{K}$ and $B\in\mathfrak{R}$ be such that $U\subset B\subset A$. We have

$$\mu(B) - \mu(A) = (\mu(B\cup K) - \mu(A)) + \mu(B\cap(K\backslash\mathring{K})) +$$
$$+ (\mu(B\cap\mathring{K}) - \mu(\mathring{K})) + (\mu(\mathring{K}) - \mu(K)) \in W + W \subset V.$$

Hence, $\mu\in\mathcal{M}(\mathfrak{R}, G; \mathfrak{T})$.

b $\Rightarrow$ c. Let $A, B\in\mathfrak{R}$ be such that $B\subset\bar{A}\backslash\mathring{A}$. Then any compact subset K of $A\cap B$ or of $B\backslash A$ is nowhere dense so by b) $\mu(K) = 0$. We get

$$\mu(B) = \mu(A\cap B) + \mu(B\backslash A) = 0.$$

c $\Rightarrow$ d and d $\Rightarrow$ b are trivial. □

Remark. Let Y be a set and μ be a positive real measure on $\mathfrak{P}(Y)$ such that $\mu(\{x\}) = 0$ for any $x\in Y$ and such that $\mu(Y) > 0$. Let us endow Y with the discrete topology, X be the Alexandrov compactification of Y, $\mathfrak{K}$ be the set of compact sets of X, $\mathfrak{T}$ be the set of open sets of X and $\mathfrak{R}$ be $\mathfrak{P}(X)$. Then the map

$$\mathfrak{R} \to \mathbb{R}, \quad A \mapsto \mu(A\cap Y)$$

belongs to $\mathcal{M}(\mathfrak{R}, \mathbb{R}; \mathfrak{T})\backslash\mathcal{M}(\mathfrak{R}, \mathbb{R}; \mathfrak{K})$. This example shows the difficulty to replace $\mathfrak{T}$ in the Proposition 4.5.26 with the set of all open relatively compact sets of X.

Theorem 4.5.27. *Let X be a Hausdorff space, $\mathfrak{T}$ be the set of open, open and σ-compact, open F_σ, or open and exact sets of X, respectively, and let $\mathfrak{R}$ be a δ-ring of subsets of X containing $\mathfrak{T}$. Then the identity map*

$$\mathscr{M}(\mathfrak{R}, G; \mathfrak{T})_{\mathfrak{T}} \to \mathscr{M}(\mathfrak{R}, G; \mathfrak{T})$$

is uniformly Φ_4-continuous. If all sets of $\mathfrak{R}$ are relatively compact, then the same assertion holds if we also take the above sets of $\mathfrak{T}$ relatively compact.

The assertion follows immediately from Proposition 4.4.4. □

§ 4.6 Measures with parameter

Throughout this section $\mathfrak{R}$ shall denote a quasi-δ-ring, $\mathfrak{K}$ a subset of $\mathfrak{R}$ closed under finite unions, and G a Hausdorff topological additive group.

Definition 4.6.1. *Let T be a set. For any map $\mu: \mathfrak{R} \to G^T$ and any $t \in T$ we denote by $\underline{\mu_t}$ the map*

$$\mathfrak{R} \to G, \quad A \mapsto (\mu(A))(t).$$

Proposition 4.6.2. *Let T be a set, Θ be a set of sequences in T, Φ be $\hat{\Phi}(\Theta)$ and $\mu \in \mathscr{E}(\mathfrak{R}, G^T_\Phi)$. If μ_t is $\mathfrak{K}$-regular for any $t \in T$, then μ is $\mathfrak{K}$-regular.*

We first remark that, by Proposition 2.1.27, G^T_Φ is a topological group. Let $A \in \mathfrak{R}$. By Proposition 4.1.16 $\mu(\mathfrak{G}(A; \mathfrak{K}))$ is a Cauchy filter in G^T_Φ and, by Proposition 4.1.26 a $\Rightarrow$ b, it converges to $\mu(A)$ in G^T. By N. Bourbaki [1] Ch II § 3 Proposition 7 $\mu(\mathfrak{G}(A, \mathfrak{K}))$ converges to $\mu(A)$ in G^T_Φ so (Proposition 4.1.26 b $\Rightarrow$ a) μ is $\mathfrak{K}$-regular. □

Theorem 4.6.3. *Let T be a uniform (topological) space, $i \in \{1, 2, 3, 4\}$ ($i \in \{1, 2\}$), $\Phi := \hat{\Phi}_i(T)$, T_0 be a Φ_i-dense (loosely Φ_i-dense) set of T and $\mathscr{F}$ be a subgroup of G^T such that any element of $\mathscr{F}$ is uniformly Φ_i-continuous (is Φ_i-continuous). Then:*

a) $\mathscr{F}_{T_0}, \mathscr{F}, \mathscr{F}_\Phi$ are Hausdorff topological additive groups:

b) the sets $\mathscr{M}(\mathfrak{R}, \mathscr{F}_{T_0}; \mathfrak{K})$, $\mathscr{M}(\mathfrak{R}, \mathscr{F}; \mathfrak{K})$, and $\mathscr{M}(\mathfrak{R}, \mathscr{F}_\Phi; \mathfrak{K})$ coincide.

If any point of T possesses a closed neighbourhood which is a Φ_3-set (Φ_1-set) and if $i = 3$ ($i = 1$), then we may choose any dense set of T as T_0.

Both assertions hold if we replace $\mathscr{M}(\mathfrak{R}, G; \mathfrak{K})$ with $\mathscr{E}(\mathfrak{R}, G; \mathfrak{K})$, 4 with 3, and 2 with 1.

We give the proof only for $\mathscr{M}(\mathfrak{R}, G; \mathfrak{K})$. If we make the corresponding changes in it, then it becomes a proof for $\mathscr{E}(\mathfrak{R}, G; \mathfrak{K})$.

a) follows immediately form Theorem 3.4.1 a).

b) By Theorem 4.2.14 a) $\mathscr{M}(\mathfrak{R}, G; \mathfrak{K})$ is a Φ_4-closed set of $G^{\mathfrak{R}}$. By Proposition 1.5.20 a) $\mathscr{M}(\mathfrak{R}, G; \mathfrak{K})$ is a strongly Φ_2-closed set of $G^{\mathfrak{R}}$. Let $\mu \in \mathscr{M}(\mathfrak{R}, \mathscr{F}_{T_0}; \mathfrak{K})$. By Corollary 2.5.3 (replacing X by $\mathfrak{R}$, Y by G, $\mathscr{F}$ by $\mathscr{M}(\mathfrak{R}, G; \mathfrak{K})$, and f by

$$\mathfrak{R} \times T \to G, \quad (A, t) \mapsto (\mu(A))(t)),$$

we deduce that

$$\{t \in T \mid \mu_t \in \mathscr{M}(\mathfrak{R}, G; \mathfrak{K})\}$$

is a Φ_i-closed (strongly Φ_i-closed) set of T. Since it contains T_0 which is Φ_i-dense (loosely Φ_i-dense) we find $\mu_t \in \mathscr{M}(\mathfrak{R}, G; \mathfrak{K})$ for any $t \in T$.

By Proposition 4.2.3 the identity map

$$\mathscr{M}(\mathfrak{R}, G; \mathfrak{K}) \to \mathscr{M}(\mathfrak{R}, G; \mathfrak{K})_{\Psi(\mathfrak{R})}$$

is uniformly Φ_4-continuous and therefore Φ_4-continuous (Proposition 1.8.3). For any $\gamma := (A_n)_{n \in \mathbb{N}} \in \Gamma(\mathfrak{R})$ let φ_γ be the map

$$\mathfrak{P}(\mathbb{N}) \to \mathfrak{R}, \quad J \mapsto \bigcup_{n \in J} A_n$$

and let us endow $\mathfrak{R}$ with the final topology with respect to the family $(\varphi_\gamma)_{\gamma \in \Gamma(\mathfrak{R})}$. By Proposition 4.1.14 a $\Rightarrow$ c any measure on $\mathfrak{R}$ is continuous with respect to this topology. By Theorem 2.5.4 a) (replacing X by $\mathfrak{R}$, Y by G, $\mathscr{F}$ by $\mathscr{M}(\mathfrak{R}, G; \mathfrak{K})$, Θ by $\Theta(\mathfrak{R})$, $\mathfrak{S}$ by $\Sigma(\mathfrak{R})$, and f by the map

$$\mathfrak{R} \times T \to G, \quad (A, t) \mapsto (\mu(A))(t)),$$

we deduce that $\mu: \mathfrak{R} \to G_\Phi^T$ is $\Psi(\mathfrak{R})$-continuous. By Proposition 4.1.14 c $\Rightarrow$ a $\mu \in \mathscr{M}(\mathfrak{R}, G_\Phi^T)$. Since $\mu_t \in \mathscr{M}(\mathfrak{R}, G; \mathfrak{K})$ for any $t \in T$ it follows that $\mu \in \mathscr{M}(\mathfrak{R}, G_\Phi^T; \mathfrak{K})$ (Proposition 4.6.2). Then $\mu \in \mathscr{M}(\mathfrak{R}, \mathscr{F}_\Phi; \mathfrak{K})$.

The last assertion follows from Proposition 1.7.12. □

Remark. The assertion b) for $\mathscr{M}$ was proved by I. Labuda (1972) for $\mathfrak{R}$ a δ-ring, $\mathfrak{K} = \mathfrak{R}$, T compact, and $T_0 = T$ ([1]. Théorème 3.2) or T compact metrizable and T_0 a dense set of T ([1] Corollaire 4.8). In the same paper I. Labuda proved b) for $\mathscr{E}$, namely for the case when T is sequentially compact and $T = T_0$ (Théorème 4.6) or T is metrizable and compact and T_0 a dense set of T (Théorème 4.7).

Corollary 4.6.4. *Let E be a Haudorff locally convex space, E' be its dual, A' be a Φ_4-dense set of E'_E and Φ be the set of $\hat{\Phi}_4$-filters on E'_E. We identify E with a set of maps of E' into $\mathbb{R}$ via the map*

$$E \to \mathbb{R}^{E'}, \quad x \mapsto \langle x, \cdot \rangle.$$

Then:

a) $E_{A'}$ and E_Φ are Hausdorff locally convex spaces;

b) $\mathscr{M}(\mathfrak{R}, E_{A'}; \mathfrak{K}) = \mathscr{M}(\mathfrak{R}, E_{E'}; \mathfrak{K}) = \mathscr{M}(\mathfrak{R}, E_\Phi; \mathfrak{K})$;

Both assertions hold if we replace $\mathscr{M}$ with $\mathscr{E}$ and 4 with 3.

E is a set of uniformly continuous real valued functions on E'_E. By Corollary 2.1.29

E_Φ is a locally convex space. The remaining assertions follows from Theorem 4.6.3 a), b). □

Remark. In particular, $\mathscr{E}(\mathfrak{R}, E_{E'}) = \mathscr{E}(\mathfrak{R}, E_{\mathfrak{A}})$, where $\mathfrak{A}$ denotes the set of Φ_3-sets of E'_E. This result was proved by L. Drewnowski and I. Labuda (1972) ([1] Théorème 1.2).

Corollary 4.6.5. *Let T be a locally compact space, T_0 be a loosely Φ_2-dense set of T and $\mathscr{F}$ be the group of continuous maps of T into G which may be extended continuously to the Alexandrov compactification of T. Then the sets $\mathscr{M}(\mathfrak{R}, \mathscr{F}_{T_0}; \mathfrak{K})$, $\mathscr{M}(\mathfrak{R}, \mathscr{F}; \mathfrak{K})$, $\mathscr{M}(\mathfrak{R}, \mathscr{F}_{\{T\}}; \mathfrak{K})$ coincide as do the sets $\mathscr{E}(\mathfrak{R}, \mathscr{F}_{T_0})$, $\mathscr{E}(\mathfrak{R}, \mathscr{F})$. The same holds for the sets $\mathscr{E}(\mathfrak{R}, \mathscr{F}_{T_0}; \mathfrak{K})$, $\mathscr{E}(\mathfrak{R}, \mathscr{F}; \mathfrak{K})$, $\mathscr{E}(\mathfrak{R}, \mathscr{F}_{\{T\}}; \mathfrak{K})$ if T is mioritic.*

We do the proof for $\mathscr{M}$ only. It also holds for $\mathscr{E}$.

We may assume that T is not compact. Let T^* be the Alexandrov compactification of T and $\mathscr{F}^*$ be the group of continuous maps of T^* into G. By Corollary 1.5.24 T_0 is a loosely Φ_2-dense set of T^*. By Theorem 4.6.3 b) the sets $\mathscr{M}(\mathfrak{R}, \mathscr{F}^*_{T_0}; \mathfrak{K})$, $\mathscr{M}(\mathfrak{R}, \mathscr{F}^*; \mathfrak{K})$ and $\mathscr{M}(\mathfrak{R}, \mathscr{F}^*_{\{T^*\}}; \mathfrak{K})$ coincide. Since the maps $\mathscr{F}^*_{T_0} \to \mathscr{F}_{T_0}$ and $\mathscr{F}^*_{\{T^*\}} \to \mathscr{F}_{\{T\}}$ defined by $f \mapsto f|T$ are isomorphisms of topological groups the sets $\mathscr{M}(\mathfrak{R}, \mathscr{F}_{T_0}; \mathfrak{K})$, $\mathscr{M}(\mathfrak{R}, \mathscr{F}; \mathfrak{K})$, and $\mathscr{M}(\mathfrak{R}, \mathscr{F}_{\{T\}}; \mathfrak{K})$ coincide. □

Theorem 4.6.6. *Let $\mathfrak{R}$ and $\mathfrak{R}'$ be δ-rings, $\mathfrak{K}'$ be a subset of $\mathfrak{R}'$ closed under finite unions and $\mathscr{M}$ be the set of maps $\mu: \mathfrak{R} \times \mathfrak{R}' \to G$* such that

$$\mu(A, \cdot) \in \mathscr{M}(\mathfrak{R}', G; \mathfrak{K}'), \quad \mu(\cdot, A') \in \mathscr{M}(\mathfrak{R}, G; \mathfrak{K})$$

for any $(A, A') \in \mathfrak{R} \times \mathfrak{R}'$. For any $\mu \in \mathscr{M}$ we denote by $\bar{\mu}$ the map

$$\mathfrak{R} \to \mathscr{M}(\mathfrak{R}', G; \mathfrak{K}'), \quad A \mapsto \mu(A, \cdot).$$

Then:

a) $\bar{\mu} \in \mathscr{M}(\mathfrak{R}, \mathscr{M}(\mathfrak{R}', G; \mathfrak{K}'); \mathfrak{K})$ for any $\mu \in \mathscr{M}$ and the map

$$\mathscr{M} \to \mathscr{M}(\mathfrak{R}, \mathscr{M}(\mathfrak{R}', G; \mathfrak{K}'); \mathfrak{K}), \quad \mu \mapsto \bar{\mu}$$

is an isomorphism of uniform spaces;

b) the sets $\mathscr{M}(\mathfrak{R}, \mathscr{M}(\mathfrak{R}', G; \mathfrak{K}')_{\mathfrak{K}'}; \mathfrak{K})$, $\mathscr{M}(\mathfrak{R}, \mathscr{M}(\mathfrak{R}', G; \mathfrak{K}'); \mathfrak{K})$ and $\mathscr{M}(\mathfrak{R}, \mathscr{M}(\mathfrak{R}', G; \mathfrak{K}')_{\Psi(\mathfrak{R}')}; \mathfrak{K})$ coincide and the identity map

$$\mathscr{M}(\mathfrak{R}, \mathscr{M}(\mathfrak{R}', G; \mathfrak{K}'); \mathfrak{K}) \to \mathscr{M}(\mathfrak{R}, \mathscr{M}(\mathfrak{R}', G; \mathfrak{K}')_{\Psi(\mathfrak{R}')}; \mathfrak{K})_{\Psi(\mathfrak{R})}$$

is uniformly Φ_4-continuous; in particular, any map $\mu: \mathfrak{R} \times \mathfrak{R}' \to G$ belongs to $\mathscr{M}$ if

$$\mu(A, \cdot) \in \mathscr{M}(\mathfrak{R}', G; \mathfrak{K}'), \quad \mu(\cdot, K') \in \mathscr{M}(\mathfrak{R}, G; \mathfrak{K})$$

for any $(A, K') \in \mathfrak{R} \times \mathfrak{K}'$.

By Proposition 4.2.2c) $\mathscr{M}(\mathfrak{R}', G; \mathfrak{K}')_{\mathfrak{K}'}$, $\mathscr{M}(\mathfrak{R}', G; \mathfrak{K}')$, and $\mathscr{M}(\mathfrak{R}', G; \mathfrak{K}')_{\Psi(\mathfrak{R}')}$ are Hausdorff topological additive groups so the expressions appearing in a) and b) are meaningful.

a) is obvious.

b) Let us endow $\mathfrak{R}'$ with the coarsest topology for which any $\mathfrak{K}'$-regular measure on $\mathfrak{R}'$ is continuous. Then any increasing sequence in $\mathfrak{R}'$ whose union A' belongs to $\mathfrak{R}'$ converges to A' (Proposition 4.1.15 a $\Rightarrow$ b). Hence, $\mathfrak{G}_{\mathfrak{R}'}(A'; \mathfrak{K}')$ is a $\Theta_1(\mathfrak{K}', \mathfrak{R}')$-filter on $\mathfrak{R}'$ for any $A' \in \mathfrak{R}'$ and by Proposition 4.1.26 a $\Rightarrow$ b it converges to A'. Hence any strongly Φ_1-dense set of $\mathfrak{R}'$ containing $\mathfrak{K}'$ is equal to $\mathfrak{R}'$ and therefore $\mathfrak{K}'$ is a loosely Φ_1-dense set of $\mathfrak{R}'$. By Proposition 4.1.14 a $\Rightarrow$ c we have $\Psi(\mathfrak{R}') \subset \hat{\Phi}_2(\mathfrak{R}')$.

By Theorem 4.6.3 b) the sets $\mathcal{M}(\mathfrak{R}, \mathcal{M}(\mathfrak{R}', G; \mathfrak{K}')_{\mathfrak{K}'}; \mathfrak{K})$, $\mathcal{M}(\mathfrak{R}, \mathcal{M}(\mathfrak{R}', G; \mathfrak{K}'); \mathfrak{K})$ and $\mathcal{M}(\mathfrak{R}, \mathcal{M}(\mathfrak{R}', G; \mathfrak{K}')_{\Psi(\mathfrak{R}')}; \mathfrak{K})$ coincide. By Proposition 4.2.3 the identity maps

$$\mathcal{M}(\mathfrak{R}', G; \mathfrak{K}') \longrightarrow \mathcal{M}(\mathfrak{R}', G; \mathfrak{K}')_{\Psi(\mathfrak{R}')}$$

$$\mathcal{M}(\mathfrak{R}, \mathcal{M}(\mathfrak{R}', G; \mathfrak{K}')_{\Psi(\mathfrak{R}')}; \mathfrak{K}) \longrightarrow \mathcal{M}(\mathfrak{R}, \mathcal{M}(\mathfrak{R}', G; \mathfrak{K}')_{\Psi(\mathfrak{R}')}; \mathfrak{K})_{\Psi(\mathfrak{R})}$$

are uniformly Φ_4-continuous. We deduce from Proposition 2.1.17 that the identity map

$$\mathcal{M}(\mathfrak{R}, \mathcal{M}(\mathfrak{R}', G; \mathfrak{K}'); \mathfrak{K}) \longrightarrow \mathcal{M}(\mathfrak{R}, \mathcal{M}(\mathfrak{R}', G; \mathfrak{K}')_{\Psi(\mathfrak{R}')}; \mathfrak{K})$$

is uniformly Φ_4-continuous. Hence, the identity map

$$\mathcal{M}(\mathfrak{R}, \mathcal{M}(\mathfrak{R}', G; \mathfrak{K}'); \mathfrak{K}) \longrightarrow \mathcal{M}(\mathfrak{R}, \mathcal{M}(\mathfrak{R}', G; \mathfrak{K}')_{\Psi(\mathfrak{R}')}; \mathfrak{K})_{\Psi(\mathfrak{R})}$$

is uniformly Φ_4-continuous (Corollary 1.8.5).

Let μ be a map $\mathfrak{R} \times \mathfrak{R}' \to G$ such that

$$\mu(A, \cdot) \in \mathcal{M}(\mathfrak{R}', G; \mathfrak{K}'), \quad \mu(\cdot, K') \in \mathcal{M}(\mathfrak{R}, G; \mathfrak{K})$$

for any $(A, K') \in \mathfrak{R} \times \mathfrak{K}'$. Then the map

$$\mathfrak{R} \longrightarrow \mathcal{M}(\mathfrak{R}', G; \mathfrak{K}')_{\mathfrak{K}'}, \quad A \longmapsto \mu(A, .)$$

is a $\mathfrak{K}$-regular measure. By the above considerations it belongs to $\mathcal{M}(\mathfrak{R}, \mathcal{M}(\mathfrak{R}', G; \mathfrak{K}'); \mathfrak{K})$ and by a) this is equivalent to $\mu \in \mathcal{M}$. □

Theorem 4.6.7. *Let E be a Hausdorff locally convex space, Φ be the set of $\hat{\Phi}_3$-filters of E'_E and $\mathfrak{A}$ be a covering of E' with Φ_4-sets of E'_E. We identify E with a subset of $\mathbb{R}^{E'}$ via the map*

$$E \longrightarrow \mathbb{R}^{E'}, \quad x \longmapsto \langle x, \cdot \rangle$$

and assume the identity map $E_\Phi \to E_\mathfrak{A}$ is sequentially continuous. Then

$$\mathscr{E}(\mathfrak{R}, E_{E'}; \mathfrak{K}) = \mathscr{E}(\mathfrak{R}, E_\mathfrak{A}; \mathfrak{K}).$$

By Corollary 4.6.4 b) $\mathscr{E}(\mathfrak{R}, E_{E'}) = \mathscr{E}(\mathfrak{R}, E_\Phi)$ and $\mathscr{E}(\mathfrak{R}, E_{E'}; \mathfrak{K}) = \mathscr{E}(\mathfrak{R}, E_\Phi; \mathfrak{K})$. By Proposition 4.1.17 $\mathscr{E}(\mathfrak{R}, E_{E'}) = \mathscr{E}(\mathfrak{R}, E_\mathfrak{A})$ so, by Corollary 4.1.31 a),

$$\mathscr{E}(\mathfrak{R}, E_{E'}; \mathfrak{K}) = \mathscr{E}(\mathfrak{R}, E_\mathfrak{A}; \mathfrak{K}) \quad \square$$

Remark. The above theorem is a kind of Orlitcz-Pettis theorem for exhaustive maps.

Examples of locally convex spaces E for which the identity map $E_{\Phi} \to E_A$ is sequentially continuous are given in § 5.9 (see Definition 5.9.11). We cannot drop this hypotheses since the map $\mathfrak{P}(\mathbb{N}) \to \ell^{\infty}$ which sends any set of $\mathfrak{P}(\mathbb{N})$ into its characteristic function on $\mathbb{N}$ is an additive exhaustive map with respect to the weak topology of ℓ^{∞}, but it is not exhaustive with respect to the norm topology of ℓ^{∞}.

§ 4.7 Bounded sets

Throughout this section G shall denote a topological additive group.

Definition 4.7.1. *A subset A of G is called* bounded *of for any 0-neigbourhood U in G there exist a finite subset P of G and an $n \in \mathbb{N}$ such that*

$$A \subset P + nU.$$

A family $(x_\iota)_{\iota \in I}$ in G is called bounded *if $\{x_\iota \mid \iota \in I\}$ is bounded.*

If G is a locally convex space then the above defined bounded sets of G coincide with the $\mathbb{R}$-bounded sets from Definition 2.1.25. But if G is merely a topological vector space, then the $\mathbb{R}$-bounded sets of G are bounded but the converse is not true. We hope that no confusion will arise from this awkward situation.

A subset A of G is precompact if for any 0-neighbourhood U in G there exists a finite subset P of G such that

$$A \subset P + U.$$

Hence, any precompact set of G is bounded. Since the two notions bounded and precompact have similar definitions it is possible to treat them simultaneously by introducing a map $\underline{\varrho}$ of $\mathbb{N}$ into itself defined by

$$\underline{\varrho}(n) := \begin{cases} n & \text{in the bounded case} \\ 1 & \text{in the precompact case} \end{cases}$$

Lemma 4.7.2. *Let U be a 0-neighbourhood in G, $(x_n)_{n \in \mathbb{N}}$ be a sequence in G and $(p_n)_{n \in \mathbb{N}}$ be an unbounded increasing sequence in $\mathbb{N}$ such that*

$$x_n \notin \{x_m \mid m \in \mathbb{N},\ m < n\} + \underline{\varrho}(p_n)U$$

for any $n \in \mathbb{N}$. Then $\{x_n \mid n \in M\}$ is not a bounded (precompact) set of G for any infinite subset M of $\mathbb{N}$.

Let M be an infinite subset of $\mathbb{N}$ and assume $\{x_n \mid n \in M\}$ is a bounded (precompact) set of G. Let V be a 0-neighbourhood in G such that $V - V \subset U$. There exist a $p \in \mathbb{N}$ and a finite subset P of G such that

$$\{x_n \mid n \in M\} \subset P + \underline{\varrho}(p)V.$$

Then there exist $x \in P$ and an infinite subset M' of M such that

$$\{x_n | n \in M'\} \subset x + \varrho(p)V.$$

Let $m, n \in M'$ such that $m < n$ and $p \leq p_n$. Then

$$x_n - x_m \in \varrho(p)V - \varrho(p)V \subset \varrho(p)U \subset \varrho(p_n)U,$$

which is a contradiction. □

Proposition 4.7.3. *Let (I, A) be a net in $\mathfrak{P}(G)$ such that for any increasing sequence $(\iota_n)_{n \in \mathbb{N}}$ in I there exists an infinite subset M of $\mathbb{N}$ such that $\bigcup_{n \in M} A_{\iota_n}$ is a bounded (precompact) set of G. Then:*

a) for any 0-neighbourhood U in G there exist $\lambda \in I$, $n \in \mathbb{N}$, and a finite subset P of G such that

$$\bigcup_{\substack{\iota \in I \\ \iota \geq \lambda}} A_\iota \subset P + \varrho(n)U;$$

b) if A_ι is a bounded (precompact) set of G for any $\iota \in I$, then $\bigcup_{n \in \mathbb{N}} A_{\iota_n}$ is a bounded (precompact) set of G for any increasing sequence $(\iota_n)_{n \in \mathbb{N}}$ in I.

a) Let U be a 0-neighbourhood in G. Assume that for any $\lambda \in I$, any $n \in \mathbb{N}$ and any finite subset P of G there exist $\iota \in I$ such that $\iota \geq \lambda$ and

$$A_\iota \not\subset P + \varrho(n)U.$$

We may construct inductively an increasing sequence $(\iota_n)_{n \in \mathbb{N}}$ in I and a sequence $(x_n)_{n \in \mathbb{N}}$ in G such that $x_n \in A_{\iota_n}$ and

$$x_n \notin \{x_m | m \in \mathbb{N}, m < n\} + \varrho(n)U$$

for any $n \in \mathbb{N}$, which contradicts Lemma 4.7.2.

b) Let $(\iota_n)_{n \in \mathbb{N}}$ be an increasing sequence in I and let U be a 0-neighbourhood in G. By a) there exist $m, p \in \mathbb{N}$ and a finite subset P of G such that

$$\bigcup_{n \geq m} A_{\iota_n} \subset P + \varrho(p)U.$$

Since $\bigcup_{n < m} A_{\iota_n}$ is bounded (precompact) there exist $q \in \mathbb{N}$ and a finite subset Q of G such that

$$\bigcup_{n < m} A_{\iota_n} \subset Q + \varrho(q)U$$

and we get

$$\bigcup_{n \in \mathbb{N}} A_{\iota_n} \subset (P \cup Q) + \varrho(p + q)U.$$

Hence, $\bigcup_{n \in \mathbb{N}} A_{\iota_n}$ is bounded (precompact). □

Proposition 4.7.4. *Let $\mathfrak{R}$ be a ring of sets and let (μ, I) be a net in $G^{\mathfrak{R}}$ such that the following hold:*

a) μ_ι is additive for any $\iota \in I$.

b) For any $A \in \mathfrak{R}$ and any increasing sequence $(\iota_n)_{n \in \mathbb{N}}$ in I there exists an infinite subset M of $\mathbb{N}$ such that $\{\mu_{\iota_n}(A) | n \in M\}$ is a bounded set of G $(\{\mu_{\iota_n}(B) | B \in \mathfrak{R}, B \subset A, n \in M\}$ is a precompact set of G).

c) For any disjoint sequence $(A_n)_{n \in \mathbb{N}}$ in $\mathfrak{R}$ and for any increasing sequence $(\iota_n)_{n \in \mathbb{N}}$ in I there exists an infinite subset M of $\mathbb{N}$ such that $\{\mu_{\iota_n}(A_n) | n \in M\}$ is a bounded (precompact) set of G.

Then for any 0-neighbourhood U in G there exist $\lambda \in I$, $n \in \mathbb{N}$, and a finite subset P of G such that

$$\{\mu_\iota(A) | \iota \in I, \iota \geq \lambda, A \in \mathfrak{R}\} \subset P + \varrho(n) U .$$

Let $A \in \mathfrak{R}$ and let U be a symmetric 0-neighbourhood in G. We want to show first that there exist $\lambda \in I$, $n \in \mathbb{N}$, and a finite subset P of G such that

$$\{\mu_\iota(B) | \iota \in I, \iota \geq \lambda, B \in \mathfrak{R}, B \subset A\} \subset P + \varrho(n) U .$$

This follows immediately from b) and Proposition 4.7.3 a) in the precompact case, so we have to prove the assertion in the bounded case only. Assume the assertion does not hold. Then for any $\lambda \in I$, any $n \in \mathbb{N}$ and any finite subset P of G there exist $\iota \in I$ and $B \in \mathfrak{R}$ such that $\iota \geq \lambda$, $B \subset A$, and

$$\mu_\iota(B) \notin P + nU .$$

We construct inductively an increasing sequence $(\iota_n)_{n \in \mathbb{N}}$ in I and a sequence $(A_n)_{n \in \mathbb{N}}$ in $\mathfrak{R}$ such that for any $n \in \mathbb{N}$:

1) $A_n \subset A \setminus \bigcup_{m < n} A_m$;

2) $\mu_{\iota_n}(A_n) \notin \{\mu_{\iota_m}(A_m) | m \in \mathbb{N}, m < n\} + nU$;

3) for any $\lambda \in I$, any $p \in \mathbb{N}$ and any finite subset P of G there exist $\iota \in I$ and $B \in \mathfrak{R}$ such that $\iota \geq \lambda$, $B \subset A \setminus \bigcup_{m \leq n} A_m$, and

$$\mu_\iota(B) \notin P + pU .$$

Let $n \in \mathbb{N}$ and assume the sequences were constructed up to $n - 1$. By b) and Proposition 4.7.3 a) there exist $\lambda \in I, p \in \mathbb{N}$, and a finite subset P of G such that

$$\{\mu_\iota(A \setminus \bigcup_{m < n} A_m) | \iota \in I, \iota \geq \lambda\} \subset P + pU .$$

By 3) (or by the hypothesis of the proof if $n = 0$) there exist $\iota_n \in I$ and $B \in \mathfrak{R}$ such that $\iota_n \geq \lambda, \iota_n \geq \iota_{n-1}, B \subset A \setminus \bigcup_{m < n} A_m$, and

$$\mu_{\iota_n}(B) \notin (P - \{\mu_{\iota_m}(A_m) | m \in \mathbb{N}, m < n\}) \cup \{\mu_{\iota_m}(A_m) | m \in \mathbb{N}, m < n\} + (n + p) U.$$

If there exist $\lambda' \in I, q \in \mathbb{N}$, and a finite subset Q of G such that

$$\{\mu_\iota(C) | \iota \in I, \iota \geq \lambda', C \in \mathfrak{R}, C \subset B\} \subset Q + qU$$

then we set $A_n := B$; otherwise we set

$$A_n := (A \setminus \bigcup_{m<n} A_m) \setminus B .$$

If $A_n = B$, then the conditions 1), 2), 3) are trivially fulfilled. If $A_n \neq B$, then 1) and 3) are fulfilled. Assume 2) is not fulfilled. Then

$$\mu_{\iota_n}(B) = \mu_{\iota_n}(A \setminus \bigcup_{m<n} A_m) - \mu_{\iota_n}(A_n) \in P + pU - \{\mu_{\iota_m}(A_m) | m \in \mathbb{N}, m < n\} - nU ,$$

which is a contradiction. Hence, 2) is also fulfilled in this case and the inductive construction is finished.

By 2) and Lemma 4.7.2 $\{\mu_{\iota_n}(A_n) | n \in M\}$ is not bounded for any infinite subset M of $\mathbb{N}$ and, by 1) this contradicts c).

We now prove the assertion of the proposition. Let U be a 0-neighbourhood in G. Assume that for any $\lambda \in I$, any $n \in \mathbb{N}$ and any finite subset P of G there exist $\iota \in I$ and $A \in \mathfrak{R}$ such that $\iota \geq \lambda$ and

$$\mu_\iota(A) \notin P + \varrho(n) U .$$

Let V be a 0-neighbourhood in G such that $V + V \subset U$. We construct an increasing sequence $(\iota_n)_{n \in \mathbb{N}}$ in I and a disjoint sequence $(A_n)_{n \in \mathbb{N}}$ in $\mathfrak{R}$ inductively such that

$$\mu_{\iota_n}(A_n) \notin \{\mu_{\iota_m}(A_m) | m \in \mathbb{N}, m < n\} + \varrho(n) V$$

for any $n \in \mathbb{N}$. Let $n \in \mathbb{N}$ and assume the sequences were constructed up to $n-1$. By the above considerations there exist $\lambda \in I, p \in \mathbb{N}$, and a finite subset P of G such that

$$\{\mu_\iota(A) | \iota \in I, \iota \geq \lambda, A \in \mathfrak{R}, A \subset \bigcup_{m<n} A_m\} \subset P + \varrho(p) V .$$

Hence, by a) there exist $\iota_n \in I$ and $A_n \in \mathfrak{R}$ such that $\iota_n \geq \iota_{n-1}$, $A_n \bigcap (\bigcup_{m<n} A_m) = \emptyset$, and

$$\mu_{\iota_n}(A_n) \notin \{\mu_{\iota_m}(A_m) | m \in \mathbb{N}, m < n\} + \varrho(n) V .$$

This finishes the inductive construction.

By Lemma 4.7.2 $\{\mu_{\iota_n}(A_n) | n \in M\}$ is not a bounded (precompact) set of G for any infinite subset M of $\mathbb{N}$, which contradicts c). □

Remark. Let μ be an additive map of $\mathfrak{R}$ into G such that $(\mu(A_n))_{n \in \mathbb{N}}$ converges to 0 for any disjoint sequence $(A_n)_{n \in \mathbb{N}}$ in $\mathfrak{R}$. Then $\{\mu(A_n) | n \in \mathbb{N}\}$ is a bounded set for any disjoint sequence $(A_n)_{n \in \mathbb{N}}$ in $\mathfrak{R}$ and by the above proposition $\{\mu(A) | A \in \mathfrak{R}\}$ is a bounded set of G. This result was proved by M. P. Katz (1972) ([1] page 1160) for μ a measure and by L. Drewnowski (1973) ([3] Corollary 1) for G locally convex.

Theorem 4.7.5. *Let X be a set, $\mathfrak{R}$ be a ring of subsets of X containing $\mathfrak{P}_f(X)$ and (μ, I) be a net in $\mathcal{M}(\mathfrak{R}, G; \mathfrak{P}_f(X))$ such that for any disjoint sequence $(K_n)_{n\in\mathbb{N}}$ in $\mathfrak{P}_f(X)$ and for any increasing sequence $(\iota_n)_{n\in\mathbb{N}}$ in I there exists an infinite subset M of $\mathbb{N}$ such that:*

a) $\{\mu_{\iota_n}(K_0)\,|\,n\in M\}$ *is a bounded (precompact) set of G;*

b) $\bigcup_{m\in M} K_m \in \mathfrak{R}$;

c) $\{\mu_{\iota_n}(\bigcup_{m\in M} K_m)\,|\,n\in M\}$ *is a bounded (precompact) set of G.*

Then for any 0-neighbourhood U in G there exist $\lambda\in I$, $n\in\mathbb{N}$, and a finite subset P of G such that

$$\{\mu_\iota(A)\,|\,\iota\in I, \iota\geq\lambda, A\in\mathfrak{R}\}\subset P+\varrho(n)U\,.$$

Let $K\in\mathfrak{P}_f(X)$ and $(\iota_n)_{n\in\mathbb{N}}$ be an increasing sequence in I. Then there exists an infinite subset M of $\mathbb{N}$ such that $\{\mu_{\iota_n}(K)\,|\,n\in M\}$ is a bounded (precompact) set of G. By Proposition 4.7.3 b) $\{\mu_{\iota_n}(L)\,|\,n\in\mathbb{N}, L\subset K\}$ is a bounded (precompact) set of G.

Let $(\iota_n)_{n\in\mathbb{N}}$ be an increasing sequence in I and $(K_n)_{n\in\mathbb{N}}$ be a disjoint sequence in $\mathfrak{P}_f(X)$. Then there exists an infinite subset M of $\mathbb{N}$ such that $\bigcup_{n\in M} K_n\in\mathfrak{R}$. Assume $\{\mu_{\iota_n}(K_n)\,|\,n\in M\}$ is not bounded (precompact). Then there exists a 0-neighbourhood U in G such that

$$\{\mu_{\iota_n}(K_n)\,|\,n\in M\}\not\subset P+\varrho(p)U$$

for any $p\in\mathbb{N}$ and for any finite subset P of G. Let V be a symmetric 0-neighbourhood in G such that $5V\subset U$. By the above considerations there exist an increasing sequense $(P_n)_{n\in\mathbb{N}}$ of finite subsets of G and an increasing sequence $(p_n)_{n\in\mathbb{N}}$ in $\mathbb{N}$ such that

$$\{\mu_{\iota_m}(K)\,|\,m\in\mathbb{N}, K\subset\bigcup_{i\leq n} K_i\}\subset P_n+\varrho(p_n)V$$

for any $n\in\mathbb{N}$. Since μ_ι is $\mathfrak{P}_f(X)$-regular for any $\iota\in I$, there exists an increasing sequence $(k_n)_{n\in\mathbb{N}}$ in $\mathbb{N}$ such that

$$\{\mu_{\iota_n}(A)\,|\,A\in\mathfrak{R}, A\subset\bigcup_{\substack{m\in M\\ m\geq k_n}} K_m\}\subset V\,.$$

for any $n\in\mathbb{N}$.

We may construct inductively an increasing sequence $(l(n))_{n\in\mathbb{N}}$ in M such that for any $n\in\mathbb{N}\backslash\{0\}$

1) $l(n)\geq k_{n-1}$;

2) $\mu_{\iota_{l(n)}}(K_{l(n)})\notin\{\mu_{\iota_{l(m)}}(K_{l(m)})\,|\,m\in\mathbb{N}, m<n\}+P_{l(n-1)}-P_{l(n-1)}+$
$+\varrho(n+2p_{l(n-1)}+2)U\,.$

Then there exists an infinite subset M_0 of $\{l(n)\,|\,n\in\mathbb{N}\}$ such that

$$A := \bigcup_{n \in M_0} K_n \in \mathfrak{R}$$

and such that $\{\mu_{\iota_n}(A) | n \in M_0\}$ is a bounded (precompact) set of G. Let $m, n \in \mathbb{N}$ with $m < n$ and $l(m), l(n) \in M_0$. Assume

$$\mu_{\iota_{l(n)}}(A) - \mu_{\iota_{l(m)}}(A) \in \varrho(n)V.$$

Then by 1)

$$\mu_{\iota_{l(n)}}(K_{l(n)}) - \mu_{\iota_{l(m)}}(K_{l(m)}) = (\mu_{\iota_{l(n)}}(A) - \mu_{\iota_{l(n)}}(\bigcup_{\substack{q \in M_0 \\ q < l(n)}} K_q) - \mu_{\iota_{l(n)}}(\bigcup_{\substack{q \in M_0 \\ q > l(n)}} K_q)) -$$

$$- (\mu_{\iota_{l(m)}}(A) - \mu_{\iota_{l(m)}}(\bigcup_{\substack{q \in M_0 \\ q < l(m)}} K_q) - \mu_{\iota_{l(m)}}(\bigcup_{\substack{q \in M_0 \\ q > l(m)}} K_q)) \in \varrho(n)V - P_{l(n-1)} -$$

$$- \varrho(p_{l(n-1)})V - V + P_{l(m-1)} + \varrho(p_{l(m-1)})V + V \subset P_{l(n-1)} - P_{l(n-1)} +$$

$$+ \varrho(n + 2p_{l(n-1)} + 2)U$$

and this contradicts 2). Hence,

$$\mu_{\iota_{l(n)}}(A) \notin \{\mu_{\iota_{l(m)}}(A) | m \in \mathbb{N}, m < n, l(m) \in M_0\} + \varrho(n)V$$

for any $n \in \mathbb{N}$, $l(n) \in M_0$ and by Lemma 4.7.2 this is a contradiction. Hence, $\{\mu_{\iota_n}(K_n) | n \in M\}$ is a bounded (precompact) set of G.

Let U be a closed 0-neighbourhood in G. By the above proof and by Proposition 4.7.4 there exist $\lambda \in I$, $n \in \mathbb{N}$, and a finite subset P of G such that

$$\{\mu_\iota(K) | \iota \in I, \iota \geq \lambda, K \in \mathfrak{P}_f(X)\} \subset P + \varrho(n)U.$$

Since U is closed and since μ_ι is $\mathfrak{P}_f(X)$-regular for any $\iota \in I$ we get

$$\{\mu_\iota(A) | \iota \in I, \iota \geq \lambda, A \in \mathfrak{R}\} \subset P + \varrho(n)U. \quad \square$$

Corollary 4.7.6. *Let I be a set and $\mathfrak{J}$ be a set of subsets of I such that:*

a) $\mathfrak{P}_f(I) \subset \mathfrak{J}$;

b) for any disjoint sequence $(J_n)_{n \in \mathbb{N}}$ in $\mathfrak{P}_f(I)$ there exists an infinite subset M of $\mathbb{N}$ such that $\bigcup_{n \in M} J_n \in \mathfrak{J}$ and $0 \in M$.

Assume G Hausdorff and let $\mathscr{F}$ be a subset of $\ell(I, G)$ such that $\{\sum_{\iota \in J} f(\iota) | f \in \mathscr{F}\}$ is a bounded (precompact) set of G for any $J \in \mathfrak{J}$. Then

$$\{\sum_{\iota \in J} f(\iota) | f \in \mathscr{F}, J \subset I\}$$

is a bounded (precompact) set of G. $\square$

Theorem 4.7.7. *Let $\mathfrak{R}$ be a quasi-σ-ring and (μ, I) be a net in $\mathscr{E}(\mathfrak{R}, G)$ such that for any increasing sequence $(\iota_n)_{n \in \mathbb{N}}$ in I there exists an infinite subset M of $\mathbb{N}$ such that*

$\{\mu_{\iota_n}(A) \mid n \in M\}$ is a bounded set of G ($\{\mu_{\iota_n}(B) \mid n \in M, B \in \mathfrak{R}, B \subset A\}$ is a precompact set of G) for any $A \in \mathfrak{R}$. Then, for any 0-neighbourhood U in G, there exist $\lambda \in I$, $n \in \mathbb{N}$ and a finite subset P of G such that

$$\{\mu_\iota(A) \mid \iota \in I, \iota \geq \lambda, A \in \mathfrak{R}\} \subset P + \varrho(n)U.$$

Let $(A_n)_{n \in \mathbb{N}}$ be a disjoint sequence in $\mathfrak{R}$, $(\iota_n)_{n \in \mathbb{N}}$ be an increasing sequence in I and U be a 0-neighbourhood in G. By Lemma 4.1.19 there exist a metrizable topological additive group H, a 0-neighbourhood V in H and a continuous surjective group homomorphism $u: G \to H$ such that $\overset{-1}{u}(V) \subset U$. By Proposition 4.1.17 $u \circ \mu_{\iota_n} \in \mathscr{E}(\mathfrak{R}, H)$ for any $n \in \mathbb{N}$. By Proposition 4.1.18 there exists an infinite subset M of $\mathbb{N}$ such that $(A_n)_{n \in M} \in \Gamma(\mathfrak{R})$ and

$$\mathfrak{P}(M) \to H, \quad J \mapsto u \circ \mu_{\iota_n}\Big(\bigcup_{m \in J} A_m\Big)$$

is a measure for any $n \in \mathbb{N}$. By Theorem 4.7.5 there exist $m_0, p' \in \mathbb{N}$ and a finite subset P' of H such that

$$\Big\{u \circ \mu_{\iota_n}\Big(\bigcup_{m \in J} A_m\Big) \,\Big|\, n \in \mathbb{N}, n \geq m_0, J \subset M\Big\} \subset P' + \varrho(p')V.$$

By Proposition 4.1.14 a $\Rightarrow$ c

$$\Big\{u \circ \mu_{\iota_n}\Big(\bigcup_{m \in J} A_m\Big) \,\Big|\, J \subset M\Big\}$$

is a compact set for any $n \in \mathbb{N}$. Hence, there exist $p'' \in \mathbb{N}$ and a finite subset P'' of H such that

$$\Big\{u \circ \mu_{\iota_n}\Big(\bigcup_{m \in J} A_m\Big) \,\Big|\, n \in \mathbb{N}, J \subset M\Big\} \subset P'' + \varrho(p'')V.$$

Let P be a finite subset of G such that $u(\mathrm{P}) = P''$. We get

$$\Big\{\mu_{\iota_n}\Big(\bigcup_{m \in J} A_m\Big) \,\Big|\, n \in \mathbb{N}, J \subset M\Big\} \subset P + U + \varrho(p'')U.$$

Since U is arbitrary the set $\{\mu_{\iota_n}(\bigcup_{m \in J} A_m) \mid n \in \mathbb{N}, J \subset M\}$ is bounded (precompact). In particular, $\{\mu_{\iota_n}(A_n) \mid n \in M\}$ is bounded (precompact).

By Proposition 4.7.4 for any 0-neighbourhood U in G there exist $\lambda \in I, n \in \mathbb{N}$, and a finite subset P of G such that

$$\{\mu_\iota(A) \mid \iota \in I, \iota \geq \lambda, A \in \mathfrak{R}\} \subset P + \varrho(n)U. \quad \square$$

Corollary 4.7.8. *Let $\mathfrak{R}$ be a quasi-σ-ring and let $\mathscr{M}$ be a set of G-valued additive exhaustive maps on $\mathfrak{R}$ such that $\{\mu(A) \mid \mu \in \mathscr{M}\}$ is a bounded set of G ($\{\mu(B) \mid \mu \in \mathscr{M}, B \in \mathfrak{R}, B \subset A\}$ is a precompact set of G) for any $A \in \mathfrak{R}$. Then $\{\mu(A) \mid \mu \in \mathscr{M}, A \in \mathfrak{R}\}$ is a bounded (precompact) set of G.* $\square$

Remark. This result was proved by O. Nikodym (1931) ([1] I), (1933) ([2] page 418) for $\mathfrak{R}$ a σ-algebra and $\mathcal{M}$ a set of real valued measures on $\mathfrak{R}$. J. Mikusińsky (1971) ([1] Theorem N) extended the bounded part of Nikodym's theorem to the case when $\mathcal{M}$ is a set of measures with values in a Banach space. R. B. Darst proved (1967) ([1] Theorem) that one may replace the measures in Nikodym's theorem by exhaustive additive maps in $\mathbb{R}$. A stronger result than the above corollary was proved by H. Weber ([1] (5.1.8)).

Corollary 4.7.9. *Let $\mathfrak{R}$ be a quasi-σ-ring, E be a locally convex space and μ be an additive map of $\mathfrak{R}$ into E. Then μ is weakly exhaustive iff $\mu(\mathfrak{R})$ is bounded.*

If μ is weakly exhaustive, then, by Corollary 4.7.8, $\mu(\mathfrak{R})$ is weakly bounded i.e. $\mu(\mathfrak{R})$ is bounded. Assume now that $\mu(\mathfrak{R})$ is bounded. Let x' be a continuous linear form on E. Then $x' \circ \mu(\mathfrak{R})$ is bounded so, by Proposition 4.1.8, $x' \circ \mu$ is exhaustive. Since x' is arbitrary μ is weakly exhaustive. □

Corollary 4.7.10. *Let $\mathfrak{R}$ be a quasi-σ-ring, E be a locally convex space and $\mathcal{M}$ be a set of additive maps of $\mathfrak{R}$ into E such that $\mu(\mathfrak{R})$ is bounded for any $\mu \in \mathcal{M}$ and $\{\mu(A) \mid \mu \in \mathcal{M}\}$ is bounded for any $A \in \mathfrak{R}$. Then*

$$\{\mu(A) \mid \mu \in \mathcal{M}, A \in \mathfrak{R}\}$$

is bounded.

By Corollary 4.7.9 any $\mu \in \mathcal{M}$ is weakly exhaustive and the assertion follows from Corollary 4.7.8. □

Remark. Let $(\mu_n)_{n \in \mathbb{N}}$ be a sequence of additive exhaustive maps of $\mathfrak{R}$ into E such that $(\mu_n(A))_{n \in \mathbb{N}}$ converges weakly for any $A \in \mathfrak{R}$. By the Corollaries 4.7.9 and 4.7.10 $\bigcup_{n \in \mathbb{N}} \mu_n(\mathfrak{R})$ is a bounded set of E. This result was proved by L. Drewnowski (1972) ([2] Theorem 2) for E a normed space and $\mathfrak{R}$ a σ-ring. Corollary 4.7.10 was proved by L. Drewnowski (1973) ([3] Theorem 1) for $\mathfrak{R}$ a σ-ring.

Proposition 4.7.11. *(L. Drewnowski, I. Labuda (1973) [1] 2.7 Lemme). Let E be a B_r-complete locally convex space, E' be its dual, A' be a dense set of E'_E, $\mathfrak{R}$ be a ring of sets and μ be a map $\mathfrak{R} \to E$ such that $x' \circ \mu$ is additive and exhaustive for any $x' \in A'$. Then μ is additive and weakly exhaustive.*

It is easy to see that μ is additive. Let us denote by F' the set of $x' \in E'$ for which $x' \circ \mu$ is exhaustive. Then F' is a dense vector subspace of E'_E. Let B' be an equicontinuous set of E'. Furthermore, let $(A_n)_{n \in \mathbb{N}} \in \Gamma(\mathfrak{R})$. Then

$$\{x' \circ \mu(\bigcup_{n \in M} A_n) \mid x' \in B'\}$$

is a bounded set for any $M \subset \mathbb{N}$. By the Corollaries 4.7.8 and 4.7.9

$$\alpha := \sup \{|x' \circ \mu(\bigcup_{n \in M} A_n)| \mid M \subset \mathbb{N}, x' \in F' \cap B'\} < \infty .$$

Let x' be a point of B' which belongs to the closure of $F' \cap B'$ in E'_E. Then

$$|x' \circ \mu(\bigcup_{n \in M} A_n)| \leq \alpha$$

for any $M \subset \mathbb{N}$. By Proposition 4.1.8

$$\lim_{n \to \infty} x' \circ \mu(A_n) = 0 .$$

Since $(A_n)_{n \in \mathbb{N}}$ is arbitrary $x' \circ \mu$ is exhaustive, i.e. $x' \in F'$. Hence, $F' \cap B'$ is a closed set of $B'_{E'}$. Since B' is arbitrary and E B_r-complete $F' = E'$ and μ is weakly exhaustive. □

Proposition 4.7.12. *Let G be Hausdorff, H be a Hausdorff sequentially complete topological additive group, $u: G \to H$ be a group homomorphism with sequentially closed graph, $\mathfrak{R}$ be a ring of sets and $\mu: \mathfrak{R} \to G$ be a measure such that $u \circ \mu$ is exhaustive. Then*

a) $u \circ \mu$ is a measure;

b) if $(u(x_n))_{n \in \mathbb{N}}$ converges to 0 for any supersummable sequence $(x_n)_{n \in \mathbb{N}}$ in G, then $u \in \mathscr{P}^c(G, H)$.

a) Let $(A_n)_{n \in \mathbb{N}} \in \Gamma(\mathfrak{R})$. By Proposition 4.1.13 a) $(u \circ \mu(A_n))_{n \in \mathbb{N}}$ is summable and since u has sequentially closed graph

$$\sum_{n \in \mathbb{N}} u \circ \mu(A_n) = \lim_{m \to \infty} \sum_{n=0}^{m} u \circ \mu(A_n) = \lim_{m \to \infty} u \circ \mu\left(\bigcup_{n=0}^{m} A_n\right) = u \circ \mu(\bigcup_{n \in \mathbb{N}} A_n) .$$

b) Let $(x_n)_{n \in \mathbb{N}}$ be a supersummable sequence in G. We set

$$v: \mathfrak{P}(\mathbb{N}) \to G, \quad M \mapsto \sum_{n \in M} x_n .$$

Then v is a measure and $u \circ v$ is exhaustive. By a) $u \circ v$ is a measure and therefore $(u(x_n))_{n \in \mathbb{N}}$ is supersummable and

$$u(\sum_{n \in \mathbb{N}} x_n) = \sum_{n \in \mathbb{N}} u(x_n) .$$

Hence, $u \in \mathscr{P}^c(G, H)$. □

Remark. b) was proved by C. Bennet and N. J. Kalton (1972) ([1] Theorem 1) for G, H locally convex spaces, H B_r-complete, and the graph of u closed.

Corollary 4.7.13. *Let G, H be Hausdorff topological additive groups having the same underlying group such that the identity map $H \to G$ is continuous and such that H is sequentially complete, let $\mathfrak{R}$ be a ring of sets and $\mu: \mathfrak{R} \to H$ be an exhaustive map such that the map $\mathfrak{R} \to G$, defined by μ, is a measure. Then $\mu: \mathfrak{R} \to H$ is a measure.*

Since the identity map $H \to G$ is continuous it has a closed graph. Hence, the identity map $G \to H$ has closed graph and the assertion follows from Proposition 4.7.12. □

Corollary 4.7.14. *Let E be a sequentially complete Hausdorff locally convex space, E' be its dual, A' be a dense set of E'_E, $\mathfrak{R}$ be a ring of sets and $\mu:\mathfrak{R}\to E$ be an exhaustive map such that $x'\circ\mu$ is a measure for any $x'\in A'$. Then μ is a measure.*

We identify E with a subset of $\mathbb{R}^{E'}$ via the map

$$E \to \mathbb{R}^{E'}, \quad x \mapsto \langle \cdot, x\rangle .$$

Then $E_{A'}$ is Hausdorff and the identity map $E\to E_{A'}$ is continuous. The assertion follows from Corollary 4.7.13. □

Lemma 4.7.15. *(R.S. Phillips (1940) [1] Corollary 7.3). Let E be a locally convex space, F be a vector subspace of E, I be a set and $u: F\to \ell^\infty(I)$ be a continuous linear map. Then there exists a continuous linear extension $E\to\ell^\infty(I)$ of u.*

Let U be a convex circled 0-neighbourhood in E such that

$$u(U\cap F)\subset\{\alpha\in\ell^\infty(I)\,|\,\|\alpha\|\le 1\}.$$

By the Hahn–Banach theorem there exists a continuous linear form x'_ι on E for any $\iota\in I$ such that

$$x\in F \Rightarrow x'_\iota(x) = u(x)_\iota,$$

$$x\in U \Rightarrow |x'_\iota(x)|\le 1.$$

We set

$$x'_\alpha: E \to \mathbb{R}, \quad x\mapsto \sum_{\iota\in I}\alpha_\iota x'_\iota(x)$$

for any $\alpha\in\ell^1(I)$; then $|x'_\alpha(x)|\le\|\alpha\|_1$ for any $x\in U$, where $\|\alpha\|_1$ denotes the norm of α in $\ell^1(I)$. For any $x\in E$ the map

$$v(x):\ell^1(I)\to\mathbb{R}, \quad \alpha\mapsto x'_\alpha(x)$$

is continuous and therefore belongs to $\ell^\infty(I)$. It is obvious that $v: E\to\ell^\infty(I)$ is linear and by the above remark $\|v(x)\|\le 1$ for any $x\in U$, i.e. v is continuous. We have

$$v(x)_\iota = x'_\iota(x) = u(x)_\iota$$

for any $x\in F$ and for any $\iota\in I$, i.e. $v|F = u$. □

Theorem 4.7.16. *Let E be a B_r-complete locally convex space and E' be its dual. Then the following assertion are equivalent:*

a) E contains no copy of ℓ^∞.

b) Any map μ of a ring of sets into E is additive and exhaustive if there exists a dense set A' of E'_E such that $x'\circ\mu$ is additive and exhaustive for any $x'\in A'$.

c) Any map μ of a ring of sets into E is a measure if there exists a dense set A' of E'_E such that $x'\circ\mu$ is a measure for any $x'\in A'$.

a ⇒ b. By Proposition 4.7.11 μ is additive and weakly exhaustive. By Corollary

4.7.9 $\{\mu(\bigcup_{n\in M} A_n) | M \subset \mathbb{N}\}$ is bounded for any $(A_n)_{n\in\mathbb{N}} \in \Gamma(\mathfrak{R})$, where $\mathfrak{R}$ denotes the ring of sets on which μ is defined, so by Corollary 4.1.44 d $\Rightarrow$ b μ is exhaustive.

b $\Rightarrow$ c follows immediately from Corollary 4.7.14.

c $\Rightarrow$ a. By α' we denote the map

$$\ell^\infty \to \mathbb{R}, \quad \beta \mapsto \sum_{n\in\mathbb{N}} \alpha_n \beta_n$$

for any $\alpha \in \ell^1$. Then $\alpha' \in (\ell^\infty)'$ and $\{\alpha' | \alpha \in \ell^1\}$ is dense in $(\ell^\infty)'_{\ell^\infty}$, where $(\ell^\infty)'$ denotes the dual of ℓ^∞. For any $A \subset \mathbb{N}$ we denote by 1_A the element of ℓ^∞ equal to 1 on A and equal to 0 on $\mathbb{N}\setminus A$ and set

$$v : \mathfrak{P}(\mathbb{N}) \to \ell^\infty, \quad A \mapsto 1_A .$$

Then $\alpha' \circ v$ is a measure for any $\alpha \in \ell^1$ but v is not a measure.

Assume a) does not hold. Then there exist a topological vector subspace F of E and an isomorphism of locally convex spaces $u : F \to \ell^\infty$. By Lemma 4.7.15 there exists a continuous linear extension $v : E \to \ell^\infty$ of u. We set $G := \overset{-1}{v}(0)$ and denote by G' the dual of G, by t the inclusion map $F \to E$, by w the map $t \circ u^{-1}$, for any $(\alpha, x') \in \ell^1 \times G'$ by $\varphi_{\alpha, x'}$ the map

$$E \to \mathbb{R}, \quad x \mapsto \alpha'(v(x)) + x'(x - w \circ v(x)),$$

and by A' the set

$$\{\varphi_{\alpha,x'} | (\alpha, x') \in \ell^1 \times G'\}.$$

It is easy to see that A' is a dense set of E'_E and $x' \circ w \circ v$ is a measure for any $x' \in A'$. By c) $w \circ v$ is a measure so $v = v \circ w \circ v$ is also a measure, which is a contradiction. □

Remark. Let E be a Fréchet space, E' be its dual, A' be a dense set of E'_E and μ be a map of a ring of sets into E such that $x' \circ \mu$ is a measure for any $x' \in A'$. By H. H. Schaefer ([1] IV 6.4 Theorem) E is B_r-complete. If E is separable, then E does not contain any copy of ℓ^∞ so by the above theorem μ is a measure; this result was proved by E. Thomas (1974) ([4] Theorem 0.2). If any sequence $(x_n)_{n\in\mathbb{N}}$ in E with

$$x' \in E' \Rightarrow \sum_{n\in\mathbb{N}} |x'(x_n)| < \infty$$

is summable, then again E contains no copy of ℓ^∞ so, by the above theorem, μ is a measure; this result was proved by E. Thomas (1974) ([4] Theorem 0.4). The above equivalence a $\Leftrightarrow$ c was proved by J. Diestel and B. Faires (1974) for E a Banach space ([1] Theorem 1.1 (i) $\Leftrightarrow$ (iii)). The equivalence a $\Leftrightarrow$ b was proved by I. Labuda (1975) ([4] (8)).

§ 4.8 Bounded sets and measures on topological spaces

Throughout this section G shall denote a topological additive group, $\mathfrak{R}$ a ring of sets and $\mathfrak{K}$ a subset of $\mathfrak{R}$ closed under finite unions.

Some results of this section were published by the author in (1981) [6].

As in the preceding section we simultaneously treat the bounded and the precompact sets by using the map $\underline{\varrho}: \mathbb{N} \to \mathbb{N}$ defined by

$$\underline{\varrho(n)} := \begin{cases} n \text{ in the bounded case} \\ 1 \text{ in the precompact case.} \end{cases}$$

Proposition 4.8.1. *Let $\mathcal{M}$ be a set of $\mathfrak{K}$-regular, additive maps of $\mathfrak{R}$ into G, $\mathfrak{T}$ be a subset of $\mathfrak{R}$ and φ be a map defined on $\mathfrak{T}$ such that the following hold:*

a) For any $K \in \mathfrak{K}$ and for any $T', T'' \in \mathfrak{T}$ with $K \subset T' \cap T''$ there exist $T \in \mathfrak{T}$ and $S \in \varphi(T)$ such that

$$K \subset T \subset S \subset T' \cap T''.$$

b) $\mathfrak{T}$ separates $\mathfrak{K}$.

c) For any sequence $(\mu_n)_{n \in \mathbb{N}}$ in $\mathcal{M}$ and any sequence $(T_n)_{n \in \mathbb{N}}$ in $\mathfrak{T}$ whose union is contained in a set of $\mathfrak{T}$ and for which there exists a disjoint sequence $(S_n)_{n \in \mathbb{N}}$ of sets such that $T_n \subset S_n \in \varphi(T_n)$ for any $n \in \mathbb{N}$, there exists an infinite subset M of $\mathbb{N}$ such that $\{\mu_n(T_n) \mid n \in M\}$ is a bounded (precompact) set of G.

Then, for any $K \in \mathfrak{K}$ and any 0-neighbourhood U in G, there exist $T \in \mathfrak{T}$, $n \in \mathbb{N}$ and a finite subset P of G such that $K \subset T$ and

$$\{\mu(A) \mid (\mu, A) \in \mathcal{M} \times \mathfrak{R}, A \subset T \backslash K\} \subset P + \varrho(n) U.$$

Assume the contrary. Then there exist $K \in \mathfrak{K}$ and a 0-neighbourhood U in G such that, for any $T \in \mathfrak{T}$ with $K \subset T$, any $n \in \mathbb{N}$ and any finite subset P of G, there exists $(\mu, A) \in \mathcal{M} \times \mathfrak{R}$ such that $A \subset T \backslash K$ and

$$\mu(A) \notin P + \varrho(n) U.$$

Let V be a 0-neighbourhood in G such that $3V \subset U$. Since $\emptyset \in \mathfrak{K}$ there exists by b) a $T'_{-1} \in \mathfrak{T}$ with $K \subset T'_{-1}$. We construct inductively a sequence $(\mu_n)_{n \in \mathbb{N}}$ in $\mathcal{M}$, two sequences $(T_n)_{n \in \mathbb{N}}$, $(T'_n)_{n \in \mathbb{N}}$ in $\mathfrak{T}$, and a sequence $(S_n)_{n \in \mathbb{N}}$ such that for any $n \in \mathbb{N}$:

1) $K \subset T'_n \subset T'_{n-1}$;
2) $T_n \subset S_n \subset T'_{n-1} \backslash T'_n$;
3) $S_n \in \varphi(T_n)$;
4) $\mu_n(T_n) \notin \{\mu_m(T_m) \mid m \in \mathbb{N}, m < n\} + \varrho(n) V$.

Let $n \in \mathbb{N}$ and assume the sequences were constructed up to $n-1$. By the hypothesis of the proof there exists $(\mu_n, A) \in \mathcal{M} \times \mathfrak{R}$ such that $A \subset T'_{n-1} \backslash K$ and

$$\mu_n(A) \notin \{\mu_m(T_m) | m \in \mathbb{N}, m < n\} + \varrho(n+1)U.$$

Since μ_n is $\mathfrak{K}$-regular there exists $L \in \mathfrak{K}$ such that $L \subset A$ and

$$\mu_n(A) - \mu_n(L) \in V.$$

By b) there exist disjoint sets $T', T'' \in \mathfrak{T}$ with $K \subset T'$, $L \subset T''$. By a), b), and Proposition 4.1.47 there exists $T \in \mathfrak{T}$ such that $L \subset T$ and

$$\mu_n(L) - \mu_n(B) \in V$$

for any $B \in \mathfrak{R}$ with $L \subset B \subset T$. By a) there exist $T_n, T_n' \in \mathfrak{T}$ and $S_n \in \varphi(T_n)$ such that

$$K \subset T_n' \subset T' \cap T_{n-1}', \quad L \subset T_n \subset S_n \subset T \cap T'' \cap T_{n-1}'.$$

The above conditions 1), 2), 3) are obviously fulfilled. We have

$$\mu_n(A) = (\mu_n(A) - \mu_n(L)) + (\mu_n(L) - \mu_n(T_n) + \mu_n(T_n)) \in \mu_n(T_n) + 2V$$

and this implies 4). This finishes the inductive construction. By c) there exists an infinite subset M of $\mathbb{N}$ such that $\{\mu_n(T_n) | n \in M\}$ is a bounded (precompact) set of G and by Lemma 4.7.2 this contradicts 4). □

Proposition 4.8.2. *Let $\mathcal{M}$ be a set of $\mathfrak{R}$-regular additive maps of $\mathfrak{R}$ into G, $\mathfrak{T}$ be a subset of $\mathfrak{R}$ and φ be a map defined on $\mathfrak{T}$ such that the following hold.*

1) For any $K \in \mathfrak{K}$ and any $T', T'' \in \mathfrak{T}$ with $K \subset T' \cap T''$ there exist $T \in \mathfrak{T}$ and $S \in \varphi(T)$ such that

$$K \subset T \subset S \subset T' \cap T'';$$

2) $\mathfrak{T}$ separates $\mathfrak{K}$;

3) for any sequence $(\mu_n)_{n \in \mathbb{N}}$ in $\mathcal{M}$, any continuous group homomorphism u of G into a metrizable topological additive group H and any sequence $(T_n)_{n \in \mathbb{N}}$ in $\mathfrak{T}$ whose union is contained in a set of $\mathfrak{T}$ and for which there exists a disjoint sequence $(S_n)_{n \in \mathbb{N}}$ of sets such that $T_n \subset S_n \in \varphi(T_n)$ for any $n \in \mathbb{N}$, there exist an infinite subset M of $\mathbb{N}$ and a map $\psi : \mathfrak{P}(M) \to \mathfrak{T}$ such that $\psi(\{n\}) = T_n$ and such that $u \circ \mu_n \circ \psi$ is an H-valued measure in $\mathfrak{P}(M)$ for any $n \in M$;

4) $\{\mu(T) | \mu \in \mathcal{M}\}$ is a bounded (precompact) set of G for any $T \in \mathfrak{T}$.

Then:

a) for any $K \in \mathfrak{K}$ and for any 0-neighbourhood U in G there exist $T \in \mathfrak{T}$, $n \in \mathbb{N}$ and a finite subset P of G such that $K \subset T$ and

$$\{\mu(A) | (\mu, A) \in \mathcal{M} \times \mathfrak{R},\ A \subset T \backslash K\} \subset P + \varrho(n)U;$$

b) $\{\mu(K) | \mu \in \mathcal{M}\}$ is a bounded (precompact) set of G for any $K \in \mathfrak{K}$.

a) Let $(\mu_n)_{n \in \mathbb{N}}$ be a sequence in $\mathcal{M}$ and $(T_n)_{n \in \mathbb{N}}$ be a sequence in $\mathfrak{T}$ whose union is contained in a set of $\mathfrak{T}$ and for which there exists a disjoint sequence $(S_n)_{n\in\mathbb{N}}$ of sets such that $T_n \subset S_n \in \varphi(T_n)$ for any $n \in \mathbb{N}$. There exist a surjective continuous group homomorphism u of G into a metrizable topological additive group H and a 0-neighbourhood V in H such that $2\overset{-1}{u}(V) \subset U$ (Lemma 4.1.19). By 3) there exist an infinite subset M of $\mathbb{N}$ and a map $\psi : \mathfrak{P}(M) \to \mathfrak{T}$ such that $\psi(\{n\}) = T_n$ and such that $u \circ \mu_n \circ \psi$ is an H-valued measure on $\mathfrak{P}(M)$ for any $n \in M$. By 4) and Theorem 4.7.5

$$\{u \circ \mu_n(T_n) \mid n \in M\}$$

is a bounded (precompact) set of H. By 1), 2), and Proposition 4.8.1 there exist $T \in \mathfrak{T}$, $n \in \mathbb{N}$ and a finite subset Q of H such that $K \subset T$, $n > 1$, and

$$\{u \circ \mu(A) \mid (\mu, A) \in \mathscr{M} \times \mathfrak{R}, A \subset T\backslash K\} \subset Q + \varrho(n-1)V.$$

Since u is surjective there exists a finite subset P of G such that $u(P) = Q$. Let $(\mu, A) \in \mathscr{M} \times \mathfrak{R}$ with $A \subset T\backslash K$. Then there exist $x \in Q$ and a family $(x_m)_{1 \le m \le \varrho(n-1)}$ in V such that

$$u \circ \mu(A) = x + \sum_{m=1}^{\varrho(n-1)} x_m .$$

Let $y \in P$ with $u(y) = x$. Since u is surjective ther exists a family $(y_m)_{1 \le m \le \varrho(n-1)}$ in $\overset{-1}{u}(V)$ such that $u(y_m) = x_m$ for any m. We get

$$u(\mu(A) - y - \sum_{m=1}^{\varrho(n-1)} y_m) = 0$$

so

$$\mu(A) - y - \sum_{m=1}^{\varrho(n-1)} y_m \in \overset{-1}{u}(V),$$

$$\mu(A) \in P + \varrho(n)U .$$

Hence,

$$\{\mu(A) \mid (\mu, A) \in \mathscr{M} \times \mathfrak{R}, A \subset T\backslash K\} \subset P + \varrho(n)U .$$

b) Let U be a 0-neighbourhood in G and $K \in \mathfrak{R}$. Further, let V be a 0-neighbourhood in G such that $V - V \subset U$. By a) there exist $T \in \mathfrak{T}$, $n \in \mathbb{N}$, and a finite subset P of G such that $K \subset T$ and

$$\{\mu(T\backslash K) \mid \mu \in \mathscr{M}\} \subset P + \varrho(n)V .$$

By 4) there exist $n' \in \mathbb{N}$ and a finite subset P' of G such that

$$\{\mu(T) \mid \mu \in \mathscr{M}\} \subset P' + \varrho(n')V .$$

We get

$$\{\mu(K) \mid \mu \in \mathscr{M}\} \subset P' - P + \varrho(n')V - \varrho(n)V \subset P' - P + \varrho(n+n')U .$$

Hence, $\{\mu(K) \mid \mu \in \mathscr{M}\}$ is a bounded (precompact) set of G. □

Proposition 4.8.3. *Let $\mathcal{M}$ be a set of $\mathfrak{K}$-regular additive maps of $\mathfrak{R}$ into G, $\mathfrak{T}$ be a subset of $\mathfrak{R}$ and φ be a map defined on $\mathfrak{T}$ such that the following hold:*

a) For any $K \in \mathfrak{K}$ and for any T', T'' with $K \subset T' \cap T''$ there exist $T \in \mathfrak{T}$ and $S \in \varphi(T)$ such that

$$K \subset T \subset S \subset T' \cap T''.$$

b) $\mathfrak{T}$ separates $\mathfrak{K}$.

c) For any $(K,T) \in \mathfrak{K} \times \mathfrak{T}$ for which there exists $S \in \varphi(T)$ with $T \subset S$ and $K \cap S = \emptyset$ there exists $T' \in \mathfrak{T}$ with $K \subset T'$ and $T \cap T' = \emptyset$.

d) Any set of $\mathfrak{R}$ is contained in a set of $\mathfrak{T}$.

e) For any sequence $(\mu_n)_{n \in \mathbb{N}}$ in $\mathcal{M}$, any continuous group homomorphism u of G into a metrizable topological additive group H and any sequence $(T_n)_{n \in \mathbb{N}}$ in $\mathfrak{T}$ whose union is contained in a set of $\mathfrak{T}$ and for which there exists a disjoint sequence $(S_n)_{n \in \mathbb{N}}$ of sets such that $T_n \subset S_n \in \varphi(T_n)$ for any $n \in \mathbb{N}$, there exist an infinite subset M of $\mathbb{N}$ and a map $\psi: \mathfrak{P}(M) \to \mathfrak{T}$ such that $\psi(\{n\}) = T_n$ and such that $u \circ \mu_n \circ \psi$ is an H-valued measure on $\mathfrak{P}(M)$ for any $n \in M$.

f) $\{\mu(T) \mid \mu \in \mathcal{M}\}$ is a bounded (precompact) set of G for any $T \in \mathfrak{T}$.

Then $\{\mu(A) \mid \mu \in \mathcal{M}\}$ is a bounded (precompact) set of G for any $A \in \mathfrak{R}$.

Assume the contrary. Then there exist A and a 0-neighbourhood U in G such that for any $n \in \mathbb{N}$ and for any finite subset P of G there exists $\mu \in \mathcal{M}$ with

$$\mu(A) \notin P + \varrho(n)U.$$

By d) there exists $T \in \mathfrak{T}$ containing A. Let V be a 0-neighbourhood in G such that $3V \subset U$. We construct inductively a sequence $(\mu_n)_{n \in \mathbb{N}}$ in $\mathcal{M}$, three sequences $(T_n)_{n \in \mathbb{N}}$, $(T_n')_{n \in \mathbb{N}}$, $(T_n'')_{n \in \mathbb{N}}$ in $\mathfrak{T}$, and two sequences $(S_n)_{n \in \mathbb{N}}$, $(S_n')_{n \in \mathbb{N}}$ such that for any $n \in \mathbb{N}$:

1) $T_n \subset S_n \in \varphi(T_n)$;

2) $S_n \subset T_n' \subset S_n' \in \varphi(T_n')$;

3) $S_n' \subset T_n'' \subset T \setminus \bigcup_{m=0}^{n-1} S_m$;

4) $\mu_n(T_n) \notin \{\mu_m(T_m) \mid m \in \mathbb{N}, m < n\} + \varrho(n)V$;

5) for any finite subset P of G and any $m \in \mathbb{N}$ there exists $\mu \in \mathcal{M}$ such that

$$\mu(A \setminus \bigcup_{m=0}^{n} T_m'') \notin P + \varrho(m)U.$$

Let $n \in \mathbb{N}$ and assume the sequences were constructed up to $n-1$. By 5) there exists $\mu_n \in \mathcal{M}$ such that

$$\mu_n(A \setminus \bigcup_{m=0}^{n-1} T_m'') \notin \{\mu_m(T_m) \mid m \in \mathbb{N}, m < n\} + \varrho(n+1)U.$$

Since μ_n is $\mathfrak{K}$-regular there exists $K \in \mathfrak{K}$ such that $K \subset A \setminus \bigcup_{m=0}^{n-1} T''_m$ and

$$\mu_n(A \setminus \bigcup_{m=0}^{n-1} T''_m) - \mu_n(K) \in V.$$

By a), b), e), f), and Proposition 4.8.2 there exist $T' \in \mathfrak{T}, p \in \mathbb{N}$, and a finite subset P of G such that $K \subset T'$ and

$$\{\mu(B) | (\mu, B) \in \mathscr{M} \times \mathfrak{R}, B \subset T' \setminus K\} \subset P + \varrho(p)V,$$
$$\{\mu(K) | \mu \in \mathscr{M}\} \subset P + \varrho(p)V.$$

By a), b) and Proposition 4.1.47 there exists $T'' \in \mathfrak{T}$ such that $K \subset T''$ and

$$\mu_n(K) - \mu_n(B) \in V$$

for any $B \in \mathfrak{R}$ with $K \subset B \subset T''$. By c), 2), and 3) there exists a family $(T'''_m)_{0 \le m < n}$ in $\mathfrak{T}$ such that

$$K \subset T'''_m, \ T'_m \cap T'''_m = \emptyset$$

for any $m \in \mathbb{N}$, $m < n$. By a) there exist $T_n, T'_n, T''_n \in \mathfrak{T}$, $S_n \in \varphi(T_n)$, and $S'_n \in \varphi(T'_n)$ such that

$$K \subset T_n \subset S_n \subset T'_n \subset S'_n \subset T''_n \subset T \cap T' \cap T'' \cap \left(\bigcap_{m=0}^{n-1} T'''_m \right)$$

Hence 1), 2), and 3) are fulfilled. Assume

$$\mu_n(T_n) \in \{\mu_m(T_m) | m \in \mathbb{N}, m < n\} + \varrho(n)V.$$

Then

$$\mu_n(A \setminus \bigcup_{m=0}^{n-1} T''_m) = (\mu_n(A \setminus \bigcup_{m=0}^{n-1} T''_m) - \mu_n(K)) + (\mu_n(K) - \mu_n(T_n)) + \mu_n(T_n) \in$$
$$\in \{\mu_m(T_m) | m \in \mathbb{N}, m < n\} + \varrho(n)V + V + V \subset$$
$$\subset \{\mu_m(T_m) | m \in \mathbb{N}, m < n\} + \varrho(n+1)U,$$

which is a contradiction. Hence, 4) holds too. We have

$$\mu(A \setminus \bigcup_{m=0}^{n-1} T''_m) = \mu(A \setminus \bigcup_{m=0}^{n} T''_m) + \mu(K) + \mu((A \setminus \bigcup_{m=0}^{n-1} T''_m) \cap (T''_n \setminus K)) \in$$
$$\in \mu(A \setminus \bigcup_{m=0}^{n-1} T''_m) + 2P + 2\varrho(p)V$$

for any $\mu \in \mathscr{M}$. By 5) if $n \neq 0$ and by the hypothesis of the proof if $n = 0$ for any finite subset Q of G and for any $m \in \mathbb{N}$ there exists $\mu \in M$ such that

$$\mu(A \setminus \bigcup_{m=0}^{n-1} T''_m) \notin Q + \varrho(m)U.$$

Hence, 5) also holds. This finishes the inductive construction.

There exist a continuous surjective group homomorphism u of G into a metrizable topological additive group H and a 0-neighbourhood W in H such that $2\overset{-1}{u}(W) \subset V$ (Lemma 4.1.19). By e), 1), 2), and 3) there exist an infinite subset M of $\mathbb{N}$ and a map $\psi : \mathfrak{P}(M) \to \mathfrak{T}$ such that $\psi(\{n\}) = T_n$ and such that $u \circ \mu_n \circ \psi$ is an H-valued measure on $\mathfrak{P}(M)$ for any $n \in M$. Since $\{u \circ \mu(T) \mid \mu \in \mathcal{M}\}$ is a bounded (precompact) set of H for any $T \in \mathfrak{T}$ we deduce by Theorem 4.7.5 that

$$\{u \circ \mu_n(T_n) \mid n \in M\}$$

is also a bounded (precompact) set of H. By Lemma 4.7.2 there exists $n \in \mathbb{N}$ such that

$$u \circ \mu_n(T_n) \in \{u \circ \mu_m(T_m) \mid m \in \mathbb{N}, m < n\} + \varrho(n)W.$$

Hence, there exist $m \in \mathbb{N}$ with $m < n$ and a family $(x_i)_{1 \le i \le \varrho(n)}$ in W such that

$$u \circ \mu_n(T_n) = u \circ \mu_m(T_m) + \sum_{i=1}^{\varrho(n)} x_i .$$

Since u is surjective there exists a family $(y_i)_{1 \le i \le \varrho(n)}$ in $\overset{-1}{u}(W)$ such that $u(y_i) = x_i$ for any $i \in \mathbb{N}$ with $0 \le i \le \varrho(n)$. We get

$$u(\mu_n(T_n) - \mu_m(T_m) - \sum_{i=1}^{\varrho(n)} y_i) = 0$$

and therefore

$$\mu_n(T_n) - \mu_m(T_m) - \sum_{i=1}^{\varrho(n)} y_i \in \overset{-1}{u}(W).$$

Hence,

$$\mu_n(T_n) \in \{\mu_m(T_m) \mid m \in \mathbb{N}, m < n\} + (\varrho(n) + 1)\overset{-1}{u}(W) \subset$$
$$\subset \{\mu_m(T_m) \mid m \in \mathbb{N}, m < n\} + \varrho(n)V,$$

which contradicts 4). □

Theorem 4.8.4. *Let X be a topological space, $\mathfrak{R}$ be a quasi-δ-ring of subsets of X, $\mathfrak{T}$ be a set of regular open sets of X belonging to $\mathfrak{R}$ and $\mathcal{M}$ be a set of $\mathfrak{K}$-regular G-valued measures on $\mathfrak{R}$ such that the followings hold:*

1) For any $K \in \mathfrak{K}$ and any $T', T'' \in \mathfrak{T}$ with $K \subset T' \cap T''$ there exists $T \in \mathfrak{T}$ with

$$K \subset T \subset \bar{T} \subset T' \cap T''.$$

2) $\mathfrak{T}$ separates $\mathfrak{K}$ and $\emptyset \in \mathfrak{T}$.

3) For any $(K,T) \in \mathfrak{K} \times \mathfrak{T}$ with $\bar{K} \cap \bar{T} = \emptyset$ there exists $T' \in \mathfrak{T}$ with $K \subset T'$ and $T \cap T' = \emptyset$.

4) Any set of $\mathfrak{R}$ is contained in a set of $\mathfrak{T}$.

5) $\overset{\circ}{\overline{\bigcup_{n \in \mathbb{N}} T_n}} \in \mathfrak{T}$ and $\bigcup_{n \in \mathbb{N}} T_n \in \mathfrak{R}$ for any sequence $(T_n)_{n \in \mathbb{N}}$ in $\mathfrak{T}$ the union of which is con-

tained in a set of $\mathfrak{T}$ *and for which there exists a disjoint sequence* $(S_n)_{n\in\mathbb{N}}$ *of open sets of* X *such that* $\bar{T}_n \subset S_n$ *for any* $n \in \mathbb{N}$.

6) $\{\mu(T) \mid \mu \in \mathcal{M}\}$ *is a bounded (precompact) set of* G *for any* $T \in \mathfrak{T}$.

Then $\{\mu(A) \mid \mu \in \mathcal{M}\}$ *is a bounded (precompact) set of* G *for any* $A \in \mathfrak{R}$. *If, in addition,* $\mathfrak{R}$ *is a quasi-*σ*-ring, then* $\{\mu(A) \mid \mu \in \mathcal{M}, A \in \mathfrak{R}\}$ *is a bounded set of* G.

We denote by φ the map

$$\mathfrak{T} \to \mathfrak{P}(\mathfrak{P}(X)), \quad T \mapsto \{S \mid \bar{T} \subset S, \quad S \text{ open set of } X\}.$$

The conditions a), b), c), d) and f) of Proposition 4.8.3 are obviously fulfilled. Let $(\mu_n)_{n\in\mathbb{N}}$ be a sequence in $\mathcal{M}$, u be a continuous group homomorphism of G into a metrizable topological additive group H and $(T_n)_{n\in\mathbb{N}}$ be a sequence in $\mathfrak{T}$ whose union is contained in a set of $\mathfrak{T}$ and for which there exists a disjoint sequence $(S_n)_{n\in\mathbb{N}}$ of sets such that $T_n \subset S_n \in \varphi(T_n)$ for any $n \in \mathbb{N}$. By 2) and 5)

$$\overset{\circ}{\overline{\bigcup_{m\in M} T_m}} \setminus \bigcup_{m\in M} T_m \in \mathfrak{R}, \quad \overset{\circ}{\overline{\bigcup_{m\in M} T_m}} \in \mathfrak{T}$$

for any $M \subset \mathbb{N}$ and therefore by Proposition 4.5.12 there exists an infinite subset M_0 of $\mathbb{N}$ such that

$$u \circ \mu_n\Big(\overset{\circ}{\overline{\bigcup_{m\in M} T_m}} \setminus \bigcup_{m\in M} T_m\Big) = 0$$

for any $n \in \mathbb{N}$ and for any $M \subset M_0$. We denote by ψ and ψ' the maps

$$\mathfrak{P}(M_0) \to \mathfrak{T}, \quad M \mapsto \overset{\circ}{\overline{\bigcup_{m\in M} T_m}}$$

and

$$\mathfrak{P}(M_0) \to \mathfrak{R}, \quad M \mapsto \bigcup_{m\in M} T_m$$

respectively. By the above relation $u \circ \mu_n \circ \psi = u \circ \mu_n \circ \psi'$ and therefore $u \circ \mu_n \circ \psi$ is an H-valued measure for any $n \in \mathbb{N}$. Hence, condition e) of Proposition 4.8.3 is also fulfilled and the first assertion follows from this proposition. The second assertion follows from the first and from Corollary 4.7.8. □

Corollary 4.8.5. *Let* X *be a regular space,* $\mathfrak{K}$ *be the set of compact sets of* X, $\mathfrak{T}$ *be the set of regular open sets of* X, $\mathfrak{R}$ *be a quasi-*σ*-ring of subsets of* X *containing any open set of* X *and* $\mathcal{M}$ *be a set of* $\mathfrak{K}$*-regular* G*-valued measures on* $\mathfrak{R}$ *such that* $\{\mu(T) \mid \mu \in \mathcal{M}\}$ *is a bounded (precompact) set of* G *for any* $T \in \mathfrak{T}$. *Then*

$$\{\mu(A) \mid (\mu, A) \in \mathcal{M} \times \mathfrak{R}\}.$$

is a bounded set of G ($\{\mu(A) \mid \mu \in \mathcal{M}\}$ *is a precompact set of* G *for any* $A \in \mathfrak{R}$). □

Remarks. 1) This result was proved by P. Gänssler (1971) ([2] Theorem 3.1) for $G = \mathbb{R}$ and $\mathfrak{R}$ a σ-ring.

2) For a similar result, which is somehow in another direction, see J. D. Stein Jr. (1972) ([1] Theorem 2).

Corollary 4.8.6. *Let X be a normal space, $\mathfrak{K}$ be the set of closed sets of X, $\mathfrak{T}$ be the set of regular open sets of X, $\mathfrak{R}$ be a quasi-σ-ring of subsets of X containing any open set of X and $\mathcal{M}$ be a set of $\mathfrak{K}$-regular G-valued measures on $\mathfrak{R}$ such that $\{\mu(T) \mid \mu \in \mathcal{M}\}$ is a bounded (precompact) set of G for any $T \in \mathfrak{T}$. Then*

$$\{\mu(A) \mid (\mu, A) \in \mathcal{M} \times \mathfrak{R}\}$$

is a bounded set of G ($\{\mu(A) \mid \mu \in \mathcal{M}\}$ is a precompact set of G for any $A \in \mathfrak{R}$). □

Corollary 4.8.7. *Let X be a locally compact paracompact space, $\mathfrak{R}$ be a σ-ring of relatively σ-compact sets of X containing any compact set of X, $\mathfrak{K}$ be the set of closed sets of X belonging to $\mathfrak{R}$, $\mathfrak{T}$ be the set of regular open sets of X belonging to $\mathfrak{R}$ and $\mathcal{M}$ be a set of $\mathfrak{K}$-regular G-valued measures on $\mathfrak{R}$ such that $\{\mu(T) \mid \mu \in \mathcal{M}\}$ is a bounded (precompact) set of G for any $T \in \mathfrak{T}$. Then $\{\mu(A) \mid (\mu, A) \in \mathcal{M} \times \mathfrak{R}\}$ is a bounded set of G ($\{\mu(A) \mid \mu \in \mathcal{M}\}$ is a precompact set of G for any $A \in \mathfrak{R}$).* □

Remark. Let X be a compact space, $\mathfrak{K}$ be the set of compact sets of X, $\mathfrak{R}$ be the σ-ring generated by $\mathfrak{K}$, E be a Banach space and $\mathcal{M}$ be a subset of $\mathcal{M}(\mathfrak{R}, E; \mathfrak{K})$ such that $\{\mu(U) \mid \mu \in \mathcal{M}\}$ is bounded for any open set U of X. By the above corollary $\{\mu(A) \mid \mu \in \mathcal{M}, A \in \mathfrak{R}\}$ is bounded. This result was proved by J. K. Brooks (1980) ([2] page 168).

Corollary 4.8.8. *Let X be a completely regular space such that the closure of any exact open set of Y is open, X be a subspace of Y, $\mathfrak{T}$ be the set of closed open sets of X, $\mathfrak{R}$ be a σ-ring of subsets of X containing $\mathfrak{T}$, $\mathfrak{K}$ be the set of compact sets of X belonging to $\mathfrak{R}$ and $\mathcal{M}$ be a set of $\mathfrak{K}$-regular G-valued measures on $\mathfrak{R}$ such that $\{\mu(T) \mid \mu \in \mathcal{M}\}$ is a bounded (precompact) set of G for any $T \in \mathfrak{T}$. Then $\{\mu(A) \mid (\mu, A) \in \mathcal{M} \times \mathfrak{R}\}$ is a bounded set of G ($\{\mu(A) \mid \mu \in \mathcal{M}\}$ is a precompact set of G for any $A \in \mathfrak{R}\}$).*

By $\mathfrak{T}'$ we denote the set of closed open sets of Y, by $\mathfrak{R}'$ the σ-ring generated by $\mathfrak{R} \cup \mathfrak{T}'$ and by μ' the map

$$\mathfrak{R}' \to G, \quad A' \mapsto \mu(A' \cap X)$$

for any $\mu \in \mathcal{M}$. It is easy to see that μ' is a $\mathfrak{K}$-regular measure for any $\mu \in \mathcal{M}$. By Theorem 4.8.4 $\{\mu'(A') \mid \mu \in \mathcal{M}, A' \in \mathfrak{R}'\}$ is a bounded set of G ($\{\mu'(A') \mid \mu \in \mathcal{M}\}$ is a precompact set of G for any $A' \in R'$). The assertion follows. □

§ 4.9 Spaces of integrals

Throughout this section E shall denote a Hausdorff locally convex space, X a completely regular space, $\mathfrak{K}$ the set of compact sets of X, $\mathfrak{R}$ the δ-ring generated by $\mathfrak{K}$, $\mathscr{C}$ the set of continuous real functions on X, $\mathscr{F}$ a dense subset of $\mathscr{C}_{\mathfrak{K}}$ such that

$$X \to \mathbb{R}, \quad x \mapsto f(x) - g(x)$$

$$X \to \mathbb{R}, \quad x \mapsto \sup(f(x), g(x))$$

belong to $\mathscr{F}$ for any $f, g \in \mathscr{F}$, and $\mathscr{M}$ shall denote a subset of $\mathscr{M}(\mathfrak{R}, E; \mathfrak{K})$.

The aim of this section is to prove Corollary 4.9.11 which is needed in the proof of Theorem 5.6.5. This theorem in its turn will be used only in the proof of Theorem 5.7.25, but at this point E will be equal to $\mathbb{R}$. So the reader who is only interested in Theorem 5.7.25 and in its consequences (i.e. the whole sections § 5.8 and § 5.9) may simplify his task by setting $E = \mathbb{R}$ throughout this section.

Definition 4.9.1. *Let Y be a set and A be a subset of Y. We denote by $\underline{1_A^Y}$ the real function on Y equal to 1 on A and equal to 0 on $Y \setminus A$. We sometimes write $\underline{1_A}$ instead of 1_A^Y. 1_A^Y is called* the characteristic function of A on Y.

Definition 4.9.2. *Let Y be a set, $\mathfrak{S}$ be a δ-ring of subsets of Y and μ be a real measure on $\mathfrak{S}$. We denote by $\underline{\mathscr{L}^1(\mu, Y)}$ the set of real functions on Y which are μ-integrable and for any $f \in \mathscr{L}^1(\mu, Y)$ by $\underline{\int f d\mu}$ its integral (C. Constantinescu [5] page 8 or C. Constantinescu, K. Weber [1] Definition 5.2.5).*

Proposition 4.9.3. *Let Y be a set, $\mathfrak{S}$ be a δ-ring of subsets of Y, $\mathfrak{L}$ be a subset of $\mathfrak{S}$ closed under finite unions, $\mu \in \mathscr{M}(\mathfrak{S}, \mathbb{R}; \mathfrak{L})$, f be a positive function of $\mathscr{L}^1(\mu, Y)$ and ε be a strictly positive real number. Then there exists a finite family $(L_\iota)_{\iota \in I}$ in $\mathfrak{L}$ and a strictly positive real number α such that*

$$\alpha \sum_{\iota \in I} 1_{L_\iota}^Y(x) < f(x)$$

for any $x \in \bigcup_{\iota \in I} L_\iota$, and

$$\Big| \sum_{\iota \in I} \alpha \mu(L_\iota) - \int f d\mu \Big| < \varepsilon .$$

We set

$$f_n := \frac{1}{2^n} \sum_{m=1}^{n 2^n} 1^Y_{\{x \in X \mid f(x) \geq \frac{m}{2^n}\}}$$

for any $n \in \mathbb{N}$. Then $(f_n)_{n \in \mathbb{N}}$ is an increasing sequence in $\mathscr{L}^1(\mu, Y)$ converging pointwise to f so

$$\lim_{n \to \infty} \int f_n d\mu = \int f d\mu .$$

Hence, we may assume f has the form $\beta \sum_{\iota \in I} 1^Y_{A_\iota}$, where β is a strictly positive real number and $(A_\iota)_{\iota \in I}$ is a finite family of subsets of Y such that $1^Y_{A_\iota} \in \mathscr{L}^1(\mu, Y)$ for any $\iota \in I$. For each $\iota \in I$ there exists an increasing sequence $(A_{\iota n})_{n \in \mathbb{N}}$ in $\mathfrak{S}$ such that $\bigcup_{n \in \mathbb{N}} A_{\iota n} \subset A_\iota$ and such that $A_\iota \setminus \bigcup_{n \in \mathbb{N}} A_{\iota n}$ is a μ-null set. Hence, we may assume $A_\iota \in \mathfrak{R}$ in the above expression. Since μ is $\mathfrak{L}$-regular there exists a family $(L_\iota)_{\iota \in I}$ in $\mathfrak{L}$ such that $L_\iota \subset A_\iota$ for any $\iota \in I$ and such that

$$\beta \left| \sum_{\iota \in I} \mu(L_\iota) - \sum_{\iota \in I} \mu(A_\iota) \right| < \varepsilon .$$

There exists a real number α such that $0 < \alpha < \beta$ and

$$\left| \alpha \sum_{\iota \in I} \mu(L_\iota) - \beta \sum_{\iota \in I} \mu(A_\iota) \right| < \varepsilon . \quad \square$$

Definition 4.9.4. *Let Y be a set, $\mathfrak{S}$ be a δ-ring of subsets of Y, $\mu \in \mathscr{M}(\mathfrak{S}, E)$, $\mathscr{N} \subset \mathscr{M}(\mathfrak{S}, E)$ and E' be the dual of E. We set*

$$\underline{\mathscr{L}^1(\mu, Y)} := \bigcap_{x' \in E'} \mathscr{L}^1(x' \circ \mu, Y),$$

$$\underline{\mathscr{L}^1(\mathscr{N}, Y)} := \bigcap_{\nu \in \mathscr{N}} \mathscr{L}^1(\nu, Y)$$

and, for any $f \in \mathscr{L}^1(\mu, Y)$, denote by $\underline{\int f d\mu}$ the linear form

$$E' \to \mathbb{R}, \quad x' \mapsto \int f d(x' \circ \mu) .$$

Let $\mathscr{G}$ be a subset of $\mathscr{L}^1(\mathscr{N}, Y)$ such that $\int g d\mu \in E$ for any $(g, \nu) \in \mathscr{G} \times \mathscr{N}$. For any $\nu \in \mathscr{N}$, denote by $\tilde{\nu}$ the map

$$\mathscr{G} \to E, \quad f \mapsto \int f d\nu$$

and by $\underline{\mathscr{N}_{\mathscr{G}}}$ the set $\mathscr{N}$ endowed with the initial uniformity with respect to the map

$$\mathscr{N} \to E^{\mathscr{G}}, \quad \nu \mapsto \tilde{\nu} .$$

Definition 4.9.5. *Let f be a map of X into an additive group. The closure of the set $\{x \in X \mid f(x) \neq 0\}$ is called the* <u>support</u> *or the* <u>carrier</u> *of f and is denoted by* <u>*Supp f*</u>.

Proposition 4.9.6. *Let $\mu \in \mathscr{M}(\mathfrak{R}, E; \mathfrak{K})$ and f be a positive function of $\mathscr{L}^1(\mu, X)$ such that $\int h d\mu \in E$ for any $h \in \mathscr{L}^1(\mu, X)$ with $0 \leq h \leq f$. Then for any 0-neighbourhood U in E there exists a positive upper semicontinuous real function g on X with compact support, taking a finite number of values only, such that $g(x) < f(x)$ for any*

$x \in Supp\, g$, and such that

$$\int h d\mu - \int f d\mu \in U$$

for any $h \in \mathscr{L}^1(\mu, X)$ with $g \leq h \leq f$.

We may assume U convex and circled. We denote by $\mathscr{G}$ the set of positive upper semicontinuous real functions g on X with compact support, taking a finite number of values only and such that $g(x) < f(x)$ for any $x \in \operatorname{Supp} g$. $\mathscr{G}$ is upper directed with respect to the usual order relation on the set of real functions on X. Let $\mathfrak{F}$ be the section filter of $\mathscr{G}$ and let φ be the map

$$\mathscr{G} \to \mathrm{E}, \quad \mathrm{g} \mapsto \int g d\mu .$$

By Proposition 4.9.3 $\varphi(\mathfrak{F})$ converges weakly to $\int f d\mu$. Let $(g_n)_{n \in \mathbb{N}}$ be an increasing sequence in $\mathscr{G}$. Then

$$\left(\int (g_{n+1} - g_n) d\mu\right)_{n \in \mathbb{N}}$$

is a weakly supersummable sequence in E. By the Orlicz-Pettis theorem (Corollary 3.5.12) this sequence is summable in E. Hence, $(\int g_n d\mu)_{n \in \mathbb{N}}$ is a convergent sequence in E. By Proposition 1.5.17 $\varphi(\mathfrak{F})$ is a Cauchy filter on E so it converges to $\int f d\mu$ (N. Bourbaki [1] Ch II § 3 Proposition 7). Hence, there exists $g \in \mathscr{G}$ such that

$$\int g' d\mu - \int f d\mu \in \tfrac{1}{2} U$$

for any $g' \in \mathscr{G}$ with $g' \geq g$. Let $h \in \mathscr{L}^1(\mu, X)$ such that $g \leq h \leq f$. Applying the above considerations to $h - g$ instead of f we find an upper semicontinuous real function g' on X with compact support taking a finite number of values only such that $0 \leq g' \leq h - g, g'(x) < h(x) - g(x)$ for any $x \in \operatorname{Supp} g'$, and

$$\int g' d\mu - \int (h - g) d\mu \in \tfrac{1}{2} U .$$

We have $g + g' \in \mathscr{G}$ so

$$\int (g + g') d\mu - \int f d\mu \in \tfrac{1}{2} U .$$

Then

$$\int h d\mu - \int f d\mu = \left(\int (h - g) d\mu - \int g' d\mu\right) + \left(\int (g + g') d\mu - \int f d\mu\right) \in$$
$$\in \tfrac{1}{2} U + \tfrac{1}{2} U = U . \quad \square$$

Corollary 4.9.7. *Let $\mu \in \mathscr{M}(\mathfrak{R}, E; \mathfrak{K})$, f_0 be a positive function of $\mathscr{F}$ belonging to $\mathscr{L}^1(\mu, X)$ such that $\int h d\mu \in E$ for any $h \in \mathscr{L}^1(\mu, X)$ with $0 \leq h \leq f_0$, let g', g'' be positive upper semicontinuous real functions in X with compact supports such that*

$$g'(x) + g''(x) < f_0(x)$$

for any $x \in Supp\,(g' + g'')$ and U be a 0-neighbourhood in E. Then there exists $f \in \mathscr{F}$ such that $g' \leq f \leq f_0 - g''$, $g'(x) < f(x) < f_0(x) - g''(x)$ for any $x \in Supp\,(g' + g'')$, and

$$\int h d\mu - \int g' d\mu \in U$$

for any $h \in \mathscr{L}^1(\mu, X)$ with $g' \leq h \leq f$.

By Proposition 4.9.6 there exists a positive upper semicontinuous real function g on X with compact support such that

$$g(x) < f_0(x) - g'(x) - g''(x)$$

for any $x \in \operatorname{Supp} g$ and such that

$$\int (f_0 - g' - g'')d\mu - \int h d\mu \in U$$

for any $h \in \mathscr{L}^1(\mu, X)$ with $g \le h \le f_0 - g' - g''$. We set

$$K := \operatorname{Supp}(g + g' + g'').$$

Then $K \in \mathfrak{K}$ and

$$g'(x) < f_0(x) - g''(x) - g(x)$$

for any $x \in K$. Since X is completely regular, K is compact, g' is upper semicontinuous, and $f_0 - g'' - g$ is lower semicontinuous there exists an $f' \in \mathscr{C}$ such that

$$g'(x) < f'(x) < f_0(x) - g''(x) - g(x)$$

for any $x \in K$. $\mathscr{F}$ being dense in $\mathscr{C}_{\mathfrak{K}}$, there exists $f'' \in \mathscr{F}$ such that

$$g'(x) < f''(x) < f_0(x) - g''(x) - g(x)$$

for any $x \in K$. We set

$$f: X \to \mathbb{R}, \quad x \mapsto \inf(f_0(x), \sup(f''(x), 0)).$$

Then $f \in \mathscr{F}$, $g' \le f \le f_0 - g'' - g$, and $g'(x) < f(x) < f_0(x) - g''(x)$ for any $x \in \operatorname{Supp}(g' + g'')$.

Let $h \in \mathscr{L}^1(\mu, X)$ such that $g' \le h \le f$. Then

$$g \le f_0 - g'' - h \le f_0 - g' - g''$$

so

$$\int h d\mu - \int g' d\mu = \int (f_0 - g' - g'')d\mu - \int (f_0 - g'' - h)d\mu \in U. \quad \square$$

Theorem 4.9.8. *Let k be a positive real function on X belonging to $\mathscr{L}^1(\mathscr{M}, X)$ such that*

$$k(x) = \sup_{\mathscr{F} \ni f \le k} f(x)$$

for any $x \in X$ and such that $\int h d\mu \in E$ for any $\mu \in \mathscr{M}$ and for any $h \in \mathscr{L}^1(\mathscr{M}, X)$ with $0 \le h \le k$, let $\bar{\mathscr{F}}$ be the set of real functions $\bar{f}$ on X such that $0 \le \bar{f} \le k$ and such that there exists an increasing sequence $(f_n)_{n \in \mathbb{N}}$ in $\mathscr{F}$ with

$$\bar{f}(x) = \sup_{n \in \mathbb{N}} f_n(x)$$

for any $x \in X$ and $\mathscr{G}$ be the set of positive upper semicontinuous real functions g on X with compact support such that $g(x) < k(x)$ for any $x \in \operatorname{Supp} g$. Then:

a) for any $\mathfrak{F} \in \hat{\Phi}_4(\mathcal{M}_{\mathcal{F}})$, for any $g \in \mathcal{G}$, and for any 0-neighbourhood U in E there exist $\mathcal{N} \in \mathfrak{F}$ and $f \in \mathcal{F}$ such that $g \leq f \leq k$ and such that

$$\int h d\mu - \int g d\mu \in U$$

for any $\mu \in \mathcal{N}$ and any $h \in \mathcal{L}^1(\mathcal{M}, X)$ with $g \leq h \leq f$;

b) the identity map $\mathcal{M}_{\mathcal{F}} \to \mathcal{M}_{\mathcal{G}}$ is uniformly Φ_4-continuous.

a) We may assume U convex and circled. By the Dini theorem there exists $f' \in \mathcal{F}$ such that $0 \leq f' \leq k$ and $g(x) < f'(x)$ for any $x \in \operatorname{Supp} g$.

Assume first that $\mathfrak{F} \in \Phi_4(\mathcal{M}_{\mathcal{F}})$ and let (I, μ) be a Θ_4-net in $\mathcal{M}_{\mathcal{F}}$ such that $\mu(\mathfrak{G}) = \mathfrak{F}$, where $\mathfrak{G}$ denotes the section filter of I. Suppose the assertion does not hold. Then for any $\iota_0 \in I$ and any $f'' \in \mathcal{F}$ such that $g \leq f'' \leq k$ there exist $\iota \in I$ and $h \in \mathcal{L}^1(\mathcal{M}, X)$ with $\iota_0 \leq \iota$, $g \leq h \leq f''$, and

$$\int h d\mu_\iota - \int g d\mu_\iota \notin U.$$

We construct inductively an increasing sequence $(\iota_n)_{n \in \mathbb{N}}$ in I and a sequence $(f_n)_{n \in \mathbb{N}}$ in $\mathcal{F}$ such that

$$\int f_n d\mu_{\iota_n} \notin \tfrac{1}{3} U, \; 0 \leq f_n \leq f' - \sum_{m=0}^{n-1} f_m,$$

$$x \in \operatorname{Supp} g \;\Rightarrow\; g(x) < f'(x) - \sum_{m=0}^{n} f_m(x)$$

for any $n \in \mathbb{N}$. Let ι_{-1} be an arbitrary element of I, $n \in \mathbb{N}$ and assume the sequences were constructed up to $n-1$. Then

$$f' - \sum_{m=0}^{n-1} f_m \in \mathcal{F}, \quad g \leq f' - \sum_{m=0}^{n-1} f_m \leq k.$$

By the above remark there exist $\iota_n \in I$ and $h \in \mathcal{L}^1(\mathcal{M}, X)$ such that

$$\iota_{n-1} \leq \iota_n, \quad g \leq h \leq f' - \sum_{m=0}^{n-1} f_m,$$

$$\int h d\mu_{\iota_n} - \int g d\mu_{\iota_n} \notin U.$$

By Proposition 4.9.6 there exists $g' \in \mathcal{G}$ such that $g' \leq h - g$,

$$g'(x) < h(x) - g(x)$$

for any $x \in \operatorname{Supp} g'$, and

$$\int g' d\mu_{\iota_n} - \int (h-g) d\mu_{\iota_n} \in \tfrac{1}{3} U.$$

We have

$$g(x) + g'(x) < f'(x) - \sum_{m=0}^{n-1} f_m(x)$$

for any $x \in \operatorname{Supp} g'$. This relation also holds for any $x \in \operatorname{Supp} g \backslash \operatorname{Supp} g'$, so it holds for any $x \in \operatorname{Supp}(g+g')$. By Corollary 4.9.7 there exists $f_n \in \mathscr{F}$ such that

$$g' \le f_n \le f' - \sum_{m=0}^{n-1} f_m - g\,,$$

$$g'(x) < f_n(x) < f'(x) - \sum_{m=0}^{n-1} f_m(x) - g(x)$$

for any $x \in \operatorname{Supp}(g+g')$, and

$$\int f_n d\mu_{\iota_n} - \int g' d\mu_{\iota_n} \in \tfrac{1}{3} U\,.$$

We have

$$U \not\ni \int h d\mu_{\iota_n} - \int g d\mu_{\iota_n} =$$
$$= \left(\int (h-g) d\mu_{\iota_n} - \int g' d\mu_{\iota_n}\right) + \left(\int g' d\mu_{\iota_n} - \int f_n d\mu_{\iota_n}\right) + \int f_n d\mu_{\iota_n} \in \int f_n d\mu_{\iota_n} + \tfrac{2}{3} U\,.$$

Hence,

$$\int f_n d\mu_{\iota_n} \notin \tfrac{1}{3} U\,.$$

This finishes the inductive construction.

Let φ be the map

$$\mathfrak{P}(\mathbb{N}) \to \bar{\mathscr{F}}, \quad M \mapsto \sum_{n \in M} f_n\,.$$

By the Orlicz-Pettis theorem (Corollary 3.5.12) the map

$$\mathfrak{P}(\mathbb{N}) \to E, \quad M \mapsto \int \varphi(M) \mathrm{d}\mu$$

is a measure for any $\mu \in \mathscr{M}$ so, by Theorem 4.1.22 i),

$$\lim_{n \to \infty} \int f_n d\mu_{\iota_n} = 0\,,$$

which is a contradiction.

Assume now $\mathfrak{F} \in \hat{\Phi}_4(\mathscr{M}_{\mathscr{F}})$. By the above proof for any $\mathfrak{G} \in \Phi_4(\mathscr{M}_{\mathscr{F}})$ there exist $\mathscr{N}_{\mathfrak{G}} \in \mathfrak{G}$ and $f_{\mathfrak{G}} \in \mathscr{F}$ such that $g \le f_{\mathfrak{G}} \le k$ and

$$\int h d\mu - \int g d\mu \in U$$

for any $\mu \in \mathscr{N}_{\mathfrak{G}}$ and any $h \in \mathscr{L}^1(\mathscr{M}, X)$ with $g \le h \le f_{\mathfrak{G}}$. There exists a finite subset Ψ of $\Phi_4(\mathscr{M}_{\mathscr{F}})$ such that $\bigcup_{\mathfrak{G} \in \Psi} \mathscr{N}_{\mathfrak{G}} \in \mathfrak{F}$. We set

$$\mathscr{N} := \bigcup_{\mathfrak{G} \in \Psi} \mathscr{N}_{\mathfrak{G}}\,,$$

$$f : X \to \mathbb{R}, \quad x \mapsto \inf_{\mathfrak{G} \in \Psi} f_{\mathfrak{G}}(x)\,.$$

$\mathscr{N}$ and f possess the required properties.

b) follows immediately from a). □

Corollary 4.9.9. *Let k be a positive real function on X belonging to $\mathscr{L}^1(\mathscr{M}, X)$ such* *that*

$$k(x) = \sup_{F \ni f \leq k} f(x)$$

for any $x \in X$ *and such that* $\int h d\mu \in E$ *for any* $\mu \in \mathcal{M}$ *and any* $h \in \mathcal{L}^1(\mathcal{M}, X)$ *with* $0 \leq h \leq k$, *let* $\mathcal{F}_k$ *be the set* $\{f \in \mathcal{F} \mid 0 \leq f \leq k\}$, $\mathcal{G}_k$ *be the set of positive upper semi-continuous real functions g on X with compact support such that* $g(x) < k(x)$ *for any* $x \in \operatorname{Supp} g$, $\bar{\mathcal{F}}$ *(resp.* $\bar{\mathcal{G}}$*) be the set of real functions on X of the form*

$$X \to \mathbb{R}, \quad x \mapsto \sup_{n \in \mathbb{N}} f_n(x),$$

where $(f_n)_{n \in \mathbb{N}}$ *is a sequence in* $\mathcal{F}_k$ *(in* $\mathcal{G}_k$*), and let* $\mathcal{N}$ *be a* Φ_4*-set of* $\mathcal{M}_{\bar{\mathcal{F}}}$. *Then:*

a) for any $\bar{g} \in \bar{\mathcal{G}}$ *and for any 0-neighbourhood U in E there exists* $\bar{f} \in \bar{\mathcal{F}}$ *such that* $\bar{g} \leq \bar{f}$ *and*

$$\int h d\mu - \int \bar{g} d\mu \in U$$

for any $\mu \in \mathcal{N}$ *and any* $h \in \mathcal{L}^1(\mathcal{M}, X)$ *with* $\bar{g} \leq h \leq \bar{f}$;

b) $\mathcal{N}$ *is a* Φ_4*-set of* $\mathcal{M}_{\bar{\mathcal{G}}}$.

a) We may assume U convex circled and closed. Let $(g_n)_{n \in \mathbb{N}}$ be a sequence in $\mathcal{G}_k$ such that

$$\bar{g}(x) = \sup_{n \in \mathbb{N}} g_n(x)$$

for any $x \in X$. By Theorem 4.9.8 a) there exists for any $n \in \mathbb{N}$ an $f_n \in \mathcal{F}_k$ such that $g_n \leq f_n$ and

$$\int h d\mu - \int g_n d\mu \in \frac{1}{2^{n+1}} U$$

for any $\mu \in \mathcal{N}$ and any $h \in \mathcal{L}^1(\mathcal{M}, X)$ with $g_n \leq h \leq f_n$. We set

$$\bar{f}: X \to \mathbb{R}, \quad x \mapsto \sup_{n \in \mathbb{N}} f_n(x).$$

Then $\bar{f} \in \bar{\mathcal{F}}$ and $\bar{g} \leq \bar{f}$. Let $h \in \mathcal{L}^1(\mathcal{M}, X)$ with $\bar{g} \leq h \leq \bar{f}$. We define inductively

$$h_n: X \to \mathbb{R}, \quad x \mapsto \inf\left(h(x) - \bar{g}(x) - \sum_{m=0}^{n-1} h_m(x),\ f_n(x) - g_n(x)\right)$$

for any $n \in \mathbb{N}$. $(g_n + h_n)_{n \in \mathbb{N}}$ is a sequence in $\mathcal{L}^1(\mathcal{M}, X)$ such that

$$g_n \leq g_n + h_n \leq f_n$$

and therefore

$$\int h_n d\mu = \int (g_n + h_n) d\mu - \int g_n d\mu \in \frac{1}{2^{n+1}} U$$

for any $\mu \in \mathcal{N}$ and for any $n \in \mathbb{N}$.

We want to prove

$$\sum_{n\in\mathbb{N}} h_n = h - \bar{g}.$$

The inequality

$$\sum_{n\in\mathbb{N}} h_n \leq h - \bar{g}$$

is obvious, so we have to prove the converse inequality only. Let $x \in X$ and let ε be a strictly positive real number. There exists $n \in \mathbb{N}$ such that

$$f_n(x) > \bar{f}(x) - \varepsilon.$$

We get

$$f_n(x) - g_n(x) \geq \bar{f}(x) - \varepsilon - \bar{g}(x) \geq h(x) - \bar{g}(x) - \varepsilon,$$

$$h_n(x) \geq h(x) - \bar{g}(x) - \sum_{m=0}^{n-1} h_m(x) - \varepsilon,$$

$$\sum_{m\in\mathbb{N}} h_m(x) \geq \sum_{m=0}^{n} h_m(x) \geq h(x) - \bar{g}(x) - \varepsilon$$

for any $n \in \mathbb{N}$. Since ε and x are arbitrary we get

$$\sum_{n\in\mathbb{N}} h_n \geq h - \bar{g}.$$

Hence,

$$\int h d\mu - \int \bar{g} d\mu = \int (h - \bar{g}) d\mu = \sum_{n=0}^{\infty} \int h_n d\mu \in \sum_{n=0}^{\infty} \frac{1}{2^{n+1}} U \subset U.$$

b) Let $\mathcal{N}_1, \mathcal{N}_2$ be Φ_4-sets of $\mathcal{M}_{\bar{\mathcal{F}}}$ and $\mathcal{N}'$ be their union. By a) the identity map $\mathcal{N}'_{\bar{\mathcal{F}}} \to \mathcal{N}'_{\mathcal{G}}$ is uniformly continuous.

Let $(\mu_n)_{n\in\mathbb{N}}$ be a sequence in $\mathcal{N}$. If it possesses a Cauchy subsequence in $\mathcal{N}_{\bar{\mathcal{F}}}$, then by the above remark it possesses a Cauchy subsequence in $\mathcal{N}_{\mathcal{G}}$ and therefore in $\mathcal{M}_{\mathcal{G}}$. Assume $(\mu_n)_{n\in\mathbb{N}}$ does not possess a Cauchy sequence in $\mathcal{N}_{\bar{\mathcal{F}}}$. Then it possesses an adherent point μ in $\mathcal{M}_{\bar{\mathcal{F}}}$. By the above remark, μ is an adherent point of $(\mu_n)_{n\in\mathbb{N}}$ in $\mathcal{M}_{\mathcal{G}}$. Hence, $(\mu_n)_{n\in\mathbb{N}}$ is a Θ_4-sequence in $\mathcal{M}_{\mathcal{G}}$ so $\mathcal{N}$ is a Φ_4-set of $\mathcal{M}_{\mathcal{G}}$ (Proposition 1.5.5 c $\Rightarrow$ a). □

Theorem 4.9.10. *Let k be a positive real function on X belonging to $\mathcal{L}^1(\mathcal{M}, X)$ such that*

$$k(x) = \sup_{F \ni f \leq k} f(x)$$

for any $x \in X$ and such that $\int h d\mu \in E$ for any $\mu \in \mathcal{M}$ and any $h \in \mathcal{L}^1(\mathcal{M}, X)$ with $0 \leq h \leq k$, let $\bar{\mathcal{F}}$ be the set of real functions $\bar{f}$ on X such that $0 \leq \bar{f} \leq k$ and such that there exists an increasing sequence $(f_n)_{n\in\mathbb{N}}$ of positive functions of $\mathcal{F}$ with

$$f(x) = \sup_{n \in \mathbb{N}} f_n(x)$$

for any $x \in X$, *and let* $\mathcal{H}$ *be the set*

$$\{h \in \mathcal{L}^1(\mathcal{M}, X) \mid 0 \leq h \leq k\}.$$

Then the identity map $\mathcal{M}_{\mathcal{F}} \longrightarrow \mathcal{M}_{\mathcal{H}}$ *is uniformly* Φ_4*-continuous.*

Assume the contrary. Then by Theorem 1.8.4 b $\Rightarrow$ a there exists a Cauchy Φ_4-filter $\mathfrak{F}$ on $\mathcal{M}_{\mathcal{F}}$ which is not a Cauchy filter on $\mathcal{M}_{\mathcal{H}}$. Let (I, μ) be a Θ_4-net in $\mathcal{M}_{\mathcal{F}}$ such that $\mu(\mathfrak{G}) = \mathfrak{F}$, where $\mathfrak{G}$ denotes the section filter of I. Since $\mathfrak{F}$ is not a Cauchy filter on $\mathcal{M}_{\mathcal{H}}$ there exists $h \in \mathcal{H}$ and a convex circled 0-neighbourhood U in E such that for any $J \in \mathfrak{G}$ there exist $\iota', \iota'' \in J$ with

$$\int h d\mu_{\iota'} - \int h d\mu_{\iota''} \notin U.$$

Let $\mathcal{G}$ be the set of positive upper semicontinuous real functions g on X with compact support such that $g(x) < k(x)$ for any $x \in \operatorname{Supp} g$. We construct inductively an increasing sequence $(\iota_n)_{n \in \mathbb{N}}$ in I, a decreasing sequence $(I_n)_{n \in \mathbb{N}}$ in $\mathfrak{G}$, and a sequence $(g_n)_{n \in \mathbb{N}}$ in $\mathcal{G}$ such that for any $n \in \mathbb{N}$:

a) $x \in \operatorname{Supp} \sum_{m=0}^{n} g_n \Rightarrow \sum_{m=0}^{n} g_m(x) < h(x)$

b) $\iota_n \in I_{n-1}$ $(I_{-1} := I)$;

c) $\iota \in I_n \Rightarrow \iota \geq \iota_n, \quad \int g_n d\mu_\iota - \int g_n d\mu_{\iota_n} \notin \frac{1}{6} U$;

d) $\iota', \iota'' \in I_n \Rightarrow \int \left(\sum_{m=0}^{n} g_m \right) d\mu_\iota - \int \left(\sum_{m=0}^{n} g_m \right) d\mu_{\iota''} \in \frac{1}{6} U$.

Let $n \in \mathbb{N}$ and assume the sequences were constructed up to $n-1$. By the above remark there exist $\iota', \iota'' \in I_{n-1}$ such that

$$\int h d\mu_{\iota'} - \int h d\mu_{\iota''} \notin U.$$

From d)

$$\int \left(h - \sum_{m=0}^{n-1} g_m \right) d\mu_{\iota'} - \int \left(h - \sum_{m=0}^{n-1} g_m \right) d\mu_{\iota''} \notin \tfrac{5}{6} U.$$

By Proposition 4.9.6 there exists $g_n \in \mathcal{G}$ such that

$$x \in \operatorname{Supp} g_n \Rightarrow g_n(x) < h(x) - \sum_{m=0}^{n-1} g_m(x),$$

$$\int \left(h - \sum_{m=0}^{n-1} g_m \right) d\mu_{\iota'} - \int g_n d\mu_{\iota'} \in \tfrac{1}{6} U,$$

$$\int \left(h - \sum_{m=0}^{n-1} g_m \right) d\mu_{\iota''} - \int g_n d\mu_{\iota''} \in \tfrac{1}{6} U.$$

We have

$$\sum_{m=0}^{n} g_m(x) < h(x) \leq k(x)$$

for any $x \in \operatorname{Supp} \sum_{m=0}^{n} g_m$ so $\sum_{m=0}^{n} g_m \in \mathscr{G}$. By Theorem 4.9.8 b) $\mathfrak{F}$ is a Cauchy filter on $\mathscr{M}_{\mathscr{G}}$. Hence, there exists $I_n \in \mathfrak{G}$ such that

$$I_n \subset \{\iota \in I_{n-1} \mid \iota \geq \iota', \iota \geq \iota''\},$$

$$\int g_n d\mu_\iota - \int g_n d\mu_\lambda \in \tfrac{1}{6} U,$$

and

$$\int \left(\sum_{m=0}^{n} g_m\right) d\mu_\iota - \int \left(\sum_{m=0}^{n} g_m\right) d\mu_\lambda \in \tfrac{1}{6} U$$

for any $\iota, \lambda \in I_n$. We deduce that either

$$\int g_n d\mu_{\iota'} - \int g_n d\mu_\iota \notin \tfrac{1}{6} U$$

for any $\iota \in I_n$ or

$$\int g_n d\mu_{\iota''} - \int g_n d\mu_\iota \notin \tfrac{1}{6} U$$

for any $\iota \in I_n$. We set $\iota_n := \iota'$ in the first case and $\iota_n := \iota''$ in the second. This finishes the inductive construction.

Let $\bar{\mathscr{G}}$ be the set of real functions $\bar{g}$ on X such that there exists a sequence $(g_n')_{n \in \mathbb{N}}$ in $\mathscr{G}$ with

$$\bar{g}(x) = \sup_{n \in \mathbb{N}} g_n'(x)$$

for any $x \in X$. We set

$$\varphi : \mathfrak{P}(\mathbb{N}) \to \bar{\mathscr{G}}, \quad M \mapsto \sum_{n \in M} g_n,$$

$$\mu' : \mathfrak{P}(\mathbb{N}) \to E, \quad M \mapsto \int \varphi(M) d\mu$$

for any $\mu \in \mathscr{M}$. Then μ' is a measure for any $\mu \in \mathscr{M}$.

$(\mu_{\iota_n})_{n \in \mathbb{N}}$ is a Θ_4-sequence in $\mathscr{M}_{\mathscr{F}}$. By Proposition 1.4.7 $\{\mu_{\iota_n} \mid n \in \mathbb{N}\}$ is a Φ_4-set of $\mathscr{M}_{\mathscr{F}}$ and therefore by Corollary 4.9.9 b) a Φ_4-set of $\mathscr{M}_{\mathscr{G}}$. We deduce $\{\mu'_{\iota_n} \mid n \in \mathbb{N}\}$ is a Φ_4-set of $\mathscr{M}(\mathfrak{P}(\mathbb{N}), E)$. Hence, by Theorem 4.1.22 i) (replacing R by $\mathbb{R}$, G by E, X by $\mathfrak{P}(\mathbb{N})$, $\mathscr{G}$ by

$$\{\mathfrak{P}(\mathbb{N}) \to \mathfrak{P}(\mathbb{N}),\ M \mapsto \bigcup_{n \in M} A_n \mid (A_n)_{n \in \mathbb{N}} \in \Gamma(\mathfrak{P}(\mathbb{N}))\},$$

$\mathfrak{F}$ by

$$\{\mathscr{N} \mid \{\mu'_{\iota_n} \mid n \in \mathbb{N}\} \subset \mathscr{N} \subset \mathscr{M}(\mathfrak{P}(\mathbb{N}), E)\},$$

and (K, h) by

$$\mathbb{N} \to \mathfrak{P}(\mathbb{N}), \quad n \mapsto \{n\})$$

$$\lim_{n \to \infty} \int g_n d\mu_{\iota_m} = 0$$

uniformly in $m \in \mathbb{N}$ and this contradicts $b \& c$. □

Corollary 4.9.11. *Let $\mathscr{K}$ be a set of positive real functions k on X belonging to $\mathscr{L}^1(\mathscr{M}, X)$ such that*

$$k(x) = \sup_{\mathscr{F} \ni f \leq k} f(x)$$

for any $x \in X$ and such that $\int h d\mu \in E$ for any $\mu \in \mathscr{M}$ and any $h \in \mathscr{L}^1(\mathscr{M}, X)$ with $0 \leq h \leq k$, let $\bar{\mathscr{F}}$ be the set of real functions $\bar{f}$ on X such that there exists $k \in \mathscr{K}$ with $0 \leq \bar{f} \leq k$ and such that there exists an increasing sequence $(f_n)_{n \in \mathbb{N}}$ of positive functions of $\mathscr{F}$ with

$$f(x) = \sup_{n \in \mathbb{N}} f_n(x)$$

for any $x \in X$, and let $\mathscr{H}$ be the set

$$\{h \in \mathscr{L}^1(\mathscr{M}, X) | \exists k \in \mathscr{K}, \ 0 \leq h \leq k\}.$$

Then the identity map $\mathscr{M}_{\bar{\mathscr{F}}} \to \mathscr{M}_{\mathscr{H}}$ is uniformly Φ_4-continuous.

We denote by φ the identity map $\mathscr{M}_{\bar{\mathscr{F}}} \to \mathscr{M}_{\mathscr{H}}$ and for any $k \in \mathscr{K}$ by $\mathscr{H}_k$ the set

$$\{h \in \mathscr{L}^1(\mathscr{M}, X) | 0 \leq h \leq k\}$$

and by φ_k the identity map $\mathscr{M}_{\mathscr{H}} \to \mathscr{M}_{\mathscr{H}_k}$. By Theorem 4.9.10 $\varphi_k \circ \varphi$ is uniformly Φ_4-continuous for any $k \in \mathscr{K}$. Since the uniformity of $\mathscr{M}_{\mathscr{H}}$ is the initial uniformity with respect to the family $(\varphi_k)_{k \in \mathscr{K}}$ the map φ is uniformly Φ_4-continuous (Proposition 1.8.13). □

§ 4.10 Supersummable families of functions and their integrals

In the applications of the results of this section to Chapter 5, E and F will always be equal to $\mathbb{R}$.

Definition 4.10.1. *Let E, F be Hausdorff locally convex spaces. We denote by $\underline{E \otimes F}$ the tensor product of these spaces endowed with the topology of bi-equicontinuous convergence and by $\underline{E \mathbin{\tilde{\tilde{\otimes}}} F}$ its completion (H. H. Schaefer [1] page 96). If E', F' denote the duals of E and F respectively, then $E \mathbin{\tilde{\tilde{\otimes}}} F$ will be identified with a set of linear forms on $E' \otimes F'$.*

Definition 4.10.2. *Let E, F be Hausdorff locally convex spaces, E', F' be the duals of E and F respectively, T be a set, $\mathfrak{R}$ be a δ-ring of subsets of T and $\mu \in \mathcal{M}(\mathfrak{R}, F)$. We set*

$$\underline{\mathcal{L}^1(\mu,T,E)} := \{f \in E^T \mid (x',y') \in E' \times F' \Rightarrow x' \circ f \in \mathcal{L}^1(y' \circ \mu, T)\}$$

and, for any $f \in \mathcal{L}^1(\mu,T,E)$, denote by $\underline{\int f d\mu}$ the linear form on $E' \otimes F'$ defined by

$$(x', y') \mapsto \int (x' \circ f) d(y' \circ \mu).$$

Definition 4.10.3. *Let T be a set and $\mathfrak{R}$ be a δ-ring of subsets of T. A subset A of T is called $\underline{\mathfrak{R}\text{-measurable}}$ if $A \cap B \in \mathfrak{R}$ for any $B \in \mathfrak{R}$. For any measure μ on $\mathfrak{R}$ and any $\mathfrak{R}$-measurable subset A of T denote by $\underline{1_A^T \cdot \mu}$ the measure $B \mapsto \mu(A \cap B)$ on $\mathfrak{R}$.*

Proposition 4.10.4. *Let E, F be Hausdorff locally convex spaces, T be a set, $\mathfrak{R}$ be a δ-ring of subsets of T, $\mathfrak{K}$ be a subset of $\mathfrak{R}$ closed under finite unions, $\mu \in \mathcal{M}(\mathfrak{R}, F; \mathfrak{K})$ and f be a map of T into E such that $1_A^T f \in \mathcal{L}^1(\mu,T,E)$ and $\int 1_A^T f d\mu \in E \tilde{\tilde{\otimes}} F$ for any $\mathfrak{R}$-measurable subset A of T. Let us order $\mathfrak{K}$ by the inclusion relation, $\mathfrak{F}$ be the section filter of $\mathfrak{K}$ and φ be the map*

$$\mathfrak{K} \to E \tilde{\tilde{\otimes}} F, \quad K \mapsto \int 1_K^T f d\mu.$$

Then $\varphi(\mathfrak{F})$ is a Φ_1-filter in $E \tilde{\tilde{\otimes}} F$ which converges to $\int f d\mu$.

Let $(K_n)_{n \in \mathbb{N}}$ be an increasing sequence in $\mathfrak{K}$. We set $K_{-1} := \emptyset$ and

$$z_n := \int 1_{K_n \setminus K_{n-1}}^T f d\mu$$

for any $n \in \mathbb{N}$. We have

$$\sum_{n \in M} (x' \otimes y')(z_n) = \sum_{n \in M} \int 1_{K_n \setminus K_{n-1}}^T (x' \circ f) d(y' \circ \mu) =$$
$$= \int 1^T_{\bigcup_{n \in M} (K_n \setminus K_{n-1})} (x' \circ f) d(y' \circ \mu) = (x' \otimes y')\left(\int 1^T_{\bigcup_{n \in M} (K_n \setminus K_{n-1})} f d\mu\right)$$

for any $M \subset \mathbb{N}$ and any $(x',y') \in E' \times F'$. Hence, $(z_n)_{n \in \mathbb{N}}$ is supersummable in $(E \tilde{\tilde{\otimes}} F)_{E' \otimes F'}$. By Theorem 3.4.1 b) $(z_n)_{n \in \mathbb{N}}$ is supersummable in $E \tilde{\tilde{\otimes}} F$ so $\left(\sum_{m=0}^{n} z_m\right)_{n \in \mathbb{N}}$ is a convergent sequence in $E \tilde{\tilde{\otimes}} F$. Hence, $(\varphi(K_n))_{n \in \mathbb{N}}$ is a convergent sequence in $E \tilde{\tilde{\otimes}} F$. Since $(K_n)_{n \in \mathbb{N}}$ is arbitrary $\varphi(\mathfrak{F})$ is a Cauchy Φ_1-filter in $E \tilde{\tilde{\otimes}} F$ (Proposition 1.5.17). Since (C. Constantinescu, K. Weber [1] Theorem 5.2.6 (i))

$$(x' \otimes y')\left(\int f d\mu\right) = \left(\int (x' \circ f) d(y' \circ \mu) = \lim_{K, \mathfrak{F}} \int 1_K^T (x' \circ f) d(y' \circ \mu) = \lim_{K, \mathfrak{F}} (x' \otimes y')(\varphi(K))\right)$$

for any $(x',y') \in E' \times F'$, the filter $\varphi(\mathfrak{F})$ converges to $\int f d\mu$ in $(E \tilde{\tilde{\otimes}} F)_{E' \otimes F'}$. Using the fact that $\varphi(\mathfrak{F})$ is a Cauchy filter on $E \tilde{\tilde{\otimes}} F$, we deduce $\varphi(\mathfrak{F})$ converges to $\int f d\mu$ in $E \tilde{\tilde{\otimes}} F$ (N. Bourbaki [1] Ch II § 3 Proposition 7). □

Proposition 4.10.5. *Let E, F be Hausdorff locally convex spaces, T be a set, $\mathfrak{R}$ be a δ-ring of subsets of T, $\mathfrak{K}$ be a subset of $\mathfrak{R}$ closed under finite unions, $\mathscr{M}$ be a subset of $\mathscr{M}(\mathfrak{R}, F; \mathfrak{K})$ such that $1_A^T \cdot \mu \in \mathscr{M}$ for any $\mu \in \mathscr{M}$ and any $\mathfrak{R}$-measurable subset A of T, and let $\mathscr{F}$ be a vector subspace of $\bigcap_{\mu \in \mathscr{M}} \mathscr{L}^1(\mu, T, E)$ such that $\int f d\mu \in E \tilde{\tilde{\otimes}} F$ for any $(f, \mu) \in \mathscr{F} \times \mathscr{M}$. We set*

$$\mathscr{N} := \{1_K^T \cdot \mu \mid K \in \mathfrak{K}, \mu \in \mathscr{M}\}.$$

Then:

a) $\mathscr{N}$ is a loosely Φ_1-dense set of $\mathscr{M}_{\mathscr{F}}$;

b) $\mathscr{F}_{\mathscr{M}}$ and $\mathscr{F}_{\mathscr{N}}$ are locally convex spaces and

$$\mathscr{G}(I, \mathscr{F}_{\mathscr{M}}) = \mathscr{G}(I, \mathscr{F}_{\mathscr{N}})$$

for any set I, where $\mathscr{F}_{\mathscr{M}}$ (resp. $\mathscr{F}_{\mathscr{N}}$) denotes the space $\mathscr{F}$ endowed with the coarsest topology for which the maps

$$\mathscr{F} \to E \tilde{\tilde{\otimes}} F, \quad f \mapsto \int f d\mu$$

are continuous for any $\mu \in \mathscr{M}$ (resp. for any $\mu \in \mathscr{N}$).

a) Let $\mu \in \mathscr{M}$. Let us order $\mathfrak{K}$ by the inclusion relation, $\mathfrak{F}$ be the section filter of $\mathfrak{K}$ and φ be the map

$$\mathfrak{K} \to \mathscr{N}, \quad K \mapsto 1_K^T \cdot \mu.$$

By Proposition 4.10.4 $\varphi(\mathfrak{F}) \in \Phi(\Theta_1(\mathscr{N}, \mathscr{M}_{\mathscr{F}}))$ and $\varphi(\mathfrak{F})$ converges to μ in $\mathscr{M}_{\mathscr{F}}$. Hence, $\mathscr{N}$ is a loosely Φ_1-dense set of $\mathscr{M}_{\mathscr{F}}$.

b) It is obvious that $\mathscr{F}_{\mathscr{M}}$ and $\mathscr{F}_{\mathscr{N}}$ are locally convex spaces. We denote by $\mathscr{C}$ the set of continuous maps of $\mathscr{M}_{\mathscr{F}}$ into $E \tilde{\tilde{\otimes}} F$ and, for any $f \in \mathscr{F}$, we denote by f' the map

$$\mathscr{M} \to E \tilde{\tilde{\otimes}} F, \quad \mu \mapsto \int f d\mu.$$

Then $f' \in \mathscr{C}$ for any $f \in \mathscr{F}$. Let $g \in \mathscr{G}(I, \mathscr{F}_{\mathscr{N}})$ and let g' be the map

$$\mathfrak{P}(I) \to \mathscr{C}, \quad J \mapsto g(J)'.$$

Then $g' \in \mathscr{G}(I, \mathscr{C}_{\mathscr{N}})$. By a) and Theorem 3.4.1 b) (and Theorem 3.3.5) $\mathscr{G}(I, \mathscr{C}_{\mathscr{N}}) = \mathscr{G}(I, \mathscr{C})$ so $g' \in \mathscr{G}(I, \mathscr{C})$. We get $g \in \mathscr{G}(I, \mathscr{F}_{\mathscr{M}})$ so

$$\mathscr{G}(I, \mathscr{F}_{\mathscr{N}}) \subset \mathscr{G}(I, \mathscr{F}_{\mathscr{M}}).$$

The converse inclusion is trivial. □

Proposition 4.10.6. *Let E, F be Hausdorff locally convex spaces, T be a set, $\mathfrak{R}$ be a δ-ring of subsets of T, $\mathfrak{K}$ be a subset of $\mathfrak{R}$ closed under finite unions, $\mathscr{M}$ be a subset of $\mathscr{M}(\mathfrak{R}, F; \mathfrak{K})$ such that $1_A^T \cdot \mu \in \mathscr{M}$ for any $\mu \in \mathscr{M}$ and for any $\mathfrak{R}$-measurable subset A of T, I be a set and $f \in \mathscr{G}(I, E_{\mathfrak{K}}^T)$ such that*

$$f(J) \in \mathscr{L}^1(\mu, T, E), \quad \int f(J) d\mu \in E \mathbin{\tilde{\tilde{\otimes}}} F$$

for any $J \subset I$ *and any* $\mu \in \mathscr{M}$. *Then* $(\int f(\{\iota\}) d\mu)_{\iota \in I}$ *is supersummable in* $E \mathbin{\tilde{\tilde{\otimes}}} F$ *for any* $\mu \in \mathscr{M}$ *and*

$$\sum_{\iota \in J} \int f(\{\iota\}) d\mu = \int f(J) d\mu$$

for any $J \subset I$.

Let $\mu \in \mathscr{M}$ such that there exists $K \in \mathfrak{K}$ with $\mu(A \setminus K) = 0$ for any $A \in \mathfrak{R}$. Let $J \subset I$ and U, V be 0-neighbourhoods in E and F respectively. By Corollary 4.7.8 there exists $\alpha \in \mathbb{R}$ such that

$$\{\mu(A) \mid A \in \mathfrak{R}\} \subset \alpha V.$$

Let J_0 be a finite subset of I such that

$$(f(J'))(t) - (f(J))(t) \in \frac{1}{|\alpha| + 1} U$$

for any $t \in K$ and any $J' \in \mathfrak{P}(I)$ with $J_0 \cap J' = J_0 \cap J$. Let U^0, V^0 be the polar sets of U, V in E', F' respectively and $(x', y') \in U^0 \times V^0$. Then

$$|(\int f(J') d\mu - \int f(J) d\mu)(x' \otimes y')| = |\int x' \circ (f(J') - f(J)) d(y' \circ \mu)| \leq 2$$

for any $J' \in \mathfrak{P}(I)$ with $J_0 \cap J' = J_0 \cap J$. Since U and V are arbitrary the map

$$\mathfrak{P}(I) \to E \mathbin{\tilde{\tilde{\otimes}}} F, \quad J \mapsto \int f(J) d\mu$$

is continuous.

Let $\mathscr{F}$ be the set of $f \in \bigcap_{\mu \in \mathscr{M}} \mathscr{L}^1(\mu, T, E)$ such that $\int f d\mu \in E \mathbin{\tilde{\tilde{\otimes}}} F$ for any $\mu \in \mathscr{M}$. $\mathscr{F}$ is a vector subspace of $\bigcap_{\mu \in \mathscr{M}} \mathscr{L}^1(\mu, T, E)$. We set

$$\mathscr{N} := \{1_K^T \cdot \mu \mid K \in \mathfrak{K}, \mu \in \mathscr{M}\}.$$

By Proposition 4.10.5 b) $\mathscr{F}_{\mathscr{M}}$ and $\mathscr{F}_{\mathscr{N}}$ are locally convex spaces and

$$\mathscr{G}(I, \mathscr{F}_{\mathscr{M}}) = \mathscr{G}(I, \mathscr{F}_{\mathscr{N}}).$$

By the first part of the proof the map

$$\mathfrak{P}(I) \to \mathscr{F}, \quad J \mapsto f(J)$$

belongs to $\mathscr{G}(I, \mathscr{F}_{\mathscr{N}})$ and therefore to $\mathscr{G}(I, \mathscr{F}_{\mathscr{M}})$. Hence, the map

$$\mathfrak{P}(I) \to E \mathbin{\tilde{\tilde{\otimes}}} F, \quad J \mapsto \int f(J) d\mu$$

belongs to $\mathscr{G}(I, E \mathbin{\tilde{\tilde{\otimes}}} F)$ for any $\mu \in \mathscr{M}$. By Theorem 3.3.5 $(\int f(\{\iota\}) d\mu)_{\iota \in I}$ is supersummable in $E\overset{\approx}{\otimes}F$ for any $\mu\in\mathcal{M}$ and

$$\sum_{\iota\in J}\int f(\{\iota\})d\mu=\int f(J)\,d\mu$$

for any $J\subset I$. □

Theorem 4.10.7. *Let E, F be Hausdorff locally convex spaces, T be a uniform (topological) space, $i\in\{1,2,3,4\}$ ($i\in\{1,2\}$), T_0 be a Φ_i-dense (loosely Φ_i-dense) set of T, $\mathfrak{R}$ be a δ-ring of subsets of T, $\mathfrak{K}$ be a subset of $\mathfrak{R}$ closed under finite unions such that any set of $\mathfrak{K}$ is a Φ_i-set of T, $\mathcal{M}$ be a subset of $\mathcal{M}(\mathfrak{R};F;\mathfrak{K})$ such that $1_A^T\cdot\mu\in\mathcal{M}$ for any $\mu\in\mathcal{M}$ and any $\mathfrak{R}$-measurable subset A of T, $\mathcal{F}$ be the vector subspace of E^T of uniformly Φ_i-continuous (Φ_i-continuous) maps and $(f_\iota)_{\iota\in I}$ be a supersummable family in $\mathcal{F}_{T_0}$ such that*

$$\sum_{\iota\in J}f_\iota\in\mathcal{L}^1(\mu,T,E),\quad \int\Big(\sum_{\iota\in J}f_\iota\Big)d\mu\in E\overset{\approx}{\otimes}F$$

for any $J\subset I$ and any $\mu\in\mathcal{M}$. Then $(\int f_\iota d\mu)_{\iota\in I}$ is supersummable in $E\overset{\approx}{\otimes}F$ for any $\mu\in\mathcal{M}$ and

$$\sum_{\iota\in J}\int f_\iota d\mu=\int\Big(\sum_{\iota\in J}f_\iota\Big)d\mu$$

for any $J\subset I$.

It is obvious that $\mathcal{F}$ is a vector subspace of E^T. Let $\mathfrak{K}'$ be the set of all Φ_i-sets of T. By Theorem 3.4.1 a) b) $\mathcal{F}_{T_0}$, $\mathcal{F}_{\mathfrak{K}'}$ are Hausdorff topological groups and

$$\ell(I,\mathcal{F}_{T_0})=\ell(I,\mathcal{F}_{\mathfrak{K}'}).$$

By Theorem 3.3.5 the map

$$\mathfrak{P}(I)\to\mathcal{F}_{\mathfrak{K}},\quad J\mapsto\sum_{\iota\in J}f_\iota$$

belongs to $\mathcal{G}(I,\mathcal{F}_{\mathfrak{K}})$ and the assertion follows from Proposition 4.10.6. □

Corollary 4.10.8. *Let T be a Hausdorff topological space, $\mathfrak{K}$ be the set of compact sets of T, $\mathfrak{R}$ be a δ-ring of subsets of T containing $\mathfrak{K}$, $\mu\in\mathcal{M}(\mathfrak{R},\mathbb{C};\mathfrak{K})$, $\mathcal{C}$ be the vector space of continuous complex functions on T, $p\in[1,\infty[$ and $(f_\iota)_{\iota\in I}$ be a supersummable family in $\mathcal{C}$ such that $\sum_{\iota\in J}f_\iota\in\mathcal{L}^p(\mu,T)$ for any $J\subset I$. Then $(\dot{f}_\iota)_{\iota\in I}$ is supersummable with respect to the norm topology of $L^p(\mu,T)$.*

Without loss of generality we may replace $\mathbb{C}$ by $\mathbb{R}$. Let $q\in]1,\infty]$ with $\frac{1}{p}+\frac{1}{q}=1$. Let $\xi\in L^q(\mu)$ if $q<\infty$ or $\xi\in\hat{L}^\infty(\mu)$ if $q=\infty$ (C. Constantinescu [5] Theorem 3.6.3). By Theorem 4.10.7 $(\int f_\iota d(\xi\cdot\mu))_{\iota\in I}$ is summable and

$$\sum_{\iota\in J}\int f_\iota d(\xi\cdot\mu)=\int\Big(\sum_{\iota\in J}f_\iota\Big)d(\xi\cdot\mu)$$

for any $J \subset I$. Hence, $(\dot{f}_\iota)_{\iota \in I}$ is weakly supersummable in $L^p(\mu)$ (C. Constantinescu [5] Theorem 3.6.3 c)). By the Orlicz-Pettits Theorem (Corollary 3.5.12) $(\dot{f}_\iota)_{\iota \in I}$ is supersummable with respect to the norm topology of $L^p(\mu)$. □

Corollary 4.10.9. *Let T be a locally compact space and $\mathscr{K}$ be the vector space of continuous complex functions on T with compact supports. Then any supersummable family in $\mathscr{K}$ is supersummable with respect to the usual inductive topology on $\mathscr{K}$ (N. Bourbaki [2] Ch 3 page 41).*

By Theorem 4.10.7 any supersummable family in $\mathscr{K}$ is weakly supersummable and the assertion follows from the Orlicz-Pettis theorem (Corollary 3.5.12). □

Definition 4.10.10. *Let T be a completely regular space, $\mathscr{C}_b$ be the vector space of bounded continuous real functions on T and $\mathscr{B}_0$ be the set of bounded real functions g on T such that for any strictly positive real number ε the set $\{t \in T \mid |g(t)| \geq \varepsilon\}$ is relatively compact. The* strict topology *of $\mathscr{C}_b$ is the topology on $\mathscr{C}_b$ generated by the seminorms*

$$\mathscr{C}_b \to \mathbb{R}, \quad f \mapsto \sup_{t \in T} |f(t) g(t)|,$$

where g runs through $\mathscr{B}_0$.

The strict topology was introduced by R.C. Buck (1958) ([1] Definition) for T locally compact and by A.C.M. van Rooij (1967) ([1], page 97) in the general case.

Corollary 4.10.11. *Let T be a completely regular space and $\mathscr{C}_b$ be the vector space of bounded continuous real functions on T. Any supersummable family in $\mathscr{C}_b$ is supersummable with respect to the strict topology of $\mathscr{C}_b$.*

Let us denote by $\mathscr{C}_0$ the vector space $\mathscr{C}_b$ endowed with the strict topology, by $\mathfrak{K}$ the set of compact sets of T, by $\mathfrak{R}$ the δ-ring generated by $\mathfrak{K}$, by $\mathscr{M}_b$ the vector subspace of bounded measures of $\mathscr{M}(\mathfrak{R}, \mathbb{R}; \mathfrak{K})$ and, for any $\mu \in \mathscr{M}_b$, by $\bar{\mu}$ the map

$$\mathscr{C}_b \to \mathbb{R}, \quad f \mapsto \int f d\mu.$$

By R. Giles ([1] Theorem 4.6) $\mu \in \mathscr{C}_0'$ for any $\mu \in \mathscr{M}_b$ and the map

$$\mathscr{M}_b \to \mathscr{C}_0', \quad \mu \mapsto \bar{\mu}$$

is an isomorphism of vector spaces. Hence, by Theorem 4.10.7 any supersummable family in $\mathscr{C}_b$ is weakly supersummable in $\mathscr{C}_0$ and by the Orlicz-Pettis theorem (Corollary 3.5.12) it is supersummable in $\mathscr{C}_0$. □

Theorem 4.10.12. *Let E, F be Hausdorff locally convex spaces, T be a set, $\mathfrak{R}$ be a δ-ring of subsets of T, $\mathscr{M}$ be a subset of $\mathscr{M}(\mathfrak{R}, F)$ such that $1_A^T \cdot \mu \in \mathscr{M}$ for any $\mu \in \mathscr{M}$ and for any $\mathfrak{R}$-measurable subset A of T and such that for any sequence $(t_n)_{n \in \mathbb{N}}$ in T which is contained in a set of $\mathfrak{R}$ there exists $y \in F \setminus \{0\}$ such that the map*

$$\mathfrak{R} \to F, \quad A \mapsto \Big(\sum_{n \in \mathbb{N}} \frac{1}{2^n} 1_A^T(t_n)\Big) y$$

belongs to $\mathscr{M}$, *and let* $\mathscr{F}$ *be a vector subspace of* $\bigcap_{\mu\in\mathscr{M}} \mathscr{L}^1(\mu,T,E)$ *such that* $\int f d\mu \in E \tilde{\tilde{\otimes}} F$ *for any* $(f,\mu)\in\mathscr{F}\times\mathscr{M}$. *Then the identity map* $\mathscr{F}\to\mathscr{F}_{\mathscr{M}}$ *belongs to* $\mathscr{P}^c(\mathscr{F},\mathscr{F}_{\mathscr{M}})$.

We denote by E', F' the duals of E and F respectively.

First, we intend to show that $f(A)$ is bounded for any $(f,A)\in\mathscr{F}\times\mathfrak{R}$. Assume the contrary. Then there exist $(f,A)\in\mathscr{F}\times\mathfrak{R}, x'\in E'$ and a sequence $(t_n)_{n\in\mathbb{N}}$ in A such that

$$x'\circ f(t_n) > 3^n\Big(1+\sum_{m=0}^{n-1}|x'\circ f(t_m)|\Big)$$

for any $n\in\mathbb{N}$. By the hypotheses there exists $y\in F\setminus\{0\}$ such that

$$\mu:\mathfrak{R}\to F,\quad B\mapsto \Big(\sum_{n\in\mathbb{N}}\frac{1}{2^n}1_B^T(t_n)\Big)y$$

belongs to $\mathscr{M}$. Let $y'\in F'$ so that $y'(y)=1$. We get $y'\circ\mu\in\mathscr{M}(\mathfrak{R},\mathbb{R})$ and $x'\circ f\in\mathscr{L}^1(y'\circ\mu,T)$ so $\left(\frac{1}{2^n}x'\circ f(t_n)\right)_{n\in\mathbb{N}}$ is summable. But

$$\sum_{m=0}^{n}\frac{1}{2^m}x'\circ f(t_m)\geq\frac{1}{2^n}x'\circ f(t_n)-\sum_{m=0}^{n-1}|x'\circ f(t_m)|>\left(\frac{3}{2}\right)^n$$

for any $n\in\mathbb{N}$, which is a contradiction. Hence, $f(A)$ is bounded for any $(f,A)\in\mathscr{F}\times\mathfrak{R}$.

Now let $(f_n)_{n\in\mathbb{N}}$ be a supersummable sequence in $\mathscr{F}$. We want to show $\bigcup_{M\in\mathfrak{P}_f(\mathbb{N})}(\sum_{n\in M}f_n)(A)$ is a bounded set of E for any $A\in\mathfrak{R}$. Assume the contrary. Then there exist $A\in\mathfrak{R}$ and $x'\in E'$ such that $\bigcup_{M\in\mathfrak{P}_f(\mathbb{N})}(x'\circ\sum_{n\in M}f_n)(A)$ is not bounded. We may even assume it is not upper bounded. We construct a sequence $(t_n)_{n\in\mathbb{N}}$ in A inductively, just as a disjoint sequence $(I_n)_{n\in\mathbb{N}}$ in $\mathfrak{P}_f(\mathbb{N})$ and an increasing sequence $(J_n)_{n\in\mathbb{N}}$ in $\mathfrak{P}(\mathbb{N})$ such that for any $n\in\mathbb{N}$:

1) $\sum_{m=0}^{n}\sum_{p\in I_m}x'\circ f_p(t_n)\geq n$,

2) $\bigcup_{m=0}^{n}I_m\subset J_n$,

3) $I_n\cap J_{n-1}=\emptyset,\quad (J_{-1}:=\emptyset)$,

4) $\sum_{m\in\mathbb{N}\setminus J_n}|x'\circ f_m(t_n)|<1$.

Let $n\in\mathbb{N}$ and assume that the sequences were constructed up to $n-1$. We set

$$\alpha:=\sum_{p\in J_{n-1}}\sup_{t\in A}|x'\circ f_p(t)|.$$

By the above considerations $\alpha<\infty$. By the hypothesis of the proof there exist $M\in\mathfrak{P}_f(\mathbb{N})$ and $t_n\in A$ such that

$$\sum_{p \in M} x' \circ f_p(t_n) > n + 2\alpha .$$

We set $I_n := M \setminus J_{n-1}$. Since $(x' \circ f_m(t_n))_{m \in \mathbb{N}}$ is summable there exists $J_n \in \mathfrak{P}_f(\mathbb{N})$ such that $J_{n-1} \subset J_n$ and 2) and 4) hold. We get (by 2) for $n-1$)

$$\sum_{m=0}^{n} \sum_{p \in I_m} x' \circ f_p(t_n) \geq \sum_{p \in I_n} x' \circ f_p(t_n) - \sum_{m=0}^{n-1} | \sum_{p \in I_m} x' \circ f_p(t_n) | \geq$$

$$\geq \sum_{p \in M} x' \circ f_p(t_n) - 2 \sum_{p \in J_{n-1}} |x' \circ f_p(t_n)| > n + 2\alpha - 2\alpha = n ,$$

which proves 1). This finishes the inductive construction. We set $I := \bigcup_{n \in \mathbb{N}} I_n$, $f := \sum_{n \in I} f_n$. By means of 1), 3), and 4)

$$x' \circ f(t_n) = \sum_{m=0}^{n} \sum_{p \in I_m} x' \circ f_p(t_n) + \sum_{m=n+1}^{\infty} \sum_{p \in I_m} x' \circ f_p(t_n) > n - 1$$

for any $n \in \mathbb{N}$ and this contradicts the fact that $f(A)$ is bounded.

We set $\mathcal{N} := \{1_A^T \cdot \mu \mid A \in \mathfrak{R}, \mu \in \mathcal{M}\}$. Let $(f_n)_{n \in \mathbb{N}}$ be a supersummable sequence in $\mathcal{F}$ and $\mu \in \mathcal{N}$. Then $\mu(\mathfrak{R})$ is a bounded set of F (Corollary 4.7.8) and by the above considerations $\bigcup_{M \in \mathfrak{P}_f(\mathbb{N})} (\sum_{n \in M} f_n)(A)$ is a bounded set of E. Hence, by Lebesgue dominated convergence theorem $(\int x' \circ f_n d(y' \circ \mu))_{n \in \mathbb{N}}$ is supersummable and

$$\sum_{n \in M} \int x' \circ f_n d(y' \circ \mu) = \int x' \circ \sum_{n \in M} f_n d(y' \circ \mu)$$

for any $(x', y') \in E' \times F'$ and for any $M \subset \mathbb{N}$. Hence, $(\int f_n d\mu)_{n \in \mathbb{N}}$ is supersummable in $(E \mathbin{\tilde{\tilde{\otimes}}} F)_{E' \times F'}$ and $\sum_{n \in M} \int f_n d\mu = \int \sum_{n \in M} f_n d\mu$ for any $M \subset \mathbb{N}$. By Theorem 3.4.1 b) $(\int f_n d\mu)_{n \in \mathbb{N}}$ is supersummable in $E \mathbin{\tilde{\tilde{\otimes}}} F$ and $\sum_{n \in M} \int f_n d\mu = \int (\sum_{n \in M} f_n) d\mu$ for any $M \subset \mathbb{N}$. By Proposition 4.10.5 b) (and Theorem 3.3.5) the same assertion holds for any $\mu \in \mathcal{M}$. Hence, the identity map $\mathcal{F} \to \mathcal{F}_{\mathcal{M}}$ belongs to $\mathcal{P}^c(\mathcal{F}, \mathcal{F}_{\mathcal{M}})$. □

§ 4.11 Measurability considerations

Definition 4.11.1. *Let X be a Hausdorff space, let $\mathfrak{K}$ be the set of compact sets of X, let $\mathfrak{R}$ be the δ-ring generated by $\mathfrak{K}$, let μ be a positive measure of $\mathcal{M}(\mathfrak{R}, \mathbb{R}; \mathfrak{K})$ (i.e. $\mu(A) \geq 0$ for any $A \in \mathfrak{R}$), let Y be a topological space, and let f be a map $X \to Y$. f is called* μ-measurable *if for any $K \in \mathfrak{K}$ and for any $\varepsilon > 0$ there exists $L \in \mathfrak{K}$ such that $f|L$ is continuous and $\mu(K \setminus L) < \varepsilon$. f is called* universally measurable *if f is ν-measurable for any positive $\nu \in \mathcal{M}(\mathfrak{R}, \mathbb{R}; \mathfrak{K})$.*

Lemma 4.11.2. *Let $\mathfrak{K}$ be the set of compact sets of $\mathfrak{P}(\mathbb{N})$ and let $\mathfrak{R}$ be the σ-ring generated by $\mathfrak{K}$. There exists $\lambda \in \mathcal{M}(\mathfrak{R}, \mathbb{R}, \mathfrak{K})$ such that:*

a) $\lambda(\mathfrak{A}) \geq 0$ *for any* $\mathfrak{A} \in \mathfrak{R}$;

b) $\lambda(\mathfrak{P}(\mathbb{N})) > 0$;

c) $\lambda(\{A \Delta B \mid A \in \mathfrak{A}\}) = \lambda(\mathfrak{A})$ *for any* $\mathfrak{A} \in \mathfrak{R}$ *and* $B \in \mathfrak{P}(\mathbb{N})$.

$\mathfrak{P}(\mathbb{N})$ endowed with the map

$$\mathfrak{P}(\mathbb{N}) \times \mathfrak{P}(\mathbb{N}) \longrightarrow \mathfrak{P}(\mathbb{N}), \quad (A, B) \longmapsto A \Delta B$$

is a compact commutative group and λ is a Haar measure on $\mathfrak{P}(\mathbb{N})$ (K. Jacobs [1] Ch. XII Theorem 3.10). □

Proposition 4.11.3. *Let G be a topological additive group, let λ be the measure introduced in Lemma 4.11.2, and let $f \in \mathcal{H}(\mathbb{N}, G)$. If f is λ-measurable then $(\{f(n)\})_{n \in \mathbb{N}}$ converges to* 0.

Let $\mathfrak{K}$ be a compact set of $\mathfrak{P}(\mathbb{N})$ such that $f|\mathfrak{K}$ is continuous and such that $\lambda(\mathfrak{K}) > \lambda(\mathfrak{P}(\mathbb{N}) \setminus \mathfrak{K})$. Since

$$\lambda(\{M \Delta \{n\} \mid M \in \mathfrak{K}\}) = \lambda(\mathfrak{K})$$

we get

$$\{M \Delta \{n\} \mid M \in \mathfrak{K}\}) \cap \mathfrak{K} \neq \emptyset$$

for any $n \in \mathbb{N}$. Hence there exist two maps $K, L \colon \mathbb{N} \to \mathfrak{K}$ such that $n \in K(n)$, $L(n) = K(n) \setminus \{n\}$ for any $n \in \mathbb{N}$.

Assume that the assertion does not hold. Then there exists a 0-neighbourhood U in G such that $M := \{n \in \mathbb{N} \mid f(\{n\}) \notin U\}$ is infinite. Let $\mathfrak{F}$ be a free ultrafilter on $\mathbb{N}$ containing M. Then $K(\mathfrak{F})$ and $L(\mathfrak{F})$ converge to the same element M_0 of $\mathfrak{K}$ and we get the contradictory relation

$$0 = f(M_0) - f(M_0) = \lim_{n, \mathfrak{F}} f(K(n)) - \lim_{n, \mathfrak{F}} f(L(n)) = \lim_{n, \mathfrak{F}} f(K(n) \setminus L(n)) = \lim_{n, \mathfrak{F}} f(\{n\}). \quad \square$$

Theorem 4.11.4. *Let I be a set, let G be a topological additive group, and let $\mu \colon \mathfrak{P}(I) \to G$ be an additive universally measurable map. Then μ is exhaustive.*

Let $(A_n)_{n \in \mathbb{N}}$ be a disjoint sequence in $\mathfrak{P}(I)$ and let φ be the map

$$\mathfrak{P}(\mathbb{N}) \longrightarrow \mathfrak{P}(I), \quad M \longmapsto \bigcup_{n \in M} A_n.$$

$\mu \circ \varphi \in \mathcal{H}(\mathbb{N}, G)$ and φ being continuous $\mu \circ \varphi$ is universally measurable. By Proposition 4.11.3 $(\mu(A_n))_{n \in \mathbb{N}}$ converges to 0. Hence μ is exhaustive. □

Remark. This theorem was proved by J.K. Pachl (1979) ([1] Proposition 3). In fact the result holds even for a more general formulation of the universal measurability; this was proved by I. Labuda (1979) ([6] Theorem 3.1) and W.H. Graves (1979) ([1] Theorem 3.5).

Theorem 4.11.5. *Let G, H be topological additive groups, G Hausdorff, let $u: G \to H$ be a universally measurable group homomorphism, and let $(x_\iota)_{\iota \in I}$ be a supersummable family in G. We have:*

a) the map

$$\mathfrak{P}(I) \to H, \quad J \mapsto u(\sum_{\iota \in J} x_\iota)$$

is exhaustive;

b) if H is Hausdorff and Φ_3-compact, then $(u(x_\iota))_{\iota \in I}$ is supersummable; if I is countable, we may replace the hypothesis Φ_3-compact by sequentially complete.

a) The map

$$\mathfrak{P}(I) \to G, \quad J \mapsto \sum_{\iota \in J} x_\iota$$

is continuous and additive (Theorem 3.3.5) so the map

$$\mathfrak{P}(I) \to H, \quad J \mapsto u(\sum_{\iota \in J} x_\iota)$$

is universally measurable and additive and the assertion follows from Theorem 4.11.4.

b) follows from a) and Corollary 4.3.6 b). □

Remark. a) was proved by J.K. Pachl (1979) ([1] Corollary 5) in the case of I countable.

Corollary 4.11.6. *Let G be a metrizable complete topological additive group. Then any universally measurable group homomorphism of G into a topological additive group is continuous.*

Let H be a topological additive group and $u: G \to H$ be a universally measurable group homomorphism. Furthermore, let $(x_n)_{n \in \mathbb{N}}$ be a supersummable sequence in G. By Theorem 4.11.5 a) $(u(x_n))_{n \in \mathbb{N}}$ converges to 0 and therefore by Proposition 3.3.28 d $\Rightarrow$ a u is continuous. □

Remark Let G, H be metrizable complete topological additive groups and $\mathscr{B}$ be the smallest sequentially closed set of H^G containing any continuous map. It is easy to see that any element of $\mathscr{B}$ is universally measurable so, by the above corollary, any group homomorphism belonging to $\mathscr{B}$ is continuous. This result was proved by S. Banach (1932) ([1] Ch. I Théorème 4). The above corollary was proved by I. Labuda (1979) ([6] Theorem 3.6) when G is a metrizable complete topological vector space and by W.H. Graves (1979) (1 Theorem 3.8.) in the general case; in both cases a more general notion of universal measurability was used.

Theorem 4.11.7. *Let G, H be Hausdorff topological additive groups having the same underlying group such that the identity map $G \to H$ is universally measurable. We have:*

a) if the identity map $H \to G$ belongs to $\mathscr{P}^c(H, G)$ (to $\mathscr{P}(H, G)$) and if H is sequentially complete (Φ_3-compact), then the identity map $G \to H$ belongs to $\mathscr{P}^c(G, H)$ (to $\mathscr{P}(G, H)$);

b) if there exists a fundamental system of 0-neighbourhoods in H which are sequentially closed (strongly Φ_1-closed) in G, then the identity map $G \to H$ belongs to $\mathscr{P}^c(G, H)$ (to $\mathscr{P}(G, H)$).

a) Let $(x_\iota)_{\iota \in I}$ be a countable (an arbitrary) supersummable family in G. By Theorem 4.11.5 b) $(x_\iota)_{\iota \in I}$ is supersummable in H. Since the identity map $H \to G$ belongs to $\mathscr{P}^c(H, G)$ (to $\mathscr{P}(H, G)$) the sums of $(x_\iota)_{\iota \in I}$ in G and H coincide. Hence, the identity map $G \to H$ belongs to $\mathscr{P}^c(G, H)$ (to $\mathscr{P}(G, H)$).

b) Let $(x_\iota)_{\iota \in I}$ be a countable (an arbitrary) supersummable family in G and x be its sum in G. By Theorem 4.11.5 a) the map

$$\mathfrak{P}(I) \to H, \quad J \mapsto \sum_{\iota \in J} x_\iota$$

is exhaustive so by Corollary 4.1.11 the image of the section filter of $\mathfrak{P}_f(I)$ with respect to this map is a Cauchy filter. Let U be a 0-neighbourhood in H which is sequentially closed (strongly Φ_1-closed) in G. There exists $J \in \mathfrak{P}_f(I)$ such that $\sum_{\iota \in K} x_\iota - \sum_{\iota \in L} x_\iota \in U$ for any $K, L \in \mathfrak{P}_f(I)$ so that $K \cap L \supset J$. U being sequentially closed (strongly Φ_1-closed) in G, we get $\sum_{\iota \in K} x_\iota - x \in U$ for any $K \in \mathfrak{P}_f(I)$ on condition that $K \supset J$. Since U is arbitrary $(x_\iota)_{\iota \in I}$ is summable in H and its sum is x. Hence, the identity map $G \to H$ belongs to $\mathscr{P}^c(G, H)$ (to $\mathscr{P}(G, H)$). □

Remark. a) was proved by I. Labuda (1979) ([6] Corollary 3.4) and b) by I. Labuda (1979) ([6] Theorem 3.3) and W. H. Graves (1979) ([1] Corollary 3.6) for a more general formulation of the universal measurability.

Lemma 4.11.8. *Let X be a topological Baire space, Y be a topological space possessing a countable base and f be a Borel measurable map $X \to Y$. Then there exists a dense G_δ-set A of X such that $f|A$ is continuous.*

Let $\mathfrak{B}$ be a countable base of Y. Then there exists a family $(U_V)_{V \in \mathfrak{B}}$ of open sets of X such that $\overset{-1}{f}(V) \Delta U_V$ is meagre for any $V \in \mathfrak{B}$ (N. Bourbaki [1] Ch. IX § 6 Lemme 8). Hence, for any $V \in \mathfrak{B}$ there exists a sequence $(F_n^V)_{n \in \mathbb{N}}$ of closed and nowhere dense sets of X such that

$$\overset{-1}{f}(V) \Delta U_V \subset \bigcup_{n \in \mathbb{N}} F_n^V .$$

We set $A := X \setminus \bigcup_{\substack{n \in \mathbb{N} \\ V \in \mathfrak{B}}} F_n^V$. Since $\mathfrak{B}$ is countable A is a G_δ-set. X being a Baire space A is dense. We want to show that $f|A$ is continuous. Let W be an open set of Y. We set $\mathfrak{W} := \{V \in \mathfrak{B} \mid V \subset W\}$. Then $W = \bigcup_{V \in \mathfrak{W}} V$ and

$$\overset{-1}{\overline{f|A}}(W) = \Big(\bigcup_{V \in \mathfrak{W}} \overset{-1}{f}(V)\Big) \cap A = \Big(\bigcup_{V \in \mathfrak{W}} U_V\Big) \cap A .$$

Hence, $\overset{-1}{\overline{f|A}}(W)$ is an open set of A and $f|A$ is continuous. □

Proposition 4.11.9. *(J.P.R. Christensen (1971) [1] Lemma). Let $\mathfrak{A}$ be a dense G_δ-set of $\mathfrak{P}(\mathbb{N})$. Then there exist $A, B, C \in \mathfrak{A}$ such that $A \cap B = \emptyset$, $A \cup B = C$ and $\{C \Delta K \mid K \in \mathfrak{P}_f(\mathbb{N})\} \subset \mathfrak{A}$.*

Let $(\mathfrak{F}_n)_{n \in \mathbb{N}}$ be a sequence of closed and nowhere dense sets of $\mathfrak{P}(\mathbb{N})$ such that $\mathfrak{A} = \mathfrak{P}(\mathbb{N}) \setminus \bigcup_{n \in \mathbb{N}} \mathfrak{F}_n$. We set

$$\mathfrak{B} := \mathfrak{P}(\mathbb{N}) \setminus \bigcup_{n \in \mathbb{N}} \{D \Delta K \mid D \in \mathfrak{F}_n,\ K \in \mathfrak{P}_f(\mathbb{N})\} .$$

Since $\mathfrak{P}_f(\mathbb{N})$ is countable $\mathfrak{B}$ is a dense G_δ-set of $\mathfrak{P}(\mathbb{N})$. We denote by φ, ψ the maps

$$\mathfrak{P}(\mathbb{N}) \times \mathfrak{P}(\mathbb{N}) \to \mathfrak{P}(\mathbb{N}), \quad (D,E) \mapsto D \cup E,$$

$$\mathfrak{P}(\mathbb{N}) \times \mathfrak{P}(\mathbb{N}) \to \mathfrak{P}(\mathbb{N}), \quad (D,E) \mapsto D \setminus E$$

respectively. It is easy to see that φ and ψ are continuous and open. Hence, $\overset{-1}{\varphi}(\mathfrak{B})$, $\overset{-1}{\psi}(\mathfrak{B})$ are dense G_δ-sets and there exists

$$(D, A) \in (\mathfrak{B} \times \mathfrak{B}) \cap \overset{-1}{\varphi}(\mathfrak{B}) \cap \overset{-1}{\psi}(\mathfrak{B}) .$$

We set $B := \psi(D,A)$, $C := \varphi(D,A)$. Then $A, B, C \in \mathfrak{B} \subset \mathfrak{A}$, $A \cap B = \emptyset$, $A \cup B = C$, and

$$\{C \Delta K \mid K \in \mathfrak{P}_f(\mathbb{N})\} \subset \mathfrak{B} \subset \mathfrak{A} . \quad \square$$

Proposition 4.11.10. *(J.P.R. Christensen (1971) [1] Theorem 1). Let X be a Hausdorff space with a countable base and $f: \mathfrak{P}(\mathbb{N}) \to X$ be a Borel measurable map such that $f(A) = f(B)$ for any $A, B \in \mathfrak{P}(\mathbb{N})$ with $A \Delta B \in \mathfrak{P}_f(\mathbb{N})$. Then there exists a dense G_δ-set of $\mathfrak{P}(\mathbb{N})$ on which f is constant.*

By Lemma 4.11.8 there exists a dense G_δ-set $\mathfrak{A}$ of $\mathfrak{P}(\mathbb{N})$ such that $f|\mathfrak{A}$ is continuous. By Proposition 4.11.9 there exists $A \in \mathfrak{A}$ such that $\{A \Delta K \mid K \in \mathfrak{P}_f(\mathbb{N})\} \subset \mathfrak{A}$. f is constant on $\{A \Delta K \mid K \in \mathfrak{P}_f(\mathbb{N})\}$ and, since this set is dense in $\mathfrak{A}$, f is constant on $\mathfrak{A}$. □

Lemma 4.11.11. *Let X be a Hausdorff space and Y be a topological space possessing a countable base. Then any Borel measurable map $X \to Y$ is universally measurable.*

Let $f: X \to Y$ be a Borel measurable map, $\mathfrak{K}$ be the set of compact sets of X, $\mathfrak{R}$ be the δ-ring generated by $\mathfrak{K}$ and μ be a positive measure of $\mathcal{M}(\mathfrak{R}, \mathbb{R}; \mathfrak{K})$. Furthermore, let $K \in \mathfrak{K}$, let ε be a strictly positive real number and $(V_n)_{n \in \mathbb{N}}$ be a sequence of open sets of Y such that $\{V_n \mid n \in \mathbb{N}\}$ is a base of Y. Take $n \in \mathbb{N}$. Since $\overset{-1}{f}(V_n)$ is a Borel set, $\overset{-1}{f}(V_n)\cap K\in\mathfrak{R}$. Hence, there exist $K_n, L_n\in\mathfrak{K}$ such that $K_n\subset\overset{-1}{f}(V_n)\cap K$, $L_n\subset K\setminus\overset{-1}{f}(V_n)$ and $\mu(K\setminus(K_n\cup L_n))<\frac{\varepsilon}{2^{n+1}}$. We set $L:=\bigcap_{n\in\mathbb{N}}(K_n\cup L_n)$. Then $L\in\mathfrak{K}$ and $\mu(K\setminus L)<\varepsilon$. We want to show that $f|L$ is continuous. Let F be a closed set of Y. Set $M:=\{n\in\mathbb{N}\mid V_n\cap F=\emptyset\}$. Then $F=\bigcap_{n\in M}(Y\setminus V_n)$ and

$$\overset{-1}{\widehat{f|L}}(F)=\bigcap_{n\in M}((X\setminus\overset{-1}{f}(V_n))\cap L)=(\bigcap_{n\in M}L_n)\cap L.$$

Hence, $\overset{-1}{\widehat{f|L}}(F)$ is a closed set of L and $f|L$ is continuous. □

Lemma 4.11.12. *Let X be a topological space, G be a topological additive group possessing a countable base and f, g be Borel measurable maps of X into G. Then $f-g$ is Borel measurable.*

We denote by φ and $f\times g$ the maps

$$G\times G\to G,\quad (x,y)\mapsto x-y,$$

$$X\to G\times G,\quad x\mapsto(f(x),g(x))$$

respectively. Let $\mathfrak{B}$ be a countable base of G and U be an open set of G. Since φ is continuous, $\overset{-1}{\varphi}(U)$ is open. We set

$$\mathfrak{W}:=\{(V,W)\in\mathfrak{B}\times\mathfrak{B}\mid V\times W\subset\overset{-1}{\varphi}(U)\}.$$

Then $\overset{-1}{\varphi}(U)=\bigcup_{(V,W)\in\mathfrak{W}}(V\times W)$ and

$$\overset{-1}{\widehat{f-g}}(U)=\overset{-1}{\widehat{f\times g}}(\overset{-1}{\varphi}(U))=\bigcup_{(V,W)\in\mathfrak{W}}\overset{-1}{\widehat{f\times g}}(V\times W)=$$

$$\bigcup_{(V,W)\in\mathfrak{W}}(\overset{-1}{f}(V)\cap\overset{-1}{g}(W)).$$

Since $\mathfrak{W}$ is countable, it follows from the above relation that $\overset{-1}{\widehat{f-g}}(U)$ is a Borel set. Hence, $f-g$ is Borel measurable. □

Lemma 4.11.13. *(Z. Frolík (1970) [1] Theorem 1). Let X be a Souslin space and φ be a Borel measurable map of X into a metrizable space. Then $\varphi(X)$ is separable.*

Assume the contrary. Let d be a metric on $\varphi(X)$ generating its topology and let ω_1 be the first uncountable ordinal number. There exists a strictly positive real number ε having the following property: for any countable subset A of $\varphi(X)$ there exists $y\in\varphi(X)$ such that $\inf_{z\in A}d(y,z)\geq\varepsilon$. We may construct inductively a family $(y_\xi)_{\xi\in\omega_1}$ such that $d(y_\xi,y_\eta)\geq\varepsilon$ for any $\xi,\eta\in\omega_1,\xi\neq\eta$. We set $A:=\{y_\xi\mid\xi\in\omega_1\}$ and denote by φ' the map $\overset{-1}{\varphi}(A)\to A$ defined by φ. A being closed $\overset{-1}{\varphi}(A)$ is a Borel set of X. Hence, there exist a Polish space Y and a continuous surjective map $\varphi'':Y\to\overset{-1}{\varphi}(A)$ (N. Bourbaki [1] Ch. IX §6 Proposition 10). Let φ''' be an injective map $A \to \mathbb{R}$. Since any subset of A is closed φ''' is continuous and $\psi := \varphi''' \circ \varphi' \circ \varphi''$ is a Borel measurable map. We deduce that $\varphi'''(A)$ is a Souslin set of $\mathbb{R}$ (K. Jacobs [1] Ch. XIII Propositions 2.6 and 2.13) so it contains a compact set K having the power of the continuum (K. Jacobs [1] Ch. XIII Theorem 2.9). Let B be a subset of K. Then $\overset{-1}{\varphi}'''(B)$ is a closed set and $\overset{-1}{\psi}(B)$ is a Borel set of Y. Hence, $B = \psi(\overset{-1}{\psi}(B))$ is also a Souslin set of K. We deduce that any subset of K is a Souslin set and therefore that any subset of K is even a Borel set of K (N. Bourbaki [1] Ch. IX §6 Corollaire 1 of Théorème 2). But the set of Borel sets of K has the power of the continuum and the set of subsets of K has a strictly greater power and this is a contradiction. □

Theorem 4.11.14. *(N. J. M. Andersen, J. P. R. Christensen (1973) [1] Theorem 1). Let G be a Hausdorff topological additive group. Any additive Borel measurable map* $\mu: \mathfrak{P}(\mathbb{N}) \to G$ *is a measure.*

By Lemma 4.1.20 there exists a family of group homomorphisms of G into metrizable topological additive groups such that the topology of G is the initial topology with respect to this family. Hence, we may assume that G is metrizable. By Lemma 4.11.13 $\mu(\mathfrak{P}(\mathbb{N}))$ is separable. Beyond that replacing G by the subgroup of G generated by $\mu(\mathfrak{P}(\mathbb{N}))$ we may suppose that G is separable and complete. By Lemma 4.11.11 μ is universally measurable and therefore by Theorem 4.11.4 μ is exhaustive. By Corollary 4.3.6 b) $(\mu(\{n\}))_{n \in \mathbb{N}}$ is supersummable. We set

$$\nu: \mathfrak{P}(\mathbb{N}) \to G, \quad M \mapsto \sum_{n \in M} \mu(\{n\})$$

and $\lambda := \mu - \nu$. ν is additive and continuous (Theorem 3.3.5) and therefore Borel measurable. λ is additive and by Lemma 4.1.12 λ is Borel measurable. We have $\lambda(A) = \lambda(B)$ for any $A, B \in \mathfrak{P}(\mathbb{N})$ so that $A \Delta B \in \mathfrak{P}_f(\mathbb{N})$. By Proposition 4.11.10 there exists a dense G_δ-set $\mathfrak{A}$ of $\mathfrak{P}(\mathbb{N})$ on which λ is constant. Let $A, B \in \mathfrak{A}$ such that $A \cap B = \emptyset$ and $A \cup B \in \mathfrak{A}$ (Proposition 4.11.9). Since

$$\lambda(A \cup B) = \lambda(A) + \lambda(B),$$

λ vanishes on $\mathfrak{A}$. The set $\{\mathbb{N} \setminus C \mid C \in \mathfrak{A}\}$ is a dense G_δ-set of $\mathfrak{P}(\mathbb{N})$. Hence, $\mathfrak{A} \cap \{\mathbb{N} \setminus C \mid C \in \mathfrak{A}\} \neq \emptyset$. Let $A \in \mathfrak{A} \cap \{\mathbb{N} \setminus C \mid C \in \mathfrak{A}\}$. Then $\mathbb{N} \setminus A \in \mathfrak{A}$ and

$$\lambda(\mathbb{N}) = \lambda(A) + \lambda(\mathbb{N} \setminus A) = 0,$$

$$\mu(\mathbb{N}) = \sum_{n \in \mathbb{N}} \mu(\{n\}).$$

Let $M \subset \mathbb{N}$. The map

$$\varphi: \mathfrak{P}(\mathbb{N}) \to \mathfrak{P}(\mathbb{N}), \quad C \mapsto M \cap C$$

being continuous, the map $\mu \circ \varphi$ is Borel measurable. Since it is additive we deduce by the above considerations that

$$\mu(M) = \mu \circ \varphi(\mathbb{N}) = \sum_{n \in \mathbb{N}} \mu \circ \varphi(\{n\}) = \sum_{n \in M} \mu(\{n\}).$$

Hence, $\mu = \nu$ and μ is a measure (Proposition 4.1.14). □

Remark. This theorem was proved by J. P. R. Christensen (1971) ([1] Theorem 2) in the case $G = \mathbb{R}$.

Corollary 4.11.15. *Let G, H be Hausdorff topological additive groups. Any Borel measurable group homomorphism $G \to H$ belongs to $\mathscr{P}^c(G, H)$.*

Let $u: G \to H$ be a Borel measurable group homomorphism and $(x_n)_{n \in \mathbb{N}}$ be a supersummable sequence in G. Then

$$\mathfrak{P}(\mathbb{N}) \to G, \quad M \mapsto \sum_{n \in M} x_n$$

is a continuous additive map (Proposition 4.1.14) and

$$\mathfrak{P}(\mathbb{N}) \to H, \quad M \mapsto u(\sum_{n \in M} x_n)$$

is a Borel measurable additive map. By Theorem 4.11.14 this map is a measure. We get $u(\sum_{n \in \mathbb{N}} x_n) = \sum_{n \in \mathbb{N}} u(x_n)$ (Proposition 4.1.14 a ⇒ e) so $u \in \mathscr{P}^c(G, H)$. □

Chapter 5: Locally convex lattices

§ 5.1 Order summable families

Definition 5.1.1. *Let G be an ordered set and $(x_\iota)_{\iota\in I}$ be a family in G. An* upper (lower) bound *of this family is an element $x\in G$ such that $x_\iota \le x$ $(x_\iota \ge x)$ for any $\iota\in I$. The family $(x_\iota)_{\iota\in I}$ is called* upper (lower) bounded *if it possesses an upper (lower) bound. If it is simultaneously upper and lower bounded, then we call it* order bounded *or simply* bounded*. The* supremum *of the family $(x_\iota)_{\iota\in I}$ in G is the smallest upper bound if it exists. In this case, it will be denoted by $\overset{G}{\bigvee\limits_{\iota\in I}} x_\iota$ or simply by $\bigvee\limits_{\iota\in I} x_\iota$. Similarly the* infimum *of the family $(x_\iota)_{\iota\in I}$ in G is the greatest lower bound if it exists. In this case, it will be denoted by $\overset{G}{\bigwedge\limits_{\iota\in I}} x_\iota$ or simply by $\bigwedge\limits_{\iota\in I} x_\iota$. If $I:=\{1,2\}$, then we set*

$$x_1\vee x_2 := x_1 \overset{G}{\vee} x_2 := \overset{G}{\bigvee_{\iota\in I}} x_\iota\,, \quad x_1\wedge x_2 := x_1 \overset{G}{\wedge} x_2 := \overset{G}{\bigwedge_{\iota\in I}} x_\iota\,.$$

We extend the above terminology for families to subsets of G by replacing any such subset A by its associated canonical family $(x)_{x\in A}$.

A filter on G is called order bounded *(or simply* bounded*) if it possesses an order bounded set. The ordered set G is called a* lattice *if any pair $x, y\in G$ possesses a supremum and an infimum. G is called* order complete *or simply* complete *if any upper bounded nonempty family in G possesses a supremum and any lower bounded nonempty family in G possesses an infimum. We say that G is* order-σ-complete *or simply* σ-complete *if the above conditions are fulfilled if we replace the word "family" by "countable family". We set*

$$[x, y] := \{z\in G \mid x\le z\le y\}$$

for any $x, y\in G$. A subset A of G is called saturated *if*

$$x, y\in A \Rightarrow [x, y]\subset A\,.$$

Let G be a lattice. A sublattice of G *is a subset F of G such that*

$$x \overset{G}{\vee} y\,, \quad x \overset{G}{\wedge} y \in F$$

for any $x, y\in F$. A σ-sublattice of G *is a sublattice F of G such that*

$$\overset{G}{\bigvee_{n\in\mathbb{N}}} x_n \in F\,, \quad \overset{G}{\bigwedge_{n\in\mathbb{N}}} y_n \in F$$

for any upper bounded sequence $(x_n)_{n\in\mathbb{N}}$ *in F for which* $\overset{G}{\bigvee\limits_{n\in\mathbb{N}}} x_n$ *exists and for any lower bounded sequence* $(y_n)_{n\in\mathbb{N}}$ *in F for which* $\overset{G}{\bigwedge\limits_{n\in\mathbb{N}}} y_n$ *exists.*

Any complete ordered set is σ-complete. Any σ-sublattice of a σ-complete lattice is σ-complete.

Definition 5.1.2. *Let G be an ordered set and* $\mathfrak{F}$ *be a filter on G. We set*

$$A := \{x \in G \mid \exists C \in \mathfrak{F},\ \ x \ \text{ is a lower bound of } \ C\},$$
$$B := \{x \in G \mid \exists C \in \mathfrak{F},\ \ x \ \text{ is an upper bound of } \ C\}.$$

If $\bigvee\limits_{x\in A} x$, $\bigwedge\limits_{x\in B} x$ *exist and are equal then we say:* $\mathfrak{F}$ *is* <u>order convergent</u>, $\mathfrak{F}$ <u>order converges to</u> $\bigvee\limits_{x\in A} x$ *and* $\bigvee\limits_{x\in A} x$ *is the* <u>order limit of $\mathfrak{F}$</u>.

Let $(x_n)_{n\in\mathbb{N}}$ *be a sequence in G,* $\mathfrak{F}$ *be the elementary filter on G generated by it and* $x \in G$. *If* $\mathfrak{F}$ *order converges to x then we say:* $(x_n)_{n\in\mathbb{N}}$ *is* <u>order convergent</u>, $(x_n)_{n\in\mathbb{N}}$ <u>order converges to x</u>, *and x is the* <u>order limit of $(x_n)_{n\in\mathbb{N}}$</u>.

If there is no danger of confusion, we drop the word "order" in the above expressions.

Let G be an ordered set, $\mathfrak{F}$ be a filter on G and $x \in G$. If $\mathfrak{F}$ converges to x, then any filter on G finer than $\mathfrak{F}$ and the filter $\{A \in \mathfrak{F} \mid A \ni x\}$ also converge to x. Any order convergent filter is order bounded.

Proposition 5.1.3. *Let G be a σ-complete ordered set,* $x \in G$ *and* $\mathfrak{F}$ *be a filter on G possessing a countable set. Then the following assertions are equivalent:*

a) $\mathfrak{F}$ *converges to x;*

b) $\mathfrak{F}$ *possesses a countable bounded set and for any such set A we have*

$$x = \bigvee_{A\supset B\in\mathfrak{F}} \Big(\bigwedge_{y\in B} y\Big) = \bigwedge_{A\supset B\in\mathfrak{F}} \Big(\bigvee_{y\in B} y\Big);$$

c) $\mathfrak{F}$ *possesses a countable bounded set A and*

$$x = \bigvee_{A\supset B\in\mathfrak{F}} \Big(\bigwedge_{y\in B} y\Big) = \bigwedge_{A\supset B\in\mathfrak{F}} \Big(\bigvee_{y\in B} y\Big).$$

a ⇒ b. Let A_0 be a countable set of $\mathfrak{F}$. By the hypothesis there exist a lower bounded set B in $\mathfrak{F}$ and an upper bounded set C in $\mathfrak{F}$. Then $A_0 \cap B \cap C$ is a countable bounded set of $\mathfrak{F}$. Now let A be a countable bounded set of $\mathfrak{F}$. Any set of $\mathfrak{F}$ contained in A is countable, nonempty and bounded, so it possesses a supremum and an infimum. It is obvious that x is an upper bound of the family $(\bigwedge\limits_{y\in B} y)_{B\in\mathfrak{F},\,B\subset A}$.
Let x_0 be another upper bound of this family. Let $z \in G$ such that there exists $B \in \mathfrak{F}$ for which z is a lower bound. Then,

$$z \leq \bigwedge_{y \in B \cap A} y \leq x_0 .$$

z being arbitrary, we get $x \leq x_0$. Hence, x is the least upper bound of the family $(\bigwedge_{y \in B} y)_{B \in \mathfrak{F}, B \subset A}$, i.e.

$$x = \bigvee_{A \supset B \in \mathfrak{F}} (\bigwedge_{y \in B} y) .$$

Similarly, we prove

$$x = \bigwedge_{A \supset B \in \mathfrak{F}} (\bigvee_{y \in B} y) .$$

b $\Rightarrow$ c and c $\Rightarrow$ a are trivial. □

Proposition 5.1.4. *Let G be a σ-complete lattice, $x \in G$ and $(x_n)_{n \in \mathbb{N}}$ be a sequence in G. The following assertions are equivalent:*

a) $(x_n)_{n \in \mathbb{N}}$ converges to x;

b) $(x_n)_{n \in \mathbb{N}}$ is bounded and

$$x = \bigvee_{n \in \mathbb{N}} \bigwedge_{m \geq n} x_m = \bigwedge_{n \in \mathbb{N}} \bigvee_{m \geq n} x_m .$$

a $\Rightarrow$ b. Let $\mathfrak{F}$ be the elementary filter on G generated by $(x_n)_{n \in \mathbb{N}}$. By Proposition 5.1.3 a $\Rightarrow$ b $\mathfrak{F}$ possesses a bounded set A. Let $n_0 \in \mathbb{N}$ such that $\{x_n \mid n \geq n_0\} \subset A$, y be a lower bound of A and z be an upper bound of A. Then $y \wedge (\bigwedge_{n \leq n_0} x_n)$ is a lower bound of $(x_n)_{n \in \mathbb{N}}$ and $z \vee (\bigvee_{n \leq n_0} x_n)$ is an upper bound of $(x_n)_{n \in \mathbb{N}}$. Hence, $(x_n)_{n \in \mathbb{N}}$ is bounded. The relation

$$x = \bigvee_{n \in \mathbb{N}} \bigwedge_{m \geq n} x_m = \bigwedge_{n \in \mathbb{N}} \bigvee_{m \geq n} x_m$$

follows immediately from Proposition 5.1.3 a $\Rightarrow$ b.

b $\Rightarrow$ a is trivial. □

Definition 5.1.5. *An* ordered additive group *is an additive group G endowed with an order relation $\leq$ such that*

$$x \leq y \Rightarrow x + z \leq y + z$$

for any $x, y, z \in G$. If G is a lattice with respect to its order relation, then we call G a lattice ordered additive group *or simply a* lattice group. *If the underlying ordered set of a lattice ordered additive groups is complete (σ-complete), we call it a* complete (σ-complete) lattice ordered additive group *or simply a* complete (σ-complete) lattice group. *A lattice ordered additive group G is called* up-down semicomplete *(F. K. Dashiell Jr. (1981) ([1]) Definition 1.1 (d)) if, for every increasing sequence $(x_n)_{n \in \mathbb{N}}$ in G and every decreasing sequence $(y_n)_{n \in \mathbb{N}}$ in G such that $\bigwedge_{n \in \mathbb{N}} (y_n - x_n) = 0$, $\bigvee_{n \in \mathbb{N}} x_n$, $\bigwedge_{n \in \mathbb{N}} y_n$,* exist and are equal.

It is obvious that every σ-complete lattice ordered additive group is up-down semicomplete.

Proposition 5.1.6. *Let G be an ordered additive group, $x, x' \in G$ and $\mathfrak{F}, \mathfrak{F}'$ be filters on G converging to x and x' respectively. Further, let φ and ψ be the maps*

$$G \times G \to G, \quad (y, z) \mapsto y + z$$

$$G \to G, \quad y \mapsto -y$$

respectively. Then $\varphi(\mathfrak{F} \times \mathfrak{F}')$ and $\psi(\mathfrak{F})$ converge to $x + x'$ and $-x$ respectively.

For any filter $\mathfrak{G}$ on G put

$$A(\mathfrak{G}) := \{y \in G \mid \exists C \in \mathfrak{G}, \quad y \quad \text{is a lower bound of} \quad C\},$$

$$B(\mathfrak{G}) := \{y \in G \mid \exists C \in \mathfrak{G}, \quad y \quad \text{is an upper bound of} \quad C\}.$$

Let $y \in A(\varphi(\mathfrak{F} \times \mathfrak{F}'))$. There exists $C'' \in \varphi(\mathfrak{F} \times \mathfrak{F}')$ such that y is a lower bound of C''. There exists $(C, C') \in \mathfrak{F} \times \mathfrak{F}'$ with $C + C' \subset C''$. We get

$$y \leq x + x'.$$

Hence, $x + x'$ is an upper bound of $A(\varphi(\mathfrak{F} \times \mathfrak{F}'))$. From

$$A(\mathfrak{F}) + A(\mathfrak{F}') \subset A(\varphi(\mathfrak{F} \times \mathfrak{F}')).$$

we get

$$\bigvee_{y \in A(\varphi(\mathfrak{F} \times \mathfrak{F}'))} y = x + x'.$$

The relation

$$\bigwedge_{y \in B(\varphi(\mathfrak{F} \times \mathfrak{F}'))} y = x + x'$$

is proved similarly. Hence, $\varphi(\mathfrak{F} \times \mathfrak{F}')$ converges to $x + x'$.

From

$$A(\psi(\mathfrak{F})) = -B(\mathfrak{F}), \quad B(\psi(\mathfrak{F})) = -A(\mathfrak{F})$$

we deduce immediately that $\psi(\mathfrak{F})$ converges to $-x$. □

Definition 5.1.7. *Let G be an ordered additive group. An element x of G is called positive, strictly positive, negative, or strictly negative if $x \geq 0$, $x > 0$, $x \leq 0$, or $x < 0$ respectively. Let $(x_\iota)_{\iota \in I}$ be a family in G, $\mathfrak{F}$ be the section filter of $\mathfrak{P}_f(I)$ and f be the map*

$$\mathfrak{P}_f(I) \to G, \quad J \mapsto \sum_{\iota \in J} x_\iota,$$

where $\sum_{\iota \in \emptyset} x_\iota = 0$. We say $(x_\iota)_{\iota \in I}$ is order summable or simply summable if the *filter* $f(\mathfrak{F})$ *order converges. Its limit is called the* order sum *or simply the* sum *of the family* $(x_\iota)_{\iota\in I}$ *and is denoted by* $\underline{\sum^{\leq}_{\iota\in I} x_\iota}$ *or simply by* $\underline{\sum_{\iota\in I} x_\iota}$. *A family in G is called* order supersummable *or simply* supersummable *if each of its subfamilies is order summable.*

We denote by $\underline{\Theta_0(G)}$ $(\underline{\Theta_e(G)})$ *the set of sequences* $(x_n)_{n\in\mathbb{N}}$ *in G for which there exists a countable supersummable family* $(y_\iota)_{\iota\in I}$ *in G such that the set*

$$\{n\in\mathbb{N}\,|\,x_n\in\{\sum_{\iota\in J} y_\iota\,|\,J\in\mathfrak{P}(I)\}\}$$

$$(\{n\in\mathbb{N}\,|\,x_n\in\{\sum_{\iota\in J} y_\iota\,|\,J\in\mathfrak{P}_f(I)\}\})$$

is infinite and set $\underline{\Phi_0(G)} := \Phi(\Theta_0(G))$, $\underline{\Phi_e(G)} := \Phi(\Theta_e(G))$, $\underline{\hat{\Phi}_0(G)} := \widehat{\Phi_0(G)}$, $\underline{\hat{\Phi}_e(G)} := \widehat{\Phi_e(G)}$.

Proposition 5.1.8. *Let G be an ordered additive group and* $(x_\iota)_{\iota\in I}$, $(y_\kappa)_{\kappa\in K}$ *be two summable families in G. We denote by L the (disjoint) sum of the sets I and K and by* $(z_\lambda)_{\lambda\in L}$ *the family defined by*

$$z_\lambda := \begin{cases} x_\lambda & \text{if } \lambda\in I \\ y_\lambda & \text{if } \lambda\in K. \end{cases}$$

Then $(z_\lambda)_{\lambda\in L}$ *and* $(-x_\iota)_{\iota\in I}$ *are summable and*

$$\sum_{\lambda\in L} z_\lambda = \sum_{\iota\in I} x_\iota + \sum_{\kappa\in K} y_\kappa, \quad \sum_{\iota\in I}(-x_\iota) = -\sum_{\iota\in I} x_\iota.$$

The assertion follows immediately from Proposition 5.1.6. □

Proposition 5.1.9. *Let G be an ordered additive group,* $(x_\iota)_{\iota\in I}$ *be a supersummable family in G and* $(I_\lambda)_{\lambda\in L}$ *be a disjoint family in* $\mathfrak{P}(I)$ *with union I. Then* $(\sum_{\iota\in I_\lambda} x_\iota)_{\lambda\in L}$ *is supersummable and its sum is* $\sum_{\iota\in I} x_\iota$.

We set

$$A := \{x\in G\,|\,\exists J\in\mathfrak{P}_f(I),\ x \text{ is a lower bound of } \{\sum_{\iota\in K} x_\iota\,|\,K\in\mathfrak{P}_f(I),\ J\subset K\}\},$$

$$B := \{x\in G\,|\,\exists J\in\mathfrak{P}_f(I),\ x \text{ is an upper bound of } \{\sum_{\iota\in K} x_\iota\,|\,K\in\mathfrak{P}_f(I),\ J\subset K\}\}.$$

Then

$$\bigvee_{x\in A} x = \sum_{\iota\in I} x_\iota = \bigwedge_{x\in B} x.$$

We have

$$A = \{x\in G\,|\,\exists J\in\mathfrak{P}_f(I),\ x \text{ is a lower bound of } \{\sum_{\iota\in K} x_\iota\,|\,K\in\mathfrak{P}(I),\ J\subset K\}\},$$

$$B = \{x\in G\,|\,\exists J\in\mathfrak{P}_f(I),\ x \text{ is an upper bound of } \{\sum_{\iota\in K} x_\iota\,|\,K\in\mathfrak{P}(I),\ J\subset K\}\}.$$

Let $\mathfrak{F}$ be the section filter of $\mathfrak{P}_f(L)$ and f be the map

$$\mathfrak{P}_f(L) \to G, \quad M \mapsto \sum_{\lambda \in M} \sum_{\iota \in I_\lambda} x_\iota .$$

Then by Proposition 5.1.8

$$A \subset \{x \in G \mid \exists J \in \mathfrak{P}_f(L),\ x \text{ is a lower bound of } \{f(M) \mid M \in \mathfrak{P}_f(L),\ J \subset M\}\},$$

$$B \subset \{x \in G \mid \exists J \in \mathfrak{P}_f(L),\ x \text{ is an upper bound of } \{f(M) \mid M \in \mathfrak{P}_f(L),\ J \subset M\}\}$$

so $f(\mathfrak{F})$ converges to $\sum_{\iota \in I} x_\iota$, i.e. $(\sum_{\iota \in I_\lambda} x_\iota)_{\lambda \in L}$ is summable and its sum is $\sum_{\iota \in I} x_\iota$. □

Proposition 5.1.10. *Let G be a σ-complete lattice group and $(x_n)_{n \in \mathbb{N}}$ be a sequence in G. Then the following assertions are equivalent;*

a) $(x_n)_{n \in \mathbb{N}}$ is supersummable ;

b) $(x_n)_{n \in \mathbb{N}}$ is summable ;

c) $\{\sum_{n \in J} x_n \mid J \in \mathfrak{P}_f(\mathbb{N})\}$ is bounded and

$$\bigvee_{J \in \mathfrak{P}_f(\mathbb{N})} \Big(\bigwedge_{J \subset K \in \mathfrak{P}_f(\mathbb{N})} \sum_{n \in K} x_n \Big) = \bigwedge_{J \in \mathfrak{P}_f(\mathbb{N})} \Big(\bigvee_{J \subset K \in \mathfrak{P}_f(\mathbb{N})} \sum_{n \in K} x_n \Big);$$

d) there exists a decreasing sequence $(y_n)_{n \in \mathbb{N}}$ in G whose infimum is 0 such that

$$-y_n \le \sum_{m \in J} x_m \le y_n$$

for any $n \in \mathbb{N}$ and any finite subset J of $\{m \in \mathbb{N} \mid m \ge n\}$.

The element defined in c) is the sum of the sequence $(x_n)_{n \in \mathbb{N}}$.

a ⇒ b is trivial.

b ⇒ c. Let $\mathfrak{F}$ be the section filter of $\mathfrak{P}_f(\mathbb{N})$ and f be the map

$$\mathfrak{P}_f(\mathbb{N}) \to G, \quad J \mapsto \sum_{n \in J} x_n .$$

Then $f(\mathfrak{F})$ converges to an element $x \in G$. By Proposition 5.1.3 a ⇒ c $f(\mathfrak{F})$ possesses a countable bounded set A and

$$x = \bigvee_{A \supset B \in f(\mathfrak{F})} \Big(\bigwedge_{t \in B} t\Big) = \bigwedge_{A \supset B \in f(\mathfrak{F})} \Big(\bigvee_{t \in B} t\Big).$$

Let $I \in \mathfrak{P}_f(\mathbb{N})$ be such that

$$\{\sum_{n \in J} x_n \mid J \in \mathfrak{P}_f(\mathbb{N}),\ I \subset J\} \subset A,$$

y be a lower bound of A and z be an upper bound of A. We set

$$u := \bigwedge_{J \subset I} \sum_{n \in J} x_n, \quad v := \bigvee_{J \subset I} \sum_{n \in J} x_n .$$

Take $J \in \mathfrak{P}_f(\mathbb{N})$. Then

$$y \leq \sum_{n \in I \cup J} x_n \leq z$$

and

$$y - v \leq y - \sum_{n \in I \setminus J} x_n \leq \sum_{n \in J} x_n \leq z - \sum_{n \in I \setminus J} x_n \leq z - u\,.$$

This shows that the set $\{\sum_{n \in J} x_n \mid J \in \mathfrak{P}_f(\mathbb{N})\}$ is bounded.

Let $B \in f(\mathfrak{F})$ with $B \subset A$. There exists $I_0 \in \mathfrak{P}_f(\mathbb{N})$ such that

$$\{\sum_{n \in K} x_n \mid K \in \mathfrak{P}_f(\mathbb{N}),\ I_0 \subset K\} \subset B$$

and we get

$$\bigwedge_{t \in B} t \leq \bigwedge_{I_0 \subset K \in \mathfrak{P}_f(\mathbb{N})} (\sum_{n \in K} x_n) \leq \bigvee_{J \in \mathfrak{P}_f(\mathbb{N})} (\bigwedge_{J \subset K \in \mathfrak{P}_f(\mathbb{N})} (\sum_{n \in K} x_n)) \leq$$
$$\leq \bigwedge_{J \in \mathfrak{P}_f(\mathbb{N})} (\bigvee_{J \subset K \in \mathfrak{P}_f(\mathbb{N})} (\sum_{n \in K} x_n)) \leq \bigvee_{I_0 \subset K \in \mathfrak{P}_f(\mathbb{N})} (\sum_{n \in K} x_n) \leq \bigvee_{t \in B} t\,.$$

From this relation and the one above we deduce immediately that

$$x = \bigvee_{J \in \mathfrak{P}_f(\mathbb{N})} (\bigwedge_{J \subset K \in \mathfrak{P}_f(\mathbb{N})} (\sum_{n \in K} x_n)) = \bigwedge_{J \in \mathfrak{P}_f(\mathbb{N})} (\bigvee_{J \subset K \in \mathfrak{P}_f(\mathbb{N})} (\sum_{n \in K} x_n))\,.$$

c $\Rightarrow$ d. For any $n \in \mathbb{N}$ put

$$J_n := \{m \in \mathbb{N} \mid m < n\}\,,$$
$$y_n := \bigvee_{J_n \subset K \in \mathfrak{P}_f(\mathbb{N})} (\sum_{m \in K} x_m) - \bigwedge_{J_n \subset K \in \mathfrak{P}_f(\mathbb{N})} (\sum_{m \in K} x_m)\,.$$

Then $(y_n)_{n \in \mathbb{N}}$ is a decreasing sequence of positive elements of G. Let y be a lower bound of this sequence and $I, J \in \mathfrak{P}_f(\mathbb{N})$. Then there exists $n \in \mathbb{N}$ with $I \cup J \subset J_n$, so

$$y \leq y_n \leq \bigvee_{I \subset K \in \mathfrak{P}_f(\mathbb{N})} (\sum_{n \in K} x_n) - \bigwedge_{J \subset K \in \mathfrak{P}_f(\mathbb{N})} (\sum_{n \in K} x_n)\,.$$

Further,

$$y \leq \bigwedge_{I, J \in \mathfrak{P}_f(\mathbb{N})} (\bigvee_{I \subset K \in \mathfrak{P}_f(\mathbb{N})} (\sum_{n \in K} x_n) - \bigwedge_{J \subset K \in \mathfrak{P}_f(\mathbb{N})} (\sum_{n \in K} x_n)) =$$
$$= \bigwedge_{I \in \mathfrak{P}_f(\mathbb{N})} (\bigvee_{I \subset K \in \mathfrak{P}_f(\mathbb{N})} (\sum_{n \in K} x_n)) - \bigvee_{J \in \mathfrak{P}_f(\mathbb{N})} (\bigwedge_{J \subset K \in \mathfrak{P}_f(\mathbb{N})} (\sum_{n \in K} x_n)) = 0\,.$$

Hence, $\bigwedge_{n \in \mathbb{N}} y_n = 0$.

d $\Rightarrow$ a. Let $M \subset \mathbb{N}$, $\mathfrak{F}$ be the section filter of $\mathfrak{P}_f(M)$ and f be the map

$$\mathfrak{P}_f(M) \to G\,, \quad J \mapsto \sum_{n \in J} x_n\,.$$

By d) the set

$$A := \{\sum_{n \in J} x_n \mid J \in \mathfrak{P}_f(M)\}$$

is bounded. Let $n \in \mathbb{N}$ and J be the set $\{m \in M \mid m \leq n\}$. We get

$$-y_n \leq \sum_{m \in K \setminus J} x_m \leq y_n,$$

$$\sum_{m \in J} x_m - y_n \leq \sum_{m \in K} x_m \leq \sum_{m \in J} x_m + y_n$$

for any $K \in \mathfrak{P}_f(M)$ with $J \subset K$, so

$$\sum_{m \in J} x_m - y_n \leq \bigwedge_{J \subset K \in \mathfrak{P}_f(\mathbb{N})} \left(\sum_{m \in K} x_m\right) \leq \bigvee_{A \supset B \in f(\mathfrak{F})} \left(\bigwedge_{y \in B} y\right) \leq$$

$$\leq \bigwedge_{A \supset B \in f(\mathfrak{F})} \left(\bigvee_{y \in B} y\right) \leq \bigvee_{J \subset K \in \mathfrak{P}_f(\mathbb{N})} \left(\sum_{m \in K} x_m\right) \leq \sum_{m \in J} x_m + y_n,$$

$$0 \leq \bigwedge_{A \supset B \in f(\mathfrak{F})} \left(\bigvee_{y \in B} y\right) - \bigvee_{A \supset B \in f(\mathfrak{F})} \left(\bigwedge_{y \in B} y\right) \leq 2y_n.$$

Since n is arbitrary and $\bigwedge_{n \in \mathbb{N}} y_n = 0$ we deduce that

$$\bigvee_{A \supset B \in f(\mathfrak{F})} \left(\bigwedge_{y \in B} y\right) = \bigwedge_{A \supset B \in f(\mathfrak{F})} \left(\bigvee_{y \in B} y\right).$$

By Proposition 5.1.3 c $\Rightarrow$ a $f(\mathfrak{F})$ is convergent. Hence, $(x_n)_{n \in M}$ is summable. □

Remark. Let us put $|x| := x \vee (-x)$ for any $x \in G$. The implication "$(x_n)_{n \in \mathbb{N}}$ supersummable $\Rightarrow (|x_n|)_{n \in \mathbb{N}}$ summable" does not hold even if G is a complete vector lattice. In order to construct such an example, let us, for any $n \in \mathbb{N}$, denote by I_n the set $\{m \in \mathbb{N} \mid 1 \leq m \leq 2n+1\}$, by E_n the vector lattice obtained by endowing the vector space $\mathbb{R}^{I_n}$ with the norm $\|(\alpha_m)_{m \in I_n}\|_n := \sum_{m \in I_n} |\alpha_m|$ and the order relation

$$(\alpha_m)_{m \in I_n} \leq (\beta_m)_{m \in I_n} \;:\Leftrightarrow\; (m \in I_n \Rightarrow \alpha_m \leq \beta_m),$$

and, for any $A \subset I_n$, denote by $x_{n,A}$ the element of E_n defined by

$$x_{n,A} : I_n \longrightarrow \mathbb{R}, \quad m \longmapsto \begin{cases} +1 & \text{if } m \in A \\ -1 & \text{if } m \notin A. \end{cases}$$

Take E as the vector space $\prod_{n \in \mathbb{N}} E_n$ endowed with the order relation

$$(x_n)_{n \in \mathbb{N}} \leq (y_n)_{n \in \mathbb{N}} \;:\Leftrightarrow\; (n \in \mathbb{N} \Rightarrow x_n \leq y_n).$$

Then E is a complete vector lattice. We set

$$G := \{(x_n)_{n \in \mathbb{N}} \in E \mid \limsup_{n \to \infty} \frac{(n!)^2 \|x_n\|_n}{(2n+1)!} < \infty\}.$$

G is a solid vector subspace of E and therefore a complete vector lattice. For any $n \in \mathbb{N}$ and any $A \subset I_n$ we denote by $y_{n,A}$ the element of G defined by

$$y_{n,A} : \mathbb{N} \to \bigcup_{m \in \mathbb{N}} E_m, \quad m \mapsto \begin{cases} x_{n,A} & \text{if } m = n \\ 0 & \text{if } m \neq n. \end{cases}$$

For any $n \in \mathbb{N}$ and any subset $\mathfrak{A}$ of $\mathfrak{P}(I_n)$ we have

$$\| \sum_{A \in \mathfrak{A}} x_{n,A} \|_n \le \frac{(2n+1)!}{(n!)^2}.$$

We set

$$I := \{(n, A) \mid n \in \mathbb{N},\ A \subset I_n\}.$$

The family $(y_{n,A})_{(n,A) \in I}$ is supersummable. We have $|y_{n,A}| = y_{n,I_n}$ for any $n \in \mathbb{N}$ and any $A \subset I_n$, so

$$\sum_{A \subset I_n} |y_{n,A}| = 2^{2n+1} y_{n,I_n}$$

for any $n \in \mathbb{N}$. By the Stirling formula it follows that $(|y_{n,A}|)_{(n,A) \in I}$ is not summable in G.

Proposition 5.1.11. *Let G be a σ-complete lattice group, $(x_\iota)_{\iota \in I}$ be a countable supersummable family in G, J be a subset of I and $(J_n)_{n \in \mathbb{N}}$ be a sequence in $\mathfrak{P}(I)$ converging to J. Then $(\sum_{\iota \in J_n} x_\iota)_{n \in \mathbb{N}}$ converges to $\sum_{\iota \in J} x_\iota$.*

We prove first that $(\sum_{\iota \in J \cap J_n} x_\iota)_{n \in \mathbb{N}}$ converges to $\sum_{\iota \in J} x_\iota$. Without loss of generality we may assume $J = \mathbb{N}$ and $J \cap J_n \neq J$ for infinitely many $n \in \mathbb{N}$. By Proposition 5.1.10 a $\Rightarrow$ d there exists a decreasing sequence $(y_n)_{n \in \mathbb{N}}$ in G whose infimum is 0 such that

$$-y_n \le \sum_{m \in K} x_m \le y_n$$

for any $n \in \mathbb{N}$ and any finite subset K of $\{m \in \mathbb{N} \mid m \ge n\}$. It is obvious that we can drop the word 'finite' in the above assertion. For any $n \in \mathbb{N}$ we denote by $k(n)$ the greatest natural number with

$$\{m \in \mathbb{N} \mid m \le k(n)\} \subset \bigcap_{p \ge n} (J \cap J_p).$$

Then $(k(n))_{n \in \mathbb{N}}$ is an increasing sequence in $\mathbb{N}$ converging to infinity. We have (Proposition 5.1.9)

$$-y_{k(n)} \le \sum_{\iota \in J \cap J_m} x_\iota - \sum_{\iota \in J} x_\iota \le y_{k(n)},$$

so

$$\sum_{\iota \in J} x_\iota - y_{k(n)} \le \sum_{\iota \in J \cap J_m} x_\iota \le \sum_{\iota \in J} x_\iota + y_{k(n)}$$

for any $m, n \in \mathbb{N}$ with $m \ge n$. We get

$$\sum_{\iota \in J} x_\iota \leq \bigvee_{n \in \mathbb{N}} \bigwedge_{m \geq n} \sum_{\iota \in J \cap J_m} x_\iota \leq \bigwedge_{n \in \mathbb{N}} \bigvee_{m \geq n} \sum_{\iota \in J \cap J_m} x_\iota \leq \sum_{\iota \in J} x_\iota .$$

By Proposition 5.1.4 b $\Rightarrow$ a $(\sum_{\iota \in J \cap J_n} x_\iota)_{n \in \mathbb{N}}$ converges to $\sum_{\iota \in J} x_\iota$.

We prove now that $(\sum_{\iota \in J_n \setminus J} x_\iota)_{n \in \mathbb{N}}$ converges to 0. Without loss of generality we may assume that $I \setminus J = \mathbb{N}$ and $\bigcap_{n \in \mathbb{N}} (\mathbb{N} \setminus J_n) \neq \emptyset$. By Proposition 5.1.10 a $\Rightarrow$ d there exists a decreasing sequence $(y_n)_{n \in \mathbb{N}}$ in G whose infimum is 0 such that

$$-y_n \leq \sum_{m \in K} x_m \leq y_n$$

for any $n \in \mathbb{N}$ and any finite subset K of $\{m \in \mathbb{N} \mid m \geq n\}$. It is obvious that we can drop the word "finite" in the above assertion. For any $n \in \mathbb{N}$ we denote by $k(n)$ the greatest natural number with

$$\{m \in \mathbb{N} \mid m \leq k(n)\} \subset \bigcap_{m \geq n} (N \setminus J_n) .$$

Then $(k(n))_{n \in \mathbb{N}}$ is an increasing sequence in $\mathbb{N}$ converging to infinity. We have

$$-y_{k(n)} \leq \sum_{\iota \in J_m \setminus J} x_\iota \leq y_{k(n)}$$

for any $m, n \in \mathbb{N}$ with $m \geq n$. Then

$$0 \leq \bigvee_{n \in \mathbb{N}} \bigwedge_{m \geq n} \sum_{\iota \in J_m \setminus J} x_\iota \leq \bigwedge_{n \in \mathbb{N}} \bigvee_{m \geq n} \sum_{\iota \in J_m \setminus J} x_\iota \leq 0 .$$

By Proposition 5.1.4 b $\Rightarrow$ a $(\sum_{\iota \in J_n \setminus J} x_\iota)_{n \in \mathbb{N}}$ converges to 0.

For any $n \in \mathbb{N}$ we have (Proposition 5.1.8)

$$\sum_{\iota \in J_n} x_\iota = \sum_{J \cap J_n} x_\iota + \sum_{J_n \setminus J} x_\iota ,$$

so (Proposition 5.1.6) $(\sum_{\iota \in J_n} x_\iota)_{n \in \mathbb{N}}$ converges to $\sum_{\iota \in J} x_\iota$. □

Proposition 5.1.12. *Let G be a lattice group and $(x_\iota)_{\iota \in I}$ be a finite family of positive elements of G. Then there exists a finite family $(y_\lambda)_{\lambda \in L}$ of strictly positive elements of G such that*

$$\sum_{\lambda \in L} y_\lambda = \bigvee_{\iota \in I} x_\iota$$

and such that for any $\iota \in I$ there exists $L_\iota \subset L$ with

$$x_\iota = \sum_{\lambda \in L_\iota} y_\lambda .$$

Let p be the cardinal number of I. For any $n \in \mathbb{N}$ with $0 < n \leq p$, denote by $\mathfrak{I}_n$ the set of subsets of I of cardinality n and set

$$z_n := \bigvee_{J \in \mathfrak{I}_n} (\bigwedge_{\iota \in J} x_\iota).$$

We set

$$z_{p+1} := 0.$$

It is obvious that $z_{n+1} \leq z_n$ for any $n \in \mathbb{N}$ with $0 < n \leq p$. For any $J \subset I$, $J \neq \emptyset$, we set

$$y_J := (\bigwedge_{\iota \in J} x_\iota - z_{n+1}) \vee 0,$$

where n denotes the cardinal number of J.

Let $n \in \mathbb{N}$ with $0 < n \leq p$ and let J, J' be two different sets of $\mathfrak{I}_n$. Then

$$y_J \wedge y_{J'} = ((\bigwedge_{\iota \in J} x_\iota - z_{n+1}) \wedge (\bigwedge_{\iota \in J'} x_\iota - z_{n+1})) \vee 0 =$$

$$= (\bigwedge_{\iota \in J \cup J'} x_\iota - z_{n+1}) \vee 0 = 0.$$

We get

$$\sum_{J \in \mathfrak{I}_n} y_J = \bigvee_{J \in \mathfrak{I}_n} y_J = (\bigvee_{J \in \mathfrak{I}_n} (\bigwedge_{\iota \in J} x_\iota - z_{n+1})) \vee 0 =$$

$$= ((\bigvee_{J \in \mathfrak{I}_n} (\bigwedge_{\iota \in J} x_\iota)) - z_{n+1}) \vee 0 = z_n - z_{n+1}.$$

We set

$$L := \{ J \subset I \,|\, J \neq \emptyset, \; y_J \neq 0 \}.$$

Then

$$\sum_{J \in L} y_J = \sum_{n=1}^{p} (\sum_{J \in \mathfrak{I}_n} y_J) = \sum_{n=1}^{p} (z_n - z_{n+1}) = z_1 = \bigvee_{\iota \in I} x_\iota.$$

Let $\iota \in I$ and set

$$L_\iota := \{ J \in L \,|\, \iota \in J \}.$$

Let $n \in \mathbb{N}$, $0 < n \leq p$. We have

$$x_\iota \wedge z_n = x_\iota \wedge (\bigvee_{J \in \mathfrak{I}_n} (\bigwedge_{\iota' \in J} x_{\iota'})) = \bigvee_{J \in \mathfrak{I}_n} (\bigwedge_{\iota' \in J \cup \{\iota\}} x_{\iota'}) =$$

$$= (\bigvee_{\iota \in J \in \mathfrak{I}_n} (\bigwedge_{\iota' \in J} x_{\iota'})) \vee (\bigvee_{\iota \in J \in \mathfrak{I}_{n+1}} (\bigwedge_{\iota' \in J} x_{\iota'})),$$

$$\bigvee_{\iota \in J \in \mathfrak{I}_n} (\bigwedge_{\iota' \in J} x_{\iota'}) \leq x_\iota \wedge z_n \leq (\bigvee_{\iota \in J \in \mathfrak{I}_n} (\bigwedge_{\iota' \in J} x_{\iota'})) \vee z_{n+1},$$

$$((\bigvee_{\iota \in J \in \mathfrak{I}_n} (\bigwedge_{\iota' \in J} x_{\iota'})) - z_{n+1}) \vee 0 \leq (x_\iota \wedge z_n - z_{n+1}) \vee 0 \leq$$

$$\leq ((\bigvee_{\iota \in J \in \mathfrak{I}_n} (\bigwedge_{\iota' \in J} x_{\iota'})) \vee z_{n+1} - z_{n+1}) \vee 0 = ((\bigvee_{\iota \in J \in \mathfrak{I}_n} (\wedge \, x_{\iota'})) - z_{n+1}) \vee 0,$$

$$x_\iota \wedge z_n - x_\iota \wedge z_{n+1} = x_\iota \wedge z_n - x_\iota \wedge z_n \wedge z_{n+1} =$$
$$= x_\iota \wedge z_n - z_{n+1} - (x_\iota \wedge z_n - z_{n+1}) \wedge 0 = (x_\iota \wedge z_n - z_{n+1}) \vee 0 =$$
$$= ((\bigvee_{\iota \in J \in \mathfrak{J}_n} (\bigwedge_{\iota' \in J} x_{\iota'})) - z_{n+1}) \vee 0 = \bigvee_{\iota \in J \in \mathfrak{J}_n} ((\bigwedge_{\iota' \in J} x_{\iota'} - z_{n+1}) \vee 0) =$$
$$= \bigvee_{\iota \in J \in \mathfrak{J}_n} y_J = \sum_{\iota \in J \in \mathfrak{J}_n} y_J = \sum_{J \in L_\iota \cap \mathfrak{J}_n} y_J .$$

Then

$$\sum_{J \in L_\iota} y_J = \sum_{n=1}^{p} \sum_{J \in L_\iota \cap \mathfrak{J}_n} y_J = \sum_{n=1}^{p} (x_\iota \wedge z_n - x_\iota \wedge z_{n+1}) = x_\iota . \qquad \square$$

Proposition 5.1.13. *Let G be a σ-complete lattice group, $(x_n)_{n \in \mathbb{N}}$ be a bounded sequence in G and*

$$x := \bigwedge_{n \in \mathbb{N}} (\bigvee_{m \geq n} x_m) .$$

Then there exists a countable supersummable family $(y_\iota)_{\iota \in I}$ of strictly positive elements of G such that, for any $n \in \mathbb{N}$, there exists $J \subset I$ with

$$x_n - x_n \wedge x = \sum_{\iota \in J} y_\iota .$$

For any $n \in \mathbb{N}$ set

$$z_n := \bigvee_{m \geq n} x_m .$$

Take $n \in \mathbb{N}$. Then, for any $y \in G$,

$$0 \leq y \wedge z_n - y \wedge z_{n+1} = y + z_n - y \vee z_n - y - z_{n+1} + y \vee z_{n+1} =$$
$$= z_n - z_{n+1} + y \vee z_{n+1} - y \vee z_n \leq z_n - z_{n+1}$$

Let $y \in G$ such that

$$x_m \wedge z_n - x_m \wedge z_{n+1} \leq y$$

for any $m \in \mathbb{N}$ with $m \leq n$. Then

$$z_n - z_{n+1} + x_m \vee z_{n+1} = z_n + x_m - x_m \wedge z_{n+1} =$$
$$= x_m \wedge z_n - x_m \wedge z_{n+1} + x_m \vee z_n \leq y + x_m \vee z_n \leq$$
$$\leq y + \bigvee_{m \leq n} (x_m \vee z_n) = y + z_0$$

for any $m \in \mathbb{N}$ with $m \leq n$. We get

$$z_n - z_{n+1} + z_0 = z_n - z_{n+1} + \bigvee_{m \leq n} (x_m \vee z_{n+1}) =$$
$$= \bigvee_{m \leq n} (z_n - z_{n+1} + x_m \vee z_{n+1}) \leq y + z_0 ,$$

so $z_n - z_{n+1} \leq y$ and

$$\bigvee_{m \leq n} (x_m \wedge z_n - x_m \wedge z_{n+1}) = z_n - z_{n+1}\,.$$

By Proposition 5.1.12 there exists a finite family $(y_\iota)_{\iota \in I_n}$ of strictly positive elements of G such that

$$\sum_{\iota \in I_n} y_\iota = z_n - z_{n+1}$$

and such that for any $m \in \mathbb{N}$ with $m \leq n$ there exists $I_{nm} \subset I_n$ with

$$x_m \wedge z_n - x_m \wedge z_{n+1} = \sum_{\iota \in I_{nm}} y_\iota\,.$$

Let I be the (disjoint) sum of $(I_n)_{n \in \mathbb{N}}$. Then $(y_\iota)_{\iota \in I}$ is a countable family of strictly positive elements of G. $(z_n - z_{n+1})_{n \in \mathbb{N}}$ being summable, we deduce that $(y_\iota)_{\iota \in I}$ is summable and therefore (Proposition 5.1.10 b $\Rightarrow$ a) supersummable.

Let $m \in \mathbb{N}$ and set

$$J := \bigcup_{n \geq m} I_{nm}\,.$$

We have

$$\sum_{\iota \in J} y_\iota = \sum_{n \geq m} \Big(\sum_{\iota \in I_{nm}} y_\iota \Big) = \sum_{n \geq m} (x_m \wedge z_n - x_m \wedge z_{n+1}) =$$

$$= x_m \wedge z_m - x_m \wedge x = x_m - x_m \wedge x\,. \quad \square$$

Proposition 5.1.14. *Let G be a σ-complete lattice group and $(x_n)_{n \in \mathbb{N}}$ be a sequence in G converging to 0. Then there exist a countable supersummable family $(y_\iota)_{\iota \in I}$ of strictly positive elements of G and two sequences $(J_n)_{n \in \mathbb{N}}$, $(K_n)_{n \in \mathbb{N}}$ in $\mathfrak{P}(I)$ converging to $\emptyset$ such that $(\bigcup_{n \in \mathbb{N}} J_n) \cap (\bigcup_{n \in \mathbb{N}} K_n) = \emptyset$ and*

$$x_n \vee 0 = \sum_{\iota \in J_n} y_\iota\,, \quad (-x_n) \vee 0 = \sum_{\iota \in K_n} y_\iota$$

for any $n \in \mathbb{N}$.

By Proposition 5.1.4 a $\Rightarrow$ b $(x_n)_{n \in \mathbb{N}}$ is bounded and

$$\bigvee_{n \in \mathbb{N}} \bigwedge_{m \geq n} x_m = \bigwedge_{n \in \mathbb{N}} \bigvee_{m \geq n} x_m = 0\,.$$

Then $(-x_n)_{n \in \mathbb{N}}$ is also bounded and

$$\bigwedge_{n \in \mathbb{N}} \bigvee_{m \geq n} (-x_m) = - \bigvee_{n \in \mathbb{N}} \bigwedge_{m \geq n} x_m = 0\,.$$

By Proposition 5.1.13 there exist two countable supersummable families $(y_\iota)_{\iota \in L}$, $(y_\iota)_{\iota \in M}$ of strictly positive elements of G and sequences $(J_n)_{n \in \mathbb{N}}$, $(K_n)_{n \in \mathbb{N}}$ of subsets of L and M respectively such that

$$\sum_{\iota \in J_n} y_\iota = x_n - x_n \wedge 0 = x_n \vee 0\,,$$

$$\sum_{\iota \in K_n} y_\iota = -x_n - (-x_n) \wedge 0 = (-x_n) \vee 0$$

for any $n \in \mathbb{N}$. Without loss of generality, we may assume $L \cap M = \emptyset$. Set $I := L \cup M$.

Let $\iota \in I$ and assume $A := \{n \in \mathbb{N} \mid \iota \in J_n\}$ is infinite. Then $y_\iota \leq x_n \vee 0$ for any $n \in A$ and $y_\iota \leq \bigvee_{m \geq n} x_m$ for any $n \in \mathbb{N}$. We get

$$y_\iota \leq \bigwedge_{n \in \mathbb{N}} \bigvee_{m \geq n} x_m = 0\,,$$

which is a contradiction, since y_ι is strictly positive. Hence, $\{n \in \mathbb{N} \mid \iota \in J_n\}$ is finite for any $\iota \in I$ and $(J_n)_{n \in \mathbb{N}}$ converges to $\emptyset$. Similarly, one shows that $(K_n)_{n \in \mathbb{N}}$ converges to $\emptyset$. □

Theorem 5.1.15. *Let G be a σ-complete lattice group, $x \in G$ and $(x_n)_{n \in \mathbb{N}}$ be a sequence in G converging to x. Then there exist a countable supersummable family $(y_\iota)_{\iota \in I}$ in G, an element $\iota_0 \in I$, and a sequence $(I_n)_{n \in \mathbb{N}}$ in $\mathfrak{P}(I)$ such that;*

a) $n \in \mathbb{N} \Rightarrow x_n = \sum_{\iota \in I_n} y_\iota$;

b) $y_{\iota_0} = x$;

c) $(I_n)_{n \in \mathbb{N}}$ *converges to* $\{\iota_0\}$.

$(x_n - x)_{n \in \mathbb{N}}$ converges to 0, so, by Proposition 5.1.14, there exist a countable supersummable family $(z_\lambda)_{\lambda \in L}$ of strictly positive elements of G and two sequences $(J_n)_{n \in \mathbb{N}}$, $(K_n)_{n \in \mathbb{N}}$ in $\mathfrak{P}(L)$ converging to $\emptyset$ such that $(\bigcup_{n \in \mathbb{N}} J_n) \cap (\bigcup_{n \in \mathbb{N}} K_n) = \emptyset$ and

$$(x_n - x) \vee 0 = \sum_{\lambda \in J_n} z_\lambda\,, \quad (x - x_n) \vee 0 = \sum_{\lambda \in K_n} z_\lambda$$

for any $n \in \mathbb{N}$. We set $I := L \cup \{\iota_0\}$, where $\iota_0 \notin L$ and

$$I_n := J_n \cup K_n \cup \{\iota_0\}$$

for any $n \in \mathbb{N}$. We put $y_\iota := z_\iota$ for any $\iota \in \bigcup_{n \in \mathbb{N}} J_n$, $y_\iota := -z_\iota$ for any $\iota \in \bigcup_{n \in \mathbb{N}} K_n$, $y_{\iota_0} := x$, and $y_\iota := 0$ for the remaining $\iota \in I$. Then $(y_\iota)_{\iota \in I}$ is a countable supersummable family in G and $(I_n)_{n \in \mathbb{N}}$ is a sequence in $\mathfrak{P}(I)$ converging to $\{\iota_0\}$. For any $n \in \mathbb{N}$ (Proposition 5.1.8)

$$\sum_{\iota \in I_n} y_\iota = \sum_{\lambda \in J_n} z_\lambda - \sum_{\lambda \in K_n} z_\lambda + x =$$
$$= (x_n - x) \vee 0 - (x - x_n) \vee 0 + x = x_n. \quad \square$$

Proposition 5.1.16. *The following assertions are equivalent for every lattice group G:*

a) G is up-down semicomplete ;

b) $(x_n - x_{n+1})_{n\in\mathbb{N}}$ *is supersummable for any convergent monotone sequence* $(x_n)_{n\in\mathbb{N}}$ *in G ;*

c) $(x_n - x_{n+1})_{n\in\mathbb{N}}$ *is supersummable for any decreasing sequence* $(x_n)_{n\in\mathbb{N}}$ *in G for which* $\bigwedge_{n\in\mathbb{N}} x_n = 0$.

a ⇒ b. Without loss of generality we may assume that $(x_n)_{n\in\mathbb{N}}$ decreases. Set $x := \bigwedge_{n\in\mathbb{N}} x_n$, take $M \subset \mathbb{N}$ and set

$$y_n := \sum_{\substack{m\in M\\ m<n}} (x_m - x_{m+1}),$$

$$z_n := x_0 - x - \sum_{\substack{m\in\mathbb{N}\setminus M\\ m<n}} (x_m - x_{m+1})$$

for any $n\in\mathbb{N}$. Then $(y_n)_{n\in\mathbb{N}}$ is an increasing sequence in G and $(z_n)_{n\in\mathbb{N}}$ is a decreasing sequence in G such that

$$z_n - y_n = x_0 - x - \sum_{m=0}^{n-1} (x_m - x_{m+1}) = x_n - x$$

for every $n\in\mathbb{N}$, so $\bigwedge_{n\in\mathbb{N}} (z_n - y_n) = 0$. By a) there exists $\bigvee_{n\in\mathbb{N}} y_n$. Since all $x_m - x_{m+1}$, $(m\in M)$, are positive we deduce that $(x_m - x_{m+1})_{m\in M}$ is summable and its sum is $\bigvee_{n\in\mathbb{N}} y_n$.

b ⇒ c is trivial.

c ⇒ a. Let $(x_n)_{n\in\mathbb{N}}$ be an increasing sequence in G and $(y_n)_{n\in\mathbb{N}}$ be a decreasing sequence in G such that $\bigwedge_{n\in\mathbb{N}} (y_n - x_n) = 0$. We set

$$z_{2n} := y_n - x_n, \quad z_{2n+1} := y_n - x_{n+1}$$

for every $n\in\mathbb{N}$. Then $(z_n)_{n\in\mathbb{N}}$ is a decreasing sequence in G with infimum 0. By c) the sequence $(x_{n+1} - x_n)_{n\in\mathbb{N}}$ is summable. We get

$$\bigvee_{n\in\mathbb{N}} x_n = x_0 + \sum_{n\in\mathbb{N}} (x_{n+1} - x_n). \quad \square$$

Corollary 5.1.17. *Let G be a σ-complete lattice group and A be an upper directed set of G whose countable subsets are upper bounded. Then the filter on G generated by the section filter of A belongs to* $\Phi_e(G)$.

Let $\mathfrak{F}$ be the section filter of A and f be the inclusion map $A \to G$. Then (A,f) is a net in G. Let $(x_n)_{n\in\mathbb{N}}$ be an increasing sequence in A. By Proposition 5.1.16 a ⇒ b the sequence $(x_{n+1} - x_n)_{n\in\mathbb{N}}$ is supersummable. Since

$$x_n = x_0 + \sum_{m=1}^{n} (x_m - x_{m-1})$$

for any $n \in \mathbb{N}$, it follows that $(x_n)_{n\in\mathbb{N}} \in \Theta_e(G)$ and $f(\mathfrak{F}) \in \Phi_e(G)$. □

Corollary 5.1.18. *Let G be an up-down semicomplete lattice group. Then any summable sequence of positive elements of G is supersummable.*

Let $(x_n)_{n\in\mathbb{N}}$ be a summable sequence of positive elements of G. Then $(\sum_{n=0}^{m} x_n)_{m\in\mathbb{N}}$ is an increasing sequence in G for which the supremum exists. By Proposition 5.1.16 a ⇒ b $(x_n)_{n\in\mathbb{N}}$ is supersummable. □

Proposition 5.1.19. *Let G be an ordered additive group, $(x_\iota)_{\iota\in I}$ be a supersummable family of positive elements of G, φ be the map*

$$\mathfrak{P}(I) \to G, \quad J \mapsto \sum_{\iota\in J} x_\iota,$$

$\mathfrak{F}$ be a convergent filter on $\mathfrak{P}(I)$ and K be the limit of $\mathfrak{F}$. Then $\varphi(\mathfrak{F})$ converges to $\varphi(K)$.

We have

$$K = \bigcup_{\mathfrak{A}\in\mathfrak{F}} \bigcap_{J\in\mathfrak{A}} J = \bigcap_{\mathfrak{A}\in\mathfrak{F}} \bigcup_{J\in\mathfrak{A}} J.$$

Let $K' \in \mathfrak{P}_f(K)$, $K'' \in \mathfrak{P}_f(I\backslash K)$. Then there exists $\mathfrak{A} \in \mathfrak{F}$ such that $K' \subset \bigcap_{J\in\mathfrak{A}} J$, $K'' \cap (\bigcup_{J\in\mathfrak{A}} J) = \emptyset$. We deduce that

$$\sum_{\iota\in K'} x_\iota \le \varphi(\bigcap_{J\in\mathfrak{A}} J) \le \varphi(J') \le \varphi(\bigcup_{J\in\mathfrak{A}} J) \le \sum_{\iota\in I} x_\iota - \sum_{\iota\in K''} x_\iota$$

for any $J' \in \mathfrak{A}$, so $\varphi(\mathfrak{F})$ converges to $\varphi(K)$. □

§ 5.2 Order continuous maps

Definition 5.2.1. *Let G be an ordered set. A map f of G into a topological space is called* <u>order continuous</u> *if, for any $x \in G$ and any filter $\mathfrak{F}$ on G converging to x, $f(\mathfrak{F})$ converges to $f(x)$. f is called* <u>order σ-continuous</u> *if, for any $x \in G$ and any sequence $(x_n)_{n\in\mathbb{N}}$ in G converging to x, $(f(x_n))_{n\in\mathbb{N}}$ converges to $f(x)$. If there is no danger of confusion, we drop the word "order" in the above expressions.*

Any order continuous map is order σ-continuous.

Theorem 5.2.2. *Let G be a σ-complete lattice group and f be a map of G into a topological space H. Then the following assertions are equivalent:*

a) f is σ-continuous;

b) for any countable supersummable family $(x_\iota)_{\iota \in I}$ *in G the map*

$$\mathfrak{P}(I) \to H, \quad J \mapsto f(\sum_{\iota \in J} x_\iota)$$

is continuous.

a ⇒ b. Let $J \in \mathfrak{P}(I)$ and $(J_n)_{n \in \mathbb{N}}$ be a sequence in $\mathfrak{P}(I)$ converging to J. By Proposition 5.1.11 $(\sum_{\iota \in J_n} x_\iota)_{n \in \mathbb{N}}$ converges to $\sum_{\iota \in J} x_\iota$. By a) $(f(\sum_{\iota \in J_n} x_\iota))_{n \in \mathbb{N}}$ converges to $f(\sum_{\iota \in J} x_\iota)$. Hence, the map

$$\mathfrak{P}(I) \to H, \quad J \mapsto f(\sum_{\iota \in J} x_\iota)$$

is continuous.

b ⇒ a. Let $x \in G$ and $(x_n)_{n \in \mathbb{N}}$ be a sequence in G converging to x. By Theorem 5.1.15 there exist a countable supersummable family $(y_\iota)_{\iota \in I}$ in G, an element $\iota_0 \in I$, and a sequence $(I_n)_{n \in \mathbb{N}}$ in $\mathfrak{P}(I)$ such that

α) $n \in \mathbb{N} \Rightarrow x_n = \sum_{\iota \in I_n} y_\iota$;

β) $y_{\iota_0} = x$;

γ) $(I_n)_{n \in \mathbb{N}}$ converges to $\{\iota_0\}$.

By b) $(f(x_n))_{n \in \mathbb{N}}$ converges to $f(x)$. Hence, f is σ-continuous. □

Proposition 5.2.3. *Let G be a σ-complete lattice group, H be a topological additive group and f be a group homomorphism of G into H. We denote by* $\mathfrak{A}$ *the set of countable sets A of positive elements of G such that* $A \setminus \{0\}$ *is not empty, is lower directed and has infimum 0. For any* $A \in \mathfrak{A}$ *we denote by* $\mathfrak{F}(A)$ *the filter on G generated by the filter base*

$$\{\{x \in A \mid x \leq y\} \mid y \in A \setminus \{0\}\}.$$

Then the following assertions are equivalent:

a) f is σ-continuous;

b) $f(\mathfrak{F}(A))$ *is a Cauchy filter for any* $A \in \mathfrak{A}$*;*

c) $f(\mathfrak{F}(A))$ *converges to 0 for any* $A \in \mathfrak{A}$*;*

d) for any decreasing sequence $(x_n)_{n \in \mathbb{N}}$ *in G with infimum 0,* $(f(x_n))_{n \in \mathbb{N}}$ *converges to 0.*

We may replace in b) and c) $\mathfrak{A}$ *by* $\{A \in \mathfrak{A} \mid A \text{ bounded}\}$. *If H is Hausdorff, all these assertions are equivalent to the following one:*

e) for any countable supersummable family $(x_\iota)_{\iota \in I}$ *in G the family* $(f(x_\iota))_{\iota \in I}$ *is supersummable and*

$$\sum_{\iota \in J} f(x_\iota) = f(\sum_{\iota \in J} x_\iota) \quad \textit{for any } J \subset I.$$

We denote by b′), c′) the assertions obtained from b) and c) respectively by replacing $\mathfrak{A}$ by $\{A\in\mathfrak{A}\mid A \text{ bounded}\}$.

a ⇒ b ⇒ b′, c ⇒ c′ ⇒ b′ are trivial.

b′ ⇒ c. Let $A\in\mathfrak{A}$ and $x\in A\setminus\{0\}$. We set

$$B:=\{y\in A\cup\{0\}\mid y\le x\}.$$

Then B is a bounded set of G belonging to $\mathfrak{A}$ and 0 is an adherent point of $f(\mathfrak{F}(B))$. By b′) $f(\mathfrak{F}(B))$ is a Cauchy filter, so it converges to 0. $f(\mathfrak{F}(A))$ being finer than $f(\mathfrak{F}(B))$, it also converges to 0.

c ⇒ d is trivial.

d ⇒ a. Let $(x_n)_{n\in\mathbb{N}}$ be a sequence of positive elements of G converging to 0. By Proposition 5.1.4 a ⇒ b $(x_n)_{n\in\mathbb{N}}$ is bounded and $\bigwedge_{n\in\mathbb{N}} y_n=0$, where

$$y_n:=\bigvee_{m\ge n} x_m$$

for any $n\in\mathbb{N}$. Assume $(f(x_n))_{n\in\mathbb{N}}$ does not converge to 0. Then there exists a closed 0-neighbourhood U in H such that

$$M:=\{n\in\mathbb{N}\mid f(x_n)\notin U\}$$

is infinite. We construct inductively a strictly increasing sequence $(k_n)_{n\in\mathbb{N}}$ in M such that

$$f(x_{k_n}-x_{k_n}\wedge y_{k_{n+1}})\notin U$$

for any $n\in\mathbb{N}$. We choose k_0 arbitrarily in M. Let $n\in\mathbb{N}$ and assume k_n has been constructed. Let $\varphi:\mathbb{N}\to M$ be an isomorphism of ordered sets. Then $(x_{k_n}\wedge y_{\varphi(m)})_{m\in\mathbb{N}}$ is a decreasing sequence in G with infimum 0, so $(f(x_{k_n}\wedge y_{\varphi(m)}))_{m\in\mathbb{N}}$ converges to 0. We get

$$\lim_{m\to\infty} f(x_{k_n}-x_{k_n}\wedge y_{\varphi(m)})=f(x_{k_n})\notin U$$

Since U is closed there exists $m\in\mathbb{N}$ with

$$f(x_{k_n}-x_{k_n}\wedge y_{\varphi(m)})\notin U.$$

We set $k_{n+1}:=\varphi(m)$. This finishes the inductive construction.

$(x_{k_n}-x_{k_n}\wedge y_{k_{n+1}})_{n\in\mathbb{N}}$ is a sequence of positive elements of G. By induction it can be seen that

$$\sum_{m=0}^{n}(x_{k_m}-x_{k_m}\wedge y_{k_{m+1}})\le y_{k_0}-y_{k_{n+1}}$$

for any $n\in\mathbb{N}$. Hence, $(x_{k_n}-x_{k_n}\wedge y_{k_{n+1}})_{n\in\mathbb{N}}$ is summable.

$$\Big(\sum_{m\ge n}(x_{k_m}-x_{k_m}\wedge y_{k_{m+1}})\Big)_{n\in\mathbb{N}}$$

is a decreasing sequence in G with infimum 0. By d)

$$(f(\sum_{m\geq n}(x_{k_m}-x_{k_m}\wedge y_{k_{m+1}})))_{n\in\mathbb{N}}$$

converges to 0. We get $(f(x_{k_n}-x_{k_n}\wedge y_{k_{n+1}}))_{n\in\mathbb{N}}$ converges to 0, which is the expected contradiction. Hence, $(f(x_n))_{n\in\mathbb{N}}$ converges to 0.

Now let $(x_n)_{n\in\mathbb{N}}$ be a sequence in G which converges to an element $x\in G$. Then $((x_n-x)\vee 0)_{n\in\mathbb{N}}$, $((x-x_n)\vee 0)_{n\in\mathbb{N}}$ are sequences of positive elements of G converging to 0. By the above proof,

$$\lim_{n\to\infty} f((x_n-x)\vee 0)=\lim_{n\to\infty} f((x-x_n)\vee 0)=0.$$

We get

$$\lim_{n\to\infty} f(x_n)=f(x).$$

Hence, f is σ-continuous.

a $\Leftrightarrow$ e. Let us denote by Δ the class of countable supersummable families in G and, for any $x:=(x_\iota)_{\iota\in I}\in\Delta$, denote by φ_x the map

$$\mathfrak{P}(I)\to G,\quad J\mapsto\sum_{\iota\in J}x_\iota.$$

By Proposition 5.1.8 $f\circ\varphi_x\in\mathscr{H}(I,H)$ for any $x\in\Delta$. By Theorem 5.2.2 f is σ-continuous iff $f\circ\varphi_x$ is continuous for any $x\in\Delta$. By Theorem 3.3.5 this is equivalent to e). □

Proposition 5.2.4. *Let G be a σ-complete lattice group, H be a topological additive group and f be a group homomorphism of G into H. We denote by $\mathfrak{A}$ the set of sets A of positive elements of G such that $A\backslash\{0\}$ is not empty, lower directed and has infimum 0. For any $A\in\mathfrak{A}$ denote by $\mathfrak{F}(A)$ the filter on G generated by the filter base*

$$\{\{x\in A\mid x\leq y\}\mid y\in A\backslash\{0\}\}.$$

Then the following assertions are equivalent:

a) f is continuous;

b) $f(\mathfrak{F}(A))$ is a Cauchy filter for any $A\in\mathfrak{A}$;

c) $f(\mathfrak{F}(A))$ converges to 0 for any $A\in\mathfrak{A}$.

We may replace in b) and c) $\mathfrak{A}$ by $\{A\in\mathfrak{A}\mid A$ bounded$\}$.

We denote by b′), c′) the assertions b), c) with $\mathfrak{A}$ replaced by

$$\{A\in\mathfrak{A}\mid A\quad\text{bounded}\}.$$

a $\Rightarrow$ b $\Rightarrow$ b′, c $\Rightarrow$ c′ $\Rightarrow$ b′ are trivial.

b′ $\Rightarrow$ c. Let $A\in\mathfrak{A}$ and $x\in A\backslash\{0\}$. We set

$$B:=\{y\in A\cup\{0\}\mid y\leq x\}.$$

Then B is a bounded set of G belonging to $\mathfrak{A}$ and 0 is an adherent point of $f(\mathfrak{F}(B))$. By b′) $f(\mathfrak{F}(B))$ is a Cauchy filter, so it converges to 0. $f(\mathfrak{F}(A))$, being finer than $f(\mathfrak{F}(B))$, it also converges to 0.

c ⇒ a. Let $\mathfrak{F}$ be a filter on G converging to 0 such that

$$G_+ := \{x \in G \mid x \geq 0\} \in \mathfrak{F}.$$

We want to show that $f(\mathfrak{F})$ converges to 0. Assume the contrary. Then there exists a closed 0-neighbourhood U in H such that for any $A \in \mathfrak{F}$ there exists $x \in A$ with $f(x) \notin U$. We set

$$B := \{x \in G \mid \exists A \in \mathfrak{F},\quad x \text{ is an upper bound of } \quad A\}.$$

Since $\mathfrak{F}$ converges to 0 we have $B \in \mathfrak{A}$. Let $y_{-1} \in B$. We construct inductively a sequence $(x_n)_{n \in \mathbb{N}}$ in G_+ and a decreasing sequence $(y_n)_{n \in \mathbb{N}}$ in B such that

$$f(x_n) \notin U, \quad x_n \leq y_{n-1},$$

$$y_n + \sum_{m=0}^{n} x_m \leq y_{-1}$$

for any $n \in \mathbb{N}$. By the definition of B there exists $A \in \mathfrak{F}$ such that y_{-1} is an upper bound of A. By the above hypothesis there exists $z \in A \cap G_+$ such that $f(z) \notin U$. Then $\{z \wedge y \mid y \in B\}$ is lower directed and 0 is its infimum. Since U is closed there exists by c) an $y_0 \in B$ with $y_0 \leq y_{-1}$ and

$$f(z - z \wedge y_0) \notin U.$$

We set

$$x_0 := z - z \wedge y_0.$$

Then

$$y_0 + x_0 = y_0 + z - z \wedge y_0 = y_0 \vee z \leq y_{-1}.$$

Let $n \in \mathbb{N}$ and assume the sequences $(x_m)_{m \in \mathbb{N}}$ and $(y_m)_{m \in \mathbb{N}}$ were constructed up to n. By the definition of B there exists $A \in \mathfrak{F}$ such that y_n is an upper bound of A. By the above hypothesis there exists $z \in A \cap G_+$ such that $f(z) \notin U$. Then $\{z \wedge y \mid y \in B\}$ is lower directed and 0 is its infimum. Since U is closed there exists by c) an $y_{n+1} \in B$ with $y_{n+1} \leq y_n$ and

$$f(z - z \wedge y_{n+1}) \notin U.$$

We set

$$x_{n+1} := z - z \wedge y_{n+1}.$$

Then

$$x_{n+1} \leq z \leq y_n$$

and

$$y_{n+1} + \sum_{m=0}^{n+1} x_m = y_{n+1} + z - z \wedge y_{n+1} + \sum_{m=0}^{n} x_m =$$

$$= y_{n+1} \vee z + \sum_{m=0}^{n} x_m \leq y_n + \sum_{m=0}^{n} x_m \leq y_{-1}.$$

This finishes the inductive construction. The sequence $(x_n)_{n \in \mathbb{N}}$ is obviously summable and therefore (Proposition 5.1.10 b $\Rightarrow$ a) supersummable. By Proposition 5.2.3 c $\Rightarrow$ a f is σ-continuous, so, by Theorem 5.2.2 a $\Rightarrow$ b, $(f(x_n))_{n \in \mathbb{N}}$ converges to 0. This is the expected contradiction.

Now let $x \in G$ and $\mathfrak{F}$ be a filter on G converging to x. We denote by φ and ψ the maps

$$G \to G, \quad y \mapsto (y - x) \vee 0,$$

$$G \to G, \quad y \mapsto (x - y) \vee 0.$$

Then $\varphi(\mathfrak{F})$ and $\psi(\mathfrak{F})$ converge to 0 and

$$\{y \in G \mid y \geq 0\} \in \varphi(\mathfrak{F}) \cap \psi(\mathfrak{F}).$$

By the above considerations $f(\varphi(\mathfrak{F}))$ and $f(\psi(\mathfrak{F}))$ converge to 0. Let $\varrho, \varrho', \varrho''$ be the maps

$$G \times G \to G, \quad (y, z) \mapsto x + y - z,$$

$$G \times G \to H \times H, \quad (y, z) \mapsto (f(y), f(z)),$$

$$H \times H \to H, \quad (y, z) \mapsto f(x) + y - z$$

respectively. Then $\varrho''(f(\varphi(\mathfrak{F})) \times f(\psi(\mathfrak{F})))$ converges to $f(x)$ and

$$\varrho(\varphi(\mathfrak{F}) \times \psi(\mathfrak{F})) \subset \mathfrak{F},$$

$$\varrho'' \circ \varrho' = f \circ \varrho.$$

We have

$$f(\mathfrak{F}) \supset f(\varrho(\varphi(\mathfrak{F}) \times \psi(\mathfrak{F}))) = \varrho''(\varrho'(\varphi(\mathfrak{F}) \times \psi(\mathfrak{F}))) =$$

$$= \varrho''(f(\varphi(\mathfrak{F})) \times f(\psi(\mathfrak{F}))),$$

so $f(\mathfrak{F})$ converges to $f(x)$, i.e. f is continuous. □

Definition 5.2.5. *Let G be an ordered additive group and H be a topological additive group. A map $f: G \to H$ will be called* <u>order exhaustive</u> *if for any countable supersummable family $(x_\iota)_{\iota \in I}$ in G the map*

$$\mathfrak{P}(I) \to H, \quad J \mapsto f(\sum_{\iota \in J} x_\iota)$$

is exhaustive.

If G is a σ-complete lattice and if f is a group homomorphism then the above notion "order exhaustive" coincides with the notion "exhaustive" defined by N.J. Kalton (1974) ([2] page 257).

Proposition 5.2.6. *Let G be an up-down semicomplete lattice group, H be a topological additive group, f be an exhaustive group homomorphism of G into H and $(x_n)_{n\in\mathbb{N}}$ be a convergent monotone sequence in G. Then $(f(x_n))_{n\in\mathbb{N}}$ is a Cauchy sequence.*

Assume the contrary. Then there exist a 0-neighbourhood U in H and an increasing sequence $(k_n)_{n\in\mathbb{N}}$ in $\mathbb{N}$ such that

$$f(x_{k_{n+1}}) - f(x_{k_n}) \notin U$$

for any $n \in \mathbb{N}$. By Proposition 5.1.16 a $\Rightarrow$ b $(x_n - x_{n+1})_{n\in\mathbb{N}}$ is supersummable. Since

$$x_{k_{n+1}} - x_{k_n} = \sum_{m=k_n}^{k_{n+1}-1} (x_{m+1} - x_m)$$

we deduce that $(f(x_{k_{n+1}} - x_{k_n}))_{n\in\mathbb{N}}$ converges to 0. This contradicts the above relation. □

Corollary 5.2.7. *Let G be a σ-complete lattice group, H be a topological group, f be a σ-continuous (an exhaustive) group homomorphism of G into H, A be a nonempty upper directed set of G whose countable subsets are upper bounded and $\mathfrak{F}$ be the filter on G generated by the section filter of A. Then $f(\mathfrak{F})$ is a Cauchy Φ_1-filter (a Cauchy Φ_3-filter).*

The result still holds if we replace "upper" by "lower" everywhere in the above hypothesis.

Let $(x_n)_{n\in\mathbb{N}}$ be an increasing sequence in A. If f is σ-continuous, then $(f(x_n))_{n\in\mathbb{N}}$ is convergent. If f is exhaustive, then by Proposition 5.2.6 $(f(x_n))_{n\in\mathbb{N}}$ is a Cauchy sequence. By Proposition 1.5.17 $f(\mathfrak{F})$ is a Cauchy filter in both cases. □

Definition 5.2.8. *Let G be an ordered additive group which is endowed with a group topology. We will call G a* Rickart group *if each sequence $(x_n)_{n\in\mathbb{N}}$ in G converges to 0 if the set $\{\sum_{n\in M} x_n \mid M \in \mathfrak{P}_f(\mathbb{N})\}$ is order bounded.*

$\mathbb{R}$ is an example of a Rickart group.

Proposition 5.2.9. *The product of an arbitary family of Rickart groups is a Rickart group. In particular $\mathbb{R}^I$ is a Rickart group for any I.*

The proof is straightforward. □

Proposition 5.2.10. *Let G be an ordered additive group and u be a group homomorphism of G into a Rickart group such that $u(x) \geq 0$ for any positive element x of G. Then u is order exhaustive.*

Let $(x_\iota)_{\iota \in I}$ be a supersummable family in G and $(A_n)_{n \in \mathbb{N}}$ be a disjoint sequence in $\mathfrak{P}(I)$. We set

$$y_n := \sum_{\iota \in A_n} x_\iota$$

for any $n \in \mathbb{N}$. Then $(y_n)_{n \in \mathbb{N}}$ is a supersummable sequence in G (Proposition 5.1.9). There exist $x, y \in G$ and a finite subset K of $\mathbb{N}$ such that

$$x \leq \sum_{n \in M} y_n \leq y$$

for any $M \in \mathfrak{P}_f(\mathbb{N} \backslash K)$. We get

$$u(x) \leq \sum_{n \in M} u(y_n) \leq u(y)$$

for any $M \in \mathfrak{P}_f(\mathbb{N} \backslash K)$, so $(u(y_n))_{n \in \mathbb{N}}$ converges to 0. Hence, u is order exhaustive. □

Proposition 5.2.11. *Let G be an ordered additive group, H be a topological additive group and u be a group homomorphism of G into H. Then the following assertions are equivalent:*

a) u is exhaustive;

b) $(u(x_n))_{n \in \mathbb{N}}$ converges to 0 for any supersummable sequence $(x_n)_{n \in \mathbb{N}}$ in G.

In particular, any σ- continuous group homomorphism is exhaustive if G is σ-complete.

a ⇒ b is trivial.

b ⇒ a. Let $(x_\iota)_{\iota \in I}$ be a countable supersummable family in G and let $(I_n)_{n \in \mathbb{N}}$ be a disjoint sequence in $\mathfrak{P}(I)$. We set

$$y_n := \sum_{\iota \in I_n} x_\iota$$

for any $n \in \mathbb{N}$. By Proposition 5.1.9 $(y_n)_{n \in \mathbb{N}}$ is supersummable and therefore by b) $(u(y_n))_{n \in \mathbb{N}}$ converges to 0. Since $(I_n)_{n \in \mathbb{N}}$ is arbitrary the map

$$\mathfrak{P}(I) \longrightarrow H, \quad J \longmapsto u\Big(\sum_{\iota \in J} x_\iota\Big)$$

is exhaustive. Hence, u is exhaustive.

The last assertion follows from b ⇒ a and from Theorem 5.2.2 a ⇒ b. □

Proposition 5.2.12. *(N. J. Kalton (1974) [2] Proposition 5). Let X be a set, $\mathfrak{R}$ be a δ-ring of subsets of X, G be the subgroup of $\mathbb{Z}^X$ generated by the characteristic functions of the sets of $\mathfrak{R}$ endowed with the order relation*

$$f \leq g \; :\Leftrightarrow (x \in X \;\Rightarrow\; f(x) \leq g(x)),$$

H be a topological additive group, u be a group homomorphism $G \longrightarrow H$ and μ be the

map

$$\mathfrak{R} \to H, \quad A \mapsto u(1_A).$$

Then the following assertions are equivalent:

a) u is order exhaustive;

b) μ is exhaustive.

a ⇒ b. Let $(A_n)_{n\in\mathbb{N}} \in \Gamma(\mathfrak{R})$. Then $(1_{A_n})_{n\in\mathbb{N}}$ is order supersummable in G, so $(u(1_{A_n}))_{n\in\mathbb{N}}$ converges to 0. Hence, $(\mu(A_n))_{n\in\mathbb{N}}$ converges to 0 and μ is exhaustive.

b ⇒ a. Let $(f_n)_{n\in\mathbb{N}}$ be an order supersummable sequence in G. Then there exist $f, g \in G$ such that

$$f \leq \sum_{n\in M} f_n \leq g$$

for any finite subset M of $\mathbb{N}$. We get

$$\bigcup_{n\in\mathbb{N}} \{x\in X \mid f_n(x) \neq 0\} \subset \{x\in X \mid f(x) \neq 0\} \cup \{x\in X \mid g(x) \neq 0\},$$

and

$$\sum_{n\in\mathbb{N}} |f_n(x)| \leq |f(x)| + |g(x)|$$

for any $x\in X$. Hence,

$$\bigcup_{n\in\mathbb{N}} \{x\in X \mid f_n(x) \neq 0\} \in \mathfrak{R}$$

and

$$m := \sup_{x\in X} \sum_{n\in\mathbb{N}} |f_n(x)| < \infty .$$

We set

$$A_{n,k} := \{x\in X \mid f_n(x) = k\}$$

for any $(n,k)\in\mathbb{N}\times\mathbb{Z}$ and, for any $n, p, q\in\mathbb{N}$ and any $k\in\mathbb{Z}$, denote by $A_{n,k}(p,q)$ the set of $x\in A_{n,k}$ such that $\{j\in\mathbb{N} \mid j<n, x\in A_{j,k}\}$ possesses exactly p elements and $\{j\in\mathbb{N} \mid j>n, x\in A_{j,k}\}$ possesses exactly q elements. Then $(A_{n,k}(p,q))_{n\in\mathbb{N}} \in \Gamma(\mathfrak{R})$ and $(\mu(A_{n,k}(p,q)))_{n\in\mathbb{N}}$ converges to 0 for any $p, q\in\mathbb{N}$ and any $k\in\mathbb{Z}\setminus\{0\}$. We have

$$f_n = \sum_{k=-m}^{m} k \sum_{p,q=0}^{m} 1_{A_{n,k}(p,q)},$$

and

$$u(f_n) = \sum_{k=-m}^{m} k \sum_{p,q=0}^{m} \mu(A_{n,k}(p,q))$$

for any $n\in\mathbb{N}$. Thus, $(u(f_n))_{n\in\mathbb{N}}$ converges to 0.

By Proposition 5.2.11 b ⇒ a u is order exhaustive. □

Proposition 5.2.13. *Let X be a set, $\mathfrak{R}$ be a δ-ring of subsets of X, E be the vector space of bounded real functions on X endowed with the order relation*

$$f \le g :\Leftrightarrow (x \in X \Rightarrow f(x) \le g(x))$$

and the norm

$$E \to \mathbb{R}, \quad f \mapsto \sup_{x \in X} |f(x)|,$$

let F be the closed vector subspace of E generated by the characteristic functions of the sets of $\mathfrak{R}$, G be a topological additive group, u be a continuous group homomorphism $F \to G$ and μ be the map

$$\mathfrak{R} \to G, \quad A \mapsto u(1_A).$$

Then the following assertions are equivalent:

a) u is order exhaustive;

b) μ is exhaustive.

a $\Rightarrow$ b. Let $(A_n)_{n \in \mathbb{N}} \in \Gamma(\mathfrak{R})$. Then $(1_{A_n})_{n \in \mathbb{N}}$ is order supersummable in F and $(u(1_{A_n}))_{n \in \mathbb{N}}$ converges to 0, i.e. $(\mu(A_n))_{n \in \mathbb{N}}$ converges to 0. Hence, μ is exhaustive.

b $\Rightarrow$ a. Let $(f_n)_{n \in \mathbb{N}}$ be an order supersummable sequence in F and U be an arbitrary 0-neighbourhood in G. Further, let V be a 0-neighbourhood in G such that $V + V \subset U$. There exists $p \in \mathbb{N} \setminus \{0\}$ such that $u(f) \in V$ for any $f \in F$ with $\|f\| < \frac{1}{p}$. We denote by H the subgroup of F generated by $\left\{\frac{1}{p} 1_A \mid A \in \mathfrak{R}\right\}$ and, for any $n \in \mathbb{N}$, we denote by g_n the map

$$X \to \mathbb{R}, \quad x \mapsto \begin{cases} \sup\left\{\frac{m}{p} \mid m \in \mathbb{N}, \ \frac{m}{p} \le f_n(x)\right\} & \text{if } f_n(x) \ge 0, \\ \inf\left\{-\frac{m}{p} \mid m \in \mathbb{N}, \ -\frac{m}{p} \ge f_n(x)\right\} & \text{if } f_n(x) < 0. \end{cases}$$

Then $\{n \in \mathbb{N} \mid g_n(x) \neq 0\}$ is finite for any $x \in X$ and

$$\bigcup_{n \in \mathbb{N}} \{x \in X \mid g_n(x) \neq 0\} \in \mathfrak{R}.$$

Hence, $(g_n)_{n \in \mathbb{N}}$ is an order supersummable sequence in H. By Proposition 5.2.12 b $\Rightarrow$ a the restriction of u to H is order exhaustive and hence $(u(g_n))_{n \in \mathbb{N}}$ converges to 0. Hence, there exists $n_0 \in \mathbb{N}$ such that $u(g_n) \in V$ for any $n \in \mathbb{N}$, $n \ge n_0$. We get

$$u(f_n) = u(f_n - g_n) + u(g_n) \in V + V \subset U$$

for any $n \in \mathbb{N}$, $n \ge n_0$. Since U is arbitrary, $(u(f_n))_{n \in \mathbb{N}}$ converges to 0.

By Proposition 5.2.11 b $\Rightarrow$ a u is order exhaustive. $\square$

Proposition 5.2.14. *Let G be an up-down semicomplete lattice group, H be a topo-*

logical additive group, and $f: G \to H$ be a group homomorphism. Then the following assertions are equivalent;

a) $(f(x_n))_{n \in \mathbb{N}}$ converges (converges to 0) for every decreasing sequence $(x_n)_{n \in \mathbb{N}}$ in G with $\bigwedge_{n \in \mathbb{N}} x_n = 0$;

b) The map

$$\mathfrak{P}(\mathbb{N}) \to H, \quad M \mapsto f(\sum_{n \in M} x_n)$$

is exhaustive (is continuous) for any supersummable sequence $(x_n)_{n \in \mathbb{N}}$ of positive elements of G.

a ⇒ b. We prove the exhaustive case first. Let $(M_m)_{m \in \mathbb{N}}$ be a disjoint sequence in $\mathfrak{P}(\mathbb{N})$ and set

$$A_n := \bigcup_{m=n}^{\infty} M_m, \quad y_n := \sum_{m \in A_n} x_m$$

for every $n \in \mathbb{N}$. Then $(y_n)_{n \in \mathbb{N}}$ is a decreasing sequence in G with infimum 0 (Proposition 5.1.19) and so $(f(y_n))_{n \in \mathbb{N}}$ converges. We get

$$\lim_{n \to \infty} f(\sum_{m \in M_n} x_m) = \lim_{n \to \infty} f(y_n) - \lim_{n \to \infty} f(y_{n+1}) = 0 .$$

Now we prove the continuous case. By Theorem 3.1.6 it is sufficient to show that the map is continuous at $\emptyset$. Assume the contrary. Then there exists a closed 0-neighbourhood U in H such that for any $K \in \mathfrak{P}_f(\mathbb{N})$ there exists $M \subset \mathbb{N} \backslash K$ such that $f(\sum_{n \in M} x_n) \notin U$. We construct a disjoint sequence $(M_n)_{n \in \mathbb{N}}$ in $\mathfrak{P}_f(\mathbb{N})$ inductively such that $f(\sum_{m \in M_n} x_m) \notin U$ for every $n \in \mathbb{N}$. Let $n \in \mathbb{N}$ and assume that the sequence was constructed up to $n-1$. By the hypothesis there exists $M \subset \mathbb{N} \backslash \bigcup_{m=0}^{n-1} M_m$ such that $f(\sum_{m \in M} x_m) \notin U$. $(\sum_{k \leq m \in M} x_m)_{k \in \mathbb{N}}$ is a decreasing sequence in G with infimum 0 (Proposition 5.1.19), so $(f(\sum_{k \leq m \in M} x_m))_{k \in \mathbb{N}}$ converges to 0 and $(f(\sum_{k > m \in M} x_m))_{k \in \mathbb{N}}$ converges to $f(\sum_{n \in M} x_n)$. Since U is closed there exists $k \in \mathbb{N}$ such that $f(\sum_{k > m \in M} x_m) \notin U$. We set

$$M_n := \{m \in M \mid m < k\} .$$

This finishes the inductive construction. The existence of the sequence $(M_n)_{n \in M}$ contradicts the first part of the proof.

b ⇒ a follows immediately from Proposition 5.1.16 a ⇒ c and Corollary 4.1.11. □

Theorem 5.2.15. *Let G be an up-down semicomplete lattice group, H be a topological additive group, $\mathfrak{A}$ be the set*

$$\{\{\sum_{n \in M} x_n \mid M \in \mathfrak{P}_f(\mathbb{N})\} \mid (x_n)_{n \in \mathbb{N}} \text{ supersummable sequence of positive elements of } G\}$$

$(\{\{\sum_{n\in M} x_n | M \in \mathfrak{P}(\mathbb{N})\}(x_n)_{n\in\mathbb{N}}$ *supersummable sequence of positive elements of* $G\})$,

$\mathcal{F}$ *be the set of group homomorphisms* $f: G \to H$ *such that* $(f(x_n))_{n\in\mathbb{N}}$ *converges (converges to 0) for every decreasing sequence* $(x_n)_{n\in\mathbb{N}}$ *in* G *with* $\bigwedge_{n\in\mathbb{N}} x_n = 0$, *and* $i = 3$ $(i = 4)$. *We have:*

a) *the identity map* $\mathcal{F} \to \mathcal{F}_{\mathfrak{A}}$ *is uniformly* Φ_i*-continuous;*

b) *if* H *is Hausdorff, then* $\mathcal{F}$ *is a* Φ_4*-closed set of* H^G*;*

c) *if* H *is Hausdorff and* Φ_j*-compact for a* $j \in \{1, 2, 3, 4\}$, *then so is* $\mathcal{F}$*;*

d) *if* H *is Hausdorff and sequentially complete, then so is* $\mathcal{F}$*;*

e) *if* $\{0\}$ *is a* G_δ*-set of* H, *then* $\mathcal{F}$ *is* Φ_2*-compact.*

Let us denote by $\mathcal{F}'$ the set of maps $f: G \to H$ such that the maps

$$\mathfrak{P}(\mathbb{N}) \to H, \quad M \mapsto f(\sum_{n\in M} x_n)$$

are exhaustive (are measures) for any supersummable sequence $(x_n)_{n\in\mathbb{N}}$ of positive elements of G. By Theorem 4.1.22 all the above assertions hold if we replace $\mathcal{F}$ by $\mathcal{F}'$. Let $\mathcal{G}$ be the set of group homomorphisms of G into H. Then by Proposition 5.2.14 (and Proposition 4.1.14 a $\Leftrightarrow$ c) $\mathcal{F} = \mathcal{F}' \cap \mathcal{G}$. If H is Hausdorff then $\mathcal{G}$ is a closed set of H^G and all assertions follow from the corresponding assertions concerning $\mathcal{F}'$ (and from Corollary 1.8.15). □

§ 5.3 Spaces of order continuous group homomorphisms

Throughout this section R shall denote a topological commutative ring, G a (resp. σ-complete) lattice group which at the same time is a left R-module, Φ_0 and Φ_e shall denote the sets $\hat{\Phi}_e(G)$ and $\hat{\Phi}_e(G)$ respectively and H a topological left R-module.

Theorem 5.3.1. *Let* $\mathcal{F}$ *be the set of order exhaustive (of σ-continuous, of continuous) R-homomorphisms of* G *into* H. *We have the following:*

a) $\mathcal{F}$ *is an R-submodule of* H^G.

b) $\mathcal{F}$ *and* $\mathcal{F}_{\Phi_e}$ *(and* $\mathcal{F}_{\Phi_0}$*) are topological left R-modules, locally convex if* H *is locally convex.*

c) *The identity map* $\mathcal{F} \to \mathcal{F}_{\Phi_e}$ $(\mathcal{F} \to \mathcal{F}_{\Phi_0})$ *is uniformly* Φ_3*-continuous (is uniformly* Φ_4*-continuous).*

d) *If* H *is Hausdorff, then* $\mathcal{F}$ *is a* Φ_4*-closed set of* H^G.

e) *If* H *is Hausdorff and sequentially complete, then* $\mathcal{F}$ *is sequentially complete. If* H *is Hausdorff and* Φ_i*-compact for an* $i \in \{1, 2, 3, 4\}$, *then so is* $\mathcal{F}$.

f) If $\{0\}$ *is a* G_δ*-set of* H*, then* $\mathscr{F}$ *is* Φ_2*-compact. If, in addition,* H *is sequentially complete, then* $\mathscr{F}$ *is* Φ_4*-compact.*

g) $f(\mathfrak{F}) \in \hat{\Phi}_3(H)$ $(f(\mathfrak{F}) \in \hat{\Phi}_1(H))$ *for any* $f \in \mathscr{F}$ *and any* $\mathfrak{F} \in \Phi_e$ $(\mathfrak{F} \in \Phi_0)$.

h) Let $\mathfrak{F} \in \hat{\Phi}_3(\mathscr{F})$ $(\mathfrak{F} \in \hat{\Phi}_4(\mathscr{F}))$, (I, h) *be a net in* G *such that* I *posseses no greatest element and such that for any strictly increasing sequence* $(\iota_n)_{n \in \mathbb{N}}$ *in* I *there exist a supersummable sequence* $(y_m)_{m \in \mathbb{N}}$ *in* G *and a sequence* $(M_n)_{n \in \mathbb{N}}$ *in* $\mathfrak{P}_f(\mathbb{N})$ *(in* $\mathfrak{P}(\mathbb{N})$*) converging to* $\emptyset$ *such that* $(\sum_{m \in M_n} y_m)_{n \in \mathbb{N}}$ *is a subsequence of* $(h(\iota_n))_{n \in \mathbb{N}}$*. Further, let* $\mathfrak{G}$ *be the section filter of* I *and* φ *be the map*

$$\mathscr{F} \times I \to H, \quad (f, \iota) \mapsto f(h(\iota)).$$

Then $\varphi(\mathfrak{F} \times \mathfrak{G})$ *converges to* 0.

Let $\mathscr{G}$ be the set of maps g of $\mathfrak{P}(\mathbb{N})$ into G such that $(g(\{n\}))_{n \in \mathbb{N}}$ is supersummable and

$$g(M) = \sum_{n \in M} g(\{n\})$$

for any $M \subset \mathbb{N}$. An R-homomorphism $f: G \to H$ is order exhaustive (is σ-continuous) iff $f \circ g \in \mathscr{E}(\mathfrak{P}(\mathbb{N}), H)$ $(f \circ g \in \mathscr{M}(\mathfrak{P}(\mathbb{N}), H))$ for any $g \in \mathscr{G}$ (Theorem 5.2.2 a $\Leftrightarrow$ b, Proposition 4.1.14 a $\Leftrightarrow$ c). Let $\mathscr{F}'$ be the set of maps f of G into H such that $f \circ g \in \mathscr{E}(\mathfrak{P}(\mathbb{N}), H)$ $(f \circ g \in \mathscr{M}(\mathfrak{P}(\mathbb{N}), H))$ for any $g \in \mathscr{G}$ and let $\mathscr{H}$ be the set of R-homomorphisms of G into H. By the above remark $\mathscr{F}' \cap \mathscr{H}$ is the set of order exhaustive (of σ-continuous) R-homomorphisms of G into H. By Theorem 4.1.22 (and Proposition 5.1.9) all assertions hold if we replace $\mathscr{F}$ by $\mathscr{F}'$.

a) is trivial.

b) follows immediately from the corresponding assertion for $\mathscr{F}'$.

c) follows from the corresponding assertion for $\mathscr{F}'$ and from Corollary 1.8.15.

d) Since $\mathscr{H}$ is a closed set of H^G if follows from the corresponding assertion for $\mathscr{F}'$ that the set of order exhaustive (of σ-continuous) R-homomorphisms of G into H is a Φ_4-closed set of H^G.

In order to prove d) in the continuous case let $\mathfrak{A}$ be the set of sets A of positive elements of G such that $A \setminus \{0\}$ is nonempty, lower directed, and has infimum 0 and for any $A \in \mathfrak{A}$ let $\mathfrak{F}(A)$ be the filter on G generated by the filter base

$$\{\{x \in A \mid x \leq y\} \mid y \in A \setminus \{0\}\}.$$

Take $\mathscr{L}$ as the set of $f \in H^G$ such that $f(\mathfrak{F}(A))$ is a Cauchy filter for any $A \in \mathfrak{A}$. By Corollary 5.1.17 $\mathfrak{F}(A) \in \Phi_e$ for any $A \in \mathfrak{A}$. Hence, by Proposition 2.1.7 $\mathscr{L}$ is a closed set of $H^G_{\Phi_e}$ and therefore a fortiori a closed set of $H^G_{\Phi_0}$.

In this part of the proof denote by $\mathscr{F}''$ (by $\mathscr{F}'''$) the set of continuous (of σ-continuous) R-homomorphisms of G into H. By the first part of the proof $\mathscr{F}'''$ is a Φ_4-closed set of H^G and therefore also a Φ_4-closed set of $H^G_{\Phi_0}$ (Proposition 1.8.17). By Proposition 5.2.4 a $\Leftrightarrow$ b $\mathscr{F}'' = \mathscr{F}''' \cap \mathscr{L}$, so $\mathscr{F}''$ is a closed set of $\mathscr{F}'''_{\Phi_0}$. We deduce that $\mathscr{F}''$ is a Φ_4-closed set of $H^G_{\Phi_0}$. By c) and Corollary 2.2.5 $\mathscr{F}''$ is a Φ_4-closed set of H^G.

e) follows from d) and Proposition 2.1.14.
f) follows from e) and Proposition 1.6.4.
g) and h) follow immediately from the corresponding assertions for $\mathscr{F}'$. □

Remark. Let H be Hausdorff and $(f_n)_{n\in\mathbb{N}}$ be a sequence of order exhaustive group homomorphisms of G into H such that $(f_n(x))_{n\in\mathbb{N}}$ converges for any $x\in G$. By d) the map

$$G \to H, \quad x \mapsto \lim_{n\to\infty} f_n(x)$$

is order exhaustive. This result was proved by N.J. Kalton (1974) ([2] Theorem 3 (ii)). In his proof G is σ-complete but the lattice property of G is replaced by the (weaker) Riesz interpolation property.

Proposition 5.3.2. *Assume G σ-complete and let F be a subgroup of G such that $\overset{G}{\bigvee\limits_{n\in\mathbb{N}}} x_n \in F$ for any sequence $(x_n)_{n\in\mathbb{N}}$ in F for which $\overset{G}{\bigvee\limits_{n\in\mathbb{N}}} x_n$ exists. For any $x\in G_+ := \{x\in G \mid x\geqq 0\}$ let $\mathfrak{F}_x$ be the filter on G generated by the section filter of $\{y\in F \mid y\leqq x\}$ and $\mathscr{F}$ be a set of σ-continuous (of order exhaustive) group homomorphisms of G into H such that $f(\mathfrak{F}_x)$ converges to $f(x)$ for any $f\in\mathscr{F}$ and any $x\in G_+$. Then the identity maps $\mathscr{F}_F\to\mathscr{F}$, $\mathscr{F}_F\to\mathscr{F}_{\Phi_0}$ ($\mathscr{F}_F\to\mathscr{F}_{\Phi_e}$) are uniformly Φ_4-continuous (uniformly Φ_3-continuous).*

Let $\mathscr{G}$ be the set of σ-continuous (of order exhaustive) group homomorphisms of F into H and Ψ be the set $\hat{\Phi}_e(F)$. Since F is a σ-complete lattice ordered additive group it follows from Theorem 5.3.1 c) that the identity map $\mathscr{G}\to\mathscr{G}_\Psi$ is uniformly Φ_4-continuous (uniformly Φ_3-continuous). Let φ be the inclusion map $F\to G$ and Ψ' be the set

$$\{\varphi(\mathfrak{G}) \mid \mathfrak{G}\in\Psi\}.$$

The set $\mathscr{F}$ may be identified with a subset of $\mathscr{G}$ via the map

$$\mathscr{F}\to\mathscr{G}, \quad f\to f|F,$$

so the identity map $\mathscr{F}_F\to\mathscr{F}_{\Psi'}$ is uniformly Φ_4-continuous (uniformly Φ_3-continuous) (Corollary 1.8.15).

Let $x\in G_+$. Then $\mathfrak{F}_x\in\Psi'$ (Corollary 5.1.17) and if we endow G with the coarsest topology for which any f of $\mathscr{F}$ is continuous, then $\mathfrak{F}_x$ converges to x. By Proposition 2.1.5 the identity map $\mathscr{F}_{\Psi'}\to\mathscr{F}_{G_+}$ is uniformly continuous. Since $\mathscr{F}_{G_+}=\mathscr{F}$ we deduce the identity map $\mathscr{F}_F\to\mathscr{F}$ is uniformly Φ_4-continuous (uniformly Φ_3-continuous). By Theorem 5.3.1 c) and Corollary 1.8.5 the identity map $\mathscr{F}_F\to\mathscr{F}_{\Phi_0}$ is uniformly Φ_4-continuous (the identity map $\mathscr{F}_F\to\mathscr{F}_{\Phi_e}$ is uniformly Φ_3-continuous). □

Theorem 5.3.3. *Let $\mathscr{F}$ be the set of order exhaustive (of σ-continuous) R-homomorphisms of G into H, $\mathfrak{F}$ be a Cauchy $\hat{\Phi}_3$-filter (Cauchy $\hat{\Phi}_4$-filter) on $\mathscr{F}$ and U be a 0-neighbourhood in H. We have the following:*

a) For any $\mathfrak{G} \in \Phi_e$ $(\mathfrak{G} \in \Phi_0)$ *and any O-neighbourhood U in H there exist* $\mathscr{A} \in \mathfrak{F}$ *and* $A \in \mathfrak{G}$ *such that*

$$f(x) - g(x) \in U$$

for any $f, g \in \mathscr{A}$ *and any* $x \in A$.

b) For any $\mathfrak{G} \in \Phi_e$ $(\mathfrak{G} \in \Phi_0)$ *for which*

$$\{f \in \mathscr{F} \mid f(\mathfrak{G}) \text{ is a Cauchy filter}\} \in \mathfrak{F}$$

the image of $\mathfrak{F} \times \mathfrak{G}$ *with respect to the map*

$$\mathscr{F} \times G \to H, \quad (f, x) \mapsto f(x)$$

is a Cauchy filter.

c) For any countable supersummable family $(x_\iota)_{\iota \in I}$ *in G there exist* $\mathscr{A} \in \mathfrak{F}$ *and* $J \in \mathfrak{P}_f(I)$ *such that*

$$f(\sum_{\iota \in K} x_\iota) - g(\sum_{\iota \in L} x_\iota) \in U, \quad f(\sum_{\iota \in M} x_\iota) - g(\sum_{\iota \in M} x_\iota) \in U$$

for any $f, g \in \mathscr{A}$ *and* $K, L, M \in \mathfrak{P}_f(I)$ $(K, L, M \in \mathfrak{P}(I))$ *with* $K \cap J = L \cap J$.

d) In the σ-continuous case for any convergent sequence $(x_n)_{n \in \mathbb{N}}$ *in G there exist* $\mathscr{A} \in \mathfrak{F}$ *and* $n \in \mathbb{N}$ *such that*

$$f(x_m) - g(x_m) \in U, \quad f(x_p) - g(x_q) \in U$$

for any $f, g \in \mathscr{A}$ *and* $m, p, q \in \mathbb{N}$ *with* $p \geq n, q \geq n$.

e) If G is σ-complete then for any upper directed, upper bounded nonempty set A of G there exist $\mathscr{A} \in \mathfrak{F}$ *and* $x \in A$ *such that*

$$f(y) - g(z) \in U$$

for any $f, g \in \mathscr{A}$ *and* $y, z \in A$ *with* $y \geq x, z \geq x$.

a) and b) follow from Theorem 5.3.1 c), Theorem 1.8.4. a ⇒ i, and Theorem 2.2.3 a), b).

c) The assertion concerning M follows immediately from a) since $\{\sum_{\iota \in J} x_\iota \mid J \in \mathfrak{P}_f(I)\}$ $(\{\sum_{\iota \in J} x_\iota \mid J \in \mathfrak{P}(I)\})$ is a Φ_e-set (a Φ_0-set). We set $\mathfrak{A}(J) := \{K \in \mathfrak{P}_f(I) \mid J \cap K = \emptyset\}$ for any $J \in \mathfrak{P}_f(I)$ and we denote by $\mathfrak{H}$ the filter on $\mathfrak{P}_f(I)$ (on $\mathfrak{P}(I)$) generated by the filter base $\{\mathfrak{A}(J) \mid J \in \mathfrak{P}_f(I)\}$ and by φ the map

$$\mathfrak{P}_f(I) \to G, \quad J \mapsto \sum_{\iota \in J} x_\iota \quad (\mathfrak{P}(I) \to G, \; J \mapsto \sum_{\iota \in J} x_\iota).$$

Then $\varphi(\mathfrak{H}) \in \Phi_e$ and by Corollary 4.1.11 (by Theorem 5.2.2 a ⇒ b) $f(\varphi(\mathfrak{H}))$ is a Cauchy filter for any $f \in \mathscr{F}$. Let V be a 0-neighbourhood in H such that $V + V \subset U$. By b) there exist $\mathscr{B} \in \mathfrak{F}$ and $J \in \mathfrak{P}_f(I)$ such that

$$f(\sum_{\iota \in K} x_\iota) - g(\sum_{\iota \in L} x_\iota) \in V$$

for any $f, g \in \mathscr{B}$ and $K, L \in \mathfrak{P}_f(I \setminus J)$. Since $\mathfrak{F}$ is a Cauchy filter on $\mathscr{F}$ there exists $\mathscr{A} \in \mathfrak{F}$ such that $\mathscr{A} \subset \mathscr{B}$ and

$$f(\sum_{\iota \in K} x_\iota) - g(\sum_{\iota \in K} x_\iota) \in V$$

for any $f, g \in \mathscr{A}$ and $K \subset J$. We get

$$f(\sum_{\iota \in K} x_\iota) - g(\sum_{\iota \in L} x_\iota) =$$

$$= (f(\sum_{\iota \in K \setminus J} x_\iota) - g(\sum_{\iota \in L \setminus J} x_\iota)) + (f(\sum_{\iota \in K \cap J} x_\iota) - g(\sum_{\iota \in L \cap J} x_\iota)) \in V + V \subset U$$

for any $f, g \in \mathscr{A}$ and $K, L \in \mathfrak{P}_f(I)$ with $K \cap J = L \cap J$.

d) By Theorem 5.1.15 there exist a countable supersummable family $(y_\iota)_{\iota \in I}$ in G and a convergent sequence $(J_n)_{n \in \mathbb{N}}$ in $\mathfrak{P}(I)$ such that

$$\{x_n \mid n \in \mathbb{N}\} \subset \{\sum_{\iota \in J} y_\iota \mid J \subset I\},$$

and such that $x_n = \sum_{\iota \in J_n} y_\iota$ for any $n \in \mathbb{N}$. The assertion follows now immediately from c).

e) Let us denote by φ the inclusion map $A \to G$ and by $\mathfrak{G}$ the section filter of A. By Corollary 5.2.7 $f(\varphi(\mathfrak{G}))$ is a Cauchy filter for any $f \in \mathscr{F}$. The assertion follows now from Corollary 5.1.17 and from b). □

Corollary 5.3.4. *Let $\mathscr{F}$ be the set of order exhaustive (of σ-continuous) R-homomorphisms of G into H, $\mathfrak{F}$ be a $\hat{\Phi}_3$-filter ($\hat{\Phi}_4$-filter) on $\mathscr{F}$ and U be a 0-neighbourhood in H. Then the following hold:*

a) for any $\mathfrak{G} \in \Phi_e$ ($\mathfrak{G} \in \Phi_0$) for which

$$\{f \in \mathscr{F} \mid f(\mathfrak{G}) \text{ is a Cauchy filter}\} \in \mathfrak{F}$$

there exist $\mathscr{A} \in \mathfrak{F}$ and $A \in \mathfrak{G}$ such that

$$f(x) - f(y) \in U$$

for any $f \in \mathscr{A}$ and any $x, y \in A$;

b) for any countable supersummable family $(x_\iota)_{\iota \in I}$ in G there exists $(\mathscr{A}, J) \in \mathfrak{F} \times \mathfrak{P}_f(I)$ such that

$$f(\sum_{\iota \in K} x_\iota) - f(\sum_{\iota \in L} x_\iota) \in U$$

for any $f \in \mathscr{A}$ and any $K, L \in \mathfrak{P}_f(I)$ ($K, L \in \mathfrak{P}(I)$) with $K \cap J = L \cap J$;

c) in the σ-continuous case for any convergent sequence $(x_n)_{n \in \mathbb{N}}$ in G there exists $(\mathscr{A}, n) \in \mathfrak{F} \times \mathbb{N}$ such that

$$f(x_p) - f(x_q) \in U$$

for any $f \in \mathscr{A}$ and any $p, q \in \mathbb{N}$ with $p \geq n, q \geq n$;

d) if G is σ-complete then for any upper directed, upper bounded nonempty set A of G there exists $(\mathscr{A}, x) \in \mathfrak{F} \times A$ *such that*

$$f(y) - f(z) \subset U$$

for any $f \in \mathscr{A}$ *and* $y, z \in A$ *with* $y \geq x$, $z \geq x$.

By Propositions 1.5.10 and 1.4.2 any ultrafilter on $\mathscr{F}$ finer than $\mathfrak{F}$ is a Cauchy filter.

a) follows from Theorem 5.3.3 b) and Lemma 2.2.7.
b) follows from Theorem 5.3.3 c) and Proposition 2.1.12.
c) follows from Theorem 5.3.3 d) and Lemma 2.2.7.
d) follows from Theorem 5.3.3 e) and Lemma 2.2.7. □

Corollary 5.3.5. *Let* $\mathscr{G}$ *be the set of order exhaustive (of σ-continuous) R-homomorphisms of G into H,* $\mathscr{F}$ *be a* Φ_3*-set (a* Φ_4*-set) of* $\mathscr{G}$ *and* $(x_\iota)_{\iota \in I}$ *be a countable supersummable family in G. Then the map*

$$\mathscr{F} \times \mathfrak{P}_f(I) \to H, \quad (f, J) \mapsto f(\sum_{\iota \in J} x_\iota)$$

$$(\mathscr{F} \times \mathfrak{P}(I) \to H, \quad (f, J) \mapsto f(\sum_{\iota \in J} x_\iota))$$

is uniformly continuous.

We set

$$f' : \mathfrak{P}_f(I) \to H, \quad J \mapsto f(\sum_{\iota \in J} x_\iota)$$

$$(f' : \mathfrak{P}(I) \to H, \quad J \mapsto f(\sum_{\iota \in J} x_\iota))$$

for any $f \in \mathscr{F}$. By Corollary 5.3.4 b) $\{f' \mid f \in \mathscr{F}\}$ is uniformly equicontinuous. Since the map

$$\mathscr{F} \to H, \quad f \mapsto f(\sum_{\iota \in J} x_\iota) \qquad (J \subset I)$$

is trivially uniformly continuous and since $\mathfrak{P}_f(I)$ ($\mathfrak{P}(I)$) is precompact the map

$$\mathscr{F} \times \mathfrak{P}_f(I) \to H, \quad (f, J) \mapsto f(\sum_{\iota \in J} x_\iota)$$

$$(\mathscr{F} \times \mathfrak{P}(I) \to H, \quad (f, J) \mapsto f(\sum_{\iota \in J} x_\iota))$$

is uniformly continuous. □

§ 5.4 Vector lattices

Definition 5.4.1. *An ordered vector space is a real vector space E endowed with an order relation $\leq$ such that*

$$x \leq y \Rightarrow x+z \leq y+z$$

$$x \leq y \Rightarrow \alpha x \leq \alpha y$$

for any $x, y, z \in E$ and any $\alpha \in \mathbb{R}$, $\alpha \geq 0$. If E is a lattice, a complete lattice or a σ-complete lattice with respect to its order, then E is called a vector lattice, a complete vector lattice or a σ-complete vector lattice respectively.

Let E be a vector lattice. A vector sublattice (vector σ-sublattice) of E is a vector subspace of E which, at the same time, is a sublattice (a σ-sublattice).

Any ordered vector space is an ordered additive group with respect to the corresponding underlying structure.

Definition 5.4.2. *Let E be a real vector space and C be a subset of E. C is called a cone of E if $\alpha C \subset C$ for any $\alpha \in \mathbb{R}$, $\alpha \geq 0$. C is called a proper cone of E if it is a cone of E and if $C \cap (-C) = \{0\}$.*

Lemma 5.4.3. *Let E be a real vector space and C be a proper convex cone of E. Then there exists a unique order relation $\leq$ on E such that*

$$C = \{x \in E \mid x \geq 0\}$$

and such that E endowed with this order relation is an ordered vector space.

We set

$$x \leq y \ :\Leftrightarrow \ y - x \in C$$

for any $x, y \in E$. It is easy to see that $\leq$ is an order relation on E and that

$$C = \{x \in E \mid x \geq 0\}.$$

Since C is a cone, E endowed with $\leq$ is an ordered vector space. The unicity is trivial. □

Definition 5.4.4. *Let E be a vector space and C be a proper convex cone of E. The order relation on E defined in Lemma 5.4.3 is called the order relation on E generated by C.*

Lemma 5.4.5. *Let E be an ordered vector space and C be the set of positive elements of E. Then the following assertions are equivalent:*

a) E is a vector lattice;

b) $E = C - C$ and any two elements of C possess an infimum in E.

a $\Rightarrow$ b is trivial.

b $\Rightarrow$ a. Let $x, y \in E$. Since $E = C - C$, x and y possess a lower bound z in E. Then $x - z$, $y - z \in C$ and $(x - z) \wedge (y - z)$ exists. It is easy to verify that

$$z + ((x - z) \wedge (y - z))$$

is the infimum of x and y and $-((-x) \wedge (-y))$ is the supremum of x and y. Hence, E is a vector lattice. □

Definition 5.4.6. *Let E be a vector lattice. We set*

$$\underline{E_+} := \{x \in E \mid x \geq 0\}$$

and

$$\underline{x^+} := x \vee 0\,, \quad \underline{x^-} := (-x) \vee 0\,, \quad \underline{|x|} := x \vee (-x)$$

for any $x \in E$. A subset A of E is called solid *if*

$$x \in E,\ y \in A,\ |x| \leq |y| \Rightarrow x \in A\,.$$

A band of *E is a solid vector subspace F of E such that the supremum in E of any family in F belongs to F, if it exists.*

Proposition 5.4.7. *Let E be a vector lattice and A be a solid set of E. Then the vector subspace of E generated by A is solid.*

Let F be the vector subspace of E generated by A and G be the solid vector subspace of E generated by A. Then $F \subset G$. Let $x \in G$. Then there exists a finite family $((\alpha_\iota, x_\iota))_{\iota \in I}$ in $(\mathbb{R}_+ \setminus \{0\}) \times A$ such that

$$|x| \leq \sum_{\iota \in I} \alpha_\iota |x_\iota|\,.$$

There also exists a family $(y_\iota)_{\iota \in I}$ in E such that

$$x = \sum_{\iota \in I} y_\iota\,, \qquad \iota \in I \Rightarrow |y_\iota| \leq \alpha_\iota |x_\iota|$$

(Ch. D. Aliprantis, O. Burkinshaw [1] Theorem 1.2). We get $\frac{1}{\alpha_\iota} y_\iota \in A$ and

$$x = \sum_{\iota \in I} \alpha_\iota \frac{1}{\alpha_\iota} y_\iota \in F\,.$$

Hence, $G \subset F$ and so $G = F$. □

Definition 5.4.8. *A convex solid set A of a vector lattice E is called* integral *if, for any $x, y \in A \cap E_+$ with $x + y \in A$, there exist $\alpha, \beta \in \mathbb{R}_+$ such that*

$$x \in \bigcap_{\gamma > \alpha} (\gamma A)\,, \quad y \in \bigcap_{\gamma > \beta} (\gamma A)\,, \quad \alpha + \beta \leq 1\,.$$

Definition 5.4.9. *Let E be a real vector space and A be a convex, circled set of E generating E as vector space. The map*

$$E \to \mathbb{R}_+, \quad x \mapsto \inf\{\alpha \in \mathbb{R}_+ \mid x \in \alpha A\}$$

is called the gauge of *A* in *E*. *It is a semi-norm on E.*

Proposition 5.4.10. *Let E be a vector lattice A be an integral set of E, F be the vector subspace of E generated by A and p be the gauge of A in F. Then $p(x+y) = p(x) + p(y)$ for any $x, y \in F_+$.*

By Proposition 5.4.7 F is solid. Let $\gamma \in \mathbb{R}_+$ such that

$$p(x+y) < \gamma .$$

Then

$$\frac{1}{\gamma} x + \frac{1}{\gamma} y \in A .$$

Since A is integral there exist $\gamma', \gamma'' \in \mathbb{R}_+$ such that

$$\frac{1}{\gamma} x \in \bigcap_{\delta > \gamma'} (\delta A), \quad \frac{1}{\gamma} y \in \bigcap_{\delta > \gamma''} (\delta A), \quad \gamma' + \gamma'' \leq 1 .$$

We get

$$p(x) \leq \gamma\gamma', \quad p(y) \leq \gamma\gamma''$$

and therefore

$$p(x) + p(y) \leq \gamma .$$

Since γ is arbitrary

$$p(x) + p(y) \leq p(x+y),$$

so

$$p(x+y) = p(x) + p(y). \quad \square$$

Proposition 5.4.11. *Let E be a vector lattice, A be an integral set of E such that $\bigcap_{\alpha > \alpha_0} (\alpha A) = \alpha_0 A$ for any $\alpha_0 \in \mathbb{R}_+$, let F be the vector subspace of E generated by A, $F \times \mathbb{R}$ be the product of the vector spaces F and $\mathbb{R}$ and C be the cone of $F \times \mathbb{R}$ generated by $(A \cap E_+) \times \{1\}$. Then:*

a) C is a proper convex cone of $F \times \mathbb{R}$;

b) $F \times \mathbb{R}$ endowed with the order relation generated by C is a vector lattice.

a) Let $a, b \in C$. There exist $x, y \in A \cap E_+$ and $\alpha, \beta \in \mathbb{R}_+$ such that

$$a = (\alpha x, \alpha), \quad b = (\beta y, \beta).$$

If $\alpha = 0$ or $\beta = 0$ then $a + b \in C$. If $\alpha > 0$ and $\beta > 0$ then

$$z := \frac{\alpha}{\alpha+\beta} x + \frac{\beta}{\alpha+\beta} y \in A \cap E_+$$

and therefore

$$a + b = (\alpha x + \beta y, \alpha + \beta) = ((\alpha+\beta) z, \alpha + \beta) \in C.$$

Hence, C is convex.

Let $c \in C \cap (-C)$. Then there exist $u, v \in A \cap E_+$ and $\gamma, \delta \in \mathbb{R}_+$ such that

$$c = (\gamma u, \gamma), \quad -c = (\delta v, \delta).$$

We get $\gamma = -\delta$, so $\gamma = \delta = 0$ and $c = 0$. Hence, C is a proper convex cone of $F \times \mathbb{R}$.

b) Let us denote by p the gauge of A in F, i.e. the function

$$F \to \mathbb{R}_+, \quad x \mapsto \inf\{\alpha \in \mathbb{R}_+ \mid x \in \alpha A\}.$$

We want to show first that

$$C = \{(x, \alpha) \in F_+ \times \mathbb{R} \mid p(x) \leq \alpha\}.$$

Let $(x, \alpha) \in F_+ \times \mathbb{R}$. Assume $(x, \alpha) \in C$. Then there exist $y \in A \cap E_+$ and $\beta \in \mathbb{R}_+$ such that $(x, \alpha) = (\beta y, \beta)$. We get $x = \beta y$, $\alpha = \beta$, so

$$p(x) = \beta p(y) \leq \beta = \alpha.$$

Assume now that $p(x) \leq \alpha$. Then

$$x \in \bigcap_{\beta > \alpha} (\beta A) = \alpha A.$$

Hence, there exists $y \in A$ with $x = \alpha y$, so $(x, \alpha) \in C$.

We now want to show that $F \times \mathbb{R} = C - C$. Let $(x, \alpha) \in F \times \mathbb{R}$. By Proposition 5.4.7 F is solid. Hence, there exist $\beta, \gamma \in \mathbb{R}_+$ such that

$$p(x^+) \leq \beta, \quad p(x^-) \leq \gamma, \quad \beta - \gamma = \alpha.$$

Then by the above considerations (x^+, β), $(x^-, \gamma) \in C$ and

$$(x, \alpha) = (x^+, \beta) - (x^-, \gamma) \in C - C.$$

Hence, $F \times \mathbb{R} = C - C$.

Let (x, α), $(y, \beta) \in C$. We set (Proposition 5.4.7)

$$z := x \wedge y, \quad \gamma := \inf(\alpha - p(x), \beta - p(y)) + p(x \wedge y).$$

By Proposition 5.4.10

$$p(x - z) = p(x) - p(x \wedge y),$$

so

$$p(x-z)+\gamma \leq p(x)-p(x \wedge y)+\alpha-p(x)+p(x \wedge y)=\alpha,$$
$$p(x-z) \leq \alpha-\gamma.$$

By the above considerations

$$(x, \alpha)-(z, \gamma)=(x-z, \alpha-\gamma) \in C.$$

Hence, $(z, \gamma) \leq (x, \alpha)$. Similarly, we can show that $(z, \gamma) \leq (y, \beta)$. Now take $(u, \delta) \in F \times \mathbb{R}$ with

$$(u, \delta) \leq (x, \alpha), \quad (u, \delta) \leq (y, \beta).$$

Then $(x-u, \alpha-\delta), (y-u, \beta-\delta) \in C$ and, by the above considerations,

$$p(x-u) \leq \alpha-\delta, \quad p(y-u) \leq \beta-\delta.$$

We have $u \leq z$, so by Proposition 5.4.10

$$p(z-u)=p(x-u)-p(x-z) \leq \alpha-\delta-p(x)+p(x \wedge y),$$
$$p(z-u) \leq \beta-\delta-p(y)+p(x \wedge y), \; p(z-u) \leq \gamma-\delta.$$

By the above considerations $(z-u, \gamma-\delta) \in C$, so $(u, \delta) \leq (z, \gamma)$. Since (u, δ) is arbitrary we get

$$(z, \gamma)=(x, \alpha) \wedge (y, \beta).$$

By the above proof and Lemma 5.4.5 $F \times \mathbb{R}$ endowed with the order generated by C is a vector lattice. □

§ 5.5 Duals of vector lattices

Definition 5.5.1. *Let E be a vector lattice. A linear form x' on E is called positive if it is positive on E_+. We denote by E^+ the set of linear forms on E which may be written as difference of positive linear forms. E^+ is a vector subspace of $\mathbb{R}^E$, the relation*

$$x' \leq y' :\Leftrightarrow y'-x' \quad \text{is positive}$$

is an order relation on E^+, and E^+ endowed with this order relation is a complete vector lattice (Ch. D. Aliprantis, O. Burkinshaw [1] Ch. 1 Theorem 3.3). We denote by E^π (E^σ) the set of order continuous (order σ-continuous) elements of E^+. E^π and E^σ are bands of E^+ (Ch. D. Aliprantis, O. Burkinshaw [1] Theorem 3.7). E^+ is called the order dual of E.

Proposition 5.5.2. *Let E be a vector lattice and Φ be the set $\hat{\Phi}_e(E)$. Then:*

a) any element of E^+ is order exhaustive;

b) the identity map $E^+ \to E_\Phi^+$ is uniformly Φ_3-continuous.

a) By Proposition 5.2.10 any positive linear form on E is order exhaustive.

b) follows from a), Theorem 5.3.1c), and Corollary 1.8.15. □

Theorem 5.5.3. *Let E be an order σ-complete vector lattice. Then the following hold:*

a) if $\mathfrak{F}$ is a convergent filter on $\mathbb{R}$ with limit α, then for any $x \in E$ the image of $\mathfrak{F}$ with respect to the map

$$\mathbb{R} \to E, \quad \beta \mapsto \beta x$$

order converges to αx;

b) any order σ-continuous group homomorphism of E into a Hausdorff topological real vector space is linear;

c) a linear form on E is order exhaustive iff it belongs to E^+;

d) any order σ-continuous linear form on E belongs to E^+ (and therefore to E^σ);

e) E^+, E^σ and E^π are Φ_4-compact spaces and Φ_4-closed sets of $\mathbb{R}^E$.

a) We set

$$y := \bigwedge_{n \in \mathbb{N}} \frac{1}{n+1} |x|.$$

Then $(ny)_{n \in \mathbb{N}}$ is an upper bounded increasing sequence in E. Set

$$z := \bigvee_{n \in \mathbb{N}} ny.$$

Then

$$z + y = \bigvee_{n \in \mathbb{N}} (n+1) y = z$$

and so $y = 0$.

Let $n \in \mathbb{N}$. There exists $A \in \mathfrak{F}$ such that $|\beta - \alpha| < \frac{1}{n+1}$ for any $\beta \in A$. Then

$$|\beta x - \alpha x| \le \frac{1}{n+1} |x|$$

and so

$$\alpha x - \frac{1}{n+1} |x| \le \beta x \le \alpha x + \frac{1}{n+1} |x|$$

for any $\beta \in A$. The assertion follows immediately from this relation and from the above considerations.

b) Let F be a Hausdorff topological real vector space, u be an order σ-continuous group homomorphism of E into F and $(\alpha, x) \in \mathbb{R} \times E$. Then

$$u(nx) = nu(x)$$

for any $n \in \mathbb{Z}$ and so

$$nu\left(\frac{m}{n}x\right) = u(mx) = mu(x),$$

$$u\left(\frac{m}{n}x\right) = \frac{m}{n}u(x)$$

for any $m, n \in \mathbb{Z}$, $n \neq 0$. Let $(\alpha_n)_{n \in \mathbb{N}}$ be a sequence of rational numbers converging to α. By a) $(\alpha_n x)_{n \in \mathbb{N}}$ order converges to αx, so, by the above considerations,

$$u(\alpha x) = \lim_{n \to \infty} u(\alpha_n x) = \lim_{n \to \infty} \alpha_n u(x) = \alpha u(x)$$

c) By Proposition 5.5.2 any element of E^+ is order exhaustive. Let x' be an order exhaustive linear form on E and let $x \in E_+$. Assume

$$\sup_{y \in [0, x]} x'(y) = \infty .$$

Then there exists a sequence $(x_n)_{n \in \mathbb{N}}$ in $[0, x]$ such that $x'(x_n) > 2^n$ for any $n \in \mathbb{N}$. We have

$$0 \leq \sum_{n \in J} \frac{1}{2^n} x_n \leq \frac{1}{2^{m-1}} x$$

for any $m \in \mathbb{N}$ and for any finite subset J of $\{n \in \mathbb{N} \mid n \geq m\}$. Since $\bigwedge_{m \in \mathbb{N}} \frac{1}{2^{m-1}} x = 0$ it follows from Proposition 5.1.10 d $\Rightarrow$ a that $\left(\frac{1}{2^n} x_n\right)_{n \in \mathbb{N}}$ is supersummable and we obtain the contradictory relation

$$1 \leq \lim_{n \to \infty} x'\left(\frac{1}{2^n} x_n\right) = 0 .$$

Hence, (Ch. D. Aliprantis, O. Burkinshaw [1] Theorem 3.3) $x' \in E^+$.

d) follows from c) and Theorem 5.2.2 a $\Rightarrow$ b.

e) follows from c) and Theorem 5.3.1 d), e). □

Remarks. 1) If E is not σ-complete, then none of the above assertions has to hold. As a counter-example for a), b), c) and d) we may take $\mathbb{R}^2$ endowed with the order relation

$$(\alpha, \beta) \leq (\gamma, \delta) :\Leftrightarrow \alpha < \gamma \quad \text{or} \quad (\alpha = \gamma \quad \text{and} \quad \beta \leq \delta) .$$

Then $\left(\frac{1}{n+1}(1, 0)\right)_{n \in \mathbb{N}}$ has no infimum and

$$\mathbb{R}^2 \to \mathbb{R}, \quad (\alpha, \beta) \mapsto \beta$$

is an order continuous linear form on $\mathbb{R}^2$ which does not belong to the order dual of $\mathbb{R}^2$. If $u : \mathbb{R} \to \mathbb{Q}$ is a group homomorphism, then

$$\mathbb{R}^2 \to \mathbb{R}, \quad (\alpha, \beta) \mapsto u(\alpha)$$

is an order continuous group homomorphism which is not linear.

We now construct a counterexample for e). Let E be the vector lattice of continuous real functions f on $\{0\} \cup \left\{\frac{1}{n+1} \,\middle|\, n \in \mathbb{N}\right\}$ for which the limit

$$\lim_{n\to\infty} (n+1)\left(f\left(\frac{1}{n+1}\right) - f(0)\right)$$

exists and let x' be the linear form on E which maps any $f \in E$ into the above limit. It is easy to see that x' does not belong to E^+. For any $n \in \mathbb{N}$ let x'_n be the linear form on E

$$f \mapsto (n+1)\left(f\left(\frac{1}{n+1}\right) - f(0)\right).$$

It is clear that $x'_n \in E^\pi$ for any $n \in \mathbb{N}$ and that $(x'_n)_{n\in\mathbb{N}}$ converges to x' in $\mathbb{R}^E$. Hence, E^π, E^σ and E^+ are not even sequentially closed in $\mathbb{R}^E$.

2) d) holds even if E is Archimedean.

Definition 5.5.4. *Let E, F, G be real vector spaces, $\varphi : E \times F \to G$ be a bilinear map, $\mathscr{L}(F, G)$ be the set of linear maps of F into G and Φ be a set of filters on F. For any $A \subset E$, we denote by $\underline{A_\Phi}$ the set A endowed with the coarsest uniformity for which the map*

$$A \to \mathscr{L}_\Phi(F, G), \quad x \mapsto \varphi(x, \cdot)$$

is uniformly continuous. $\underline{A_\mathfrak{A}}$ and $\underline{A_B}$ are defined similarly for any set $\mathfrak{A}$ of subsets of F and any subset B of F. If $B = F$ we sometimes drop the index. We say φ separates E if the above map is injective for $A = E$.

If $G = \mathbb{R}$, then φ is called a duality and the maps

$$E \to \mathscr{L}(F, \mathbb{R}), \quad x \mapsto \varphi(x, \cdot),$$
$$F \to \mathscr{L}(E, \mathbb{R}), \quad y \mapsto \varphi(\cdot, y)$$

are called evaluation maps. In this case, we denote by $\underline{A^0}$ the set

$$\{y \in F \mid x \in A \Rightarrow |\varphi(x, y)| \leq 1\}$$

for any $A \subset E$ and call it the polar of A with respect to φ or the polar of A in F or simply the polar of A. The bipolar of A (with respect to φ) is the polar of A^0 with respect to φ. This is denoted by $\underline{A^{00}}$.

Definition 5.5.5. *Let E, F be vector lattices. A linear map $u : E \to F$ is called a* homomorphism of vector lattices *if*

$$u(x \vee y) = u(x) \vee u(y),$$
$$u(x \wedge y) = u(x) \wedge u(y)$$

for any $x, y \in E$.

Proposition 5.5.6. *Let E be a vector lattice, F be a solid vector subspace of E^+ and, for any $x \in E$, let $\tilde{x}$ be the map*

$$F \to \mathbb{R}, \quad x' \mapsto x'(x).$$

Then $\tilde{x} \in F^\pi$ for any $x \in E$ and the map

$$E \to F^\pi, \quad x \mapsto \tilde{x}$$

is a homomorphism of vector lattices.

The assertions follow from Ch. D. Aliprantis, O. Burkinshaw [1] Theorem 3.11. □

Proposition 5.5.7. *Let E be a vector lattice, F be a solid subspace of E^+, A be a subset of F and $\mathscr{G}$ be a set of maps g of $\mathfrak{P}(\mathbb{N})$ into A such that $(g(\{n\}))_{n\in\mathbb{N}}$ is order supersummable in F and*

$$g(M) = \sum_{n \in M} g(\{n\})$$

for any $M \subset \mathbb{N}$. Further, let Θ be the set of sequences $(x_n)_{n\in\mathbb{N}}$ in A for which there exists $g \in \mathscr{G}$ such that the set

$$\{n \in \mathbb{N} \mid x_n \in g(\mathfrak{P}(\mathbb{N}))\}$$

is infinite and Φ be the set $\hat{\Phi}(\Theta)$. Then the identity map $E_A \to E_\Phi$ is uniformly Φ_4-continuous.

Let $\mathscr{F}$ be the set of real functions f on A such that $f \circ g \in \mathscr{M}(\mathfrak{P}(\mathbb{N}), \mathbb{R})$ for any $g \in \mathscr{G}$. By Theorem 4.1.22 c) the identity map $\mathscr{F}_A \to \mathscr{F}_\Phi$ is uniformly Φ_4-continuous. For any $x \in E$ denote by $\tilde{x}$ the map

$$A \to \mathbb{R}, \quad y \mapsto y(x).$$

By Theorem 5.2.2 a ⇒ b and Proposition 5.5.6 $\tilde{x} \in \mathscr{F}$ for any $x \in E$. We set $\mathscr{F}' := \{\tilde{x} \mid x \in E\}$. By Corollary 1.8.15 the identity map $\mathscr{F}'_A \to \mathscr{F}'_\Phi$ is uniformly Φ_4-continuous. Hence, the identity map $E_A \to E_\Phi$ is uniformly Φ_4-continuous too. □

Corollary 5.5.8. *Let E be a vector lattice, F be a solid vector subspace of E^+ and Φ be the set $\hat{\Phi}_0(F)$. Then the identity map $E_F \to E_\Phi$ is uniformly Φ_4-continuous.* □

Proposition 5.5.9. *Let E be a vector lattice, F be a solid vector subspace of E^+, φ be the evaluation map $E \longrightarrow F^+$, and G be a solid vector subspace of F^+ containing $\varphi(E)$. We set*

$$X := \{\bigvee_{n\in\mathbb{N}} \varphi(x_n) \mid (x_n)_{n\in\mathbb{N}} \text{ increasing sequence in } E_+\} \cap G,$$

$$Y := \{\bigvee_{x\in A} \varphi(x) \mid A \subset E_+\} \cap G.$$

Then the identity map $F_X \longrightarrow F_Y$ is uniformly Φ_4-continuous.

Let $\mathscr{G}$ be the set of maps g of $\mathfrak{P}(\mathbb{N})$ into X such that $(g(\{n\}))_{n\in\mathbb{N}}$ is order supersummable in G and such that

$$g(M) = \sum_{n\in M} g(\{n\})$$

for any $M \subset \mathbb{N}$, let Θ be the set of sequences $(x_n)_{n\in\mathbb{N}}$ in X for which there exists $g \in \mathscr{G}$ such that the set

$$\{n \in \mathbb{N} \mid x_n \in g(\mathfrak{P}(\mathbb{N}))\}$$

is infinite, and take Φ as the set $\hat{\Phi}(\Theta)$. By Proposition 5.5.7 the identity map $F_X \longrightarrow F_\Phi$ is uniformly Φ_4-continuous.

Let $\mathfrak{F}$ be a Cauchy $\hat{\Phi}_4$-filter on F_X and take $y \in Y$. There exists a subset A of E_+ such that $y = \bigvee_{x\in A} \varphi(x)$. We may assume A upper directed (Proposition 5.5.6). Let ψ be the map

$$A \longrightarrow X, \quad x \longmapsto \varphi(x)$$

and $\mathfrak{G}$ be the section filter of A. It is easy to see that $\psi(\mathfrak{G}) \in \Phi$. Hence, by the above considerations, for any $\varepsilon > 0$ there exists $\mathscr{F} \in \mathfrak{F}$ such that for any $x'\, y' \in \mathscr{F}$ there exists $x \in A$ with

$$|x'(z) - y'(z)| < \varepsilon$$

for any $z \in A$, $x \leq z$. Then (Proposition 5.5.6)

$$|y(x') - y(y')| \leq \varepsilon$$

for any $x', y' \in \mathscr{F}$. Since y is arbitrary $\mathfrak{F}$ is a Cauchy filter on F_Y. Hence, the identity map $F_X \longrightarrow F_Y$ is uniformly Φ_4-continuous. □

Proposition 5.5.10. *Let E be a vector lattice and A be a solid set of E. Then the polar A^0 of A in E^+ is solid.*

Let $x' \in E^+$, $y' \in A^0$ such that $|x'| \leq |y'|$. We have (Ch. D. Aliprantis, O. Burkinshaw [1] Theorem 3.3)

$$|x'(x)| \leq |x'|(|x|) \leq |y'|(|x|) = \sup_{\substack{y\in E \\ |y| \leq |x|}} |y'(y)| \leq 1$$

for any $x \in A$, so $x' \in A^0$ and A^0 is solid. □

Proposition 5.5.11. *Let E be a vector lattice, F be a vector sublattice of E^+ and A be a solid set of E. If A is integral (directed), then its polar in F is directed (integral).*

Suppose first that A is integral and let x', y' be elements of A^0. Further, let z be an element of A and x, y be positive elements of E such that

$$x + y = |z|.$$

Then, by the hypothesis, there exist positive real numbers α, β for which

$$x \in \bigcap_{\gamma > \alpha} (\gamma A), \quad y \in \bigcap_{\gamma > \beta} (\gamma A), \quad \alpha + \beta \leq 1$$

and (Proposition 5.5.10)

$$|x'|(x) + |y'|(y) \leq \alpha + \beta \leq 1.$$

Hence,

$$|(x' \vee y')(z)| \leq (|x'| \vee |y'|)(|z|) =$$
$$= \sup_{\substack{x, y \in E_+ \\ x + y = |z|}} (|x'|(x) + |y'|(y)) \leq 1$$

and $x' \vee y' \in A^0$. Since x', y' are arbitrary, A^0 is upper directed. Being circled, it is also lower directed.

Now suppose A is directed. A^0 is convex. By Proposition 5.5.10 A^0 is solid. Let x', y' be positive elements of F such that $x' + y' \in A^0$. We set

$$\alpha := \sup_{x \in A} x'(x), \quad \beta := \sup_{x \in A} y'(x).$$

Then

$$\alpha + \beta = \sup_{x \in A} (x' + y')(x) \leq 1,$$

$$x' \in \bigcap_{\gamma > \alpha} (\gamma A^0), \quad y' \in \bigcap_{\gamma > \beta} (\gamma A^0). \quad \square$$

Definition 5.5.12. *Let E be a real vector space and A be a convex set of E. An* <u>extreme point of A</u> *is a point $x \in A$ such that*

$$y, z \in A, \alpha, \beta \in]0, 1[, \alpha + \beta = 1, \alpha y + \beta z = x \Rightarrow y = z = x.$$

We denote by <u>$\partial_e A$</u> *the set of extreme points of A.*

Definition 5.5.13. *Let E be a real vector space. A* <u>ray of E</u> *is a subset of E of the form $\{\alpha x \mid \alpha \in \mathbb{R}_+\}$ where $x \in E$. An* <u>extreme ray of a proper convex cone C</u> *of E is a ray $D \subset C$ such that*

$$x, y \in C, \quad x + y \in D \Rightarrow x, y \in D.$$

Proposition 5.5.14. *Let E be a vector lattice and A be an integral set of E such that*

$$\bigcap_{\alpha>\alpha_0}(\alpha A)=\alpha_0 A$$

for any $\alpha_0\in\mathbb{R}_+$. Then any extreme point of $A\cap E_+$ belongs to an extreme ray of E_+.

Let x be an extreme point of $A\cap E_+$ and y, z be points of E_+ with $y+z=x$. Then there exist $\alpha,\beta\in\mathbb{R}_+$ such that

$$y\in\bigcap_{\gamma>\alpha}(\gamma A),\quad z\in\bigcap_{\gamma>\beta}(\gamma A),\quad \alpha+\beta\le 1.$$

Assume first that $\alpha>0$, $\beta>0$. Then

$$\frac{\alpha+\beta}{\alpha}y,\quad \frac{\alpha+\beta}{\beta}z\in A.$$

From

$$\frac{\alpha}{\alpha+\beta}\left(\frac{\alpha+\beta}{\alpha}y\right)+\frac{\beta}{\alpha+\beta}\left(\frac{\alpha+\beta}{\beta}z\right)=y+z=x$$

we get

$$\frac{\alpha+\beta}{\alpha}y=\frac{\alpha+\beta}{\beta}z=x.$$

If $\alpha=0$, then $y=0$ and $z=x$. Putting together the above results, we see that x belongs to an extreme ray of E_+. □

Corollary 5.5.15. *Let E be a vector lattice, F be a vector subspace of E^+ separating E and K be an integral compact set of E_F. Then any extreme point of $E_+\bigcap K$ belongs to an extreme ray of E_+.*

Let $\alpha_0\in\mathbb{R}_+$ and $x\in\bigcap_{\alpha>\alpha_0}(\alpha K)$. Then $\frac{1}{\alpha}x\in K$ for any $\alpha>\alpha_0$. Since K is a compact set of E_F, x belongs to $\alpha_0 K$. Hence,

$$\bigcap_{\alpha>\alpha_0}(\alpha K)=\alpha_0 K$$

and the assertion follows from Proposition 5.5.14. □

Lemma 5.5.16. *Let E be a vector lattice and F be a solid vector subspace of E. A ray of F is an extreme ray of F_+ iff it is an extreme ray of E_+.*

Let D be an extreme ray of F_+ and take $x,y\in E_+$ with $x+y\in D$. Since F is solid $x,y\in F_+$, so $x,y\in D$. Hence, D is an extreme ray of E_+. The converse implication is trivial. □

Proposition 5.5.17. *Let E be a vector lattice, F be a solid vector subspace of E^+, K' be a compact integral set of F_E, φ be the evaluation map $E\to F^+$ and $(x_n)_{n\in\mathbb{N}}$ be a Cauchy* *sequence in $E_{K'}$, which is a bounded sequence in $E_{\{K'\}}$ and has the property that $(\varphi(x_n))_{n\in\mathbb{N}}$ is order bounded. We set*

$$y_n := \bigwedge_{m\geq n} \varphi(x_m), \quad z_n := \bigvee_{m\geq n} \varphi(x_m)$$

for any $n\in\mathbb{N}$. Then

$$\inf_{n\in\mathbb{N}} (z_n - y_n)(|x'|) = 0$$

for any $x'\in K'$.

Let X' be the set of elements of F lying on extreme rays of F_+. By Lemma 5.5.16 X' is the set of elements of F lying on extreme rays of $(E^+)_+$. By Ch. D. Aliprantis, O. Burkinshaw [1] Theorem 3.13 X' is the set of homomorphisms of vector lattices of E into $\mathbb{R}$ belonging to F, so it is a closed set of F_E. Hence, $X'\cap K'$ is a compact set of F_E.

Let $\mathscr{F}$ be the set of universally measurable bounded functions on $X'\bigcap K'$. $\mathscr{F}$ is a vector σ-sublattice of $\mathbb{R}^{X'\cap K'}$. For any $x\in E$ let $\tilde{x}$ be the map

$$X'\cap K' \to \mathbb{R}, \quad x' \mapsto x'(x).$$

Then $\tilde{x}\in\mathscr{F}$ for any $x\in E$. We have

$$y_n|X'\cap K' = \bigwedge_{m\geq n} \tilde{x}_m, \quad z_n|X'\cap K' = \bigvee_{m\geq n} \tilde{x}_m$$

for any $n\in\mathbb{N}$. Since $(x_n)_{n\in\mathbb{N}}$ is a bounded sequence in $E_{\{K'\}}$, $y_n|X'\cap K', z_n|X'\cap K'\in\mathscr{F}$ for any $n\in\mathbb{N}$. Since $(x_n)_{n\in\mathbb{N}}$ is a Cauchy sequence in $E_{K'}$,

$$((z_n - y_n)|X'\cap K')_{n\in\mathbb{N}}$$

is a decreasing sequence in $\mathscr{F}$, which converges to 0.

Let $x'\in K'_+$. By Corollary 5.5.15 and Choquet's theorem (E. M. Alfsen [1] Theorem I 4.8 and Proposition I 4.6) there exists a measure μ on $X'\cap K'$ such that

$$x'(x) = \int \tilde{x}\, d\mu$$

for any $x\in E$. Then

$$\begin{aligned}(z_n - y_n)(x') &= \sup_{p\in\mathbb{N}}\left(\bigvee_{m=n}^{n+p} \varphi(x_m) - \bigwedge_{m=n}^{n+p} \varphi(x_m)\right)(x') = \sup_{p\in\mathbb{N}}\left(\varphi\left(\bigvee_{m=n}^{n+p} x_m - \bigwedge_{m=n}^{n+p} x_m\right)\right)(x')\\ &= \sup_{p\in\mathbb{N}} x'\left(\bigvee_{m=n}^{n+p} x_m - \bigwedge_{m=n}^{n+p} x_m\right) = \sup_{p\in\mathbb{N}} \int\left(\bigvee_{m=n}^{n+p} \tilde{x}_m - \bigwedge_{m=n}^{n+p} \tilde{x}_m\right) d\mu\\ &= \int (z_n - y_n)|X'\cap K'\, d\mu\end{aligned}$$

for any $n\in\mathbb{N}$, so

$$\lim_{n\to\infty} (z_n - y_n)(x') = 0. \quad \square$$

Proposition 5.5.18. *Let E be an order σ-complete vector lattice. Then the map*

$$E^+ \longrightarrow E^+, \quad x' \longmapsto |x'|$$

preserves the bounded sets of E_E^+.

Let A' be a bounded set of E_E^+ and $x \in E_+$. Assume

$$\sup_{x' \in A'} |x'|(x) = \infty .$$

We may construct inductively a sequence $(x_n)_{n \in \mathbb{N}}$ in $[-x, x]$ and a sequence $(x_n')_{n \in \mathbb{N}}$ in A' such that

$$x_n'(x_n) > \frac{3}{4} |x_n'|(x) > 3^{n+2} \left(n + \sup_{x' \in A'} |x' \left(\sum_{m=0}^{n-1} \frac{1}{3^m} x_m \right)| \right).$$

Since E is order σ-complete the sequence $\left(\frac{1}{3^n} x_n \right)_{n \in \mathbb{N}}$ is order summable. We have

$$| \sum_{m=n+1}^{\infty} \frac{1}{3^m} x_m | \leq \frac{1}{2.3^n} x ,$$

so

$$| x_n' \left(\sum_{m=n+1}^{\infty} \frac{1}{3^m} x_m \right) | \leq \frac{1}{2.3^n} |x_n'|(x) < \frac{2}{3^{n+1}} x_n'(x_n) ,$$

$$x_n' \left(\sum_{m \in \mathbb{N}} \frac{1}{3^m} x_m \right) \geq x_n' \left(\frac{1}{3^n} x_n \right) - |x_n' \left(\sum_{m=0}^{n-1} \frac{1}{3^m} x_m \right)| - |x_n' \left(\sum_{m=n+1}^{\infty} \frac{1}{3^m} x_m \right)| \geq$$

$$\geq \frac{1}{3^n} x_n'(x_n) - \frac{1}{3^{n+2}} x_n'(x_n) - \frac{2}{3^{n+1}} x_n'(x_n) > n$$

for any $n \in \mathbb{N}$, which yields the contradictory relation

$$\sup_{x' \in A'} |x' \left(\sum_{m \in \mathbb{N}} \frac{1}{3^m} x_m \right)| = \infty .$$

Hence,

$$\sup_{x' \in A'} |x'|(x) < \infty$$

for any $x \in E_+$ and so $\{|x'| \,|\, x' \in A'\}$ is a bounded set of E_E^+. □

Proposition 5.5.19. *(C. Constantinescu (1969) [1] Proposition 3.1). Let E be an order σ-complete vector lattice, A be a lower directed subset of E_+ and A' be a subset of E^+ such that any sequence in A' either a) possesses a convergent subsequence in E_E^+ or b) possesses an adherent point in E_E^+ which belongs to E^σ. If*

$$\inf_{x \in A} |x'|(x) = 0$$

for any $x' \in A'$, *then*

$$\inf_{x \in A} (\sup_{x' \in A'} |x'|(x)) = 0 .$$

We set

$$\alpha(x) := \sup_{x' \in A'} |x'|(x)$$

for any $x \in E_+$. By Proposition 5.5.18 $\alpha(x)$ is finite. Assume

$$\alpha := \inf_{x \in A} \alpha(x) > 0 .$$

We may construct inductively a decreasing sequence $(x_n)_{n \in \mathbb{N}}$ in A and a sequence $(x'_n)_{n \in \mathbb{N}}$ in A' such that

$$\alpha(x_0) < \alpha + \frac{\alpha}{8}$$

and

$$|x'_n|(x_n - x_{n+1}) > \alpha - \frac{\alpha}{8}$$

for any $n \in \mathbb{N}$. If the alternative a) holds for $(x'_n)_{n \in \mathbb{N}}$, then we may suppose, in addition, that this sequence converges in E_E^+ to x'. For any $n \in \mathbb{N}$ there exists

$$y_n \in [x_{n+1} - x_n, x_n - x_{n+1}]$$

with

$$x'_n(y_n) = \alpha - \frac{\alpha}{8}.$$

We have

$$\sum_{n \in \mathbb{N} \setminus \{m\}} |x'_m|(|y_n|) \leq \sum_{n \in \mathbb{N}} |x'_m|(x_n - x_{n+1}) - |x'_m|(x_m - x_{m+1}) <$$
$$< \alpha + \frac{\alpha}{8} - \alpha + \frac{\alpha}{8} = \frac{2\alpha}{8}$$

for any $m \in \mathbb{N}$.

Set

$$x := \bigwedge_{n \in \mathbb{N}} x_n$$

and

$$a_{n,p} := \sum_{m=n}^{n+p} (-1)^m y_m ,$$

$$b_{n,p} := a_{n,p} - (x_{n+p+1} - x) ,$$

$$c_{n,p} := a_{n,p} + (x_{n+p+1} - x)$$

for any $n, p \in \mathbb{N}$. Let $n \in \mathbb{N}$. The sequence $(b_{n,p})_{p\in\mathbb{N}}$ is increasing, the sequence $(c_{n,p})_{p\in\mathbb{N}}$ is decreasing, and

$$b_{n,p} \le a_{n,p} \le c_{n,p}, \quad c_{n,p} - b_{n,p} = 2(x_{n+p+1} - x).$$

Hence,

$$a_n := \bigvee_{p\in\mathbb{N}} b_{n,p} = \bigwedge_{p\in\mathbb{N}} c_{n,p}$$

and $(a_{n,p})_{p\in\mathbb{N}}$ order converges to a_n.

We have

$$\alpha + \frac{\alpha}{8} > \alpha(x_m) \ge |x'_m|(x_m) = |x'_m|(x_m - x_{m+1}) + |x'_m|(x_{m+1}) >$$

$$> \alpha - \frac{\alpha}{8} + |x'_m|(x_{n+p+1}),$$

so

$$|x'_m|(x_{n+p+1}) < \frac{2\alpha}{8}$$

for any $m, n, p \in \mathbb{N}$ with $n + p \ge m$. Let $m, n \in \mathbb{N}$ and $n \le m$ and choose $p \in \mathbb{N}$ such that $n + p \ge m$. Then

$$|x'_m(a_n) - (-1)^m \frac{7\alpha}{8}| \le$$

$$\le |x'_m(a_n) - x'_m(a_{n,p})| + |x'_m(a_{n,p}) - (-1)^m x'_m(y_m)| \le$$

$$\le 2|x'_m|(x_{n+p+1}) + \sum_{p\in\mathbb{N}\setminus\{m\}} |x'_m|(|y_p|) < \frac{4\alpha}{8} + \frac{2\alpha}{8} = \frac{6\alpha}{8}.$$

If the alternative a) holds for $(x'_n)_{n\in\mathbb{N}}$, then we obtain the contradictory relations

$$|x'(a_0) - \frac{7\alpha}{8}| = \lim_{m\to\infty} |x'_{2m}(a_0) - \frac{7\alpha}{8}| \le \frac{6\alpha}{8},$$

$$|x'(a_0) + \frac{7\alpha}{8}| = \lim_{m\to\infty} |x'_{2m+1}(a_0) + \frac{7\alpha}{8}| \le \frac{6\alpha}{8}.$$

If the alternative a) does not hold for $(x'_n)_{n\in\mathbb{N}}$, then $(x'_{2m})_{m\in\mathbb{N}}$ has an adherent point y' in E^+_E which belongs to E^σ. We have

$$|y'(a_n) - \frac{7\alpha}{8}| \le \limsup_{m\to\infty} |x'_{2m}(a_n) - \frac{7\alpha}{8}| \le \frac{6\alpha}{8}$$

for any $n \in \mathbb{N}$ and

$$|y'(a_{n,p})| \le \sum_{m=n}^{n+p} |y'|(|y_m|) \le \sum_{m=n}^{n+p} |y'|(x_m - x_{m+1}) \le |y'|(x_n - x)$$

for any $n, p \in \mathbb{N}$. Since $y' \in E^\sigma$ and $(a_{n,p})_{p \in \mathbb{N}}$ order converges to a_n

$$|y'(a_n)| = \lim_{p \to \infty} |y'(a_{n,p})| \leq |y'|(x_n - x)$$

for any $n \in \mathbb{N}$, so

$$\lim_{n \to \infty} y'(a_n) = 0.$$

This contradicts the relation

$$\frac{7\alpha}{8} = \lim_{n \to \infty} |y'(a_n) - \frac{7\alpha}{8}| \leq \frac{6\alpha}{8}. \quad \square$$

Definition 5.5.20. *A vector lattice E is said to be* Archimedean *if $x, y \in E_+$ and $nx \leq y$ for all $n \in \mathbb{N}$ imply $x = 0$. A vector lattice E is called* uniformly complete *if $\bigvee_{n \in \mathbb{N}} x_n$ exists for any increasing sequence $(x_n)_{n \in \mathbb{N}}$ in E for which there is $x \in E$ such that*

$$x_{m+n} - x_n \leq \frac{1}{n+1} x$$

for any $m, n \in \mathbb{N}$.

Any σ-complete vector lattice is uniformly complete. If X is a topological space and $\mathscr{C}$ is the vector lattice of continuous real functions on X then any solid vector subspace of $\mathscr{C}$ is uniformly complete.

Proposition 5.5.21. *The following assertions are equivalent for every vector lattice E:*

a) E is Archimedean;

b) $\bigwedge_{n \in \mathbb{N}} \left(\frac{1}{n+1} x\right) = 0$ for all $x \in E_+$;

c) $\bigwedge_{n \in \mathbb{N}} \left(\frac{1}{n+1} x\right)$ exists for all $x \in E_+$.

In particular, every uniformly complete vector lattice is Archimedean and every Archimedean up-down semicomplete vector lattice is uniformly complete.

a $\Rightarrow$ b. Let y be a lower bound of $\left(\frac{1}{n+1} x\right)_{n \in \mathbb{N}}$ in E. Then $n(y \vee 0) \leq x$ for all $n \in \mathbb{N}$ and so $y \vee 0 = 0$. Hence, $\bigwedge_{n \in \mathbb{N}} \left(\frac{1}{n+1} x\right) = 0$.

b $\Rightarrow$ c is trivial.

c $\Rightarrow$ a. Let $x, y \in E_+$ such that $nx \leq y$ for all $n \in \mathbb{N}$. We set $z := \bigwedge_{n \in \mathbb{N}} \left(\frac{1}{n+1} y\right)$. Then

$$2(n+1)(x+z) \leq 2y$$

for all $n \in \mathbb{N}$, so

$$x + z \leq \bigwedge_{n \in \mathbb{N}} \left(\frac{1}{n+1} y \right) = z .$$

Hence, $x = 0$.

Assume now E is uniformly complete and let $x \in E_+$. Then $\left(\frac{n}{n+1} x \right)_{n \in \mathbb{N}}$ is an increasing sequence in E and

$$\frac{m+n}{m+n+1} x - \frac{n}{n+1} x \leq \frac{1}{n+1} x$$

and so $\bigvee_{n \in \mathbb{N}} \left(\frac{n}{n+1} x \right)$ exists. We deduce $\bigwedge_{n \in \mathbb{N}} \left(\frac{1}{n+1} x \right)$ also exists and by c ⇒ a E is Archimedean.

Assume now E is Archimedean and up-down semicomplete. Let $(x_n)_{n \in \mathbb{N}}$ be an increasing sequence in E and let $x \in E_+$ such that

$$x_{m+n} - x_n \leqq \frac{1}{n+1} x ,$$

for any $m, n \in \mathbb{N}$. Then $\left(x_{n^2} + \frac{1}{n^2+1} x \right)_{n \in \mathbb{N}}$ is a decreasing sequence in E and

$$0 \leq \left(x_{n^2} + \frac{1}{n^2+1} x \right) - x_n \leq \frac{1}{n+1} (2x)$$

for any $n \in \mathbb{N}$. By b ⇒ a we get

$$\bigwedge_{n \in \mathbb{N}} \left(\left(x_{n^2} + \frac{1}{n^2+1} x \right) - x_n \right) = 0$$

and so $\bigvee_{n \in \mathbb{N}} x_n$ exists. Hence E is uniformly complete. □

Proposition 5.5.22. *Let E be an Archimedean vector lattice and x' be a linear form on E. Then the following assertions are equivalent:*

a) $x' \in E^+$;

b) $(x'(x_n))_{n \in \mathbb{N}}$ *converges for any bounded monotone sequence* $(x_n)_{n \in \mathbb{N}}$ *in E ;*

c) $(x'(x_n))_{n \in \mathbb{N}}$ *converges for any decreasing sequence* $(x_n)_{n \in \mathbb{N}}$ *in E such that* $\bigwedge_{n \in \mathbb{N}} x_n = 0$.

a ⇒ b ⇒ c is trivial.

c ⇒ a. Let $x \in E_+$. Assume

$$\sup \{ x'(y) \mid y \in E_+, y \leq x \} = \infty .$$

We construct a decreasing sequence $(x_n)_{n \in \mathbb{N}}$ in E_+ inductively such that

$$|x'(x_n)| \geq n^2, \quad \sup\{x'(y) \mid y \in E_+, \; y \leq x_n\} = \infty$$

for any $n \in \mathbb{N}$. Set $x_0 := x$. Let $n \in \mathbb{N}$ and assume that the sequences were constructed up to n. There exists $y \in E_+$ such that $y \leq x_n$ and

$$x'(y) \geq (n+1)^2 + |x'(x_n)|.$$

Then

$$|x'(x_n - y)| \geq x'(y) - |x'(x_n)| \geq (n+1)^2.$$

Let $z \in E_+$ and $z \leq x_n$. Then

$$z - y \wedge z = y \vee z - y \leq x_n - y,$$

so

$$\infty = \sup\{x'(z) \mid z \in E_+, \; z \leq x\} \leq$$
$$\leq \sup\{x'(z) \mid z \in E_+, \; z \leq y\} + \sup\{x'(z) \mid z \in E_+, \; z \leq x_n - y\}.$$

Hence, either

$$\sup\{x'(z) \mid z \in E_+, \; z \leq y\} = \infty$$

or

$$\sup\{x'(z) \mid z \in E_+, \; z \leq x_n - y\} = \infty.$$

We set $x_{n+1} := y$ in the first case and $x_{n+1} := x_n - y$ in the second case. This finishes the inductive construction.

The sequence $\left(\frac{1}{n+1} x_n\right)_{n \in \mathbb{N}}$ is decreasing and $\frac{1}{n+1} x_n \leq \frac{1}{n+1} x_0$ for all $n \in \mathbb{N}$. By Proposition 5.5.21 a $\Rightarrow$ b we obtain $\bigwedge_{n \in \mathbb{N}} \left(\frac{1}{n+1} x_n\right) = 0$. By c) $\left(\frac{1}{n+1} x'(x_n)\right)_{n \in \mathbb{N}}$ converges, which contradicts the relation

$$n \in \mathbb{N} \Rightarrow \left|\frac{1}{n+1} x'(x_n)\right| \geq \frac{n^2}{n+1}. \quad \square$$

Proposition 5.5.23. *Let E be a uniformly complete vector lattice and x' be a linear form on E. Then the following assertions are equivalent:*

a) $x' \in E^+$;

b) the map

$$\mathfrak{P}(\mathbb{N}) \to \mathbb{R}, \quad M \mapsto x'\Big(\sum_{n \in M} x_n\Big)$$

is exhaustive for any supersummable sequence $(x_n)_{n \in \mathbb{N}}$ in E;

c) the map

$$\mathfrak{P}(\mathbb{N}) \to \mathbb{R}, \quad M \mapsto x'(\sum_{n \in M} x_n)$$

is exhaustive for every supersummable sequence $(x_n)_{n \in \mathbb{N}}$ of positive elements of E.

a ⇒ b follows from Proposition 5.2.10.
b ⇒ c is trivial.
c ⇒ a. Let $x \in E_+$. Assume

$$\sup \{x'(y) \mid y \in E_+, y \leq x\} = \infty .$$

There exists a sequence $(x_n)_{n \in \mathbb{N}}$ in E_+ such that $x_n \leq x$ and $x'(x_n) > 2^n$ for any $n \in \mathbb{N}$. Let $(k_n)_{n \in \mathbb{N}}$ be a strictly increasing sequence in $\mathbb{N}$. We set

$$y_n := \sum_{m=0}^{n} \frac{1}{2^{k_n}} x_{k_n}$$

for any $n \in \mathbb{N}$. Then $(y_n)_{n \in \mathbb{N}}$ is an increasing sequence in E and

$$y_{m+n} - y_n \leq \frac{1}{n+1} x$$

for any $m, n \in \mathbb{N}$. Since E is uniformly complete $\bigvee_{n \in \mathbb{N}} y_n$ exists. We deduce $\left(\frac{1}{2^{k_n}} x_{k_n}\right)_{n \in \mathbb{N}}$ is order summable and $\left(\frac{1}{2^n} x_n\right)_{n \in \mathbb{N}}$ is order supersummable. By c) we get the contradictory relation

$$1 \leq \lim_{n \to \infty} x'\left(\frac{1}{2^n} x_n\right) = 0 .$$

Hence

$$\sup \{x'(y) \mid y \in E_+, y \leqq x\} < \infty$$

and $x' \in E^+$ (Ch. D. Aliprantis, O. Burkinshaw [1] Theorem 3.3). □

Theorem 5.5.24. *Let E be a uniformly complete vector lattice. Then E^+ is a Φ_4-closed set of $\mathbb{R}^E$ and a Φ_4-compact space.*

By Proposition 5.5.23 a ⇔ b E^+ is the set of all order exhaustive linear forms on E and the assertion follows from Theorem 5.3.1 d), e). □

Remark. The assertion of Theorem 5.5.24 does not hold any more if we replace the hypothesis "uniformly complete" by the weaker one "Archimedean". Indeed let E be the set of continuous real functions on $[0,1]$ having a right derivative at 0, for any $f \in E$ let $f'(0)$ be the right derivative of f at 0 and for $n \in \mathbb{N}$ let x'_n be the real function $f \mapsto n\left(f\left(\frac{1}{n}\right) - f(0)\right)$ on E. Then E is a vector sublattice of $\mathbb{R}^{[0,1]}$

and so Archimedean, and $(x'_n)_{n\in\mathbb{N}}$ is a Cauchy sequence in E^+ converging in $\mathbb{R}^E$ to the map $f\mapsto f'(0)$, which does not belong to E^+. Hence in this case E^+ is not even sequentially closed in $\mathbb{R}^E$.

Corollary 5.5.25. *Let X be a topological space, $\mathscr{C}$ be the vector lattice of continuous real functions on X and $\mathscr{F}$ be a solid vector subspace of $\mathscr{C}$. Then $\mathscr{F}^+$ is a Φ_4-closed set of $\mathbb{R}^{\mathscr{F}}$ and a Φ_4-compact space.* □

§ 5.6 Spaces of vector valued measures

Throughout this section X shall denote a set, $\mathfrak{R}$ a δ-ring of subsets of X, $\mathscr{M}$ a band of $\mathscr{M}(\mathfrak{R},\mathbb{R})$, E a Hausdorff locally convex space, E' the dual of E, and E'^* the algebraic dual of E'. We also use the notation and terminology of C. Constantinescu [5].

Proposition 5.6.1. *Let $\mathscr{F}$ be the set of bounded $\mathfrak{R}$-measurable real functions on X which vanish outside a set of $\mathfrak{R}$. Then*

$$\xi=\bigvee_{\substack{f\in\mathscr{F}\\ \dot f\le\xi}}\dot f$$

for any $\xi\in\mathscr{M}_+^{\varrho}$.

By C. Constantinescu [5] Theorem 2.3.8 and Proposition 3.1.1

$$\xi=\bigvee_{\substack{g\in\mathscr{L}\\ \dot g\le\xi}}\dot g$$

and the assertion follows from the relation

$$\dot g=\bigvee_{\substack{f\in\mathscr{F}\\ f\le g}}\dot f$$

for any $g\in\mathscr{L}_+$. □

Theorem 5.6.2. *Let $\mathscr{F}$ be the vector lattice of $\mathfrak{R}$-measurable real functions on X, $\mathscr{G}$ be a solid vector subspace of $\mathscr{F}$ such that*

$$\{1_A^X\mid A\in\mathfrak{R}\}\subset\mathscr{G},$$

$\mathscr{H}$ be the solid vector subspace of $\mathscr{M}^{\varrho}$ generated by $\{\dot g\mid g\in\mathscr{G}\}$, Φ be the set $\hat{\Phi}_0(\mathscr{H})$ and $\mathscr{N}$ be a subset of $\mathscr{M}(E)$ such that $\xi\in\mathscr{L}^1(\mu)$ and $\int\xi\,d\mu\in E$ for any $(\xi,\mu)\in\mathscr{H}\times\mathscr{N}$. Then the identity map $\mathscr{N}_{\mathscr{G}}\to\mathscr{N}_{\Phi}$ is uniformly Φ_4-continuous.

For any $\xi\in\mathscr{H}_+$ let $\mathfrak{F}_\xi$ be the filter on $\mathscr{H}$ generated by the section filter of

$\{\dot{g} \mid g \in \mathcal{G}, \dot{g} \leq \xi\}$. By Proposition 5.6.1 and C. Constantinescu [5] Proposition 4.1.6 $\mu(\mathfrak{F}_\xi)$ converges to $\int \xi d\mu$ for any $(\xi, \mu) \in \mathcal{H}_+ \times \mathcal{N}$. By Proposition 5.3.2 the identity map $\mathcal{N}_{\mathcal{G}} \to \mathcal{N}_{\Phi}$ is uniformly Φ_4-continuous. □

Proposition 5.6.3. *Let $\mathfrak{K}$ be a subset of $\mathfrak{R}$ closed under finite unions. Then $\mathcal{M}(\mathfrak{R}, \mathbb{R}; \mathfrak{K})$ is a band of $\mathcal{M}(\mathfrak{R}, \mathbb{R})$. If we set $\mathcal{M} := \mathcal{M}(\mathfrak{R}, \mathbb{R}; \mathfrak{K})$ then $\mathcal{M}(E) = \mathcal{M}(\mathfrak{R}, E; \mathfrak{K})$.*

By Proposition 4.2.1 $\mathcal{M}(\mathfrak{R}, \mathbb{R}; \mathfrak{K})$ is a vector subspace of $\mathcal{M}(\mathfrak{R}, \mathbb{R})$; we want to show that it is also a solid set of $\mathcal{M}(\mathfrak{R}, \mathbb{R})$. Let $\mu \in \mathcal{M}(\mathfrak{R}, \mathbb{R})$, $\nu \in \mathcal{M}(\mathfrak{R}, \mathbb{R}; \mathfrak{K})$ such that $|\mu| \leq |\nu|$, let $A \in \mathfrak{R}$ and ε be a strictly positive real number. Then there exists $K \in \mathfrak{K}$ such that $K \subset A$ and

$$|\nu(B) - \nu(A)| < \frac{\varepsilon}{2}$$

for any $B \in \mathfrak{R}$ with $K \subset B \subset A$. We get (K. Jacobs [1] Ch VIII Theorem 1.3)

$$|\mu(B) - \mu(A)| \leq |\mu|(A \backslash B) \leq |\nu|(A \backslash B) < \varepsilon$$

for any $B \in \mathfrak{R}$ with $K \subset B \subset A$. Hence, $\mu \in \mathcal{M}(\mathfrak{R}, \mathbb{R}; \mathfrak{K})$, so $\mathcal{M}(\mathfrak{R}, \mathbb{R}; \mathfrak{K})$ is a solid vector subspace of $\mathcal{M}(\mathfrak{R}, \mathbb{R})$.

We want to show now that $\mathcal{M}(\mathfrak{R}, \mathbb{R}; \mathfrak{K})$ is a band of $\mathcal{M}(\mathfrak{R}, \mathbb{R})$. Let $(\mu_\iota)_{\iota \in I}$ be a family in $\mathcal{M}(\mathfrak{R}, \mathbb{R}; \mathfrak{K})_+$ possessing a supremum μ in $\mathcal{M}(\mathfrak{R}, \mathbb{R})$. We have to show $\mu \in \mathcal{M}(\mathfrak{R}, \mathbb{R}; \mathfrak{K})$. By the above considerations we may assume $(\mu_\iota)_{\iota \in I}$ an upper directed family in $\mathcal{M}(\mathfrak{R}, \mathbb{R}; \mathfrak{K})_+$. Let $A \in \mathfrak{R}$ and ε be a strictly positive real number. There exists $\iota \in I$ such that

$$\mu_\iota(A) > \mu(A) - \frac{\varepsilon}{2}$$

(K. Jacobs [1] Ch VIII Theorem 1.4). There also exists $K \in \mathfrak{K}$ such that $K \subset A$ and

$$\mu_\iota(B) > \mu_\iota(A) - \frac{\varepsilon}{2}$$

for any $B \in \mathfrak{R}$ with $K \subset B \subset A$. Then

$$\mu(B) \geq \mu_\iota(B) > \mu_\iota(A) - \frac{\varepsilon}{2} > \mu(A) - \varepsilon$$

for any $B \in \mathfrak{R}$ with $K \subset B \subset A$. Hence, $\mu \in \mathcal{M}(\mathfrak{R}, \mathbb{R}; \mathfrak{K})$ and $\mathcal{M}(\mathfrak{R}, \mathbb{R}; \mathfrak{K})$ is a band of $\mathcal{M}(\mathfrak{R}, \mathbb{R})$.

We prove now the last assertion. Let $\mu \in \mathcal{M}(E)$ and let E_σ be the vector space underlying E endowed with the weak topology of E. Then $\mu \in \mathcal{M}(\mathfrak{R}, E_\sigma; \mathfrak{K})$ and by Corollary 4.1.31 a) $\mu \in \mathcal{M}(\mathfrak{R}, E; \mathfrak{K})$. Hence $\mathcal{M}(E) \subset \mathcal{M}(\mathfrak{R}, E; \mathfrak{K})$. The converse inclusion is trivial. □

Proposition 5.6.4. *If $\mathcal{F}$ denotes the set of bounded $\mathfrak{R}$-measurable real functions on X which vanish outside a set of $\mathfrak{R}$ and if E is sequentially complete, then $\int f d\mu \in E$ for* *any* $(f, \mu) \in \mathscr{F} \times \mathscr{M}(E)$ *and the identity map*

$$\mathscr{M}(E) \to \mathscr{M}(E)_{\mathscr{F}}$$

is uniformly Φ_5*-continuous.*

Let $(f, \mu) \in \mathscr{F}_+ \times \mathscr{M}(E)$. We set

$$f_n := \frac{1}{2^n} \sum_{m=1}^{n 2^n} 1^X \left\{ f \geq \frac{m}{2^n} \right\}$$

for any $n \in \mathbb{N}$. Then $(f_n)_{n \in \mathbb{N}}$ is an increasing sequence in $\mathscr{F}$ converging uniformly to f. Hence, $(\int f_n d\mu)_{n \in \mathbb{N}}$ is a Cauchy sequence (Corollary 4.7.8) and therefore a convergent sequence in E. Then $\int f d\mu \in E$.

Let $\mathfrak{F}$ be a Cauchy Φ_5-filter on $\mathscr{M}(E)$, $f \in \mathscr{F}$, $A \in \mathfrak{R}$ such that f vanishes on $X \setminus A$ and U be an absolutely convex 0-neighbourhood in E. By Theorem 4.7.7 there exist $\alpha \in \mathbb{R}$ and $\mathscr{M}' \in \mathfrak{F}$ such that

$$\{\mu(B) \mid \mu \in \mathscr{M}', \ B \in \mathfrak{R}, \ B \subset A\} \subset \alpha U .$$

Let g be a step function on X with respect to $\mathfrak{R}$ vanishing on $X \setminus A$ such that $|f - g| < \dfrac{1}{6|\alpha| + 1}$. There exists $\mathscr{M}'' \in \mathfrak{F}$ such that

$$\int g d\mu' - \int g d\mu'' \in \tfrac{1}{3} U$$

for any $\mu', \mu'' \in \mathscr{M}''$. Then

$$\int f d\mu' - \int f d\mu'' = \int (f - g) d\mu' + \left(\int g d\mu' - \int g d\mu''\right) + \int (g - f) d\mu'' \in$$
$$\in \tfrac{1}{3} U + \tfrac{1}{3} U + \tfrac{1}{3} U = U$$

for any $\mu', \mu'' \in \mathscr{M}' \cap \mathscr{M}''$. Hence, $\mathfrak{F}$ is a Cauchy filter on $\mathscr{M}(E)_{\mathscr{F}}$. By Theorem 1.8.4 b $\Rightarrow$ a the identity map

$$\mathscr{M}(E) \to \mathscr{M}(E)_{\mathscr{F}}$$

is uniformly Φ_5-continuous. □

Theorem 5.6.5. *Let X be a completely regular space, $\mathfrak{K}$ be the set of compact sets of X, $\mathfrak{R}$ be the δ-ring generated by $\mathfrak{K}$, $\mathscr{M}$ be $\mathscr{M}(\mathfrak{R}, \mathbb{R}; \mathfrak{K})$, $\mathscr{N}$ be a subset of $\mathscr{M}(E) = \mathscr{M}(\mathfrak{R}, E; \mathfrak{K})$, $\mathscr{C}$ be the vector lattice of continuous real functions on X, $\mathscr{F}$ be a dense vector sublattice of $\mathscr{C}_{\mathfrak{K}}$, $\mathscr{K}$ be a set of positive real functions k on X belonging to $\mathscr{L}^1(\mathscr{N}, X)$ such that*

$$k(x) = \sup_{\mathscr{F} \ni f \leq k} f(x)$$

for any $x \in X$ and $\int h d\mu \in E$ for any $\mu \in \mathscr{N}$ and for any $h \in \mathscr{L}^1(\mathscr{N}, X)$ with $0 \leq h \leq k$, *let* $\overline{\mathscr{F}}$ *be the set* $\left\{ \bigvee_{n\in\mathbb{N}}^{\mathbb{R}^X} f_n \mid (f_n)_{n\in\mathbb{N}} \text{ being an increasing sequence in } \mathscr{F}_+, \exists k\in\mathscr{K}, \bigvee_{n\in\mathbb{N}}^{\mathbb{R}^X} f_n \le k \right\}$, $\mathscr{G}$ *be the solid vector subspace of* $\mathscr{M}^{\varrho}$ *generated by* $\{\dot{k} \mid k\in\mathscr{K}\}$, Φ *be the set* $\hat{\Phi}_0(\mathscr{G})$, $\tilde{\mathscr{N}}$ *be the set* $\mathscr{N}$ *endowed with a uniformity such that the identity maps* $\mathscr{N}_{\Phi} \to \tilde{\mathscr{N}} \to \mathscr{N}_{\overline{\mathscr{F}}}$ *are uniformly* Φ_4*-continuous and* $\mathscr{P}$ *be the set of* $\mu\in\mathscr{M}(E)$ *such that* $\bar{f}\in\mathscr{L}^1(\mu,X)$ *and* $\int \bar{f}d\mu\in E$ *for any* $\bar{f}\in\overline{\mathscr{F}}$. *We assume 1)* $\int \xi d\mu \in E$ *for any* $(\xi,\mu)\in\mathscr{G}\times\mathscr{N}$ *and 2) for any* $K\in\mathfrak{K}$ *there exists* $k\in\mathscr{K}$ *such that* $k\ge 1_K$. *Then:*

a) the identity map $\mathscr{N}_{\overline{\mathscr{F}}} \to \mathscr{N}_{\Phi}$ *is uniformly* Φ_4*-continuous ;*

b) if, in addition, 3) E is Φ_4*-compact, 4)* $\mathscr{N}$ *is a* Φ_4*-closed set of* $\mathscr{P}_{\overline{\mathscr{F}}}$, *5) any function of* $\overline{\mathscr{F}}$ *is bounded on the compact sets of X, then* $\tilde{\mathscr{N}}$ *is* Φ_4*-compact.*

By Proposition 5.6.3 $\mathscr{M}(\mathfrak{R},\mathbb{R};\mathfrak{K})$ is a band of $\mathscr{M}(\mathfrak{R},\mathbb{R})$ and $\mathscr{M}(E) = \mathscr{M}(\mathfrak{R},E;\mathfrak{K})$. The functions of $\mathscr{K}$ and $\overline{\mathscr{F}}$ are lower semicontinuous and therefore $\mathfrak{R}$-measurable; hence $\overline{\mathscr{F}}\subset\mathscr{L}_\infty$, $\mathscr{K}\subset\mathscr{L}_\infty$ and $\mathscr{G}$ and $\mathscr{N}_{\overline{\mathscr{F}}}$ are well-defined.

a) We denote by $\mathscr{H}$ the set

$$\mathscr{H} := \{f\in\mathbb{R}^X \mid f \ \mathfrak{R}\text{-measurable}, \exists k\in\mathfrak{K}, f\le k\}.$$

By Corollary 4.9.11 the identity map $\mathscr{N}_{\overline{\mathscr{F}}} \to \mathscr{N}_{\mathscr{H}}$ is uniformly Φ_4-continuous, by Theorem 5.6.2 the identity map $\mathscr{N}_{\mathscr{H}} \to \mathscr{N}_{\Phi}$ is uniformly Φ_4-continuous, and the assertion follows from Corollary 1.8.5.

b) Let $\mathfrak{F}$ be a Φ_4-ultrafilter on $\mathscr{N}_{\overline{\mathscr{F}}}$ and φ be the inclusion map $\mathscr{N}_{\overline{\mathscr{F}}} \to \mathscr{M}(E)$. Since the identity map $\mathscr{N}_{\overline{\mathscr{F}}} \to \mathscr{N}_{\Phi}$ is uniformly Φ_4-continuous, φ is also uniformly Φ_4-continuous, so $\varphi(\mathfrak{F})$ is a Φ_4-ultrafilter on $\mathscr{M}(E)$ (Theorem 1.8.4 a $\Rightarrow$ g). By Theorem 4.2.14 b) $\varphi(\mathfrak{F})$ converges in $\mathscr{M}(E)$. We denote its limit by μ.

Let $\bar{f}\in\overline{\mathscr{F}}$. We denote by $\mathfrak{B}$ the set of Borel sets of X and for any $\nu\in\mathscr{P}$ let ν' be the map

$$\mathfrak{B} \to E, \quad A \mapsto \int 1_A \bar{f} d\nu.$$

Then $\nu'\in\mathscr{M}(\mathfrak{B},E;\mathfrak{K})$ for any $\nu\in\mathscr{P}$. Take ψ as the map

$$\mathscr{N}_{\overline{\mathscr{F}}} \to \mathscr{M}(\mathfrak{B},E;\mathfrak{K}), \quad \nu \mapsto \nu'.$$

Since the map

$$\mathscr{N}_{\Phi} \to \mathscr{M}(\mathfrak{B},E;\mathfrak{K}), \quad \nu \mapsto \nu'$$

is uniformly continuous ψ is uniformly Φ_4-continuous so $\psi(\mathfrak{F})$ is a Φ_4-ultrafilter on $\mathscr{M}(\mathfrak{B},E;\mathfrak{K})$ (Theorem 1.8.4 a $\Rightarrow$ g). By Theorem 4.2.14 b), $\psi(\mathfrak{F})$ converges in $\mathscr{M}(\mathfrak{B},E;\mathfrak{K})$. We denote its limit by λ. By Proposition 5.6.4 and Proposition 1.8.3 we have

$$\int 1_A \bar{f} d\mu = \lambda(A)$$

for any $A\in\mathfrak{R}$. Let x' be a continuous linear form on E. Then

$$\int^{*} \bar{f} d|x' \circ \mu| = \sup\{\int 1_A \bar{f} d(x' \circ \mu) - \int 1_B \bar{f} d(x' \circ \mu) \mid A, B \in \mathfrak{R}\} =$$
$$= \sup\{x' \circ \lambda(A) - x' \circ \lambda(B) \mid A, B \in \mathfrak{R}\} = |x' \circ \lambda|(X) < \infty .$$

Hence, $\bar{f} \in \mathscr{L}^1(x' \circ \mu)$ and

$$\int \bar{f} d(x' \circ \mu) = \lim_{A, \mathfrak{G}} \int 1_A \bar{f} d(x' \circ \mu) = \lim_{A, \mathfrak{G}} (x' \circ \lambda)(A) = x' \circ \lambda(X),$$

where $\mathfrak{G}$ denotes the section filter of $\mathfrak{R}$ (ordered by the inclusion relation). Since x' is arbitrary we get $\bar{f} \in \mathscr{L}^1(\mu)$ and

$$\int \bar{f} d\mu = \lambda(X) = \lim_{\nu, \mathfrak{F}} \int \bar{f} d\nu .$$

Since $\bar{f}$ is arbitrary $\mu \in \mathscr{P}$ and, if ϱ denotes the inclusion map $\mathscr{N} \to \mathscr{P}$, $\varrho(\mathfrak{F})$ converges to μ in $\mathscr{P}_{\bar{\mathscr{F}}}$. Since $\mathscr{N}$ is a Φ_4-closed set of $\mathscr{P}_{\bar{\mathscr{F}}}$ we get $\mu \in \mathscr{N}$ and $\mathscr{N}_{\bar{\mathscr{F}}}$ is Φ_4-compact.

By Corollary 1.8.5 the identity map $\mathscr{N}_{\bar{\mathscr{F}}} \to \tilde{\mathscr{N}}$ is uniformly Φ_4-continuous so, by Proposition 1.8.11, $\tilde{\mathscr{N}}$ is Φ_4-compact. □

Theorem 5.6.6. *Let $\mathscr{G}$ be the solid vector subspace of $\mathscr{M}^\pi$ generated by $\{\dot{1}_A \mid A \in \mathfrak{R}\}$, Φ be the set $\hat{\Phi}_0(\mathscr{G})$, $i \in \{1, 2, 3, 4\}$ and $\mathfrak{A}$ be a subset* of $\mathfrak{R}$ *such that the identity map $\mathscr{M}(E)_{\mathfrak{A}} \to \mathscr{M}(E)$ is uniformly Φ_i-continuous. If E is complete, then $\int \xi d\mu \in E$ for any $(\xi, \mu) \in \mathscr{G} \times \mathscr{M}(E)$ and the identity map $\mathscr{M}(E)_{\mathfrak{A}} \to \mathscr{M}(E)_{\Phi}$ is uniformly Φ_i-continuous.*

By C. Constantinescu [5] Corollary 4.2.17 a) $\int \xi d\mu \in E$ for any $(\xi, \mu) \in \mathscr{G} \times \mathscr{M}(E)$. Let $\mathscr{F}$ be the set of bounded $\mathfrak{R}$-measurable real functions on X which vanish outside a set of $\mathfrak{R}$. By Proposition 5.6.4 the identity map $\mathscr{M}(E) \to \mathscr{M}(E)_{\mathscr{F}}$ is uniformly Φ_5-continuous. By Theorem 5.6.2 the identity map $\mathscr{M}(E)_{\mathscr{F}} \to \mathscr{M}(E)_{\Phi}$ is uniformly Φ_4-continuous, and the assertion follows from Corollary 1.8.5. □

Corollary 5.6.7. *Let X be a Hausdorff space, $\mathfrak{K}$ be the set of compact sets of X, $\mathfrak{T}$ be the set of open sets of X, $\mathfrak{R}$ be the δ-ring generated by $\mathfrak{K}$ (by $\mathfrak{T}$), $\mathscr{M}$ be $\mathscr{M}(\mathfrak{R}, \mathbb{R}; \mathfrak{K})$ and $\mathscr{G}$ be the solid vector subspace of $\mathscr{M}^\pi$ generated by $\{\dot{1}_A \mid A \in \mathfrak{R}\}$. If E is complete, then $\int \xi d\mu \in E$ for any $(\xi, \mu) \in \mathscr{G} \times \mathscr{M}(E)$ and the identity map $\mathscr{M}(E)_{\mathfrak{K}} \to \mathscr{M}(E)_{\mathscr{G}}$ ($\mathscr{M}(E)_{\mathfrak{T}} \to \mathscr{M}(E)_{\mathscr{G}}$) is uniformly Φ_4-continuous.*

By Proposition 5.6.3 $\mathscr{M}(\mathfrak{R}, \mathbb{R}; \mathfrak{K})$ is a band of $\mathscr{M}(\mathfrak{R}, \mathbb{R})$ and $\mathscr{M}(E) = \mathscr{M}(\mathfrak{R}, E; \mathfrak{K})$. Hence, by Corollary 4.5.11 (by Corollary 4.5.3), the identity map $\mathscr{M}(E)_{\mathfrak{K}} \to \mathscr{M}(E)$ ($\mathscr{M}(E)_{\mathfrak{T}} \to \mathscr{M}(E)$) is uniformly Φ_4-continuous, and the assertion immediately follows from Theorem 5.6.6. □

Corollary 5.6.8. *Let Y be a completely regular space (a locally compact space) such that the closure of any exact open set (relatively compact exact open set) of Y is open, let X be a subspace (a closed subspace) of Y, $\mathfrak{T}$ be the set of closed open sets (of compact open sets) of X, $\mathfrak{R}$ be a δ-ring of subsets of X (of relatively compact sets of X) containing $\mathfrak{T}$. Further, let $\mathfrak{K}$ be the set of compact sets of X belonging to $\mathfrak{R}$, $\mathscr{M}$ be $\mathscr{M}(\mathfrak{R}, \mathbb{R}; \mathfrak{K})$*

and $\mathscr{G}$ be the solid vector subspace of $\mathscr{M}^\pi$ generated by $\{1_A \mid A \in \mathfrak{R}\}$. If E is complete, then $\int \xi d\mu \in E$ for any $(\xi, \mu) \in G \times \mathscr{M}(E)$ and the identity map $\mathscr{M}(E)_{\mathfrak{T}} \to \mathscr{M}(E)_{\mathscr{G}}$ is uniformly Φ_3-continuous.

By Proposition 5.6.3 $\mathscr{M}(\mathfrak{R}, \mathbb{R}; \mathfrak{K})$ is a band of $\mathscr{M}(\mathfrak{R}, \mathbb{R})$ and $\mathscr{M}(E) = \mathscr{M}(\mathfrak{R}, E; \mathfrak{K})$. Hence, by Corollary 4.5.17 (by Corollary 4.5.18), the identity map $\mathscr{M}(E)_{\mathfrak{T}} \to \mathscr{M}(E)$ is uniformly Φ_3-continuous and the assertion immediately follows from Theorem 5.6.6. □

Remark. Let Y be a completely regular space such that the closure of any exact open set of Y is open, X be a subspace of Y, $\mathfrak{R}$ be the σ-ring generated by the open sets of X, $\mathfrak{K}$ be the set of compact sets of X and $(\mu_n)_{n \in \mathbb{N}}$ be a sequence in $\mathscr{M}(\mathfrak{R}, \mathbb{R}; \mathfrak{K})$ such that

$$\lim_{n \to \infty} \int f d\mu_n = 0$$

for any continuous bounded real function f on X. If we endow $\mathscr{M}(\mathfrak{R}, \mathbb{R}; \mathfrak{K})$ with the norm

$$\mathscr{M}(\mathfrak{R}, \mathbb{R}; \mathfrak{K}) \longrightarrow \mathbb{R}, \quad \mu \longmapsto |\mu|(X),$$

then, by the above corollary (and by Proposition 1.8.3 and C. Constantinescu [5] Proposition 3.4.2 b)), $(\mu_n)_{n \in \mathbb{N}}$ converges weakly to 0. This result was proved by A. Grothendieck (1953) ([2] Theorem 9) when Y is a compact stonian space and $X = Y$ and by Z. Semadeni (1964) ([1] Theorem (i) ⇒ (iv)) when Y is compact and X is a closed subspace of Y. G.L. Seever (1968 ([1] Theorem B) and F.K. Dashiell Jr. (1981) [(1] page 408) obtained improvements of these results of Grothendieck and Semadeni which do not follow from Corollary 5.6.8.

Proposition 5.6.9. *Assume that $\mathfrak{R}$ is a σ-ring and let $\mathscr{N}$ be a compact convex nonempty set of $\mathscr{M}$. Then there exists $\mu \in \mathscr{N}$ such that any $\nu \in \mathscr{N}$ is absolutely continuous with respect to μ.*

Let E be the Banach space obtained by endowing the real vector space of affine continuous real functions on $\mathscr{N}$ with the supremum norm and for any $A \in \mathfrak{R}$ let $\lambda(A)$ be the element

$$\mathscr{N} \longrightarrow \mathbb{R}, \quad \nu \longmapsto \nu(A)$$

of E. By Theorem 4.2.16 d) $\lambda \in \mathscr{M}(\mathfrak{R}, E)$. By the theorem of Rybakov (V.I. Rybakov (1970) [1] Teorema or B. Walsh [1] page 509) there exists $x' \in E'$ such that λ is absolutely continuous with respect to $x' \circ \lambda$. By the Hahn-Banach theorem and by the Riesz representation theorem there exists a real Radon measure ϱ on $\mathscr{N}$ such that

$$x'(f) = \int f d\varrho$$

for any $f \in E$. Since any $f \in E$ is affine there exist $\mu \in \mathscr{N}$ and $\alpha \in \mathbb{R}$ such that

$$\int f d\varrho = \alpha f(\mu)$$

for any $f \in E$. Then $\alpha\mu = x' \circ \lambda$ and any $\nu \in \mathcal{N}$ is absolutely continuous with respect to μ. $\square$

Theorem 5.6.10. *Let $\mathfrak{S}$ be the σ-ring generated by $\mathfrak{R}$, $\mathcal{F}$ be the set $\{\dot{1}_A \mid A \in \mathfrak{S}\}$ and $\mathcal{N}$ be a compact convex set of $(\mathcal{M}_c)_{\mathcal{F}}$. Then there exists $A \in \mathfrak{R}$ such that $X \backslash A$ is a μ-null set for any $\mu \in \mathcal{N}$ and such that*

$$\sup_{\mu \in \mathcal{N}} |\mu|(A) < \infty .$$

We may assume $\mathcal{N}$ nonempty. Set

$$\mu' : \mathfrak{S} \to \mathbb{R}, \quad A \mapsto \int 1_A d\mu$$

for any $\mu \in \mathcal{M}_b$ and $\mathcal{N}' := \{\mu' \mid \mu \in \mathcal{N}\}$. Then $\mathcal{N}'$ is a compact convex set of $\mathcal{M}(\mathfrak{S}, \mathbb{R})$. By Proposition 5.6.9 there exists $\nu \in \mathcal{N}$ such that μ' is absolutely continuous with respect to ν' for any $\mu \in \mathcal{N}$. Since $\mathcal{N} \subset \mathcal{M}_c$ there exists $A \in \mathfrak{R}$ such that $X \backslash A$ is a ν-null set. Then $X \backslash A$ is *a* μ-null set for any $\mu \in \mathcal{N}$. By Corollary 4.7.8 $\{\mu'(B) \mid \mu \in \mathcal{N}, B \in \mathfrak{S}\}$ is a bounded set of $\mathbb{R}$ so

$$\sup_{\mu \in \mathcal{N}} |\mu|(A) < \infty . \quad \square$$

Corollary 5.6.11. *The Mackey topology $\tau(\mathcal{M}_c^\pi, \mathcal{M}_c)$ on $\mathcal{M}_c^\pi$ is coarser than the topology of $(\mathcal{M}_c^\pi, \mathcal{F})$, where $\mathcal{F} := \{\dot{1}_A \mid A \in \mathfrak{R}\}$.*

Let $\mathcal{N}$ be a convex set of $\mathcal{M}_c$ compact with respect to the $\sigma(\mathcal{M}_c, \mathcal{M}_c^\pi)$-topology. By Theorem 5.6.10 there exists $A \in \mathfrak{R}$ such that $X \backslash A$ is a μ-null set for any $\mu \in \mathcal{N}$ and such that

$$\alpha := \sup_{\mu \in \mathcal{N}} |\mu|(A) < \infty$$

We set $\xi := \dot{1}_A$. Then

$$\left\{\eta \in \mathcal{M}_c^\pi \mid p_\xi(\eta) < \frac{1}{\alpha + 1}\right\} \subset \mathcal{N}^0 ,$$

which proves the assertion. $\square$

Remark. The above result was tacitly used in the proof of C. Constantinescu [5] Corollary 3.8.6.

Lemma 5.6.12. *Let X be a paracompact space and $(F_\iota)_{\iota \in I}$ be a locally finite family of closed sets of X. Then there exists a locally finite family $(U_\iota)_{\iota \in I}$ of open sets of X such that $F_\iota \subset U_\iota$ for any $\iota \in I$.*

For any $x \in X$ there exists an open V_x of X such that $x \in V_x$ and

$$I_x := \{\iota \in I \mid F_\iota \cap V_x \neq \emptyset\}$$

is finite. Let $(W_\lambda)_{\lambda \in L}$ be a locally finite open covering of X finer than the open covering $(V_x)_{x \in X}$ of X. We set

$$L_\iota := \{\lambda \in L \mid F_\iota \cap W_\lambda \neq \emptyset\}, \quad U_\iota := \bigcup_{\lambda \in L_\iota} W_\lambda$$

for any $\iota \in I$. U_ι is an open set of X and $F_\iota \subset U_\iota$ for any $\iota \in I$. We have to show that $(U_\iota)_{\iota \in I}$ is locally finite.

Let $y \in X$. There exists a neighbourhood A of y such that

$$J := \{\lambda \in L \mid A \cap W_\lambda \neq \emptyset\}$$

is finite. For any $\lambda \in J$ there exists $x_\lambda \in X$ with $W_\lambda \subset V_{x_\lambda}$. Then

$$\{\iota \in I \mid \lambda \in L_\iota\} \subset I_{x_\lambda}$$

for any $\lambda \in J$ and

$$\{\iota \in I \mid A \cap U_\iota \neq \emptyset\} \subset \bigcup_{\lambda \in J} I_{x_\lambda},$$

so $\{\iota \in I \mid A \cap U_\iota\} \neq \emptyset$ is finite. Hence, $(U_\iota)_{\iota \in I}$ is a locally finite family. □

Proposition 5.6.13. *Let X be a paracompact space, $\mathscr{C}$ be the vector lattice of continuous real functions on X, l be a linear form of $\mathscr{C}^+$ and K be the set of $x \in X$ such that for any neighbourhood U of x in X there exists $f \in \mathscr{C}$ with* Supp $f \subset U$, $l(f) \neq 0$. *Then K is a compact set of X.*

We may assume without loss of generality that l positive. Let $(x_n)_{n \in \mathbb{N}}$ be a sequence in K. Assume $(x_n)_{n \in \mathbb{N}}$ has no adherent point. By Lemma 5.6.12 there exists a locally finite sequence $(U_n)_{n \in \mathbb{N}}$ of open sets of X such that $x_n \in U_n$ for any $n \in \mathbb{N}$. For any $n \in \mathbb{N}$ let $f_n \in \mathscr{C}_+$ such that $\operatorname{Supp} f_n \subset U_n$ and $l(f_n) > 0$. Then $\sum_{n \in \mathbb{N}} \frac{1}{l(f_n)} f_n \in \mathscr{C}_+$ and

$$p < \sum_{n=0}^{p} \frac{1}{l(f_n)} l(f_n) = l\left(\sum_{n=0}^{p} \frac{1}{l(f_n)} f_n\right) \leq l\left(\sum_{n \in \mathbb{N}} \frac{1}{l(f_n)} f_n\right) < \infty$$

for any $p \in \mathbb{N}$, which is a contradiction. Hence, $(x_n)_{n \in \mathbb{N}}$ possesses adherent points so K is a Φ_2-set of X (Proposition 1.5.5 c $\Rightarrow$ a). By Theorem 1.5.8 K is relatively compact. Since K is closed it is compact. □

Definition 5.6.14. *Let X be a Hausdorff space, $\mathfrak{K}$ be the set of compact sets of X, $\mathfrak{R}$ be a δ-ring of subsets of X containing $\mathfrak{K}$ and $\mu \in \mathscr{M}(\mathfrak{R}, \mathbb{R}; \mathfrak{K})$. Then there exists a greatest open set of X, which is a μ-null set. The complement of this set in X is called the* carrier of μ *and is denoted by* Supp μ.

Theorem 5.6.15. *Let X be a paracompact space, $\mathscr{C}$ be the vector lattice of continuous real functions on X, l be a positive linear form on $\mathscr{C}$, $\mathfrak{K}$ be the set of compact sets of X and $\mathfrak{R}$ be the δ-ring generated by $\mathfrak{K}$. Then the following assertions are equivalent:*

a) we have $\inf_{\iota\in I} l(f_\iota) = 0$ *for any lower directed family* $(f_\iota)_{\iota\in I}$ *in* $\mathscr{C}$ *with* $\overset{\mathbb{R}^X}{\bigwedge_{\iota\in I}} f_\iota = 0$;

b) there exists $\mu\in\mathscr{M}(\mathfrak{R},\mathbb{R};\mathfrak{K})_+$ *with* $\operatorname{Supp}\mu\in\mathfrak{K}$ *and* $\int f d\mu = l(f)$ *for any* $f\in\mathscr{C}$;

c) there exists $\mu\in\mathscr{M}(\mathfrak{R},\mathbb{R};\mathfrak{K})$ *such that* $f\in\mathscr{L}^1(\mu)$ *and* $\int f d\mu = l(f)$ *for any* $f\in\mathscr{C}$.

The measure μ *in c) is unique.*

a ⇒ b. We denote by K the set of $x\in X$ such that for any neighbourhood U of x in X there exists $f\in\mathscr{C}$ with $\operatorname{Supp} f\subset U$ and $l(f)\neq 0$. By Proposition 5.6.13 K is compact.

Let $f\in\mathscr{C}_+$ and $f=0$ on K. We set

$$\mathscr{G} := \{g\in\mathscr{C}_+ \mid g\leq f,\ l(g)=0\}.$$

$\mathscr{G}$ is upper directed. Let $x\in X\backslash K$. There exists a neighbourhood U of x such that $l(h)=0$ for any $h\in\mathscr{C}$ with $\operatorname{Supp} h\subset U$. Let $h\in\mathscr{C}_+$, $\operatorname{Supp} h\subset U$ and $h(x)>0$. We set

$$g := f\wedge \frac{f(x)}{h(x)}h.$$

Then $g\in\mathscr{G}$ and $g(x)=f(x)$. Hence, $\overset{\mathbb{R}^X}{\bigvee_{g\in\mathscr{G}}} g = f$, so $(f-g)_{g\in\mathscr{G}}$ is a lower directed family in $\mathscr{C}$ with $\overset{\mathbb{R}^X}{\bigwedge_{g\in\mathscr{G}}}(f-g)=0$. By a)

$$0 = \inf_{g\in\mathscr{G}} l(f-g) = \inf_{g\in\mathscr{G}} (l(f)-l(g)) = l(f).$$

Let $\mathscr{C}(K)$ be the vector lattice of continuous real functions on K and, for any $f\in\mathscr{C}(K)$, let $f'\in\mathscr{C}$ and $f'|K=f$. By the above considerations $l(f')$ does not depend on the choice of f'. Hence,

$$\mathscr{C}(K)\to\mathbb{R},\quad f\mapsto l(f')$$

is a positive linear form on $\mathscr{C}(K)$. By the Riesz representation theorem there exists $\mu\in\mathscr{M}(\mathfrak{R},\mathbb{R};\mathfrak{K})_+$ with $\operatorname{Supp}\mu\subset K$ such that $\int f d\mu = l(f)$ for any $f\in\mathscr{C}$.

b ⇒ c is trivial.

c ⇒ a. By Proposition 5.6.13 the carrier of μ is compact and the assertion follows from the Dini theorem.

The unicity of μ in c) is trivial. □

Corollary 5.6.16. *Let* X *be a paracompact space,* $\mathscr{C}$ *be the set of continuous real functions on* X, $\mathfrak{K}$ *be the set of compact sets of* X, $\mathfrak{R}$ *be the* δ*-ring generated by* $\mathfrak{K}$ *and* $\mu\in\mathscr{M}(\mathfrak{R},\mathbb{R};\mathfrak{K})$. *Then the following assertions are equivalent:*

a) $\operatorname{Supp}\mu\in\mathfrak{K}$;

b) $\mathscr{C}\subset\mathscr{L}^1(\mu)$.

a $\Rightarrow$ b is trivial.

b $\Rightarrow$ a follows from Theorem 5.6.15 c $\Rightarrow$ b via the unicity of μ. □

Remark. The implication a $\Rightarrow$ b holds for arbitrary Hausdorff spaces while the implication b $\Rightarrow$ a fails if we replace paracompact by locally compact in the above result. Indeed let $\beta\mathbb{N}$ be the Stone-Čech compactification of $\mathbb{N}$ and let $x \in \beta\mathbb{N}\setminus\mathbb{N}$. We set $X := \beta\mathbb{N}\setminus\{x\}$. Then X is locally compact and any continuous real function on X is bounded. Hence, $\mathscr{C} \subset \mathscr{L}^1(\mu)$ for any bounded $\mu \in \mathscr{M}(\mathfrak{R}, \mathbb{R}; \mathfrak{K})$ but not any such μ has compact carrier.

Proposition 5.6.17. *Let X be a completely regular space, $\mathscr{C}$ be the set of continuous real functions on X, $\mathfrak{K}$ be the set of compact sets of X, $\mathfrak{R}$ be the δ-ring generated by $\mathfrak{K}$ and $\mathscr{N}$ be a subset of $\mathscr{M}(\mathfrak{R}, \mathbb{R}; \mathfrak{K})$ such that any measure of $\mathscr{N}$ has compact carrier and such that $\{\int f d\mu \mid \mu \in \mathscr{N}\}$ is bounded for any $f \in \mathscr{C}$. Then $\{\int f d|\mu| \mid \mu \in \mathscr{N}\}$ is bounded for any $f \in \mathscr{C}$.*

Assume the contrary. Then there exists a positive function f of $\mathscr{C}$ such that $\{\int f d|\mu| \mid \mu \in \mathscr{N}\}$ is not bounded. We construct a sequence $(\mu_n)_{n \in \mathbb{N}}$ in $\mathscr{N}$ and a sequence $(g_n)_{n \in \mathbb{N}}$ in $\mathscr{C}$ inductively such that we have

a) $\int f d|\mu_n| > 6 \cdot 3^n (n + \sup_{\mu \in \mathscr{N}} |\int f \sum_{m=0}^{n-1} \frac{1}{3^m} g_m d\mu|)$,

b) $\int f g_n d\mu_n > \frac{2}{3} \int f d|\mu_n|$,

c) $|g_n| \leq 1$

for any $n \in \mathbb{N}$. Let $n \in \mathbb{N}$ and assume that the sequences were constructed up to $n-1$. Since $\{\int f d|\mu| \mid \mu \in \mathscr{N}\}$ is not bounded there exists $\mu_n \in \mathscr{N}$ such that a) is fulfilled. Let K be the carrier of μ_n and $\mathfrak{R}' := \{A \in \mathfrak{R} \mid A \subset K\}$. Then $K \in \mathfrak{K}$. By Hahn's theorem there exists $A \in \mathfrak{R}$ such that $\mu_n(B) \geq 0$ for any $B \in \mathfrak{R}$ so that $B \subset A$ and $\mu_n(B) \leq 0$ for any $B \in \mathfrak{R}$ with $A \cap B = 0$. We have

$$\int f d|\mu_n| = \int f(1_A - 1_{X \setminus A}) d\mu_n = \int_K (1_A - 1_{X \setminus A}) d(f \cdot \mu_n).$$

There exists a continuous real function g on K such that $|g| \leq 1$ and

$$\int g d((f \cdot \mu_n) | \mathfrak{R}') > \tfrac{2}{3} \int_K (1_A - 1_{X \setminus A}) d(f \cdot \mu_n).$$

Since X is completely regular there exists $g_n \in \mathscr{C}$ such that $|g_n| \leq 1$ and $g_n | K = g$. This finishes the inductive construction.

We set $g := \sum_{n \in \mathbb{N}} \frac{1}{3^n} g_n$. Then $g \in \mathscr{C}$ and

$$\int f g d\mu_n = \int f \sum_{m=0}^{n-1} \frac{1}{3^m} g_m d\mu_n + \frac{1}{3^n} \int f g_n d\mu_n + \sum_{m=n+1}^{\infty} \frac{1}{3^m} \int f g_m d\mu_n >$$

$$> \frac{1}{3^n} \int f g_n d\mu_n - \sup_{\mu \in \mathscr{N}} |\int f \sum_{m=0}^{n-1} \frac{1}{3^m} g_m d\mu| - \sum_{m=n+1}^{\infty} \frac{1}{3^m} \int f d|\mu_n| >$$

$$> \frac{2}{3 \cdot 3^n} \int f d|\mu_n| - \sup_{\mu \in \mathcal{N}} |\int f \sum_{m=0}^{n-1} \frac{1}{3^m} g_m d\mu| - \frac{1}{2 \cdot 3^n} \int f d|\mu_n| =$$

$$= \frac{1}{6 \cdot 3^n} \int f d|\mu_n| - \sup_{n \in \mathcal{N}} |\int f \sum_{m=0}^{n-1} \frac{1}{3^m} g_m d\mu| > n$$

for any $n \in \mathbb{N}$, which is the expected contradiction. □

Theorem 5.6.18. *Let X be a paracompact space, $\mathfrak{K}$ be the set of compact sets of X and $\mathscr{C}$ be the vector space of continuous real functions on X. Then $\mathscr{C}_{\mathfrak{K}}$ is barreled.*

Let $\mathfrak{R}$ be the δ-ring generated by $\mathfrak{K}$ and $\mathscr{M}$ be $\mathscr{M}(\mathfrak{R}, \mathbb{R}; \mathfrak{K})$. By Proposition 5.6.3 $\mathscr{M}$ is a band of $\mathscr{M}(\mathfrak{R}, \mathbb{R})$. For any $\mu \in \mathscr{M}_c$ we denote by μ' the map

$$\mathscr{C} \to \mathbb{R}, \quad f \mapsto \int f d\mu .$$

Then $\mu' \in (\mathscr{C}_{\mathfrak{K}})'$ for any $\mu \in \mathscr{M}_c$ and by Riesz representation theorem the map

$$\mathscr{M}_c \to (\mathscr{C}_{\mathfrak{K}})', \quad \mu \mapsto \mu'$$

is an isomorphism of vector spaces. We identify $(\mathscr{C}_{\mathfrak{K}})'$ with $\mathscr{M}_c$ via this isomorphism.

Let $\mathscr{U}$ be a barrel of $\mathscr{C}_{\mathfrak{K}}$ and $\mathscr{U}^0$ be its polar set in $\mathscr{M}_c$. First we wish to show that $\bigcup_{\mu \in \mathscr{U}^0} \operatorname{Supp} \mu$ is a relatively compact set of X. By Theorem 1.5.8 it is sufficient to show that $\bigcup_{\mu \in \mathscr{U}^0} \operatorname{Supp} \mu$ is a Φ_2-set. Assume the contrary. By Proposition 1.5.5 c $\Rightarrow$ a there exists a sequence $(x_n)_{n \in \mathbb{N}}$ in $\bigcup_{\mu \in \mathscr{U}^0} \operatorname{Supp} \mu$ which possesses no adherent points in X. Let $(\mu_n)_{n \in \mathbb{N}}$ be a sequence in $\mathscr{U}^0$ such that $x_n \in \operatorname{Supp} \mu_n$ for any $n \in \mathbb{N}$. Since $\mathscr{U}$ is a barrel of $\mathscr{C}_{\mathfrak{K}}$ the set $\{\int f d\mu_n \mid n \in \mathbb{N}\}$ is bounded for any $f \in \mathscr{C}$. By Proposition 5.6.17 $\{\int f d|\mu_n| \mid n \in \mathbb{N}\}$ is bounded for any $f \in \mathscr{C}$. Hence, the map

$$l: \mathscr{C} \to \mathbb{R}, \quad f \mapsto \sum_{n \in \mathbb{N}} \frac{1}{2^n} \int f d|\mu_n|$$

is well-defined and it is obvious that it is a positive linear form on $\mathscr{C}$. Let K be the set of $x \in X$ such that for any neighbourhood U of x there exists $f \in \mathscr{C}$ so that $\operatorname{Supp} f \subset U$ and $l(f) \neq 0$. By Proposition 5.6.13 K is compact. Since $\{x_n \mid n \in \mathbb{N}\} \subset K$ this is a contradiction. Hence, $\bigcup_{\mu \in \mathscr{U}^0} \operatorname{Supp} \mu$ is relatively compact.

We denote by Y the compact space $\overline{\bigcup_{\mu \in \mathscr{U}^0} \operatorname{Supp} \mu}$ and by $\mathscr{C}(Y)$ the Banach space of continuous real functions on Y (provided with the supremum norm). For any $f \in \mathscr{C}(Y)$ take f' as an arbitrary function of $\mathscr{C}$ with $f' | Y = f$ and, for any $\mu \in \mathscr{U}^0$, take μ' as the map

$$\mathscr{C}(Y) \to \mathbb{R}, \quad f \mapsto \int f' d\mu .$$

Then $\mu' \in \mathscr{C}(Y)'$ for any $\mu \in \mathscr{U}^0$ and $\{\mu'(f) \mid \mu \in \mathscr{U}^0\}$ is a bounded set for any $f \in \mathscr{C}(Y)$. Since $\mathscr{C}(Y)$ is barreled $\{\mu' \mid \mu \in \mathscr{U}^0\}$ is equicontinuous. We deduce that

$\{\int f d\mu \mid \mu \in \mathscr{U}^0, f \in \mathscr{C}, |f| \leq 1\}$ is a bounded set so $\mathscr{U}^0$ is equicontinuous. Hence, $\mathscr{U}$ is a 0-neighbourhood in $\mathscr{C}_{\mathfrak{K}}$ and $\mathscr{C}_{\mathfrak{K}}$ is barreled. □

Proposition 5.6.19. *Let X be a topological space, $\mathscr{C}$ be the vector lattice of continuous real functions on X and $l \in \mathscr{C}^+$. If $l(f) = 0$ for any bounded $f \in \mathscr{C}$, then $l = 0$.*

Let $f \in \mathscr{C}_+$. Then $\sum_{n \in \mathbb{N}} (f - n 1_X) \vee 0 \in \mathscr{C}$. It is easy to see that $|l|(g) = 0$ for any bounded $g \in \mathscr{C}$. For any $n \in \mathbb{N}$ the function $f - (f - n 1_X) \vee 0$ is bounded so

$$|l|(f) = |l|((f - n 1_X) \vee 0).$$

Then

$$m|l|(f) = \sum_{n=0}^{m-1} |l|((f - n 1_X) \vee 0) = |l|(\sum_{n=0}^{m-1} (f - n 1_X) \vee 0) \leq$$
$$\leq |l|(\sum_{n \in \mathbb{N}} (f - n 1_X) \vee 0)$$

for any $m \in \mathbb{N}$, so $|\ell|(f) = 0$ and $l = 0$. □

Theorem 5.6.20. *Let X be a completely regular space, βX be the Stone-Čech compactification of X, $\mathscr{C}$ be the vector latice of continuous real functions on X and, for any $f \in \mathscr{C}$, let βf be the continuous map $\beta X \to \overline{\mathbb{R}}$ extending f. Moreover, let Y be the subspace $\bigcap_{f \in \mathscr{C}} \overset{-1}{\beta f}(\mathbb{R})$ of βX, $\mathfrak{K}$ be the set of compact sets of Y, $\mathfrak{R}$ be the δ-ring generated by $\mathfrak{K}$, $\mathscr{M}$ be the band $\mathscr{M}(\mathfrak{R}, \mathbb{R}; \mathfrak{K})$ of $\mathscr{M}(\mathfrak{R}, \mathbb{R})$ and $\mathscr{C}'$ be the vector lattice of continuous real functions on Y. For any $f \in \mathscr{C}$ let f' be the map*

$$Y \to \mathbb{R}, \quad x \mapsto \beta f(x)$$

and, for any $\mu \in \mathscr{M}_c$, let μ' be the map

$$\mathscr{C} \to \mathbb{R}, \quad f \mapsto \int f' d\mu.$$

Then the following hold:

a) $f' \in \mathscr{C}'$ for any $f \in \mathscr{C}$ and the map

$$\mathscr{C} \to \mathscr{C}', \quad f \mapsto f'$$

is an isomorphism of vector lattices;

b) Y is Φ_2-compact;

c) $\mu' \in \mathscr{C}^+$ for any $\mu \in \mathscr{M}_c$ and the map

$$\mathscr{M}_c \to \mathscr{C}^+, \quad \mu \mapsto \mu'$$

is an isomorphism of vector lattices;

d) the following assertions are equivalent for any $x \in \beta X$:

d_1) $x \in Y$;

d_2) $X \cap \bigcap_{n \in \mathbb{N}} U_n \neq \emptyset$ for any sequence $(U_n)_{n \in \mathbb{N}}$ of neighbourhoods of x in βX;

d_3) $(\beta f_n(x))$ *is summable and* $\sum_{n\in\mathbb{N}} \beta f_n(x) = (\beta \sum_{n\in\mathbb{N}} f_n)(x)$ *for any summable sequence* $(f_n)_{n\in\mathbb{N}}$ *in* $\mathscr{C}$;

d_4) *the map* $\mathscr{C} \to \mathbb{R}$, $f \mapsto \beta f(x)$ *belongs to* $\mathscr{P}^c(\mathscr{C}, \mathbb{R})$;

e) $\mathscr{C}^+ = \mathscr{P}^c(\mathscr{C}, \mathbb{R}) = \{\varphi \mid \varphi$ *order exhaustive linear form on* $\mathscr{C}\}$;

f) if X *is normal, then the following hold:*

f_1) *any sequence in* X *possessing no adherent point in* X *does not possess any adherent point in* Y;

f_2) X *is a sequentially closed set of* Y;

f_3) X *is a* Φ_2*-set of the topological space* $X \cup (\beta X \setminus Y)$;

f_4) *if* $\mathfrak{F}$ *is a* Φ_2*-filter of* Y *such that* $A \cap X \neq \emptyset$ *for any* $A \in \mathfrak{F}$, *then the filter induced by* $\mathfrak{F}$ *on* X *is a* Φ_2*-filter of* X;

f_5) $A \cap X$ *is a* Φ_2*-set of* X *for any* Φ_2*-set* A *of* Y;

g) if X *is normal and* Φ_2*-compact (e.g. paracompact), then* $X \cap K$ *is a compact set of* X *for any compact set* K *of* Y;

h) if X *is discrete and* Φ *denotes the set of* δ*-stable ultrafilters on* X, *then* $\lim \varphi(\mathfrak{F}) \in Y$ *for any* $\mathfrak{F} \in \Phi$, *the map*

$$\Phi \to Y, \quad \mathfrak{F} \mapsto \lim \varphi(\mathfrak{F})$$

is a bijection, and any Φ_2*-set of* Y *is finite, where* φ *denotes the inclusion map* $X \to \beta X$.

By Proposition 5.6.3 $\mathscr{M}(\mathfrak{R}, \mathbb{R}; \mathfrak{K})$ is a band of $\mathscr{M}(\mathfrak{R}, \mathbb{R})$.

a) is trivial.

b) Let $\mathfrak{F}$ be a Φ_2-ultrafilter of Y and $f \in \mathscr{C}$. Then $f'(\mathfrak{F})$ is a Φ_2-ultrafilter and therefore a convergent ultrafilter on $\mathbb{R}$ (Theorem 1.5.8). Hence, the ultrafilter on βX defined by $\mathfrak{F}$ converges to a point of Y, so $\mathfrak{F}$ is convergent. It follows that Y is Φ_2-compact (Proposition 1.4.2).

c) Let l be a positive linear form on $\mathscr{C}$. Moreover, let $\tilde{\mathscr{C}}$ be the vector lattice of continuous real functions on βX, $\tilde{\mathfrak{K}}$ be the set of compact sets of βX and $\tilde{\mathfrak{R}}$ be the σ-ring generated by $\tilde{\mathfrak{K}}$. Then

$$\tilde{\mathscr{C}} \to \mathbb{R}, \quad f \mapsto l(f \mid X)$$

is a positive linear form so there exists $\tilde{\mu} \in \mathscr{M}(\tilde{\mathfrak{R}}, \mathbb{R}; \tilde{\mathfrak{K}})$ such that $l(f \mid X) = \int f d\tilde{\mu}$ for any $f \in \tilde{\mathscr{C}}$. Let $f \in \mathscr{C}_+$. Then

$$\int \beta(f \wedge n 1_X) d\tilde{\mu} = l(f \wedge n 1_X) \leq l(f)$$

for any $n \in \mathbb{N}$. We get $\beta f \in \mathscr{L}^1(\tilde{\mu})$ and $\int \beta f d\tilde{\mu} \leq l(f)$. Hence,

$$\mathscr{C} \to \mathbb{R}, \quad f \mapsto l(f) - \int \beta f d\tilde{\mu}$$

is a positive linear form that vanishes at any bounded function of $\mathscr{C}$. By Proposition 5.6.19 $l(f) = \int \beta f d\tilde{\mu}$ for any $f \in \mathscr{C}$.

Let $f \in \mathscr{C}_+$ and $(\alpha_n)_{n\in\mathbb{N}}$ be a sequence of positive real numbers. We set

$$g := \sum_{n\in\mathbb{N}} \alpha_n (f - n 1_X) \vee 0.$$

Then $g \in \mathscr{C}_+$, $\beta g = \sum_{n\in\mathbb{N}} \alpha_n \beta(f - n 1_X) \vee 0$, and

$$\int \beta g d\tilde{\mu} = \int (\sum_{n\in\mathbb{N}} \alpha_n \beta(f - n 1_X) \vee 0) d\tilde{\mu} =$$
$$= \sum_{n\in\mathbb{N}} \alpha_n \int (\beta f - n 1_{\beta X}) \vee 0 \, d\tilde{\mu} \geq \sum_{n\in\mathbb{N}} \alpha_n \tilde{\mu}(\overset{-1}{\widehat{\beta f}}(]n+1, n+2])).$$

Since $(\alpha_n)_{n\in\mathbb{N}}$ is arbitrary the set

$$\{n \in \mathbb{N} \mid \tilde{\mu}(\overset{-1}{\widehat{\beta f}}(]n+1, n+2]) > 0\}$$

is finite. Hence, $\operatorname{Supp} \tilde{\mu} \subset \overset{-1}{\widehat{\beta f}}(\mathbb{R})$. Since f is arbitrary $\operatorname{Supp} \tilde{\mu} \subset Y$.

We set $\mu := \tilde{\mu} \mid \mathfrak{R}$. Then $\mu \in \mathscr{M}$ and $\operatorname{Supp} \mu = \operatorname{Supp} \tilde{\mu}$ and therefore $\mu \in \mathscr{M}_c$. Moreover,

$$\mu'(f) = \int f' d\mu = \int \beta f d\tilde{\mu} = l(f)$$

for any $f \in \mathscr{C}$, i.e. $\mu' = l$. Hence, the map

$$\mathscr{M}_c \to \mathscr{C}^+, \quad \mu \mapsto \mu'$$

is surjective. It is now easy to see that this map is an isomorphism of vector lattices.

d) $d_1 \Rightarrow d_2$. Assume $X \cap \bigcap_{n\in\mathbb{N}} U_n = \emptyset$. We construct a decreasing sequence $(f_n)_{n\in\mathbb{N}}$ of continuous positive real functions on βX inductively such that $f_n(x) = 1$ and $\operatorname{Supp} f_n \subset \bigcap_{m=0}^{n} U_m$ and set $f := \sum_{n\in\mathbb{N}} (f_n \mid X)$. Then $f \in \mathscr{C}$ and $\beta f(x) = \infty$, which is a contradiction.

$d_2 \Rightarrow d_3$. Assume the contrary. Then there exists $\alpha \in \overline{\mathbb{R}}$, $\alpha \neq (\beta \sum_{n\in\mathbb{N}} f_n)(x)$ such that for any neighbourhood V of α in $\overline{\mathbb{R}}$ and for any $K \in \mathfrak{P}_f(\mathbb{N})$ there exists $L \in \mathfrak{P}_f(\mathbb{N})$ with $K \subset L$ and $\sum_{n\in L} \beta f_n(x) \in V$. Let U, V be open neighbourhoods of $(\beta \sum_{n\in\mathbb{N}} f_n)(x)$ and of α in $\overline{\mathbb{R}}$ respectively such that $\bar{U} \cap \bar{V} = \emptyset$. We may construct an increasing sequence $(K_n)_{n\in\mathbb{N}}$ in $\mathfrak{P}_f(\mathbb{N})$ inductively such that $n \in K_n$ and $\sum_{m\in K_n} \beta f_m(x) \in V$ for any $n \in \mathbb{N}$. We set

$$V_n := \{y \in \beta X \mid (\beta \sum_{n\in\mathbb{N}} f_n)(y) \in U, \ \sum_{m\in K_n} \beta f_m(y) \in V\}$$

for any $n \in \mathbb{N}$. Then $(V_n)_{n\in\mathbb{N}}$ is a sequence of neighbourhoods of x in βX with $X \cap \bigcap_{n\in\mathbb{N}} V_n = \emptyset$. This is a contradiction.

$d_3 \Rightarrow d_4$ is trivial.

$d_4 \Rightarrow d_1$. Let $f \in \mathscr{C}_+$. Then $(f \wedge (n+1)1_X - f \wedge n 1_X)_{n\in\mathbb{N}}$ is a supersummable sequence in $\mathscr{C}$ and its sum is f. Hence, $\beta f(x) \in \mathbb{R}$. Since f is arbitrary if follows that $x \in Y$.

e) Let $l \in \mathscr{C}^+$ and $(f_n)_{n\in\mathbb{N}}$ be a supersummable sequence in $\mathscr{C}$. By c) there exists $\mu \in \mathscr{M}_c$ such that $\mu' = l$ and by $d_1 \Rightarrow d_4$ $(f_n')_{n\in\mathbb{N}}$ is supersummable in $\mathscr{C}'$ and its sum is $(\sum_{n\in\mathbb{N}} f_n)'$. By Theorem 4.10.7 $(l(f_n))_{n\in\mathbb{N}}$ is summable and

$$l(\sum_{n\in\mathbb{N}} f_n) = \int (\sum_{n\in\mathbb{N}} f_n') d\mu = \sum_{n\in\mathbb{N}} \int f_n' d\mu = \sum_{n\in\mathbb{N}} l(f_n).$$

Hence $l \in \mathscr{P}^c(\mathscr{C}, \mathbb{R})$ and $\mathscr{C}^+ \subset \mathscr{P}^c(\mathscr{C}, \mathbb{R})$.

Let now $\varphi \in \mathscr{P}^c(\mathscr{C}, \mathbb{R})$. By proposition 3.3.19 φ is linear. Let $f \in \mathscr{C}_+$. Assume

$$\sup\{\varphi(g) \mid g \in \mathscr{C}_+, g \leq f\} = \infty.$$

Then there exists a sequence $(f_n)_{n\in\mathbb{N}}$ in $\mathscr{C}_+$ such that $f_n \leqq f$ and $\varphi(f_n) > 2^n$ for any $n \in \mathbb{N}$. Since $\left(\frac{1}{2^n} f_n\right)_{n\in\mathbb{N}}$ is a supersummable sequence in $\mathscr{C}$, we get the contradictory relation

$$1 \leq \lim_{n\to\infty} \varphi\left(\frac{1}{2^n} f_n\right) = 0.$$

Hence

$$\sup\{\varphi(g) \mid g \in \mathscr{C}_+, g \leq f\} < \infty,$$

$\varphi \in \mathscr{C}^+$ (Ch. D. Aliprantis, O. Burkinshaw [1] Theorem 3.3) and $\mathscr{P}^c(\mathscr{C}, \mathbb{R}) \subset \mathscr{C}^+$.

The equality

$$\mathscr{C}^+ = \{\varphi \mid \varphi \text{ order exhaustive linear form on } \mathscr{C}\}$$

follows from Proposition 5.5.23 a $\Leftrightarrow$ b. $\square$

f_1) Let $(x_n)_{n\in\mathbb{N}}$ be a sequence in X which possesses no adherent point. We may assume that $x_m \neq x_n$ for any different $m, n \in \mathbb{N}$. Since X is normal there exists $f \in \mathscr{C}$ such that $f(x_n) = n$ for any $n \in \mathbb{N}$. Let x be an adherent point of $(x_n)_{n\in\mathbb{N}}$ in βX. Then $\beta f(x) = \infty$. Hence, $x \notin Y$ and $(x_n)_{n\in\mathbb{N}}$ possesses no adherent point in Y.

f_2) Let $(x_n)_{n\in\mathbb{N}}$ be a sequence in X converging to a point x of Y. By f_1) $x \in X$ so X is a sequentially closed set of Y.

f_3) Let $(x_n)_{n\in\mathbb{N}}$ be a sequence in X. f_1) asserts that if $(x_n)_{n\in\mathbb{N}}$ possesses no adherent point in X, then it possesses an adherent point in $\beta X \setminus Y$. By Proposition 1.5.5 a $\Rightarrow$ c X is a Φ_2-set of the topological space $X \cup (\beta X \setminus Y)$.

f_4) Let (I, f) be a Θ_2-net in Y such that $f(\mathfrak{G}) = \mathfrak{F}$, where $\mathfrak{G}$ denotes the section filter of I. We set $J := \overset{-1}{f}(X)$ and

$$g: J \to X, \quad \iota \mapsto f(\iota).$$

If we endow J with the preorder relation induced by I, then (J, g) is a net in X such that $g(\mathfrak{H})$ is the filter induced by $\mathfrak{F}$ on X, where $\mathfrak{H}$ is the section filter of J. We have to show that (J, g) is a Θ_2-net in X. Let $(\iota_n)_{n\in\mathbb{N}}$ be an increasing sequence in J. Then $(f(\iota_n))_{n\in\mathbb{N}}$ possesses an adherent point in Y. By f_1) $(g(\iota_n))_{n\in\mathbb{N}}$ possesses an adherent point in X so (J, g) is a Θ_2-net in X.

f_5) follows immediately from f_4).

g) By Proposition 1.5.5 c $\Rightarrow$ a K is a Φ_2-set of Y so by f_5) $K \cap X$ is a Φ_2-set of X. By Proposition 1.2.10 it is a relatively compact set of X. Since $X \cap K$ is a closed set of X it is even a compact set of X.

By Theorem 1.5.8 any paracompact space is Φ_2-compact.

h) Let $x \in \beta X$ and $\mathfrak{F}$ be the filter induced on X by the neighbourhood filter of x. It is easy to see that $\mathfrak{F}$ is an ultrafilter. Assume $\mathfrak{F} \in \Phi$ and let $f \in \mathscr{C}$. Let $(U_n)_{n\in\mathbb{N}}$ be a sequence of subsets of $\overline{\mathbb{R}}$ forming a fundamental system of neighbourhoods for $\lim f(\mathfrak{F})$. Then $\overset{-1}{f}(U_n) \in \mathfrak{F}$ for any $n \in \mathbb{N}$ and so $\bigcap_{n\in\mathbb{N}} \overset{-1}{f}(U_n) \in \mathfrak{F}$. We have $f = \lim f(\mathfrak{F})$ on $\bigcap_{n\in\mathbb{N}} \overset{-1}{f}(U_n)$ so $\lim f(\mathfrak{F}) \in \mathbb{R}$ and $x \in Y$. Now assume $\mathfrak{F} \notin \Phi$. There exists a sequence $(A_n)_{n\in\mathbb{N}}$ in $\mathfrak{F}$ such that $\bigcap_{n\in\mathbb{N}} A_n \notin \mathfrak{F}$. Let f be the real function on X equal to n on $\bigcap_{m=0}^{n} A_m \setminus \bigcap_{m=0}^{n+1} A_m$ for any $n \in \mathbb{N}$ and equal to 0 otherwise. Then $\lim f(\mathfrak{F}) = \infty$ and $x \notin Y$. Hence, the map

$$\Phi \to Y, \quad \mathfrak{F} \mapsto \lim \varphi(\mathfrak{F})$$

is a bijection.

Let $(x_n)_{n\in\mathbb{N}}$ be a sequence in Y such that $x_m \neq x_n$ for different $m, n \in \mathbb{N}$ and for any $n \in \mathbb{N}$ let $\mathfrak{F}_n \in \Phi$ so that $\lim \varphi(\mathfrak{F}_n) = x_n$. Furthermore, let $(A_n)_{n\in\mathbb{N}}$ be a sequence of subsets of X such that $A_n \in \mathfrak{F}_n \setminus \bigcup_{m=1}^{n-1} \mathfrak{F}_m$ for any $n \in \mathbb{N}$. We denote by f the real function on X equal to n on $A_n \setminus \bigcup_{m=n+1}^{\infty} A_m$ for any $n \in \mathbb{N}$ and equal to 0 otherwise. Then $f'(x_n) = n$ for any $n \in \mathbb{N}$ so $(x_n)_{n\in\mathbb{N}}$ possesses no adherent point in Y. Hence, any Φ_2-set of Y in finite. $\square$

Remarks. 1) The first uncountable ordinal number endowed with the usual topology furnishes an example of a normal space X which is not a Φ_1-closed set of Y.

2) Let X be a set, Φ be the set of δ-stable ultrafilters on X and $l \in (\mathbb{R}^X)^+$. By c) and h) there exists a real function g on Φ such that $\{\mathfrak{F} \in \Phi \mid g(\mathfrak{F}) \neq 0\}$ is finite and such that $l(f) = \sum_{\mathfrak{F}\in\Phi} g(\mathfrak{F}) \lim f(\mathfrak{F})$ for any $f \in \mathbb{R}^X$. This result was proved by G. Choquet (1967) ([1] Théorème 8).

Theorem 5.6.21. *Let X be a paracompact space, $\mathfrak{K}$ be the set of compact sets of X, $\mathfrak{R}$ be the δ-ring generated by $\mathfrak{K}$, $\mathscr{M}$ be $\mathscr{M}(\mathfrak{R}, \mathbb{R}; \mathfrak{K})$, $\mathscr{C}$ be the vector space of continuous real functions on X, $\mathscr{F}$ be the set of $\mathfrak{R}$-measurable real functions on X which are bounded on each compact set of X, $\mathscr{G}$ be the solid subspace of $\mathscr{M}_c^\pi$ generated by*

$\{\dot{f} \mid f \in \mathscr{F}\}$, Φ be $\hat{\Phi}_0(\mathscr{G})$, and assume $\int \xi d\mu \in E$ for any $(\xi, \mu) \in \mathscr{G} \times \mathscr{M}_c(E)$. Then the identity map $\mathscr{M}_c(E)_{\mathfrak{K} \cup \mathscr{C}} \to \mathscr{M}_c(E)_{\Phi}$ is uniformly Φ_4-continuous. If, in addition, X is locally compact, then the identity map $\mathscr{M}_c(E)_{\mathfrak{L} \cup \mathscr{C}} \to \mathscr{M}_c(E)_{\Phi}$ is uniformly Φ_4-continuous, where $\mathfrak{L}$ denotes the set of compact G_δ-set of X.

We prove both assertions simultaneously by writing the locally compact case in parentheses. Let us denote by $\mathscr{H}$ the set of functions of $\mathscr{F}$ with compact carrier and by Ψ the set $\hat{\Phi}_0(\mathscr{H})$. By Corollary 4.5.11 (Corollary 4.5.8) the identity map $\mathscr{M}(E)_{\mathfrak{K}} \to \mathscr{M}(E)$ ($\mathscr{M}(E)_{\mathfrak{L}} \to \mathscr{M}(E)$) is uniformly Φ_4-continuous so by Theorem 5.6.6 the identity map $\mathscr{M}(E)_{\mathfrak{K}} \to \mathscr{M}(E)_{\Psi}$ ($\mathscr{M}(E)_{\mathfrak{L}} \to \mathscr{M}(E)_{\Psi}$) is uniformly Φ_4-continuous. Hence, by Corollary 1.8.15 the identity map $\mathscr{M}_c(E)_{\mathfrak{K}} \to \mathscr{M}_c(E)_{\Psi}$ ($\mathscr{M}_c(E)_{\mathfrak{L}} \to \mathscr{M}_c(E)_{\Psi}$) is uniformly Φ_4-continuous. By Theorem 5.6.2 the identity map $\mathscr{M}_c(E)_{\mathscr{F}} \to \mathscr{M}_c(E)_{\Phi}$ is uniformly Φ_4-continuous. Hence, in order to prove that the identity map $\mathscr{M}_c(E)_{\mathfrak{K} \cup \mathscr{C}} \to \mathscr{M}_c(E)_{\Phi}$ ($\mathscr{M}_c(E)_{\mathfrak{L} \cup \mathscr{C}} \to \mathscr{M}_c(E)_{\Phi}$) is uniformly Φ_4-continuous, it is sufficient, by Corollary 1.8.5, to show the identity map $\mathscr{M}_c(E)_{\mathscr{H} \cup \mathscr{C}} \to \mathscr{M}_c(E)_{\mathscr{F}}$ is uniformly Φ_4-continuous.

Let $\mathfrak{F}$ be a Cauchy Φ_4-filter on $\mathscr{M}_c(E)_{\mathscr{H} \cup \mathscr{C}}$ and (I, μ) be a Φ_4-net in $\mathscr{M}_c(E)_{\mathscr{H} \cup \mathscr{C}}$ such that $\mu(\mathfrak{G}) = \mathfrak{F}$, where $\mathfrak{G}$ denotes the section filter of I. Assume $\mathfrak{F}$ is not a Cauchy filter on $\mathscr{M}_c(E)_{\mathscr{F}}$. Then there exists $f \in \mathscr{F}$ and a convex closed 0-neighbourhood U *in* E such that for any $J \in \mathfrak{G}$ there exist $\iota', \iota'' \in J$ with

$$\int f d\mu_{\iota'} - \int f d\mu_{\iota''} \notin U.$$

We construct inductively an increasing sequence $(\iota_n)_{n \in \mathbb{N}}$ in I, a decreasing sequence $(I_n)_{n \in \mathbb{N}}$ in $\mathfrak{G}$, an increasing sequence $(K_n)_{n \in \mathbb{N}}$ in $\mathfrak{K}$, and a sequence $(x'_n)_{n \in \mathbb{N}}$ in U^0 such that for any $n \in \mathbb{N}$:

a) $\iota_n \in I_{n-1}$, $(I_{-1} := I)$

b) $\iota \in I_n \Rightarrow |\int f 1_{K_n} d(x'_n \circ \mu_\iota) - \int f 1_{K_n} d(x'_n \circ \mu_{\iota_n})| > \frac{1}{3}$.

Let $n \in \mathbb{N}$ and assume the sequences were constructed up to $n-1$. By the above remark there exist $\iota', \iota'' \in I_{n-1}$ and $x'_n \in U^0$ such that

$$|\int f d(x'_n \circ \mu_{\iota'}) - f d(x'_n \circ \mu_{\iota''})| > 1.$$

We set

$$K_n := K_{n-1} \cup \operatorname{Supp}(x'_n \circ \mu_{\iota'}) \cup \operatorname{Supp}(x'_n \circ \mu_{\iota''}) \qquad (K_{-1} := \emptyset).$$

Then $K_n \in \mathfrak{K}$ and $K_{n-1} \subset K_n$. Since $f 1_{K_n} \in \mathscr{H}$ and since $\mathfrak{F}$ is a Cauchy filter on $\mathscr{M}_c(E)_{\mathscr{H}}$ there exists $I_n \in \mathfrak{G}$ such that $I_n \subset I_{n-1} \cap \{\iota \in I \mid \iota \geq \iota_{n-1}\}$ and

$$\int f 1_{K_n} d\mu_{\lambda'} - \int f 1_{K_n} d\mu_{\lambda''} \in \tfrac{1}{3} U$$

for any $\lambda', \lambda'' \in I_n$. Then either

$$|\int f 1_{K_n} d(x'_n \circ \mu_\iota) - \int f 1_{K_n} d(x'_n \circ \mu_{\iota'})| > \tfrac{1}{3}$$

for any $\iota \in I_n$ or

$$|\int f 1_{K_n} d(x'_n \circ \mu_\iota) - \int f 1_{K_n} d(x'_n \circ \mu_{\iota''})| > \tfrac{1}{3}$$

for any $\iota \in I_n$. We set $\iota_n := \iota'$ in the first case and $\iota_n := \iota''$ in the second. This finishes the inductive construction.

For any $(\nu, x') \in \mathcal{M}_c(E) \times E'$ let $\varphi(\nu, x')$ be the map

$$\mathscr{C} \to \mathbb{R}, \quad g \mapsto \int g\, d(x' \circ \nu)$$

which belongs to $(\mathscr{C}_{\mathfrak{K}})' \cdot (\mu_{\iota_n})_{n \in \mathbb{N}}$ is a Θ_4-sequence in $\mathcal{M}_c(E)_{\mathscr{C}}$ so $\{\varphi(\mu_{\iota_n}, x') \mid n \in \mathbb{N},\ x' \in U^0\}$ is a bounded set of $(\mathscr{C}_{\mathfrak{K}})'_{\mathscr{C}}$. By Theorem 5.6.18 $\mathscr{C}_{\mathfrak{K}}$ is barreled. Hence, $\{\varphi(\mu_{\iota_n}, x') \mid n \in \mathbb{N}, x' \in U^0\}$ is equicontinuous and there exists $K \in \mathfrak{K}$ with

$$\bigcup_{\substack{n \in \mathbb{N} \\ x' \in U^0}} \operatorname{Supp}(x' \circ \mu_{\iota_n}) \subset K.$$

$(\mu_{\iota_n})_{n \in \mathbb{N}}$ is a Θ_4-sequence in $\mathcal{M}_c(E)_{\mathscr{H}}$. By the above considerations the identity map $\mathcal{M}_c(E)_{\mathscr{H}} \to \mathcal{M}_c(E)_{\Psi}$ is uniformly Φ_4-continuous so, by Theorem 1.8.4 a $\Rightarrow$ e, $(\mu_{\iota_n})_{n \in \mathbb{N}}$ is a Θ_4-sequence in $\mathcal{M}_c(E)_{\Psi}$. $(f 1_{K \cap (K_n \setminus K_{n-1})})_{n \in \mathbb{N}}$ is order supersummable in $\mathscr{H}$ so $(f 1_{K \cap K_n})_{n \in \mathbb{N}} \in \Theta_0(\mathscr{H})$. Hence, there exist $m, n \in \mathbb{N}$ such that $m < n$ and

$$\int f 1_{K \cap K_p} d\mu_{\iota_m} - \int f 1_{K \cap K_p} d\mu_{\iota_n} \in \tfrac{1}{3} U$$

for any $p \in \mathbb{N}$. Then

$$|\int f 1_{K_m} d(x'_m \circ \mu_{\iota_m}) - \int f 1_{K_m} d(x'_m \circ \mu_{\iota_n})| \leq \tfrac{1}{3},$$

which contradicts a & b.

Hence, $\mathfrak{F}$ is a Cauchy filter on $\mathcal{M}_c(E)_{\mathscr{F}}$ and, by Theorem 1.8.4 b $\Rightarrow$ a, the identity map $\mathcal{M}_c(E)_{\mathscr{H} \cup \mathscr{C}} \to \mathcal{M}_c(E)_{\mathscr{F}}$ is uniformly Φ_4-continuous. □

Theorem 5.6.22. *Let X be a paracompact space, $\mathfrak{K}$ be the set of compact sets of X, $\mathfrak{R}$ be the δ-ring generated by $\mathfrak{K}$, $\mathcal{M}$ be $\mathcal{M}(\mathfrak{R}, \mathbb{R}; \mathfrak{K})$, $\mathscr{C}$ be the vector space of continuous real functions on X and $\varphi : \mathscr{C} \to E$ be a linear map. Then the following assertions are equivalent:*

a) $\varphi : \mathscr{C}_{(\mathscr{C}_{\mathfrak{K}})'} \to E_{E'}$ is continuous;

b) $\varphi : \mathscr{C}_{\mathfrak{K}} \to E$ is continuous;

c) there exists $\mu \in \mathcal{M}(E'^*)$ such that $f \in \bigcap_{x' \in E'} \mathscr{L}^1(x' \circ \mu)$ and $\int f d\mu = \varphi(f)$ for any $f \in \mathscr{C}$;

d) there exists $\mu \in \mathcal{M}_c(E'^*)$ such that $f \in \bigcap_{x' \in E'} \mathscr{L}^1(x' \circ \mu)$ and $\int f d\mu = \varphi(f)$ for any $f \in \mathscr{C}$, and such that $\bigcup_{x' \in A'} \operatorname{Supp} x' \circ \mu$ is a relatively compact set of X and $\sup_{x' \in A'} \|x' \circ \mu\| < \infty$ for any bounded set A' of E'_E.

By Proposition 5.6.3 $\mathcal{M}$ is a band of $\mathcal{M}(\mathfrak{R}, \mathbb{R})$.

a $\Rightarrow$ b follows from the fact that $\mathscr{C}_{\mathfrak{K}}$ is barreled (Theorem 5.6.18).

b ⇒ d. By the Riesz representation theorem there exists a $\mu_{x'} \in \mathcal{M}_c$ for any $x' \in E'$ such that $x' \circ \varphi(f) = \int f d\mu_{x'}$ for any $f \in \mathcal{C}$. For any $A \in \mathfrak{R}$ put

$$\mu(A) : E' \to \mathbb{R}, \quad x' \mapsto \mu_{x'}(A).$$

Then $\mu \in \mathcal{M}_c(E'^*)$ and $x' \circ \mu = \mu_{x'}$ for any $x' \in E'$. We have $\mathcal{C} \subset \bigcap_{x' \in E'} \mathcal{L}^1(x' \circ \mu)$ and $\int f d\mu = \varphi(f)$ for any $f \in \mathcal{C}$.

Let A' be a bounded set of E'_E. Then $\{x' \circ \varphi \mid x' \in A'\}$ is a bounded set of $(\mathcal{C}_{\mathfrak{K}})'_{\mathcal{C}}$. By Theorem 5.6.18 there exists $K \in \mathfrak{K}$ such that

$$\sup \{|x' \circ \varphi(f)| \mid x' \in A', f \in \mathcal{C}, \sup_{x \in K} |f(x)| \leq 1\} < \infty .$$

We get $\bigcup_{x' \in A'} \operatorname{Supp} x' \circ \mu \subset K$ and $\sup_{x' \in A'} \|x' \circ \mu\| < \infty$.

d ⇒ c is trivial.

c ⇒ a. Let $x' \in E'$. Then $x' \circ \mu \in \mathcal{M}$. Since $\mathcal{C} \subset \mathcal{L}^1(x' \circ \mu)$ we get $x' \circ \mu \in \mathcal{M}_c$ by Corollary 5.6.16 b ⇒ a. Hence, $x' \circ \varphi \in (\mathcal{C}_{\mathfrak{K}})'$. Since x' is arbitrary the map $\psi : \mathcal{C}_{(\mathcal{C}_{\mathfrak{K}})'} \to E_{E'}$ is continuous. □

Theorem 5.6.23. *Let $\mathfrak{S}$ be the σ-ring generated by $\mathfrak{R}$ and $\mathcal{N}$ be a subset of $\mathcal{M}_b(E)$ such that $\int \xi d\mu \in E$ for any $(\xi, \mu) \in \mathcal{M}_b^\pi \times \mathcal{N}$. Then the identity map $\mathcal{N}_{\mathfrak{S}} \to \mathcal{N}_{\mathcal{M}_b^\pi}$ is uniformly Φ_4-continuous.*

Let us denote by $\bar{\mathcal{M}}$ the set of $\mu \in \mathcal{M}(\mathfrak{S}, \mathbb{R})$ such that $\mu | \mathfrak{R} \in \mathcal{M}$. Then $\bar{\mathcal{M}}$ is a band of $\mathcal{M}(\mathfrak{S}, \mathbb{R})$, $\mu | \mathfrak{R} \in \mathcal{M}_b$ for any $\mu \in \bar{\mathcal{M}}$, and the map

$$\bar{\mathcal{M}} \to \mathcal{M}_b, \quad \mu \mapsto \mu | \mathfrak{R}$$

is an isomorphism of vector lattices. For any $\xi \in \mathcal{M}_b^\pi$ let $\bar{\xi}$ be the map

$$\bar{\mathcal{M}} \to \mathbb{R}, \quad \mu \mapsto \int \xi d(\mu | \mathfrak{R}).$$

Then $\bar{\xi} \in \bar{\mathcal{M}}^\pi$ and the map

$$\mathcal{M}_b^\pi \to \bar{\mathcal{M}}^\pi, \xi \mapsto \bar{\xi}$$

is an isomorphism of vector lattices.

Let $\mathcal{F}$ be the solid subspace of $\mathcal{M}_b^\pi$ generated by $\{\dot{1}_A \mid A \in \mathfrak{S}\}$. Then $\bar{\mathcal{F}} := \{\bar{\xi} \mid \xi \in \mathcal{F}\}$ is the solid subspace of $\bar{\mathcal{M}}^\pi$ generated in the same way. We may assume E complete. By Theorem 5.6.6 $\int \xi d\mu \in E$ for any $(\xi, \mu) \in \bar{\mathcal{M}}^\pi \times \bar{\mathcal{M}}(E)$ and the identity map $\bar{\mathcal{M}}(E) \to \bar{\mathcal{M}}(E)_{\bar{\mathcal{F}}}$ is uniformly Φ_4-continuous.

For any $\mu \in \mathcal{N}$ let $\bar{\mu}$ be the map

$$\mathfrak{S} \to E, \quad A \mapsto \int 1_A d\mu.$$

Then $\bar{\mu} \in \bar{\mathcal{M}}(E)$ for any $\mu \in \mathcal{N}$ by virtue of Orlicz-Pettis theorem (Corollary 3.5.12). The map

$$\mathcal{N} \to \bar{\mathcal{M}}(E), \quad \mu \mapsto \bar{\mu}$$

is obviously injective. We identify $\mathcal{N}$ with a subset of $\bar{\mathcal{M}}(E)$ via this injection. By the above considerations and by Corollary 1.8.15 the identity map $\mathcal{N}_{\mathfrak{S}} \to \mathcal{N}_{\mathcal{F}}$ is uniformly Φ_4-continuous.

Let $\mathcal{G}$ be the set of order continuous linear maps of $\mathcal{M}_b^\pi$ into E and, for any $\mu \in \mathcal{N}$, let μ' be the map

$$\mathcal{M}_b^\pi \to E, \quad \xi \mapsto \int \xi \, d\mu .$$

By C. Constantinescu [5] Theorem 4.2.11 a $\Rightarrow$ d, $\mu' \in \mathcal{G}$ for any $\mu \in \mathcal{N}$. The map

$$\mathcal{N} \to \mathcal{G}, \quad \mu \mapsto \mu'$$

is obviously an injection. We identify $\mathcal{N}$ with a subset of $\mathcal{G}$ via this injection. It is obvious that $\mathcal{F}$ is a solid subspace of $\mathcal{M}_b^\pi$ such that $\bigvee_{n \in \mathbb{N}} \xi_n \in \mathcal{F}$ for any sequence $(\xi_n)_{n \in \mathbb{N}}$ in $\mathcal{F}$ which is upper bounded in $\mathcal{M}_b^\pi$. By Proposition 5.3.2 the identity map $\mathcal{G}_{\mathcal{F}} \to \mathcal{G}_{\mathcal{M}_b^\pi}$ is uniformly Φ_4-continuous so by Corollary 1.8.15 the identity map $\mathcal{N}_{\mathcal{F}} \to \mathcal{N}_{\mathcal{M}_b^\pi}$ is uniformly Φ_4-continuous too. By Corollary 1.8.5 the identity map $\mathcal{N}_{\mathfrak{S}} \to \mathcal{N}_{\mathcal{M}_b^\pi}$ is uniformly Φ_4-continuous. □

Corollary 5.6.24. *Let $\mathfrak{S}$ be the σ-ring generated by $\mathfrak{R}$, $\mu \in \mathcal{M}$, $L^1(\mu)$ be the Banach space of equivalence classes of μ-integrable real functions on X, and $L^1(\mu)_{\mathfrak{S}}$ be the set $L^1(\mu)$ endowed with the uniformity of $\mathfrak{S}$-convergence (i.e. the coarsest uniformity for which the maps*

$$L^1(\mu) \to \mathbb{R}, \quad \mathcal{F} \mapsto \int_A \mathcal{F} \, d\mu$$

are uniformly continuous, where A runs through $\mathfrak{S}$). Further let $\tilde{L}^1(\mu)$ be the set $L^1(\mu)$ endowed with a uniformity such that the identity maps $L^1(\mu)_{L^1(\mu)'} \to \tilde{L}^1(\mu) \to L^1(\mu)_{\mathfrak{S}}$ are uniformly Φ_4-continuous. Then the identity map $L^1(\mu)_{\mathfrak{S}} \to L^1(\mu)_{L^1(\mu)'}$ is uniformly Φ_4-continuous and $\tilde{L}^1(\mu)$ is Φ_4-compact. In particular, $L^1(\mu)$ is weakly Φ_4-compact.

For any $\mathcal{F} \in L^1(\mu)$ let $\mathcal{F} \cdot \mu$ be the map

$$\mathfrak{R} \to \mathbb{R}, \quad A \mapsto \int_A \mathcal{F} \, d\mu .$$

Then $\mathcal{F} \cdot \mu \in \mathcal{M}_b$ for any $\mathcal{F} \in L^1(\mu)$ and the map

$$L^1(\mu) \to \mathcal{M}_b, \quad \mathcal{F} \mapsto \mathcal{F} \cdot \mu$$

is an injection. We identify $L^1(\mu)$ with a subset of $\mathcal{M}_b$ via this injection. Then $L^1(\mu)_{\mathfrak{S}}$ is a subspace of $(\mathcal{M}_b)_{\mathfrak{S}}$ and by C. Constantinescu [5] Theorem 3.6.3 c) $L^1(\mu)_{L^1(\mu)'}$ is a subspace of $(\mathcal{M}_b)_{\mathcal{M}_b^\pi}$. By Theorem 5.6.23 the identity map $(\mathcal{M}_b)_{\mathfrak{S}} \to (\mathcal{M}_b)_{\mathcal{M}_b^\pi}$ is uniformly Φ_4-continuous so by Corollary 1.8.15 the identity map $L^1(\mu)_{\mathfrak{S}} \to L^1(\mu)_{L^1(\mu)'}$ is also uniformly Φ_4-continuous.

Let $\mathfrak{F}$ be a Φ_4-ultrafilter on $L^1(\mu)_{\mathfrak{S}}$ and ν be the map

$$\mathfrak{S} \to \mathbb{R}, \quad A \mapsto \lim_{\mathcal{F}, \mathfrak{F}} \int_A \mathcal{F} \, d\mu .$$

By Theorem 4.2.14 a) ν is a measure on $\mathfrak{S}$ and by Corollary 4.7.8 it is bounded. ν vanishes on any μ-null set so by Radon-Nikodym's theorem there exists $\mathscr{F} \in L^1(\mu)$ such that $\nu(A) = \int_A \mathscr{F} d\mu$ for any $A \in \mathfrak{S}$. Hence, $L^1(\mu)_{\mathfrak{S}}$ is Φ_4-compact. By the above considerations and Corollary 1.8.5 the identity map $L^1(\mu)_{\mathfrak{S}} \to \tilde{L}^1(\mu)$ is uniformly Φ_4-continuous so by Proposition 1.8.11 $\tilde{L}^1(\mu)$ is Φ_4-compact. □

Proposition 5.6.25. *Let $\mathfrak{S}$ be a set of $\mathfrak{R}$-measurable subsets of X, $\mu \in \mathscr{M}(E)$ such that for any equicontinuous set A' of E' there exists $A \in \mathfrak{S}$ such that $X \backslash A$ is an $x' \circ \mu$-null set for any $x' \in A'$, and let $\xi \in \hat{L}^1(\mu)$ with $\int \xi 1_A d\mu \in E$ for any $A \in \mathfrak{S}$. If E is complete, then $\int \xi d\mu \in E$.*

Let A' be an equicontinuous set of E'. By the hypothesis there exists $A \in \mathfrak{S}$ such that $X \backslash A$ is an $x' \circ \mu$-null set for any $x' \in A'$. Then

$$\int \xi d(x' \circ \mu) = \int \xi 1_A d(x' \circ \mu) = \langle \int \xi 1_A d\mu, x' \rangle$$

for any $x' \in A'$ so the map

$$A'_E \to \mathbb{R}, \quad x' \mapsto \int \xi d(x' \circ \mu)$$

is continuous. Since E is complete Grothendieck's completeness criterium (H. H. Schaefer [1] IV 6.2 Corollary 2 (a) ⇒ (b)) yields $\int \xi d\mu \in E$. □

§ 5.7 Quasi *M*-spaces

Definition 5.7.1. *For any locally convex space E denote its dual by* <u>E'</u> *and by* <u>E''</u> *its bidual, i.e. the dual of the strong dual of E, (the strong dual of E is $E'_{\mathfrak{A}}$, where $\mathfrak{A}$ denotes the set of bounded sets of E).*

E' is a set of maps defined on E so, according to our general conventions, it will be considered automatically endowed with the uniformity (and topology) of pointwise convergence if not other uniformity or topology is mentioned explicitly. The same convention holds for the subsets of E'.

Definition 5.7.2. *A* <u>locally convex lattice</u> *is a vector lattice E endowed with a Hausdorff locally convex topology such that its topological dual space is a solid vector subspace of E^+.*

The expressions *complete*, *convergence*, *continuous*, *bounded*, *summable*, and *supersummable* used on locally convex lattices will always be understood with respect to its underlying locally convex space. The corresponding expressions referring to its underlying vector lattice will always be preceded by the word *order*.

Let E be a locally convex lattice and $\mathfrak{A}$ be the set of closed intervals of E'. Any set of $\mathfrak{A}$, being a compact set of E'_E, it follows that $E' = (E_{\mathfrak{A}})'$. The topology of $E_{\mathfrak{A}}$ is locally convex-solid, (i.e. there exists a fundamental system of solid convex 0-neighbourhoods in $E_{\mathfrak{A}}$). This allows us to apply to locally convex lattices many results concerning the vector lattices endowed with locally convex-solid topologies.

The dual of a locally convex lattice endowed with the strong topology is also a locally convex lattice (Ch. D. Aliprantis, O. Burkinshaw [1] page 59).

Definition 5.7.3. *A* quasi-M-space *is a locally convex lattice E such that for any weak 0-neighbourhood U in E there exists a solid directed 0-neighbourhood for the Mackey topology of E which is enclosed by U.*

Proposition 5.7.4. *Let E be a quasi-M-space, $\varphi : E \longrightarrow E''$ be the evaluation map and A be a bounded set of E. Then:*

a) the solid directed set of E generated by A is bounded;
b) $\varphi(A)$ is order bounded.

a) Let B be the solid directed set of E generated by A and $\mathfrak{U}$ be the set of solid directed 0-neighbourhoods for the Mackey topology of E. There exists, for any $U \in \mathfrak{U}$, an $\alpha_U \in \mathbb{R}_+$ such that $A \subset \alpha_U U$. Then $\bigcap_{U \in \mathfrak{U}} \alpha_U U$ is a solid directed set of E containing A so

$$B \subset \bigcap_{U \in \mathfrak{U}} \alpha_U U .$$

Let V be a weak 0-neighbourhood in E. There exists $U \in \mathfrak{U}$ with $U \subset V$. Then $B \subset \alpha_U V$. Hence, B is a bounded set of E.

b) Let $\mathfrak{F}$ be the section filter of $\mathfrak{P}_f(A)$ and y'', z'' be the maps

$$E' \longrightarrow \mathbb{R}, \quad x' \longmapsto \lim_{B, \mathfrak{F}} x'(\bigwedge_{x \in B} x),$$

$$E' \longrightarrow \mathbb{R}, \quad x' \longmapsto \lim_{B, \mathfrak{F}} x'(\bigvee_{x \in B} x)$$

respectively. By a) $y'', z'' \in E''$. It is obvious that y'' and z'' are lower and upper bounds of $\varphi(A)$ respectively so $\varphi(A)$ is order bounded. □

Definition 5.7.5. *Let E be a vector lattice and $\mathfrak{F}$ be a filter on E. We set*

$$\underline{|A|} := \{x \in E \mid \exists y \in A, \ |x| \leq |y|\}$$

for any $A \subset E$ and denote by $\underline{|\mathfrak{F}|}$ the filter on E generated by the filter base

$$\{\underline{|A|} \mid A \in \mathfrak{F}\} .$$

Put

$$\underline{|\Phi|} := \{\underline{|\mathfrak{G}|} \mid \mathfrak{G} \in \Phi\}$$

for any set Φ of filters on E.

Theorem 5.7.6. *Let E be a quasi-M-space, $\varphi : E \to E''$ be the evaluation map and F be a vector σ-sublattice of E'' containing $\varphi(E)$ such that $\bigvee^{E''}_{n\in\mathbb{N}} \varphi(x_n) \in F$ for any sequence $(x_n)_{n\in\mathbb{N}}$ in E for which $\bigvee^{E''}_{n\in\mathbb{N}} \varphi(x_n)$ exists. We set*

$$\Phi := \hat{\Phi}_4(E'_F), \quad \Psi := |\Phi_4(E'_F)|.$$

Then:

a) the identity map $E_{E'} \to E_\Phi$ is uniformly Φ_4-continuous;

b) if, in addition, F is a solid subspace of E'', then the identity map $E_{E'} \to E_{\hat{\Psi}}$ is uniformly Φ_4-continuous.

Let $(x'_n)_{n\in\mathbb{N}}$ be a Θ_4-sequence in E'_F. For any $n\in\mathbb{N}$ let U_n be a solid directed 0-neighbourhood for the Mackey topology of E such that $x'_n \in U_n^0$. We set

$$X := \bigcup_{n\in\mathbb{N}} U_n^0$$

and denote by $\mathscr{C}$ the set of continuous real functions on X. For any $x\in E$ let $\tilde{x}$ be the map

$$X \to \mathbb{R}, \quad x' \mapsto x'(x).$$

Then $\tilde{x}\in\mathscr{C}$ for any $x\in E$. We denote by ψ the map

$$E \to \mathscr{C}, \quad x \mapsto \tilde{x}.$$

Let $\mathfrak{F}$ be a Cauchy Φ_4-filter on $E_{E'}$ and (I,f) be a Θ_4-net in $E_{E'}$ such that $f(\mathfrak{G}) = \mathfrak{F}$, where $\mathfrak{G}$ denotes the section filter of I. Then $(I, \psi\circ f)$ is a Θ_4-net in $\mathscr{C}$. By Theorem 2.3.14 and Proposition 1.7.3 a $\Rightarrow$ c $(I, \psi\circ f)$ is a Θ_3-net in $\mathscr{C}$.

Let $(\iota_n)_{n\in\mathbb{N}}$ be an increasing sequence in I. By the above considerations there exists a subsequence $(\iota_{k_n})_{n\in\mathbb{N}}$ of $(\iota_n)_{n\in\mathbb{N}}$ such that $(\psi(f(\iota_{k_n})))_{n\in\mathbb{N}}$ is a Cauchy sequence in $\mathscr{C}$. $(f(\iota_{k_n}))_{n\in\mathbb{N}}$ is a Θ_4-sequence in $E_{E'}$ and therefore a bounded sequence in E. Hence, $\{\varphi(f(\iota_{k_n}))\,|\,n\in\mathbb{N}\}$ is order bounded (Proposition 5.7.4 b)). Set

$$z''_n := \bigvee_{m=n}^{\infty} \varphi(f(\iota_{k_m})), \quad y''_n := \bigwedge_{m=n}^{\infty} \varphi(f(\iota_{k_m}))$$

for any $n\in\mathbb{N}$. Then $y''_n, z''_n \in F$ for any $n\in\mathbb{N}$.

Let $m\in\mathbb{N}$. $(f(\iota_{k_n}))_{n\in\mathbb{N}}$ is a bounded sequence in $E_{\{U_m^0\}}$. By Proposition 5.5.11 U_m^0 is an integral set of E'. By Proposition 5.5.17

$$\inf_{n\in\mathbb{N}} (z''_n - y''_n)(|x'|) = 0$$

for any $x'\in U_m^0$. Since m is arbitrary

$$\inf_{n\in\mathbb{N}} (z''_n - y''_n)(|x'|) = 0$$

for any $x'\in X$.

Let G be the solid subspace of E' generated by X. For any $x' \in E'$ let $\bar{x}'$ be the map

$$F \to \mathbb{R}, \quad x'' \mapsto x''(x').$$

Then $\bar{x}' \in F^\sigma$ for any $x' \in E'$.

Let $x' \in G$ and $x'' \in F_+$. Then

$$|\bar{x}'(y'')| = |y''(x')| \le |y''|(|x'|) \le x''(|x'|) = \overline{|x'|}(x'')$$

for any $y'' \in F$ with $|y''| \le x''$ so

$$|\bar{x}'|(x'') \le \overline{|x'|}(x'').$$

Hence, $|\bar{x}'| \le \overline{|x'|}$ and

$$0 \le \inf_{n \in \mathbb{N}} |\bar{x}'_m|(z''_n - y''_n) \le \inf_{n \in \mathbb{N}} \overline{|x'_m|}(z''_n - y''_n) =$$
$$= \inf_{n \in \mathbb{N}} (z''_n - y''_n)(|x'_m|) = 0$$

so

$$\inf_{n \in \mathbb{N}} |\bar{x}'_m|(z''_n - y''_n) = 0$$

for any $m \in \mathbb{N}$.

The map

$$E'_F \to F^\sigma, \quad x' \mapsto \bar{x}'$$

being uniformly continuous, $(\bar{x}'_m)_{m \in \mathbb{N}}$ is a Θ_4-sequence in F^σ. By Theorem 5.5.3 e) F^σ is sequentially complete so $(\bar{x}'_m)_{m \in \mathbb{N}}$ is a Θ_2-sequence in F^σ. By Proposition 1.4.7 $\{\bar{x}'_m \mid m \in \mathbb{N}\}$ is a Φ_2-set in F^σ. By Proposition 5.5.19

$$\inf_{n \in \mathbb{N}} \sup_{m \in \mathbb{N}} |\bar{x}'_m|(z''_n - y''_n) = 0.$$

Let $\varepsilon > 0$. There exists $n \in \mathbb{N}$ such that

$$\sup_{m \in \mathbb{N}} |\bar{x}'_m|(z''_n - y''_n) < \varepsilon.$$

Then

$$\sup_{m \in \mathbb{N}} |x'_m(f(\iota_{k_n}) - f(\iota_{k_p}))| =$$
$$= \sup_{m \in \mathbb{N}} |(\varphi(f(\iota_{k_n})) - \varphi(f(\iota_{k_p})))(x'_m)| =$$
$$= \sup_{m \in \mathbb{N}} |\bar{x}'_m(\varphi(f(\iota_{k_n})) - \varphi(f(\iota_{k_p})))| \le$$
$$\le \sup_{m \in \mathbb{N}} |\bar{x}'_m|(|\varphi(f(\iota_{k_n})) - \varphi(f(\iota_{k_p}))|) \le \sup_{m \in \mathbb{N}} |\bar{x}'_m|(z''_n - y''_n) < \varepsilon$$

for any $p \in \mathbb{N}$, $p \ge n$. Hence, $(f(\iota_{k_n}))_{n \in \mathbb{N}}$ is a Cauchy sequence in $E_{\{A'\}}$, where A'

denotes the set $\{x'_m \mid m \in \mathbb{N}\}$. We deduce further that $\mathfrak{F}$ is a Φ_3-filter in $E_{\mathfrak{A}}$, where

$$\mathfrak{A} := \{\{y'_n \mid n \in \mathbb{N}\} \mid (y'_n)_{n \in \mathbb{N}} \text{ is a } \Theta_4\text{-sequence in } E'_F\}.$$

By Theorem 2.2.3 a) $\mathfrak{F}$ is a Cauchy filter on E_Φ. Hence, the identity map $E_{E'} \to E_\Phi$ is uniformly Φ_4-continuous (Theorem 1.8.4 b $\Rightarrow$ a).

Assume now that F is a solid subspace of E''. Then, with the above notation, $|\overline{x}'| = \overline{|x'|}$ for any $x' \in G$ (Proposition 5.5.6) so

$$\inf_{n \in \mathbb{N}} \sup_{m \in \mathbb{N}} (z''_n - y''_n)(|x'_m|) = \inf_{n \in \mathbb{N}} \sup_{m \in \mathbb{N}} \overline{|x'_m|}(z''_n - y''_n) =$$

$$= \inf_{n \in \mathbb{N}} \sup_{m \in \mathbb{N}} |\overline{x}'_m|(z''_n - y''_n) = 0$$

Let $\varepsilon > 0$. There exists $n \in \mathbb{N}$ such that

$$\sup_{m \in \mathbb{N}} (z''_n - y''_n)(|x'_m|) < \varepsilon.$$

Then (Proposition 5.5.6)

$$\sup_{m \in \mathbb{N}} |x'_m|(|f(\iota_{k_n}) - f(\iota_{k_p})|) =$$

$$= \sup_{m \in \mathbb{N}} |\varphi(f(\iota_{k_n})) - \varphi(f(\iota_{k_p}))|(|x'_m|) \leq$$

$$\leq \sup_{m \in \mathbb{N}} (z''_n - y''_n)(|x'_m|) < \varepsilon$$

for any $p \in \mathbb{N}$ with $p \geq n$. Hence, $(f(\iota_{k_n}))_{n \in \mathbb{N}}$ is a Cauchy sequence in $E_{\{|A'|\}}$, where $|A'|$ denotes the set

$$\{x' \in E' \mid \exists m \in \mathbb{N},\ |x'| \leq |x'_m|\}.$$

Further, $\mathfrak{F}$ is a Φ_3-filter in $E_{|\mathfrak{A}|}$, where $|\mathfrak{A}|$ denotes the set of subsets of E' of the above form $|A'|$ with $(x'_m)_{m \in \mathbb{N}}$ running through the set of all Θ_4-sequences in E'_F.

Let Θ be the set of sequences $(x'_n)_{n \in \mathbb{N}}$ in E' for which there exists $A \in |\mathfrak{A}|$ such that the set $\{n \in \mathbb{N} \mid x'_n \in A\}$ is infinite. Furthermore, let $\mathfrak{H} \in \Phi_4(E'_F)$ and (J, g) be a Θ_4-net in E'_F such that $\mathfrak{H} = g(\mathfrak{H}')$, where $\mathfrak{H}'$ denotes the section filter of J. We set

$$K := \{(\iota, x') \in J \times E' \mid |x'| \leq |g(\iota)|\},$$

$$h : K \to E', \quad (\iota, x') \mapsto x'$$

and endow K with the upper directed preorder relation defined by

$$(\iota, x') \leq (\lambda, y') :\Leftrightarrow \iota \leq \lambda.$$

Then (K, h) is a Θ-net in E' and $h(\mathfrak{H}'') = |\mathfrak{H}|$, where $\mathfrak{H}''$ denotes the section filter of K. Hence, $|\mathfrak{H}| \in \Phi(\Theta)$ and $\hat{\Psi} \subset \hat{\Phi}(\Theta)$. By Theorem 2.2.3 a) $\mathfrak{F}$ is a Cauchy filter in $E_{\hat{\Psi}}$. Hence, the identity map $E_{E'} \to E_{\hat{\Psi}}$ is uniformly Φ_4-continuous (Theorem 1.8.4 b $\Rightarrow$ a). $\square$

Corollary 5.7.7. *Let E be a quasi-M-space, $\varphi : E \longrightarrow E''$ be the evaluation map, F be a vector σ-sublattice of E'' containing $\varphi(E)$ such that $\bigvee^{E''}_{n\in\mathbb{N}} \varphi(x_n) \in F$ for any sequence $(x_n)_{n\in\mathbb{N}}$ in E for which $\bigvee^{E''}_{n\in\mathbb{N}} \varphi(x_n)$ exists and let u be a continuous linear map of E into a locally convex space G such that the biadjoint map of u maps F into G. We set*

$$\Phi := \hat{\Phi}_4(G').$$

Then the map

$$u : E_{E'} \longrightarrow G_{\Phi}$$

is uniformly Φ_4-continuous.

Let $u' : G' \longrightarrow E'$ be the adjoint map of u. Put $\Psi := \hat{\Phi}_4(E'_F)$. By the hypothesis the map $u' : G'_G \longrightarrow E'_F$ is continuous so $u'(\mathfrak{F}) \in \Psi$ for any $\mathfrak{F} \in \Phi$. Hence, the map $u : E_\Psi \longrightarrow G_\Phi$ is continuous. By Theorem 5.7.6 a) the identity map $E_{E'} \longrightarrow E_\Psi$ is uniformly Φ_4-continuous so the map $u : E_{E'} \longrightarrow G_\Phi$ is uniformly Φ_4-continuous. □

Definition 5.7.8. *Let E be a locally convex space and Φ be the set $\hat{\Phi}_4(E'_{E''})$. If the identity map $E_{E'} \longrightarrow E_\Phi$ is uniformly Φ_4-continuous, then we say that* <u>E possesses the strong D. P.-property</u>. *A locally convex lattice* <u>possesses the strong D. P.-property</u> *if its underlying locally convex space possesses this property.*

Let E be a locally convex space and $\mathfrak{K}$ be the set of convex circled equicontinuous compact sets of $E'_{E''}$. E possesses the D. P.-property (the strict D. P.-property) if the identity map $E_{E'} \longrightarrow E_{\mathfrak{K}}$ preserves the compact convex sets (the Cauchy sequences), (A. Grothendieck (1953) [2] Définitions 1, 2; D. P. is an abbreviation for Dunford-Pettis). It is obvious that the strong D. P.-property implies the D. P.-property and the strict D. P.-property (Theorem 1.8.4 a $\Rightarrow$ j).

Corollary 5.7.9. *Any quasi-M-space possesses the strong D. P.-property.*

The assertion follows immediately from Theorem 5.7.6 a). □

Remark. Let E be a quasi-M-space, $\mathfrak{K}$ be the set of equicontinuous sets of E' which are compact sets in $E'_{E''}$. By the above result and Theorem 1.8.4 a $\Rightarrow$ h the identity map $E_{E'} \longrightarrow E_{\mathfrak{K}}$ preserves the Φ_4-sets. This result was proved by C. Constantinescu (1969) ([1] Theorem 4.9).

Proposition 5.7.10. *Let E be a locally convex lattice and X' be the set of elements of E' which are homomorphisms of vector lattices. Then the following hold:*

a) X' is a closed set of E'_E;

b) for any $x', y' \in X'$, which do not belong to the same ray of E', there exist $x, y \in E_+$ such that

$$x'(x) = 1, \quad x'(y) = 0,$$
$$y'(x) = 0, \quad y'(y) = 1, \quad x \wedge y = 0;$$

c) let K' be a compact set of $X' \setminus \{0\}$ such that any two different points of K' lie on different rays of E', let f be a continuous real function on K' and ε be a strictly positive real number; then there exists $x \in E$ such that

$$|x'(x) - f(x')| < \varepsilon$$

for any $x' \in K'$.

a) follows from

$$X' = \{x' \in E' \mid x, y \in E \Rightarrow x'(x \vee y) = \sup(x'(x), x'(y))\}.$$

b) We obviously have $x' \wedge y' = 0$ (Ch. D. Aliprantis, O. Burkinshaw [1] Theorem 3.13). Let $z \in E_+$ with $x'(z) > 0$ and $y'(z) > 0$. There exist $z_1, z_2 \in E$ such that

$$z_1 + z_1 = z, \quad x'(z_1) + y'(z_2) < \tfrac{1}{2} \inf(x'(z), y'(z))$$

(Ch. D. Aliprantis, O. Burkinshaw [1] Theorem 3.3 (i)). Then $x'(z_1) < x'(z_2)$, $y'(z_2) < y'(z_1)$ so

$$x'((z_2 - z_1)^+) = \sup(x'(z_2) - x'(z_1), 0) > 0,$$
$$x'((z_1 - z_2)^+) = \sup(x'(z_1) - x'(z_2), 0) = 0,$$
$$y'((z_1 - z_2)^+) = \sup(y'(z_1) - y'(z_2), 0) > 0,$$
$$y'((z_2 - z_1)^+) = \sup(y'(z_2) - y'(z_1), 0) = 0.$$

Put

$$x := \frac{1}{x'((z_2 - z_1)^+)} (z_2 - z_1)^+, \qquad y := \frac{1}{y'((z_1 - z_2)^+)} (z_1 - z_2)^+.$$

x, y possess the required properties.

c) Let $x', y' \in K'$, $x' \neq y'$. By b) there exist $x, y \in E$ such that

$$x'(x) = 1, \quad x'(y) = 0,$$
$$y'(x) = 0, \quad y'(y) = 1.$$

Then $f(x')x + f(y')y \in E$ and

$$x'(f(x')x + f(y')y) = f(x'),$$
$$y'(f(x')x + f(y')y) = f(y').$$

The assertion now follows immediately from N. Bourbaki [1] Ch. X § 4 Proposition 2. □

Definition 5.7.11. *Let E be a locally convex space, $\mathfrak{R}$ be a ring of subsets of E and $\mu \in \mathcal{M}(\mathfrak{R}, \mathbb{R})$ such that $E' \subset \mathcal{L}^1(\mu, E)$. A point $x \in E$ such that*

$$x'(x) = \int x' d\mu$$

for any $x' \in E'$ is called the barycenter of μ in E.

If it exists, the barycenter of a measure is unique. Let X be a subset of E containing $\bigcup_{A \in \mathfrak{R}} A$. Then

$$f \in \mathcal{L}^1(\mu, E) \Leftrightarrow f | X \in \mathcal{L}^1(\mu, X) \Rightarrow \int f d\mu = \int (f | X) d\mu$$

for any real function f on E.

Definition 5.7.12. *Let E be a real vector space and A be a convex set of E. A* convex function on A *is a real function f on A such that*

$$f(\alpha x + \beta y) \leq \alpha f(x) + \beta f(y)$$

for any $x, y \in A$ and any $\alpha, \beta \in \mathbb{R}_+$ with $\alpha + \beta = 1$.

Definition 5.7.13. *Let E be a locally convex space, K be a weakly compact convex set of E, $\mathfrak{K}$ be the set of weakly compact subsets of K and $\mathfrak{R}$ be the σ-ring generated by $\mathfrak{K}$ (i.e. the set of weak Borel sets of K). We define an order relation $\preceq$ on $\mathcal{M}(\mathfrak{R}, \mathbb{R}; \mathfrak{K})_+$ by setting $\mu \prec \nu$ for any $\mu, \nu \in \mathcal{M}(\mathfrak{R}, \mathbb{R}; \mathfrak{K})_+$ such that*

$$\int f d\mu \leq \int f d\nu$$

for any weakly continuous convex function f on K. The measures $\mu \in \mathcal{M}(\mathfrak{R}, \mathbb{R}; \mathfrak{K})$ for which $|\mu|$ is a maximal element with respect to the order relation $\prec$ will be called boundary measures on K. *If any two different positive boundary measures μ, ν on K with $\mu(K) = \nu(K)$ have different barycenters, then K is called a* simplex of E.

Theorem 5.7.14. *Let E be a locally convex lattice, X' be the set of elements of E' which are homomorphisms of vector lattices, U be a directed solid 0-neighbourhood for the Mackey topology of E, F' be the vector subspace of E' generated by U^0 and p be the gauge of U^0 in F'. Write $K' := U^0 \cap E'_+$. Then the following hold:*

a) F' is the solid vector subspace of E' generated by U^0;

b) for any $x' \in F'_+$

$$p(x') = \sup_{x \in U \cap E_+} x'(x);$$

c) p is a norm and its restriction to F'_+ is lower semi-continuous;

d) there exists $x'' \in (F')^\pi$ such that $x'' = p$ on F'_+;

e) K' is a simplex of E'_E;

f) $K' \cap X'$ is a compact set of K';

g) $\partial_e K' = \{x' \in K' \cap X' \mid p(x') \in \{0, 1\}\}$; *in particular,* $\partial_e K'$ *is a Borel set of both* K' *and* X';

h) $|\mu|(K' \backslash \partial_e K') = 0$ *for any boundary measure* μ *on* K';

i) for any compact set L' *of* $\partial_e K' \backslash \{0\}$, *any continuous real function* f *on* L' *such that* $0 \leq f \leq 1$ *and any strictly positive real number* ε *there exists* $x \in U \cap E_+$ *such that*

$$|x'(x) - f(x')| < \varepsilon$$

for any $x' \in L'$.

a) By Proposition 5.5.10 U^0 is a solid set of E' and, by Proposition 5.4.7, F' is solid.

b) Take $x \in U \cap E_+$ and $x'(x) > 0$ and let $\alpha \in \mathbb{R}$ such that $0 < \alpha < x'(x)$. Then $\frac{1}{\alpha} x' \notin U^0$ and $p(x') \geq \alpha$. Since α and x are arbitrary

$$p(x') \geq \sup_{x \in U \cap E_+} x'(x).$$

Assume $x' \neq 0$ and set

$$\beta := \sup_{x \in U \cap E_+} x'(x).$$

Then

$$\left|\frac{1}{\beta} x'(x)\right| \leq \frac{1}{\beta} x'(|x|) \leq 1$$

for any $x \in U$ so $\frac{1}{\beta} x' \in U^0$. Hence, $p(x') \leq \beta$ and

$$p(x') = \sup_{x \in U \cap E_+} x'(x).$$

c) follows immediately from b) and $\bigcap_{\alpha > 0} \alpha U^0 = \{0\}$.

d) Since $U \cap E_+$ is upper directed the assertion follows immediately from b) and Proposition 5.5.6.

e) By Proposition 5.5.11 U^0 is an integral set of E'. We have $\bigcap_{\alpha > \alpha_0} \alpha U^0 = \alpha_0 U^0$ for any $\alpha_0 \in \mathbb{R}_+$. From

$$E'_+ = \{x' \in E' \mid x \in E_+ \Rightarrow x'(x) \geq 0\}$$

we see that E'_+ is a closed set of E'_E so K' is a compact set of E'_E. By Proposition 5.4.11 and Choquet's theorem (G. Choquet, P.A. Meyer [1] Théorème 11 1 ⇒ 5 or E.M. Alfsen [1] Theorem II 3.6) K' is a simplex.

f) By Proposition 5.7.10 a) X' is a closed set of E' and therefore $K' \cap X'$ is a compact set of K'.

g) By Corollary 5.5.15 $\partial_e K' \subset K' \cap X'$. We have $p(K') \subset [0, 1]$. Let $x' \in \partial_e K'$ and assume $p(x') > 0$. Then $p\left(\frac{1}{p(x')} x'\right) = 1$ and $\frac{1}{p(x')} x' \in K'$. Since

$$x' = p(x')\left(\frac{1}{p(x')} x'\right) + (1 - p(x'))0$$

we get $p(x') = 1$. Hence,

$$\partial_e K' \subset \{x' \in K' \cap X' \mid p(x') \in \{0, 1\}\}.$$

Now let $x' \in K' \cap X' \backslash \partial_e K'$ with $p(x') = 1$. Then there exist $y', z' \in K'$, $\alpha, \beta \in \mathbb{R}_+ \backslash \{0\}$ such that $y' \neq z'$, $\alpha + \beta = 1$, $x' = \alpha y' + \beta z'$. It is easy to see that $y', z' \in X'$. We have by d) $p(y') = p(z') = 1$ so y' and z' belong to different rays of E'. By Proposition 5.7.10b) there exist $y, z \in E_+$ such that $y \wedge z = 0$,

$$y'(y) = 1, \quad y'(z) = 0,$$
$$z'(y) = 0, \quad z'(z) = 1.$$

Then

$$x'(y \vee z) = \sup(x'(y), x'(x)) = \sup(\alpha, \beta) < 1,$$
$$x'(y \vee z) = \alpha y'(y \vee z) + \beta z'(y \vee z) = \alpha + \beta = 1,$$

which is a contradiction. Hence,

$$\partial_e K' = \{x' \in K' \cap X' \mid p(x') \in \{0, 1\}\}.$$

h) Let $\mathscr{C}$ be the vector lattice of continuous real functions on K'. The map

$$l : \mathscr{C} \to \mathbb{R},$$
$$f \mapsto \int_{K' \backslash \{0\}} p(x') f\left(\frac{1}{p(x')} x'\right) d|\mu|(x') + \int_{K'} (1 - p(x')) f(0)\, d|\mu|(x')$$

is a positive linear form on $\mathscr{C}$ so by Riesz representation theorem there exists a positive Radon measure on K' such that

$$l(f) = \int f\, dv$$

for any $f \in \mathscr{C}$. Let f be a convex function of $\mathscr{C}$. Then

$$f(x') \leq p(x') f\left(\frac{1}{p(x')} x'\right) + (1 - p(x')) f(0)$$

for any $x' \in K' \setminus \{0\}$ so

$$\int f d|\mu| = f(0)|\mu|(\{0\}) + \int_{K'\setminus\{0\}} f d|\mu| \leq$$

$$\leq f(0)|\mu|(\{0\}) + \int_{K'\setminus\{0\}} p(x') f\left(\frac{1}{p(x')} x'\right) d|\mu|(x') +$$

$$+ \int_{K'\setminus\{0\}} (1 - p(x')) f(0) d|\mu|(x') =$$

$$= \int_{K'\setminus\{0\}} p(x') f\left(\frac{1}{p(x')} x'\right) d|\mu|(x') + \int_{K'} (1 - p(x')) f(0) d|\mu|(x') = \int f d\nu .$$

Since f is arbitrary we obtain $|\mu| \prec \nu$ and since μ is a boundary measure on K', $|\mu| = \nu$. Since $U \cap E_+$ is upper directed, from b) it follows that

$$\int_{K'} p^2 d|\mu| = \sup_{x \in U \cap E_+} \int_{K'} x^2 d|\mu| = \sup_{x \in U \cap E_+} \int_{K'} x^2 d\nu =$$

$$= \sup_{x \in U \cap E_+} \int_{K'\setminus\{0\}} p(x') \left(\frac{x'(x)^2}{p(x')}\right) d|\mu|(x') =$$

$$= \int_{K'\setminus\{0\}} p(x') d|\mu|(x') = \int_{K'} p d|\mu| .$$

Hence,

$$\int_{K'} (p - p^2) d|\mu| = 0$$

so

$$|\mu|(\{x' \in K' \mid 0 < p(x') < 1\}) = 0 .$$

By f), g), and E. M. Alfsen [1] Proposition I 4.6 $|\mu|(K' \setminus X') = 0$ so, by g), $|\mu|(K' \setminus \partial_e K') = 0$.

i) By g) and Proposition 5.7.10c) there exists $y \in E$ such that

$$|x'(y) - f(x')| < \varepsilon$$

for any $x' \in L'$. By b) and g) there exists $z \in U \cap E_+$ such that $x'(z) > 1 - \varepsilon$ for any $x' \in L'$. We set $x := (y \vee 0) \wedge z$. Then $x \in U \cap E_+$ and

$$|x'(x) - f(x')| < \varepsilon$$

for any $x' \in L'$. □

Theorem 5.7.15. *Let E be a locally convex lattice, U be a directed solid 0-neighbourhood for the Mackey topology of E and F' be the band of E' generated by U^0. Put*

$$K' := U^0 \cap E'_+, \quad X' := \partial_e K' \setminus \{0\},$$

denote by $\mathfrak{K}$ the set of compact sets of X', by $\mathfrak{R}$ the δ-ring generated by $\mathfrak{K}$, by $\mathcal{M}$ the set of measures of $\mathcal{M}(\mathfrak{R}, \mathbb{R}; \mathfrak{K})$ which possess a barycenter in E'_E, and, for any $\mu \in \mathcal{M}$, *denote by x'_μ its barycenter in E'_E. Then the following hold:*

a) $\mathscr{M}$ is a solid vector subspace of the vector lattice $\mathscr{M}(\mathfrak{R}, \mathbb{R}; \mathfrak{K})$ containing any bounded measure of $\mathscr{M}(\mathfrak{R}, \mathbb{R}; \mathfrak{K})$ and $x'_\mu \in F'$ for any $\mu \in \mathscr{M}$;

b) the map

$$\mathscr{M} \to F', \quad \mu \mapsto x'_\mu$$

is an isomorphism of vector lattices;

c) x'_μ belongs to the solid vector subspace of E' generated by U^0 iff μ is bounded; $x'_\mu \in U^0$ iff $\sup_{A \in \mathfrak{R}} |\mu|(A) \leq 1$.

By Proposition 5.6.3 $\mathscr{M}(\mathfrak{R}, \mathbb{R}; \mathfrak{K})$ is a complete vector lattice.

Take φ as the evaluation map $E \to E''$, G' as the solid vector subspace of E' generated by U^0, $\tilde{\mathscr{M}}$ as the set of measures of $\mathscr{M}(\mathfrak{R}, \mathbb{R}; \mathfrak{K})$ which possess a barycenter in E^+. For any $\mu \in \tilde{\mathscr{M}}$ let x'_μ be its barycenter in E^+. It is obvious that $\tilde{\mathscr{M}}$ and $\mathscr{M}$ are vector subspaces of $\mathscr{M}(\mathfrak{R}, \mathbb{R}; \mathfrak{K})$, that the map

$$\tilde{\mathscr{M}} \to E^+, \quad \mu \mapsto x'_\mu$$

is linear, and that $\mu \geq 0$ implies $x'_\mu \geq 0$. Let $\mu \in \mathscr{M}(\mathfrak{R}, \mathbb{R}; \mathfrak{K})$ and $\nu \in \tilde{\mathscr{M}}$ such that $|\mu| \leq |\nu|$. Then

$$\varphi(E) \subset \mathscr{L}^1(\nu, E') = \mathscr{L}^1(|\nu|, E') \subset \mathscr{L}^1(\mu, E')$$

and

$$-\int \varphi(x) d|\nu| \leq \int \varphi(x) d\mu \leq \int \varphi(x) d|\nu|$$

for any $x \in E_+$. Since the map

$$E \to \mathbb{R}, \quad x \mapsto \int \varphi(x) d|\nu|$$

is linear and positiv the map

$$E \to \mathbb{R}, \quad x \mapsto \int \varphi(x) d\mu$$

belongs to E^+. This is the barycenter of μ in E^+. Hence, $\mu \in \tilde{\mathscr{M}}$ and $\tilde{\mathscr{M}}$ is a solid vector subspace of $\mathscr{M}(\mathfrak{R}, \mathbb{R}; \mathfrak{K})$.

Let $(\mu_\iota)_{\iota \in I}$ be an upper directed family in $\mathscr{M}$ which possesses a supremum μ in $\tilde{\mathscr{M}}$. Then (K. Jacobs [1] Ch IX Theorem 2.3)

$$x'_\mu(x) = \int \varphi(x) d\mu = \sup_{\iota \in I} \int \varphi(x) d\mu_\iota = \sup_{\iota \in I} x'_{\mu_\iota}(x)$$

for any $x \in E_+$ so (Proposition 5.5.6) $x'_\mu = \bigvee_{\iota \in I} x'_{\mu_\iota}$.

We denote by $\mathscr{M}_b$ the set of bounded measures of $\mathscr{M}(\mathfrak{R}, \mathbb{R}; \mathfrak{K})$, by $\|\cdot\|$ the norm

$$\mathscr{M}_b \to \mathbb{R}, \quad \mu \mapsto \sup_{A \in \mathfrak{R}} |\mu|(A)$$

on $\mathscr{M}_b$, and by $\mathscr{N}$ the set

$$\mathscr{N} := \{\mu \in \mathscr{M}_{b+} \mid \|\mu\| \leq 1\}.$$

Let $x' \in K'$. By G. Choquet, P.A. Meyer [1] Théorème 3 there exists a positive boundary measure μ' on K' having x' as barycenter. By Theorem 5.7.14 h) $\mu'(K' \setminus \partial_e K') = 0$. We set $\mu := \mu' \mid \mathfrak{R}$. Then $\mu \in \mathscr{N}$ and $x'_\mu = x'$. Hence, the map

$$\mathscr{N} \to K', \quad \mu \to x'_\mu$$

is surjective. Let $\mu, \nu \in \mathscr{N}$ with $x'_\mu = x'_\nu$. Let $\mathfrak{S}$ be the set of Borel sets of K'. We set

$$\alpha := \sup_{A \in \mathfrak{R}} \mu(A), \quad \beta := \sup_{A \in \mathfrak{R}} \nu(A),$$

$$\mu' : \mathfrak{S} \to \mathbb{R}, \quad A \mapsto \sup_{A \supset K \in \mathfrak{K}} \mu(K) + (1 - \alpha) 1_A(0),$$

$$\nu' : \mathfrak{S} \to \mathbb{R}, \quad A \mapsto \sup_{A \supset K \in \mathfrak{K}} \nu(K) + (1 - \beta) 1_A(0).$$

Then μ', ν' are positive Radon measures on K' having the same barycenter. Since

$$\mu'(K' \setminus \partial_e K') = \nu'(K' \setminus \partial_e K') = 0,$$

μ' and ν' are boundary measures on K' (G. Choquet, P.A. Meyer [1] Proposition 15). Hence, by Theorem 5.7.14 e) $\mu' = \nu'$ so $\mu = \nu$. Hence, the map

$$\mathscr{N} \to K', \quad \mu \mapsto x'_\mu$$

is a bijection. Let $\mu, \nu \in \mathscr{N}$ with $\mu \wedge \nu = 0$. By the Hahn Theorem there exists a subset A' of X' such that

$$A' \cap B' \in \mathfrak{R}, \quad \mu(B' \setminus A') = 0, \quad \nu(A' \cap B') = 0$$

for any $B' \in \mathfrak{R}$. Let $x \in E_+$ and ε be a strictly positive real number. There exists $\alpha > 0$ such that $\alpha x \in U$. Let L', M' be compact sets of X' such that

$$L' \subset A', \quad \sup_{\substack{B' \in \mathfrak{R} \\ B' \subset A' \setminus L'}} \mu(B') < \frac{\varepsilon}{4},$$

$$M' \subset X' \setminus A', \quad \sup_{\substack{B' \in \mathfrak{R} \\ B' \subset (X' \setminus A') \setminus M'}} \nu(B') < \frac{\varepsilon}{4}.$$

By Theorem 5.7.14 i) there exist $y, z \in U \cap E_+$ such that

$$|\varphi(\alpha x) - \varphi(y)| < \frac{\varepsilon}{8}, \quad |\varphi(z)| < \frac{\varepsilon}{8}$$

on L' and

$$|\varphi(y)| < \frac{\varepsilon}{8}, \quad |\varphi(\alpha x) - \varphi(z)| < \frac{\varepsilon}{8}$$

on M'. We set

$$u := (y-z)^+ \wedge \alpha x, \quad v := (y-z)^- \wedge \alpha x.$$

Then $\varphi(u) \geq \varphi(\alpha x) - \frac{\varepsilon}{4}$ on L' so

$$x'_\mu(u) = \int \varphi(u) d\mu \geq \int_{L'} \varphi(u) d\mu \geq \int_{L'} (\varphi(\alpha x) - \frac{\varepsilon}{4})\, d\mu =$$

$$= \int \varphi(\alpha x) d\mu - \int_{X' \setminus L'} \varphi(\alpha x) d\mu - \frac{\varepsilon}{4}\, \mu(L') \geq$$

$$\geq x'_\mu(\alpha x) - \frac{\varepsilon}{4} - \frac{\varepsilon}{4} = x'_\mu(\alpha x) - \frac{\varepsilon}{2}.$$

Similarly,

$$x'_\nu(v) \geq x'_\nu(\alpha x) - \frac{\varepsilon}{2}.$$

We have $u \wedge v = 0$ so

$$u + v = u \vee v \leq \alpha x.$$

Hence,

$$x'_\mu \vee x'_\nu(\alpha x) \geq x'_\mu \vee x'_\nu(u+v) \geq x'_\mu(u) + x'_\nu(v) \geq$$

$$\geq x'_\mu(\alpha x) - \frac{\varepsilon}{2} + x'_\nu(\alpha x) - \frac{\varepsilon}{2} = (x'_\mu + x'_\nu)(\alpha x) - \varepsilon.$$

Since ε and x are arbitrary

$$x'_\mu \vee x'_\nu \geq x'_\mu + x'_\nu$$

so $x'_\mu \wedge x'_\nu = 0$. By the above considerations, $\mathscr{M}_b$ is the solid vector subspace of $\tilde{\mathscr{M}}$ generated by $\mathscr{N}$, $x'_\mu \in G'$ for any $\mu \in \mathscr{M}_b$, and $x'_\mu \wedge x'_\nu = 0$ for any $\mu, \nu \in \mathscr{M}_b$ with $\mu \wedge \nu = 0$.

We set

$$\mathscr{M}_\mu := \{\nu \in \mathscr{M}_{b+} \mid \nu \leq \mu\}$$

for any $\mu \in \tilde{\mathscr{M}}_+$. Then $\mu = \bigvee_{\nu \in \mathscr{M}_\mu} \nu$ so by a remark made above

$$x'_\mu = \bigvee_{\nu \in \mathscr{M}_\mu} x'_\nu$$

for any $\mu \in \tilde{\mathscr{M}}_+$. In particular, $x'_\mu \in F'$ for any $\mu \in \tilde{\mathscr{M}}_+ \cap \mathscr{M}$.

Let μ, ν be two elements of $\tilde{\mathscr{M}}_+$ such that $\mu \wedge \nu = 0$. Then $\mu' \wedge \nu' = 0$ for any $(\mu', \nu') \in \mathscr{M}_\mu \times \mathscr{M}_\nu$ so (W.A.J. Luxembourg, A.C. Zaanen [1] Theorem 12.2)

$$x'_\mu \wedge x'_{\nu'} = (\bigvee_{\mu' \in \mathscr{M}_\mu} x'_{\mu'}) \wedge x'_{\nu'} = \bigvee_{\mu' \in \mathscr{M}_\mu} (x'_{\mu'} \wedge x'_{\nu'}) = 0$$

for any $\nu' \in \mathcal{M}_\nu$ and

$$x'_\mu \wedge x'_\nu = x'_\mu \wedge \Big(\bigvee_{\nu' \in \mathcal{M}_\nu} x'_{\nu'} \Big) = \bigvee_{\nu' \in \mathcal{M}_\nu} (x'_\mu \wedge x'_{\nu'}) = 0 .$$

Let $\mu \in \mathcal{M}$. Then $\mu^+, \mu^- \in \tilde{\mathcal{M}}$ and $\mu^+ \wedge \mu^- = 0$. Hence, $x'_{\mu+} \wedge x'_{\mu-} = 0$. Since $x'_\mu = x'_{\mu+} - x'_{\mu-}$

$$x'^+_\mu = x'_{\mu+}, \quad x'^-_\mu = x'_{\mu-} .$$

In particular, $x'_{\mu+}, x'_{\mu-} \in E'$ so $\mu^+, \mu^- \in \mathcal{M}$. Hence, $\mathcal{M}$ is a solid vector subspace of $\mathcal{M}(\mathfrak{R}, \mathbb{R}; \mathfrak{K})$. Assume now that $x'_\mu \geq 0$. Then $x'_{\mu-} = 0$ and

$$\bigvee_{\nu \in \mathcal{M}_{\mu-}} x'_\nu = x'_{\mu-} = 0$$

so $x'_\nu = 0$ for any $\nu \in \mathcal{M}_{\mu-}$. Since the map

$$\mathcal{N} \to F', \quad \nu \mapsto x'_\nu$$

is injective $\nu = 0$ for any $\nu \in \mathcal{M}_{\mu-}$ so $\mu^- = 0$ and $\mu = \mu^+ \geq 0$. It follows immediately that the map

$$\mathcal{M} \to F', \quad \mu \mapsto x'_\mu$$

is injective.

It remains only to show that this map is surjective. Let $x' \in F'_+$. We set

$$C' := \{ y' \in G'_+ \mid y' \leq x' \}, \quad \mathcal{P} := \{ \mu \in \mathcal{M}_b \mid x'_\mu \in C' \}$$

By the above considerations for any $y' \in C'$ there exists $\mu \in \mathcal{P}$ such that $x'_\mu = y'$. In particular, $\mathcal{P}$ is upper directed. Let $L' \in \mathfrak{K}$. By Theorem 5.7.14 i) there exists $x \in E_+$ such that $\varphi(x) > 1$ on L'. We get

$$\sup_{\mu \in \mathcal{P}} \mu(L') \leq \sup_{\mu \in \mathcal{P}} \int \varphi(x) d\mu \leq$$

$$\leq \sup_{y' \in C'} y'(x) \leq x'(x) < \infty .$$

Hence, $\mathcal{P}$ is upper bounded in $\mathcal{M}(\mathfrak{R}, \mathbb{R}; \mathfrak{K})$ (K. Jacobs [1] Ch VIII Theorem 1.4). We set $\nu := \bigvee_{\mu \in \mathcal{P}} \mu$. By the above considerations $\nu \in \mathcal{M}$ and

$$x'_\nu = \bigvee_{\mu \in \mathcal{P}} x'_\mu = \bigvee_{y' \in C'} y' = x' .$$

Hence, the map

$$\mathcal{M} \to F', \quad \mu \mapsto x'_\mu$$

is surjective. □

Corollary 5.7.16. *Let E be a quasi-M-space, X' be the set of continuous linear forms on E which are homomorphisms of vector lattices, $\mathfrak{K}$ be the set of compact sets of $X'\backslash\{0\}$, $\mathfrak{R}$ be the δ-ring generated by $\mathfrak{K}$, $\mathfrak{U}$ be the set of directed solid 0-neighbourhoods for the Mackey topology of E and $\mathscr{M}$ be the set of bounded measures $\mu \in \mathscr{M}(\mathfrak{R}, \mathbb{R}; \mathfrak{K})$ for which there exists $U \in \mathfrak{U}$ such that the carrier of μ in $X'\backslash\{0\}$ lies in U^0. Then any $\mu \in \mathscr{M}$ possesses a barycenter x'_μ in E'_E and the map*

$$\mathscr{M} \to E', \quad \mu \mapsto x'_\mu$$

is surjective.

It is obvious that any $\mu \in \mathscr{M}$ possesses a barycenter in E'_E. Let $x' \in E'$. Since E is a quasi-M-space there exists $U \in \mathfrak{U}$ such that $x' \in U^0$. By Theorem 5.7.14 f), g) $U^0 \cap X'$ is compact and $\partial_e(U^0 \cap E'_+) \subset U^0 \cap X'$. By Theorem 5.7.15 b), c) x' is the barycenter of a bounded measure of $\mathscr{M}(\mathfrak{R}, \mathbb{R}; \mathfrak{K})$ whose carrier in $X'\backslash\{0\}$ lies in $U^0 \cap X'$. □

Corollary 5.7.17. *Let E be a quasi-M-space and X' be the set of continuous linear forms on E which are homomorphisms of vector lattices. Then any supersummable family in $E_{X'}$ is supersummable in E.*

Let $\mathfrak{K}$ be the set of compact sets of X', $\mathfrak{R}$ be the δ-ring generated by $\mathfrak{K}$ and $\mathscr{M}$ be the set of measures of $\mathscr{M}(\mathfrak{R}, \mathbb{R}; \mathfrak{K})$ for which

$$X' \to \mathbb{R}, \quad x' \mapsto x'(x)$$

is integrable for any $x \in E$. Further, let $(x_\iota)_{\iota \in I}$ be a supersummable family in $E_{X'}$ and $x' \in E'$. By Corollary 5.7.16 x' is the barycenter in E' of a $\mu \in \mathscr{M}$. By Theorem 4.10.7

$$x'(\sum_{\iota \in J} x_\iota) = \int y'(\sum_{\iota \in J} x_\iota) d\mu(y') =$$

$$= \sum_{\iota \in J} \int y'(x_\iota) d\mu(y') = \sum_{\iota \in J} x'(x_\iota)$$

for any $J \subset I$. Hence, $(x_\iota)_{\iota \in I}$ is weakly supersummable and by the Orlicz-Pettis theorem (Corollary 3.5.12) it is supersummable in E. □

Proposition 5.7.18. *Let E be a locally convex lattice, A be an upper directed set of E, $\mathfrak{F}$ be the filter on E generated by the section filter of A and $\varphi : E \to E''$ be the evaluation map. Then the following assertions are equivalent:*

a) $\varphi(A)$ is upper bounded;

b) $\mathfrak{F}$ is a weak Cauchy filter.

a ⇒ b. Let x' be a positive element of E'. From

$$\sup_{x \in A} x'(x) = \sup_{x'' \in \varphi(A)} x''(x') < \infty,$$

it follows that $\mathfrak{F}$ is a weak Cauchy filter.

b ⇒ a. Let x'' be the linear form

$$E' \to \mathbb{R}, \quad x' \mapsto \lim x'(\mathfrak{F}),$$

and $x_0 \in A$. Then $B := \{x \in A \mid x_0 \le x\}$ is a bounded set of E and

$$x' \in B^0 \Rightarrow |x''(x')| \le 1.$$

Hence, $x'' \in E''$. It is obvious that x'' is an upper bound of $\varphi(A)$. □

Proposition 5.7.19. *Let E be a locally convex lattice and $\varphi : E \to E''$ be the evaluation map. Then the following assertions are equivalent:*

a) any increasing weak Cauchy sequence in E_+ possesses a supremum;

b) E is order σ-complete and any increasing weak Cauchy sequence in E is upper bounded;

c) any increasing sequence $(x_n)_{n \in \mathbb{N}}$ in E_+ possesses a supremum if $(\varphi(x_n))_{n \in \mathbb{N}}$ is upper bounded;

d) E is order σ-complete and any increasing sequence $(x_n)_{n \in \mathbb{N}}$ in E is upper bounded if $(\varphi(x_n))_{n \in \mathbb{N}}$ is upper bounded.

a ⇔ c and b ⇔ d follow from Proposition 5.7.18.

c ⇒ d. Let $(x_n)_{n \in \mathbb{N}}$ be an increasing sequence in E for which $(\varphi(x_n))_{n \in \mathbb{N}}$ is upper bounded. Then $(x_n - x_0)_{n \in \mathbb{N}}$ is an increasing sequence in E_+ for which $(\varphi(x_n - x_0))_{n \in \mathbb{N}}$ is upper bounded. By c) $(x_n - x_0)_{n \in \mathbb{N}}$ possesses a supremum. Hence, $(x_n)_{n \in \mathbb{N}}$ possesses a supremum and d) follows.

d ⇒ c is trivial. □

Proposition 5.7.20. *Let E be an order σ-complete locally convex lattice, φ be the evaluation map $E \to E''$ and X'' be the set*

$$\{\bigvee_{n \in \mathbb{N}} \varphi(x_n) \mid (x_n)_{n \in \mathbb{N}} \text{ upper bounded sequence in } E\}.$$

Then the identity map $E'_E \to E'_{X''}$ is uniformly Φ_3-continuous.

Let $x'' \in X''$. There exists an upper bounded sequence $(x_n)_{n \in \mathbb{N}}$ in E such that $x'' = \bigvee_{n \in \mathbb{N}} \varphi(x_n)$. By Proposition 5.5.6 we may assume $(x_n)_{n \in \mathbb{N}}$ increasing. By Proposition 5.1.16 $(x_{n+1} - x_n)_{n \in \mathbb{N}}$ is order supersummable so $(x_n)_{n \in \mathbb{N}} \in \Theta_e(E)$. We set $\Phi := \hat{\Phi}_e(E)$. By the above considerations and by Propositions 5.5.6, 2.1.5, the identity map $E'_\Phi \to E'_{X''}$ is uniformly continuous. By Proposition 5.5.2 b) and Corollary 1.8.15, the identity map $E'_E \to E'_\Phi$ is uniformly Φ_3-continuous. Hence, the identity map $E'_E \to E'_{X''}$ is uniformly Φ_3-continuous. □

Proposition 5.7.21. *Let E be a locally convex lattice for which any bounded set is contained in an upper directed bounded set (this occurs e.g. if E is a quasi-M-space) and let φ be the evaluation map $E \to E''$. Then:*

a) for any $x'' \in E''$ *there exists an upper directed bounded set* A *of* E *such that* $\varphi(A)$ *is upper bounded and* $|x''| \leq \bigvee_{x \in A} \varphi(x)$;

b) E'' *is the band of* E'' *generated by* $\varphi(E)$;

c) $E'' \subset E'^{\pi}$.

By Proposition 5.7.4 a) any bounded set of a quasi-M-space is contained in an upper directed bounded set of E.

a) There exists a bounded set B of E such that $x'' \in B^{00}$. We may assume B solid. Let A be an upper directed solid bounded set of E containing B. We set

$$l : E'_+ \to \mathbb{R}, \quad x' \mapsto \sup_{x \in A} x'(x).$$

Then l is additive and positive so (Ch. D. Aliprantis, O. Burkinshaw [1] Lemma 3.1) there exists a positive linear form y'' on E' such that $l = y'' | E'_+$. We have $|y''(x')| \leq 1$ for any $x' \in A^0$ and this implies $y'' \in E''$. Moreover, $x' \in y''(x')B^0$ so $|x''|(x') \leq y''(x')$ for any $x' \in E'_+$. Hence,

$$|x''| \leq y'' = \bigvee_{x \in A} \varphi(x).$$

b) follows immediately from a).

c) follows immediately from b) and Proposition 5.5.6. □

Proposition 5.7.22. *Let* E *be a locally convex lattice,* F' *be a solid vector subspace of* E', X' *be the set of elements of* $E' \setminus \{0\}$ *which are homomorphisms of vector lattices,* Y' *be a subset of* X' *such that any two different points of* Y' *lie on different rays of* E', $\mathfrak{K}$ *be the set of compact sets of* Y', $\mathfrak{R}$ *be the* δ*-ring generated by* $\mathfrak{K}$, $\mathcal{M}$ *be the set of measures of* $\mathcal{M}(\mathfrak{R}, \mathbb{R}; \mathfrak{K})$ *which possess barycenters in* E'_E *which lie in* F' *and, for any* $\mu \in \mathcal{M}$, *let* x'_μ *be its barycenter in* E'_E. *Then* $\mathcal{M}$ *is a solid vector subspace of* $\mathcal{M}(\mathfrak{R}, \mathbb{R}; \mathfrak{K})$ *and the map*

$$\mathcal{M} \to F', \quad \mu \mapsto x'_\mu$$

is an injective homomorphism of vector lattices.

It is obvious that $\mathcal{M}$ is a vector subspace of $\mathcal{M}(\mathfrak{R}, \mathbb{R}; \mathfrak{K})$ and that the map

$$\mathcal{M} \to F', \quad \mu \mapsto x'_\mu$$

is linear.

Let $\mu, \nu \in \mathcal{M}(\mathfrak{R}, \mathbb{R}; \mathfrak{K})$ with $\mu \wedge \nu = 0$ and

$$\{x | Y' \mid x \in E\} \subset \mathcal{L}^1(\mu, Y') \cap \mathcal{L}^1(\nu, Y').$$

We set

$$y' : E \to \mathbb{R}, \quad x \mapsto \int_{Y'} x'(x) d\mu(x'),$$

$$z' : E \to \mathbb{R}, \quad x \mapsto \int_{Y'} x'(x) d\nu(x').$$

Then $y', z' \in E^+$. We want to show $y' \wedge z' = 0$. Let $x \in E_+$. Then there exists a σ-compact set A' of Y' such that $x'(x) = 0$ for $\mu + \nu$-almost all $x' \in Y' \setminus A'$. By Hahn's theorem there exists an $\mathfrak{R}$-measurable set B' of Y' such that $B' \subset A'$ and

$$\mu(C' \setminus B') = 0, \quad \nu(C' \cap B') = 0$$

for any $C' \in \mathfrak{R}$, $C' \subset A'$. Let ε be a strictly positive real number. There exist $K', L' \in \mathfrak{K}$ such that $K' \subset B'$, $L' \subset A' \setminus B'$ and

$$\int_{Y' \setminus K'} x'(x) d\mu(x') < \frac{\varepsilon}{2}, \quad \int_{Y' \setminus L'} x'(x) d\nu(x') < \frac{\varepsilon}{2}.$$

Let ε' be another strictly positive real number. By Proposition 5.7.10 c) there exists $y \in E$ such that

$$x' \in K' \Rightarrow |x'(y) - x'(x) - \varepsilon'| < \varepsilon',$$

$$x' \in L' \Rightarrow |x'(y) + \varepsilon'| < \varepsilon'.$$

We set $x_1 := x \wedge y^+$, $x_2 := x - x_1$. Then $x_1, x_2 \in E_+$, $x_1 + x_2 = x$, $x_1 | L' = 0$ and $x_2 | K' = 0$. We get

$$y'(x_2) = \int_{Y'} x'(x_2) d\mu(x') \le \int_{Y' \setminus K'} x'(x) d\mu(x') < \frac{\varepsilon}{2},$$

$$z'(x_1) = \int_{Y'} x'(x_1) d\nu(x') \le \int_{Y' \setminus L'} x'(x) d\nu(x') < \frac{\varepsilon}{2},$$

$$y'(x_2) + z'(x_1) < \varepsilon.$$

By Ch. D. Aliprantis, O. Burkinshaw [1] Theorem 3.3 $y' \wedge z' = 0$.

Now let $\mu \in \mathscr{M}$. Then

$$\{x | Y' \mid x \in E\} \subset \mathscr{L}^1(\mu, Y') \subset \mathscr{L}^1(\mu^+, Y') \cap \mathscr{L}^1(\mu^-, Y').$$

We set

$$y' : E \to \mathbb{R}, \quad x \mapsto \int_{Y'} x'(x) d\mu^+(x'),$$

$$z' : E \to \mathbb{R}, \quad x \mapsto \int_{Y'} x'(x) d\mu^-(x').$$

By the above considerations $y', z' \in E^+$ and $y' \wedge z' = 0$. We obviously have $x'_\mu = y' - z'$ so

$$y' = (x'_\mu)^+, \quad z' = (x'_\mu)^-.$$

Then $y', z' \in F'$ and so $\mu^+, \mu^- \in \mathscr{M}$.

Now let $\mu \in \mathscr{M}(\mathfrak{R}, \mathbb{R}; \mathfrak{K})$, $\nu \in \mathscr{M}$ with $|\mu| \le |\nu|$. By the above considerations $|\nu| \in \mathscr{M}$. Then $\mu^+ \le |\nu|$, $\mu^- \le |\nu|$, so

$$\{x | Y' \mid x \in E\} \subset \mathscr{L}^1(\mu^+, Y') \cap \mathscr{L}^1(\mu^-, Y')$$

and

$$0 \leq \int_{Y'} x'(x)d\mu^+(x') \leq x'_{|\nu|}(x), \quad 0 \leq \int_{Y'} x'(x)d\mu^-(x') \leq x'_{|\nu|}(x)$$

for any $x \in E_+$. Hence, $\mu^+, \mu^- \in \mathcal{M}$ and $\mu \in \mathcal{M}$. Hence, $\mathcal{M}$ is a solid set of $\mathcal{M}(\mathfrak{R}, \mathbb{R}; \mathfrak{K})$.
Let $\mu, \nu \in \mathcal{M}$. We have

$$(\mu - \mu \wedge \nu) \wedge (\nu - \mu \wedge \nu) = 0\,.$$

By the above considerations $\mu - \mu \wedge \nu \in \mathcal{M}$, $\nu - \mu \wedge \nu \in \mathcal{M}$ and

$$x'_{\mu - \mu \wedge \nu} \wedge x'_{\nu - \mu \wedge \nu} = 0\,.$$

We get

$$x'_\mu \wedge x'_\nu = (x'_{\mu - \mu \wedge \nu} + x'_{\mu \wedge \nu}) \wedge (x'_{\nu - \mu \wedge \nu} + x'_{\mu \wedge \nu}) = x'_{\mu \wedge \nu}$$

so the map

$$\mathcal{M} \to F', \quad \lambda \mapsto x'_\lambda$$

is a homomorphism of vector lattices.

Let $\mu \in \mathcal{M}$ such that $x'_\mu = 0$. By the above proof $x'_{|\mu|} = |x'_\mu| = 0$. Let $K \in \mathfrak{K}$. By Proposition 5.7.10 c) there exists $x \in E$ such that $x'(x) \geq 1$ for any $x' \in K$. We get

$$|\mu|(K) \leq \int_{Y'} x'(x^+)d|\mu|(x') = x'_{|\mu|}(x^+) = 0\,.$$

Since K is arbitrary $|\mu| = 0$ so $\mu = 0$. Hence, the map

$$\mathcal{M} \to F', \quad \mu \mapsto x'_\mu$$

is injective. □

Proposition 5.7.23. *Let E be a locally convex lattice, U be a directed solid 0-neighbourhood for the Mackey topology of E, F' be the band of E' generated by U^0 and x' be an element of F' which is a homomorphism of vector lattices. Then there exists $n \in \mathbb{N}$ such that $x' \in nU^0$.*

We set $G' := \bigcup_{n \in \mathbb{N}} (nU^0)$. G' is the solid vector subspace of E' generated by U^0. Let $y' \in G'$, $y' \leq x'$. By Ch. D. Aliprantis, O. Burkinshaw [1] Theorem 3.13 x' lies on an extremal ray of E'_+ so there exists $\alpha \in \mathbb{R}_+$ such that $y' = \alpha x'$. We have

$$x' = \bigvee_{\substack{y' \in G'_+ \\ y' \leq x'}} y'$$

so $x' \in G'$. Hence, there exists $n \in \mathbb{N}$ such that $x' \in nU^0$. □

Theorem 5.7.24. *Let E be a locally convex lattice, X' be the set of elements of $E' \backslash \{0\}$ which are homomorphisms of vector lattices, $\mathfrak{U}$ be a countable set of directed solid* 0-

neighbourhoods for the Mackey topology of E and F′ be the band of E′ generated by $\bigcup_{U\in\mathfrak{U}} U^0$. *Then there exists a subset Y′ of $X'\cap F'$ such that, if we denote by $\mathfrak{K}$ the set of compact sets of Y′, by $\mathfrak{R}$ the δ-ring generated by $\mathfrak{K}$, by $\mathcal{M}$ the set of measures of $\mathcal{M}(\mathfrak{R},\mathbb{R};\mathfrak{K})$ which possess barycenters in E'_E and for any $\mu\in\mathcal{M}$ by x'_μ its barycenter in E'_E, then $\mathcal{M}$ is a solid vector subspace of $\mathcal{M}(\mathfrak{R},\mathbb{R};\mathfrak{K})$, $x'_\mu\in F'$ for any $\mu\in\mathcal{M}$, and the map*

$$\mathcal{M}\to F',\quad \mu\mapsto x'_\mu$$

is an isomorphism of vector lattices.

Let $(U_n)_{n\in\mathbb{N}}$ be a decreasing sequence of directed solid 0-neighbourhoods for the Mackey topology of E such that F' is the band generated by $\bigcup_{n\in\mathbb{N}} U_n^0$. For any $n\in\mathbb{N}$ we denote by F'_n the band of E' generated by U_n^0 and set

$$K'_n:=U_n^0\cap E'_+,\quad X'_n:=\partial_e K'_n\setminus\{0\},\quad Y'_n:=X'_n\setminus F'_{n-1},\quad (F'_{-1}:=\emptyset),$$
$$Y':=\bigcup_{n\in\mathbb{N}} Y'_n.$$

By Theorem 5.7.14 g) $X'_n\subset X'$ for any $n\in\mathbb{N}$ so $Y'\subset X'\cap F'$. Let $n\in\mathbb{N}$. By Theorem 5.7.14 g) X'_n is a Borel set of E' and by Proposition 5.7.23

$$Y'_n=X'_n\setminus\bigcup_{m\in\mathbb{N}} mU^0_{n-1},\quad (U^0_{-1}:=\emptyset).$$

Hence, Y'_n is a Borel set of E' and so it is $\mathfrak{R}$-measurable.

By Proposition 5.7.22 $\mathcal{M}$ is a solid vector subspace of $\mathcal{M}(\mathfrak{R},\mathbb{R};\mathfrak{K})$ and the map

$$\mathcal{M}\to E',\quad \mu\mapsto x'_\mu$$

is an injective homomorphism of vector lattices. Let $\mu\in\mathcal{M}_+$. We set

$$\mu_n:\mathfrak{R}\to\mathbb{R},\quad A'\mapsto\mu(A'\cap Y'_n)$$

for any $n\in\mathbb{N}$. Then $\mu_n\in\mathcal{M}_+$ and, by Theorem 5.7.15 a), $x'_{\mu_n}\in F'_n\subset F'$ for any $n\in\mathbb{N}$. We have

$$x'_\mu(x)=\int_{Y'} x'(x)d\mu(x')=\sup_{n\in\mathbb{N}}\sum_{m=0}^{n}\int_{Y'_m} x'(x)d\mu(x')=$$
$$=\sup_{n\in\mathbb{N}}\sum_{m=0}^{n}\int_{Y'} x'(x)d\mu_m(x')=\sup_{n\in\mathbb{N}}\sum_{m=0}^{n} x'_{\mu_m}(x)$$

for any $x\in E_+$ and therefore (Ch. D. Aliprantis, O. Burkinshaw [1] Theorem 3.3)

$$x'_\mu=\bigvee_{n\in\mathbb{N}}\sum_{m=0}^{n} x'_{\mu_m}\in F'.$$

We deduce that $x'_\mu\in F'$ for any $\mu\in\mathcal{M}$.

It only remains to show that the map

$$\mathcal{M}\to F',\quad \mu\mapsto x'_\mu$$

is surjective. Let $x' \in F'_+$. For any $n \in \mathbb{N}$ we denote by x'_n the component of x' on the band F'_n, by $\mathfrak{K}_n$ the set of compact sets of X'_n and by $\mathfrak{R}_n$ the δ-ring generated by $\mathfrak{K}_n$. Let $n \in \mathbb{N}$. By Theorem 5.7.15 b) there exists a $\mu_n \in \mathscr{M}(\mathfrak{R}_n, \mathbb{R}; \mathfrak{K}_n)_+$ such that $x'_n - x'_{n-1}$ ($x'_{-1} := 0$) is the barycenter of μ_n in E'_E. Let $m \in \mathbb{N}$. We set

$$\nu_m : \mathfrak{R}_n \to \mathbb{R}, \quad A \mapsto \mu_n(A \cap m U^0_{n-1}).$$

Then $\nu_m \in \mathscr{M}(\mathfrak{R}_n, \mathbb{R}; \mathfrak{K}_n)$ and its barycenter y'_m in E'_E lies in F'_{n-1} and $(x'_n - x'_{n-1}) \wedge y'_m = 0$. By Proposition 5.7.22

$$\nu_m = \mu_n \wedge \nu_m = 0.$$

Hence,

$$\mu_n(A \cap m U^0_{n-1}) = 0$$

for any $A \in \mathfrak{R}_n$. By Proposition 5.7.23 we get

$$\mu_n(A \cap F'_{n-1}) = \sup_{m \in \mathbb{N}} \mu_n(A \cap m U^0_{n-1}) = 0$$

for any $A \in \mathfrak{R}_n$.

Let $K' \in \mathfrak{K}$. We denote for any $n \in \mathbb{N}$ and for any $A' \in \mathfrak{R}$ by $f_n^{A'}$ the characteristic function of $A' \cap Y'_n$ on Y'_n. By Proposition 5.7.10c) there exists $x \in E_+$ such that $x'(x) \geq 1$ for any $x' \in K'$. We get $f_n^{K'} \in \mathscr{L}^1(\mu_n)$ for any $n \in \mathbb{N}$ and

$$\sum_{n \in \mathbb{N}} \int f_n^{K'} d\mu_n \leq \sum_{n \in \mathbb{N}} \int_{Y'_n} y'(x) d\mu_n(y') =$$

$$= \sum_{n \in \mathbb{N}} (x'_n - x'_{n-1})(x) \leq x'(x) < \infty .$$

Hence the map

$$\mu : \mathfrak{R} \to \mathbb{R}, \quad A' \mapsto \sum_{n \in \mathbb{N}} \int f_n^{A'} d\mu_n$$

is well defined. It is easy to see that $\mu \in \mathscr{M}(\mathfrak{R}, \mathbb{R}; \mathfrak{K})$.

Let $x \in E_+$. Then (Ch. D. Aliprantis, O. Burkinshaw [1] Theorem 3.3)

$$x'(x) = \sup_{n \in \mathbb{N}} x'_n(x) = \sup_{n \in \mathbb{N}} \sum_{m=0}^{n} (x'_m - x'_{m-1})(x) =$$

$$= \sum_{n \in \mathbb{N}} (x'_n - x'_{n-1})(x) = \sum_{n \in \mathbb{N}} \int_{Y'_n} y'(x) d\mu_n(y').$$

Hence

$$Y' \to \mathbb{R}, \quad y' \mapsto y'(x)$$

is μ-integrable and

$$\int y'(x) d\mu(y') = \sum_{n \in \mathbb{N}} \int_{Y'_n} y'(x) d\mu_n(y') = x'(x).$$

We deduce $\mu \in \mathcal{M}$ and $x'_\mu = x'$. Hence the map

$$\mathcal{M} \longrightarrow F', \quad \mu \longmapsto x'_\mu$$

is surjective. □

Theorem 5.7.25. *Let E be a quasi-M-space, $\mathfrak{U}$ be a countable set of directed solid 0-neighbourhoods for the Mackey topology of E, F' be the band of E' generated by $\bigcup_{U \in \mathfrak{U}} U^0$, φ be the evaluation map $E \to E''$, F'' be a solid vector subspace of E'' containing $\varphi(E)$ such that for any $x'' \in F''_+$ there exists a bounded set A of E with $x'' \leq \bigvee_{x \in A} \varphi(x) \in F''$, and let X'' be the set*

$$\{x'' \in F'' \mid \exists (x_n)_{n \in \mathbb{N}} \text{ an increasing sequence in } E_+, x'' = \bigvee_{n \in \mathbb{N}} \varphi(x_n)\}.$$

Then the identity map $F'_{X''} \to F'_{F''}$ is uniformly Φ_4-continuous and $U^0_{F''}$ is Φ_4-compact for any $U \in \mathfrak{U}$. If in addition E is order σ-complete and if X'' is contained in the solid vector subspace of E'' generated by $\varphi(E)$, then the identity map $F'_E \to F'_{F''}$ is uniformly Φ_3-continuous and the Cauchy sequences of E'_E and of $E'_{E''}$ coincide.

Let X' be the set of elements of $E' \setminus \{0\}$ which are homomorphisms of vector lattices. By Theorem 5.7.24 there exists a subset Y' of $X' \cap F'$ such that, if $\mathfrak{K}$ is the set of compact sets of Y', $\mathfrak{R}$ is the δ-ring generated by $\mathfrak{K}$, $\mathcal{M}$ is the set of measures of $\mathcal{M}(\mathfrak{R}, \mathbb{R}; \mathfrak{K})$ which possess barycenters in E'_E, and for any $\mu \in \mathcal{M}$ x'_μ is its barycenter in E'_E, then $\mathcal{M}$ is a solid vector subspace of $\mathcal{M}(\mathfrak{R}, \mathbb{R}; \mathfrak{K})$, $x'_\mu \in F'$ for any $\mu \in \mathcal{M}$, and the map

$$\psi : \mathcal{M} \longrightarrow F', \quad \mu \longmapsto x'_\mu$$

is an isomorphism of vector lattices.

By Proposition 5.7.21 c) $E'' \subset E'^\pi$ so $(x'' \mid F') \circ \psi \in \mathcal{M}^\pi$ for any $x'' \in E''$. We denote by ψ' the map

$$E'' \longrightarrow \mathcal{M}^\pi, \quad x'' \longmapsto (x'' \mid F') \circ \psi .$$

We want to prove that the identity map $\mathcal{M}_{\psi'(X'')} \longrightarrow \mathcal{M}_{\psi'(F'')}$ is uniformly Φ_4-continuous and that, if E is barreled, then $\overset{-1}{\psi}(U^0)_{\psi'(F'')}$ is Φ_4-compact for any $U \in \mathfrak{U}$.

Let $\mathscr{C}$ be the set of continuous real functions on Y', $\mathscr{F}$ be the set of functions on Y' of the form

$$Y' \longrightarrow \mathbb{R}, \quad x' \longmapsto x'(x),$$

where x runs through E and $\mathscr{K}$ be the set of positive real functions k on Y' for which there exist a bounded set A of E and an $x'' \in F''$ such that

$$k(x') = \sup_{x \in A} x'(x) \leq x''(x')$$

for any $x' \in Y'$. $\mathscr{F}$ is a dense vector sublattice of $\mathscr{C}_{\mathfrak{K}}$ (Theorem 5.7.10 c) and, for any $x'' \in X''$, there exists an increasing sequence $(f_n)_{n \in \mathbb{N}}$ in $\mathscr{F}$ and a $k \in \mathscr{K}$ such that

$$x''(x') = \sup_{n \in \mathbb{N}} f_n(x') \leq k(x')$$

for any $x' \in Y'$. We want to prove $\mathscr{K} \subset \mathscr{L}^1(\mathscr{M}, Y')$. Let $(k, \mu) \in \mathscr{K} \times \mathscr{M}_+$. There exists a bounded set A of E such that

$$k(x') = \sup_{x \in A} x'(x)$$

for any $x' \in Y'$. By Proposition 5.7.4 a) there exists a bounded upper directed set B of E such that $A \subset B$. We have

$$\int^* k \, d\mu \leq \sup_{x \in B} \int x'(x) \, d\mu(x') = \sup_{x \in B} x'_\mu(x) < \infty .$$

Hence, $k \in \mathscr{L}^1(\mu, Y')$ and $\mathscr{K} \subset \mathscr{L}^1(\mathscr{M}, Y')$.

Let $x'' \in F''_+$. By the hypothesis there exists a bounded set A of E such that $x'' \leq \bigvee_{x \in A} \varphi(x) \in F''$. We set

$$k : Y' \to \mathbb{R}, \quad x' \mapsto \sup_{x \in A} x'(x) .$$

Then $k \in \mathscr{K}$. We set

$$\dot{k} : \mathscr{M} \to \mathbb{R}, \quad \mu \mapsto \int k \, d\mu .$$

Then $\dot{k} \in \mathscr{M}^\pi$. Let $\mu \in \mathscr{M}_+$ and

$$\alpha := \sup_{x \in A} x'_\mu(x) < \infty .$$

Then $x'_\mu \in \alpha A^0$ and so

$$(\psi'(x''))(\mu) = x''(x'_\mu) \leq \alpha = \sup_{x \in A} x'_\mu(x) = \sup_{x \in A} \int x'(x) \, d\mu(x') \leq$$
$$\leq \int k \, d\mu = \dot{k}(\mu) .$$

Since μ is arbitrary $\psi'(x'') \leq \dot{k}$. Hence, $\psi'(F'')$ is contained in the solid vector subspace of $\mathscr{M}^\pi$ generated by $\{\dot{k} \mid k \in \mathscr{K}\}$.

By the above considerations and by Theorem 5.6.5 a), the identity map $\mathscr{M}_{\psi'(X'')} \to \mathscr{M}_{\psi'(F'')}$ is uniformly Φ_4-continuous.

Since the maps

$$\mathscr{M}_{\psi'(X'')} \to F'_{X''}, \quad \mu \mapsto x'_\mu ,$$
$$\mathscr{M}_{\psi'(F'')} \to F'_{F''}, \quad \mu \mapsto x'_\mu$$

are isomorphisms of uniform spaces the identity map $F'_{X''} \to F'_{F''}$ is uniformly Φ_4-continuous.

Let $U \in \mathfrak{U}$. We want to show $U^0_{F''}$ is Φ_4-compact. For this part of the proof we may assume $\mathfrak{U} := \{U\}$. We set $\mathcal{N} := \overset{-1}{\psi}(U_0)$,

$$\bar{\mathscr{F}} := \{x''|Y \ |x'' \in X''\},$$

$$\mathscr{P} := \{\mu \in \mathscr{M}(\mathfrak{R}, \mathbb{R}, \mathfrak{K}) \ |\bar{\mathscr{F}} \subset \mathscr{L}^1(\mu)\}.$$

Then $\mathcal{N}_{\mathscr{F}}$ is homeomorph to U^0_E and so it is compact. We deduce $\mathcal{N}$ is a closed set of $\mathscr{P}_{\mathscr{F}}$ and a fortiori a closed set of $\mathscr{P}_{\bar{\mathscr{F}}}$. By the Corollary 1.5.11 $\mathbb{R}$ is Φ_4-compact and any $x'' \in E''$ is bounded on U^0, so, by Theorem 5.6.5 b), $\mathcal{N}_{\psi'(F'')}$ is Φ_4-compact. We deduce that $U^0_{F''}$ is Φ_4-compact.

Now, in addition, assume that E is order σ-complete and that X'' is contained in the solid vector subspace of E'' generated by $\varphi(E)$. Then, by Proposition 5.7.20 and Corollary 1.8.15, the identity map $F'_E \to F'_{X''}$ is uniformly Φ_3-continuous. By the above result and by Corollary 1.8.5 the identity map $F'_E \to F'_{F''}$ is uniformly Φ_3-continuous.

Let $(x'_n)_{n \in \mathbb{N}}$ be a Cauchy sequence in E'_E. There exists a sequence $(U_n)_{n \in \mathbb{N}}$ of directed solid 0-neighbourhoods for the Mackey topology of E such that $x'_n \in U^0_n$ for any $n \in \mathbb{N}$. Let G' be the band of E' generated by $\bigcup_{n \in \mathbb{N}} U^0_n$. By the above considerations the identity map $G'_E \to G'_{F''}$ is uniformly Φ_3-continuous. It is obvious that $(x'_n)_{n \in \mathbb{N}}$ is a Cauchy sequence in G'_E so, by Theorem 1.8.4 a $\Rightarrow$ e, it is a Cauchy sequence in $G'_{F''}$. In particular, $(x'_n)_{n \in \mathbb{N}}$ is a Cauchy sequence in $E'_{F''}$. □

Remarks. 1) Let c_0 be the Banach lattice of sequences of real numbers converging to 0 and U be the unit ball of c_0. Then c_0 is an order complete quasi-M-space but the identity map $c'_0 \to (c'_0)_{c''_0}$ is not sequentially continuous and therefore a fortiori not uniformly Φ_3-continuous. This example shows that, in the preceding proposition, we cannot drop the hypothesis that X'' is contained in the solid vector subspace of E'' generated by $\varphi(E)$.

2) Let E be the vector lattice of continuous real functions on $\mathbb{R}$ which vanish at the infinity endowed with the norm

$$E \to \mathbb{R}_+, \quad f \mapsto \sup_{x \in \mathbb{R}} |f(x)|$$

and let F'' be the solid vector subspace of E'' generated by the bounded Borel functions f on $\mathbb{R}$ such that the interior of the set

$$\{x \in \mathbb{R} \,|\, |f(x)| \geq \alpha\}$$

is relatively compact for any strictly positive real number α. Then the identity map $E'_{X''} \to E'_{F''}$ is not even sequentially continuous. This example shows that we cannot drop the supplementary hypothesis on F''.

§ 5.8 M-spaces

Throughout this section we shall write locally convex space instead of Hausdorff locally convex space.

Some results of this and of the next section were announced by the author in (1983) [8].

Definition 5.8.1. *An* M-space *is a locally convex lattice possessing a fundamental system of solid directed* 0*-neighbourhoods.*

Any M-space is a quasi-M-space. Any vector sublattice of an M-space is an M-space with respect to the induced topology. The product of any family of M-spaces is an M-space. The M-space was defined by S. Kakutani (1941) ([1] page 994) in the case of a Banach lattice.

Proposition 5.8.2. *Let E be a metrizable M-space, φ be the evaluation map $E \longrightarrow E''$, F'' be a solid vector subspace of E'' containing $\varphi(E)$ such that for any $x'' \in F''_+$ there exists a bounded set A of E with $x'' \leq \bigvee_{x \in A} \varphi(x) \in F''$, and let X'' be the set*

$$\{x'' \in F'' \mid \exists (x_n)_{n \in \mathbb{N}} \text{ an increasing sequence in } E_+,\, x'' = \bigvee_{n \in \mathbb{N}} \varphi(x_n)\}.$$

Then the identity map $E'_{X''} \longrightarrow E'_{F''}$ is uniformly Φ_4-continuous. If, in addition, E is order σ-complete and if X'' is contained in the solid vector subspace of E'' generated by $\varphi(E)$, then the identity map $E'_F \longrightarrow E'_{F''}$ is uniformly Φ_3-continuous.

Let $(U_n)_{n \in \mathbb{N}}$ be a sequence of directed solid 0-neighbourhoods in E forming a fundamental system of 0-neighbourhoods in E. Then $E' = \bigcup_{n \in \mathbb{N}} U_n^0$. The assertion follows from Theorem 5.7.25 now. □

Proposition 5.8.3. *Let E be a Fréchet M-space, φ be the evalutation map $E \longrightarrow E''$, F'' be a solid vector subspace of E'' containing $\varphi(E)$ such that for any $x'' \in F''_+$ there exists a bounded set A of E with $x'' \leq \bigvee_{x \in A} \varphi(x) \in F''$, let X'' be the set*

$$\{x'' \in F'' \mid \exists (x_n)_{n \in \mathbb{N}} \text{ an increasing sequence in } E_+,\, x'' = \bigvee_{n \in \mathbb{N}} \varphi(x_n)\},$$

and $\tilde{E}'$ be the set E' endowed with a uniformity such that the identity maps $E'_{F''} \longrightarrow \tilde{E}' \longrightarrow E'_{X''}$ are uniformly Φ_4-continuous. Then $\tilde{E}'$ is Φ_4-compact.

Let $\mathfrak{F}$ be a Φ_4-ultrafilter on $E'_{F''}$. By Proposition 2.1.30 b) there exists a directed solid 0-neighbourhood U in E such that $U^0 \in \mathfrak{F}$. Let $\mathfrak{G}$ be the ultrafilter induced by $\mathfrak{F}$ on U^0. By Theorem 5.7.25 $U^0_{F''}$ is Φ_4-compact so $\mathfrak{G}$ is a Φ_4-ultrafilter on $U^0_{F''}$ and therefore converges in $U^0_{F''}$. We deduce that $\mathfrak{F}$ converges in $E'_{F''}$ and $E'_{F''}$ is Φ_4-compact.

By Proposition 5.8.2 the identity map $E'_{X''} \to E'_{F''}$ is uniformly Φ_4-continuous so, by Corollary 1.8.5, the identity map $\tilde{E}' \to E'_{F''}$ is also uniformly Φ_4-continuous. By Proposition 1.8.11 $\tilde{E}'$ is Φ_4-compact. □

Remark. In the above proposition we have F'' at our disposal. We can take $F'' = E''$ for instance or F'' may be the solid vector subspace generated by $\varphi(E)$. This can create the impression that E', endowed with any uniformity between these extremes, is Φ_4-compact. This is not the case, as the following example shows. As E we choose the Banach lattice c_0. Then ℓ^1 is its dual and ℓ^∞ its bidual. If we denote by c the set of convergent sequences of ℓ^∞, then ℓ^1_c is not even sequentially complete.

Proposition 5.8.4. *Let E be a barreled M-space, φ be the evalutation map $E \to E''$, F'' be a solid vector subspace of E'' containing $\varphi(E)$ such that for any $x'' \in F''_+$ there exists a bounded set A of E with $x'' \leq \bigvee_{x \in A} \varphi(x) \in F''$, let X'' be the set*

$$\{x'' \in F'' \mid \exists (x_n)_{n \in \mathbb{N}} \text{ an increasing sequence in } E_+,\ x'' = \bigvee_{n \in \mathbb{N}} \varphi(x_n)\},$$

A' be a Φ_4-set of $E'_{X''}$ and $\bar{A}'$ be its closure in $E'_{X''}$. Then $\bar{A}'$ is the closure of A' in $E'_{F''}$ and $\bar{A}'_{X''}$, $\bar{A}'_{F''}$ are equal and compact. In particular, the Θ_4-sequences, the Θ_4-nets, the Φ_4-filters and the $\hat{\Phi}_4$-filters on $E'_{X''}$ and on $E'_{F''}$ coincide.

A' is a bounded set of E'_E. Since E is a barreled M-space there exists a directed solid 0-neighbourhood U in E such that $A' \subset U^0$. By Theorem 5.7.25, Proposition 1.8.11, and Corollary 1.8.15 $U^0_{X''}$, $U^0_{F''}$ are Φ_4-compact and the identity map $U^0_{X''} \to U^0_{F''}$ is uniformly Φ_4-continuous so, by Proposition 1.8.8, $\bar{A}'$ is the closure of A' in $E'_{F''}$ and $\bar{A}'_{X''} = \bar{A}'_{F''}$. By Corollary 1.2.10 $\bar{A}'_{X''}$ is compact. The last assertion follows immediately from the first by Theorem 1.8.4. □

Lemma 5.8.5. *Let E, F be locally convex spaces, $\varphi: E \to E''$, $\psi: F \to F''$ be the evaluation maps, X'' be a subset of E'' containing $\varphi(E)$, $\mathfrak{K}$ be the set of convex circled equicontinuous compact sets of $E'_{X''}$, $u: E \to F$ be a continuous linear map and u'' be its biadjoint map. Then:*

a) if $u''(X'') \subset \psi(F)$, then $u: E_{\mathfrak{K}} \to F$ is continuous;

b) if $u: E_{\mathfrak{K}} \to F$ is continuous and F is quasicomplete, then $u''(X'') \subset \psi(F)$.

a) Let V be a convex circled closed 0-neighbourhood in F and let u' be the adjoint map of u. V^0_F being compact and $u': F'_F \to E'_{X''}$ continuous we get $u'(V^0) \in \mathfrak{K}$ so $u'(V^0)^0$ is a 0-neighbourhood in $E_{\mathfrak{K}}$. From $u'(V^0)^0 \subset \overset{-1}{u}(V)$ it follows that $\overset{-1}{u}(V)$ is also a 0-neighbourhood in $E_{\mathfrak{K}}$. Since V is arbitrary, $u: E_{\mathfrak{K}} \to F$ is continuous.

b) Since for any point $x'' \in E''$ there exists a bounded set A of E such that x'' belongs to the closure of $\varphi(A)$ in $E''_{E'}$ we may assume that F is complete. Let $x'' \in X''$ and V be a 0-neighbourhood in F. Then $\overset{-1}{u}(V)^0 \in \mathfrak{K}$ so $x'' \mid \overset{-1}{u}(V)^0_E$ is continuous. Since $u': F'_F \to E'_E$ is continuous and $u'(V^0) \subset \overset{-1}{u}(V)^0$, $x'' \circ u' \mid V^0_F$ is continuous. V

being arbitrary and F complete, we get $x'' \circ u' \in \psi(F)$ (H.H. Schaefer [1] Corollary 2 of Theorem IV 6.2). □

Proposition 5.8.6. *Let E be an M-space, φ be the evaluation map $E \to E''$, G'' be a solid vector subspace of E'' containing $\varphi(E)$ such that for any $x'' \in G''_+$ there exists a bounded set A of E with $x'' \le \bigvee_{x \in A} \varphi(x) \in G''$, let F be a quasicomplete locally convex space, ψ be the evaluation map $F \to F''$, u be a continuous linear map of E into F such that $(u(x_n))_{n \in \mathbb{N}}$ converges weakly for any increasing sequence $(x_n)_{n \in \mathbb{N}}$ in E_+ for which $\bigvee_{n \in \mathbb{N}} \varphi(x_n)$ exists and belongs to G'', and let u'' be the biadjoint map of u. Then:*

a) $u''(G'') \subset \psi(F)$:

b) $u(A)$ is a weakly relatively compact set of F for any subset A of E for which $\varphi(A)$ is order bounded in G''.

a) Let X'' be the set

$$\{x'' \in G'' \mid \exists (x_n)_{n \in \mathbb{N}} \text{ an increasing sequence in } E_+,\ x'' = \bigvee_{n \in \mathbb{N}} \varphi(x_n)\}$$

and $\mathfrak{K}$ be the set of convex circled equicontinuous compact sets of $E'_{X''}$. By the hypothesis $u''(X'') \subset \psi(F)$ and therefore by Lemma 5.8.5a) the map $E_{\mathfrak{K}} \to F$ is continuous. By Theorem 5.7.25 and Corollary 1.8.15 the identity map $K'_{X''} \to K'_{G''}$ is uniformly Φ_4-continuous so, by Proposition 1.8.8, K' is a compact set of $E'_{G''}$ for any $K' \in \mathfrak{K}$. Lemma 5.8.5b) $u''(G'') \subset \psi(F)$.

b) $\varphi(A)$ is a relatively compact set of $G''_{E'}$, for it is order bounded in G'' (Ch. D. Aliprantis, O. Burkinshaw [1] Theorem 19.14). By a) the map $G''_{E'} \to F_{F'}$ defined by u'' is continuous so $u(A)$ is weakly relatively compact. □

Remark. The example given in the second remark of Theorem 5.7.25 shows that we cannot drop the supplementary condition about F'' (about G'') in Propositions 5.8.2, 5.8.3 and 5.8.4 (in Proposition 5.8.6).

Definition 5.8.7. *Let E, F be locally convex spaces. A continuous linear map $u: E \to F$ is called* boundedly weakly compact (weakly compact) *if $u(A)$ is a weakly relatively compact set of F for any bounded set (for some 0-neighbourhood) A of E.*

Let F be a quasicomplete locally convex space, $\tilde{F}$ be its completion, $\varphi: F \to \tilde{F}$ be the inclusion map and $u: E \to F$ be a continuous linear map. Then u is boundedly weakly compact iff $\varphi \circ u$ is. Indeed, assume that $\varphi \circ u$ is boundedly weakly compact and let A be a bounded set of E. Then $\overline{\varphi \circ u(A)}$ is weakly compact. Since F is quasicomplete, the inclusion map $\overline{u(A)} \to \overline{\varphi \circ u(A)}$ is surjective so it is a homeomorphism, i.e., $\overline{u(A)}$ is weakly compact and u is boundedly weakly compact. The converse implication is trivial.

Any weakly compact map is boundedly weakly compact and these two notions coincide if the domain of the map is normed.

Definition 5.8.8. *(A. Grothendieck (1953) [2] Définition 4.) A locally convex space E* possesses the D-property *(D is an abbreviation of Dieudonné) if any continuous linear map u of E into a quasicomplete locally convex space is boundedly weakly compact if $(u(x_n))_{n\in\mathbb{N}}$ is weakly convergent for any weak Cauchy sequence $(x_n)_{n\in\mathbb{N}}$ in E.*

Theorem 5.8.9. *Let E be an M-space, F be a quasicomplete locally convex space, Φ be the set $\hat{\Phi}_4(F'_F)$ and u be a continuous linear map of E into F such that $(u(x_n))_{n\in\mathbb{N}}$ is a weakly convergent sequence for any increasing weak Cauchy sequence $(x_n)_{n\in\mathbb{N}}$ in E_+. Then:*

a) u is boundedly weakly compact;

b) the map $u: E_{E'} \to F_\Phi$ is uniformly Φ_4-continuous;

c) E possesses the D-property.

Let $\varphi: E \to E''$, $\psi: F \to F''$ be the evaluation maps and u'' be the biadjoint map of u.

a) Let A be a bounded set of E. By Proposition 5.7.4 b) $\varphi(A)$ is order bounded so $u(A)$ is a weakly relatively compact set of F by Proposition 5.8.6 b).

b) By Proposition 5.8.6 a) $u''(E'') \subset \psi(F)$ and, by Corollary 5.7.7, the map $u: E_{E'} \to F_\Phi$ is uniformly Φ_4-continuous.

c) follows from a). □

Corollary 5.8.10. *Let E be an M-space, F be a quasicomplete and weakly sequentially complete locally convex space and u be a continuous linear map $E \to F$. Then u is boundedly weakly compact and $u(A)$ is a relatively compact set of F for any weak Φ_4-set A of E.*

Since F is weakly sequentially complete, $(u(x_n))_{n\in\mathbb{N}}$ is weakly convergent for any weak Cauchy sequence $(x_n)_{n\in\mathbb{N}}$ in E. By Theorem 5.8.9 a) u is boundedly weakly compact. By Theorem 5.8.9 b) and Theorem 1.8.4 a $\Rightarrow$ h, $u(A)$ is a Φ_4-set of F and therefore a precompact set of F (Proposition 1.5.6 a $\Rightarrow$ c). Since F is quasicomplete $u(A)$ is a relatively compact set of F. □

Remark. Let X be a compact space, $\mathscr{C}$ be the Banach lattice of continuous real functions on X and u be a continuous linear map from $\mathscr{C}$ into a Banach space of type $L^1(\mu)$. By the above result u is boundedly weakly compact and $u(A)$ is a relatively compact set of $L^1(\mu)$ for any weakly compact set A of $\mathscr{C}$ and for any bounded weakly metrizable set A of $\mathscr{C}$. This result was proved by A. Grothendieck (1953) ([2] Theorem 7).

Lemma 5.8.11. *Let E be a locally convex lattice, $(x_n)_{n\in\mathbb{N}}$ be a Cauchy sequence in E and U be a directed solid 0-neighbourhood in E. Then there exists $k \in \mathbb{N}$ such that*

$$\bigvee_{m=n}^{n+p} x_m - x_q, \quad \bigwedge_{m=n}^{n+p} x_m - x_q, \quad \bigvee_{m=j}^{n} x_m - \bigvee_{m=j}^{n'} x_m, \quad \bigwedge_{m=j}^{n} x_m - \bigwedge_{m=j}^{n'} x_m$$

belong to U for any $j, n, n', p, q \in \mathbb{N}$ with $k \le n$, $k \le n'$, $j \le n$, $j \le n'$, $k \le q$.

There exists $k \in \mathbb{N}$ such that $x_m - x_n \in U$ for any $m, n \in \mathbb{N}$ with $m \geq k$, $n \geq k$. Since U is directed and solid, we get

$$\bigvee_{m=n}^{n+p} x_m - x_q \in U, \quad \bigwedge_{m=n}^{n+p} x_m - x_q \in U$$

for any $n, p, q \in \mathbb{N}$ with $k \leq n$, $k \leq q$. Once again, using the fact that U is directed and solid, we also find that

$$\bigvee_{m=n}^{n+p} x_m - \bigvee_{m=n}^{n+p'} x_m \in U, \quad \bigwedge_{m=n}^{n+p} x_m - \bigwedge_{m=n}^{n+p'} x_m \in U$$

for any $n, p, p' \in \mathbb{N}$ with $k \leq n$. We have (Ch. D. Aliprantis, O. Burkinshaw [1] Theorem 1.1 (12))

$$\left| \bigvee_{m=j}^{n} x_m - \bigvee_{m=j}^{n'} x_m \right| \leq \left| \bigvee_{m=k}^{n} x_m - \bigvee_{m=k}^{n'} x_m \right|,$$

$$\left| \bigwedge_{m=j}^{n} x_m - \bigwedge_{m=j}^{n'} x_m \right| \leq \left| \bigwedge_{m=k}^{n} x_m - \bigwedge_{m=k}^{n'} x_m \right|$$

for any $j, n, n' \in \mathbb{N}$ with $j < k \leq n$, $k \leq n'$. Since U is solid

$$\bigvee_{m=j}^{n} x_m - \bigvee_{m=j}^{n'} x_m \in U, \quad \bigwedge_{m=j}^{n} x_m - \bigwedge_{m=j}^{n'} x_m \in U$$

for any $j, n, n' \in \mathbb{N}$ with $k \leq n$, $k \leq n'$, $j \leq n$, $j \leq n'$. □

Proposition 5.8.12. *Let E be a sequentially complete M-space and $(x_n)_{n \in \mathbb{N}}$ be a convergent sequence in E. Then $(x_n)_{n \in \mathbb{N}}$ order converges to $\lim_{n \to \infty} x_n$.*

By Lemma 5.8.11 $\left(\bigvee_{m=n}^{n+p} x_m \right)_{p \in \mathbb{N}}$ and $\left(\bigwedge_{m=n}^{n+p} x_m \right)_{p \in \mathbb{N}}$ are Cauchy sequences and therefore convergent sequences in E for any $n \in \mathbb{N}$. By Ch. D. Aliprantis, O. Burkinshaw [1], Theorem 5.6 (iii)

$$\bigvee_{m \geq n} x_m = \lim_{p \to \infty} \bigvee_{m=n}^{n+p} x_m, \quad \bigwedge_{m \geq n} x_m = \lim_{p \to \infty} \bigwedge_{m=n}^{n+p} x_m.$$

Let V be an arbitrary closed 0-neighbourhood in E and U be a directed solid 0-neighbourhood in E contained in V. By Lemma 5.8.11 there exists $k \in \mathbb{N}$ such that $\bigvee_{m=n}^{n+p} x_m - x_q \in U$, $\bigwedge_{m=n}^{n+p} x_m - x_q \in U$ for any $n, p, q \in \mathbb{N}$ with $k \leq n$, $k \leq q$. We get $\bigvee_{m \geq n} x_m - x_q \in V$, $\bigwedge_{m \geq n} x_m - x_q \in V$ for any $n, q \in \mathbb{N}$ with $k \leq n$, $k \leq q$. Since V is arbitrary

$$\lim_{n \to \infty} \bigvee_{m \geq n} x_m = \lim_{n \to \infty} \bigwedge_{m \geq n} x_m = \lim_{n \to \infty} x_n.$$

Once again, using Ch. D. Aliprantis, O. Burkinshaw [1] Theorem 5.6 (iii)

$$\bigwedge_{n\in\mathbb{N}}\bigvee_{m\geq n} x_m = \lim_{n\to\infty} x_n = \bigvee_{n\in\mathbb{N}}\bigwedge_{m\geq n} x_m .$$

Hence, $(x_n)_{n\in\mathbb{N}}$ order converges to $\lim_{n\to\infty} x_n$. □

Proposition 5.8.13. *Let E be a locally convex space. We consider the following assertions:*

a) E contains no copy of ℓ^*;

b) any group homomorphism of a lattice ordered additive group G into E is order exhaustive if it maps the order bounded sets of G into bounded sets of E;

c) any continuous linear map of a locally convex lattice into E is order exhaustive;

d) E contains no copy of ℓ^∞.

Then $a \Rightarrow b \Rightarrow c \Rightarrow d$ *and, if E is sequentially complete, all these assertions are equivalent.*

a ⇒ b. Let $u: G \to E$ be a group homomorphism which maps the order bounded sets of G into bounded sets of E. Moreover, let $(x_\iota)_{\iota\in I}$ be an order supersummable family in G and μ be the map

$$\mathfrak{P}(I) \to E, \quad J \mapsto u(\sum_{\iota\in J} x_\iota).$$

Then $\{\sum_{\iota\in J} x_\iota \mid J \subset I\}$ is an order bounded set of G and μ is a bounded additive map. By Corollary 4.1.44 a ⇒ c μ is exhaustive, i. e., u is order exhaustive.

b ⇒ c follows from the remark that any order bounded set of a locally convex lattice is bounded (Ch D. Aliprantis, O. Burkinshaw [1] Theorem 5.4 (i)).

c ⇒ d. Assume that d) does not hold. Then there exists a topological vector subspace F of E and an isomorphism of topological vector spaces $u: \ell^\infty \to F$. We endow ℓ^∞ with the order relation induced by $\mathbb{R}^{\mathbb{N}}$ and denote by v the inclusion map $F \to E$ and, for any $n \in \mathbb{N}$, denote by a_n the element of ℓ^∞ which is equal to 1 at n and equal to 0 otherwise. Then ℓ^∞ is a locally convex lattice, $v \circ u$ is continuous, and $(a_n)_{n\in\mathbb{N}}$ is order supersummable in ℓ^∞. Hence, by c) $(v \circ u(a_n))_{n\in\mathbb{N}}$ converges to 0 in E, i. e., $(u(a_n))_{n\in\mathbb{N}}$ conveges to 0 in F and $(a_n)_{n\in\mathbb{N}}$ converges to 0 in ℓ^∞, which is a contradiction.

d ⇒ a follows from Lemma 4.1.42. □

Theorem 5.8.14. *Let E be an order σ-complete and sequentially complete M-space containing no copy of* ℓ^∞. *Then the order convergence in E and the topological convergence in E coincide for sequences.*

Because of Proposition 5.8.12 we only have to show that an order convergent sequence converges topologically.

First, let $(x_n)_{n\in\mathbb{N}}$ be an increasing sequence in E which possesses a supremum. Since E is order σ-complete $(\alpha_n(x_{n+1} - x_n))_{n\in\mathbb{N}}$ is order summable for any $(\alpha_n)_{n\in\mathbb{N}} \in \ell^\infty$. We set

$$u: \ell^\infty \to E, \quad (\alpha_n)_{n\in\mathbb{N}} \mapsto \sum_{n\in\mathbb{N}}^{\leq} \alpha_n(x_{n+1}-x_n)$$

and endow ℓ^∞ with the order relation induced by $\mathbb{R}^{\mathbb{N}}$. Then ℓ^∞ is a vector lattice and u is a linear map which sends the order bounded sets into topologically bounded sets. By Proposition 5.8.13 d $\Rightarrow$ b u is order exhaustive. For any $n \in \mathbb{N}$ we denote by a_n the element of ℓ^∞ which is equal to 1 on $\{m \in \mathbb{N} \mid m \leq n\}$ and equal to 0 otherwise. By Proposition 5.2.6 $(u(a_n))_{n\in\mathbb{N}}$ converges, i.e., $(x_n)_{n\in\mathbb{N}}$ converges.

Now let $(x_n)_{n\in\mathbb{N}}$ be an arbitrary sequence in E which order converges to an $x \in E$. Then by Proposition 5.1.4 a $\Rightarrow$ b $(x_n)_{n\in\mathbb{N}}$ is order bounded and

$$x = \bigvee_{n\in\mathbb{N}} \bigwedge_{m\geq n} x_m = \bigwedge_{n\in\mathbb{N}} \bigvee_{m\geq n} x_m .$$

By the first part of the proof

$$\lim_{n\to\infty} \Big(\bigvee_{m\geq n} x_m\Big) = \lim_{n\to\infty} \Big(\bigwedge_{m\geq n} x_m\Big) = x .$$

Let U be a solid 0-neighbourhood in E. Then there exists $n \in \mathbb{N}$ such that

$$\bigvee_{m\geq n} x_m - \bigwedge_{m\geq n} x_m \in U .$$

We have

$$|x_m - x| \leq \bigvee_{m\geq n} x_m - \bigwedge_{m\geq n} x_m$$

so $x_m - x \in U$ for any $m \in \mathbb{N}$ with $m \geq n$. Hence,

$$\lim_{n\to\infty} x_n = x . \quad \square$$

Remark Let E be the vector subspace of ℓ^∞ formed by the convergent sequences endowed with the order relation induced by $\mathbb{R}^{\mathbb{N}}$. Then E is a Banach space containing no copy of ℓ^∞ and it is an M-space. For any $n \in \mathbb{N}$ we denote by x_n the element of E equal to 1 on $\{m \in \mathbb{N} \mid m \leq n\}$ and equal to 0 otherwise. Then $(x_n)_{n\in\mathbb{N}}$ is an increasing sequence in E which possesses a supremum but does not converge to this supremum. Hence, we cannot drop the hypothesis that E is order σ-complete in the above theorem.

Proposition 5.8.15. *The following assertions are equivalent for any sequentially complete locally convex space E:*

a) E contains no copy of c_0;

b) for any continuous linear map u of an M-space F into E the map $F_{F'} \to E$ defined by u is uniformly Φ_4-continuous;

c) any continuous linear map $c_0 \to E$ maps the weak convergent sequences into Cauchy sequences.

a $\Rightarrow$ b. Let $(x_n)_{n\in\mathbb{N}}$ be an increasing weak Cauchy sequence in F_+. Then $\{\sum_{n\in M} (x_{n+1}-x_n) \mid M \in \mathfrak{P}_f(\mathbb{N})\}$ is a bounded set of F

$\{\sum_{n\in M}(u(x_{n+1})-u(x_n))\,|\,M\in\mathfrak{P}_f(\mathbb{N})\}$ is a bounded set of E. By Theorem 4.1.45 d $\Rightarrow$ e $(u(x_{n+1})-u(x_n))_{n\in\mathbb{N}}$ is summable so $(u(x_n))_{n\in\mathbb{N}}$ is convergent. By Theorem 5.8.9 b) the map $F_{F'}\to E$ is uniformly Φ_4-continuous.

b $\Rightarrow$ c. If we endow c_0 with the order relation induced by $\mathbb{R}^{\mathbb{N}}$ then c_0 becomes an M-space.

c $\Rightarrow$ a. The property described in c) holds for any topological vector subspace of E and does not hold for c_0 so E contains no copy of c_0. □

Theorem 5.8.16. *Let E be a sequentially complete M-space and $\tilde{E}$ be its completion. Then the following assertions are equivalent:*

a) E contains no copy of c_0;

b) the identity map $E_{E'}\to E$ is uniformly Φ_4-continuous;

c) the weak Cauchy sequences of E are Cauchy sequences;

d) the inclusion map $E\to\tilde{E}$ is boundedly weakly compact.

a $\Rightarrow$ b follows from Proposition 5.8.15 a $\Rightarrow$ b.

b $\Rightarrow$ c is trivial.

c $\Rightarrow$ d follows from Theorem 5.8.9 a).

d $\Rightarrow$ a. The property described in d) holds for any topological vector subspace of E but does not hold for c_0. Hence, E contains no copy of c_0. □

Theorem 5.8.17. *Let E be a sequentially complete M-space containing no copy of c_0. Then:*

a) E is weakly sequentially complete;

b) E is order σ-complete;

c) the order convergence and the topological convergence in E coincide for sequences.

a) follows immediately from Theorem 5.8.16 a $\Rightarrow$ c.

b) Let $(x_n)_{n\in\mathbb{N}}$ be an upper bounded increasing sequence in E. By Proposition 5.7.18 a $\Rightarrow$ b $(x_n)_{n\in\mathbb{N}}$ is a weak Cauchy sequence so, by a), it converges. By Proposition 5.8.12 $(x_n)_{n\in\mathbb{N}}$ possesses a supremum. Hence, E is order σ-complete.

c) follows immediately from b) and Theorem 5.8.14. □

Theorem 5.8.18. *The following assertions are equivalent for any quasicomplete locally convex space E:*

a) E contains no copy of c_0;

b) any continuous linear map u of an M-space F into E is boundedly weakly compact and the map $F_{F'}\to E$ defined by u is uniformly Φ_4-continuous;

c) any continuous linear map of an M-space into E is boundedly weakly compact;

d) any continuous linear map $c_0\to E$ is boundedly weakly compact.

a ⇒ b. By Proposition 5.8.15 a ⇒ b the map $F_{F'} \to E$ defined by u is uniformly Φ_4-continuous. In particular $(u(x_n))_{n \in \mathbb{N}}$ converges for any weak Cauchy sequence. By Theorem 5.8.9 a) u is boundedly weakly compact.

b ⇒ c is trivial.

c ⇒ d. c_0 endowed with the order relation induced by $\mathbb{R}^{\mathbb{N}}$ is an M-space.

d ⇒ a. The property described by d) holds for any closed topological vector space of E but does not hold for c_0. Hence, E contains no copy of c_0. □

Remarks. 1) Let X be a compact space, $\mathscr{C}$ be the usual Banach space of continuous real functions on X and E be a Banach space containing no copy of c_0. By the above theorem a ⇒ c any continuous linear map of $\mathscr{C}$ into E is weakly compact. This result was proved by A. Pelczynski (1960) ([1] Theorem 5).

2) E. Thomas called a locally convex space E "faiblement Σ-complete" if, for any sequence $(x_n)_{n \in \mathbb{N}}$ in E, the following assertions are equivalent: a) $(x_n)_{n \in \mathbb{N}}$ is weakly supersummable; b) $\sum_{n \in \mathbb{N}} |x'(x_n)| < \infty$ for any $x' \in E'$ (1968) ([1] Définition). If E is sequentially complete, then by Theorem 4.1.45 d ⇔ e (and by the Orlicz-Pettis theorem) E is faiblement Σ-complete iff E contains no copy of c_0. For any locally compact space X let us denote by $\mathscr{C}_0(X)$ the usual Banach lattice of continuous real functions on X that vanish at the infinity. Then $\mathscr{C}_0(X)$ is an M-space. Assume that E is quasicomplete. By the above considerations and by Theorem 5.8.18 a⇔c⇔d E is faiblement Σ-complete iff the following property holds for any locally compact space X: any continuous linear map $\mathscr{C}_0(X) \to E$ is weakly compact. This result was proved by E. Thomas (1968) ([1] Théorème 1).

Lemma 5.8.19. *Let E be a locally convex space for which the identity map $E \to E$ is boundedly weakly compact and the identity map $E_{E'} \to E$ is Φ_2-continuous. Then any bounded set of E is relatively compact.*

Let A be a bounded set. Since the identity map $E \to E$ is boundedly weakly compact, the closure $\bar{A}$ of A in $E_{E'}$ is a compact and therefore a Φ_2-set of $E_{E'}$ (Proposition 1.5.5 c ⇒ a). By Proposition 1.8.9 e) the inclusion map $\bar{A}_{E'} \to E$ is continuous so $\bar{A}$ is a compact set of E. □

Theorem 5.8.20. *The following assertions are equivalent for any quasicomplete M-space E:*

a) E contains no copy of c_0;

b) any bounded set of E is relatively compact;

c) E is semi-reflexive;

d) E is weakly sequentially complete.

If E is complete, then all these assertions are equivalent to the following:

e) E is weakly Φ_4-compact.

a ⇒ b. By Theorem 5.8.18 a ⇒ b the identity map $E \to E$ is boundedly weakly compact and the identity map $E_{E'} \to E$ is uniformly Φ_4-continuous and therefore Φ_2-continuous (Proposition 1.8.3). By Lemma 5.8.19 any bounded set of E is relatively compact.

b ⇒ c ⇒ d ⇒ a and e ⇒ a are trivial.

d ⇒ e. By Corollary 2.3.19 E is weakly Φ_2-compact and by d) the weak Φ_2-filters and the weak Φ_4-filters coincide. □

Remark. The example given in Proposition 5.8.34 shows that E need not be reflexive. The example given in Proposition 5.8.35 shows that we cannot replace quasicomplete by sequentially complete in the above theorem even if the space is weakly sequentially complete.

Corollary 5.8.21. *Let E be a quasicomplete M-space containing no copy of c_0. Then:*

a) E is order complete;

b) $E' \subset E^\pi$.

Let us denote the evaluation map $E \to E''$ by φ. By Theorem 5.8.20 a ⇒ c, φ is bijective and therefore an isomorphism of vector lattices (Ch. D. Aliprantis, O. Burkinshaw [1] Theorem 3.11 (ii)). Hence, E is order complete. Let A be a lower directed set of E with infimum 0 and $x' \in E'_+$. Then (Proposition 5.5.6)

$$\inf_{x \in A} x'(x) = \inf_{x'' \in \varphi(A)} x''(x') = 0$$

so $x' \in E^\pi$ and $E' \subset E^\pi$ (Proposition 5.2.4 c ⇒ a). □

We now give some examples of M-spaces.

Proposition 5.8.22. *Let E be an M-space and F be a solid closed vector subspace of E. Then E/F endowed with the quotient structure is an M-space.*

Let $\varphi : E \to E/F$ be the canonical map. The order relation on E/F is defined by setting $(E/F)_+ = \varphi(E_+)$. Then E/F is a vector lattice and φ is a homomorphism of vector lattices. Let $\mathfrak{U}$ be the set of solid directed 0-neighbourhoods in E and $U \in \mathfrak{U}$. Furthermore, let $x', y' \in \varphi(U)$. There exist $x, y \in U$ so that $x' = \varphi(x)$, $y' = \varphi(y)$. Then $x \vee y \in U$ and

$$x' \vee y' = \varphi(x \vee y) \in \varphi(U).$$

Hence, $\varphi(U)$ is directed. Now let $u' \in E/F$, $v' \in \varphi(U)$ and $|u'| \leq |v'|$. There exist $u \in E$, $v \in U$ so that $u' = \varphi(u)$, $v' = \varphi(v)$. Then $(u \wedge |v|) \vee (-|v|) \in U$ and

$$\begin{aligned} u' &= (u' \wedge |v'|) \vee (-|v'|) = (\varphi(u) \wedge |\varphi(v)|) \vee (-|\varphi(v)|) = \\ &= \varphi((u \wedge |v|) \vee (-|v|)) \in \varphi(U). \end{aligned}$$

Hence, $\varphi(U)$ is solid. Since $\{\varphi(U) \mid U \in \mathfrak{U}\}$ is a fundamental system of 0-neighbourhoods in E/F, E/F is an M-space. □

Proposition 5.8.23. *Let E be a vector lattice, p be a seminorm on E and U be the set*

$$\{x \in E \,|\, p(x) \leq 1\}.$$

Then the following assertions are equivalent:

a) U is solid and directed;

b) $x, y \in E \Rightarrow p(|x| \vee |y|) = \sup(p(x), p(y))$.

a $\Rightarrow$ b. Let x, $y \in E$. Without loss of generality we may assume $p(x) \leq p(y)$. Let $\alpha \in \mathbb{R}$ with $\alpha > p(y)$. Then $\frac{1}{\alpha}x, \frac{1}{\alpha}y \in U$ and

$$\frac{1}{\alpha}(|x| \vee |y|) = \left|\frac{1}{\alpha}x\right| \vee \left|\frac{1}{\alpha}y\right| \in U, \quad p\left(\frac{1}{\alpha}(|x| \vee |y|)\right) \leq 1,$$
$$p(|x| \vee |y|) \leq \alpha.$$

Since α is arbitrary

$$p(|x| \vee |y|) \leq \sup(p(x), p(y)).$$

Let $\beta \in \mathbb{R}$ with $\beta > p(|x| \vee |y|)$. Then

$$p\left(\frac{1}{\beta}(|x| \vee |y|)\right) \leq 1, \quad \left|\frac{1}{\beta}x\right| \vee \left|\frac{1}{\beta}y\right| = \frac{1}{\beta}(|x| \vee |y|) \in U$$

so $\frac{1}{\beta}y \in U$ and $p(y) \leq \beta$. Since β is arbitrary

$$\sup(p(x), p(y)) \leq p(|x| \vee |y|).$$

b $\Rightarrow$ a. Let $x \in E$, $y \in U$ so that $|x| \leq |y|$. Then $|x| \vee |y| = |y|$ and

$$p(x) \leq \sup(p(x), p(y)) = p(|x| \vee |y|) = p(|y|) \leq 1,$$

so $x \in U$. Hence, U is solid.

Let $x, y \in U$. Then

$$p(|x| \vee |y|) = \sup(p(x), p(y)) \leq 1$$

and $|x| \vee |y| \in U$. We have

$$|x \vee y| \leq |x| \vee |y|, \quad |x \wedge y| \leq |x| \vee |y|.$$

Since U is solid $x \vee y, x \wedge y \in U$. Hence, U is directed. □

Corollary 5.8.24. *A locally convex lattice E is an M-space iff its topology may be generated by a set $\mathscr{P}$ of seminorms such that*

$$p(|x| \vee |y|) = \sup(p(x), p(y))$$

for any $p \in \mathscr{P}$ and any $x, y \in E$. □

Corollary 5.8.25. *A normable locally convex lattice E is an M-space iff its topology may be generated by a norm p such that*

$$p(|x| \vee |y|) = \sup(p(x), p(y))$$

for any $x, y \in E$. □

Corollary 5.8.26. *Let X be a completely regular space and $\mathscr{C}_b$ be the vector lattice of bounded continuous real functions on X. Then $\mathscr{C}_b$ endowed with the strict topology is an M-space.* □

Remark. In particular $\mathscr{C}_b$ has the strong D.P.-property (Corollary 5.7.9) and so the strict D.P.-property. This last result was proved by S.S. Khurana (1978) ([2] Theorem 3).

Corollary 5.8.27. *Let $\mathfrak{R}$ be a δ-ring, $\mathscr{M}$ be a band of $\mathscr{M}(\mathfrak{R}, \mathbb{R})$, $\mathscr{G}$ be a solid vector subspace of $\mathscr{M}_c^\pi$ (with the notation of C. Constantinescu [5]), and let $\mathscr{F}$ be a subset of $\tilde{\mathscr{G}}$. Then $(\mathscr{G}, \mathscr{F})$ is an M-space if it is Hausdorff.* □

Proposition 5.8.28. *If E is an M-space, then the maps*

$$E \to E, \quad x \mapsto |x|,$$

$$E \times E \to E, \quad (x, y) \mapsto x \vee y$$

are uniformly continuous.

By Corollary 5.8.24 there exists a set $\mathscr{P}$ of seminorms on E generating its topology such that

$$p(|x| \vee |y|) = \sup(p(x), p(y))$$

for any $p \in \mathscr{P}$ and any $x, y \in E$.

Let $p \in \mathscr{P}$ and $x, y \in E$ such that $|x| \leq |y|$. Then

$$p(x) \leq \sup(p(x), p(y)) = p(|x| \vee |y|) = p(|y|) = p(y).$$

Let $x, y \in E$ and $p \in \mathscr{P}$. We have

$$||x| - |y|| \leq |x - y|$$

so, by the above consideration,

$$p(|x| - |y|) \leq p(|x - y|) = p(x - y).$$

Hence, the map

$$E \to E, \quad x \mapsto |x|$$

is uniformly continuous. Since

$$x \vee 0 = \tfrac{1}{2}(|x| + x)$$

holds for any $x \in E$, the map

$$E \to E, \quad x \mapsto x \vee 0$$

is also uniformly continuous. Since

$$x \vee y = (x - y) \vee 0 + y$$

for any $x, y \in E$ we see that the map

$$E \times E \to E, \quad (x, y) \mapsto x \vee y$$

is uniformly continuous. □

Theorem 5.8.29. *The completion of an M-space is an M-space.*

Let E be an M-space and $\tilde{E}$ be the completion of the locally convex space underlying E. By Proposition 5.8.28 the lattice operations in E may be extended continuously on $\tilde{E}$ and $\tilde{E}$ endowed with these operations is a vector lattice. By Corollary 5.8.24 there exists a set $\mathscr{P}$ of seminorms on E generating its topology such that

$$p(|x| \vee |y|) = \sup(p(x), p(y))$$

for any $p \in \mathscr{P}$ and any $x, y \in E$. If we extend the seminorms of $\mathscr{P}$ continuously on $\tilde{E}$ we get a set of seminorms on $\tilde{E}$ generating its topology. By Proposition 5.8.28 these seminorms satisfy the conditions of Corollary 5.8.24 so, by Corollary 5.8.24, $\tilde{E}$ is an M-space. □

Definition 5.8.30. *Let X be a set and E be a vector lattice. We endow E^X with the order relation defined by*

$$\underline{f \leq g}: \Leftrightarrow (x \in X \Rightarrow f(x) \leq g(x)).$$

E^X is a vector lattice. For any $f, g \in E^X$ the map

$$X \to E, \quad x \mapsto f(x) \vee g(x)$$

$$(X \to E, \quad x \mapsto f(x) \wedge g(x))$$

is the supremum (infimum) of f and g in E^X.

The above definition contradicts Definition 5.5.1 but we hope it will lead to no confusion.

Theorem 5.8.31. *Let E be an M-space, T be a locally compact σ-compact space, $\mathscr{K}(T, E)$ be the set of continuous maps of T into E with compact carrier and, for any compact set K of T, let $\mathscr{K}_0(K, E)$ be the set of $f \in \mathscr{K}(T, E)$ with carrier contained in K endowed with the topology of uniform convergence. We endow $\mathscr{K}(T, E)$ with the order relation induced by E^T and with the finest locally convex topology for which the inclusion map $\mathscr{K}_0(K, E) \to \mathscr{K}(T, E)$ is continuous for any compact set K of T. Then $\mathscr{K}(T, E)$ is a vector sublattice of E^T and an M-space so it possesses the strong D.P.-property and the D-property.*

By Proposition 5.8.28 the map

$$E \times E \to E, \quad (x, y) \mapsto x \vee y$$

is continuous so $\mathscr{K}(T, E)$ is a vector sublattice of E^T.

By Corollary 5.8.24 there exists a set $\mathscr{P}$ of semi-norms on E generating its topology such that

$$p(|x| \vee |y|) = \sup(p(x), p(y))$$

for any $x, y \in E$ and any $p \in \mathscr{P}$. Let $\mathscr{G}$ be the set of positive real functions g on $\mathscr{P} \times T$ such that

1) $g(p, \cdot)$ is upper semi-continuous for any $p \in \mathscr{P}$,
2) any $t \in T$ possesses a neighbourhood U such that the set

$$\{p \in \mathscr{P} \mid \sup_{s \in U} g(p, s) > 0\}$$

is finite.

Put

$$r_g : \mathscr{K}(T, E) \to \mathbb{R}, \quad f \mapsto \sup_{p \in \mathscr{P}} \sup_{t \in T} g(p, t) p(f(t))$$

for any $g \in \mathscr{G}$. Then r_g is a semi-norm and, by Theorem 3.5.20 g), the family of these semi-norms generates the topology of $\mathscr{K}(T, E)$.

Let $g \in \mathscr{G}, p \in \mathscr{P}$ and $f', f'' \in \mathscr{K}(T, E)$. Since the map

$$T \to \mathbb{R}, \quad t \mapsto g(p, t) p(|f'(t)| \vee |f''(t)|)$$

is upper semi-continuous and has a compact carrier it takes its supremum at a point $t_0 \in T$. Without loss of generality we may assume

$$p(f'(t_0)) \leq p(f''(t_0)).$$

We have

$$\begin{aligned} &\sup(g(p,t)p(f'(t)), g(p,t)p(f''(t))) = \\ &= g(p, t) \sup(p(f'(t)), p(f''(t))) = g(p,t)p(|f'(t)| \vee |f''(t)|) \leq \\ &\leq g(p, t_0)p(|f'(t_0)| \vee |f''(t_0)|) = g(p,t_0) \sup(p(f'(t_0)), p(f''(t_0))) = \\ &= g(p,t_0)p(f''(t_0))) \leq r_g(f'') \end{aligned}$$

for any $t \in T$ so

$$\sup(r_g(f'), r_g(f'')) \leq r_g(|f'| \vee |f''|) \leq r_g(f''),$$
$$r_g(|f'| \vee |f''|) = \sup(r_g(f'), r_g(f'')).$$

By Corollary 5.8.24 $\mathscr{K}(T, E)$ is an M-space and so it possesses the strong D. P.-property (Corollary 5.7.9) and the D-property (Theorem 5.8.9 c)). □

Remarks. 1) A. Grothendieck showed (1953) that $\mathscr{K}(T, \mathbb{R})$ possesses the D-property ([2] Théorème 6) the D. P.-property, and the strict D. P.-property ([2] Théorème 1 and Corollaire of Lemme 3) for T compact. C. Constantinescu extended these last two results to the case when T is locally compact and paracompact and showed that $\mathscr{K}(T, \mathbb{R})$ is an M-space when T is locally compact and σ-compact (1969) ([1] Corollary 4.14).

2) $\mathscr{K}(T, \mathbb{R})$ is not always a quasi-M-space if T is locally compact and paracompact but not σ-compact. Indeed, let T be an uncountable set endowed with the discrete topology. Then

$$U := \{f \in \mathscr{K}(T, \mathbb{R}) \mid \sum_{t \in T} f(t) < 1\}$$

is a weak 0-neighbourhood in $\mathscr{K}(T, \mathbb{R})$. Assume that $\mathscr{K}(T, \mathbb{R})$ is a quasi-M-space. Then there exists a solid directed 0-neighbourhood $\mathscr{U}$ for the Mackey topology of $\mathscr{K}(T, \mathbb{R})$ such that $\mathscr{U} \subset U$. For every $t \in T$ there exists a strictly positive real number α_t such that $\alpha_t 1_{\{t\}} \in \mathscr{U}$. We get

$$\sum_{t \in A} \alpha_t 1_{\{t\}} = \bigvee_{t \in A} \alpha_t 1_{\{t\}} \in \mathscr{U}$$

so $\sum_{t \in A} \alpha_t < 1$ for any finite subset A of T. This implies T is countable, which is a contradiction.

Proposition 5.8.32. *Let E be an M-space, T be a topological space, Φ be a set of filters on T and $\mathscr{F}$ be the set of continuous maps f of T into E such that $f(\mathfrak{F})$ is an $\mathbb{R}$-bounded filter on E for any $\mathfrak{F} \in \Phi$. Then*

a) $\mathscr{F}$ is a vector sublattice of E^T;

b) if $\mathscr{F}_\Phi$ is Hausdorff, then it is an M-space so it possesses the strong D.P.-property and the D-property.

By Proposition 2.1.27 $\mathscr{F}$ is a vector subspace of E^T. By Corollary 5.8.24 there exists a set $\mathscr{P}$ of seminorms on E generating its topology such that

$$p(|x| \vee |y|) = \sup(p(x), p(y))$$

for any $p \in \mathscr{P}$ and any $x, y \in E$. For any $(p, \mathfrak{F}) \in \mathscr{P} \times \Phi$ we denote by $p_{\mathfrak{F}}$ the map

$$\mathscr{F} \to \mathbb{R}, \quad f \mapsto \inf_{A \in \mathfrak{F}} \sup_{t \in A} p(f(t)).$$

By Proposition 2.1.28 $\{p_{\mathfrak{F}} \mid (p, \mathfrak{F}) \in \mathscr{P} \times \Phi\}$ is a set of seminorms on $\mathscr{F}_\Phi$ generating its topology.

Let $f, g \in \mathscr{F}$ and $(p, \mathfrak{F}) \in \mathscr{P} \times \Phi$. By Proposition 5.8.28 $|f| \vee |g|$ is continuous. We have

$$\begin{aligned}\sup_{t \in A} p((|f| \vee |g|)(t)) = \sup_{t \in A} p(|f(t)| \vee |g(t)|) = \sup_{t \in A} \sup(p(f(t)), p(g(t))) = \\ = \sup(\sup_{t \in A} p(f(t)), \sup_{t \in A} p(g(t)))\end{aligned}$$

for any $A \in \mathfrak{F}$. First, we deduce that $|f| \vee |g| \in \mathscr{F}$ so $\mathscr{F}$ is a sublattice of E^T. Furthermore,

$$p_{\mathfrak{F}}(|f| \vee |g|) = \sup(p_{\mathfrak{F}}(f), p_{\mathfrak{F}}(g)).$$

By Corollary 5.8.24 $\mathscr{F}_\Phi$ is an M-space. By Corollary 5.7.9 it possesses the strong D. P.-property and, by Theorem 5.8.9 c), it possesses the D-property. □

Remark. Let T be a normal space, $\mathfrak{A}$ be a set of subsets of T and $\mathscr{F}$ be the set of continuous real functions on T which are bounded on any $A \in \mathfrak{A}$. By the above proposition $\mathscr{F}_{\mathfrak{A}}$ possesses the D. P.-property and the strict D. P.-property. This result was proved by A. Grothendieck (1953) ([2] Proposition 6 (a)).

Theorem 5.8.33. *Let T be a paracompact space, $\mathfrak{K}$ be the set of compact sets of T, $\mathfrak{R}$ be the δ-ring generated by $\mathfrak{K}$, $\mathscr{C}$ be the vector space of continuous real functions on T, $\mathfrak{L}$ be the set $\{\{f = 0\} \mid f \in \mathscr{C}\}$, E be a locally convex space, Φ be the set $\hat{\Phi}_4(E'_E)$ and $\mu \in \mathscr{M}(\mathfrak{R}, E'^*; \mathfrak{K})$ such that $f \in \bigcap_{x' \in E'} \mathscr{L}^1(x' \circ \mu)$ and $\int f d\mu \in E$ for any $f \in \mathscr{C}$, where $\int g d\mu$ denotes the map*

$$E' \to \mathbb{R}, \quad x' \mapsto \int g d(x' \circ \mu)$$

for any $g \in \bigcap_{x' \in E'} \mathscr{L}^1(x' \circ \mu)$. Then the following hold:

a) $1_L \in \bigcap_{x' \in E'} \mathscr{L}^1(x' \circ \mu)$ for any $L \in \mathfrak{L}$;

b) if $\int 1_L d\mu \in E$ for any $L \in \mathfrak{L}$ and E is sequentially complete, then $(\int f_n d\mu)_{n \in \mathbb{N}}$ is weakly convergent for any increasing weak Cauchy sequence $(f_n)_{n \in \mathbb{N}}$ in $\mathscr{C}_{\mathfrak{K}}$;

c) if $(\int f_n d\mu)_{n \in \mathbb{N}}$ is weakly convergent for any increasing weak Cauchy sequence $(f_n)_{n \in \mathbb{N}}$ in $\mathscr{C}_{\mathfrak{K}}$, then the map

$$\mathscr{C}_{(\mathscr{C}_{\mathfrak{K}})'} \to E, \quad f \mapsto \int f d\mu$$

is uniformly Φ_4-continuous and $\int 1_L d\mu \in E$ for any $L \in \mathfrak{L}$;

d) if $\int 1_L d\mu \in E$ for any $L \in \mathfrak{L}$ and E is quasicomplete, then $\{\int f d\mu \mid f \in \mathscr{F}\}$ is weakly relatively compact for any bounded set $\mathscr{F}$ of $\mathscr{C}_{\mathfrak{K}}$, $\mu \in \mathscr{M}(\mathfrak{R}, E; \mathfrak{K})_c$, the map

$$\mathscr{C}_{(\mathscr{C}_{\mathfrak{K}})'} \to E_\Phi, \quad f \mapsto \int f d\mu$$

is uniformly Φ_4-continuous, and $\xi \in \hat{\mathscr{L}}^1(\mu)$ and $\int \xi d\mu \in E$ for any $\xi \in \mathscr{M}(\mathfrak{R}, \mathbb{R}; \mathfrak{K})_c^\pi$.

a) By Corollary 5.6.16 b ⇒ a $x' \circ \mu$ has a compact carrier for any $x' \in E'$ so $1_L \in \bigcap_{x' \in E'} \mathscr{L}^1(x' \circ \mu)$.

b) We may assume $f_0 \geq 0$. Set $f := \bigvee_{n \in \mathbb{N}}^{\mathbb{R}^T} f_n$ and let A' be an equicontinuous set of E'. By Theorem 5.6.22 c ⇒ d, $\alpha := \sup_{x' \in A'} \|x' \circ \mu\| < \infty$ and there exists $K \in \mathfrak{K}$ such that

$$\bigcup_{x' \in A'} \operatorname{Supp} x' \circ \mu \subset K.$$

Assume that f is not bounded on K. Then there exists a sequence $(t_n)_{n \in \mathbb{N}}$ in K such that $f(t_n) > 2^n$ for any $n \in \mathbb{N}$. If δ_t denotes the Dirac measure at t on T for any $t \in T$, then

$$\nu := \sum_{n \in \mathbb{N}} \frac{1}{2^n} \delta_{t_n} \in \mathcal{M}(\mathfrak{R}, \mathbb{R}; \mathfrak{K})$$

and

$$\lim_{n \to \infty} \int f_n d\nu = \sum_{n \in \mathbb{N}} \frac{1}{2^n} f(t_n) = \infty ,$$

which is a contradiction. Hence, f is bounded on K. We set

$$g_n := \frac{1}{2^n} \sum_{m=1}^{n 2^n} 1_{\{f > \frac{m}{2^n}\}}$$

for any $n \in \mathbb{N}$. $(g_n)_{n \in \mathbb{N}}$ is an increasing sequence in $\mathbb{R}^T$ which converges uniformly on K to f. It is easy to see that $\{f \leq \beta\} \in \mathfrak{L}$ for any $\beta \in \mathbb{R}$. We deduce that $\int g_n d\mu \in E$ for any $n \in \mathbb{N}$. There exists $n_0 \in \mathbb{N}$ such that $|g_m - g_n| < \dfrac{1}{1+\alpha}$ on K for any $m, n \in \mathbb{N}$ so that $m \geq n_0, n \geq n_0$. Then

$$|< \int g_m d\mu - \int g_n d\mu, x' >| = |\int g_m d(x' \circ \mu) - \int g_n d(x' \circ \mu)| \leq$$

$$\leq \int |g_m - g_n| d|x' \circ \mu| < \frac{\alpha}{1+\alpha} < 1$$

for any $m, n \in \mathbb{N}$ with $m \geq n_0, n \geq n_0$ and any $x' \in A'$. Hence,

$$\int g_m d\mu - \int g_n \, d\mu \in A'^0$$

for any $m, n \in \mathbb{N}$ so that $m \geq n_0, n \geq n_0$. Since A' is arbitrary $(\int g_n d\mu)_{n \in \mathbb{N}}$ is a Cauchy sequence and therefore a convergent sequence in E. Put

$$x := \lim_{n \to \infty} \int g_n d\mu .$$

We have

$$\lim_{n \to \infty} < \int f_n d\mu, x' > = \lim_{n \to \infty} \int f_n d(x' \circ \mu) = \int f d(x' \circ \mu) = \lim_{n \to \infty} \int g_n d(x' \circ \mu) =$$
$$= \lim_{n \to \infty} < \int g_n d\mu, x' > = \langle x, x' \rangle$$

for any $x' \in E'$ so $(\int f_n d\mu)_{n \in \mathbb{N}}$ converges weakly to x in E.

c) The second assertion follows immediately from a) and from the hypothesis of c). In order to prove the first assertion, we may assume that E is complete. $\mathscr{C}_{\mathfrak{K}}$ is an M-space (Proposition 5.8.32 b)) and the assertion follows from Theorem 5.8.9 b).

d) $\mathscr{C}_{\mathfrak{K}}$ is an M-space (Proposition 5.8.32 b)). By b) and Theorem 5.8.9 a), b) $\{\int f d\mu \mid f \in \mathscr{F}\}$ is weakly relatively compact for any bounded set $\mathscr{F}$ of $\mathscr{C}_{\mathfrak{K}}$ and the map

$$\mathscr{C}_{(\mathscr{C}_{\mathfrak{R}})'} \to E_{\Phi}, \quad f \mapsto \int f d\mu$$

is uniformly ϕ_4-continuous. Let $K \in \mathfrak{K}$, put

$$\mathscr{G} := \{f \in \mathscr{C} \mid 1 \geq f \geq 1_K^X\},$$

endow $\mathscr{G}$ with the converse order relation, and denote by $\mathfrak{F}$ the section filter of $\mathscr{G}$. By the above result $\{\int f d\mu \mid f \in \mathscr{G}\}$ is a weakly relatively compact set of E. Since

$$\mathscr{G} \to E, \quad f \mapsto \int f d\mu$$

maps $\mathfrak{F}$ into a weak Cauchy filter $\mathfrak{G}$ on E we deduce that $\mathfrak{G}$ weakly converges to a point x of E. We have $\langle x, x' \rangle = \lim_{f,\mathfrak{F}} \langle \int f d\mu, x' \rangle = \lim_{f,\mathfrak{F}} \int f d(x' \circ \mu) = x' \circ \mu(K)$ for any $x' \in E'$. Hence, $\mu(K) = x \in E$. By C. Constantinescu [5] Theorem 4.4.1 $\mu(\mathfrak{R}) \subset E$. We get $\mu \in \mathscr{M}(\mathfrak{R}, E_{E'}; \mathfrak{K})$. By Corollary 4.6.4 b) $\mu \in \mathscr{M}(\mathfrak{R}, E; \mathfrak{K})$. By Theorem 5.6.22 c $\Rightarrow$ d for any equicontinuous set A' of E' there exists $K \in \mathfrak{K}$ such that $\bigcup_{x' \in A'} \operatorname{Supp}(x' \circ \mu) \subset K$. Let $\xi \in \mathscr{M}(\mathfrak{R}, \mathbb{R}; \mathfrak{K})_c^{\pi}$. We have $\xi \in \mathscr{L}^1(\mu)$. By C. Constantinescu [5] Theorem 4.4.1 $\int \xi 1_A d\mu \in E$ for any $A \in \mathfrak{R}$ so, by Proposition 5.6.25, $\int \xi d\mu$ belongs to the completion of E. Since $\{\int f d\mu \mid f \in \mathscr{F}\}$ is weakly relatively compact for any bounded set of $\mathscr{C}_{\mathfrak{K}}$, $\int \xi d\mu \in E$. □

Remark. C. Constantinescu proved (1980) ([5] Theorem 4.4.7) that if T is locally compact and σ-compact, if $\int 1_L d\mu \in E$ for any $L \in \mathfrak{L}$ and if E is complete, then $\{\int f d\mu \mid f \in \mathscr{F}\}$ is weakly relatively compact for any bounded set $\mathscr{F}$ of $\mathscr{C}_{\mathfrak{K}}$ and $\{\int f d\mu \mid f \in \mathscr{F}\}$ is relatively compact for any weak Φ_4-set $\mathscr{F}$ of $\mathscr{C}_{\mathfrak{K}}$. These results follow from d).

Proposition 5.8.34. *Let E be the solid vector subspace of $\mathbb{R}^{\mathbb{N}}$ formed by the bounded functions, $\mathscr{F}$ be the set of strictly positive real functions on $\mathbb{N}$ so that $\lim_{n\to\infty} f(n) = \infty$, and for any $f \in \mathscr{F}$ let $\bar{f}$ be the semi-norm*

$$E \to \mathbb{R}, \quad g \mapsto \inf\{\alpha \in \mathbb{R}_+ \mid |g| \leq \alpha f\}.$$

Then E endowed with the topology generated by the set $\{\bar{f} \mid f \in \mathscr{F}\}$ of semi-norms is a complete, separable, compactly generated, order complete and non-reflexive M-space, possessing no copy of c_0.

By Corollary 5.8.24 E is an M-space. It is obviously a separable complete and order complete space. A subset A of E is bounded iff there exists $\alpha \in \mathbb{R}_+$ such that $|g| \leq \alpha$ for any $g \in A$. Hence, any bounded set of E is relatively compact and E is compactly generated. By Theorem 5.8.20 b $\Rightarrow$ a E contains no copy of c_0. The set $\{g \in E \mid |g| \leq 1\}$ is a barrel which is not a 0-neighbourhood, consequently E cannot be reflexive (it is easy to see that the bidual of E is isomorphic to ℓ^{∞}). □

Remark. If we endow $\mathbb{N}$ with the discrete topology then the above space E is the vector lattice of continuous bounded real functions on $\mathbb{N}$ endowed with the strict topology.

Proposition 5.8.35. *Let X be a topological space and $\mathscr{C}$ be the vector lattice of continuous real functions on X. Then the following assertions are equivalent:*

a) $\mathscr{C}$ is a weakly sequentially complete and order σ-complete M-space;

b) $\mathscr{C}$ is sequentially complete;

c) for any sequence $(f_n)_{n\in\mathbb{N}}$ in $\mathscr{C}$ for which $\sup\limits_{n\in\mathbb{N}} f_n(x) < \infty$ for any $x \in X$, the map

$$X \to \mathbb{R}, \quad x \mapsto \sup_{n\in\mathbb{N}} f_n(x)$$

is continuous;

d) for any upper bounded sequence $(f_n)_{n\in\mathbb{N}}$ in $\mathscr{C}$ the map

$$X \to \mathbb{R}, \quad x \mapsto \sup_{n\in\mathbb{N}} f_n(x)$$

is continuous;

e) the intersection of any sequence of exact open sets of X is an exact open set;

f) the intersection of any sequence of exact open sets of X is open.

a $\Rightarrow$ b is trivial.

b $\Rightarrow$ c. We set $g_n := \bigvee\limits_{m=0}^{n} f_m$ for any $n \in \mathbb{N}$. Then $(g_n)_{n\in\mathbb{N}}$ is a Cauchy sequence in $\mathscr{C}$. Let f be the limit point of $(g_n)_{n\in\mathbb{N}}$ in $\mathscr{C}$. Then

$$f(x) = \lim_{n\to\infty} g_n(x) = \sup_{n\in\mathbb{N}} f_n(x)$$

for any $x \in X$ so the map

$$X \to \mathbb{R}, \quad x \mapsto \sup_{n\in\mathbb{N}} f_n(x)$$

is continuous.

c $\Rightarrow$ d is trivial.

d $\Rightarrow$ e. Let $(U_n)_{n\in\mathbb{N}}$ be a sequence of exact open sets of X. Without loss of generality we may assume that $(U_n)_{n\in\mathbb{N}}$ decreases. For any $n \in \mathbb{N}$ there exists $f_n \in \mathscr{C}$ such that

$$U_n = \{x \in X \mid f_n(x) \neq 0\}.$$

We set

$$g_n : X \to \mathbb{R}, \quad x \mapsto \sup_{m\in\mathbb{N}} \inf(|f_1(x)|, m|f_n(x)|)$$

for any $n \in \mathbb{N}$ and

$$f : X \to \mathbb{R}, \quad x \mapsto \inf_{n\in\mathbb{N}} g_n(x).$$

By d) $g_n \in \mathscr{C}$ for any $n \in \mathbb{N}$ and $f \in \mathscr{C}$. From

$$\bigcap_{n\in\mathbb{N}} U_n = \{x \in X \mid f(x) \neq 0\}$$

it follows that $\bigcap_{n \in \mathbb{N}} U_n$ is an exact open set of X.

e ⇒ f is trivial.

f ⇒ a. $\mathscr{C}$ is obviously an M-space. Let $(f_n)_{n \in \mathbb{N}}$ be a weak Cauchy sequence in $\mathscr{C}$ and f be the map

$$X \to \mathbb{R}, \quad x \mapsto \lim_{n \to \infty} f_n(x).$$

We have to show that f is continuous. Let $x \in X$ and $\alpha, \beta \in \mathbb{R}$ such that $\alpha < f(x) < \beta$. Then

$$U := \bigcup_{n \in \mathbb{N}} \bigcap_{m \geq n} \overset{-1}{f_m}(]\alpha, \beta[)$$

is a neighbourhood of x and $f(y) \in [\alpha, \beta]$ for any $y \in U$. Hence, f is continuous and $\mathscr{C}$ is weakly sequentially complete. We deduce that $\mathscr{C}$ contains no copy of c_0. By Theorem 5.8.17 b) $\mathscr{C}$ is order σ-complete. □

Remark. There exist nondiscrete paracompact spaces possessing the above properties (L. Gillman – M. Jerison [1] 4 N).

§ 5.9 Strict M-spaces

Throughout this section we shall write locally convex space instead of Hausdorff locally convex space.

Definition 5.9.1. *A* strict M-space *is an M-space E for which any increasing weak Cauchy sequence in E_+ possesses a supremum.*

The Banach lattice c_0 of sequences of real numbers converging to 0 is an order complete M-space but not a strict M-space.

Proposition 5.9.2. *Let E be an M-space, φ be the evaluation map $E \to E''$ and X'' be the set*

$$\{\bigvee_{n \in \mathbb{N}} \varphi(x_n) \mid (x_n)_{n \in \mathbb{N}} \text{ is an increasing sequence in } E_+\}.$$

Then the following assertions are equivalent:

a) E is a strict M-space;

b) E is order σ-complete and X'' is contained in the solid vector subspace of E'' generated by $\varphi(E)$.

The assertion follows immediately from Proposition 5.7.19 a ⇔ d. □

Remark. There exists a strict M-space which is neither order complete nor topologically complete. Let $\mathscr{F}$ be the set of real functions on $\mathbb{R} \times \mathbb{R}$ such that $f(\cdot, y)$ is a bounded Borel function for any $y \in \mathbb{R}$ and such that the set $\{y \in \mathbb{R} \mid f(\cdot, y) \neq 0\}$ is countable. $\mathscr{F}$ is a vector sublattice of $\mathbb{R}^{\mathbb{R} \times \mathbb{R}}$ and

$$\mathscr{F} \to \mathbb{R}, \quad f \mapsto \sup_{x \in \mathbb{R}} |f(x,y)|$$

is a semi-norm on $\mathscr{F}$ for any $y \in \mathbb{R}$. It is easy to see that $\mathscr{F}$ endowed with the above structures is a strict M-space which is neither order complete nor topologically complete.

Corollary 5.9.3. *Let E be a strict M-space and $\mathfrak{A}$ be the set of bounded sets of E. Then:*

a) the Cauchy sequences and the Φ_3-sets of E'_E and of $E'_{E''}$ coincide;

b) any supersummable sequence in E'_E is supersummable in $E'_{\mathfrak{A}}$.

a) Let $(x'_n)_{n \in \mathbb{N}}$ be a Cauchy sequence in E'_E. There exists a countable set $\mathfrak{U}$ of directed solid 0-neighbourhoods in E such that $\{x'_n \mid n \in \mathbb{N}\} \subset \bigcup_{U \in \mathfrak{U}} U^0$. By Proposition 5.9.2 a $\Rightarrow$ b, Theorem 5.7.25, and Proposition 5.7.21 a) $(x'_n)_{n \in \mathbb{N}}$ is a Cauchy sequence in $E'_{E''}$. By Theorem 1.8.4 e $\Rightarrow$ h the Φ_3-sets of E'_E and of $E'_{E''}$ coincide.

b) By a) the identity map $E'_E \to E'_{E''}$ is sequentially continuous. By Proposition 3.3.20 any supersummable sequence in E'_E is supersummable in $E'_{E''}$ and the assertion follows from Orlicz-Pettis theorem (Corollary 3.5.12). □

Proposition 5.9.4. *Let E be a metrizable strict M-space and $\mathfrak{A}$ be the set of bounded sets of E. Then*

a) the identity map $E'_E \to E'_{E''}$ is uniformly Φ_3-continuous;

b) any supersummable family in E'_E is supersummable in $E'_{\mathfrak{A}}$.

a) follows immediately from Proposition 5.9.2 a $\Rightarrow$ b, Proposition 5.7.21 a), and Proposition 5.8.2.

b) By a) and the Propositions 1.8.3 and 3.3.20 any supersummable family in E'_E is supersummable in $E'_{E''}$ and the assertion follows from Orlicz-Pettis theorem (Corollary 3.5.12). □

Corollary 5.9.5. *Let E be a Frechet strict M-space and $\tilde{E}'$ be the set E' endowed with a uniformity such that the identity maps $E'_{E''} \to \tilde{E}' \to E'_E$ are uniformly Φ_3-continuous. Then $\tilde{E}'$ is Φ_3-compact.*

By Proposition 2.1.30 E'_E is Φ_5-compact and, by Proposition 5.9.4, the identity map $E'_E \to E'_{E''}$ is uniformly Φ_3-continuous. We deduce that the identity map $E'_E \to \tilde{E}'$ is uniformly Φ_3-continuous (Proposition 1.8.5) and the assertion follows from Proposition 1.8.11. □

Proposition 5.9.6. *Let E be a barreled strict M-space, A' be a Φ_3-set of E'_E and $\bar{A}'$ be its closure in E'_E. Then $\bar{A}'$ is the closure of A' in $E'_{E''}$ and $\bar{A}'_E = \bar{A}'_{E''}$.*

A' is a bounded set of E'_E. Since E is a barreled M-space there exists a solid directed 0-neighbourhood U in E such that $A' \subset U^0$. We get $\bar{A}' \subset U^0$. By Theorem 5.7.25, Proposition 5.7.21 a), and Corollary 1.8.15 the identity map $U^0_E \to U^0_{E''}$ is uniformly Φ_3-continuous so, by Proposition 1.8.8, $\bar{A}'$ is the closure of A' in $E'_{E''}$ and $\bar{A}'_E = \bar{A}'_{E''}$. □

Proposition 5.9.7. *Let $\mathfrak{R}$ be a ring of sets, $\mathfrak{K}$ be a subset of $\mathfrak{R}$ which is closed with respect to finite unions and E be a strict M-space. Then:*

a) $\mathscr{M}(\mathfrak{R}, E'_E) = \mathscr{M}(\mathfrak{R}, E'_{E''})$, $\mathscr{E}(\mathfrak{R}, E'_E) = \mathscr{E}(\mathfrak{R}, E'_{E''})$;

b) if $\mathfrak{R}$ is a quasi-δ-ring and E is metrizable or barreled, then

$$\mathscr{M}(\mathfrak{R}, E'_E; \mathfrak{K}) = \mathscr{M}(\mathfrak{R}, E'_{E''}; \mathfrak{K}), \quad \mathscr{E}(\mathfrak{R}, E'_E; \mathfrak{K}) = \mathscr{E}(\mathfrak{R}, E'_{E''}; \mathfrak{K}).$$

a) By Corollary 5.9.3 a) the identity map $E'_E \to E'_{E''}$ is sequentially continuous and the assertion follows from Proposition 4.1.17.

b) Let $\mu \in \mathscr{M}(\mathfrak{R}, E'_E; \mathfrak{K})$ ($\mu \in \mathscr{E}(\mathfrak{R}, E'_E; \mathfrak{K})$) and $A \in \mathfrak{R}$. By Proposition 4.1.26 a $\Rightarrow$ b and Proposition 4.1.16 $\mu(\mathfrak{G}(A, \mathfrak{K}))$ is a Φ_3-filter, which converges to $\mu(A)$ in E'_E.

If E is metrizable, then by Proposition 5.9.4 the identity map $E'_E \to E'_{E''}$ is uniformly Φ_3-continuous so $\mu(\mathfrak{G}(A, \mathfrak{K}))$ converges to $\mu(A)$ in $E'_{E''}$ (Proposition 1.8.3). By Proposition 4.1.26 b $\Rightarrow$ a, μ is $\mathfrak{K}$-regular with respect to $E'_{E''}$.

Now assume that E is barreled. By Corollary 4.7.8 $A' := \{\mu(B) \mid B \in \mathfrak{R}, B \subset A\}$ is a bounded set of E'_E. Let $\mathfrak{F}$ be the filter induced by $\mu(\mathfrak{G}(A, \mathfrak{K}))$ on A'. By Proposition 5.9.6 $A'_E = A'_{E''}$ so $\mathfrak{F}$ converges to $\mu(A)$ in $A'_{E''}$. We deduce that $\mu(\mathfrak{G}(A, \mathfrak{K}))$ converges to $\mu(A)$ in $E'_{E''}$. By Proposition 4.1.26 b $\Rightarrow$ a, μ is $\mathfrak{K}$-regular with respect to $E'_{E''}$. □

Proposition 5.9.8. *Let E be a strict M-space, F be a locally convex space and $u: E \to F$ be a continuous order exhaustive linear map. Then the map $E_{E'} \to F$ defined by u is uniformly Φ_4-continuous. If, in addition, F is quasicomplete, then u is boundedly weakly compact.*

We may assume that F is quasicomplete. Let $(x_n)_{n \in \mathbb{N}}$ be an increasing weak Cauchy sequence in E. Then $(x_n)_{n \in \mathbb{N}}$ is upper bounded and by Proposition 5.9.2 a $\Rightarrow$ b and Proposition 5.2.6, $(u(x_n))_{n \in \mathbb{N}}$ is convergent. Since $(x_n)_{n \in \mathbb{N}}$ is arbitrary the assertions follows from Theorem 5.8.9 a), b). □

Theorem 5.9.9. *Let E be a sequentially complete locally convex space. Then the following assertions are equivalent:*

a) E contains no copy of ℓ^∞;

b) for any continuous linear map u of a strict M-space F into E the map $F_{F'} \to E$ defined by u is uniformly Φ_4-continuous;

c) for any continuous linear map $u:\ell^\infty \to E$ the map $(\ell^\infty)_{(\ell^\infty)'} \to E$ defined by u preserves the Cauchy sequences.

If E is quasicomplete, then the above relations are equivalent to the following;

d) any continuous linear map u of a strict M-space into E is boundedly weakly compact;

e) any continuous linear map $\ell^\infty \to E$ is weakly compact.

a ⇒ b and a ⇒ d. By Proposition 5.8.13 d ⇒ c u is order exhaustive and the assertions follows from Proposition 5.9.8.

b ⇒ c and d ⇒ e are trivial since ℓ^∞ endowed with the order relation induced by $\mathbb{R}^{\mathbb{N}}$ is a strict M-space.

c ⇒ a and e ⇒ a. Assume a) does not hold. Then there exist a topological vector subspace F of E and an isomorphism of topological vector spaces $u:\ell^\infty \to F$. If v denotes the inclusion map $F \to E$, then $v \circ u$ is a continuous linear map of ℓ^∞ into E.

First, assume that c) holds. For any $n \in \mathbb{N}$ we denote by x_n the element of ℓ^∞ which is equal to 1 on $\{m \in \mathbb{N} \mid m \leq n\}$ and equal to 0 otherwise. Then $(x_n)_{n \in \mathbb{N}}$ is a weak Cauchy sequence in ℓ^∞ but $(v \circ u(x_n))_{n \in \mathbb{N}}$ is not a Cauchy sequence in E, which is a contradiction.

Now assume that e) holds. Then F possesses a weakly compact 0-neighbourhood and the same has to be valid for ℓ^∞, which is an obvious contradiction. □

Remark. H. P. Rosenthal proved the implication a ⇒ e in the case when E is a Banach space (1970) ([1] Corollary 1.4).

Corollary 5.9.10. *The following assertions are equivalent for any quasicomplete strict M-space E:*

a) E contains no copy of c_0;

b) E contains no copy of ℓ^∞;

c) any bounded set of E is relatively compact;

d) E is semi-reflexive;

e) E is weakly sequentially complete;

If E is complete then these assertions are equivalent to the following:

f) E is weakly Φ_4-compact.

a ⇒ b is trivial.

b ⇒ c. By Theorem 5.9.9 a ⇒ b & d the identity map $E \to E$ is boundedly weakly compact and the map $E_{E'} \to E$ is uniformly Φ_4-continuous and therefore Φ_2-continuous (Proposition 1.8.3). By Lemma 5.8.19 any bounded set of E is relatively compact.

c ⇒ d ⇒ e is trivial.

e ⇒ f. By Corollary 2.3.19 E is weakly Φ_2-compact and by e) the weak Φ_2-filters and the weak Φ_4-filters coincide.

The deductions e ⇒ a and f ⇒ a are trivial. □

Remark. For any set I the locally convex lattice $\mathbb{R}^I$ is a reflexive complete strict M-space; consequently the above equivalences do not imply that E is finite dimensional. The example given in Proposition 5.8.34 shows that E need not be reflexive.

Definition 5.9.11. *A* G-space *is a locally convex space E for which the identity map $E_{\mathfrak{A}} \to E$ is uniformly Φ_4-continuous, where $\mathfrak{A}$ denotes the set of equicontinuous Φ_1-sets of E'_E.*

G is an abbreviation of Grothendieck, who considered a similar topology (1953) namely in [2] Théorème 5. It will be proved later (Corollary 5.9.26) that any G-space possesses no copy of ℓ^∞.

Proposition 5.9.12. *Let E, F be locally convex spaces, u be a continuous linear map of E into F, $i, j \in \{1, 2, 3, 4, 5\}$ and $\mathfrak{A}$ ($\mathfrak{B}$ resp.) be the set of equicontinuous Φ_i-sets of E'_E (F'_F resp.). If the identity map $E_{E'} \to E_{\mathfrak{A}}$ is uniformly Φ_j-continuous, then the map*

$$E_{E'} \to F_{\mathfrak{B}}, \quad x \mapsto u(x)$$

is uniformly Φ_j-continuous. If, in addition, F is a G-space and $j \leq 4$, then the map

$$E_{E'} \to F, \quad x \mapsto u(x)$$

is uniformly Φ_j-continuous.

Let u' be the adjoint map of u. Then $u'(B) \in \mathfrak{A}$ for any $B \in \mathfrak{B}$ so the map

$$E_{\mathfrak{A}} \to F_{\mathfrak{B}}, \quad x \mapsto u(x)$$

is uniformly continuous. Hence, the map

$$E_{E'} \to F_{\mathfrak{B}}, \quad x \mapsto u(x)$$

is uniformly Φ_j-continuous. If F is a G-space, then the identity map $F_{\mathfrak{B}} \to F$ is uniformly Φ_4-continuous so (Corollary 1.8.5) the map

$$E_{E'} \to F, \quad x \mapsto u(x)$$

is uniformly Φ_j-continuous. □

Corollary 5.9.13. *Let E be a strict M-space, F be a locally convex space, $\mathfrak{A}$ be the set of equicontinuous Φ_3-sets of F'_F, and u be a continuous linear map of E into F. Then:*

a) the map $u: E_{E'} \to F_{\mathfrak{A}}$ is uniformly Φ_4-continuous;

b) if F is a G-space, then the map $u: E_{E'} \to F$ is uniformly Φ_4-continuous;

c) if F is a quasicomplete G-space, then u is boundedly weakly compact.

a & b. Let us denote by $\mathfrak{B}$ the set of equicontinuous Φ_3-sets of E'_E. By Theorem 5.7.6 a) and Corollary 5.9.3 a) the identity map $E_{E'} \to E_{\mathfrak{B}}$ is uniformly Φ_4-continuous and the assertions follow from Proposition 5.9.12.

c) Let $(x_n)_{n \in \mathbb{N}}$ be a weak Cauchy sequence in E. By b) and Theorem 1.8.4 a $\Rightarrow$ e,

$(u(x_n))_{n\in\mathbb{N}}$ is a convergent sequence in F. By Theorem 5.8.9 a), u is boundedly weakly compact. □

We now give some examples of strict M-spaces.

Lemma 5.9.14. *Let X be a topological space, $\mathscr{C}$ be the vector lattice of continuous real functions on X and $(f_\iota)_{\iota\in I}$ be a nonempty family in $\mathscr{C}$ such that: a) any $x\in X$ possesses a neighbourhood V such that $\sup_{(\iota,x)\in I\times V} f_\iota(x)<\infty$; b) the set*

$$\overline{\bigcup_{\iota\in I}\{x\in X \mid f_\iota(x)>\alpha\}}$$

is open for any $\alpha\in\mathbb{R}$. We set

$$f: X\to\mathbb{R},\quad x\mapsto \sup_{\iota\in I} f_\iota(x),$$

$$g: X\to\mathbb{R},\quad x\mapsto \limsup_{y\to x} f(y).$$

Then $g=\bigvee_{\iota\in I}^{\mathscr{C}} f_\iota$.

Let $x\in X$ and α be a real number such that $\alpha<g(x)$. We set

$$U:=\bigcup_{\iota\in I}\{x\in X \mid f_\iota(x)>\alpha\}.$$

Let V be an arbitrary neighbourhood of x. Then there exists $y\in V$ such that $f(y)>\alpha$. We have $y\in U$ so $U\cap V\neq\emptyset$. Since V is arbitrary we get $x\in\bar{U}$. Since $f>\alpha$ on U we have $g\geq\alpha$ on $\bar{U}$. By b) $\bar{U}$ is open so g is lower semi-continuous. But g is also upper semi-contrinuous so $g\in\mathscr{C}$.

g is obviously an upper bound of $(f_\iota)_{\iota\in I}$ in $\mathscr{C}$. Let h be another upper bound of $(f_\iota)_{\iota\in I}$ in $\mathscr{C}$. We have $f\leq h$ so $g\leq h$. Hence, $g=\bigvee_{\iota\in I}^{\mathscr{C}} f_\iota$. □

Lemma 5.9.15. *Let X be a completely regular topological space, $\mathscr{C}$ be the vector lattice of continuous real functions on X, $\mathscr{C}_b$ be the solid vector subspace of $\mathscr{C}$ formed by the bounded functions of $\mathscr{C}$, βX be the Stone-Čech compactification of X and $\beta\mathscr{C}$ be the vector lattice of continuous real functions on βX. Then the following assertions are equivalent:*

a) $\mathscr{C}$ is order σ-complete;

b) $\mathscr{C}_b$ is order σ-complete;

c) the closure in X of any exact open set of X is an open set of X;

d) $\beta\mathscr{C}$ is order σ-complete;

e) the closure of any open σ-compact set of βX is open.

All these conditions are satisfied if the intersection of any sequence of exact open sets of X is open.

a ⇒ b is trivial.

b ⇒ c. Let U be an exact open set of X. There exists $f \in \mathscr{C}$ such that

$$U = \{x \in X \mid f(x) \neq 0\}.$$

We set

$$f_n := (n|f|) \wedge 1_X$$

for any $n \in \mathbb{N}$. $(f_n)_{n \in \mathbb{N}}$ is an upper bounded sequence in $\mathscr{C}_b$ so it possesses a supremum g in $\mathscr{C}_b$. Then $g = 1$ on U so $g = 1$ on $\bar{U}$. Since X is completely regular $g = 0$ on $X \setminus \bar{U}$. Hence, $\bar{U}$ is open.

c ⇒ a follows from Lemma 5.9.14.

b ⇔ d follows from the fact that $\mathscr{C}_b$ and $\beta\mathscr{C}$ are isomorphic in their role of vector lattices.

d ⇔ e follows from a ⇔ c and the fact that on βX the exact open sets are exactly the open σ-compact sets.

The last assertion follows from Proposition 5.8.35 f ⇒ d. □

Remark. H. Nakano proved the equivalence a ⇔ c for the case when X is normal (1941) ([1] Satz 2 and Satz 4).

Theorem 5.9.16. *Let X be a completely regular space such that the closure of any exact open set of X is open, $\mathscr{C}_b$ be the vector lattice of bounded continuous real functions on X endowed with the strict topology, $\mathfrak{K}$ be the set of compact sets of X, $\mathfrak{R}$ be the σ-ring generated by the closed sets of X and, for any $\mu \in \mathscr{M}(\mathfrak{R}, \mathbb{R}; \mathfrak{K})$, let μ' be the map*

$$\mathscr{C}_b \to \mathbb{R}, \quad f \mapsto \int f \, d\mu.$$

Then the following hold:

a) $\mathscr{C}_b$ is a strict M-space;

b) $\mu' \in \mathscr{C}_b'$ for any $\mu \in \mathscr{M}(\mathfrak{R}, \mathbb{R}; \mathfrak{K})$ and the map

$$\mathscr{M}(\mathfrak{R}, \mathbb{R}; \mathfrak{K}) \to \mathscr{C}_b', \quad \mu \mapsto \mu'$$

is an isomorphism of vector lattices;

c) the identity map

$$\mathscr{M}(\mathfrak{R}, \mathbb{R}; \mathfrak{K})_{\mathscr{C}_b} \to \mathscr{M}(\mathfrak{R}, \mathbb{R}; \mathfrak{K})_{\mathscr{C}_b''}$$

is uniformly Φ_3-continuous.

d) for any quasicomplete locally convex space E containing no copy of ℓ^∞ and any continuous linear map $u: \mathscr{C}_b \to E$ is boundedly weakly compact and there exists $\mu \in \mathscr{M}(\mathfrak{R}, E: \mathfrak{K})$ such that $\mathscr{C}_b \subset \mathscr{L}^1(\mu, X)$ and $u(f) = \int f \, d\mu$ for any $f \in \mathscr{C}_b$.

a) By Corollary 5.8.26 $\mathscr{C}_b$ is an M-space. Let $(f_n)_{n \in \mathbb{N}}$ be an increasing weak Cauchy sequence in $(\mathscr{C}_b)_+$. Assume that $(f_n)_{n \in \mathbb{N}}$ is not upper bounded. Then there exists a sequence $(x_n)_{n \in \mathbb{N}}$ in X such that $\sup_{m \in \mathbb{N}} f_m(x_n) > 3^n$ for any $n \in \mathbb{N}$ and

$$l: \mathscr{C}_b \to \mathbb{R}, \quad f \mapsto \sum_{n \in \mathbb{N}} \frac{1}{2^n} f(x_n)$$

is a continuous linear form but

$$\sup_{n \in \mathbb{N}} l(f_n) = \infty,$$

which is a contradiction. Hence, $(f_n)_{n \in \mathbb{N}}$ is upper bounded. By Lemma 5.9.15 c $\Rightarrow$ b the supremum of $(f_n)_{n \in \mathbb{N}}$ in $\mathscr{C}_b$ exists so $\mathscr{C}_b$ is a strict M-space.

b) follows form R. Giles ([1] Theorem 4.6).

c) By Proposition 5.6.3 $\mathscr{M}(\mathfrak{R}, \mathbb{R}; \mathfrak{K})$ is a band of $\mathscr{M}(\mathfrak{R}, \mathbb{R})$; we denote it by $\mathscr{M}$ and use the notations of C. Constantinescu [5] in the sequel. By Corollary 5.6.8 the identity map $\mathscr{M}_{\mathscr{C}_b} \to \mathscr{M}_{\mathscr{M}^\pi}$ is uniformly Φ_3-continuous. A subset $\mathscr{F}$ of $\mathscr{C}_b$ is bounded iff

$$\sup \{|f(x)| \mid (f, x) \in \mathscr{F} \times X\} < \infty$$

so $\mathscr{M}^\pi = \mathscr{C}_b''$ (C. Constantinescu [5] Proposition 3.6.2 b)). Hence, the identity map $\mathscr{M}_{\mathscr{C}_b} \to \mathscr{M}_{\mathscr{C}_b''}$ is uniformly Φ_3-continuous.

d) By a) and Theorem 5.9.9 a $\Rightarrow$ d u is boundedly weakly compact and the assertion follows from C. Constantinescu [5] Propositions 4.2.2 and 4.3.9 a). □

Remark. For a historical commentary concerning the assertion c) of the above theorem, see the remark following Corollary 5.6.8. The fact that the map u of d) is boundedly weakly compact was proved by H. P. Rosenthal (1970) ([1] Theorem 3.7).

Theorem 5.9.17. *Let X be a locally compact σ-compact space such that the closure of any relatively compact σ-compact open set of X is open, let $\mathscr{K}$ be the vector lattice of continuous real functions on X with compact carrier endowed with the topology defined in Theorem 5.8.31, $\mathfrak{K}$ be the set of compact sets of X, $\mathfrak{R}$ be the δ-ring generated by $\mathfrak{K}$ and, for any $\mu \in \mathscr{M}(\mathfrak{R}, \mathbb{R}; \mathfrak{K})$, let μ' be the map*

$$\mathscr{K} \to \mathbb{R}, \quad f \mapsto \int f \, d\mu.$$

Then the following hold:

a) $\mathscr{K}$ is a strict M-space;

b) $\mu' \in \mathscr{K}'$ for any $\mu \in \mathscr{M}(\mathfrak{R}, \mathbb{R}; \mathfrak{K})$ and the map

$$\mathscr{M}(\mathfrak{R}, \mathbb{R}; \mathfrak{K}) \to \mathscr{K}', \quad \mu \mapsto \mu'$$

is an isomorphism of vector lattices;

c) the identity map

$$\mathscr{M}(\mathfrak{R}, \mathbb{R}; \mathfrak{K})_{\mathscr{K}} \to \mathscr{M}(\mathfrak{R}, \mathbb{R}; \mathfrak{K})_{\mathscr{K}''}$$

is uniformly Φ_3-continuous;

d) for any quasicomplete locally convex space E containing no copy of ℓ^∞ and any continuous linear map $u: \mathscr{K} \to E$ there exists $\mu \in \mathscr{M}(\mathfrak{R}, E; \mathfrak{K})$ such that $\mathscr{K} \subset \mathscr{L}^1(\mu, X)$ and $u(f) = \int f\, d\mu$ for any $f \in \mathscr{K}$.

a) By Theorem 5.8.31 $\mathscr{K}$ is an M-space. Let $(f_n)_{n\in\mathbb{N}}$ be an increasing weak Cauchy sequence in $\mathscr{K}_+$ and f be its supremum in $\mathbb{R}^X$. If f is not bounded, then there exists a sequence $(x_n)_{n\in\mathbb{N}}$ in X such that $f(x_n) > 2^n$ for any $n \in \mathbb{N}$. The map

$$l: \mathscr{K} \to \mathbb{R}, \quad g \mapsto \sum_{n\in\mathbb{N}} \frac{1}{2^n} g(x_n)$$

is continuous and linear but

$$\lim_{n\to\infty} l(f_n) = \infty,$$

which is a contradiction. Hence, f is bounded. If the carrier of f is not compact then there exists a sequence $(x_n)_{n\in\mathbb{N}}$ in X with no adherent point such that $f(x_n) > 0$ for any $n \in \mathbb{N}$. The map

$$l: \mathscr{K} \to \mathbb{R}, \quad g \mapsto \sum_{n\in\mathbb{N}} \frac{g(x_n)}{f(x_n)}$$

is linear and continuous but

$$\lim_{n\to\infty} l(f_n) = \infty,$$

which is a contradiction. Hence, f has a compact carrier. By Lemma 5.9.14 $(f_n)_{n\in\mathbb{N}}$ possesses a supremum in $\mathscr{K}$ so $\mathscr{K}$ is a strict M-space.

b) is a classical result.

c) By Proposition 5.6.3 $\mathscr{M}(\mathfrak{R}, \mathbb{R}; \mathfrak{K})$ is a band of $\mathscr{M}(\mathfrak{R}, \mathbb{R})$; we denote it by $\mathscr{M}$ and use the notation of C. Constantinescu [5] in the sequel. Since X is locally compact and paracompact $\mathscr{M}^\pi$ is the solid vector subspace of $\mathscr{M}^\pi$ generated by $\{1_A | A \in \mathfrak{R}\}$ and, by b) and Proposition 5.7.21 c), $\mathscr{M}^\pi = \mathscr{K}''$. Hence, by Corollary 5.6.8, the identity map $\mathscr{M}_{\mathscr{K}} \to \mathscr{M}_{\mathscr{K}''}$ is uniformly Φ_3-continuous.

d) By a) and Theorem 5.9.9 a $\Rightarrow$ d u is boundedly weakly compact and the assertion follows from C. Constantinescu [5] Proposition 4.2.2 and Theorem 4.4.7 a), c). □

Proposition 5.9.18. *Let X be a completely regular space such that the closure of any exact open set is open, $\mathscr{C}$ be the vector lattice of continuous real functions on X, Θ be a set of sequences in X such that the neighbourhood filter of any point of X belongs to $\hat{\Phi}(\Theta)$, and let Φ be a set of filters on X such that $f(\mathfrak{F})$ is an $\mathbb{R}$-bounded filter for any $(f, \mathfrak{F}) \in \mathscr{C} \times \Phi$ (this occurs, e.g., if any filter of Φ is a Φ_2-filter on X) and such that*

$$\{A \,|\, \{x_n \,|\, n \in \mathbb{N}\} \subset A \subset X\} \in \Phi$$

for any sequence $(x_n)_{n\in\mathbb{N}}$ belonging to Θ. Then $\mathscr{C}_\Phi$ is a strict M-space.

By Corollary 1.5.12 and Proposition 2.1.26 $f(\mathfrak{F})$ is an $\mathbb{R}$-bounded filter for any $(f, \mathfrak{F}) \in \mathscr{C} \times \Phi_2(X)$. By Proposition 5.8.32 $\mathscr{C}_\Phi$ is an M-space.

Let $(f_n)_{n\in\mathbb{N}}$ be an increasing weak Cauchy sequence in $\mathscr{C}_\Phi$ and let $x\in X$. We intend to show that $(f_n)_{n\in\mathbb{N}}$ is upper bounded in a neighbourhood of x. Assume the contrary and put

$$f: X \to \overline{\mathbb{R}}, \quad y \mapsto \sup_{n\in\mathbb{N}} f_n(y),$$

$$A_\alpha := \{y\in A \mid f(y) > \alpha\}$$

for any $A\subset X$ and any $\alpha\in\mathbb{R}$ and denote by $\mathfrak{B}$ the neighbourhood filter of x. Then

$$\{V_\alpha \mid (V,\alpha)\in\mathfrak{B}\times\mathbb{R}\}$$

is a filter base. Let $\mathfrak{F}$ be an ultrafilter on X finer than this filter base. $\mathfrak{F}$ is finer than $\mathfrak{B}$ so it belongs to $\Phi(\Theta)$ (Proposition 1.4.2). Let (I, g) be a Θ-net in X such that $g(\mathfrak{G}) = \mathfrak{F}$, where $\mathfrak{G}$ denotes the section filter of I. There exists an increasing sequence $(\iota_n)_{n\in\mathbb{N}}$ in I such that $(f(g(\iota_n)))_{n\in\mathbb{N}}$ converges to ∞. Then $(g(\iota_n))_{n\in\mathbb{N}}$ is a Θ-sequence and we may assume it belongs to Θ. Since $(f_n)_{n\in\mathbb{N}}$ is bounded in $\mathscr{C}_\Phi$ and

$$\{A \mid \{g(\iota_n) \mid n\in\mathbb{N}\}\subset A\subset X\}\in\Phi$$

we get

$$\infty = \sup_{n\in\mathbb{N}} f(g(\iota_n)) = \sup_{n\in\mathbb{N}}\sup_{m\in\mathbb{N}} f_m(g(\iota_n)) = \sup_{m\in\mathbb{N}}\sup_{n\in\mathbb{N}} f_m(g(\iota_n)) < \infty,$$

which is a contradiction. Hence, $(f_n)_{n\in\mathbb{N}}$ is locally upper bounded so, by Lemma 5.9.14, it possesses a supremum in $\mathscr{C}$. We get $\mathscr{C}_\Phi$ is a strict M-space. □

Corollary 5.9.19. *Let X be a completely regular space such that the closure of any exact open set of X is open, $\mathscr{C}$ be the vector lattice of continuous real functions on X and $\mathfrak{K}$ be a set of subsets of X such that any point of X belongs to the interior of a set of $\mathfrak{K}$ and such that any function of $\mathscr{C}$ is bounded on any set of $\mathfrak{K}$, (as $\mathfrak{K}$ one may choose the set of compact sets of X if X is locally compact). Then $\mathscr{C}_\mathfrak{K}$ is a strict M-space.*

The assertion follows immediately from Proposition 5.9.18 (and Proposition 1.5.5) by replacing Φ by $\{\{A \mid B\subset A\subset X\} \mid \exists K\in\mathfrak{K}, B\subset K\}$ and Θ by the set of sequences $(x_n)_{n\in\mathbb{N}}$ in X for which the set $\{x_n \mid n\in\mathbb{N}\}$ is included in a set of $\mathfrak{K}$. □

Corollary 5.9.20. *Let X be a locally compact σ-compact space such that the closure of any exact open set of X is open, $\mathscr{C}$ be the vector lattice of continuous real functions on X, $\mathfrak{K}$ be the set of compact sets of X, $\mathfrak{R}$ be the δ-ring generated by $\mathfrak{K}$, E be a quasi-complete locally convex space containing no copy of ℓ^∞ and $u: \mathscr{C}_\mathfrak{K}\to E$ be a continuous linear map. Then there exists $\mu\in\mathscr{M}(\mathfrak{R}, E; \mathfrak{K})$ such that $\mathscr{C}\subset\mathscr{L}^1(\mu, X)$ and $u(f) = \int f\,d\mu$ for any $f\in\mathscr{C}$.*

By Corollary 5.9.19 $\mathscr{C}_\mathfrak{K}$ is a strict M-space and, by Theorem 5.9.9 a ⇒ d, u is boundedly weakly compact. Let $f\in\mathscr{C}$. Then $\{g\in\mathscr{C} \mid |g|\le|f|\}$ is a bounded set of $\mathscr{C}_\mathfrak{K}$ so $\{u(g) \mid g\in\mathscr{C}, |g|\le|f|\}$ is a weakly relatively compact set of E and the assertion follows from C. Constantinescu [5], Proposition 4.2.2 and Theorem 4.4.7 a), d). □

Corollary 5.9.21. *Let X be a set, $\mathfrak{R}$ be a δ-ring of subsets of X, $\mathscr{M}$ be a band of $\mathscr{M}(\mathfrak{R}, \mathbb{R})$ and $\mathscr{F}$ be a subset of $\mathscr{M}^{\pi}$ containing $\{\dot{1}_A | A \in \mathfrak{R}\}$ (with the notation of C. Constantinescu [5]). Then $(\mathscr{M}_c^{\pi}, \mathscr{F})$ and $(\mathscr{M}_b^{\pi}, \mathscr{M}_b^{\pi})$ are strict M-spaces.*

By C. Constantinescu [5] Theorem 2.1.3 and Proposition 3.1.1 $(\mathscr{M}_b^{\pi}, \mathscr{M}_b^{\pi})$ is isomorphic to $\mathscr{C}_{\{X\}}$, where X is a compact stonian space and $\mathscr{C}$ is the vector lattice of continuous real functions on X. By Corollary 5.9.19 $(\mathscr{M}_b^{\pi}, \mathscr{M}_b^{\pi})$ is a strict M-space.

By C. Constantinescu [5] Theorem 2.3.8 and Proposition 3.1.1 there exist a locally compact stonian space X and a set $\mathfrak{K}$ of subsets of X such that: a) any point of X is an interior point of a set of $\mathfrak{K}$; b) any continuous real function on X is bounded on any set of $\mathfrak{K}$; c) $(\mathscr{M}_c^{\pi}, \mathscr{F})$ is isomorphic to $\mathscr{C}_{\mathfrak{K}}$, where $\mathscr{C}$ denotes the vector lattice of continuous real functions on X. By Corollary 5.9.19 $(\mathscr{M}_c^{\pi}, \mathscr{F})$ is a strict M-space. □

Proposition 5.9.22. *Let X be a set, $\mathfrak{R}$ be a δ-ring of subsets of X, $\mathfrak{S}$ be a σ-ring of subsets of X, $\mathfrak{A}$ be a set of sets such that $\bigcup_{A \in \mathfrak{A}} A = X$ and E be the vector lattice of $\mathfrak{R}$-measurable real functions on X which are bounded on the sets of $\mathfrak{A}$ and which vanish outside of a set of $\mathfrak{S}$. Then $E_{\mathfrak{A}}$ is a strict M-space.*

By Proposition 5.8.24 $E_{\mathfrak{A}}$ is an M-space. Let $(f_n)_{n \in \mathbb{N}}$ be an increasing weak Cauchy sequence in $E_{\mathfrak{A}}$ and let f be the map

$$X \to \overline{\mathbb{R}}, \quad x \mapsto \sup_{n \in \mathbb{N}} f_n(x).$$

It is obvious that f vanishes outside a set of $\mathfrak{S}$ and is $\mathfrak{R}$-measurable. We want to show that f is bounded on any set of $\mathfrak{A}$. Let $A \in \mathfrak{A}$. Assume that f is not bounded on A. Then there exists a sequence $(x_n)_{n \in \mathbb{N}}$ in A such that $f(x_n) > 3^n$ for any $n \in \mathbb{N}$. We denote by φ the map

$$E \to \mathbb{R}, \quad g \mapsto \sum_{n \in \mathbb{N}} \frac{1}{2^n} g(x_n).$$

Then $\varphi \in (E_{\mathfrak{A}})'$ and $\sup_{n \in \mathbb{N}} \varphi(f_n) = \infty$ and this is a contradiction. Hence, f is bounded on any set of $\mathfrak{A}$ so f belongs to E. $(f_n)_{n \in \mathbb{N}}$ therefore possesses a supremum in E. Thus, $E_{\mathfrak{A}}$ is a strict M-space. □

Proposition 5.9.23. *Let $\mathfrak{R}$ be a δ-ring and μ be a real measure on $\mathfrak{R}$. Then $L^{\infty}(\mu)$ and the dual of $L^1(\mu)$ are strict M-spaces.*

It is obvious that $L^{\infty}(\mu)$ is an M-space. Since the unit ball of $L^1(\mu)$ is an integral set the unit ball of its dual $L^1(\mu)'$ is a solid directed set (Propositions 5.510 and 5.5.11) so $L^1(\mu)'$ is also an M-space. Let $(\mathscr{F}_n)_{n \in \mathbb{N}}$ be an increasing weak Cauchy sequence in $L^{\infty}(\mu)$ or $L^1(\mu)'$. Then $(\mathscr{F}_n)_{n \in \mathbb{N}}$ is a (topologically) bounded and therefore an order bounded sequence. Since the spaces $L^{\infty}(\mu)$, $L^1(\mu)'$ are order σ-complete $(\mathscr{F}_n)_{n \in \mathbb{N}}$ possesses a supremum. Hence, $L^{\infty}(\mu)$ and $L^1(\mu)'$ are strict M-spaces. □

We give some examples of G-spaces.

Proposition 5.9.24. *The completion of any metrizable G-space is a G-space.*

Let E be a metrizable G-space, $\tilde{E}$ be its completion and $\mathfrak{A}$ be the set of equicontinuous Φ_1-sets of E'_E. Then the identity map $E_{\mathfrak{A}} \to E$ is uniformly Φ_4-continuous. $\tilde{E}$ is metrizable so, by Theorem 2.3.22, $\tilde{E}_{\mathfrak{A}}$ is mioritic. By Proposition 1.8.21 the identity map $\tilde{E}_{\mathfrak{A}} \to \tilde{E}$ is uniformly Φ_4-continuous. Since $\bar{A}_E = \bar{A}_{\tilde{E}}$ for any $A \in \mathfrak{A}$ we deduce that any set of $\mathfrak{A}$ is a Φ_1-set of $E'_{\tilde{E}}$ so $\tilde{E}$ is a G-space. □

Proposition 5.9.25. *Let E be a locally convex space and $(u_\iota)_{\iota \in I}$ be a family of linear maps of E into G-spaces such that the topology of E is the initial topology with respect to $(u_\iota)_{\iota \in I}$. Then E is a G-space.*

Let us denote by $\mathfrak{A}$ the set of equicontinuous Φ_1-sets of E'_E and by v the identity map $E_{\mathfrak{A}} \to E$. For any $\iota \in I$, let E_ι be the target of u_ι, $\mathfrak{A}_\iota$ be the set of equicontinuous Φ_1-sets of $(E'_\iota)_{E_\iota}$, v_ι be the identity map $(E_\iota)_{\mathfrak{A}_\iota} \to E_\iota$, and w_ι be the map

$$E_{\mathfrak{A}} \to (E_\iota)_{\mathfrak{A}_\iota}, \quad x \mapsto u_\iota(x).$$

It is easy to see that w_ι is uniformly continuous so $u_\iota \circ v = v_\iota \circ w_\iota$ is uniformly Φ_4-continuous for any $\iota \in I$. By Proposition 1.8.13 v is uniformly Φ_4-continuous. □

Corollary 5.9.26. *The vector subspaces of a G-space and the product of a family of G-spaces are G-spaces. In particular, a G-space possesses no copy of ℓ^*.*

The first assertion follows immediately from Proposition 5.9.25 and the last one from Corollary 5.9.13c), Proposition 5.9.24, and Lemma 4.1.42. □

Proposition 5.9.27. *Any locally convex space possessing a weakly σ-compact and weakly dense set is a G-space. In particular, any reflexive Banach space is a G-space.*

Let E be a locally convex space possessing a weakly σ-compact and weakly dense set. By Proposition 2.3.23 any equicontinuous set of E' is a Φ_1-set of E'_E so $E = E_{\mathfrak{A}}$, where $\mathfrak{A}$ denotes the set of equicontinuous Φ_1-sets of E'. □

Lemma 5.9.28. *Let E be a locally convex space. The following assertions are equivalent:*

a) E is semi-separable;

b) the topology of E is the initial topology with respect to a family of linear maps of E into separable Banach spaces;

c) the topology of E is the initial topology with respect to a family of linear maps of E into semi-separable locally convex spaces.

a ⇒ b. Let us denote by $\mathfrak{U}$ the set of convex circled 0-neighbourhoods in E and, for any $U \in \mathfrak{U}$, let $E(U)$ be the associated Banach space and u_U be the canonical map $E \to E(U)$. As E is semi-separable $E(U)$ is separable. It is obvious that the topology of E is the initial topology with respect to the family $(u_U)_{U \in \mathfrak{U}}$.

b ⇒ c is trivial.

c ⇒ a. Let $(u_\iota)_{\iota \in I}$ be a family of linear maps of E into semi-separable locally convex spaces such that the topology of E is the initial topology with respect to this family. Let U be a 0-neighbourhood in E. We may assume there exist $\iota \in I$ and a 0-neighbourhood V in the target F of u_ι such that $\overset{-1}{u_\iota}(V) \subset U$. Since F is semi-separable there exists a countable covering $(A_\lambda)_{\lambda \in L}$ of F such that $A_\lambda - A_\lambda \subset V$ for any $\lambda \in L$. $(\overset{-1}{u_\iota}(A_\lambda))_{\lambda \in L}$ is a countable covering of E such that $\overset{-1}{u_\iota}(A_\lambda) - \overset{-1}{u_\iota}(A_\lambda) \subset U$ for any $\lambda \in L$. Since U is arbitrary E is semi-separable. □

Corollary 5.9.29. *Any semi-separable locally convex space is a G-space.*

The assertion follows immediately from Lemma 5.9.28 a ⇒ b and from Propositions 5.9.25 and 5.9.27. □

Remarks. 1) Let X be a compact stonian space, $\mathscr{C}$ be the Banach space of continuous real functions on X, E be a separable complete locally convex space and u be a continuous linear map $\mathscr{C} \to E$. By the above corollary E is a G-space and by Theorem 5.9.16a) $\mathscr{C}$ is a strict M-space. Hence, by Corollary 5.9.13c), u is weakly compact. This result was proved by A. Grothendieck (1953) ([2] Corollaire 1 of Théorème 9).

2) Let X be a set, $\mathfrak{R}$ be a σ-algebra on X, $\mathscr{F}$ be the Banach space of $\mathfrak{R}$-measurable bounded real valued functions on X, E be a separable Banach space and $u : \mathscr{F} \to E$ be a continuous linear map. By the above result E is a G-space and, by Proposition 5.9.22, $\mathscr{F}$ is a strict M-space. Hence, by Corollary 5.9.13c), u is weakly compact. This result was proved by J. Diestel (1973) ([2] Theorem 2.4).

3) The locally convex lattice defined in Proposition 5.8.34 is a strict M-space whose underlying locally convex space is a G-space.

Proposition 5.9.30. *Let X be a Hausdorff topological space, $\mathfrak{K}$ be the set of compact metrizable sets of X, E be a semi-separable locally convex space and $\mathscr{C}$ be the vector space of continuous maps of X into E. Then $\mathscr{C}_{\mathfrak{K}}$ is semi-separable and therefore a G-space.*

First assume $X \in \mathfrak{K}$ and E is a separable Banach space. Let $\mathscr{B}$ be the vector space of continuous real functions on X endowed with the norm

$$\mathscr{B} \to \mathbb{R}, \quad f \mapsto \sup_{x \in X} \|f(x)\|.$$

Let $\mathscr{F}$ be a countable dense subset of $\mathscr{B}$ and A be a countable dense set of E. Let $g \in \mathscr{C}$ and $\varepsilon > 0$. There exists a finite open covering $(U_\iota)_{\iota \in I}$ of X such that

$$\|g(x) - g(y)\| < \frac{\varepsilon}{3}$$

for any $x, y \in U_\iota$ and any $\iota \in I$. Let $(f_\iota)_{\iota \in I}$ be a partition of the unity on X such that $\operatorname{Supp} f_\iota \subset U_\iota$ for any $\iota \in I$. For any $\iota \in I$ let $z_\iota \in A$ such that

$$\|g(x) - z_\iota\| < \frac{\varepsilon}{2}$$

for any $x \in U_\iota$ and $f'_\iota \in \mathscr{F}$ such that

$$\|f'_\iota - f_\iota\| < \frac{\varepsilon}{2(1 + \sum_{\iota \in I} \|z_\iota\|)}.$$

We set

$$g' : X \to E, \quad x \mapsto \sum_{\iota \in I} f'_\iota(x) z_\iota .$$

Let $x \in X$. We have

$$\begin{aligned} \|g'(x) - g(x)\| &\leq \|\sum_{\iota \in I} f'_\iota(x) z_\iota - \sum_{\iota \in I} f_\iota(x) z_\iota\| + \|\sum_{\iota \in I} f_\iota(x) z_\iota - g(x)\| \leq \\ &\leq \sum_{\iota \in I} |f'_\iota(x) - f_\iota(x)| \, \|z_\iota\| + \sum_{\iota \in I} f_\iota(x) \|z_\iota - g(x)\| < \\ &< \frac{\varepsilon}{2} + \frac{\varepsilon}{2} = \varepsilon . \end{aligned}$$

Hence the maps of the form g' are dense in $\mathscr{C}_{\{X\}}$. Since there are countable many maps of the form g', $\mathscr{C}_{\{X\}}$ is separable.

By Lemma 5.9.28 a $\Rightarrow$ b there exists a family $(u_\iota)_{\iota \in I}$ of linear maps of E into separable Banach spaces such that the topology of E is the initial topology with respect to this family. Let $(\iota, K) \in I \times \mathfrak{K}$. We denote by E_ι the target of u_ι, by $\mathscr{C}(\iota, K)$ the vector space of continuous maps of K into E_ι, and by $\varphi_{\iota, K}$ the map

$$\mathscr{C} \to \mathscr{C}(\iota, K)_{\{K\}}, \quad g \mapsto u_\iota \circ g | K .$$

Then the topology of $\mathscr{C}_\mathfrak{K}$ is the initial topology with respect to the family $(\varphi_{\iota, K})_{(\iota, K) \in I \times \mathfrak{K}}$. By the above considerations $\mathscr{C}(\iota, K)_{\{K\}}$ is a separable Banach space for any $(\iota, K) \in I \times \mathfrak{K}$ so, by Lemma 5.9.28 b $\Rightarrow$ a, $\mathscr{C}_\mathfrak{K}$ is semi-separable. By Corollary 5.9.29 $\mathscr{C}_\mathfrak{K}$ is a G-space. $\square$

Proposition 5.9.31. *Let E be a locally convex space, $\mathscr{P}$ be a set of semi-norms on E generating its topology, φ be a positive real function on $\mathscr{P}$ and $(u_\iota)_{\iota \in I}$ be a family of continuous linear maps of E into itself such that:*

a) $u_\iota(E)$ is a G-space for any $\iota \in I$;

b) $u_\iota \circ u_\iota = u_\iota$ for any $\iota \in I$;

c) $p(x - u_\iota(x)) \leq \varphi(p) p(x)$ for any $(x, p, \iota) \in E \times \mathscr{P} \times I$;

d) for any $\iota', \iota'' \in I$ there exists $\iota \in I$ such that

$$u_{\iota'}(E) \cup u_{\iota''}(E) \subset u_\iota(E) ;$$

e) for any countable subset A of E there exists $\iota \in I$ such that $A \subset u_\iota(E)$.

Then E is a G-space.

Let $\mathfrak{A}$ be the set of equicontinuous Φ_1-sets of E'_E and let $\mathfrak{F}$ be a Cauchy Φ_4-filter on $E_{\mathfrak{A}}$. Assume that $\mathfrak{F}$ is not a Cauchy filter on E. Then we may assume there exist $p \in \mathscr{P}$ and $\varepsilon > 0$ such that for any $A \in \mathfrak{F}$ there exist $x, y \in A$ with $p(x-y) > \varepsilon$.

Let $\iota \in I$. By a) $u_\iota(\mathfrak{F})$ is a Cauchy filter on $u_\iota(E)$. Assume there exists $A \in \mathfrak{F}$ such that

$$p(x - u_\iota(x)) \le \frac{\varepsilon}{3}$$

for any $x \in A$. Since there exists $B \in \mathfrak{F}$ such that

$$p(u_\iota(x) - u_\iota(y)) \le \frac{\varepsilon}{3}$$

for any $x, y \in B$ we get

$$p(x-y) \le p(x - u_\iota(x)) + p(y - u_\iota(y)) + p(u_\iota(x) - u_\iota(y)) \le \varepsilon$$

for any $x, y \in A \cap B$, which is a contradiction. Hence for any $A \in \mathfrak{F}$ there exists $x \in A$ such that

$$p(x - u_\iota(x)) > \varepsilon .$$

Let (L, f) be a Θ_4-net in $E_{\mathfrak{A}}$ such that $f(\mathfrak{G}) = \mathfrak{F}$, where $\mathfrak{G}$ denotes the section filter of L. We construct an increasing sequence $(\lambda_n)_{n \in \mathbb{N}}$ in L inductively together with a sequence $(\iota_n)_{n \in \mathbb{N}}$ in I such that for any $n \in \mathbb{N}$:

1) $f(\lambda_{n+1}) \in u_{\iota_{n+1}}(E)$;

2) $u_{\iota_n}(E) \subset u_{\iota_{n+1}}(E)$;

3) $p(f(\lambda_{n+1}) - u_{\iota_n}(f(\lambda_{n+1}))) > \varepsilon$.

We choose λ_0 and ι_0 arbitrarily. Let $n \in \mathbb{N}$ and assume the sequences were constructed up to n. By the above considerations there exists $\lambda_{n+1} \in L$ such that $\lambda_{n+1} \ge \lambda_n$ and such that 3) holds. By d) and e) there exists $\iota_{n+1} \in I$ such that 1) and 2) hold. This finishes the inductive construction.

Let $m, n \in \mathbb{N}$ with $0 < m < n$. By virtue of b), c), 1), 2) and 3)

$$\varphi(p)p(f(\lambda_m) - f(\lambda_n)) \ge p(f(\lambda_m) - f(\lambda_n) - u_{\iota_{n-1}}(f(\lambda_m) - f(\lambda_n))) =$$
$$= p(f(\lambda_n) - u_{\iota_{n-1}}(f(\lambda_n))) > \varepsilon .$$

Hence, (L, f) is not a Θ_5-net in E.

By a), b), e) and Proposition 1.8.20 the identity map $E_{\mathfrak{A}} \to E$ preserves the Φ_4-sets so by Theorem 1.8.4 h $\Rightarrow$ f (L, f) is a Θ_4-net in E, which is a contradiction. Hence, $\mathfrak{F}$ is a Cauchy filter on E and, by Theorem 1.8.4 b $\Rightarrow$ a, the identity map $E_{\mathfrak{A}} \to E$ is uniformly Φ_4-continuous, i.e., E is a G-space. □

Proposition 5.9.32. *Let $\mathfrak{R}$ be a δ-ring and μ be a real measure on $\mathfrak{R}$. Then:*

a) if there exists a sequence $(A_n)_{n \in \mathbb{N}}$ in $\mathfrak{R}$ such that $\mu(A \setminus \bigcup_{n \in \mathbb{N}} A_n) = 0$ for any $A \in \mathfrak{R}$ then $L^1(\mu)$ possesses a weakly σ-compact dense set;

b) $L^1(\mu)$ is a G-space.

a) We may assume $(A_n)_{n\in\mathbb{N}}$ is an increasing sequence. We set $X := \bigcup_{A\in\mathfrak{R}} A$ and put

$$\mathscr{F}_n := \{\mathscr{F} \in L^1(\mu) \mid \exists f \in \mathscr{F}, \quad |f| \le n 1^X_{A_n}\}$$

for any $n \in \mathbb{N}$. $\mathscr{F}_n$ is also a subset of $L^\infty(\mu)$ and it is easy to see that it is a $\sigma(L^\infty(\mu), L^1(\mu))$-compact subset. Since the weak topology on $L^1(\mu)$ induces a coarser topology on $\mathscr{F}_n$ than the $\sigma(L^\infty(\mu), L^1(\mu))$-topology, $\mathscr{F}_n$ is a weakly compact set of $L^1(\mu)$. Hence, $\bigcup_{n\in\mathbb{N}} \mathscr{F}_n$ is a weakly σ-compact set of $L^1(\mu)$ and it is obviously dense.

b) Let $\mathfrak{A}$ be the set of sets of the form $\bigcup_{n\in\mathbb{N}} A_n$, where $(A_n)_{n\in\mathbb{N}}$ runs through the set of all sequences in $\mathfrak{R}$, and for any $A \in \mathfrak{A}$ let u_A be the map

$$L^1(\mu) \to L^1(\mu), \quad \mathscr{F} \mapsto \mathscr{F} 1_A.$$

By a) and Proposition 5.9.27 $u_A(L^1(\mu))$ is a G-space for any $A \in \mathfrak{A}$. By Proposition 5.9.31 $L^1(\mu)$ is a G-space. □

Corollary 5.9.33. *Let $\mathfrak{R}$ be a δ-ring and $\mathscr{M}$ be the vector space of bounded real measures on $\mathfrak{R}$ endowed with the norm*

$$\mathscr{M} \to \mathbb{R}, \quad \mu \mapsto \sup_{A\in\mathfrak{R}} |\mu|(A).$$

Then $\mathscr{M}$ is a G-space.

For any $\mu \in \mathscr{M}$ we denote by u_μ the map of $\mathscr{M}$ into itself which maps any measure of $\mathscr{M}$ into its component on the band of $\mathscr{M}$ generated by μ. By Radon-Nikodym theorem $u_\mu(\mathscr{M})$ is isomorphic to $L^1(\mu)$ so by Proposition 5.9.32 b), it is a G-space. By Proposition 5.9.31 $\mathscr{M}$ is also a G-space. □

Corollary 5.9.34. *Let $\mathfrak{R}$ be a δ-ring, $\mathscr{N}$ be a fundamental solid subspace of $\mathscr{M}(\mathfrak{R}, \mathbb{R})$ and $\mathscr{F}$ be a subset of $\mathscr{N}^\pi$. Then $(\mathscr{N}, \mathscr{F})$ (C. Constantinescu [5], Definition 3.4.1) is a G-space.*

Let $\mathscr{M}$ be the vector space of bounded real measures on $\mathfrak{R}$ endowed with the norm

$$\mathscr{M} \to \mathbb{R}, \quad \mu \mapsto \sup_{A\in\mathfrak{R}} |\mu|(A)$$

and for any $\xi \in \mathscr{F}$ let u_ξ be the map

$$\mathscr{N} \to \mathscr{M}, \quad \mu \mapsto \xi \cdot \mu.$$

Then u_ξ is linear for any $\xi \in \mathscr{F}$ and the topology of $(\mathscr{N}, \mathscr{F})$ is the initial topology with respect to the family $(u_\xi)_{\xi\in\mathscr{F}}$. By Corollary 5.9.33 $\mathscr{M}$ is a G-space so, by Proposition 5.9.25, $(\mathscr{N}, \mathscr{F})$ is also a G-space. □

Proposition 5.9.35. *Let E be a G-space and I be a set. Then $\ell^1(I, E)$ and $c_0(I, E)$ are G-spaces.*

It is obvious that $\ell^1(I, E)$ is a locally convex space. Let $\mathfrak{A}$ be the set of equicontinuous Φ_1-sets of $\ell^1(I, E)'_{\ell^1(I,E)}$, $\mathfrak{F}$ be a Cauchy Φ_4-filter on $\ell^1(I, E)_{\mathfrak{A}}$ and (L, f) be a Θ_4-net in $\ell^1(I, E)_{\mathfrak{A}}$ such that $f(\mathfrak{G}) = \mathfrak{F}$, where $\mathfrak{G}$ denotes the section filter of I. For any $\iota \in I$ we denote by φ_ι the map

$$\ell^1(I, E) \to E, \quad (x_\lambda)_{\lambda \in L} \mapsto x_\iota .$$

Then φ_ι is a continuous linear map so, E being a G-space, $\varphi_\iota(\mathfrak{F})$ is a Cauchy filter on E for any $\iota \in I$.

Assume $\mathfrak{F}$ is not a Cauchy filter on $\ell^1(I, E)$. Then there exists a continuous seminorm p on E such that for any $A \in \mathfrak{F}$ there exist $(x_\iota)_{\iota \in I}, (y_\iota)_{\iota \in I} \in A$ with

$$\sum_{\iota \in I} p(x_\iota - y_\iota) > 7 .$$

We set $L_{-1} := L$ and construct a disjoint sequence $(I_n)_{n \in \mathbb{N}}$ in $\mathfrak{P}_f(I)$, a decreasing sequence $(L_n)_{n \in \mathbb{N}}$ in $\mathfrak{G}$, and an increasing sequence $(\lambda_n)_{n \in \mathbb{N}}$ in L inductively such that for any $n \in \mathbb{N}$:

a) $\lambda_n \in L_{n-1}$;

b) $\lambda', \lambda'' \in L_n \Rightarrow \sum_{\iota \in \bigcup_{m=0}^{n} I_m} p(\varphi_\iota(f(\lambda') - f(\lambda''))) < 1$

c) $\lambda \in L_n \Rightarrow \sum_{\iota \in I_n} p(\varphi_\iota(f(\lambda) - f(\lambda_n))) > 2 .$

Let $n \in \mathbb{N}$ and assume the sequences were constructed up to $n - 1$. By the above considerations there exist $\lambda', \lambda'' \in L_{n-1}$ such that

$$\sum_{\iota \in I} p(\varphi_\iota(f(\lambda') - f(\lambda''))) > 7 .$$

By b)

$$\sum_{\iota \in I \setminus \bigcup_{m=0}^{n-1} I_m} p(\varphi_\iota(f(\lambda') - f(\lambda''))) =$$

$$= \sum_{\iota \in I} p(\varphi_\iota(f(\lambda') - f(\lambda''))) - \sum_{\iota \in \bigcup_{m=0}^{n-1} I_m} p(\varphi_\iota(f(\lambda') - f(\lambda''))) > 7 - 1 = 6$$

Hence, there exists $I_n \in \mathfrak{P}_f\left(I \setminus \bigcup_{m=0}^{n-1} I_m\right)$ such that

$$\sum_{\iota \in I_n} p(\varphi_\iota(f(\lambda') - f(\lambda''))) > 6 .$$

Since $\varphi_\iota(\mathfrak{F})$ is a Cauchy filter for any $\iota \in I$ there exists $L_n \in \mathfrak{G}$ such that $L_n \subset L_{n-1}$ and such that b) holds. Then either

$$\sum_{\iota \in I_n} p(\varphi_\iota(f(\lambda) - f(\lambda'))) > 2$$

for any $\lambda \in L_n$ or

$$\sum_{\iota \in I_n} p(\varphi_\iota(f(\lambda) - f(\lambda''))) > 2$$

for any $\lambda \in L_n$. We set $\lambda_n := \lambda'$ in the first case and $\lambda_n := \lambda''$ in the second. This finishes the inductive construction.

Put

$$V := \{x \in E \mid p(x) \leq 1\}$$

and denote by V^0 the polar set of V in E'. By a) and c) there exists a family $(x'_\iota)_{\iota \in I_n}$ in V^0 for any $n \in \mathbb{N}$ such that

$$\sum_{\iota \in I_n} x'_\iota(\varphi_\iota(f(\lambda_{n+1}) - f(\lambda_n))) > 2\,.$$

By a) and b)

$$\sum_{\iota \in I_n} x'_\iota(\varphi_\iota(f(\lambda_m) - f(\lambda_n))) =$$

$$= \sum_{\iota \in I_n} x'_\iota(\varphi_\iota(f(\lambda_{n+1}) - f(\lambda_n))) - \sum_{\iota \in I_n} x'_\iota(\varphi_\iota(f(\lambda_{n+1}) - f(\lambda_m))) > 2 - 1 = 1$$

for any $m \in \mathbb{N}$ so that $m > n$. We set

$$A := \{\sum_{\iota \in I_n} x'_\iota \circ \varphi_\iota \mid n \in \mathbb{N}\}\,.$$

A is an equicontinuous set of $\ell^1(I, E)'$. Since

$$\lim_{n \to \infty} \sum_{\iota \in I_n} x'_\iota \circ \varphi_\iota = 0$$

in $\ell^1(I, E)'_{\ell^1(I,E)}$, A is a Φ_1-set of $\ell^1(I, E)'_{\ell^1(I,E)}$ (Proposition 1.4.7). Hence, $A \in \mathfrak{A}$ and so there exist $m, n \in \mathbb{N}$ such that $m > n$ and

$$\sup_{q \in \mathbb{N}} \sum_{\iota \in I_q} x'_\iota(\varphi_\iota(f(\lambda_m) - f(\lambda_n))) < 1\,,$$

which is a contradiction. We deduce that $\mathfrak{F}$ is a Cauchy filter on $\ell^1(I, E)$. By Theorem 1.8.4 b $\Rightarrow$ a the identity map

$$\ell^1(I, E)_{\mathfrak{A}} \to \ell^1(I, E)$$

is uniformly Φ_4-continuous so $\ell^1(I, E)$ is a G-space.

The proof that $c_0(I, E)$ is a G-space proceeds similarly. □

Proposition 5.9.36. *Let $\mathfrak{R}$ be a quasi-δ-ring of sets, $\mathfrak{K}$ be a subset of $\mathfrak{R}$ closed under finite unions and E be a G-space. Then*

$$\mathscr{E}(\mathfrak{R}, E_{E'}; \mathfrak{K}) = \mathscr{E}(\mathfrak{R}, E; \mathfrak{K}).$$

The assertion follows immediately from Theorem 4.6.7. □

Theorem 5.9.37. *Let $\mathfrak{R}$ be a σ-ring, E be a quasicomplete locally convex space and μ be an additive map of $\mathfrak{R}$ into E. Then the following assertions are equivalent:*

a) μ is exhaustive;

b) the circled convex closed hull of $\mu(\mathfrak{R})$ is weakly compact;

c) $\mu(\mathfrak{R})$ is bounded and there exists a countable family $(K_\iota)_{\iota \in I}$ of weakly compact sets of E such that $\mu(\mathfrak{R})$ is enclosed by the closed vector subspace of E generated by $\bigcup_{\iota \in I} K_\iota$;

d) $\mu(\mathfrak{R})$ is bounded and there exists a closed topological vector subspace of E containing $\mu(\mathfrak{R})$, which is a G-space;

e) $\mu(\mathfrak{R})$ is bounded and there exists a topological vector subspace of E containing $\mu(\mathfrak{R})$, which is a G-space.

a ⇒ b. We set $X := \bigcup_{A \in \mathfrak{R}} A$ and denote by F the vector space of bounded $\mathfrak{R}$-measurable real functions on X that vanish outside a set of $\mathfrak{R}$ endowed with the order relation induced by $\mathbb{R}^X$ and with the norm

$$F \to \mathbb{R}, \quad f \mapsto \sup_{x \in X} |f(x)|.$$

By Proposition 5.9.22 F is a strict M-space. By Corollary 4.7.8 $\mu(\mathfrak{R})$ is a bounded set of E so there exists a continuous linear map $u: F \to E$ such that $u(1_A) = \mu(A)$ for any $A \in \mathfrak{R}$. By Proposition 5.2.13 b ⇒ a u is order exhaustive so, by Proposition 5.9.8, u is boundedly weakly compact. Hence, the circled convex closed hull of $\mu(\mathfrak{R})$ is weakly compact.

b ⇒ c is trivial.

c ⇒ d. There exists a closed topological vector subspace F of E containing $\mu(\mathfrak{R})$ such that F possesses a weakly σ-compact dense set. By Proposition 5.9.27 F is a G-space.

d ⇒ e is trivial.

e ⇒ a. Let F be a topological vector subspace of E containing $\mu(\mathfrak{R})$, which is a G-space. Since $\mu(\mathfrak{R})$ is bounded $\mu \in \mathscr{E}(\mathfrak{R}, F_{F'})$. By Proposition 5.9.36

$$\mu \in \mathscr{E}(\mathfrak{R}, F) \subset \mathscr{E}(\mathfrak{R}, E). \quad \square$$

Remarks. 1) The fact that $\mu(\mathfrak{R})$ is weakly relatively compact was proved by R.G. Bartle, N. Dunford, and J. Schwartz (1955) ([1] Theorem 2.9) in the case $\mathfrak{R}$ is a σ-algebra, E a Banach space, and μ a measure. The implication a ⇒ b was proved by I. Tweddle (1969) ([1] Theorem 3) in the case μ is a measure. The equivalence a ⇔ b was proved by J. Diestel (1973) ([1] Corollary 1 and [2] Theorem 2.2) in the case E is a Banach space.

2) Assume $\mu(\mathfrak{R})$ is bounded and separable. Let F be the vector subspace of E generated by $\mu(\mathfrak{R})$. By Corollary 5.9.29 F is *a* G-space so, by e $\Rightarrow$ a, μ is exhaustive. This result was proved by L. Drewnowski (1973) ([5] 2.16 Theorem (a)) and J. Diestel (1973) ([2] Theorem 1.6) in the case E is a Banach space.

Proposition 5.9.38. *Let I be a set, let $\aleph$ be an infinite cardinal number which is not the union of a sequence of cardinal numbers strictly smaller than it, and let E be the solid vector subspace of $\ell^\infty(I)$ formed by those $f \in \ell^\infty(I)$ for which*

$$\operatorname{Card}\{f \neq 0\} < \aleph .$$

Then E is a strict M-space.

The assertion follows immediately from Proposition 5.9.22 by replacing X, $\mathfrak{R}$, $\mathfrak{S}$, and $\mathfrak{A}$ with I, $\mathfrak{P}(I)$, $\{A \subset X \mid \operatorname{Card} A < \aleph\}$, and $\{I\}$ respectively. □

Proposition 5.9.39. *Let X be a metrizable locally compact space and let $\mathscr{F}$ be a vector space of continuous real functions on X having a limit at the Alexandrov point of X if X is not compact, endowed with the norm*

$$\mathscr{F} \to \mathbb{R}, \quad f \mapsto \sup_{x \in X} |f(x)| .$$

Then $\mathscr{F}$ is a G-space.

Assume first X σ-compact. Then its Alexandrov compactification is metrizable and therefore $\mathscr{F}$ is separable. By Corollary 5.9.29 $\mathscr{F}$ is a G-space.

Assume now X is not σ-compact and let $\mathfrak{A}$ be the set of open closed σ-compact sets of X. For each $f \in \mathscr{F}$ and for each $A \in \mathfrak{A}$ let f_A be the real function on X equal to f on A and equal to the limit of f at the Alexandrov point of X on $X \backslash A$. Then $f_A \in \mathscr{F}$ for every $f \in \mathscr{F}$ and for every $A \in \mathfrak{A}$. We set for each $A \in \mathfrak{A}$

$$u_A : \mathscr{F} \to \mathscr{F}, \quad f \mapsto f_A .$$

It is easy to check that the family $(u_A)_{A \in \mathfrak{A}}$ possesses the properties $a - e$ of Proposition 5.9.31 for $\varphi = 2$ (a) follows from the first part of the proof). Hence $\mathscr{F}$ is a G-space. □

Theorem 5.9.40. *Let I be a set, let $\aleph$ be an infinite cardinal number which is not the union of a sequence of cardinal numbers strictly smaller than it, and let E be the Banach subspace of $\ell^\infty(I)$ formed by those $f \in \ell^\infty(I)$ for which*

$$\operatorname{Card}\{f \neq 0\} < \aleph .$$

Further, let X be a metrizable locally compact space and let $\mathscr{F}$ be the vector space of continuous real functions on X having a limit at the Alexandrov point of X if X is not compact, endowed with the norm

$$\mathscr{F} \to \mathbb{R}, \quad f \mapsto \sup_{x \in X} |f(x)| .$$

Then any continuous linear map $E \to \mathscr{F}$ is weakly compact and the map $E_{E'} \to \mathscr{F}$ defined by it is uniformly Φ_4-continuous.

By Proposition 5.9.38 E is a strict-M-space and by Proposition 5.9.39 $\mathscr{F}$ is a G-space and the assertions follow from Corollary 5.9.13 b), c). □

Corollary 5.9.41. *Let I be an infinite set endowed with the discrete topology, let $\aleph$ be an infinite cardinal number which is not the union of a sequence of cardinal numbers strictly smaller than it, let E be the Banach subspace of $\ell^\infty(I)$ formed by those $f \in \ell^\infty(I)$ for which*

$$\mathrm{Card}\,\{f \neq 0\} < \aleph,$$

let U be a ball of E, let $c(I)$ be the Banach subspace of $\ell^\infty(I)$ formed by the functions having a limit at the Alexandrov point of I, and let $u: E \to c(I)$, $v: \ell^\infty(I) \to c(I)$ (resp. $u: E \to c_0(I)$, $v: E \to c_0(I)$) be continuous linear maps. Then $v(u(U))$ is a relatively compact set of $c(I)$.

By Theorem 5.9.40 u is weakly compact, i.e. $u(U)$ is a weakly relatively compact set of $c(I)$ (resp. $c_0(I)$). By Proposition 1.5.5 c $\Rightarrow$ a $u(U)$ is a weak Φ_4-set of $\ell^\infty(I)$ (resp. E). By Theorem 5.9.40 and Theorem 1.8.4 a $\Rightarrow$ h $v(u(U))$ is a Φ_4-set and therefore a precompact set (Proposition 1.5.6 a $\Rightarrow$ c) of $c(I)$ (resp. $c_0(I)$). Since $c(I)$ (resp. $c_0(I)$) is complete $v(u(U))$ is a relatively compact set of $c(I)$ (resp. $c_0(I)$). □

Corollary 5.9.42. *Let I be an infinite set endowed with the discrete topology, let $\aleph$ be an infinite cardinal number which is not the union of a sequence of cardinal numbers strictly smaller than it, let E be the Banach subspace of $\ell^\infty(I)$ formed by those $f \in \ell^\infty(I)$ for which*

$$\mathrm{Card}\,\{f \neq 0\} < \aleph,$$

and let $c(I)$ be the Banach subspace of $\ell^\infty(I)$ formed by the functions having a limit at the Alexandrov point of I. Further, let U be a ball of E (of $\ell^\infty(I)$) and let $u: E \to E$ ($u: \ell^\infty(I) \to \ell^\infty(I)$) be a continuous linear map such that $u \circ u = u$ and $u(E) \subset c_0(I)$ ($u(\ell^\infty(I)) \subset c(I)$). Then $u(U)$ is a relatively compact set of $c_0(I)$ (of $c(I)$). In particular, there exist no projections $E \to c_0(I)$, $\ell^\infty(I) \to c(I)$. □

Remark. R. S. Phillips proved (1940) [1] (7.5) that there exists no projection $\ell^\infty \to c(\mathbb{N})$.

Theorem 5.9.43. *Let X be a metrizable topological space, let $\mathscr{B}$ be the vector space of bounded Borel real functions on X endowed with the norm*

$$\mathscr{B} \to \mathbb{R}, \quad f \mapsto \sup_{x \in X} |f(x)|,$$

let $\mathscr{C}_b$ be the vector space of continuous bounded real functions on X endowed with the strict topology, and let $u, v: \mathscr{B} \to \mathscr{C}_b$ be continuous linear maps. Then:

a) $\mathscr{C}_b$ is complete;

b) $\mathscr{C}_b$ *is semi-separable and therefore it is a G-space;*

c) u *is weakly compact and the map* $\mathscr{B}_{\mathscr{B}'} \to \mathscr{C}_b$ *defined by* u *is uniformly* Φ_4-*continuous;*

d) if the map $\mathscr{C}_b \to \mathscr{C}_b$ *defined by* v *is continuous then it is boundedly weakly compact, the map* $(\mathscr{C}_b)_{\mathscr{C}_b'} \to \mathscr{C}_b$ *defined by* v *is uniformly* Φ_4-*continuous, and* $v(u(U))$ *is a relatively compact set of* $\mathscr{C}_b$ *for every ball* U *of* $\mathscr{B}$.

a) follows from the fact that any real function on X is continuous if its restrictions to the compact sets of X are continuous.

b) Let g be a bounded real function on X such that $\{x\in X \mid |g(x)| \geq \varepsilon\}$ is relatively compact for every strictly positive real number ε. For every $n \in \mathbb{N}$ there exists a compact set K_n of X such that $n|g| < 1$ on $X \setminus K_n$ and, since K_n is metrizable, a countable subset $\mathscr{F}_n$ of $\mathscr{C}_b$ such that for every $f \in \mathscr{C}_b$ there exists $f' \in \mathscr{F}_n$ with $n|f-f'| < 1$ on K_n and such that every function of $\mathscr{F}_n$ reaches its supremum on K_n. Let $f \in \mathscr{C}_b$ and let ε be a strictly positive real number. Let $n \in \mathbb{N}$ such that

$$n > \frac{1}{\varepsilon}\left(\sup_{x\in X}|g(x)| + 2\sup_{x\in X}|f(x)| + 1\right).$$

There exists $f' \in \mathscr{F}_n$ such that $n|f-f'| < 1$ on K_n. We get

$$|(f-f')g| \leq \frac{1}{n}\sup_{x\in X}|g(x)| < \varepsilon$$

on K_n,

$$|f'| \leq |f| + |f-f'| \leq \sup_{x\in X}|f(x)| + 1$$

on K_n and therefore on X, and

$$|(f-f')g| \leq (|f| + |f'|)|g| \leq \left(2\sup_{x\in X}|f(x)| + 1\right)\frac{1}{n} < \varepsilon$$

on $X \setminus K_n$. Hence $|(f-f')g| < \varepsilon$ on X. We deduce $\mathscr{C}_b$ is semi-separable and by Corollary 5.9.29 it is a G-space.

c) By Proposition 5.9.22 $\mathscr{B}$ is a strict M-space and the assertions follow from a), b), and Corollary 5.9.13 b), c).

d) Let $(f_n)_{n\in\mathbb{N}}$ be an increasing weak Cauchy sequence in $\mathscr{C}_b$. Then $\{f_n \mid n \in \mathbb{N}\}$ is a bounded set of $\mathscr{B}$ and by c) $\{v(f_n) \mid n \in \mathbb{N}\}$ is a weakly relatively compact set of $\mathscr{C}_b$. Moreover, $(v(f_n))_{n\in\mathbb{N}}$ is a weak Cauchy sequence in $\mathscr{C}_b$ and therefore a weakly convergent sequence in $\mathscr{C}_b$. By Corollary 5.8.26 $\mathscr{C}_b$ is an M-space. By a) and by Theorem 5.8.9 a), b) the map $\mathscr{C}_b \to \mathscr{C}_b$ defined by v is boundedly weakly compact and the map $(\mathscr{C}_b)_{\mathscr{C}_b} \to \mathscr{C}_b$ defined by v is uniformly Φ_4-continuous. Let U be a ball of $\mathscr{B}$. By c) $u(U)$ is a weakly relatively compact set of $\mathscr{C}_b$. By Proposition 1.5.5 c $\Rightarrow$ a $u(U)$ is a weak Φ_4-set of $\mathscr{C}_b$ and therefore by the above considerations and by Theorem 1.8.4 a $\Rightarrow$ h $v(u(U))$ is a Φ_4-set and therefore a precompact set (Proposition 1.5.6 a $\Rightarrow$ c) of $\mathscr{C}_b$. By a) $v(u(U))$ is a relatively compact set of $\mathscr{C}_b$. □

Corollary 5.9.44. *Let X be a set endowed with the discrete topology and let* $\mathscr{C}_b$ *be the vector space of continuous bounded real functions on X endowed with the strict topology. Then* $\{f \in \mathscr{C}_b \mid |f| \leq 1\}$ *is a compact set of* $\mathscr{C}_b$.

The assertion follows immediately from Theorem 5.9.43 d). □

References

(3.1.1., 3.5.9., 4.3.5., 5.7.7.) written at the end of a paper means that the corresponding paper was cited in or after Definition 3.1.1, Proposition 3.5.9, Theorem 4.3.5, and Corollary 5.7.7.

Alfsen, E.M.:

[1] *Compact convex sets and boundary integrals.* (Ergebnisse der Mathematik und ihrer Grenzgebiete **57**). Berlin – Heidelberg – New York: Springer 1971, (*5.5.17, 5.7.14*).

Aliprantis, Ch.D., O. Burkinshaw:

[1] *Locally solid Riesz spaces.* New York – San Francisco – London: Academic Press 1978, (*4.1.51, 5.4.7, 5.5.1, 5.5.3, 5.5.6, 5.5.10, 5.5.17, 5.5.23, 5.7.2, 5.7.10, 5.7.21, 5.7.22, 5.7.23, 5.7.24, 5.8.6, 5.8.11, 5.8.12, 5.8.13, 5.8.21*).

Andersen, N.J.M., J.P.R. Christensen:

[1] *Some results on Borel structures with applications to subseries convergence in abelian topological groups.* Israel J. Math. **15** (1973), 414–420, (*4.11.14*).

Andô, T.:

[1] *Convergent sequences of finitely additive measures.* Pacific J. Math. **11** (1961), 395–404, (*4.2.14, 4.3.3*).

Antosik, P.:

[1] *Mappings from L-groups into topological groups I.* Bull. Acad. Pol. Sci., Série Sci. Math. Astr. Phys. **21** (1973), 145–152, (*3.3.14*).

Banach, S.:

[1] *Théorie des opérations linéaires.* (Monografje matematiczne I) Warszawa 1932, (*3.5.12, 4.11.6*).

Bartle, R.G., N. Dunford, J. Schwartz:

[1] *Weak compactness and vector measures.* Canadian J. Math. **7** (1955) 289–305, (*4.2.13, 5.9.37*).

Bennet, C., N.J. Kalton:

[1] *Addendum to "FK-spaces containing c_0".* Duke Math. J. **39** (1972), 819–821, (*3.3.18, 4.7.12*).

Bessaga, C., A. Pełczyński:

[1] *On bases and unconditional convergence of series in Banach spaces.* Studia Math. **17** (1958), 151–164, (*4.1.45*).

Birkhoff, G.:

[1] *Integration of functions with values in Banach spaces.* Trans. Amer. Math. Soc. **38** (1935), 357–378, (*3.3.2*).

Bourbaki, N.:

[1] *Topologie générale*. Paris: Hermann 1974, (*3.5.2, 4.6.2, 4.9.6, 4.10.4, 4.11.8, 4.11.13, 5.7.10*).

[2] *Intégration* Ch. I–IV. Paris: Hermann, 1973, (*4.10.9*).

Brooks, J.K.:

[1] *Equicontinuous sets of measures and applications to Vitali's integral convergence theorem and control measures*. Adv. in Math. **10** (1973), 165–171, (*4.3.3*).

[2] *On a theorem of Dieudonné*. Adv. in Math. **36** (1980), 165–168, (*4.8.7*).

Brooks, J.K., N. Dinculeanu:

[1] *Critères de compacité dans les espaces de mesures vectorielles*. C.R. Acad. Sci., Paris **274** (1972), A 1627–A 1628, (*4.2.13*).

Brooks, J.K., R.S. Jewett:

[1] *On finitely additive vector measures*. Proc. Nat. Acad. Sci. USA **67** (1970), 1294–1298, (*4.2.14, 4.3.3, 4.3.7*).

Buck, R.C.:

[1] *Bounded continuous functions on a locally compact space*. Michigan Math. J. **5** (1958), 95–104, (*4.10.10*).

Choquet,G.:

[1] *Cardinaux 2-measurables et cônes faiblement complets*. Ann. Inst. Fourier **17**, 2 (1967), 383–393, (*5.6.20*).

Choquet, G., P.A. Meyer:

[1] *Existence et unicité des représentations intégrales dans les convexes compacts quelconques*. Ann. Inst. Fourier **13** (1963) 139–154, (*5.7.14, 5.7.15*).

Christensen, J.P.R.:

[1] *Borel structures and a topological zero-one law*. Math. Scand. **29** (1971), 245–255, (*4.11.9, 4.11.10, 4.11.14*),

Constantinescu, C.:

[1] *Weakly compact sets in locally convex vector lattices*. Rev. Roumaine Math. Pures Appl. **14** (1969), 325–351, (*5.5.19, 5.7.9, 5.8.31*).

[2] *Šmulian – Eberlein spaces*. Comment. Math. Helv. **48** (1973), 254–317, (*1.5.8, 2.3.10, 2.3.14, 2.3.19, 2.3.22*).

[3] *Familles multipliables dans les groupes topologiques séparés*. C.R. Acad. Sci., Paris **282** (1976), A 191–A 193, (*3.5.1, 3.5.8*).

[4] *Familles multipliables dans les groupes topologiques séparés*. C.R. Acad. Sci, Paris **282** (1976), A 271–A 274, (*3.5.2, 3.5.18, 3.5.19*).

[5] *Duality in measure theory*. (Lecture Notes in Math. **796**). Berlin – Heidelberg – New York: Springer Verlag 1980, (*4.9.2, 4.10.8, 5.6.1, 5.6.2, 5.6.6, 5.6.11, 5.6.23, 5.6.24, 5.8.27, 5.8.33, 5.9.16, 5.9.17, 5.9.20, 5.9.21, 5.9.34*).

[6] *On Nikodym's boundedness theorem*. Libertas Math. **1** (1981), 51–73 (§*4.8*).

[7] *Spaces of measures on topological spaces*. Hokkaido Math. J. **10** (1981), 89–156 (§*4.5*).

[8] *Sur les espaces du type* M.C.R. Acad. Sci., Paris, Série I, 296 (1983) 303–305 (§*5.8*).

Constantinescu, C., K. Weber (in collaboration with A. Sontag):

[1] *Integration Theory I*, New York – Chichester – Brisbane – Toronto: John Wiley & Sons (to appear) (*4.9.2, 4.10.4*).

Darst, R. B.:

[1] *On a theorem of Nikodym with applications to weak convergence and von Neumann algebras*. Pacific J. Math. **23** (1967), 473–477, (*4.7.8*).

Dashiell, F. K. Jr.:

[1] *Nonweakly compact operators from order-Cauchy complete C(S) lattices, with application to Baire classes*. Trans. Amer. Math. Soc. **266** (1981), 397–413, (*5.1.5, 5.6.8*).

Dierolf, P.:

[1] *Summierbare Familien und assoziierte Orlicz-Pettis Topologien*. Dissertation zur Erlangung des Doktorgrades (1975), (*3.3.18, 3.3.29, 3.5.11*).

[2] *Theorems of the Orlicz-Pettis-type for locally convex spaces*. Manuscripta Math. **20** (1977), 73–94, (*3.5.11*).

[3] *Summable sequences and associated Orlicz-Pettis-topologies*. Commentationes Math. Tomus specialis in honorem Ladislai Orlicz II, Warszawa (1979), 71–88, (*3.3.29, 3.5.10*).

Diestel, J.:

[1] *Grothendieck spaces and vector measures*. In: Vector and operator valued measures and applications (edited by D. H. Tucker and H. B. Maynard), 97–108, New York – London: Academic Press 1973, (*4.2.20, 5.9.37*).

[2] *Applications of weak compactness and bases to vector measures and vectorial integration*. Rev. Roumaine Math. Pures Appl. **18** (1973), 211–224, (*4.1.46, 5.9.29, 5.9.37*).

Diestel, J., B. Faires:

[1] *On vector measures*. Trans. Amer. Math. Soc. **198** (1974), 253–271, (*4.1.44, 4.7.16*).

Dieudonné, J.:

[1] *Sur la convergence des suites de mesures de Radon*. An. Acad. Brasil. Ciencias **23** (1951), 21–38, (*4.5.3*).

Dieudonné, J., L. Schwartz:

[1] *La dualité dans les espaces (F) et (LF)*. Ann. Inst. Fourier **1** (1950), 61–101, (*2.3.19, 2.3.22*).

Dini, U.:

[1] *Fondamenti per la teorica delle funzioni di variabili reali*. Pisa 1878, (*2.1.32*).

Drewnowski, L.:

[1] *Topological rings of sets, continuous set functions, integration* II. Bull. Acad. Pol. Sci., Série Sci. Math. Astr. Phys. **20** (1972), 277–286, (*4.3.3*).

[2] *Equivalence of Brooks-Jewett, Vitali-Hahn-Saks and Nikodym theorems*. Bull. Acad. Pol. Sci., Série Sci. Math. Astr. Phys. **20** (1972), 725–731, (*4.1.18, 4.2.14, 4.3.3, 4.7.10*).

[3] *Uniform boundedness principle for finitely additive vector measues*. Bull. Acad. Pol. Sci., Série Sci. Math. Astr. Phys. **21** (1973), 115–118, (*4.7.4, 4.7.10*).
[4] *On the Orlicz-Pettis type theorems of Kalton*. Bull. Acad. Pol. Sci., Série Sci. Math. Astr. Phys. **21** (1973), 515–518, (*3.3.25, 3.3.27, 4.1.35*).
[5] *Decomposition of set functions*. Studia Math. **48** (1973), 23–48, (*5.9.37*).
[6] *Another note on Kalton's theorem*. Studia Math. **52** (1975), 233–237, (*3.3.25, 4.1.34, 4.1.35*).
[7] *Un théorème sur les opérateurs de l^∞ (Γ)*. C.R. Acad. Sci., Paris **281** (1975), A 967–A 969, (*4.1.37, 4.1.39, 4.1.41*).

Drewnowski, L., I. Labuda:
[1] *Sur quelques théorèmes du type d'Orlicz-Pettis* II. Bull. Acad. Pol. Sci., Série Sci. Math. Astr. Phys. **21** (1973), 119–125, (*4.6.4, 4.7.11*).

Dunford, N., J.T. Schwartz:
[1] *Linear operators* I. (Pure and applied Math. 7). New York – London – Sydney – Toronto, John Wiley: Interscience Publishers 1957, (*4.2.14, 4.3.3*).

Dvoretzky, A., C.A. Rogers:
[1] *Absolute and unconditional convergence in normed linear spaces*. Proc. Nat. Acad. Sci. USA **36** (1950), 192–197, (*3.5.6*).

Eberlein, W.F.:
[1] *Weak compactness in Banach spaces*. Proc. Nat. Acad. Sci. USA **33** (1947), 51–53, (*2.3.19*).

Frolik, Z.:
[1] *A measurable map with analytic domain and metrizable range is quotient*. Bull. Amer. Math. Soc. **76** (1970), 1112–1117, (*4.11.13*).

Gänssler, P.:
[1] *Compactness and sequential compactness in spaces of measures*. Z. Wahrscheinlichkeitstheorie Verw. Gebiete **17** (1971), 124–146, (*4.2.14, 4.5.14*).
[2] *A convergence theorem for measures in regular Hausdorff spaces*. Math. Scand. **29** (1971), 237–244, (*4.5.14, 4.8.5*).

Garnir, H.G., M. de Wilde, J. Schmets:
[1] *Analyse fonctionelle* I. Basel – Stuttgart: Birkhäuser 1968, (*3.1.11*).

Giles, R.:
[1] *A generalization of the strict topology*. Trans. Amer. Math. Soc. **161** (1971), 467–474, (*4.10.11, 5.9.16*).

Gillman, L., M. Jerison:
[1] *Rings of continuous functions* (Graduate texts in mathematics 43) New York – Heidelberg – Berlin, Springer Verlag 1976, (*5.8.35*).

Graves, W.H.:
[1] *Universal Lusin measurability and subfamily summable families in abelian topological groups*. Proc. Amer. Math. Soc. **73** (1979), 45–50, (*4.11.4, 4.11.6, 4.11.7*).

Graves, W.H., W. Ruess:
[1] *Compactness in spaces of vector-valued measures and a natural Mackey topology in spaces of bounded measurable functions*. Contemporary Math. **2** (1980), 189–203 (*4.2.13*).

Grothendieck, A.:

[1] *Critères de compacité dans les espaces fonctionnels généraux*. Amer. J. Math. **74** (1952), 168–186, (*2.1.16, 2.3.10, 2.3.22*).

[2] *Sur les applications linéaires faiblement compactes d'espace du type* C(K). Canadian J. Math. **5** (1953), 129–173, (*3.5.12, 4.5.3, 5.6.8, 5.7.8, 5.8.8, 5.8.10, 5.8.31, 5.8.32, 5.9.11, 5.9.29*).

Gusel'nikov, N.S.:

[1] *An analog of the Vitali-Hahn-Saks theorem*. Mat. Zametki **19** (1976), 641–652 or Math. Notes Acad. Sci. USSR **19** (1976), 387–392, (*4.3.3*).

Hahn, H.:

[1] *Über Folgen linearer Operationen*. Monatshefte Math. Phys. **32** (1922), 3–88, (*3.3.7, 4.3.3*).

Jacobs, K.:

[1] *Measure and integral*. New York - San Francisco – London: Academic Press 1978, (*4.11.2, 4.11.13, 5.6.3, 5.7.15*).

Kakutani, S.:

[1] *Concrete representation of abstract* (M)–*spaces*. Ann. of Math. **42** (1941), 994–1024 (*5.8.1*).

Kalton, N.J.:

[1] *Subseries convergence in topological groups and vector spaces*. Israel J. Math. **10** (1971), 402–412, (*3.3.25, 3.3.27, 4.1.33, 4.1.35*).

[2] *Topologies on Riesz groups and applications to measure theory*. Proc. London Math. Soc. (3) **28** (1974), 253–273, (*4.1.33, 4.1.35, 5.2.5, 5.2.12, 5.3.1*).

Katz, M.P.:

[1] *On extension of measures* (Russian). Sibirskii Math. J. **13** (1972), 1158–1168, (*4.7.4*).

Khurana, S.S.:

[1] *Convergent sequences of regular measures*. Bull. Acad. Pol. Sci., Série Sci. Math. Astr. Phys. **24** (1976) 37–42 (*4.5.14*).

[2] *Dunford-Pettis property*. J. Math. Anal. App. **65** (1978) 361–364 (*5.8.26*).

Kluvanek, I.:

[1] *On the theory of vector measures*. Mat. Fyz. Časopis SAV **11** (1961), 173–191, (*3.3.15*).

Labuda, I.:

[1] *Sur quelques généralisations des théorèmes de Nikodym et de Vitali-Hahn-Saks*. Bull. Acad. Pol. Sci., Série Sci. Math. Astr. Phys. **20** (1972), 447–456, (*3.4.4, 4.2.14, 4.3.3, 4.6.3*).

[2] *A generalization of Kalton's theorem*. Bull. Acad. Pol. Şci., Série Sci. Math. Astr. Phys. **21** (1973), 509–510, (*4.1.33*).

[3] *Sur quelques théorèmes du type d'Orlicz – Pettis* III. Bull. Acad. Pol. Sci., Série Sci. Math. Astr. Phys. **21** (1973), 559–605, (*3.5.20*).

[4] *A note on exhaustive measures*. Commentationes math. **18** (1975), 217–221, (*4.1.44, 4.7.16*).

[5] *Sur les measures exhaustives et certaines classes d'espaces vectoriels topologiques considérés par W. Orlicz et L. Schwartz*. C.R. Acad. Sci., Paris **280** (1975), A 997–A 999, (*4.1.38*).
[6] *Universal measurability and summable families in tvs*. Proc. Kon. Ned. Acad. v Wet. Series A **82** (1979), 27–34, (*4.11.4, 4.11.6, 4.11.7*).

Landers, D., L. Rogge:
[1] *The Hahn-Vitali-Saks and the uniform boundedness theorem in topological groups*. Manuscirpta Math. **4** (1971), 351–359, (*4.2.14*).
[2] *Cauchy convergent sequences of regular measures with values in a topological group*. Z. Wahrscheinlichkeitstheorie Verw. Gebiete **21** (1972), 188–196, (*4.5.2, 4.5.3, 4.5.5*).

Luxemburg, W.A.J., A.C. Zaanen:
[1] *Riesz spaces* I. Amsterdam – London: North Holland Publishing Company 1971, (*5.7.15*).

Meyer-Nieberg, P.:
[1] *Zur schwachen Kompaktheit in Banachverbänden*. Math. Zeitschrift **134** (1973), 303–315, (*1.5.4*).

Mikusiński, J.:
[1] *On a theorem of Nikodym on bounded measures*. Bull. Acad. Pol. Sci., Série Sci. Math. Astr. Phys. **19** (1971), 441–444, (*4.7.8*).

Moore, E.H.:
[1] *General Analysis, Part* II. Mem. Amer. Phil. Soc. **1**, 2 (1939), 1–255, (*3.3.1*).

Nakano, H.:
[1] *Über das System aller stetigen Funktionen auf einem topologischen Raum*. Proc. Imperial Acad. Tokyo **17** (1941), 308–310, (*5.9.15*).

Nikodym, O.:
[1] *Sur les suites de fonctions parfaitement additives d'ensembles abstraits*. C.R. Acad. Sci., Paris **192** (1931), 727, (*4.2.14, 4.7.8*).
[2] *Sur les familles bornées de fonctions parfaitement additives d'ensemble abstrait*. Monatshefte Math. **40** (1933), 418–426, (*4.7.8*).
[3] *Sur les suites convergentes de fonctions parfaitement additives d'ensemble abstrait*. Monatshefte Math. **40** (1933), 427–432, (*4.2.14*).

Orlicz, W.:
[1] *Beiträge zur Theorie der Orthogonalentwicklungen* II. Studia Math. **1** (1929), 241–255, (*3.5.12*).

Osgood, W.F.:
[1] *Non-uniform convergence and the integration of series term by term*. Amer. J. Math. **19** (1897), 155–190, (*2.1.8*).

Pachl, J.K.:
[1] *A note on Orlicz-Pettis theorem*. Proc. Kon. Ned. Akad. v Wet. Series A **82** (1979), 35–37, (*4.11.4, 4.11.5*).

Pettis, B.J.:
[1] *Uniformity on linear spaces*. Trans. Amer. Math. Soc. **44** (1938), 277–304, (*3.5.12*).

Pełczyński, A.:
[1] *Projections in certain Banach spaces*. Studia Math. **19** (1960), 209–228, (*5.8.18*).
Pfanzagl, J.:
[1] *Convergent sequences of regular measures*. Manuscripta Math. **4** (1971), 91–98, (*4.5.3*).
Phillips, R.S.:
[1] *On linear transformations*. Trans. Amer. Math. Soc. **48** (1940), 516–541, (*4.1.50, 4.3.7, 4.7.15, 5.9.42*).
Pryce, J.D.:
[1] *A device of R.J. Whitley's applied to pointwise compactness in spaces of continuous functions*. Proc. London Math. Soc. (3) **23** (1971), 532–546, (*2.3.14, 2.3.19*).
Rickart, C.E.:
[1] *Decomposition of additive set functions*. Duke Math. J. **10** (1943), 653–665, (*4.1.7, 4.3.4*).
Robertson, A.P.:
[1] *On unconditional convergence in topological vector spaces*. Proc. Royal Soc. Edinburgh **68** (1970), 145–157, (*3.5.17*).
[2] *Unconditional convergence and the Vitali-Hahn-Saks theorem*. Bull. Soc. Math. France **31–32** (1972), 335–341, (*3.3.9, 3.4.3, 3.4.4*).
Rogge, L.:
[1] *The convergence determining class of regular open sets*. Proc. Amer. Math. Soc. **37** (1973), 581–585, (*4.5.15*).
Rooij, A.C.M. van:
[1] *Tight functionals and the strict topology*. Kyungpook Math. J. **7** (1967), 41–43, (*4.10.10*).
Rosenthal, H.P.:
[1] *On relatively disjoint families of measures, with some applications to Banach space theory*. Studia Math. **37** (1970), 13–36, (*5.9.9, 5.9.16*).
Rybakov, V.I.:
[1] *K teoreme Bartla-Danforda-Švarca o vektornyh merah*. Mat. Zametki **7** (1970), 247–254, (*5.6.9*).

Saks, S.:
[1] *Addition to the note On some functionals*. Trans. Amer. Math. Soc. **35** (1933), 967–970, (*4.3.3*).

Schaefer, H.H.:
[1] *Topological vector spaces*. New York – Heidelberg – Berlin: Springer-Verlag 1971, (*2.3.19, 3.5.13, 3.5.15, 3.5.16, 4.7.16, 4.10.1, 5.6.25, 5.8.5*).

Schur, J.:
[1] *Über lineare Transformationen in der Theorie der unendlichen Reihen*. J. Reine Angew. Math. **151** (1920), 79–111, (*3.3.7*).

Seever, G.L.:
[1] *Measures on F-spaces*. Trans. Amer. Math. Soc. **133** (1968), 267–280, (*5.6.8*).

Semadeni, Z.:

[1] *On weak convergence of measures and σ-complete Boolean algebra*. Colloq. Math. **12** (1964), 229–233, (*5.6.8*).

Sierpiński, W.:

[1] *Sur une décomposition d'ensembles*. Monatshefte für Math. Phys. **35** (1928), 239–242, (*3.2.9*).

Šmulian, V.:

[1] *Über lineare topologische Räume*. Mat. Sbornik NS **7** (1940), 425–448, (*2.3.22*).

Stein, J.D. Jr.:

[1] *A uniform boundedness theorem for measures*. Michigan Math. J. **19** (1972), 161–165 (*4.8.5*).

Swartz, C.:

[1] *A generalized Orlicz-Pettis theorem and applications*. Math. Zeitschrift **163** (1978), 283–290, (*3.4.1, 3.5.8*).

[2] *The Schur lemma for bounded multiplier convergent series*. Math. Ann. **263** (1983), 283–288 (*3.3.33, 3.3.34*).

Thomas, E.:

[1] *Sur les mesures vectorielles à valeurs dans les espaces d'un type particulier*. C.R. Acad. Sci., Paris **266** (1968), A 1135–A 1137, (*5.8.18*).

[2] *Sur le théorème d'Orlicz et un problème de M. Laurent Schwartz*. C.R. Acad. Sci., Paris **267** (1968), A 7–A 10, (*3.5.8, 3.5.16*).

[3] *L'integration par rapport à une mesure de Radon vectorielle*. Ann. Int. Fourier **20.2** (1970), 55–191, (*3.4.4, 3.5.1, 3.5.8, 3.5.16*).

[4] *The Lebesgue-Nikodym theorem for vector valued Radon measures*. Mem. Amer. Math. Soc. **139** (1974), 101 pages, (*3.5.16, 4.7.16*).

Tumarkin, Ju.B.:

[1] *On locally convex spaces with basis*. Soviet Math. Dokl. **11** (1970), 1672–1675 or Dokl. Acad. Nauk SSSR **195**, 6 (1970), (*4.1.45*).

Tweddle, I.:

[1] *Weak compactness in locally convex spaces*. Glasgow Math. J. **9** (1968), 123–127, (*5.9.37*).

[2] *Vector-valued measures*. Proc. London Math. Soc. (3) **20** (1970), 469–485, (*3.5.11*).

[3] *Unconditional convergence and vector-valued measures*. J. London Math. Soc. (2) **2** (1970), 603–610, (*3.5.10*).

Vitali, G.:

[1] *Sull' integrazione per serie*. Rendiconti Circolo Mat. Palermo **23** (1907), 137–155, (*4.3.3*).

Walsh, B.:

[1] *Mutual absolute continuity of sets of measures*. Proc. Amer. Math. Soc. **29** (1971), 506–510, (*5.6.9*).

Weber, H.:
[1] *Kompaktheit in Räumen von gruppen- und vektor-wertigen Inhalten, der Satz von Vitali-Hahn-Saks und der Beschränktheitssatz von Nikodym* (to appear) (*4.2.13*, *4.7.8*).
Weil, A.:
[1] *Sur les espaces à structure uniforme et sur la topologie générale*. (Actualités Sci. Ind. **551**), Paris: Hermann 1938, (*1.5.6*).
Wells, B. B. Jr.:
[1] *Weak compactness of measures*. Proc. Amer. Math. Soc. **20** (1969), 124–130, (*4.5.14*).
Weston, J. D.:
[1] *On the comparison of topoligies*. J. London Math. Soc. **32** (1957), 342–354, (*3.3.26*).

Index

Notations